How Does Earth Work?

How Does Earth Work?

Physical Geology *and the* Process of Science

Gary A. Smith
University of New Mexico

Aurora Pun
University of New Mexico

Upper Saddle River, New Jersey 07458

Library of Congress Cataloging-in-Publication Data

Smith, Gary A. (Gary Allen)
   How does Earth work? : physical geology and the process of science / Gary A. Smith, Aurora Pun.
     p. cm.
   Includes index.
   ISBN 0-13-034129-0 (alk. paper)—ISBN 0-13-186747-4 (alk. paper)
   1. Physical geology—Textbooks.   I. Pun, Aurora. II. Title.
   QE28.2.S59 2006
   551--dc22

                    2005004918

Executive Editor: *Patrick Lynch*
Development Editor: *Patricia Nealon*
Editor-in-Chief, Development: *Carol Trueheart*
Executive Managing Editor: *Kathleen Schiaparelli*
Assistant Managing Editor: *Beth Sweeten*
Editorial Assistant: *Sean Hale*
Production Editor: *Kim Dellas*
Project Manager: *Dorothy Marrero*
Director of Marketing, Science: *Linda Taft-MacKinnon*
Managing Editor, AV Production & Management: *Patty Burns*
Director of Creative Services: *Paul Belfanti*
Manager, Composition: *Allyson Graesser*
Electronic Production Specialist/Electronic Page Makeup: *Joanne Del Ben*
Art Director: *John Christiana*
Interior Design: *Jonathan Boylan*

AV Editor: *Jess Einsig*
Illustrator: *Precision Graphics/Mapquest.com*
Director, Image Resource Center: *Melinda Reo*
Manager, Rights and Permissions: *Zina Arabia*
Manager, Visual Research: *Beth Brenzel*
Manager, Cover Visual Research & Permissions: *Karen Sanatar*
Image Permission Coordinator: *Cathy Mazzuca*
Photo Researcher: *Truitt & Marshall and Yvonne Gerin*
Manufacturing Buyer: *Alan Fischer*
Senior Media Editor: *Chris Rapp*
Managing Editor, Science Media: *Nicole M. Jackson*
Media Production Editor: *Elizabeth Wright*
Assistant Managing Editor, Science Supplements: *Becca Richter*
Cover Design: *John Christiana*
Cover Photo: *El Capitan in Yosemite, Marc Muench/CORBIS*

PEARSON
Prentice Hall

© 2006 by Pearson Education, Inc.
Pearson Prentice Hall
Pearson Education, Inc.
Upper Saddle River, NJ 07458

+
DH
<S>
copy1

Printed in the United States of America

10 9 8 7 6 5 4 3 2 1

ISBN 0-13-034129-0

Pearson Education LTD., *London*
Pearson Education Australia PTY, Limited, *Sydney*
Pearson Education *Singapore,* Pte. Ltd.
Pearson Education North Asia Ltd., *Hong Kong*
Pearson Education Canada, Ltd., *Toronto*
Pearson Educación de Mexico, S.A. de C.V.
Pearson Education—Japan, *Tokyo*
Pearson Education Malaysia, Pte. Ltd.

About the Authors

Gary A. Smith

Gary A. Smith is a Professor of Earth and Planetary Sciences at the University of New Mexico and Fellow of the Geological Society of America. He has an undergraduate geology degree with a specialty in geophysics from Bowling Green State University and a Ph.D. in geology from Oregon State University. He has taught first-year geology courses, primarily to nonmajors, for 14 of the last 22 years. His research, reported in more than 100 publications, and upper-division-teaching experience range widely across the geologic sciences, including sedimentology, volcanology, geomorphology, and hydrology. Gary's research has a strong field emphasis, and he has contributed to 12 published geologic map quadrangles in Oregon and New Mexico. Nonacademic employment experiences in the oil industry and with Department of Energy facilities engaged in environmental remediation further enhance his background. Gary has strong interests in science education through his membership in the National Association of Geoscience Teachers and the National Science Teachers Association and as chair of the University of New Mexico's Faculty Senate Teaching Enhancement Committee.

Aurora Pun

Aurora Pun is an Adjunct Assistant Professor, Research Scientist, and instructor in the Department of Earth and Planetary Sciences at the University of New Mexico. She holds an undergraduate degree in paleontology from the University of California, Berkeley and M.S. and Ph.D. degrees in Geology from the University of New Mexico, Institute of Meteoritics. Aurora has taught physical geology for the last 10 years. Also a member of the National Association of Geoscience Teachers, Aurora taught, with Gary and an education instructor, a course for teachers on developing inquiry-based K–12 curricula in the earth and space sciences. Her research focuses on meteorite geochemistry with an emphasis on planetary igneous systems, and the petrology, mineralogy, and chemistry of ultrafine-grained sub-millimeter particles, from both terrestrial and extraterrestrial environments. Aurora's research utilizes a variety of laboratory instruments (electron microprobe, ion-microprobe, and transmission electron microscope) that provide in-place chemical analyses of microscopic minerals.

Gary and Aurora have a variety of nonacademic interests. They enjoy hiking, traveling, and xeric gardening around their home in Albuquerque.

Thank You

We would like to express our deep gratitude to more than 80 of our colleagues from schools across the country who provided us with hundreds of hours of their time over the past four years. They reviewed manuscript chapters, participated in focus groups, reviewed animations, class tested an early chapter, and patiently answered our questions.

How Does Earth Work? has benefited immeasurably from your effort, insights, objections, encouragements, and your selfless willingness to share your expertise as scientists and teachers.

Patrick Abbott, San Diego State University
Gerry Adams, Columbia College
John Anderson, Georgia Perimeter College
Abdolali Babaei, Cleveland State University
Kathy Baldwin, Washington State University
Lisa Barlow, University of Colorado – Boulder
Jerry Bartholomew, University of Memphis
Raymond Beiersdorfer, Youngstown State University
Glen Blaylock, Utah Valley State College
Robert Blodgett, Austin Community College
James Brophy, Indiana University
Lloyd Burckle, Columbia University
Patrick Burkhart, Slippery Rock University
Susan Cashman, Humboldt State University
Beth Christensen, Georgia State University
Jane Dawson, Iowa State University
John Dembosky, University of Akron
Halilang Dong, Miami University
William Dupre, University of Houston
Roy Ewing, Aims Community College – Greeley
Andrew Frank, Chemeketa Community College
Daniel Frederick, Austin Peay State University
Sharon Gabel, State University of New York – Oswego
Jacqueline Gallagher, Florida Atlantic University
Francisco Gomez, University of Missouri – Columbia
Yuri Gorokhovich, State University of New York – Purchase
Joseph Gould, University of South Florida
Nathan Green, University of Alabama
Lisa Greer, Pennsylvania State University
Paul Grogger, University of Colorado – Colorado Springs
Roy Hagerty, Oregon State University
John Hanchar, George Washington University
William Heins, Lewis-Clark State College
Rosemary Hickey-Vargas, Florida International University
Scott Hippensteel, University of North Carolina – Charlotte
Lincoln Hollister, Princeton University
Daniel Holm, Kent State University
Mary Hubbard, Kansas State University
Scott Hughes, Idaho State University
Robert Jorstad, Eastern Illinois University
Richard Josephs, University of North Dakota
David King, Auburn University
Nicholas Lancaster, Desert Research Institute

Patricia Lee, University of Hawaii
Robert Leighty, Mesa Community College
Alan Lester, University of Colorado – Boulder
Steven Lev, Towson University
Mian Liu, University of Missouri – Columbia
I. Peter Martini, University of Guelph
Maureen McCurdy, Louisiana Tech University
Dan McNally, Bryant College
Robert Meeks, St. Philip's College
Stanley Mertzman, Franklin & Marshall College
Barry Metz, Delaware County Community College
William Minarik, University of Maryland
LeeAnn Munk, University of Alaska – Anchorage
Kathryn Nagy, University of Illinois – Chicago
Donald Neal, East Carolina University
Clive Neal, University of Notre Dame
Yuet-Ling O'Connor, Pasadena City College
Raymond Pheifer, Eastern Illinois University
Nicholas Pinter, Southern Illinois University
Roy Plotnick, University of Illinois – Chicago
Jay Quade, University of Arizona
Monica Ramirez, Aims Community College – Greeley
John Renton, West Virginia University
Carl Richter, University of Louisiana – Lafayette
Jennifer Rivers, Northeastern University
Michael Roden, University of Georgia
Sarah Rogers, Arapahoe Community College
Laura Sanders, Northeastern Illinois University
William Sanford, Colorado State University
William Cullen Sherwood, James Madison University
John Paul Stimac, Eastern Illinois University
Karen Swanson, William Paterson University
Martha Sykes, Hawaii Pacific University
Charles Trupe, Georgia Southern University
Dorothy Vesper, West Virginia University
Chunzeng Wang, Hunter College of the City University of New York
David Warburton, Florida Atlantic University
James Wittke, Northern Arizona University
Ellen Wohl, Colorado State University
Wan Yang, Wichita State University
Aaron Yoshinobu, Texas Tech University
Brent Zaprowski, Salisbury University
Sally Zellers, Mississippi State University

Brief Contents

Contents

Chapter 7   Earth Materials as Time Keepers   166

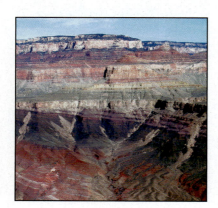

Part II   Earth's Internal Processes   196

Chapter 8   Journey to the Center of Earth   196

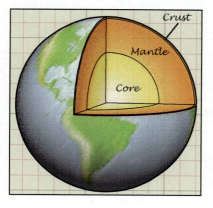

Chapter 9   **Making Earth   222**

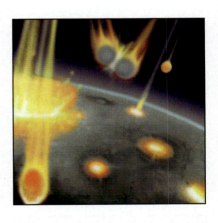

Part III   Earth Deformation   268

Building collapse. Yulin, Taiwan, 1999. Moment magnitude 7.8.

Chapter 12 Global Tectonics: Plates and Plumes 306

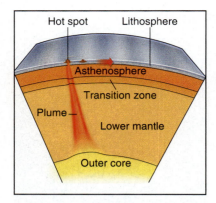

Part IV   Surface and Near Surface Processes   382

Chapter 14   Soil Formation and Landscape Stability   382

Chapter 16   Streams: Flowing Water Shapes the Landscape   440

Chapter 18   **Glaciers: Cold-Climate Sculptors of Continents   516**

Preface

> Science is built up of facts, as a house is built of stones; but an accumulation of facts is no more science than a heap of stones is a house.
> –Henri Poincaré, Science and Hypothesis, 1905

Why Now?

We undertook to write this book because we find that no existing introductory text adequately supports our primary objective as introductory geology teachers: Helping the average non-science student to truly understand that science is not simply a body of knowledge, but a method of acquiring knowledge. It is a way of thinking that once internalized will serve them their entire lives. The current textbooks—such a crucial component of our introductory courses—focus on presenting "what we know." (This is no small burden, to be sure, and we do truly admire the achievements of these authors.) We find that this mile-wide-inch-deep approach, however, only further buttresses the students' erroneous preconception of science as a revealed set of facts, Poincaré's "heap of stones." We set out instead to write a book that better furthers that primary objective as introductory teachers, a book about *how we know* what we know.

This text is designed for use in introductory physical geology courses, generally taught with no prerequisites, and commonly taken by a diverse student population. For many—if not most—of these students, this course will be their sole college-level experience in science.

How Does Earth Work? emphasizes conceptual learning of processes, rather than immersion in facts, and adopts the credo "how we know" rather than only "what we know". The organization focuses on asking and then seeking answers to questions so as to:

1. Further engage the student to learn from reading, and
2. Encourage instructors to use a classroom pedagogy that facilitates interaction with students, regardless of class size.

The pedagogical spirit is:

1. Consistent with recently emphasized learning paradigms, rather than simply as a companion to traditional teaching-paradigm lecturing, and
2. Intended to enhance science literacy through better appreciation of how scientific knowledge is obtained, and how it is relevant to societal concerns, needs, and decision-making.

What Makes this Different?

These are the principal features of the text, some of which differ significantly from most comparable texts:

— **The presentation leans away from encyclopedic presentation of facts toward emphasizing active-process concepts**, and an understanding how earth scientists acquire fundamental knowledge about our planet.

— **Each chapter opens with an essay that places a curious observer in realistic outdoor (field) or laboratory circumstances to make observations and ask questions about geological phenomena.** To the extent possible, these hypothetical experiences are of the type that students can readily relate to or may have experienced themselves. The chapter presentation then seeks to explain phenomena that piqued the interest of the hypothetical curious observer.

— **The text uses questions as devices for learning.** Section headings are in the form of questions that intend to maintain reader interest, and to persistently illustrate relevancy. Students also encounter many questions in the text that are designed to cause them to stop, contemplate, and integrate what has been learned. The section-heading titles could serve as examination questions.

— **Key points are summarized at the end of each major section ("Putting it Together") rather than at the end of the chapter.** These bullets (1) relate directly to answering the question stated in the section heading, and (2) provide a learning check for students before moving on in the text.

— **Wherever appropriate, topics are linked to issues of societal concern, real-world experience, or both**, so as to increase relevancy and appreciation of the value of science learning. Topics of societal relevancy are not relegated to sidebars, boxes or separate chapters, but are integrated throughout the text.

— **Illustrations are chosen/constructed to emphasize learning enhancement.** In particular, illustrations are as simple as possible, spatially associated with applicable text, and highly annotated in accordance with research demonstrating distinctions between effective versus ineffective use of textbook illustrations.

— **Animation is used to illustrate active processes.** The majority of the 60 "Active Art" animations use the same base illustration as the static art in the text, providing a clear link between text illustration and CD animation.

— General concepts about Earth processes are presented with introduction of the absolute, bare minimum number of unfamiliar terms. **Classifications, where necessary, are inductively "derived" from "observation" rather than simply produced on the page as facts to be memorized.** More mineral and rock names are used than in some comparable texts in the spirit of understanding the diversity of Earth materials. Instructors sharing the pedagogical spirit of the text will not, however, use this as an opportunity to make students memorize more mineral and rock terms.

— **There is always consideration of "how it is that we know what we know," and a context for why it is important to understand these processes.** This focus ties to societal relevance in many cases, but also includes satisfaction of innate curiosity (i.e., the distinction, but equal importance, of strategic- and curiosity-based science).

— **Each chapter has a "How do we Know?" section that provides more detail on the scientific accomplishments that provided a piece of key information in the topic for the chapter.** The *"How do we Know?"* sections explicitly show how concepts about geological processes are developed, and implicitly serve (1) to be a basis for appreciating how all concepts in the book have become scientific knowledge, and (2) to illustrate how geoscientists apply scientific methodology.

— **To facilitate text brevity, some background material (including that which we consider to be remedial), detailed examples, and supplemental information (which an instructor may or may not require students to know) are provided on compact disc as Extension Modules.** Unlike pasted-in-the-back-of-the-book CD's that are commonly treated as supplements or optional elements, and ignored by many instructors and students, we designed the extension modules as elements to be fully integrated with the text at the discretion of instructor assignments or student curiosity. Instructors may assign selected CD modules, and hold students responsible for the material just as if it were a text section. The text itself reads continuously without the necessity of consulting the CD, and the modules can be assigned, or not, to supplement textbook content. The modules are of two types:

1. Coverage of information that we assume students to have previously learned in high-school chemistry or physical science. The modules serve as a refresher for students who previously learned this material but are also written to suffice as first-time treatments for the less thoroughly prepared student.
2. Coverage of information that extends beyond the text. These include topics found in some physical geology texts, but which we feel are optional, in addition to topics that some books treat in "boxes." Some extension modules provide substantially more depth or mathematical treatment of subjects than in most comparable textbooks; these features can be selected for courses designed for science majors. Other modules will be deemed essential for instructors teaching in regions where the subject matter is highly relevant, or will be ignored by instructors who feel that their students will not relate to the subjects.

How Does the Book Work?

Following, we detail our approach to the core topics of the course and the ways, if any, we depart from the standard treatment in most texts.

Overall organization

The parts of the text follow a logical, progressive development of geological knowledge. Part I focuses on Earth materials seen at the surface; what they are and how they are used and studied. Part II moves on to what is known or inferred about Earth's interior, how this knowledge is acquired, and the origin of Earth's internal structure. Part III takes what was learned about Earth's internal processes to understand descriptions of rock deformation, including plate tectonics. Part IV integrates knowledge of Earth materials and deformation with hydrosphere and atmosphere processes to determine the origin of diverse landscape features.

Plate tectonics

Plate tectonics is an underlying theme that can be threaded throughout the course. In writing a concept-oriented text, which encourages inquiry and application of scientific reasoning, we recognize that we must first establish an understanding of the nature of geologic materials and phenomena before fully developing the theory by which they are explained: plate tectonics. Observations and fundamental understanding are first in scientific methodology. We introduce the theory in Chapter 1. Emphasis at this early stage is placed on the general phenomena at plate boundaries, and this background permits integration of plate tectonics into the chapters following. A thorough development of the theory is placed in Chapter 12, after students have a firm grasp of the geological materials and structural features required to understand the theory, rather than simply having its salient features presented as undisputed fact.

Rocks and rock forming processes

The dedication of an entire chapter to an overview of rocks is atypical for physical geology texts. We believe that this overview in Chapter 3 is important, however, for the following reasons:

- The inter-relationship of the three rock types (the popular rock cycle) is an important concept that provides a context for students to understand what rocks are and how they form.
- The classification of rocks, although typically encountered in grades 5-8 (and in some states in high school), is a nebulous concept to many students, many of whom are unclear on what classification schemes are or why they are developed.
- An overview of the three rock classes, and an understanding of the general concepts of the genetic classification, help students put rocks in perspective before considering each group in detail. Otherwise, students learn about each of the three classes separately, and rarely know where to begin identifying a rock unless they already know that it is igneous, sedimentary, or metamorphic to begin with.

We appreciate that many instructors may prefer to skip this chapter and sail onward into igneous rocks. For those inclined to do so, Chapters 4-6 are written independently of Chapter 3 so as to not impede its omission.

Weathering and soil

Many physical geology textbooks insert a chapter on weathering and soil formation ahead of the chapter on sedimentary rocks. *How Does Earth Work?* takes a different approach. Weathering processes are included in the sedimentary rock chapter to emphasize how the components of sedimentary rock form. We find the consideration of soil to be a digression in the understanding of rocks, and provide a separate short chapter on soil in the context of landscape evolution in Part IV of the text.

Earth's interior

Earth's interior is dealt with in a single short chapter in many texts. We believe that this brief treatment does an injustice to the vast wealth of information obtained about the interior of the planet. We dedicate chapters 8-10 to the interior, one chapter to each of these concepts: what it is made of, how it got to be that way, and why it is in motion. Convection within Earth is worthy of more description and explanation than is found in most physical geology texts. Recent advances in geodynamics support the idea that plate tectonics can mostly be described as being mantle convection, rather than seeing convection as some nebulous, and unpopular, driving force. Convection is, therefore, given a prominent place in this text, but done at a level appropriate for the introductory student.

Earthquakes

The subject of earthquakes is divided into parts of two chapters rather than lumped into one. This differs from most physical geology texts where there is a single earthquake chapter, which deals with two completely different subjects: (a) The inelastic deformation mechanism of earthquakes and resulting effects and hazards; and (b) the use of elastic waves to deduce the internal structure of the planet. We treat the surface effects and deformation caused by earthquakes along with the discussion of faults and faulting in Chapter 11. Chapter 8 deals only with the properties of seismic waves that are fundamental to investigating the planetary interior.

Global tectonics

Global tectonics (plate motions and plumes) is treated in a single chapter. This construct is similar to many physical geology textbooks, whereas others break the material into two chapters. In contrast to most text treatments of the subject, Chapter 12 does not provide significant historical development of the plate tectonics theory and plume hypothesis. Instead, our emphasis is on "showing the evidence." We think that it is more important for students to be able to explain why plate tectonics works, than to recite the history of critical developments. This approach is most consistent with the "how do we know" theme of the text, and keeps students focused on how scientific data successfully test hypotheses and theories. An underlying, subdued theme in this chapter is the distinction between plate tectonics as a theory and the mantle plume hypothesis; these two concepts do not enjoy the same level of supporting data and confirmation. We hope that students will come away with confidence in understanding why plate tectonics is entrenched as a fundamental concept in geoscience, while at the same time seeing the mantle-plume hypothesis as an area of continued fundamental testing, a continued frontier of exciting, ongoing science.

Surface relief

We replace the typical chapter on mountain building with Chapter 13 on the origin of surface relief. While giving considerable attention to the formation of mountains, this chapter deals more generally with the tectonic foundation for the irregular surface of the planet as a fundamental feature that any introductory geology student should be able to explain. This chapter fully develops concepts of isostasy by making extensive use of simple analogy for complex Earth features, and the scientific modeling approach of starting simple and adding more complexity only as necessary to explain observed features.

Surficial processes

Part IV, which emphasizes surficial processes, is written with an emphasis on understanding these processes and their visible results, rather than on developing an encyclopedic glossary of landform names. Our approach is somewhat more in depth than in most texts, because we think that students relate to studying and understanding landscapes that they see and have experienced more than any other topic in physical geology. We strive to develop understanding of how observed processes create observed landforms, resources, and hazards. A significant component of this understanding comes from understanding how surface processes impact humans, and how human modifications of the environment impact the surface processes in predictable ways.

Modern soil treatment

The soil chapter (Chapter 14) uses modern soil descriptions and genetic interpretations. The archaic terms *pedalfer* and *pedocal* are not used. This chapter should be particularly engaging for students studying geology in those large regions of the country where agriculture plays an important role in the economy. This chapter also introduces the role of soils in evaluating landscape stability, which is integrated through the succeeding chapters.

Mass movements

Our discussion of mass movements in Chapter 15 emphasizes the characteristics of these phenomena that lead to a natural classification with genetic significance. As with most texts, we focus on the hazards of mass movements, but also provide insights into how these processes contribute to overall landscape development.

Rivers, streams, and lakes

Chapter 16 provides many of the familiar elements of learning about streams, but within the context of understanding how streams do work. With an understanding of how streams pick up, transport, and deposit sediment, it is possible to deduce the origins of a wide variety of landforms, and to understand why streams change along their courses and through time, in response to climate change, tectonic activity, sea-level fluctuation, and human activities. This chapter also includes a treatment of the origin and landscape-significance of lakes, which are rarely mentioned in physical geology texts.

Ground water

Chapter 17 discusses ground water flow and its effects. Emphasis is placed on the simple mechanics of ground water movement and particularly aspects of ground water mechanics that influence the exploitation of this water resource. Ground water also plays a role in shaping landscapes. The formation of karst topography is discussed without undue emphasis on cave features, which are elaborated in an extension module. The role of ground water sapping, rarely mentioned in other texts, to form canyon landscapes is also explained.

Glaciers

Chapter 18 considers the domain of frozen water—glaciers. Introducing a minimum of unfamiliar-sounding landform terms, the chapter first explains the nature of glacier flow and then relates glacier dynamics to erosional and depositional landforms. Glaciers are responsible for a large variety of landscapes and resources in North America, even though they are unseen by most students. Consideration of the late Pleistocene ice ages and the climatic driving forces for their occurrence is key to understanding these landscapes and resources. This chapter provides critical examples of how geologists interpret landscape origins by reference to vanished agents of landscape modification.

Shoreline processes

Chapter 19 explains the role of waves, tides, and sediment supply to shape ocean shorelines. As with other chapters in this section, the shoreline chapter begins with grounding in the operative processes and then shows how these processes account for familiar shoreline features. Rapid, historical change of shorelines, by both natural and anthropogenic forces, is also a key part of the chapter and links to how scientists document and explain historical sea-level rise.

Wind

Chapter 20 evaluates wind as a geological process. Most introductory geological texts discuss wind within a chapter on deserts. *How does Earth Work?* does not have a deserts chapter. The river chapter develops concepts of stream dynamics in both humid and arid landscapes and includes much of the information traditionally found in desert chapters. Wind is a critical geological agent everywhere on Earth, and not just in deserts—for this reason the text treats wind in a separate chapter. Although including significant information on desert dunes, wind is also featured as an effective landscape-forming agent along coastlines and as a globally important sediment-transport process as revealed by global dust movement.

Instructor and Student Resources

Instructors and students using *How Does Earth Work?* have at their disposal a tremendous range of resources.

The centerpiece of the instructor resources is the **Instructor Resource Center (IRC) on DVD**, which aggregates in one, conveniently-accessed place: an **Instructor's Manual** with test questions, lecture outlines, and suggestions for the classroom based on the latest pedagogical research; a **library of PowerPoint®️ presentations**, including "art only" and "lecture outline" versions; a library of **60 Active-Art animations** set up for classroom projection, pre-loaded into

PowerPoint® slides; and an additional set of photos. A complete set of transparencies is also available.

The student resources are designed to help focus effort and maximize results. The resources include: the **Online Study Guide** (www.prenhall.com/smithpun), with multiple quizzes for each chapter; the **Student Lecture Notebook**, which reproduces the illustrations from the text with space for note taking.

For a complete description of these resources or to request a desk copy, please contact your local Prentice Hall representative or visit www.prenhall.com/geology.

Acknowledgments

The execution of this project was possible only through the assistance, advice, encouragement, and hard work of many people. We are especially grateful to Executive Editor Patrick Lynch for his devotion to assisting us accomplish the vision for the book, his valuable advice, his persistent efforts to help us however possible, and for being the head cheerleader at all times to encourage us to keep battling downfield. Development editor Patricia Nealon was invaluable for helping us mold a unique and readable style, which was then fine-tuned by the copy editing of Pat Daly and Mary Eberle. Production Editor Kim Dellas masterfully coordinated a large production team that completed the project on an accelerated schedule. Sean Hale, Editorial Assistant, efficiently tracked all of the manuscripts. We are deeply indebted to photo researchers Jerry Marshall and Yvonne Gerin for a remarkable job in locating photos to match our very demanding specifications. Our hats go off to Precision Graphics for executing the art program consistent with our themes for the text. Chris Rapp, Media Editor, designed and developed the student CD, which exceeded all of our expectations, and all the while kept the project on a complicated schedule while deflecting as much work as possible from our desk to his. The quality of the CD is largely because of the team at Cadre Design, especially Ian Shakeshaft with his gift for translating scientific processes into efficient animations. Many reviewers helped us to mold the book into form, but we are particularly grateful to the team that reviewed virtually the entire final draft of the book: Gerry Adams, Beth Christensen, Ray Beiersdorfer, Sharon Gabel, Scott Hippensteel, David King, Jennifer Rivers, Karen Swanson, and Aaron Yoshinobu. Ray and Beth also co-authored most of the end-of-chapter questions.

Many others were instrumental in our successful development of the text. At an early stage, high school geoscience teachers Doug Earick and Alex Castrounis provided critical assistance to help us find a style and approach consistent with the tools students bring to the college classroom. The NSF-CETP program at the University of New Mexico supported early development and class testing of course materials. Linda Doran helped with background research, and Theresa Hoyt provided helpful accuracy checks on the text. Frans Rietmeijer and Lysa Chyzmadia offered constructive advice on presenting topics that were unfamiliar to us. We are grateful to student Cody Wiley who reviewed nearly the entire book out of pure interest, and provided unique criticism and insight from a student's perspective. We also appreciate the encouragement of our colleagues in the Department of Earth and Planetary Sciences, University of New Mexico, during the entire four-year project and especially the support of department chair, Les McFadden.

We greatly appreciate the generosity of geoscientists who eagerly responded to our personal requests to provide unique images that appear in the text. From long-cherished colleagues to new acquaintances, we express our special thanks to the following people: Robert S. Anderson (University of Colorado), David Barber (University of Essex), James Bartolino (USGS), Ronald C. Blakey (Northern Arizona University), John Bloch (University of New Mexico), Mike Blum (Louisiana State University), Roderick Brown (University of Glasgow), Susan H. Cannon (USGS), Michael Clynne (USGS), Yildirim Dilek (Miami University), Michael Dolan (Michigan Technological University), Peter J. Fawcett (University of New Mexico), Tobias Fischer (University of New Mexico), Virginia H. Garrison (USGS), John W. Geissman (University of New Mexico), Gary Glatzmaier (University of California-Santa Cruz), Fraser Goff (Los Alamos National Laboratory), Andrea Holland Sears (USFS), Rhian Jones (University of New Mexico), Karl E. Karlstrom (University of New Mexico), Shari Kelley (New Mexico Institute of Mining and Technology), Jerry D. Kudenov (University of Alaska-Anchorage), Stephen P Leatherman (Florida International University), David Love (New Mexico Bureau of Geology and Mineral Resources), Brian H. Luckman (University of Western Ontario), Patrick Lynch (Prentice Hall), Leslie D. McFadden (University of New Mexico), Grant Meyer (University of New Mexico), O. Richard Norton (Science Graphics), Tom Pierson (USGS), Gary Muckel (Natural Resources Conservation Service), Henry Pollack (University of Michigan), David A. Schmidt (University of Oregon), Michael N. Spilde (University of New Mexico), Susie Staples (Chickasaw National Recreation Area), Paul J. Tackley (University of California-Los Angeles), Larry W. Thomason (NASA), Keith J. Tinkler (Brock University), Laura Wilson, Aaron Yoshinobu (Texas Tech University), and Richard Young (SUNY-Geneseo).

Gary Smith and Aurora Pun
Albuquerque, New Mexico

1 Why Study Earth?

Chapter Outline

Why Study Earth?

HERE ARE SOME REASONS TO CONSIDER.

- Earth is the planet upon which you live.

- Earth is an active planet that experiences cataclysmic earthquakes, volcanic eruptions, floods, landslides, and other processes that constantly reshape the surface and sometimes threaten human lives and property.

- Earth is a source of energy resources that provide for your quality of life: oil, natural gas, coal, and uranium.

- Earth is the host for essential water supplies for human consumption, irrigation, industrial use, power generation, and recreation.

- Earth is a source of raw materials for metals, plastics, building materials, and precious gemstones.

- Earth exhibits vastly varying surface features that are the foundation for countless ecosystems on continents and in oceans.

- Earth's landmasses have a thin layer of soil that supports natural plant diversity and agricultural richness.

- Earth is a template with which to compare and contrast the evolution of other planets and the processes that shaped them.

- Earth provides awesome, beautiful landscapes and seascapes, which inspire the mind to wonder how the planet came to be, what it is made up of, and how it works.

You are about to embark on a study of **geology**—the science of Earth. The goal is to learn how your planet works.

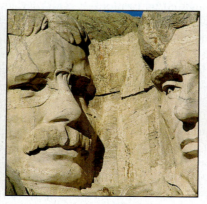
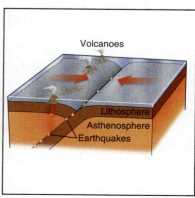

Geology includes the study of all the processes that shape Earth's surface and that take place beneath the surface. These processes produce dramatic scenery, provide Earth resources, and are sometimes highly destructive. ▶

Along the way, you will learn the origins of essential resources and the forces driving the processes that are hazardous to lives and livelihoods. This text is not simply a collection of facts to memorize. Instead, you will ponder how scientists have carefully, and cleverly, arrived at the current body of knowledge about Earth.

This opening chapter lays the groundwork for your geologic studies by addressing the following objectives:

✔ To learn the breadth of topics and studies that compose geology

✔ To see how geology is relevant to you and our society

✔ To understand how geologists undertake scientific studies

✔ To understand the basic elements of a key principle and an integrating theory that explain how Earth works
The questions to answer along the way are as follows:

1.1 What is geology?
1.2 Why study geology?
1.3 *How do we know . . .* how to study Earth?
1.4 What does the principle of uniformitarianism mean?
1.5 What is the theory of plate tectonics?
1.6 How does the concept of work apply to Earth?

In the FIELD

Your geologic inquiry begins with three imaginary field outings—excursions into the great outdoors with a focus on observing, describing, and asking questions about Earth processes and materials. Each chapter opens with such an experience, either in the field or in a hypothetical laboratory.

Your first stop is the island of Sicily, the "football" located off the "toe of the boot" of Italy. The view appears in **Figure 1.1**a. Fountains of lava light up the night sky on the flank of the great volcano Mount Etna, and molten-rock cinders fall back upon the slopes of the growing conical hill. What is happening and why is it happening? How safe is it to be where you are standing to take in this view? Are the communities at the base of the erupting volcano at risk of destruction? How can the risks to lives and properties be determined, and can they possibly be reduced?

The active processes on and within Earth—erupting volcanoes, shattering earthquakes, widespread inundation by floodwater, devastating landslides, or destructive rock falls—always make the news. These events clearly illustrate that the planet is dynamic, not stationary and static. As a student of geology, you are probably curious to know what processes lead to these hazardous events and the extent to which geologists can forecast or pinpoint their occurrence.

Your second stop is a summer morning along the Snake River in Grand Tetons National Park, Wyoming, portrayed in Figure 1.1b. If you are an outdoors enthusiast, you can envision countless opportunities in this landscape—hiking, rock climbing, canoeing, and fishing. The diverse landscapes and seascapes of Earth provide widely different challenges, uses, and inspirational views.

Ever wonder why there is so much diversity in the appearance of Earth's surface? Why are mountains, such as the Grand Tetons, not found everywhere? How long have the mountains existed and how, when, and how fast did they form? Why do rivers, such as the Snake River, flow where they do? What determines how much water the river carries, how fast it moves, how much sediment is carried by the flow, and how often the river expands beyond its banks to flood the adjacent countryside? Geologists strive to answer such questions.

Your third stop overlooks a vast, open-pit coal mine in West Virginia. Gargantuan shovels scoop out huge bites of the black rock (Figure 1.1c). The coal, carried by oversized trucks to conveyors, moves to equipment that crushes and sorts the fragments by size and then delivers the fuel to waiting train cars. In the ground for hundreds of millions of years, the coal will soon burn in a power plant to generate electricity.

About half of the electricity generated in the United States comes from coal. What is coal? How did it form? Why is coal found only in particular locations? The same questions can be asked about a wide variety of Earth resources, from rock processed for iron, gold, or gemstones, to rock hosting oil, natural gas, and even water.

How does Earth work? You are embarking on an in-depth inquiry about your planet. Along the way you will ask and answer questions about Earth that are important to societal concerns—hazards and resources—in addition to the questions that simply pique human curiosity.

Figure 1.1 Why study how Earth works? ▶
(a) Volcanic eruptions are examples of dynamic Earth processes that are visually stunning and extremely hazardous. (b) Scenic landscapes result from geologic processes and have long fascinated humans and inspired curiosity to know more about how Earth works. (c) This coalmine reminds us that most of the resources we depend on for everyday life are directly or indirectly derived from the solid Earth at our feet.

A. *Field observations: Mt. Etna, Italy*

City possibly threatened by erupting volcano

Molten cinders ejected from crater in small volcanic cone

B. *Field observations: Grand Tetons, Wyoming*

Scenic high mountains →

Snake River flows in valley at base of mountains →

C. *Field observations: Coal mine in West Virginia*

Light-colored rock is stripped away to expose the coal

Black rock is a layer of coal

1.1 **What Is Geology?**

Geology is the study of Earth. What types of studies compose geological science? Geology is commonly introduced in college classrooms as two broad fields—physical and historical geology.

Physical geology, the subject of this text, originated as the study of the appearance of Earth's surface and the processes that form surface features. Sweeping scenic landscapes, like those seen in **Figure 1.2**, are expressions of the processes that sculpt Earth's surface, from the high peaks of the Himalayan Mountains in Nepal to the deep Grand Canyon in Arizona. Geologists later concluded that heat energy within Earth drives many of these processes, so physical geology expanded to include the physical and chemical interior workings of the planet.

Historical geology applies an understanding of physical, chemical, and biological processes to interpret the history of Earth. Historical geology includes the ordering and timing of events in Earth history such as major mountain-building periods, changes in sea level and climate through time, and the evolution of organisms preserved in rocks as fossils.

(a)

(b)

(c)

▲ **Figure 1.2 Earth's surface is varied.**
Dynamic geologic processes deform the surface and sculpt it into varied landscapes. (a) Mount Everest is the highest point on Earth, rising to 8857 m above sea level. Spectacular mountains adorn the landscape of all continents and also reside out of sight on the floors of all of the oceans. (b) The Grand Canyon of the Colorado River is but one of many examples of deep gorges eroded by rivers. The canyon is more than 1675 m deep and carved through rock that formed as much as 1.7 billion years ago. (c) This Illinois landscape results from a combination of geologic processes acting over half a billion years of Earth history. These processes account not only for the flatness of the landscape but also for the fertile soils that support economically important agriculture.

This text emphasizes the physical geology of Earth. The objective is not only to describe how Earth looks but also why it looks the way it does and how geologists acquired this knowledge.

Geologists Work in Many Disciplines of Study

Geology integrates aspects of other sciences and molds these diverse concepts to the study of Earth. Each geologic discipline has its own name, but all of these studies fall broadly within the geologic sciences. Here are some examples:

- Geochemists bring rigorous chemical methods and theory to the study of rocks and water and also determine the ages of Earth materials.
- Geophysicists apply fundamental physics and measurement of physical properties of rock to understand how interior and exterior processes shape surface features.
- Paleontologists study fossilized organisms and are sometimes referred to as paleobiologists because they study the biology of past life.
- Planetary geologists work with astronomers to apply terrestrial geologic knowledge to explain materials from, and surface features visible on, other planets and moons.
- Geomorphologists study the origin of Earth's landscapes.
- Mineralogists study minerals that are the building blocks of rocks.
- Petrologists study rocks and the processes that form them.
- Stratigraphers study the common layered rocks that preserve a sequential archive of Earth history.
- Structural geologists study how rocks deform and are uplifted into mountains or depressed into lowlands, whereas the field of tectonics endeavors to understand the underlying causes for this deformation.
- Hydrogeologists and hydrologists study surface and ground water.
- Resource geologists explore for energy resources (e.g., coal, oil, gas, uranium) and other important resources, ranging from the constituents of concrete to metal ores and gemstones.
- Environmental geologists work to protect the environment by determining how human-caused contaminants move through soil and rock and how to clean up contaminated areas.

During the course of study with this text, you will learn a bit about each of these disciplines within geology.

Examples of Geologic Knowledge

Among the critical insights that await in chapters to come, you will learn how geologists know that the interior of the Earth, depicted in **Figure 1.3**, is made up of three concentric layers—the **crust**, **mantle**, and **core**—that differ from one another in composition and physical properties. The continental crust also differs from the crust beneath oceans, and this difference explains the contrasting high and low elevations of these two large-scale regions of Earth's surface. The mantle is Earth's thickest layer, and it consists of rocks that are distinct from the crust and mantle. The crust and uppermost mantle, down to about 100 kilometers, compose a strong outer lid on Earth's surface called the **lithosphere** (rocky sphere). The remaining upper mantle, down to about 440 kilometers, is the less rigid, squishier but mostly solid layer called the **asthenosphere** (weak layer). The core consists of metal and is divided into the inner, solid core and the outer, liquid core (Figure 1.3).

▼ **Figure 1.3 What the interior looks like.**
Although geologists do not voyage through Earth like fictional Jules Verne characters, they have determined the internal structure of the planet in surprising detail. The Earth consists of concentric shells beginning with a solid inner core and followed successively by molten outer core and solid mantle and crust. The composition and physical properties of these layers, and subordinate layers within them, vary considerably. You regularly see only the upper part of the relatively thick crust of continents, which differs from the thin crust of the oceans. The lithosphere is the firm, rigid outer skin of the Earth consisting of the crust and uppermost mantle, which overlies a weaker layer of upper mantle called the asthenosphere.

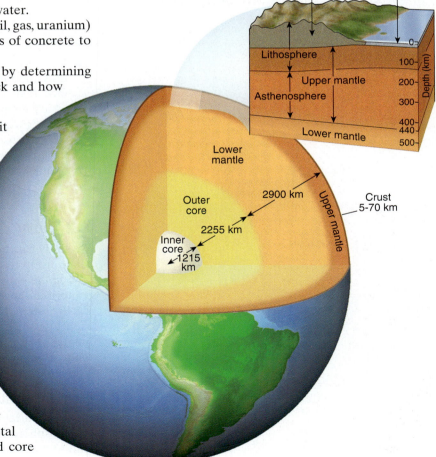

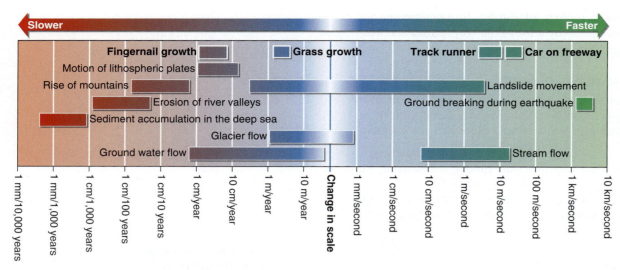

▲ Figure 1.4   Rates of geologic processes.
Many processes affecting Earth's surface take place rapidly, such as the flow of water in a stream or ground breaking during an earthquake (which is faster than the speed of sound). Most of the processes that have an enduring effect take place very slowly. Each vertical bar represents a tenfold increase in the rate of a process from left to right. Rates are a distance of movement per specified interval of time. In the metric measurement system there are 10 millimeters in 1 centimeter, 100 centimeters in 1 meter, and 1000 meters in a kilometer. Many of the listed processes occur at rates much slower than the growth of human fingernails. To account for the geologic features resulting from such slow but persistent processes, it is necessary to remember that Earth is 4.56 billion years old.

Geochemical studies of rocks show that Earth is approximately 4.56 billion years old. Geologists strive to understand the variety of processes active on the planet at present and how these processes varied over the enormous expanse of Earth's history.

The great antiquity of the planet requires an appreciation of the slow rates of many geologic processes, as illustrated in **Figure 1.4**. The rates of mountain building and processes that wear down mountainous landscapes are so slow that most aspects of these processes are not observed over a human lifetime. The slow rates of these geologic processes invite us to contemplate the vast length of time necessary to produce geologic features such as the Himalayan Mountains (Figure 1.2a), the Grand Canyon (Figure 1.2b), the Grand Tetons (Figure 1.1b), and the very rapid rates at which humans consume Earth resources (Figure 1.1c).

Not all geologic processes are so imperceptibly slow (see Figure 1.4). The volcanic eruption in Sicily (Figure 1.1a) altered the landscape as people watched, and large earthquakes can heave the ground surface many meters in an instant. This means that geologists need to understand the dynamics of both fast and slow Earth processes.

---

*Putting It Together–**What Is Geology?***

■ Geology is the study of Earth. Geology includes the study of Earth materials, the inner workings of the planet, and the origin and changes of surface features. Geology includes narrowly focused disciplines, many of which integrate knowledge from other fields of science, each addressing specific aspects of the planet.

■ Geologists not only strive to understand the cause and distribution of geologic processes on and within Earth today but also to interpret the 4.56-billion-year-long history of these processes on the planet.

1.2 Why Study Geology?

We study Earth partly out of curiosity about our planet and also to examine our relationship to the Earth system, which encompasses the dynamic interactions among the geosphere, hydrosphere, biosphere, and atmosphere. The geosphere comprises the solid Earth, the sturdy though ever-changing foundation of the Earth system. On a more practical level geologic topics impact virtually every aspect of our daily lives, including such diverse issues as the economy, environmental health, and climate. The following discussion considers just a few of the many reasons to study geology.

Geologists Study Earth Out of Curiosity

People invariably ask questions about their natural surroundings, and because the geosphere forms the foundation of our environment, even basic questions lead to contemplating geologic processes. Consider the question, "Why does a particular type of tree grow only in certain places?" Answering this question requires more than biologic knowledge because the growth of specific trees partly depends on climate (e.g., temperature and precipitation) and the available nutrients in the soil. The climate, in turn, partly depends on how moisture from the ocean moves landward in the atmosphere and interacts with the topography of continents molded by geologic processes. Soil forms by the complex breakdown of rocks involving biologic processes, action of water, and chemical reactions with atmospheric gases. The disintegration of different rock types leads to the formation of a variety of soils, each with different nutrient properties. The slope of the land influences the thickness of the soil and its suitability as a viable foundation for trees, because water runoff erodes soil in steep areas. The slope of the land, in turn, results from geologic processes that form and modify the shape of the landscape. Answering the question "Why do these trees grow here?" requires an understanding of multiple aspects of the Earth system. The geosphere is central to the Earth system, so geologic studies are also central to human curiosity about our surroundings.

Geologists Study Earth to Find Essential Resources

Geologists are motivated to understand the workings of the planet in order to locate and develop essential resources. Everything you use comes from planet Earth, from the food we grow to eat to the metal we use to build cars and the fuel that a car consumes. As humans need specific resources, such as iron ore for making steel, geologists apply their knowledge to find economically viable deposits. The resources bound in rocks and the geologic processes that concentrate them are illustrated in Chapters 2 through 6.

Water for drinking and irrigating crops is the most fundamental resource derived from the planet. We cannot survive without clean drinking water. How do we find vast amounts of water for a growing population? How do we know the water is drinkable? Geologists use knowledge of how water is stored and moves on the surface and in the shallow subsurface to locate useable supplies and to protect them from contamination. You will learn more about the geology of water in Chapters 16 and 17.

Geologists Study Earth to Reduce Hazards

Geologists work to diminish the detrimental impact of hazardous geologic processes. For example, an understanding of how different types of volcanoes erupt allows geologists to gauge potential hazards and respond to them. By examining the geochemistry of the fuming gases and the location and abundance of earthquakes below a volcano, it is sometimes possible to forecast the onset of an eruption and to determine the areas at risk in order to plan evacuations.

Geologists routinely map the distribution of different materials at Earth's surface. They also use maps to determine the suitability of a site for building or highway foundations or the instability of hillsides that could fail in disastrous landslides. **Figure 1.5** shows how geologists use their knowledge to produce maps that indicate the risk of damage from earthquakes and other hazardous phenomena. Geologic information is essential to land-use planning and responses to hazardous circumstances.

All natural geologic disasters are expressions of the dynamic nature of the planet. Understanding the geologic processes that cause these events reduces negative outcomes. This is especially important as current population growth makes it more and more difficult to keep humans out of harm's way.

▼ **Figure 1.5 Geologists map the risk of geological hazards.** Many geologic processes are hazardous to humans, and as cities grow it becomes ever more difficult to assure that people are safe from harm. Geologists use their knowledge of these processes to assess the hazards from different phenomena. This map outlines the severity of ground shaking as a hazard during a possible future strong earthquake in San Francisco, California.

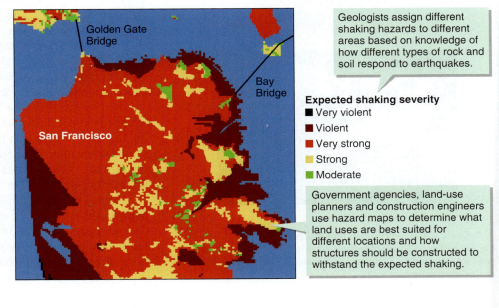

Geologists assign different shaking hazards to different areas based on knowledge of how different types of rock and soil respond to earthquakes.

Golden Gate Bridge

Bay Bridge

San Francisco

Expected shaking severity
- ■ Very violent
- ■ Violent
- ■ Very strong
- ■ Strong
- ■ Moderate

Government agencies, land-use planners and construction engineers use hazard maps to determine what land uses are best suited for different locations and how structures should be constructed to withstand the expected shaking.

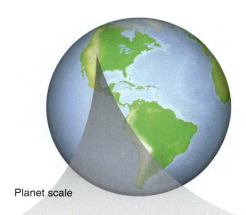

Planet scale

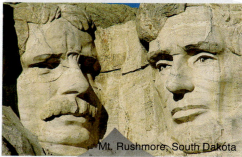

Mt. Rushmore, South Dakota

Landscape scale

Rock sample scale

1 cm

Microscopic scale

▲ **Figure 1.6 Geology at different scales.**
Geologic observations range from visualizing the whole planet with satellite images to observing the minute, such as the microscopic components of rocks.

Putting It Together—Why Study Geology?
■ Geologic studies are at the center of interdisciplinary efforts to understand the Earth system, which is composed of the geosphere, hydrosphere, atmosphere, and biosphere.
■ Geologic knowledge is required to locate and develop essential natural resources and to avoid or diminish the effects of hazardous phenomena.

1.3 How Do We Know ... How to Study Earth?

UNDERSTANDING GEOLOGIC SCIENCE
How Do We Know What We Know?
Science is the activity that furthers understanding of why things happen the way they do in the natural world by proposing new ideas and then putting them to the test. The tests involve making measurements and observations, which constitute data, to see if they are consistent with the new ideas. Geologists offer explanations of geologic phenomena, including the application of these explanations, to make informed decisions to reduce hazards and locate resources.

As shown in **Figure 1.6**, geologic studies range from the minute, such as a view of the arrangements of atoms within minerals composing rocks, to immense, such as visualizing whole continents with satellite images. Collection of these data at these different scales of observation employs a sophisticated battery of tools in addition to simple visual observation.

Data do not by themselves comprise science. Data collection is one step toward explaining natural phenomena in a process that also includes asking the right questions, seeking the answers, and making predictions about the phenomena that are tested by additional study.

SCIENTIFIC METHOD
How Is Science Done?
The **scientific method** is the inquiring process that examines and explains a problem or observed phenomenon. The scientific method is not a step-by-step, cookbook recipe for doing science but is a description of how scientists measure natural phenomena and rigorously test new ideas about how some part of the natural world works.

There is no single method to do good science. Instead, as depicted in **Figure 1.7**, scientists undertake many activities that collectively further our knowledge about the natural world. As the arrows indicate in Figure 1.7, certain activities repeat during the course of a research project. The process of acquiring knowledge requires asking many questions, which is central to the scientific activities depicted in Figure 1.7 and also reflected in the questions that serve as section titles throughout this text.

Scientists start by asking a question about something they do not understand. The question results from, or leads to, data collection. The data may include number of occurrences, location of occurrences, appearance of features or processes, and changes of the features over time.

Reviewing the data leads to stating a problem that defines the limits of what we need to know to answer the question. Library research is essential during the early stages of a project to find out what scientists already know about the problem. Not only does investigation of prior knowledge prevent duplicating already completed efforts, but it commonly leads to a modification or refinement of the problem that the new study focuses on.

After, or sometimes while, data are collected, scientists propose a **hypothesis**, which is an explanation of the problem that accounts for existing data and predicts additional phenomena that should exist if the hypothesis is correct. A hypothesis is always testable because the hypothesis makes predictions about a natural process that can be checked by collecting more data. If these tests refute the hypothesis, then scientists modify or abandon the hypothesis in favor of a new hypothesis that also explains the new results. If tests undertaken by different scientists repeatedly support the hypothesis, then it gains recognition as the logical tested explanation for the studied phenomena.

Communicating results is an important part of how scientific knowledge accumulates and is put to use (Figure 1.7). Most scientists present hypotheses, preliminary results, and initial conclusions at conferences convened just for these purposes. Not only does this allow scientists to stay abreast of what others are doing, but researchers get feedback from colleagues on whether the hypothesis is sound or whether data could be interpreted in a different way. Unless guarded for reasons of national security or proprietary interests of companies doing the research, most results are reported in journal articles and books. Other scientists typically review and critique these written communications prior to publication to help assure dissemination of only well-tested results and high-quality data. Nonetheless, the results may still be disputed or interpreted in other ways, such that later papers refute some published results or modify them into more robust hypotheses.

Scientific research does not always lead immediately to correct explanations of natural phenomena. Changes in widely accepted hypotheses do not represent the failure of science to provide the "right" result but rather the success of science to constantly improve the approach and eventually provide a meaningful solution.

APPLYING THE SCIENTIFIC METHOD IN GEOLOGY

How Do Geologists Do Science?

Geologists do not apply the scientific method the same way to all geologic problems. Many traditional scientific hypotheses are tested by laboratory experiments conducted under easily controlled conditions over periods of minutes to perhaps a few years. For many geologic questions, however, the number of variables is typically too large and the rates of processes much too slow (Figure 1.4) for direct laboratory experimentation. Time plays a critical role in geologic processes. Clearly, the experimenter must outlive the experiment, and geologists cannot reasonably run experiments that precisely duplicate processes that require centuries or millions of years in nature.

Earth itself is the laboratory for many geoscience research objectives. Rather than testing hypotheses by laboratory experiments, geologists assess critical questions through careful observation of what transpired on or within the planet. Hypothesis testing may include observing or measuring features in rocks or landscapes that the hypothesis predicts should be present.

Geologists use a variety of tools and methods to gather information about Earth. The simplest data are visual observations, commonly made in the field—such as the observations you made at the beginning of this chapter. Geologists apply sophisticated instruments in analytical laboratories to measure abundances of chemical elements in rocks. Laboratory and field measurements quantify physical properties of rocks, such as density and magnetic characteristics. Other experimental labs simulate the elevated temperature and pressure conditions within Earth's interior.

Computer simulations are important in geologic studies. Time can be "sped up" to reproduce Earth processes and variables can change one at a time to examine their effects on the outcome. Necessary assumptions or simplifications limit the reality of some simulations, and many are constrained by computational memory or the uncertainty of mathematical formulations of natural phenomena.

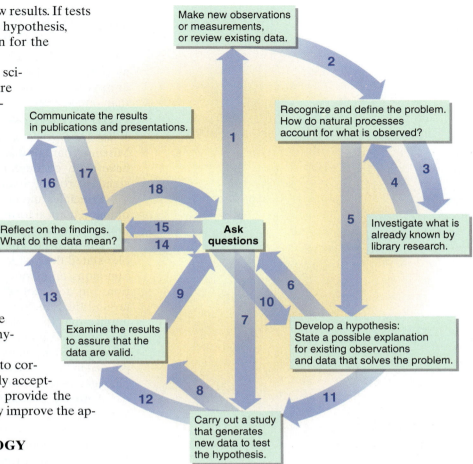

▲ **Figure 1.7 Activities reflect the scientific method.**
This chart lists the activities that scientists undertake in their research. The scientific method emphasizes the development and testing of hypotheses. The numbered arrows show one of many possible routes that a scientist might follow among these activities. Asking questions is central to scientific investigation.

Many geologic investigations apply multiple working hypotheses whereby, instead of formulating a single hypothesis, the geologist simultaneously considers all conceivable alternatives. Then the geologist identifies observations or data required to support a particular hypothesis while simultaneously refuting all other hypotheses. This approach reduces bias, and hypothesis testing progresses rapidly because many possibilities are evaluated simultaneously rather than one at a time.

UNDERSTANDING THE BIG PICTURE

What Are Theories and the Laws of Nature?

Missing from this discussion of scientific method are some other terms that you have likely encountered in your past science classes. It is important to understand the role of theories, laws, and principles to advance scientific knowledge.

Principles and **laws** of nature are synonymous terms referring to statements or mathematical formulas that always succeed in generalizing data and observations about data without necessarily offering explanations for why things work the way they do. For example, seventeenth-century astronomer Johannes Kepler stated the first law of planetary motion, which states that planets orbit the Sun along an elliptical orbit. The law derives from many undisputed observations but does not offer an explanation for why planets follow elliptical paths around the Sun. Some scientists dislike the phrase "laws of nature" because it implies that these are inviolate concepts. Laws, or principles, simply describe how humans observe nature to behave. This book describes many principles and laws applied by geologists, including the principle of uniformitarianism, explained in Section 1.4, which guides researchers to use observed processes to explain geologic features.

A **theory** is a widely applicable and generally accepted explanation for natural phenomena that explains all of the relevant data. A scientific theory is a rigorously scrutinized and tested concept and not a tentative explanation or opinion, as is commonly misunderstood in statements like "it's only a theory." Where laws and principles are generalizations of how nature is observed to work, theories are explanations of why nature works this way. For example, the theory describing gravity forces between objects offers an explanation for Kepler's law of planetary motion. Theories sometimes arise from successfully tested hypotheses, and many theories incorporate several such hypotheses. All theories undergo further testing by application in new studies and may eventually, although rarely, be found as unsatisfactory. Unless data disprove or lead to modification of a theory, that theory serves as the valid accepted explanation of the observed phenomenon. Section 1.5 introduces the theory of plate tectonics, which is the most important unifying idea in geology and is, therefore, threaded as a common theme through this entire book.

In general, principles and theories call for seeking the simplest explanation for natural phenomena without bringing in unneeded explanations if the tested ideas at hand are adequate for the job. Scientific theories also exclude supernatural explanations, as they cannot be observed or measured by scientific procedures—this is what *supernatural* means.

INSIGHT

How Do We Know?

The knowledge of Earth conveyed in the following chapters results from applying the scientific method. This body of knowledge results from two centuries of observation, experimentation, critique, challenge, reformulation, and testing of hypotheses documented in countless pages of reviewed scientific journals and books. What each generation of scientists takes for granted as established "fact" is actually the collective essential contribution of previous generations of hardworking researchers who proposed, tested, and established these ideas as firm knowledge. Throughout the book we endeavor to convey an appreciation for how geologists have come to know what they know. It is impractical to assess every concept, but each chapter includes the scientific background for at least one important idea.

Putting It Together–How Do We Know … How to Study Earth?
■ The scientific method integrates asking questions, proposing explanatory hypotheses, and testing hypotheses to understand natural phenomena.
■ Geologists use laboratory experiments less often to test hypotheses than do other scientists because the variables governing Earth processes are usually too many and the rates of the processes are typically too slow for direct experimental analysis.
■ Principles (or laws) are generalizations about how nature is observed to work, whereas theories offer well-tested and accepted explanations, not offhand hunches, of why the natural system works this way.

1.4 What Does the Principle of Uniformitarianism Mean?

There are many guiding principles in geology, including those from other disciplines that are integrated within geology, such as Sir Isaac Newton's laws of motion and gravity. Geologic studies use a concept that originated in geology but applies to all science. This very important concept is the **principle of uniformitarianism**. The principle states that observations of both processes and results in case studies apply to interpret other results where the process was not observed.

The Original View of Uniformitarianism

Historical context is important to understand uniformitarianism. The idea of long-term uniformity in natural processes originated at least 2000 years ago among Greek natural philosophers. The idea became a focus in geology in the late eighteenth century with the work of an extraordinarily curious and observant Scottish physician and gentleman farmer, James Hutton. Englishman Charles Lyell formalized the concept in the first physical geology textbook (*Principles of Geology*, three volumes published 1830–1835), although the term *uniformitarianism* did not appear until 1875. Initially, the principle argued for explaining geologic features only in terms of observed processes and at rates that are measured today. In other words, if you can understand the geologic processes responsible for Earth materials and features you presently see, then you infer that similar materials and features found in ancient rocks are the result of these same processes.

Hutton, Lyell, and geologists following in their footsteps today appreciated that no extraordinary means are required to explain geologic features. For example, great thicknesses of sand, gravel, and mud later consolidated into rock are explained by the observed slow, but persistent, processes of erosion, transport, and deposition of sediment by flowing water and wind operating over immense geologic time and do not require rapid cataclysms.

As another example, compare rippled sand shown in **Figure 1.8**, formed by the to-and-fro swash of water on a sandy beach, with the identical feature seen in a consolidated rock. Clearly, it is possible to infer the origin of features in the ancient rock by a process readily observed today on the beach. Therefore, you are applying the principle of uniformitarianism to infer the origin of the ripple marks in the ancient rock without having observed the process that formed them.

Likewise, although nobody has witnessed the formation of a mountain range thousands of meters high, there are many observations of ground surfaces heaved upward by many meters during single earthquakes as rock layers break and crumble. **Figure 1.9** illustrates an example of such instantaneous uplift during a historic earthquake. With many of these observations, it becomes readily accepted that mountains of broken and otherwise mangled rocks are the result of rock movement during innumerable earthquakes over long intervals of time.

▲ **Figure 1.8 An application of uniformitarianism.**
Surging waves on a South Carolina ocean beach move sand grains back and forth to form conspicuous ripple marks. The surface of a rock layer hundreds of millions of years old shows similar ripple marks, and the rock consists of cemented sand grains similar to those found on the modern beach. The similarities of the modern and ancient features indicate that the ancient rock formed in an environment similar to a beach. These photographs illustrate an application of the principle of uniformitarianism, whereby geologists interpret features preserved in ancient rocks in terms of observed processes.

Mountains uplifted during many earthquakes over a long time interval

Land surface broken and raised up 2 meters during earthquake

▲ **Figure 1.9 Uniformitarianism explains mountains by uplift during repeated earthquakes.**
Uplift during an earthquake in Taiwan in 1999 raised part of an athletic field and running track by two meters. Adding up the uplift during many earthquakes over long intervals of geologic time explains the elevations of mountains.

The Modern View of Uniformitarianism

Modern geologists do not apply uniformitarianism in the same fashion as Lyell and his contemporary scientists. Twentieth-century geologists questioned if uniformity of process should also require that the rate of processes be uniform through time and limited to the values measured during the history of geologic study. Is it possible for processes to have been active in the past that humans have not witnessed or for such processes to have operated at different rates?

The immensity of geologic time plays an important role in answering these questions because geologic studies of active processes have been ongoing for less than 4 ten-millionths of Earth history. Certainly many geologic processes occur episodically (e.g., collisions of meteors and comets with Earth) or over a wide range of scale (e.g., volcanic eruptions, floods). Are two centuries of geologic investigation and five millennia of recorded history sufficient for observing the rare or extreme event? If the primordial Earth was hotter, could processes driven by thermal energy have taken place at faster rates? Would not rates of erosion have been higher before the appearance of rooted plants on land beginning 400 million years ago? With these questions in mind, one should not place too many restrictions on applying uniformitarianism, especially with regard to the rates and conditions of processes that might change over time.

The principle of uniformitarianism really states a more fundamental concept in natural science. Scientists use examined cases, where they observe both process and result, as a guide to explain features where only the result, and not the active process of formation, is seen. An ancient rippled sandstone (Figure 1.8) is only one of a multitude of examples where geologists see the result of a process without actually seeing the process at work. Examined causes are not limited to events witnessed in nature. The results of laboratory experiments and mathematical calculations based on known natural phenomena also guide our interpretations of unseen processes concealed from view beneath the surface of Earth or that occurred in the ancient past. These experiments and calculations permit scientific simulation of processes that cannot be directly observed or that operate too slowly to study in nature.

If uniformitarianism guides interpretation of geologic history, then can it also be used to forecast future events? In a general way, yes, because scientists infer that the same natural processes will operate in the future as they understand to be active today and in the geologic past. This does not mean, however, that scientists can predict exactly when or where a geologic event will occur. For example, geologists study the behavior of past eruptions of a volcano from studying the resulting deposits. The geologists then present hazard assessments that future eruptions will feature similar behavior as documented by the deposits of past eruptions. This application of uniformitarianism does not, however, mean that the time and severity of the next eruption can be forecast far in advance.

*Putting It Together–**What Does the Principle of Uniformitarianism Mean?***

■ The principle of uniformitarianism states that examined cases of process and result guide interpretation of results where the process was not witnessed. Ancient geologic features are interpreted by understanding active processes that are readily studied.

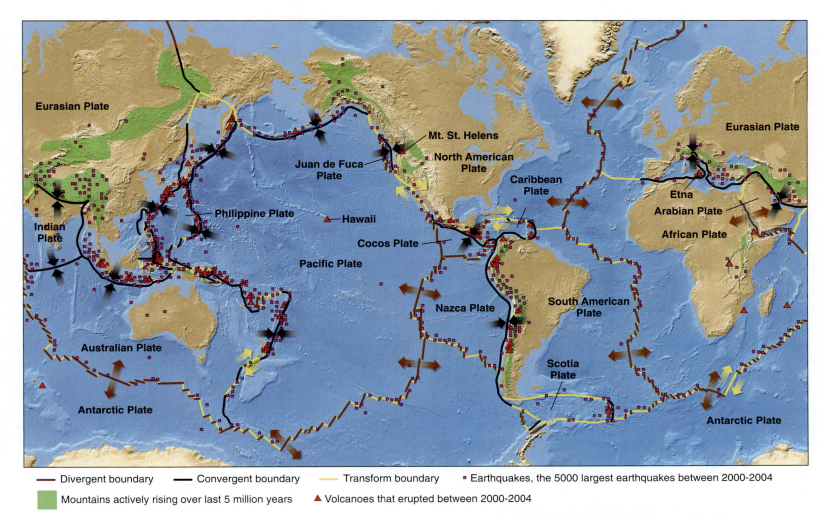

— Divergent boundary   — Convergent boundary   — Transform boundary   ▪ Earthquakes, the 5000 largest earthquakes between 2000-2004

🟩 Mountains actively rising over last 5 million years   ▲ Volcanoes that erupted between 2000-2004

▲ **Figure 1.10 The world composed of plates.**
This world map shows the outlines of the lithospheric plates whose motion is explained by the plate tectonics theory. Figure 1.11 explains the three types of boundaries between plates, which depend on the type of motion between the neighboring plates. Notice that nearly all historically recent active volcanoes, earthquakes, and geologically recent mountain building occur near plate boundaries.

1.5 What Is the Theory of Plate Tectonics?

Every scientific field profits from revolutionary advances that simultaneously explain and integrate many observed phenomena while also redirecting the focus of new research. Examples are the theory of relativity in the field of physics, resolving the structure of the invisible atom in chemistry, and deciphering the genetic code in biology. The equivalent revolution in geology occurred relatively recently, during the 1960s, and is enveloped in the most important geologic theory, the theory of **plate tectonics**, which explains a vast array of geologic processes and features by the motion of separate plates of lithosphere.

To appreciate the evidence for plate tectonics, you first need to develop sufficient geologic knowledge to understand the strength of the supporting arguments. Here you will learn just the broad features of the theory, whereas Chapter 12 fully develops the theory using knowledge gained from the intervening chapters.

Describing a Plate

Plate tectonics theory explains that the outer shell of Earth, the lithosphere (see Figure 1.3), is not seamlessly continuous but is broken into discrete, roughly 100-km-thick slabs, called **plates**. The lithospheric plates contain both continental and oceanic crust and part of the immediately underlying mantle. **Figure 1.10** depicts the outlines of plates on Earth today. Plates move slowly relative to one another and change in size over geologic time. The plates are rigid and strong, in contrast to the underlying weaker asthenosphere (Figure 1.3).

Plates move away from one another at a divergent boundary, resulting in shallow earthquakes and abundant volcanic activity where partly melted asthenosphere rises and solidifies to form new lithosphere in the gap. Most divergent plate boundaries coincide with mid-ocean ridges, which snake through the ocean.

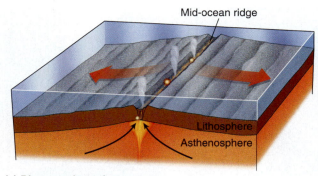

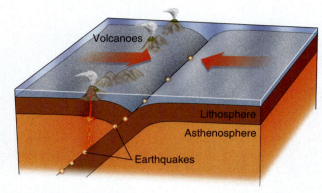

(a) Divergent boundary

Plates move toward each other at convergent boundaries and one plate subducts beneath the other. Earthquakes occur over a wide range of depths where the rigid lithosphere thrusts downward into the weaker asthenosphere. Volcanic activity is triggered above the downward-moving slab to produce lines of volcanoes.

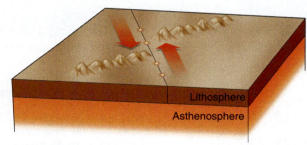

(b) Convergent boundary

Plates slide past one another at transform boundaries. The shearing of one plate past its neighbor generates abundant earthquakes but is not typically associated with volcanic activity.

(c) Transform boundary

▲ **Figure 1.11 What happens at plate boundaries.**
The motion between adjacent plates describes the type of boundary between plates.

Motion at Plate Boundaries: *See how the plates move along their boundaries.*

Types of Plate Boundaries

Plates collide, move apart, or slide past one another at plate boundaries. These motions cause deformation of the lithosphere described by the term **tectonics**, the study of the causes of rock deformation. **Figure 1.11** shows that there are three types of conceivable relative motions between two plates at their mutual boundary.

Plate edges stretch and separate along **divergent plate boundaries** (Figure 1.11a). The Mid-Atlantic Ridge, for example, marks the divergence of the North American and Eurasian plates in the center of the Atlantic Ocean (Figure 1.10). As the plates separate, the resulting open gash simultaneously fills with molten material rising from the asthenosphere. The molten material solidifies to form new seafloor lithosphere along a line of submarine volcanoes. This process causes the Atlantic Ocean to widen by about 5 centimeters each year as new lithosphere forms along the mid-ocean ridge.

Plates collide nearly head on into one another at **convergent plate boundaries** (Figure 1.11b). Where the plates converge, one plunges into the deeper mantle in a process called **subduction**, while the overriding plate experiences volcanic activity and buckling that uplifts tall mountain ranges. Earthquakes, mountains, and active volcanoes in the Pacific Northwest result from a convergent boundary just offshore of Oregon and Washington (Figure 1.10). On the global scale, the creation of lithosphere at divergent boundaries is balanced by its destruction at convergent boundaries, so that the overall size of the planet does not change.

Plates slide past one another along **transform plate boundaries**, without creation or destruction of lithosphere (Figure 1.11c). An example of a transform boundary is the San Andreas fault, which runs along nearly the entire length of California and marks where the Pacific plate slides past the North American plate (Figure 1.10).

Plate tectonics explains the tendency for earthquakes and active volcanoes to concentrate in long, narrow belts (Figure 1.10). The breaking and folding of rocks are focused at plate boundaries while there is much less deformation within plates. The largest and most damaging earthquakes happen at or near convergent and transform boundaries, whereas divergent-boundary earthquakes are small and mostly occur in the center of oceans away from cities and therefore rarely cause damage. Convergent and divergent plate boundaries are associated with processes that promote melting in the mantle, thus focusing volcanoes in narrow belts (Figures 1.11a, b). You will examine these melting processes in Chapter 4.

Hot Spots

If volcanoes relate to plate-boundary processes, how do geologists explain the volcanic Hawaiian Islands found near the center of the Pacific plate (Figure 1.10)? Midplate volcanoes are **hot spots**, where molten material rises from deep in the mantle below the moving lithosphere by processes not explained by plate tectonics. **Figure 1.12** shows how large volcanic islands build up as the Pacific

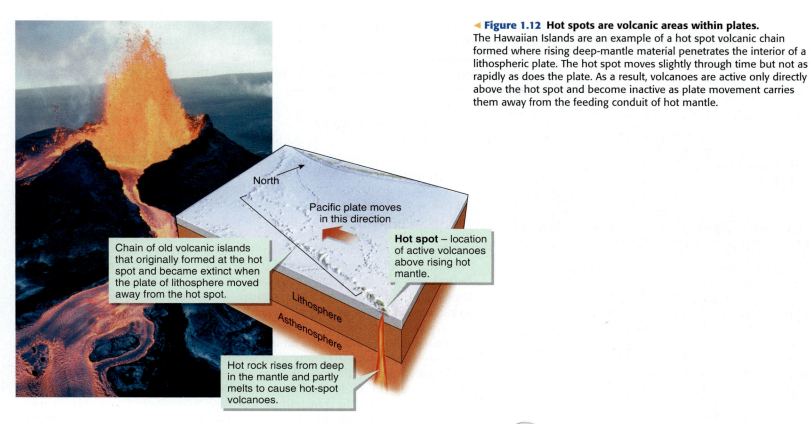

◀ **Figure 1.12 Hot spots are volcanic areas within plates.**
The Hawaiian Islands are an example of a hot spot volcanic chain
formed where rising deep-mantle material penetrates the interior of a
lithospheric plate. The hot spot moves slightly through time but not as
rapidly as does the plate. As a result, volcanoes are active only directly
above the hot spot and become inactive as plate movement carries
them away from the feeding conduit of hot mantle.

North

Pacific plate moves
in this direction

Hot spot – location
of active volcanoes
above rising hot
mantle.

Chain of old volcanic islands
that originally formed at the hot
spot and became extinct when
the plate of lithosphere moved
away from the hot spot.

Lithosphere

Asthenosphere

Hot rock rises from deep
in the mantle and partly
melts to cause hot-spot
volcanoes.

plate moves over the Hawaiian hot spot. A volcano becomes extinct as plate
motion carries it away from the hot spot of rising hot mantle, and a new volcano
appears in its wake. Over a long period of time, the plate movement over the
hot spot generates a chain of volcanic islands and submerged volcanoes.

Hot Spot Volcano Tracks: *See how an island chain is formed by
plate motion across a hot spot.*

Applying Plate Tectonics

Perhaps the most compelling evidence before you at this point to support plate
tectonics is the ability of the theory to explain the restricted distribution of most
earthquakes, active volcanoes, and young mountain belts (look at Figure 1.10).
As you progress through your study of rocks, the internal workings of the planet, rock deformation, and the evolution of landscapes in later sections of the
book, you will see how plate tectonics provides the basis for understanding far
more than simply earthquakes and volcanoes. It is the ability of the theory to
satisfy all of these conditions simultaneously that makes it the centerpiece of
geologic studies.

Putting It Together–What Is the Theory of Plate Tectonics?
- The plate tectonics theory states that the outer shell of Earth, the lithosphere, is divided into plates that move toward, away, or past one another.
- The interaction of plates at their boundaries accounts for the distribution of
  earthquakes, volcanoes, and actively growing mountain belts and provides a
  basis for interpreting most geologic processes and products.

1.6 How Does the Concept of Work Apply to Earth?

Earth is an active place. Water flows in streams and erodes rock particles that
are carried to the ocean, where crashing waves move the particles onto sandy
beaches. The slow shifting of the lithospheric plates across the entire surface of
Earth leads to ground-shaking earthquakes, violent volcanic eruptions, and

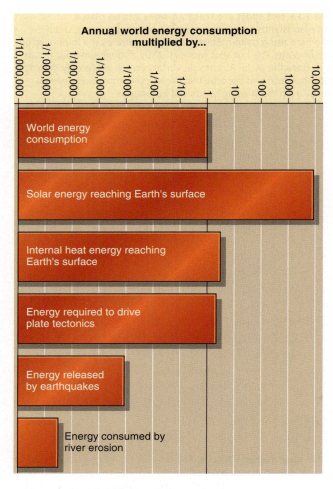

Annual world energy consumption multiplied by...

1/10,000,000
1/1,000,000
1/100,000
1/10,000
1/1000
1/100
1/10
1
10
100
1000
10,000

World energy consumption

Solar energy reaching Earth's surface

Internal heat energy reaching Earth's surface

Energy required to drive plate tectonics

Energy released by earthquakes

Energy consumed by river erosion

▲ **Figure 1.13 Visualizing energy numbers.**
This bar graph illustrates geologically important energy sources and uses by comparison to the annual energy consumption by humans. For example, solar energy reaching Earth's surface is almost 10,000 times greater than world energy consumption, whereas internal heat energy reaching Earth's surface is only three times greater.

gradually rising mountains. These are but a few of the observed Earth processes where material mass moves from one place to another. These motions are evidence of work, just as you do work to move an object from one place to another. This text, as suggested by the title, explains the many types of work that occur within Earth and on its surface.

Work Requires Energy

Energy is the measure of the ability to do work. This idea may be apparent from everyday experiences. Movement of an automobile is evidence of work. Energy required for that work is provided by burning gasoline. If the automobile runs out of gas, no more work is done and the vehicle stops moving. You obtain energy from the food that you eat, which biochemical processes in your body convert to energy. Your body requires food as "fuel" and the accumulated body energy "burns" during exercise, which is a form of work.

Energy is most obvious when motion takes place, but energy is also stored. A battery is a familiar example of stored energy. Chemical reactions inside the battery produce electricity. If wires are connected to the battery, then the electricity lights a bulb, causes a toy to move, or produces other kinds of work. The work consumes the battery energy until it drains completely.

There is stored energy, motion energy, and there is radiant energy like the visible light or detectable heat emitted by a light bulb. Energy transfers from one place to another and from one form to another. The stored chemical energy in the battery transfers through wires as electrical current, which may transform to motion energy in the toy or be converted to heat and light energy in the bulb. The heat also serves to transport energy away from the bulb, causing air molecules to vibrate and produce the warmth that you feel when placing your hand close to the light.

Work on and within Earth requires energy. Energy is stored, it transfers to cause motion and do work, and it moves as heat. **Figure 1.13** compares natural energy sources and energy expenditures with human energy consumption. While examining and understanding the active work represented by geologic processes described in this text, you will also consider the energy required to do that work.

Heat Drives Geologic Processes

Heat represents the most important form of energy transfer for geologic processes. Heat from the Sun fuels Earth-surface processes, like blowing wind, ocean waves, and the evaporation and precipitation of water. Solar heat results from reactions within the Sun, so in a sense the Sun is like a huge battery of stored energy that constantly radiates heat and light energy through space.

The total energy output of the Sun each second is equivalent to the energy in about 4×10^{24} (4 followed by 24 zeros) 100-watt light bulbs. Not all of that energy reaches Earth, because the heat moves outward in all directions from the Sun and Earth is just a small speck in space about 150 million kilometers away from the Sun. Not all of the energy radiating from the Sun is heat. Some of the energy is visible light and some of it is ultraviolet radiation that causes sunburn. Still, the warmth you feel on a sunny day clearly indicates the arrival of heat energy from the Sun. The amount of solar energy received by each square meter of Earth's surface is about one-tenth the amount of energy that emerges as light and heat from a 100-watt bulb. Some of the energy is consumed by work on Earth's surface and in the atmosphere, but most of it radiates back into space.

Natural radioactivity produces heat inside Earth's interior. Chapters 7 through 10 thoroughly examine this internal heat and the work it fuels. The heat received from the Sun is about 4000 times larger than the heat energy emitted from within Earth through an average area of the surface. This is why heat from within Earth is rarely obvious unless you visit an unusually hot place, like a volcano. Nonetheless, this energy source is sufficient to power the work done by plate tectonics for more than 20 billion years, which is far longer than the history of the planet.

How Heat Causes Motion

Heat is a form of energy that you feel, but how does it cause the motion that defines work? To answer this question requires further consideration of how heat transfers from place to place, as illustrated in **Figure 1.14**. The heat you feel on your skin on a sunny day moves as waves of energy called **radiation**. Light is another example of radiation. The painful burn felt when touching a hot stove is conduction. **Conduction** is the transfer of heat from a hot object to a cold object. Heat in the hot object is caused by energetic collisions of rapidly vibrating molecules. These rapidly vibrating molecules in the hot object collide with the less active molecules in the cold substance, setting them in rapid motion, too. The rapid molecular vibrations are detected as heat, which transfers from the hot object to the cold one. **Convection** is the simultaneous transfer of heat and mass.

Figure 1.15a illustrates how conduction of heat from a stove burner into a pot of liquid causes the liquid to move by convection. The liquid heated by conduction at the bottom of the pot expands as its molecules vibrate more energetically. The expanded hot liquid is less dense than overlying, cooler, and unexpanded liquid. The denser liquid, therefore, sinks toward the bottom of the pot, which displaces the less dense hot liquid to the top of the pot. The motion results from transfer of heat energy into the liquid and the moving mass of liquid carries heat with it. Once exposed to colder air, the liquid at the surface cools by conduction, becomes denser, and sinks while more hot liquid in the bottom of the pot rises to replace it.

Convection in nature resembles convection on the stovetop. Where solar heating warms the atmosphere the air expands and rises while cooler air sweeps in to replace it, as shown in Figure 1.15b. This motion in the atmosphere is wind. Wind moving over the oceans produces waves. Liquid water evaporates where the Sun warms air, ground, and water surfaces. The warm air then moves upward by convection, where the air cools and the water condenses back to liquid. The liquid water falls as rain or snow that flows on the surface as streams, which do work eroding surface rock and soil. Atmosphere convection, therefore, is an important observed process for shaping Earth's surface.

Convection also causes the motion of plates, as shown in Figure 1.15c. Cold, dense rock near Earth's surface sinks downward, and less dense rock expanded by heating in the interior rises upward. Convection-driven motion of the mostly solid mantle is extremely slow compared to the convection revealed by a bubbling pot on the stove or a windy day. The motion is sufficient, however, to move the rigid lithosphere in much the same way as you may notice patches of grease moving about on the upper surface of hot soup on the stove. While convection of rock may seem implausible, you will learn the scientific rationale for this phenomenon in Chapter 10 along with convection in the molten outer core, which generates Earth's magnetic field.

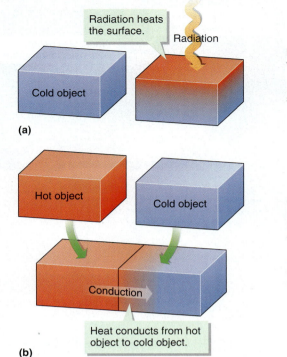

(a)

(b)

Radiation: Energy transferred as waves.

Example: Heat radiates through space from the Sun to Earth.

Conduction: Heat transferred by rapidly vibrating molecules when a hot object is placed against a cold object.

Example: Heat conducts from a stovetop burner to the bottom of a pan.

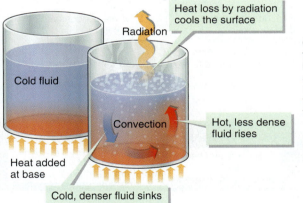

(c)

Convection: Simultaneous transport of mass and heat caused by sinking of cold, dense material and rise of hot, less dense material. Radiation and conduction typically convey heat in or out of the convecting material.

Example: Heat convects when fluid turns over in a liquid heated from below on a stove.

▲ **Figure 1.14 How heat transfers from place to place.**

Heat conducts from a stovetop burner to a pot of soup. The heated liquid is less dense and rises upward as colder, denser liquid sinks to replace it. This circulation is convection and not only moves the soup up and down in the pot, but also transports heat upward with the warmer liquid.

Soup conducts and radiates heat into the air.

Warm, less dense soup rises to the surface.

Soup cools at the surface, becomes denser and sinks to the bottom.

Heat source

(a)

Where solar heating warms the atmosphere, the air expands and rises while cooler air sweeps in to replace it, producing a convection cell of air circulation. The air movement near the surface is wind. Water moves with the convecting air and evaporates from liquid into vapor and then condenses from vapor into liquid as the temperature changes.

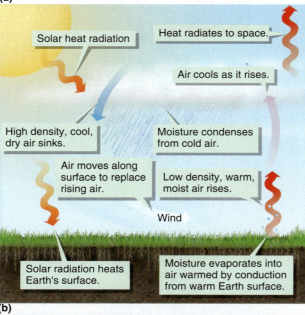

Solar heat radiation

Heat radiates to space.

Air cools as it rises.

High density, cool, dry air sinks.

Moisture condenses from cold air.

Air moves along surface to replace rising air.

Low density, warm, moist air rises.

Wind

Solar radiation heats Earth's surface.

Moisture evaporates into air warmed by conduction from warm Earth surface.

(b)

Rocks also convect. Cold, dense lithosphere sinks at convergent plate boundaries. Sinking lithosphere pulls the plates apart at divergent plate boundaries. Hot spots are places where hot, low-density mantle convectively rises to the surface. Plate tectonics and hot spots, therefore are convection.

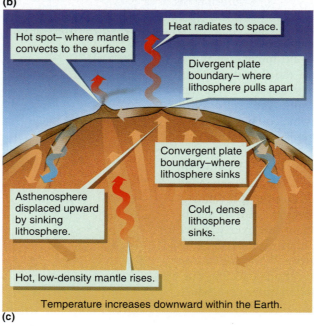

Hot spot– where mantle convects to the surface

Heat radiates to space.

Divergent plate boundary– where lithosphere pulls apart

Convergent plate boundary–where lithosphere sinks

Asthenosphere displaced upward by sinking lithosphere.

Cold, dense lithosphere sinks.

Hot, low-density mantle rises.

Temperature increases downward within the Earth.

(c)

◀ **Figure 1.15 Visualizing how convection works.** Convection is a process involving simultaneous transport of heat and matter. Convection is witnessed on a stovetop, occurs in atmospheric circulation, and occurs within Earth.

Landscapes Have Potential Energy

Convection within Earth drives plate tectonics and plate tectonics, in turn, causes uplift of mountains. Considerable work is done to raise mountains because the force of gravity pulls all mass toward Earth's interior. The higher rocks are uplifted, the greater the work required in overcoming the downward gravitational pull.

The effect is similar to lifting a heavy box from the floor to a shelf, as illustrated in **Figure 1.16**. The higher the shelf you chose to lift the box to, the more work you do. Once placed on the shelf, the box just sits there but it actually has a form of stored energy called **potential energy**. Potential energy is the energy an object possesses because of its elevation and weight. The potential energy of the box on a high shelf is greater than the energy of the same box on a lower shelf or on the floor. The difference in energy between the box on the floor and the box on a high shelf is equal to the work done to move the box from the floor to the shelf. The downward pull of gravity tends to move objects from positions of high potential energy to positions of low potential energy. If the box is not secured on the shelf, then it may fall downward to the floor. The potential energy stored when the box was stationary on the shelf converts to motion energy as the box falls and also to sound-wave energy and a minor amount of heat when the box hits the floor.

Differences in potential energy between high elevations and low elevations in landscapes drive processes on Earth's surface. Rocks high on mountain peaks have higher potential energy than those on valley bottoms. The energy difference causes rocks to fall or slide downslope, sometimes in impressive movement of large masses of material that are commonly called landslides. The mass movement represents the conversion of the potential energy of elevation to motion, sound, and heat energy analogous to a box falling from the shelf. Likewise, water on Earth's surface has a relatively high potential energy at high elevations, and this energy decreases as the water flows downhill in stream channels to the ocean. The potential energy converts to motion energy, the noise of the flowing water, a little bit of heat, and also allows the stream to do work. Erosion and transport of rock material to form the stream channel are the evidence of the work done by the stream. Each year streams erode about 14 cubic kilometers of sediment from the land surface, which is equivalent to filling up railroad cars on a train that would encircle Earth 34 times at the equator. Clearly, considerable work is done to move this much mass for distances as great as several thousand kilometers. The annual energy consumed by stream work

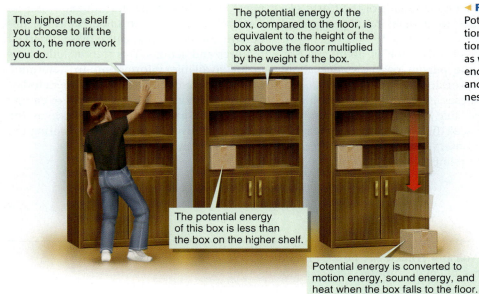

The higher the shelf you choose to lift the box to, the more work you do.

The potential energy of the box, compared to the floor, is equivalent to the height of the box above the floor multiplied by the weight of the box.

The potential energy of this box is less than the box on the higher shelf.

Potential energy is converted to motion energy, sound energy, and heat when the box falls to the floor.

◄ **Figure 1.16 Visualize potential energy.**
Potential energy is the energy an object possesses because of its elevation and weight. The downward pull of gravity moves objects from locations of high potential energy to positions of low potential energy, such as when an unstable box falls from the shelf onto the floor. The difference in potential energy of the box between its positions on the shelf and on the floor is equal to the motion, sound, and heat energy witnessed when the box falls.

to erode Earth's surface is, however, less than one-half of one percent of the energy released each year by earthquakes, which continually deform Earth's surface to create greater potential energy differences that cause even more stream erosion.

Putting It Together–*How Does the Concept of Work Apply to Earth?*

- Movement of mass within Earth, on Earth's surface, and in the atmosphere is evidence of work.
- Work requires energy.
- Geologically important energy sources are heat from the Sun and internal heat resulting from natural radioactivity.
- Heat energy is transferred from place to place by radiation, conduction, and convection. Convection involves simultaneous movement of mass and heat and powers geologically important motion in the gaseous atmosphere and in the solid and liquid parts of Earth's interior.
- Potential energy is an important form of stored energy that objects possess because of their elevation. Conversion of potential energy to motion energy occurs when masses move from high elevation to low elevation.

Where Are You and Where Are You Going?

You now have a basis for studying geology. You know the types of study encompassed within geology and how these studies help us understand the Earth system and some of its beneficial resources and perilous hazards. You have also been introduced to the tools that geologists use and can anticipate seeing the outcome of research efforts throughout the chapters that follow.

Scientists neither randomly collect data and observations nor make up explanations for natural phenomena. Scientists design their work to test specific hypotheses and develop guiding theories and principles. You have been introduced to the principle of uniformitarianism and the plate tectonics theory. Uniformitarianism states that the observed link between processes and results in a few cases can be used to explain results in cases where the process was not observed. Plate tectonics describes the slow motion of surface slabs to produce earthquakes, volcanoes, mountain ranges, and other geologic features. In the chapters ahead, you will develop an even stronger appreciation of how these concepts explain geologic phenomena.

Earth is a dynamic, active planet. Active motion requires energy. Geologic processes are powered by heat energy from the Sun and from within the planet. These energy-fueled processes reveal how Earth works.

The chapters that follow are arranged into four parts. Before delving deeply into understanding routinely observed geologic features, you must first understand what the planet is made of. The first part of the book (Chapters 2–7) emphasizes the solid Earth materials that are seen at the surface. The second part (Chapters 8–10) turns attention to sleuthing out the materials hidden from view within the planet. This knowledge will bring to light appreciation of internal processes that drive external deformation through plate tectonics. Deformation at and near the surface and formulation of the plate tectonics theory compose the third part of the book (Chapters 11–13). The final part (Chapters 14–20) combines knowledge of Earth materials with dynamic deformation processes and integrates this knowledge with the other components of the Earth system in order to understand the origin of landscapes.

Above all, this book is about science. Science is the pursuit of knowledge about the natural world, using observation to propose hypotheses and to test them. Geology is the science that seeks knowledge about Earth. We seek first to understand what is there—the observed rocks, soils, mountains, streams, shorelines, glaciers, erupting volca-noes, land shaken by earthquakes, and so forth. Next, we seek to understand the underlying processes that cause these observed features and actions. This step includes analyzing available information to determine how things came to be as we see them today. We must also consider the sources of energy to drive these processes. In essence, as the title states, this book is about answering the question "How does Earth work?"

Active Art

Motion at Plate Boundaries. See how the plates move along their boundaries.

Hot Spot Volcano Tracks. See how an island chain forms by plate motion across a hot spot.

Confirm Your Knowledge

1. What is geology? What is the main difference between physical and historical geology?

2. Give two examples of fast Earth processes and two examples of slow Earth processes.

3. In many scientific disciplines, hypotheses are tested by direct laboratory experiments. List two reasons why this is not always possible in geology.

4. What type of explanation is not included in scientific theories? Why?

5. Uniformitarianism has changed since it was originally developed. Explain its original meaning, its modern meaning, and how it has changed.

6. List and describe the three types of plate boundaries. Give an example of each.

7. List and describe the three types of energy. Give an example of each.

8. List the three types of heat transfer. Give an example of each.

9. Each year streams erode about 14 cubic kilometers of sediment. This amount is equivalent to filling up a train of railroad cars that would encircle Earth 34 times at the equator. If this amount of sediment was spread over your home state, how thick would it be?

10. Earth is composed of three concentric layers. What are they, and how do they differ?

11. How old is Earth? What are some of the changes it has experienced?

12. Geologists study Earth to find essential resources. What are some of these resources?

13. Many students learn that the scientific method consists of formulating a hypothesis, testing that question, and developing a theory based on results. In actual practice, the scientific method is much more complex. Explain how.

14. What is the difference between a law and a theory? List an example of each.

15. Energy drives earth processes and when utilized, gives off heat. What are the 2 major sources for energy in and on Earth? How do they differ from potential energy?

Confirm Your Understanding

1. Write out an answer for each question in the Chapter Outline for the chapter sections assigned by your instructor.

2. The text illustrates how answering a question like "why do these trees grow here?" requires an understanding of climate, soil formation and slope development. What understandings would you need to answer the question, "Where should we build a dam to provide water for our town?" List at least three items you would need to understand.

3. If you were going to explore a planet orbiting around a star other than our Sun and you wanted to determine if plate tectonics was occurring there, what would you look for? What instruments would you bring?

4. Many geologic processes operate on such long time scales, beyond the lifetime of the investigator and this complicates the scientific method. What are some of the ways that geologists accommodate time in their research?

5. You find fossil horseshoe crabs in a rock. When you take it to a local geology museum you find out they are 400 million years old. What can the principle of uniformitarianism tell you about their lifestyle and the environment of deposition?

6. How do hot spots, which are not explained by plate tectonics, help to confirm that lithospheric plates are moving?

7. Given what you know about convection, consider how ocean circulation must work if the deep waters are known to be cold and surface waters are warm in the equatorial regions but cool down when transported to cold polar regions.

8. From the nine reasons to consider as to why you should study Earth, which is the most important to you? Why is it personally important?

Key Terms

asthenosphere (p. 5)
conduction (p. 17)
convection (p. 17)
convergent plate boundaries
   (p. 14)
core (p. 5)
crust (p. 5)
divergent plate boundaries
   (p. 14)

energy (p. 16)
geology (p. 1)
hot spots (p. 14)
hypothesis (p. 8)
laws (p. 10)
lithosphere (p. 5)
mantle (p. 5)
plates (p. 13)

plate tectonics (p. 13)
potential energy (p. 18)
principle of uniformitarianism
   (p. 11)
principles (p. 10)
radiation (p. 17)
scientific method (p. 8)
subduction (p. 14)

tectonics (p. 14)
theory (p. 10)
transform plate boundaries
   (p. 14)

2 Minerals:
Building Blocks of the Planet

Chapter Outline

Why Study Minerals?

MINERALS ARE THE FUNDAMENTAL BUILDING BLOCKS OF OUR rocky planet. Geologists study minerals both to understand the formation of rocks and to locate and extract resources that you use every day. Minerals form under a wide range of physical and chemical conditions that provide clues to the processes that generate Earth's lithosphere and deeper layers. Minerals containing significant iron, aluminum, lead, zinc, copper, silver, gold, platinum, and many more elements are valuable economic resources, supplying all of our industrial and precious metals. Other minerals fulfill important needs in the manufacture of useful commodities from powder for cosmetics, to filtering agents for water purification, to graphite for pencils.

There are two primary objectives in this chapter:

✔ To understand that each mineral has a definitive chemical composition and internal atomic structure. These determine physical properties and relate to where and how minerals form and how they are used

✔ To learn the names, compositions, and properties of an important handful of the more than 4000 known minerals

In the chapters that follow, you will see how these key minerals combine to form most of Earth's rocks. To achieve these objectives, consider the following questions:

2.1 What are the properties of minerals?
2.2 What are minerals composed of?
2.3 *How do we know ...* the atomic structure of minerals?
2.4 How do elements combine to make minerals?
2.5 What is a mineral?
2.6 What determines the physical properties of minerals?
2.7 What are the most important minerals?

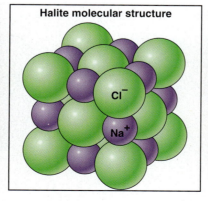

Halite molecular structure

Cl⁻

Na⁺

Mineral crystals are among the most beautiful natural features on Earth. ▶

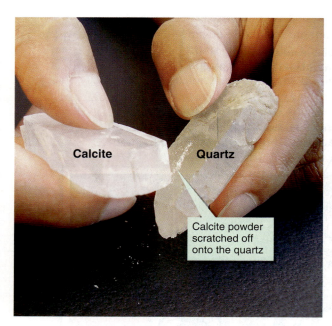

▲ **Figure 2.3** **Comparing the hardness of minerals.**
Scraping two minerals against one another tests their relative hardness. The quartz crystal remains unblemished while the calcite is scratched and powdered, indicating that calcite is softer than quartz.

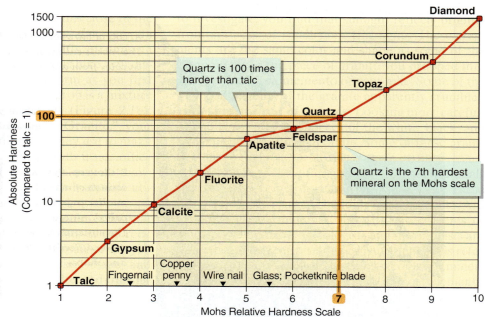

▲ **Figure 2.4** **How mineral hardness is defined.**
Mohs hardness scale is a standardized *relative* scale with values between 1 and 10 simply ranking the minerals from softest to hardest. More quantitative measurement methods determine the *absolute* hardness of minerals. Calcite and quartz are four steps apart on Mohs' scale, but quartz is more than 10 times harder than calcite. The black triangles show Mohs hardnesses of common objects for comparison.

when a thermometer measures temperature. Sophisticated methods more accurately quantify hardness rather than simply ranking minerals based on hardness. Results of these experiments reveal that quartz is about ten times harder than calcite (Figure 2.4).

If hardness, like external shape, distinguishes minerals, then do geologists carry around pieces of the ten minerals on the Mohs scale in order to measure relative hardness? They could, but instead field geologists use a few common objects, such as a fingernail, a penny, and the blade of a pocketknife to conduct scratch tests (Figure 2.4).

Some Minerals Have Characteristic Cleavage

The next observation, illustrated in **Figure 2.5**, is the way that quartz and calcite break differently when struck by a hammer. Quartz breaks into very irregularly

▼ **Figure 2.5** **How minerals break.**
Broken mineral fragments have a different shape than mineral crystals. Broken quartz fractures like glass into irregular, sharp fragments. Calcite breaks into regularly shaped pieces that have six planar sides with rhomb shapes. The planes defining the pieces are cleavage planes and the pieces of calcite are cleavage fragments.

shaped pieces that resemble broken glass. The broken surfaces display an indented, rounded pattern but never smooth, flat planes (Figure 2.5a). On the other hand, the calcite easily breaks into similar six-sided pieces where each side has the shape of a rhomb (Figure 2.5b). Each rhomb-shaped piece can be broken into still smaller pieces of similar shape. The broken surfaces of calcite are flat, smooth, planes that reflect light. The term **cleavage** describes the flat, smooth planes along which some minerals break and the shape of the resulting fragments. Calcite has three directions of cleavage that form the rhomb-shaped fragments. As another example, **Figure 2.6** illustrates mica minerals that split into very thin transparent sheets along a single cleavage direction. Quartz breaks along unpredictable, irregular surfaces so it does not have cleavage (Figure 2.5a). This nonuniform breakage is **fracture**, rather than cleavage.

Mineral Color and Shape Can Vary

Some ways to distinguish calcite and quartz seem clear at this point. Do *all* specimens of calcite and quartz have the same properties? Specific gravity, luster, hardness, and cleavage are diagnostic in all cases for quartz and calcite but other properties are more variable, as shown in **Figure 2.7**. Quartz (Figure 2.7a) is found in pink hues (rose quartz), cloudy white (milky quartz), black (smoky quartz), purple (amethyst) and yellow (citrine). Calcite (Figure 2.7b) also comes in a variety of shades; cloudy white, gray, orange, red, and pink are the most common. Color is not always a reliable property for identifying a mineral because of these variations, even though many minerals have less variable color than quartz and calcite.

A property related to color is **streak**, the color of the residue produced by scratching a mineral on a nonglazed porcelain plate, as illustrated in **Figure 2.8**. A mineral may vary in color, but its streak color is always the same.

▲ **Figure 2.6 Micas cleave into thin sheets.**
The common mica minerals, silvery muscovite and nearly black biotite, have a single cleavage direction. This means that a knife blade or fingernail easily separates samples into thin, transparent sheets.

(a)

(b)

▲ **Figure 2.7 Variations in mineral color.**
Samples of quartz and calcite come in a variety of external forms and colors. One type of mineral can occur in a variety of colors, and various minerals can have the same color.

Red hematite streak

White calcite streak

▲ **Figure 2.8 Mineral streak colors.**
Minerals softer than porcelain (about 6.5 on Mohs scale) leave a powdery residue, called the streak, when scraped across the porcelain plate. Calcite streak color typically matches the crystal color. The iron mineral hematite always leaves a red-brown streak on the plate, even when the mineral specimen is a different color.

TABLE 2.1 Physical Properties of Minerals	
Description of property	Factors that determine the property
Luster describes how mineral surfaces reflect light.	Luster depends on the smoothness of the mineral surface at the atomic scale, or how mobile electrons are within the crystal.
Specific gravity is the ratio of the mass of a substance compared to the mass of the same volume of water.	Specific gravity depends on the atomic weights of atoms and how the atoms are arranged in the crystal structure. The more tightly together the atoms are, the higher the specific gravity.
Density is the measure of the mass of a substance contained within a particular volume of the substance.	Density and specific gravity are different measures of essentially the same property, so density also depends on how the atoms are arranged in the crystal structure. The more tightly together the atoms are, the higher the density.
Crystal faces are flat, smooth surfaces on mineral exteriors with regular geometric forms.	Crystal faces reflect the atomic arrangement of atoms within the crystal structure and produce exterior geometric shapes during growth unless the crystal grows against another crystal.
Hardness is the resistance to scratching on a smooth surface of the mineral.	Hardness reflects the bond strength, which depends on bond type and the spacing of atoms within the crystal.
Cleavage describes planes along which a mineral breaks and the shape of the resulting fragments.	Cleavage forms along regularly spaced internal planes where bonds are weakest in minerals.
Color results from the interaction of light with the mineral.	Atomic arrangement and composition determine how light passes through or interacts with the atoms through the crystal, determining the color.
Streak is the color of the residue remaining from scratching a mineral on a nonglazed porcelain plate.	The fine-grained nature of the powdered residue results in a more reliable observed color than the whole crystal, which depends on the types of atoms and their arrangement.

Not all specimens of each mineral are identical in shape, either. All calcite crystals have six sides, but the shapes and arrangement of crystal faces differ. Furthermore, some specimens of both calcite and quartz lack obvious crystal faces (Figure 2.7). External crystal form, therefore, is also not always helpful for identification.

Table 2.1 summarizes the physical properties that distinguish minerals and that serve as tests to identify mineral specimens. The table also provides information on the causes of these properties, which are examined later in the chapter.

> *Putting It Together—**What Are the Properties of Minerals?***
> ■ Minerals have observed or easily measured physical properties, some of which are more diagnostic than others for recognizing a specific mineral.
> ■ The principal physical properties used to describe minerals are color, luster, streak, hardness, cleavage (or fracture), specific gravity, and external crystal form.

2.2 What Are Minerals Composed Of?

Minerals, like all substances, are composed of atoms of chemical elements. Atoms consist of even smaller particles—positively charged protons, negatively charged electrons, and uncharged neutrons. The central nucleus houses the protons and neutrons, and electrons surround the nucleus. The number of protons defines a particular element. Most minerals (e.g., quartz, calcite) are compounds, composed of two or more elements, although a few contain atoms of only one element (e.g., diamond, contains only carbon). Mineralogists perform laboratory chemical analyses to determine mineral compositions. **Appendix A** provides the periodic table of the elements for reference.

Calcite and quartz have distinct compositions. Calcite is mostly calcium (Ca), carbon (C), and oxygen (O), whereas quartz primarily consists of silicon (Si) and oxygen (O). The chemical formula for calcite is $CaCO_3$ and quartz is SiO_2.

Figure 2.9 lists the chemical compositions of three quartz varieties identified by different colors. The analyses report trace amounts of elements other than silicon and oxygen. The abundances of these other elements may be insufficient to affect properties such as specific gravity, luster, hardness, and cleavage, which are qualities shared by all specimens of the same mineral. Color, however, may depend on these trace constituents. For example, Figure 2.9 shows that rose quartz contains a tiny amount of titanium, and amethyst contains a tiny amount of iron.

Chemical composition is a reasonably definitive characteristic, within narrow ranges, of individual minerals. Other than to some extent for color, it is not yet obvious how composition determines the physical properties that distinguish calcite from quartz. These properties are determined not only by what elements are present but also by how these elements combine, and this is the subject of the next two sections.

Colorless quartz

99.998% SiO_2

Rose quartz

0.003% Titanium

99.996% SiO_2

Amethyst

0.020% Iron

99.978% SiO_2

▲ **Figure 2.9 Why quartz has many colors.**
All quartz specimens are almost entirely composed of silica (SiO_2), but even minutely small amounts of other elements determine mineral color. The purest quartz is colorless, but traces of titanium account for the pink color of rose quartz, whereas iron colors purple amethyst.

---

*Putting It Together—**What Are Minerals Composed Of?***

- Minerals are chemical compounds consisting of combinations of atoms of one or more elements.
- Each mineral has a definitive, but possibly slightly varying, chemical composition.

2.3 **How Do We Know** ... The Atomic Structure of Minerals?

UNDERSTAND THE TOOL
How Do Geologists Visualize Atoms Inside Minerals?
Chemical analyses reveal that atoms of particular elements are present in minerals, but the analyses do not show where the atoms are located. To understand the arrangements of atoms within minerals requires tools that allow geologists to "see" atoms.

One way to see small objects is to magnify them with a microscope. A common optical microscope uses glass lenses to focus light that reflects off or passes through a mineral specimen. Under the best circumstances, an optical microscope magnifies a mineral about 1000 times and permits recognition of features about one-thousandth of a millimeter across. While this dimension seems very tiny, it is still much larger than atoms.

The much more sophisticated transmission electron microscope (TEM) uses magnets, rather than lenses, to focus streams of tiny electrons, rather than light, through a small part of a mineral sample. Some electrons pass through the sample without encountering any obstacles. Other electrons bounce off atoms composing the mineral. A detector beneath the sample records the arrival of electrons that passed through unhindered and reveals electron paths that encountered atomic obstacles. Mineralogists use data from the detector to produce images magnified up to 10 million times and resolve features less than one-millionth of a millimeter across. This very high level of magnification is sufficient to see the fuzzy outlines of most atoms.

VISUALIZE THE RESULT
What Does the Atomic Structure Look Like?
Figure 2.10 shows how a TEM produces an image of atoms in a mineral. The interaction of the streams of electrons fired inside the microscope produces an image of black and white dots. White areas are where electrons struck the

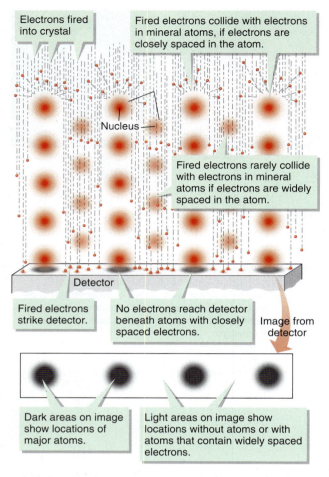

Electrons fired into crystal

Fired electrons collide with electrons in mineral atoms, if electrons are closely spaced in the atom.

Nucleus

Fired electrons rarely collide with electrons in mineral atoms if electrons are widely spaced in the atom.

Detector

Fired electrons strike detector.

No electrons reach detector beneath atoms with closely spaced electrons.

Image from detector

Dark areas on image show locations of major atoms.

Light areas on image show locations without atoms or with atoms that contain widely spaced electrons.

▲ **Figure 2.10 How a TEM makes images of atoms inside minerals.**
Electrons are fired at thin, carefully oriented slices of minerals inside the electron microscope. Those electrons that transmit all of the way through the mineral, without encountering obstacles, reach and brightly illuminate a detector on the other side. Fired electrons that encounter electron clouds surrounding mineral atoms do not reach the detector, leaving a shadow on the image. Not all atoms leave a dark shadow, however, because they contain too few orbiting electrons that are too widely spaced apart to intercept all of the electrons fired inside the microscope.

detector beneath the mineral sample, and black areas are where no electrons reached the detector. The black areas are like shadows formed where an object blocks light. In the case of a TEM image, however, the shadow forms because electrons encountered atomic obstacles when passing through the mineral and never reached the detector.

Most of the volume and diameter of an atom consist of the cloud of electrons that rapidly orbit the nucleus. If there are many electrons in a small area around the nucleus, then there is a good chance that electrons fired in the TEM will collide with an orbiting electron and bounce off elsewhere in the microscope rather than continuing through to the detector. If the electrons are widely spaced around a nucleus, however, then the fired electrons may pass through the atom without hitting any obstacles. This means that the shadows in the TEM image may not represent all of the types of atoms in a mineral, but rather only the atoms with closely clustered electrons. Figure 2.10 also shows that white and dark areas appear in the image only where the atoms are arranged in neat columns. If the atoms were randomly spaced in the sample, then in all likelihood none of the electrons would succeed in reaching the detector without encountering an obstacle.

Figure 2.11 shows a TEM image of dolomite $(CaMg[CO_3]_2)$, a mineral similar to calcite $(CaCO_3)$ but with magnesium atoms replacing half of the calcium atoms. The image has a resolution smaller than one-millionth of a millimeter. Clearly, this magnification is enough to see the shadows caused by the calcium and magnesium atoms in repetitive rows of black dots. Calcium atoms are much bigger than magnesium atoms, so the dots in the calcium rows are larger than in the magnesium rows. Where are the carbon and oxygen atoms? The TEM image does not clearly show these elements because compared to calcium and magnesium there are very few electrons spinning around carbon and oxygen and they are also very widely spaced.

INSIGHTS

What Do TEM Images Reveal about Mineral Structures?

Transmission electron microscope images provide insights into the internal arrangement of atoms within minerals. Atoms in minerals form systematic, repetitive patterns, as revealed by the image of dolomite (Figure 2.11). Minerals are defined not only by the types and abundances of atoms but also their arrangement. Other types of laboratory experiments (using x-rays, for example) also reveal the orderly internal arrangement of atoms within minerals, so geologists knew this aspect of minerals long before development of the electron microscope. Nonetheless, the TEM was the first instrument that actually permitted geologists to "see" atoms in minerals.

▶ **Figure 2.11 Visualizing atoms inside a mineral.**
This TEM image of the mineral dolomite resembles a wallpaper of regularly spaced black and white dots. The black dots reveal locations of atoms that interfered with electrons fired inside the TEM. The image interpretation shows that rows of large dots are calcium atoms and rows of small dots are magnesium atoms. Dolomite also contains oxygen and carbon atoms that the TEM does not detect, but which must occupy spaces between the calcium and magnesium atoms. All minerals feature regular repeated patterns of atoms, which define an orderly internal atomic structure.

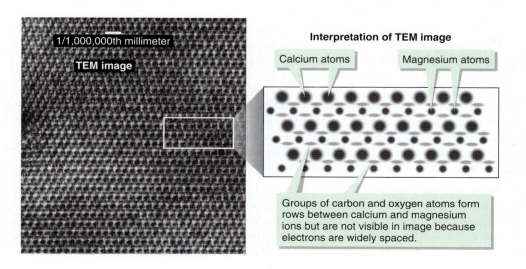

1/1,000,000th millimeter

TEM image

Interpretation of TEM image

Calcium atoms

Magnesium atoms

Groups of carbon and oxygen atoms form rows between calcium and magnesium ions but are not visible in image because electrons are widely spaced.

Putting It Together—How Do We Know ... *The Atomic Structure of Minerals?*
- Geologists use transmission electron microscopes to visualize arrangements of atoms inside minerals.
- TEM images, made by firing electrons into a mineral sample, show the orderly internal arrangement of atoms unique to each mineral.

2.4 How Do Elements Combine to Make Minerals?

We have seen that each mineral consists of atoms of particular elements arranged in an orderly pattern. How do atoms join to make minerals, what element combinations are possible in minerals, and what determines the orderly arrangement of the atoms?

Chemical Bonds Form Mineral Compounds

The distribution of electrons around atomic nuclei determines how atoms **bond**, which is the process of combining atoms into compounds. Three bond types, and a fourth weak force that attracts atoms to one another, are most common in minerals. Electrons involved in bonding in most minerals orbit atomic nuclei in concentric spherical layers. Only two electrons occupy the innermost layer, but most of the other layers will accommodate eight electrons. Atoms transfer or share electrons in order to fill their outermost layer with eight electrons. Transfer and share processes are responsible for the most common bonds.

Most elements tend to shed or gain electrons in order to end up with a filled outer electron layer. This tendency produces an excess or deficiency of electrons compared to protons, so that the atoms become positively or negatively charged particles called **ions**. Negative ions have more electrons than protons whereas positive ions have fewer electrons than protons. **Appendix B** lists the ions for the elements that are most abundant in rocks. **Ionic bonds** form by the attraction of negative and positive ions in order to balance their charges. **Figure 2.12** illustrates ionic bonding in table salt, which is also the mineral halite.

In a different case, two or more atoms mutually share electrons to fill the outer electron layer. Sharing creates a **covalent bond**, as shown in **Figure 2.13**. Ionic bonds, covalent bonds, or a mixture of both dominate most minerals.

Less commonly, electrons freely roam around a number of different atoms, typically of the same element, to form **metallic bonds**, as shown in **Figure 2.14**. The mobility of the electrons within metallic substances accounts for their ability to conduct electricity. Minerals with freely roaming electrons, as expected with metallic bonds and some covalent bonds, cause bright reflection of light; this quality results in metallic luster because of the similarity to bright, shiny metal.

Electrons are not always equally distributed around molecules or atoms. This means that the neutrally charged molecule or atom behaves like a weak miniature magnet. There is a weak negative charge on the side with more electrons and a weak positive charge on the side with an electron deficiency. This causes a **van der Waals force**, which is a weak attraction of neutrally charged particles.

Bonds Break when Minerals Dissolve

Water (H_2O) is a covalently bonded compound of considerable geologic importance. **Figure 2.15** shows that the single electron encircling each of two hydrogen atoms in a water molecule is shared with a single oxygen atom. The water molecule is electrically neutral, but it is lopsided, giving the molecule a slight positive tendency on the side with hydrogen atoms, and electrons spend more time on the oxygen side of the molecule, which has a slightly negative tendency.

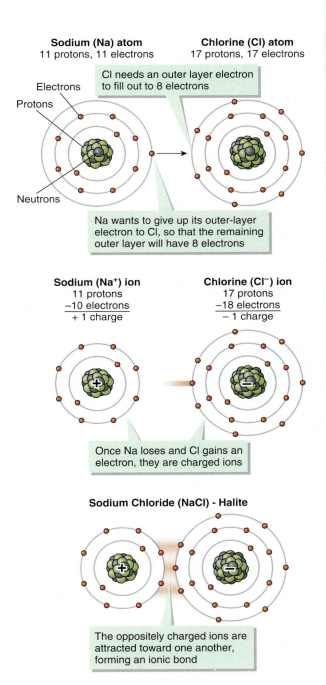

Sodium (Na) atom
11 protons, 11 electrons

Chlorine (Cl) atom
17 protons, 17 electrons

Cl needs an outer layer electron to fill out to 8 electrons

Electrons
Protons
Neutrons

Na wants to give up its outer-layer electron to Cl, so that the remaining outer layer will have 8 electrons

Sodium (Na⁺) ion
11 protons
−10 electrons
+ 1 charge

Chlorine (Cl⁻) ion
17 protons
−18 electrons
− 1 charge

Once Na loses and Cl gains an electron, they are charged ions

Sodium Chloride (NaCl) - Halite

The oppositely charged ions are attracted toward one another, forming an ionic bond

▲ **Figure 2.12 How ionic bonding works.**
A sodium (Na) atom has 11 electrons, with one electron in its outermost electron layer. Chlorine (Cl) has 17 electrons, with 7 electrons in its outermost layer. The transfer of an electron away from Na to Cl leaves each atom with 8 electrons in its outer layer. The transfer of an electron creates a positive Na⁺ ion and a negative Cl⁻ ion. The two oppositely charged ions attract to each other to form an ionic bond. The resulting compound, NaCl (sodium chloride), is table salt, and this same compound is also the mineral halite.

Carbon

6 Electrons

6 Protons
6 Neutrons

Four vacant spaces
in outer layer

Covalent bond forms where each carbon
atom shares 4 electrons with
neighboring carbon atoms so that all
atoms have 8 electrons in outer layer.

▲ **Figure 2.13** **How covalent bonding works.**
Two or more uncharged atoms may bond by sharing electrons in
their outermost layers. The outer electron layer surrounding a
carbon nucleus has four electrons and needs eight to be full.
A carbon atom, therefore, can share electrons with four other surrounding carbon atoms so that all of them have full outer layers.
Sharing electrons defines covalent bonds.

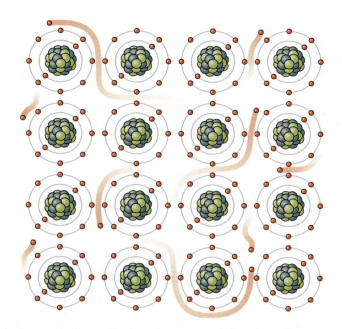

▲ **Figure 2.14** **How metallic bonding works.**
This diagram illustrates a cross section across the atomic structure of
a metal. Metallic bonds form where electrons freely roam between
nuclei. At any one time, each atom experiences a neutral charge
because as one electron leaves the proximity of a nucleus another
replaces it.

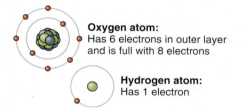

Oxygen atom:
Has 6 electrons in outer layer
and is full with 8 electrons

Hydrogen atom:
Has 1 electron

Two hydrogen atoms covalently bond with one
oxygen atom to produce a water molecule:

Protons unshielded by electrons
produce a weak positive charge
at this end of the molecule.

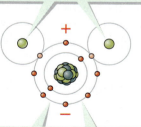

When orbiting electrons are on this side of the
molecule they are not close to protons, so a slight
negative charge exists at this end of the molecule.

▲ **Figure 2.15** **What the water molecule looks like.**
Two hydrogen atoms form a covalent bond with oxygen to make a
water molecule. Water is a neutral molecule but it has mild electrical
charges at opposite ends. The mild charges occur because the water
molecule is asymmetric. The hydrogen atoms are skewed toward
one end, which produces a weak positive charge. Electrons are more
numerous on the other end of the molecule, resulting in a weak
negative charge. These weak charges cause water molecules to
attract one another and other charged molecules and ions.

Figure 2.16 illustrates how the uneven distribution of electrical charge around a neutral water molecule allows the molecule to tug at ions in compounds such as halite. The positive ends of the water molecules surround the negative chloride ions (Cl^-), and the negative ends of water surround the positive sodium (Na^+) ions. When the ionic attraction between Na^+ and Cl^- is broken, the halite dissolves in the water. As long as enough water molecules are present to separate the sodium and chloride ions, the salt remains dissolved. If the water evaporates or if more Na^+ and Cl^- are added to the solution, it becomes more difficult for the water molecules to keep the attracting ions apart. Then the Na^+ and Cl^- ions combine, and crystals of halite reappear, or precipitate, from the solution.

All minerals with ionic bonds dissolve to some extent in water. Mineral solubility in water depends on the strength of the ionic bonds to resist the charge attraction of the lopsided water molecule. Minerals dominated by ionic bonds more readily dissolve in water than those with covalent bonds. Adding acid to water further enhances solubility, because acids contain ions that draw atoms in minerals apart from one another.

Internal Structure of Minerals

Figure 2.17 shows the presence of both ionic and covalent bonds in $CaCO_3$ molecules of calcite. Three oxygen atoms surround and share electrons with each carbon atom to form covalent bonds. This group of carbon and oxygen atoms does, however, have two more electrons than protons, so the entire group of atoms behaves like a negative ion, CO_3^{2-}, which is called carbonate. The oppositely charged carbonate and Ca^{2+} form ionic bonds (Figure 2.17). Like other ionic compounds, calcite dissolves in water and acid, although much less so than halite.

When calcite dissolves in water, the Ca^{2+} ions separate from the CO_3^{2-} ions but acid is required to break the strong covalent bond between carbon and oxygen. A common test for calcite, and other carbonate minerals, is to apply dilute hydrochloric acid (sold as muriatic acid in hardware and pool-supply stores) to the

Halite molecular structure

Water molecule

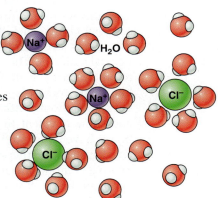

H_2O

◀ Figure 2.16 **How bonds break when halite dissolves in water.**

Halite (rock salt) consists of ionic bonds between sodium (Na) and chloride (Cl) ions.

The covalent bonding of two hydrogen atoms to one oxygen atom causes a lopsided shape for the water molecule. One end of the molecule has a weak positive charge while the other has a weak negative charge, as explained in Figure 2.15.

When halite is placed in water, the charged water molecules align with negative ends close to the sodium ions and positive ends next to the chloride ions. The weak electrical forces in the water molecules tug at the relatively weak ionic bonds and pull the mineral molecule apart.

The halite completely dissolves in water when there are sufficient water molecules surrounding each ion to keep ionic bonds from forming between sodium and chloride.

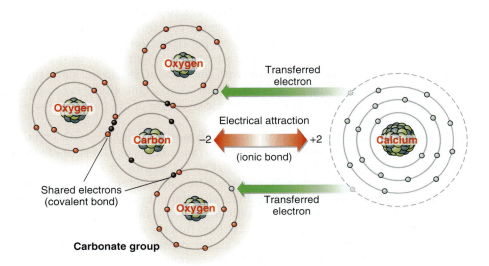

Carbonate group

◀ Figure 2.17 **How atoms bond in calcite.**
Calcite has both ionic and covalent bonds. Covalent bonding of carbon and oxygen atoms forms carbonate groups. Calcium ions (Ca^{2+}) form when electrons transfer to the carbonate groups, which also forms carbonate ions (CO_3^{2-}). The oppositely charged calcium and carbonate ions attract to form ionic bonds.

▲ **Figure 2.18 Calcite reacts with hydrochloric acid.**
A violent bubbling reaction occurs when hydrochloric acid (HCl) drops land on this calcite specimen. The bubbles are carbon dioxide gas released when the CO_3^{2-} groups in the carbonate mineral break down. This chemical reaction test distinguishes carbonate minerals from other, similar appearing minerals.

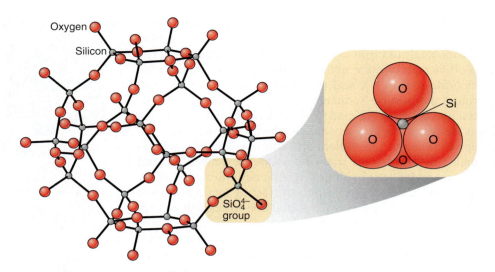

▲ **Figure 2.19 How atoms bond in quartz.**
Quartz has both ionic and covalent bonds. Four large oxygen atoms surround and covalently bond to each smaller silicon atom to make the basic SiO_4^{4-} group. This negatively charged group easily attracts positively charged Si ions. In quartz, each oxygen atom bonds to two different silicon atoms in a combination of covalent and ionic bonds. Oxygen atoms form the corners of a pyramid-like structure with a silicon atom in the center, as shown in the blow-up view.

sample. The mineral not only dissolves but, as shown in **Figure 2.18**, violent bubbling of carbon dioxide (CO_2) gas reveals the breakdown of the CO_3^{2-} group.

Bonding in quartz, depicted in **Figure 2.19**, is more complex than in calcite. The silicon (Si) and oxygen (O) atoms share electrons, but the geometry of the covalent bond is such that each Si atom shares electrons with four adjacent oxygen atoms, even though two oxygen atoms suffice to produce an electrically neutral molecule. The resulting negatively charged SiO_4^{4-} groups ionically bond to other nearby Si atoms so that the quartz structure becomes a linked, scaffold-like framework of Si–O groupings with each oxygen atom bonded to two different silicon atoms. Although early researchers predicted some ionic Si–O bonds and some covalent Si–O bonds, laboratory experiments show that all of the bonds have equal strength, indicating that each bond is partly ionic and partly covalent. The relative sizes of the atoms determine the four-to-one arrangement of oxygen atoms around each silicon atom. The much smaller silicon atom fits neatly in the small space between four oxygen atoms (Figure 2.19, inset).

Calcite illustrates the importance of understanding not only what atoms are present in a mineral and how they bond together, but also how the atoms are positioned relative to one another. **Figure 2.20** shows how you can use grapefruits, oranges, and grapes to reproduce the repetitive pattern of the calcite atomic structure. These three fruits have approximately the same relative sizes as oxygen, calcium, and carbon, respectively. Figure 2.20 shows that there are two ways to build the models with the same atoms, in the same proportions and bonded so that there is no overall charge. One model represents calcite. The other model illustrates the atomic structure of aragonite, the mineral constituent of pearls.

Calcite and aragonite are examples of **polymorphs**, minerals with identical chemical composition but with different arrangements of atoms. Shelly invertebrate organisms, like clams, secrete calcite, aragonite, or both minerals. Why do the same atoms arrange themselves differently to form different minerals? When considering *nonbiologic* mineral-forming processes, experiments show that pressure and temperature influence the atomic structure. Experimental results summarized in **Figure 2.21** show that the calcite atomic configuration is preferred at low pressure. At higher pressure, however, the same atoms pack more closely together and form aragonite.

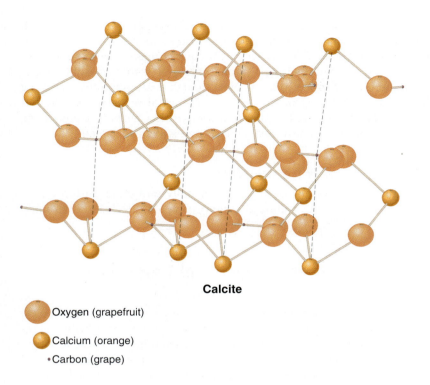

Calcite

- Oxygen (grapefruit)
- Calcium (orange)
- Carbon (grape)

Aragonite

▲ **Figure 2.20 Is there more than one way to configure atoms?**
In these diagrams, familiar fruits of different sizes represent atoms of different sizes and show two ways to make a repeated pattern of Ca^{2+} and CO_3^{2-} ions. Although both depict the same chemical composition, they represent different structural arrangements of the elements, so each is a different mineral. In the calcite pattern, calcium ions and carbonate ions are in alternating rows. The other pattern, representing the mineral aragonite, has two staggered carbonate rows between each row of more closely spaced calcium ions. Notice that the atoms composing the aragonite structure crowd more closely together than in the calcite structure.

Putting It Together–*How Do Elements Combine to Make Minerals?*

- Minerals are atoms of elements combined by ionic, covalent, and less common metallic bonds and weak van der Waals forces.
- Ionic bonds are weaker than covalent bonds, where atoms share electrons.
- The neutral, but lopsided, water molecule has a weak positive charge at one end and a negative charge at the other. These charges pull apart some weakly bonded ions, causing some minerals to dissolve in water.
- Many minerals exhibit combinations of ionic and covalent bonds or bonds that have both ionic and covalent traits.
- The atomic structure of minerals is determined not only by the type of bond but also by how the atoms are positioned relative to one another. Different arrangements of the same atoms represent different minerals.

2.5 What Is a Mineral?

Pause to reflect upon what you know about minerals. Each mineral has a composition, which varies only slightly (e.g., Figure 2.9). Each mineral has a repetitive ordered internal structure of atoms that even distinguishes minerals that have the same composition but different atomic arrangements (e.g., Figure 2.20). Minerals have different physical properties, and you are beginning to see how the properties relate to the composition and internal structure. Minerals compose natural rocks formed by geologic processes, and biologic processes also form minerals, as is the case of clam shells and pearls (and also teeth and bone).

Definition of a Mineral

These observations lead to a comprehensive definition of a mineral. A **mineral** is a naturally occurring solid, usually inorganic, with a definite, only slightly variable chemical composition and an ordered atomic structure. Minerals include only naturally formed substances to separate them from manufactured materials such as synthetic gemstones (e.g., cubic zirconia). A mineral must be a solid with an orderly internal arrangement of atoms or it otherwise is a gas or liquid. At first

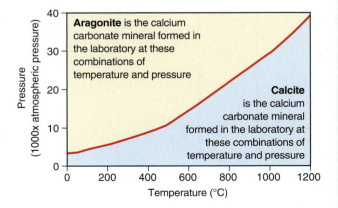

Aragonite is the calcium carbonate mineral formed in the laboratory at these combinations of temperature and pressure

Calcite is the calcium carbonate mineral formed in the laboratory at these combinations of temperature and pressure

Pressure (1000x atmospheric pressure)

Temperature (°C)

▲ **Figure 2.21 What conditions determine the formation of aragonite rather than calcite?**
Experimental crystallization of calcium carbonate under conditions of changing temperature and pressure indicate that calcite preferentially forms at lower pressure compared to aragonite.

glance, naturally occurring glass (e.g., volcanic obsidian) appears to be a mineral, but transmission electron microscope studies and other analyses reveal a lack of orderly atomic arrangement. Glass is actually a liquid. Organic compounds, containing mostly carbon and hydrogen atoms, are not minerals. Table sugar ($C_{12}H_{22}O_{11}$), therefore, is not a mineral, even though it is a naturally occurring solid with an orderly atomic structure. A mineral must also have a definite but not fixed chemical composition. The chemical composition can vary slightly (e.g., Figure 2.9), but the principal constituents are common to all specimens.

> *Putting It Together—What Is a Mineral?*
> ■ A mineral is a naturally occurring solid, usually inorganic, with a definite, only slightly variable chemical composition and an ordered atomic structure.

2.6 What Determines the Physical Properties of Minerals?

When comparing quartz and calcite (Section 2.1), physical properties of color, external or crystal form, cleavage, hardness, specific gravity, and luster of reflected light are, to varying degrees, diagnostic of each mineral. These identifying physical properties are a result of the chemical composition and internal structure of the minerals. The causes of these properties, at the atomic level, are summarized in Table 2.1 and explained in the following paragraphs.

Composition Commonly Determines Color

The substitution of a few atoms for dominant atoms in the crystal structure of a mineral has minimal effect on its composition or most of its physical properties but can have considerable effect on its color. Data depicted in Figure 2.9 show that trace amounts of some elements can affect color, with purple quartz containing minute amounts of iron and the pink variety containing a bit of titanium. **Figure 2.22** shows that the iron (Fe^{3+}) and titanium (Ti^{4+}) ions are larger than

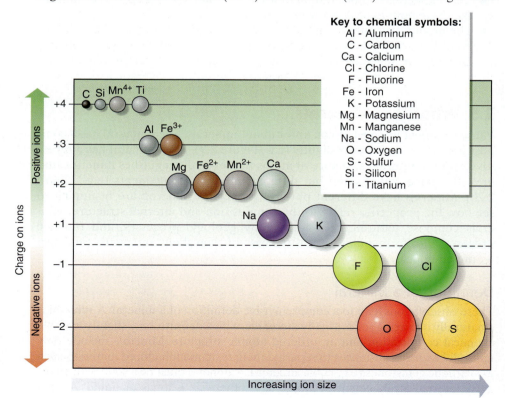

▶ **Figure 2.22 Comparing the sizes of ions.** This chart schematically shows the electrical charges and diameters of several ions that are common in rock-forming minerals. Ions can substitute for one another in a mineral atomic structure when their charges and sizes are similar.

Key to chemical symbols:
Al - Aluminum
C - Carbon
Ca - Calcium
Cl - Chlorine
F - Fluorine
Fe - Iron
K - Potassium
Mg - Magnesium
Mn - Manganese
Na - Sodium
O - Oxygen
S - Sulfur
Si - Silicon
Ti - Titanium

the silicon (Si^{4+}) ion, but Fe^{3+} and Ti^{4+} can still nestle between the four oxygen atoms in the quartz structure in place of Si^{4+}. The Fe^{3+}, Ti^{4+}, and Si^{4+} ions also have the same or nearly the same number of electrons to share.

Similarly, calcite is clear or white when pure, but its colors vary with elemental substitutions (Figure 2.7b). Varying amounts of iron (Fe^{2+}) provide color to some dark blue, green, and brown calcites. Manganese (Mn^{2+}) incorporated in the calcite structure produces colors from pale purple to deep red. The Fe^{2+} and Mn^{2+} ions have the same charge and similar size as the Ca^{2+} ion (Figure 2.22), permitting them to substitute for Ca^{2+} in the calcite atomic structure.

Not all color variations result from element substitutions. White, milky quartz is clouded by tiny droplets of water, trapped between atoms in the crystal structure, but not bonded to them. Black, smoky quartz is dark to opaque because of damage in the crystal structure caused by natural radiation released by traces of radioactive elements enclosed in the quartz.

Atomic Structure Determines Most Physical Properties

Other diagnostic physical properties, especially hardness and cleavage, are more closely related to the arrangement of atoms within the crystal structure than to the types of atoms that define mineral composition. A comparison of diamond and graphite, illustrated in **Figure 2.23**, provides an example of the importance of crystal structure to determine physical properties. Both minerals contain only carbon, but diamond is the hardest mineral on Earth, and graphite, with a Mohs hardness of 2, is soft enough to mark paper and to lubricate machinery. Graphite readily cleaves into thin scaly sheets, similar to mica (Figure 2.6) whereas diamond cleaves only with great difficulty into eight-sided fragments.

Cleavage planes form where bonds are weakest in minerals. Figure 2.23 shows how different arrangements of carbon atoms in diamond and graphite account for their contrasting physical properties. Diamond forms at very high pressure and consists of a tightly packed three-dimensional framework of covalently bonded carbon atoms. Graphite forms at low pressure and consists of covalently bonded carbon atoms arranged in sheets. Weak van der Waals forces hold the sheets together. Graphite breaks easily between these weakly linked layers to form thin scaly plates, whereas carbon bonds strongly in all directions within diamond, so the mineral is much more difficult to break.

Scratching a mineral during a hardness test breaks bonds. Bond type and spacing of atoms determines bond strength. Covalent bonds are stronger than ionic bonds, which are in turn

▼ **Figure 2.23 Why diamond and graphite have different physical properties.**
Diamond and graphite consist only of carbon atoms but their physical properties are very different. Carbon atoms in diamond are closely spaced and strongly bonded in all directions; this configuration produces the hardest mineral on Earth. The smooth crystal faces defining diamond's external form coincide with planes of carbon atoms in the crystal structure. Carbon atoms are more widely spaced in graphite than in diamond and are strongly bonded only in two dimensions. The covalently bonded carbon sheets are weakly held together by van der Waals forces. The weakly linked sheets readily separate from each other, accounting for the softness of graphite and its tendency to cleave readily into thin scaly plates.

Diamond

Graphite

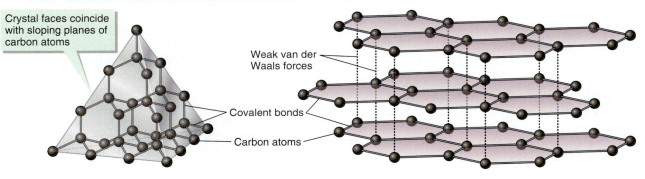

Crystal faces coincide with sloping planes of carbon atoms

Weak van der Waals forces

Covalent bonds

Carbon atoms

stronger than van der Waals forces. Hardness of ionic substances increases with increasing electrical charge on the ions. This trend means that an ionically bonded mineral with a +2 ion is harder than a mineral with a +1 ion. For example, calcite ($CaCO_3$) is harder than halite ($NaCl$). The calcium ion in calcite has a +2 charge whereas the sodium ion in halite has a +1 charge. The closer atoms are bonded within a molecule, in any fashion, the stronger the bonds. The weak electrical attraction between relatively distant carbon sheets renders graphite far softer than the closely spaced, covalently bonded carbon atoms in diamond.

The atomic arrangements of carbon in diamond and graphite also affect other physical properties. The contrasting specific gravity of the two minerals, 3.51 for diamond compared to 2.23 for graphite, is consistent with the more compact arrangement of the carbon atoms in diamond (Figure 2.23). The arrangement of the carbon atoms also determines their crystal forms. The covalent bonding of carbon in diamond commonly produces octahedron crystals, which resemble two four-sided pyramids stuck together at their bases and pointing in opposite directions (Figure 2.23). Luster depends on the smoothness of the mineral surface at the atomic scale or on how mobile electrons are within the atomic crystal. Diamond has an adamantine luster—a brilliant, almost oily appearance—related to a very smooth surface and strongly fixed electrons. Graphite has metallic luster that results from some mobility of electrons within the covalent bonds. When light strikes the surface of graphite, most of the energy is absorbed, but some of it reflects as visible light that provides the metallic luster. No light transmits through the graphite, so it is opaque. In contrast, light passes through the diamond crystal, so it appears transparent.

Think back to your initial experience with calcite and quartz—can you now explain their different physical properties? The most notable differences are hardness and cleavage. Covalent bonding dominates in quartz, whereas the ionic bond between Ca^{2+} and CO_3^{2-} is the "weak link" in the calcite structure. The ionically bonded calcite (Mohs hardness of 3), therefore, is softer than the more covalently bonded quartz (Mohs hardness of 7) and is also much more soluble in water or dilute acid. Calcite is harder and less soluble than halite (hardness of 2.5), another ionic mineral you have considered, because the ions in calcite have +2 and −2 charges that hold them together more strongly than the +1 and −1 charges of the ions in halite. The interlocking three-dimensional framework of equally spaced SiO_4^{4-} groups in quartz features no planes of preferential weakness; therefore, specimens fracture irregularly rather than cleave when broken. Calcite, however, fails across ionic bonds, whose arrangement in the crystal structure produces rhomb-shaped cleavage fragments of various sizes. Calcite and aragonite have the same composition, so the closer atomic packing of aragonite (Figure 2.20) reasonably accounts for the higher specific gravity compared to calcite (2.95 compared to 2.71) and for greater hardness (3.5–4 on the Mohs hardness scale).

Putting It Together–**What Determines the Physical Properties of Minerals?**

- The physical properties of minerals correspond to their composition and structure.
- Very low abundances of some elements cause occurrence of minerals with different colors but with otherwise identical physical properties.
- The type of bond and the distance between bonded atoms or ions determine hardness and cleavage. Covalent bonds are stronger than ionic bonds, and van der Waals forces are the weakest. Covalent bonds are strongest where atoms are closer together, and ionic bonds are relatively stronger where charges on the ions are higher.
- The specific gravity of minerals is greater for minerals composed of larger and more closely packed atoms.

2.7 **What Are the Most Important Minerals?**

With more than 4000 minerals known on Earth and more discovered and named every year, how essential is it to know the details of all of them? The importance of minerals can be gauged in two ways. On one hand, a few dozen minerals comprise the vast majority of rocks exposed at Earth's surface, which earns them important geologic status as the "rock-forming minerals." On the other hand, many minerals, although not abundant, are essential economic resources. Minerals like talc, sulfur, gold, and graphite are used almost as they are when removed from the ground. Others, called **ore minerals**, contain important metallic elements extracted from the minerals by metallurgical processes that break the mineral bonds. Among the economically important elements obtained from ore minerals are iron (Fe), lead (Pb), zinc (Zn), copper (Cu), titanium (Ti), nickel (Ni), chromium (Cr), and uranium (U). Even combined, these essential elements form far less than one percent of Earth's crust. This means that although many of these useful elements are found in rock-forming minerals, they are typically present in such minute concentration that they cannot be extracted at acceptable cost. Fortunately, there are geologic processes (discussed in Chapters 4 through 6) that concentrate the economically important elements into mineral deposits that can be mined.

Just 12 out of 89 naturally occurring elements account for 99.7 percent of the mass of the crust, which explains why there are relatively few common rock-forming minerals (see Appendix A). **Figure 2.24** depicts geologists' estimates of the average composition of the whole planet and just the crust. The composition of the crust is known from analyses of actual rocks and estimates of their various proportions. The whole-Earth composition is clearly more difficult to determine but can be accomplished using methods explained in Chapter 8. The two dominant elements in the whole Earth are iron and oxygen. Although iron composes approximately 35 percent of the total Earth, it represents only 6 percent of the crust. The two dominant elements within the crust are oxygen (46 wt percent) and silicon (28 wt percent). Minerals containing silicon and oxygen, such as quartz, form the most common mineral group, known as silicates.

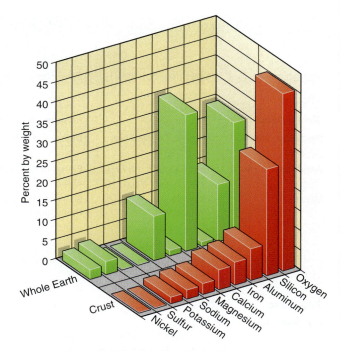

▲ **Figure 2.24** **Visualize Earth's composition.**
This graph depicts the abundances of the ten most common elements within the whole Earth and within just the crust. Iron is the most abundant element in the whole Earth, whereas oxygen, silicon, and aluminum are most abundant in the crust.

Silicate Minerals Are the Primary Rock-Forming Minerals in the Crust

Silicate minerals are those containing silicon and oxygen bonded in SiO_4^{4-} groups (Figure 2.19). Quartz is the chemically simplest silicate mineral, but most silicates contain one to four additional elements in sufficient abundances to produce different crystal structures. The silicates are further classified primarily on the basis of their crystal structures and secondarily on the basis of composition. Some of the dominant silicates found in the Earth's crust are listed and described in **Table 2.2**.

Many rock-forming minerals, including silicates, differ from one another only slightly as a result of element substitutions within the crystal structure. You have already seen how substitution of a small number of atoms in place of a more dominant element causes variations in mineral color without sufficient effect on composition or crystal structure to warrant definition of a new mineral (Figure 2.9). In many other cases, however, wholesale substitution of one element for another leads to the formation of a series of minerals that differ by gradual changes in composition and structure. These gradual compositional variations correspond to equally gradual variations in some physical properties, most commonly color and specific gravity.

Certain conditions must be met in order for element substitutions to occur. For one thing, the ions should be of comparable size so that a substituting ion fits into the same site in the crystal structure (Figure 2.22). Substitution of Fe^{3+} and Ti^{4+} for Si^{4+} in quartz is very minor (Figure 2.9) because the substituting ions are nearly 50 percent larger than Si^{4+} and the fit is not very good. Consider, however, the common ion of the third most abundant crustal element (Figure

EXTENSION MODULE 2.2
Silicate Mineral Structures.
See how the SiO_4^{4-} groups, called silica tetrahedra, bond together to form a variety of silicate-mineral structures.

TABLE 2.2 Descriptions and Generalized Chemical Compositions of Important Minerals

Mineral	Chemical Formula	Brief Description	Chapter(s) where you will learn more
Silicates			
Olivine Group	$(Mg, Fe)_2SiO_4$	Rounded equidimensional, green to brown crystals that fracture. Gem form is peridot.	3, 4, 5, 6, 8, 9, 11
Pyroxene Group	$(Mg, Fe, Ca, Na)(Mg, Fe, Al)Si_2O_6$	Blocky, black to green crystals with cleavage in two directions at almost 90°.	3, 4, 5, 6, 8, 9
Amphibole Group	$(Na, Ca)_2(Mg, Al, Fe)_5 (Si, Al)_8O_{22}(OH)_2$	Elongate to needle-shaped, black, green, and brown crystals with cleavage in two directions at 60° and 120°. Includes some forms of asbestos.	4, 5, 6, 8
Micas	$KAl_2(AlSi_3O_{10})(OH)_2$ (Muscovite)	Silvery muscovite and dark dark-brown to black biotite both have one direction of cleavage that produces thin sheets.	4, 5, 6, 7, 8
	$K(Mg, Fe)_3(AlSi_3O_{10}) (OH)_2$ (Biotite)		
	$(Mg, Fe)_3(Si, Al)_4O_{10} (OH)_2(Mg, Fe)_3(OH)_6$ (Chlorite)	Green chlorite usually forms very fine-grained aggregates.	
Serpentine	$Mg_3Si_2O_5(OH)_4$	Fibrous, green crystals. Includes the chrysotile variety of asbestos.	6
Feldspar Group	$KAlSi_3O_8$ (Potassium feldspar)	Rectangular, blocky crystals in various colors from colorless to black. Feldspars have cleavage in two directions at 90°.	3, 4, 5, 6, 7, 8, 9, 14, 19
	$(Ca, Na)Al_2Si_2O_8$ (Plagioclase feldspar)		
Quartz	SiO_2	Quartz has six-sided crystal faces and is found in a variety of colors. Quartz fractures but does not cleave.	3, 4, 5, 6, 8, 9, 11, 14, 19
Garnet	$A_3B_2(SiO_4)_3$ where "A" can be Ca, Mg, Fe^{2+} or Mn^{2+} and "B" can be Al^{3+}, Fe^{3+}, or Cr^{3+}	Garnet is variable in color, from green, to yellow, to red, and fractures rather than cleaves.	6, 8, 19
Nonsilicates			
Calcite/aragonite	$CaCO_3$ Carbonate	Calcite varies in color, has cleavage in three directions at less than 90°, and reacts with dilute hydrochloric acid.	3, 5, 6, 7, 13, 14, 17, 19, 20
Dolomite	$CaMg(CO_3)_2$ Carbonate	Colorless to white and sometimes pink crystals with cleavage in three directions at less than 90°. Powdered dolomite reacts with dilute hydrochloric acid.	3, 5, 6, 17
Magnetite	Fe_3O_4 Oxide	Black with metallic luster; strong magnetism.	3, 4, 10, 19
Hematite	Fe_2O_3 Oxide	Black, red, or silvery gray with a red streak. Hematite is a major iron ore.	3, 5
Corundum	Al_2O_3 Oxide	Variable in color; fractures rather than cleaves; insoluble in all acids. Commonly used in abrasives because of its hardness (9 on Mohs hardness scale).	6
Halite	$NaCl$ Halide	Salty tasting, colorless to white cubic crystals with cleavage in three directions at 90°.	3, 5, 7, 14, 16, 17, 19, 20
Gypsum	$CaSO_4 \cdot 2H_2O$ Sulfate	Colorless to white crystals with cleavage in three directions at less than 90°. Commonly used in the manufacture of plaster and wallboard.	3, 5, 14, 16, 17, 19, 20
Gold	Au Native element	Various shades of yellow, very malleable with a high specific gravity.	4, 11
Copper	Cu Native element	Copper-red to dark brown-green from tarnish, very malleable with a high specific gravity.	4, 6, 11, 17
Chalcopyrite	$CuFeS_2$ Sulfide	Brass yellow often tarnished with iridescent colors. Common copper ore.	4
Pyrite	FeS_2 Sulfide	Pale brass yellow cubic crystals with fracture. Also known as "fool's gold."	5, 6, 17
Sphalerite	ZnS Sulfide	Various colors of yellow, brown, and black with cleavage in six directions. Chief zinc ore.	17
Galena	PbS Sulfide	Lead-gray, commonly cubic crystals with cleavage in three directions at 90°. Major lead ore.	17

2.24) aluminum (Al^{3+}), which is only about 22 percent larger than Si^{4+} (Figure 2.22). Aluminum more readily substitutes for Si^{4+} between four oxygen atoms than do iron or titanium. The aluminum substitution, however, clearly suffers from an unequal charge substitution, Al^{3+} for Si^{4+}. To maintain a neutral charge, other positive ions are added simultaneously, which produces crystal structures that quickly depart from that of quartz.

Feldspars, diagrammatically illustrated in **Figure 2.25**, are the most common group of rock-forming minerals in the crust; they result from aluminum-for-silicon substitution. Potassium (K^+), sodium (Na^+), and calcium (Ca^{2+}), which are also significant constituents of the crust (Figure 2.24), are the most commonly added positively charged ions that offset the imbalance of the Al^{3+} for Si^{4+} substitution. Na^+ and K^+ ions are the same charge, but the potassium ion is significantly larger than the sodium ion. Substitution of K^+ and Na^+ does not easily occur; therefore, most potassium feldspars contain very little sodium, and sodium feldspar contains very little potassium. The calcium (Ca^{2+}) ion is close in size to the Na^+ ion, permitting both ions to occupy the same sites in the feldspar structure, but this substitution also requires balancing charges. This balancing is accomplished by simultaneous additional substitution of Al^{3+} for Si^{4+} as Ca^{2+} replaces Na^+ in the atomic structure. The complete, gradual variation between relatively calcium- and aluminum-rich feldspars and sodium- and silicon-rich feldspars defines a mineral group called the plagioclase feldspars (Figure 2.25).

Magnesium and iron are two other significant components of the crust (Figure 2.24) that substitute for one another in silicate minerals pictured in **Figure 2.26**. This substitution readily occurs because Mg^{2+} and Fe^{2+} ions have the same charge and are very similar in size (Figure 2.22). Four groups of silicate minerals—olivine, pyroxene, amphibole, and biotite—illustrate magnesium and iron substitutions. Although geologists casually refer to each of these four as single minerals, each is really a group of minerals whose crystal structures and physical properties vary depending on the proportions of magnesium and iron

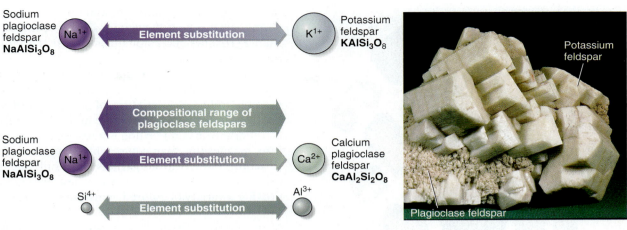

▲ **Figure 2.25** **How composition varies among feldspars.** Feldspar minerals are aluminum-silicate minerals containing various amounts of sodium, potassium, and calcium. In some feldspars, sodium and potassium substitute for one another since they have the same ion charge (+1), even though they are different sizes. A complete gradation from sodium rich to calcium rich compositions defines the plagioclase feldspars. Sodium and calcium ions are similar in size but have different charges. To account for charge imbalance, aluminum substitutes for silicon at the same time that calcium substitutes for sodium.

▼ **Figure 2.26** **Examples of iron- and magnesium-rich silicate minerals.** Iron and magnesium substitute for one another, and in some cases also with calcium, within the crystal structures of these dark silicate minerals—olivine, pyroxene, and amphibole.

▲ **Figure 2.27 Colorful garnet gemstones.**
Garnets vary in color from green to red depending on the elements incorporated into the crystal structure. These color variations account for a wide variety of popular gemstones in jewelry.

that are present. In addition, some pyroxene and amphibole crystal structures permit substitution of calcium, and less commonly sodium, for iron or magnesium. Pyroxenes and amphiboles consist, therefore, of calcium-rich and calcium-poor or absent varieties.

Substitutions also produce silicate mineral groups with widely varying composition but only slight differences in crystal structure. The term "garnet," for example, refers to a mixture of six groups of common silicate minerals with nearly identical atomic structure and indistinguishable crystal form. They differ considerably in composition because of complex substitutions between calcium, iron, magnesium, manganese, chromium, and aluminum. Although the crystal structures of the various garnets differ only slightly, there are wide variations in color from wine-red to green that not only aid in their identification but also provide a variety of colored gemstones as illustrated in **Figure 2.27**.

Nonsilicate Minerals Are Also Important

Table 2.2 shows that not all major rock-forming minerals are silicates. Another important group are the carbonates, which contain CO_3^{2-}. These include the calcium carbonate minerals calcite, aragonite, and dolomite, which you have already learned about. Carbonate minerals typically precipitate from water. The uses of carbonate minerals range from formulation of concrete to application as a gentle abrasive in toothpaste.

Many other nonsilicate minerals precipitate from water besides carbonates. Halite, or rock salt, is a halide mineral shown in **Figure 2.28** that forms from evaporation of water. Sulfate minerals all contain the SO_4^{2-} group, and the most common example is a calcium-sulfate mineral called gypsum. Gypsum is extensively mined for the production of plaster used in the manufacture of wallboard.

Oxides contain elements, other than silicon, bonded to oxygen (Table 2.2). The related hydroxide minerals also contain the hydroxyl ion (OH^-). The most common of these minerals are iron oxides such as hematite (Fe_2O_3) and magnetite (Fe_3O_4). The iron oxides are the primary iron ores extracted for making steel. Hematite is a source of red pigment. Magnetite, as the name suggests, is strongly magnetic, as shown in **Figure 2.29**.

Other oxide and hydroxide minerals are common ores of aluminum, chromium, uranium, tin, and titanium. Common oxide minerals other than iron oxides, such as rutile (TiO_2), ilmenite $(FeTiO_3)$, corundum (Al_2O_3), and uraninite (UO_2), also have industrial uses. Ilmenite and rutile are resources for making alloys where titanium (Ti) is combined with other metals. Titanium provides the alloy with high strength but low weight, which is ideal for aircraft engines, electricity-generating turbines, and a variety of other industrial equipment. Titanium from rutile is also used to make titanium oxide pigment, which is the white

▶ **Figure 2.28 Salt is a mineral.**
Common table salt is the mineral halite, also called rock salt. Halite crystallizes as cubes and commonly forms from the evaporation of water saturated with sodium and chloride.

color in paint, paper, and plastic. The principal source of aluminum is bauxite, the name applied to a mixture of very soft aluminum hydroxides. Aluminum has a wide variety of applications where low weight is essential, such as car parts, packaging and containers, and building products. Corundum, an aluminum oxide, is used as a polishing abrasive because of its high hardness (9 on the Mohs scale) and is also the source of the colorful, hard gemstones sapphire and ruby.

Atoms of many metallic elements are incompatible with silicate crystal structures but readily bond to sulfur (S) to form sulfide minerals illustrated in **Figure 2.30** (Table 2.2). Examples include galena (PbS), chalcopyrite (CuFeS$_2$), and sphalerite (ZnS). Pyrite (FeS$_2$), commonly called "fool's gold" for its resemblance to the precious metal, is a very common iron sulfide. Although the lead, copper, iron, and zinc atoms are easily separated from the sulfur atoms, there is a fundamental drawback to the process. The processing of sulfide minerals produces sulfuric acid as a by-product and it is quite damaging to the environment, including the effects of acid rain. To limit these harmful by-products, pyrite, one of the most common mineral occurrences of iron, is not commonly used as an ore for iron. Most iron is extracted instead from oxide minerals, such as hematite.

Lead (Pb), obtained from galena, has many industrial uses because it can be easily cast, molded, and shaped into pipes and storage vessels as well as used in ceramic glazes, glassware, car batteries, munitions (small bullets and shot), and even gasoline to improve engine performance. With mounting evidence that toxic levels of lead have accumulated in the environment, the use of lead has decreased as indicated by its removal from gasoline, paint, and glassware.

Zinc (Zn), obtained from sphalerite, is primarily used for galvanizing, a process that puts a protective coating on steel and iron to prevent corrosion. Zinc is also commonly used to make white pigment and in ointments and lotions to prevent sunburn and infections. U.S. pennies are mostly made of zinc with only a thin copper coat so as to conserve copper resources for other uses.

Copper (Cu) was one of the very first metals used by humans because it is present in some places as a native metal that, like gold shown in Figure 2.30, does not require chemical processing to separate the metal from other elements. Most copper production in the last century has, however, relied primarily on the extraction of copper sulfide minerals, such as chalcopyrite. Copper is also alloyed with tin to make bronze and with zinc to make brass. Copper and its alloys are attractive in appearance, very malleable, and corrosion resistant. As a consequence, copper and its alloys have a long history of use in weapons, tools, jewelry, pipes, and even utensils. With the great expansion in the use of electricity beginning in the late nineteenth century, the abundance and highly conductive nature of copper made it ideally suited for the manufacture of electrical wiring.

▲ **Figure 2.29 Magnetite is magnetic.**
Magnetite is an oxide mineral that is strongly magnetic. This particular property distinguishes magnetite from other, similar looking minerals.

▼ **Figure 2.30 A collection of metallic minerals.**
Galena and pyrite are examples of sulfide minerals in which metallic minerals bond with sulfur. Galena, a lead sulfide, is the most abundant source for industrial lead. Pyrite is iron sulfide and is sometimes called "fool's gold." Gold is an example of a mineral consisting of only a single element, similar to graphite and diamond.

Galena   Pyrite   Gold

Putting It Together–What Are the Most Important Minerals?

■ Minerals are most important if they are common, provide essential resources, or both.

■ Although there are more than 4000 known minerals, only a few dozen are important as rock-forming minerals, because only 12 of the 89 naturally occurring elements comprise 99.7 percent of the crust.

■ Silicon and oxygen are the most abundant elements in the crust and, bonded with other elements, form silicates, the principal rock-forming minerals.

■ Elements with similar ionic charge, size, or both substitute for each other within mineral structures. These substitutions cause minor changes in crystal structure that define groups of related minerals. Among the silicate minerals the most important substitutions are Al^{3+} for Si^{4+} and related interchange of K^+, Na^+, and Ca^{2+} in the feldspar group, and the substitution of Mg^{2+} and Fe^{2+} in the olivine, pyroxene, amphibole, and biotite groups.

■ Most economically valuable metals are processed from nonsilicate ore minerals, especially oxide and sulfide minerals.

Where Are You and Where Are You Going?

Minerals are the fundamental building blocks of the solid Earth. A mineral is a naturally occurring solid with a definite, only slightly variable chemical composition and an ordered atomic structure usually, but not always, formed by inorganic processes.

Each mineral possesses a unique set of observed or easily measured physical properties that directly relate to the elemental composition and internal atomic structure of the mineral. Color is commonly a reflection of composition and can vary significantly with only very small changes in composition. Specific gravity, closely akin to density, is determined by the mass of the atoms and their spacing. Scratching a mineral or breaking it into pieces both involve breaking chemical bonds. If a mineral breaks along predetermined planes to form flat, shiny surfaces, we say it has cleavage; otherwise, it exhibits fracture. Both hardness and cleavage, therefore, vary depending on the type of bonds, the proximity of bonded atoms, and the locations of repeated relatively weak bonds within the crystal's atomic structure.

Some groups of minerals differ only slightly in composition and structure because they are related to one another by the substitution of elements within the crystalline atomic structure. Elements of similar charge, size, or both can replace each other to cause minor changes in the structure, and in related physical properties, so as to produce distinctly different minerals.

Although there are more than 4000 known minerals on Earth, only a handful are found to predominate in most rocks, because just 12 elements account for 99.7 percent of the mass of the crust. The principal rock-forming-mineral class is the silicates. Quartz and feldspar are the most common silicate minerals in the crust, although others are also found in most rocks. A number of minerals precipitate from water, including most carbonates, sulfates, and some halides, such as halite (rock salt). Oxide and sulfide minerals are the primary ores for essential metals.

Now you are ready to embark on the study of rocks—what they are, how they form, and why they have so many different appearances. Rocks consist of one or more than one, and usually several, different minerals. Clearly, part of your understanding of rocks will need to include the processes that determine which minerals form together, why and how they form, and how the many different mineral grains become consolidated into a coherent rock. To pursue these questions geologists begin by examining rocks in their natural settings to look for clues. You will begin this avenue of inquiry by considering the full array of rocks and the general relationships that exist among the three rock groups in Chapter 3. Then you will consider the rock-forming processes for each group in greater detail in Chapters 4 through 6.

Extension Modules

Extension Module 2.1: Basics of an Atom. Learn about the basic components of an atom.

Extension Module 2.2: Silicate Mineral Structures. Learn how the SiO_4^{4-} groups, called silica tetrahedra, bond together to form a variety of silicate-mineral structures.

Extension Module 2.3: Gemstones. Learn about the features valued in gems and the types of minerals and rocks that form gemstones.

Confirm Your Knowledge

1. Calcite and quartz are two minerals. What properties do they have in common? How do they differ?

2. Why is color not a useful property in identifying a mineral?

3. Use the absolute hardness scale (Figure 2.4) to determine how many times harder fluorite is than gypsum.

4. Why don't all elements in a mineral show up on a TEM image?

5. List and describe the three types of bonds exhibited in minerals. For each bond type, give an example of a mineral that exhibits that type of bond.

6. What are three reasons why some minerals exhibit a variety of colors? Give an example for each explanation.

7. What is the implication of iron representing almost 35 percent of the entire Earth's composition yet less than 6 percent of Earth's crust?

8. Why are silicates the dominant rock-forming minerals?

9. Give two uses for each of the following metals.

   Iron   Lead
   Titanium   Zinc
   Aluminum   Copper

10. List and define the physical properties used to identify minerals.

11. What is the difference between density and specific gravity? What factors determine the specific gravity of a mineral?

12. How does cleavage differ from fracture? How does cleavage relate to mineral structure?

13. How does a geologist "see" atoms?

14. How does the type of bond affect the luster of a mineral?

15. Silicates are the most common mineral group, and differences between some silicates are due to only slight variations in their compositions. Explain the conditions necessary for substitutions to occur for Si and Mg.

16. List 6 minerals that are important economic ores.

Confirm Your Understanding

1. Write out an answer for each question in the Chapter Outline for the chapter sections assigned by your instructor.

2. Calcite and aragonite are mineral polymorphs. They have the same chemical composition but are different minerals because their atoms are arranged differently. Research two other pairs of minerals that are polymorphs.

3. In order for a substance to be a mineral it must be a naturally occurring, usually inorganic solid with a definite chemical composition and an ordered atomic arrangement. For each of the following substances determine if it is a mineral or not. If it is not a mineral list all of the properties that don't apply.
   a. amber
   b. beer
   c. calcite
   d. diamond
   e. emerald
   f. plagioclase feldspar
   g. glacial ice
   h. halite
   i. ice cubes from an ice machine
   j. rock candy
   k. kryptonite
   l. cubic zirconia
   m. mineral oil
   n. muscovite
   o. obsidian

4. Minerals have a wide variety of applications. For five of the following substances identify what minerals they contain.
   Baby powder
   Antacid
   Toothpaste
   Toothpaste with sparkles
   Epsom salt
   Red lipstick
   Red blush
   Table salt
   Kitty litter

5. Determine if your state has a state gemstone or state mineral. Research your state gemstone or state mineral and list its chemical formula, two physical properties and the reason why it is your state gemstone or state mineral.

6. What are some possible reasons that color and streak of a mineral are not necessarily the same?

7. Consider the van der Waals force and dissolution of halite (NaCl) in water. Why does quartz not dissolve readily in water?

Key Terms

bond (p. 31)
cleavage (p. 27)
covalent bond (p. 31)
crystal faces (p. 24)
density (p. 24)

fracture (p. 27)
hardness (p. 24)
ionic bonds (p. 31)
ions (p. 31)
luster (p. 24)

metallic bonds (p. 31)
mineral (p. 35)
Mohs hardness scale (p. 24)
ore minerals (p. 39)
polymorphs (p. 34)

specific gravity (p. 24)
streak (p. 27)
van der Waals force (p. 31)

3 Rocks and Rock-Forming Processes

Chapter Outline

3.1 How and where do rocks form?
3.2 Can rocks be classified according to the processes that form them?
3.3 How do we know ... how to determine rock origins?
3.4 How are the rock classes related to one another?

Why Study Rocks?

ASK A NONGEOLOGIST WHAT GEOLOGY IS ALL ABOUT AND THERE is good chance that he or she will say, "Geology is the study of rocks." Rocks certainly are central to most geologic studies. Most of Earth is solid rock, so understanding what rocks are and how they form are important aspects of learning geology.

Almost all natural rocks are consolidated aggregates of minerals. How do minerals combine to make rocks? What do rock-forming processes reveal about how Earth works? Rocks form over a wide range of conditions. You learned in the previous chapter that some minerals form by precipitation from solution (e.g., calcite) or are left over after evaporation (e.g., halite). Minerals also solidify from cooling molten mixtures and transform from one to another under different geologic conditions. Rock-forming processes must be understood to classify rocks in a meaningful way. In this chapter you will learn that observations, information about rock origins, and basic scientific classification principles all

work together in geology to yield a working classification system for rocks. This classification provides a framework for understanding rock-forming processes in the chapters that follow.

There are two primary objectives in this chapter:

✔ To learn the basics of rock formation and the relationships among rock types

✔ To learn how to classify rocks and the factors that contribute to geologic classification systems

To accomplish these objectives, the following four questions are addressed:

3.1 How and where do rocks form?
3.2 Can rocks be classified according to the processes that form them?
3.3 *How do we know ...* how to determine rock origins?
3.4 How are the rock classes related to one another?

   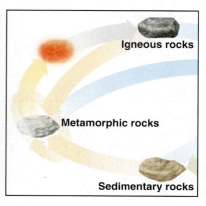

Rock climber reaches for a hold on a New York cliff. ▶

In the FIELD

Imagine yourself on a rafting field trip to see the geology of the Grand Canyon. **Figure 3.1** shows some of the views you might encounter. Few localities in the world expose so much bare rock continuously for more than 200 kilometers. No wonder geologists flock to this place to study Earth processes and history.

Your eye quickly discerns differences among the rocks (Figures 3.1a and b). They vary considerably in color: Many are red, quite a few outcrops are black, other rocks are off-white to gray, and a few have a greenish tint. You assume that these differences somehow relate to the combinations of minerals within the rocks. Some rocks form vertical cliffs, whereas others underlie more easily climbed slopes of loose rubble. These differences in slope must have something to do with the durability of the different rocks when they are exposed to weather—rain, snow, ice, and wind—over long periods of time.

The rocks are also striking in their patterns across the landscape. Most of the rocks form very even, nearly horizontal layers that are continuous as far as your eye can see (Figure 3.1a). In the dark depths at the bottom of the canyon, however, are more chaotic bands of color (Figure 3.1b). What processes account for these patterns?

Up close, some of the rocks contain large mineral crystals (Figure 3.1c). Others contain no clearly visible minerals and will require a microscope to determine what composes them. Some contain remarkable fossils of ancient animals (Figure 3.1d).

The raft trip is memorable for the spectacular scenery and for whetting your appetite to learn more about rocks. Why do rocks exhibit such a huge variety of colors, patterns, durabilities, mineral compositions, and mineral textures? Do these variations relate to an equally diverse set of processes that form rocks? What are these processes and how can you hope to understand them from studying rocks? Where do you begin in classifying and naming this great variety of rock types? To answer these questions, you need to expand your experience in looking at rocks.

Figure 3.1 **Rocks in the Grand Canyon.** ▶
Scenes from a raft trip in the Grand Canyon reveal a wide variety of rocks that are distinguished by color, presence or absence of persistent layers, identifiable garnet crystals, and even fossils in some cases.

3.1 How and Where Do Rocks Form?

Scientific explanations of natural processes require observations in the natural world. Observations of rocks provide insights into rock-forming processes. Three imaginary, but realistic, field trips can provide the necessary observations and insights.

Observations on a Beach

Walking along an ocean beach, you feel gritty sand under your bare feet, punctuated by sharp-edged shell fragments. Waves crash against the shore. A small stream crosses the beach and enters the sea. A low cliff of solid, layered rock rises above the landward side of the beach. **Figure 3.2**a depicts this location.

On your walk, you first wonder, what is the sand composed of? Where did the sand come from? Figure 3.2b shows a handful of sand grains, seen with the aid of a magnifying glass. Most grains are colorless and glassy-looking quartz. Other, darker grains are present, too, including some that move toward the magnet attached to your key chain; this indicates the presence of magnetite. You guess that the broken shell fragments in the sand (Figure 3.2b) are composed of calcite or aragonite. If handy, adding a drop of dilute acid could test this idea. Identifying all of the tiny grains is challenging, but the sand is clearly composed of mineral grains.

The mineral grains and most of the shell fragments are smooth and round, with polished edges (Figure 3.2b). Breaking one of the shells produces a contrasting sharp, jagged edge. Clearly, the edges were somehow modified to form the smooth margins observed. A possible solution arises while watching the movement of sand and shell fragments as waves surge across the beach—the waves and surf grind the grains against one another, abrading away sharp edges and smoothing the outlines. Could it be, as well, that the mineral grains began as angular fragments separated from a large rock outcrop and then abraded into the smoothly rounded forms seen today?

A stream crossing the beach clearly delivers water to the sea, but a closer look, shown in Figure 3.2c indicates that the stream also transports sand, small pebbles, and suspended silt. Some of these moving particles appear well rounded, like the sand grains on the beach. Transport along the streambed may also reshape originally angular grains before they reach the ocean. Gusts of wind sting your face with blowing sand and are a reminder that wind, as well as water, moves sediment particles and likely contributes to the abrasion process. Ocean waves crash into the mouth of the stream and spread the recently delivered sediment along the beach. At least some of the beach sand, therefore, does not originate in the ocean but is fed into it by streams and simply redistributed by the ocean waves and currents, which add in local shell

<u>Field observations in the Grand Canyon</u>

A. These rocks form persistent layers. Some layers are apparently stronger than others, with strong layers forming cliffs and weaker layers making rubbly slopes.

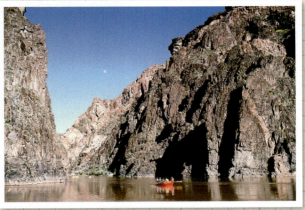

B. These rocks in the bottom of the canyon have a few discontinuous bands that seem to swirl up and down the canyon wall.

Lots of shiny mica crystals reflect the sunlight

Garnet crystals

C. Some of the dark rocks in the bottom of the canyon have visible crystals, like this rock that has little garnets surrounded by shiny mica.

D. This rock has fossils in it. According to the river guide, this is an ancient chambered nautilus. Lens cap shows the scale.

▶ **Figure 3.2** **Geologic processes at a beach.**

(a)

Waves move sand to and fro on a beach. The nearby sea cliff consists of sand grains consolidated into rock.

(b)

A close look at the beach sand shows that it consists of rounded mineral grains, with the glassy appearance of quartz, and colorful, polished pieces of broken shells.

(c)

A stream crosses the beach and flows into the ocean. The current slowly rolls pebbles along the streambed. The water is cloudy with suspended grains of sand and silt.

(d)

The rock composing the nearby sea cliff is a consolidated mixture of sand grains and fossil shell fragments that resembles the modern beach.

fragments. The mineral and rock grains have, therefore, come from higher inland areas through which the stream flows. Somehow, the mineral grains composing most of the sand break off from rocks by processes that are not obvious from this vantage point.

Next, your attention turns to the low cliff of layered rock at the landward edge of the beach. Examining a piece of rock dislodged from the cliff, you notice that it shares some similarities with the beach sand. As seen in Figure 3.2d, the rock also consists of rounded sand grains of quartz and fragments of broken shells. Using the principle of uniformitarianism, you infer that the grains in this rock were originally deposited in a setting like the modern beach. The most obvious difference is that the sand grains in the rock are not loose but have somehow been stuck together (compare Figures 3.2b and 3.2d). It seems reasonable to apply the name **sandstone** to a rock composed of sand grains, to distinguish the rock from the loose sand on the beach. Using a magnifying glass, you see a pale film along the margins of the grains in the sandstone that appears to act like glue to hold them together. Applying dilute acid causes the film to fizz and dissolve, suggesting that it is a carbonate mineral, such as calcite.

Spring water emerges from the top of a mound of white rock, stained orange by living bacteria. Notice the dead trees entombed in the rock.

The rock around the spring opening consists of thin layers of tiny crystals and sponge-like holes.

You leave the beach with some important knowledge of rock-forming processes:

- For some rocks, at least, the mineral constituents are pieces of other rocks with the addition of some biologically produced mineral matter.
- An abrasion process—the grinding of pieces together during transport by moving water or wind—reshapes the fragments.
- Fragments of older rock consolidate into new rock by the addition of mineral cement.

Observations at a Spring

The next stop is at a spring, illustrated in **Figure 3.3**a. Water issues from an opening atop a mound of light-colored rock. Examining this rock more closely, there are thin layers and very small, clear to white crystals, as shown in Figure 3.3b. The rock reacts with mild acid, which again suggests a carbonate mineral, perhaps calcite. The layers in the rock slope uniformly outward from the spring opening, and the calcite-rich rock is found only immediately surrounding the spring, suggesting a relationship between the rock and the exiting spring water. Another striking feature visible in Figure 3.3a is that the rock encloses dead trees. This rock, therefore, seems to form from the spring water by mineral growth at the surface. As the rock builds outward it incorporates plants. You hypothesize that the rock forms by a chemical reaction that precipitates calcite, forming solids out of a watery solution as spring water flows away from its exit. This is similar to halite (table salt) crystals forming along the edge of a glass as salty water evaporates and mineral to crusts clogging water pipes.

► **Figure 3.4 Rock-forming processes at a volcano.**

(a)

An erupting volcano puts on a dazzling display as lava fountains into the air and then falls back to the ground to flow away in molten rivers.

(b)

Red-hot, fluid lava solidifies into hard black rock around a geologist's rock hammer.

Olivine
Plagioclase feldspar
Pyroxene
(c)   10 mm   0.5 mm

The volcanic rock contains a few visible crystals of green olivine and white plagioclase feldspar set against a dull, gray background.
A microscopic view shows that the gray background consists of interlocking crystals.

Mosaic of feldspar, pyroxene, & olivine
Bands of biotite, quartz, & feldspar
(d)   CENTIMETER

The volcano ejected these blocks, which differ from the solidified lava. The rock on the left is a coarser-grained version of the lava rock, consisting entirely of visible crystals. The rock on the right contains different minerals in distinct bands. These rock types underlie the volcano and the pieces were blown out during the eruption.

Remembering the sandstone at the beach composed of mineral grains cemented by calcite, you can now hypothesize that the cement precipitated from water percolating through what was originally loose sediment. The rock formed at the spring resulted almost entirely from such precipitation, and the precipitated minerals are so closely intergrown that they form a coherent rock. You have observed a second rock formation process:

• Some rocks form by precipitation of minerals from water that cements loose sediment together, or by intergrowth of precipitated minerals.

Observations at a Volcano

Perhaps no geologic phenomenon simultaneously causes awe and fear so dramatically as the erupting volcano at the third location, portrayed in **Figure 3.4**a. Fiery jets of molten material shoot into the sky and flow away in rivers of lava. Unlike rivers of water, these molten streams solidify—they literally turn to stone. Figure 3.4b shows that the fluid lava turns to rock as it cools, much as liquid water chills to form solid ice. Clearly, the temperature at which the lava transforms from liquid to solid is much higher than the temperature for the conversion of water to ice. Most water on Earth's surface is liquid but lava "freezes" into rock.

Looking very carefully at the safely cooled volcanic rock, shown in Figure 3.4c, it is difficult to see what composes it. A few visible blocky green olivine and white, rectangular prisms of feldspar are surrounded by a nondescript black background. Most of a volcanic rock consists of minerals too small to see with a magnifying glass; you need to specially prepare the rock and view it under a microscope, where minute crystals of olivine, plagioclase feldspar, and pyroxene are visible (Figure 3.4c). It is not surprising that these are all silicate minerals, given the dominance of the silicate minerals on Earth (Section 2.7). The crystals intergrow with one another to form a coherent rock, much as the interlocking of ice crystals produces coherent ice cubes.

What distinguishes this volcanic rock in appearance from those observed at the beach and the spring? These rocks clearly formed by different processes:

• A major difference is the shape of the mineral grains in the sandstone compared to those in the volcanic rock. The mineral grains in the volcanic rock have sharp edges and flat surfaces representing the crystal faces (Figure 3.4c) whereas those in the sandstone are rounded and do not preserve crystal outlines. Minerals in the volcanic rock crystallized in place from the original molten material and were not tumbled and abraded during transport by moving water and wind.

• Intergrowth of newly formed minerals explains the solid nature of the rock formed at both the spring and the volcano. These rock types differ, however, in that the rock at the spring formed by precipitation of minerals out of water that remained liquid. The volcanic rock formed by the complete crystallization of a molten-rock liquid.

Volcano craters are rimmed not only by solidified lava but also by rock fragments ripped from the walls of the volcanic conduit. These fragments reveal the kinds of rocks hidden below the volcano.

One group of ripped-up rock fragments, shown in Figure 3.4d, contains the same minerals seen in the solidified lava flows, but in this case they are readily visible to the naked eye. The intergrown mineral texture, like the lava rock, suggests solidification by freezing of a liquid. The larger crystal sizes, however, suggest some different conditions during mineral and rock formation.

A second group of ripped-up rocks seen in Figure 3.4d contains angular minerals in distinct layers or bands. The minerals are intergrown in a fashion vaguely similar to the calcite-rich spring rock and the lava rock, but the minerals are curiously arranged in alternating bands of dark and light crystals unlike anything you have seen forming at the surface.

Apparently, these two groups of rocks ejected from the volcano came from some depth below Earth's surface because they are quite different from rocks observed forming at the surface. These observations imply that rock-forming processes occur within, as well as on, Earth.

*Putting It Together–*How and Where Do Rocks Form?
- Most natural rocks are aggregates of mineral grains.
- Many rocks originate from observable processes that take place at Earth's surface.
- The presence of rocks that are not related to observable surface processes suggests that they relate to processes active within Earth.

3.2 Can Rocks Be Classified According to the Processes that Form Them?

Classification is an important part of natural science. Scientists commonly group features or phenomena according to similarities in order to seek explanations for their origins. *Descriptive classifications* group items of similar appearance. *Genetic classifications* group features or phenomena by noting similarities in the processes that cause or create them. Chapter 2 utilized a descriptive classification to group minerals based on chemical compositions and, in some cases, crystal structure. You may be familiar with other descriptive classifications of natural objects, such as animals and plants. Historically, the principal groupings of rocks have emphasized rock origin.

External and Internal Processes

Field observations (Section 3.1) suggest that some rocks form by processes that leave visible evidence at the surface, whereas other rocks likely originate below the surface. Might the distinction between internal and external processes serve as a means to classify rocks? The sandstone near the beach, the rock formed by precipitation of calcite from the spring water, and the rock formed by crystallization of cooling molten lava erupted from the volcano could be viewed as rocks forming by external processes. Rocks that contain minerals similar to volcanic rocks, but with much larger crystals, and others with bands of different kinds of minerals could be grouped as those formed by interior processes.

Although this genetic division seems workable at first, it is not wholly satisfying. Consider the rocks composed of similar minerals intergrown in a fashion

TABLE 3.1 The Genetic Classification of Rocks						
Origin	Rocks formed from the products of the breakdown of preexisting rock		Rocks formed by the solidification of molten rock material (magma)		Rocks formed by the solid-state transformation of minerals in a preexisting rock under the influence of elevated temperature, pressure, hot fluids, or all three variables		
Rock Class	Sedimentary Rocks		Igneous Rocks		Metamorphic Rocks		
Principal Subdivisions	*Clastic Rocks:* Formed from particles of preexisting rocks liberated by physical weathering or formed during chemical weathering	*Chemical and Biogenic Rocks:* Formed of mineral precipitates originating by dissolution of rock; includes biologic materials, mostly formed of mineral precipitates	*Volcanic (Extrusive) Rocks:* Formed by solidification of magma (lava) at Earth's surface	*Plutonic (Intrusive) Rocks:* Formed by solidification of magma below Earth's surface	*Regional Metamorphic Rocks:* Formed over large areas, typically by the combined influence of increasing temperature and pressure	*Contact Metamorphic Rocks:* Formed in small areas and resulting from locally elevated temperature adjacent to magma intrusions	*Hydrothermal Metamorphic Rocks:* Formed by reactions with hot fluids circulating through rock

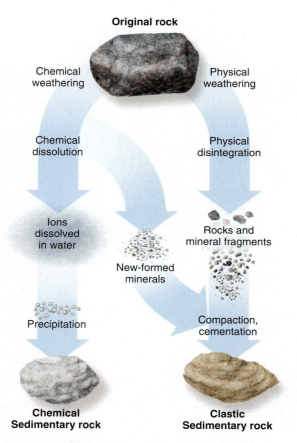

▲ **Figure 3.5 Sedimentary rocks contain the weathering products of other rocks.**
Rocks break down at the surface through the weathering processes of physical disintegration and chemical dissolution. Clastic sedimentary rocks contain fragments set loose through physical weathering or newly formed by chemical weathering. Changing chemical conditions can cause ions dissolved in water to precipitate as interlocking mineral grains to form chemical sedimentary rocks.

suggesting crystallization of a freezing liquid (Figures 3.4c and 3.4d). Descriptively, these rocks differ only in the size of the mineral grains composing them, which suggests that a great similarity in process should link them, rather than separate them, into distinct genetic categories. Rocks formed from solidification of molten material, therefore, could be grouped together. The rocks with large crystals are the product of crystallization below ground, whereas the volcanic rock with smaller crystals forms after the melt breaches the surface. The distinction of internal and external processes is further blurred because although the volcanic rock formed on the surface, the molten liquid it solidified from clearly originated below the surface. Volcanic activity is merely one of the more obvious of many links to be explored between the Earth's interior workings and its surface features. True, some processes are wholly interior and some are wholly exterior to the planet, but the formation of rocks exposed at the surface is the result of complex relationships between different processes in these two locations. It seems unwise, therefore, to use the site of formation as the primary criterion for classification.

The Three Rock Classes

Geologists use a three-category rock classification scheme, based on rock-forming processes rather than on where rocks form. The three rock types are **sedimentary**, **igneous**, and **metamorphic** and are described in **Table 3.1**. A brief overview of these rock classes and their origins and relationships to one another is presented here and is followed with three chapters, one devoted to each rock type. The genetic classification emphasizes the processes responsible for forming rocks in each class. Nonetheless, you will see in later chapters that further divisions within each rock class are descriptive, with an emphasis on visible features rather than inference on the origins of these features.

Sedimentary Rocks

Sedimentary rocks form by deposition and precipitation of products formed by the breakdown of older rocks under the ordinary conditions found in Earth's surface environment. Rocks of any kind exposed at the surface slowly deteriorate by the processes of **weathering**, schematically illustrated in **Figure 3.5**. Physical weathering causes rocks to disintegrate into smaller rocks or into even smaller mineral grains. Chemical weathering involves reactions between minerals and water to produce dissolved ions and new minerals. Mineral or rock fragments produced by weathering, such as the sand you observed on the beach (Section 3.1), are called **clastic sediment** (from the Greek *clastos*, meaning "broken") (Table 3.1; Figure 3.5).

Mineral constituents dissolved in water by chemical weathering may later precipitate as new mineral crystals if conditions of water temperature, pressure, or element composition change appropriately. These precipitates form **chemical sedimentary rocks** (Table 3.1), such as the rock you observed forming at the spring in Section 3.1. The minerals precipitated from water are those whose chemical constituents most readily dissolve in water. Although silicate minerals are the most abundant constituents of Earth's crust, they do not dissolve as easily as other more strongly ionic compounds, such as carbonates, sulfates, common salts, and some oxides (see Section 2.4). Chemical sedimentary rocks are, therefore, most commonly composed of minerals such as calcite, dolomite, gypsum, halite, and hematite rather than silicate minerals. Quartz is the only silicate mineral to form a common mineral precipitate from water.

Precipitated minerals also cement particles of loose clastic sediment to form clastic sedimentary rocks, as observed in the beachside sandstone in Section 3.1. Clastic sedimentary rocks consist of particles generated by weathering and then transported to a final site of deposition by flowing water, blowing wind, or glaciers. Compaction under the weight of accumulating sediment and precipitation of cementing minerals between grains convert loose sediment into consolidated sedimentary rocks. This transformation is **lithification**, which loosely translates as "making of rock" (Greek *lithos* means "rock").

The twofold division of sedimentary rocks into clastic and chemical types is not without some minor problems. For instance, in Chapter 4 you will appreciate that volcanic explosions produce fragments, such as volcanic ash and pumice. When these fragments of volcanic origin wash away and ultimately show up in sedimentary rocks, it is clear that they did not originate from the physical disintegration of an older rock by weathering but rather formed by igneous processes. Also, you will need to allow for the presence of biologically produced fragments, such as the shells in the sandstone (Figure 3.2d), within a more complete classification of sedimentary rocks in Chapter 5.

Sedimentary rocks are most readily recognized from a distance by their distinctive layering, called **bedding**. Bedding was a conspicuous feature seen on your Grand Canyon raft trip (Figure 3.1a). Bedding originates from the sequential deposition of different sedimentary materials, one above the other, at Earth's surface.

Igneous Rocks

Igneous rocks crystallize from molten material called **magma**, which originates from melting rock within Earth. The root for "igneous" is the Latin word *igneus*, meaning "fiery," which is an appropriate label for these rocks. **Figure 3.6** shows that magma may erupt at volcanoes, as lava and volcanic ash, or remain underground to solidify. **Volcanic rocks** solidify at the surface whereas **plutonic rocks** (named for Pluto, the Greek god of the underworld) solidify beneath the surface. In some cases, their synonyms **extrusive** (or volcanic) and **intrusive** (or plutonic) are applied and refer to, respectively, magma extruded onto the surface at volcanoes and intruded into preexisting rocks below the surface.

Experiments show that fast magma solidification usually produces small crystals and slower crystallization favors formation of larger crystals. Although there are exceptions to this relationship between cooling rate and crystal size, it helps to account for the different appearances of the two types of rock that originate from magma (Table 3.1). The generally fine-grained rocks result from rapid cooling following eruptions from volcanoes, as you encountered in Section 3.1 (see Figures 3.4b and c). The coarser rocks form from the slow, subterranean solidification of magma (an example is the rock on the left in Figure 3.4d). Magma is overwhelmingly dominated by the elements composing silicate minerals. Igneous rocks, therefore, consist mostly of silicate minerals, a composition that also helps to distinguish them from chemical sedimentary rocks, in most cases.

The accumulation of successive lava flows or other volcanic materials on the surface produces layers that superficially resemble sedimentary-rock bedding,

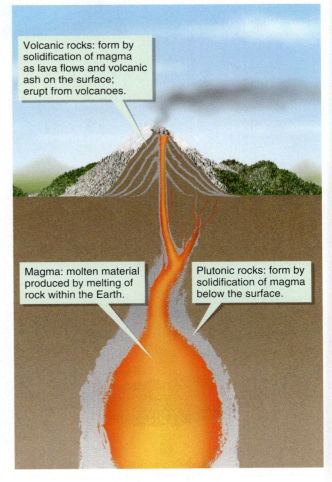

Volcanic rocks: form by solidification of magma as lava flows and volcanic ash on the surface; erupt from volcanoes.

Magma: molten material produced by melting of rock within the Earth.

Plutonic rocks: form by solidification of magma below the surface.

▲ **Figure 3.6 Igneous rocks originate from magma.**

▶ **Figure 3.7 What igneous rocks look like in the field.**

Successive lava flows stack up in layers, as in this view along the Columbia River in Washington. Notice how thick the lava layers are compared to the train.

Plutonic igneous rocks are usually massive without layers. Plutonic igneous rocks compose many bold mountain landscapes, like the Wind River Mountains in Wyoming.

Original rock

Increasing temperature and, in most cases, pressure

Metamorphism: new minerals form and align perpendicular to the applied pressure, if present.

Metamorphic rock

Applied pressure

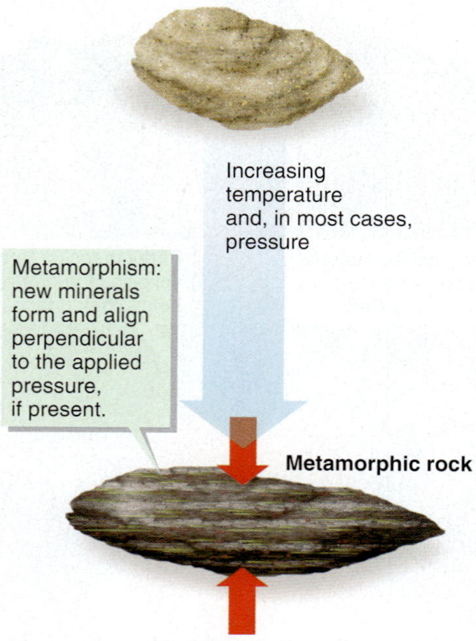

▲ **Figure 3.8 Metamorphic rocks form from preexisting rocks.**

as shown in **Figure 3.7**a. Plutonic rocks, however, rarely accumulate as successive layers (see Figure 3.7b), and they seldom contain features that might be mistaken as the characteristic bedding of sedimentary rocks. Geologists use the term **massive** to describe rocks that lack layering.

Metamorphic Rocks

Metamorphic rocks result from changes of preexisting rocks (Greek *meta* means "change" and *morphe* means "form"). The changes typically occur at temperatures in excess of 200°C and therefore beyond the typical realm of rock weathering at the surface and lithification of sedimentary rock. Notably, metamorphism does not include very high temperatures where melting occurs. A change in the appearance of a rock requires a change in the minerals that compose it, or a rearrangement of the existing minerals, or both. In nearly all cases the processes causing these changes, referred to as **metamorphism**, take place at temperatures and pressures found only within Earth and not at the surface.

Field observations and laboratory experiments summarized in **Figure 3.8** show that metamorphism almost always occurs at elevated temperatures. High pressure and chemical reactions between minerals and high-temperature fluid are also important metamorphic agents. Pressure increases with depth in Earth because of the increasing weight of overlying rock. Temperatures measured in deep wells and mines show that Earth's interior is hotter than the surface and that temperature increases progressively with increasing depth. Another clue to internal heat is magma that forms by melting rock at temperatures exceeding 1000°C and rises from somewhere below to be erupted at volcanoes. Metamorphic transformations, therefore, require conditions found inside Earth.

Geologists simulate metamorphic temperatures and pressures in the laboratory. These experiments reveal that minerals within rocks of any origin change under conditions present at great depths within the crust or mantle, especially in the presence of watery fluids. Changes range from modifications in the shape and orientation of the existing crystals to wholesale changes in the minerals found in the rock. The new minerals are the products of chemical reactions between the original minerals and between the minerals and hot water rich in dissolved ions that passes through the rock.

Metamorphic changes affect vast regions, particularly at convergent plate boundaries (Figure 1.11b), where rocks originating at or near the surface are forced down to great depth and have high, near-horizontal pressure exerted on them in addition to the vertically directed weight of overlying rock. These **regional metamorphic rocks** (Table 3.1) include those with minerals oriented preferentially in layers or bands (right side of Figure 3.4d and Figure 3.8). The rocks seen deep in the Grand Canyon in Figure 3.1b are dark, regional metamorphic rocks cut through by pink intrusive igneous rocks.

In other cases, metamorphism is a very local phenomenon, a relatively simple case of the intrusion or extrusion of magma cooking the adjacent rocks. The resulting **contact metamorphic rocks** are seen at the bottom of some lava flows, but their occurrence is most impressive adjacent to large igneous intrusions. Here, the great magmatic heat together with the naturally elevated temperature at depth and large volumes of hot water that circulate through rocks near the intrusions cause dramatic metamorphic changes.

Hydrothermal metamorphic rocks form where hot-fluid reactions with rock are the primary cause of metamorphism. These rocks may be found with regional and contact metamorphic rocks or separately, as in the case of metamorphism of igneous rocks at divergent plate boundaries, where seawater flows through rock heated by the rising magma (Figure 1.11a). Hydrothermally metamorphosed rocks commonly host important metal-ore deposits.

The minerals found in metamorphic rocks depend on the minerals in the original rock and the nature of the metamorphic processes. Metamorphic rocks commonly contain abundant silicate minerals, but many of these silicate minerals are minor or nonexistent in rocks that crystallize from magma. This fact helps to distinguish metamorphic and igneous rocks.

The growth of metamorphic minerals in bands, illustrated in **Figure 3.9**, only superficially resembles sedimentary bedding, and the bands are neither continuous nor uniform in thickness over large distances. In contrast, sedimentary beds are almost always readily traced across whole outcrop faces hundreds of meters long and typically can be traced for many kilometers. The banding of many metamorphic rocks may be apparent from great distances, but rarely is it as distinct and continuous as sedimentary bedding (compare Figure 3.9 to Figure 3.1a).

(a)

Bands of different minerals are common in metamorphic rocks, such as the swirling dark and light-colored bands in this rock outcrop in Scotland.

Putting It Together—*Can Rocks Be Classified According to the Processes that Form Them?*

- Rocks can be classified descriptively or genetically. The three principal rock classes are defined genetically, according to the processes that form them.
- Igneous rocks crystallize as intergrown mineral aggregates from molten lava erupted at volcanoes (volcanic, or extrusive, rocks) or from magma that remains to solidify below ground (plutonic, or intrusive, rocks).
- Sedimentary rocks consist of minerals derived from the physical disintegration or chemical dissolution of preexisting rocks. Clastic sedimentary rocks contain particles of weathered rock cemented by minerals that precipitate from water between the sediment grains. Chemical sedimentary rocks are intergrown mineral aggregates that precipitate from water. Biological components are significant constituents of some sedimentary rocks.
- Metamorphic rocks form by reactions of minerals in preexisting rocks where changing temperature, pressure, fluid composition, or all three produce new minerals and rock characteristics.

3.3 **How Do We Know** ... How to Determine Rock Origins?

PICTURE THE PROBLEM
How Do Geologists Use Observations of Rocks to Infer Process?
Do you feel confident that you could now classify a rock into one of the three principal classes? Whatever confusion you may experience at this point is inherent to the problem of comprehending a genetic classification scheme. Descriptive classifications require only your powers of observation, whereas genetic divisions require drawing interpretations from those observations. The imaginary experiences at the beach, spring, and volcano are similar to those of early geologists when they formulated questions and made inferences from their observations of rock-forming processes. The current threefold genetic classification scheme is less than 200 years old and represents the efforts of geologists of the late eighteenth and early nineteenth centuries, who wrestled to understand how rocks form.

(b)

The sheer cliffs of the Black Canyon of the Gunnison River, in Colorado, consist mostly of metamorphic rocks. The rocks appear massive from a distance but contain fine-scale mineral banding like that seen in photo (a).

▲ **Figure 3.9 What metamorphic rocks look like in the field.**

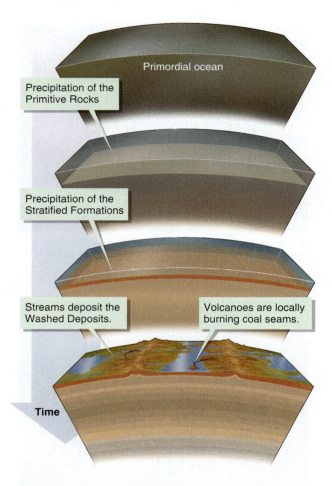

Primordial ocean

Precipitation of the Primitive Rocks

Precipitation of the Stratified Formations

Streams deposit the Washed Deposits.

Volcanoes are locally burning coal seams.

Time

▲ **Figure 3.10 The Neptunist view of Earth.**
Werner envisioned concentric shells of different types of rocks and minerals that precipitated in sequence from a "primordial ocean" solution that was extremely rich in dissolved ions. The interior of the planet was cold. Volcanoes were rare, relatively recent phenomena related to the natural burning of subterranean coal beds.

Two competing views of rock-forming processes arose in the geologic community of the late 1700s. Members of both intellectual camps appreciated the wearing down of rocky landscapes to make loose sand, gravel, and mud, just as you readily observed those processes today. When it came, however, to explaining how consolidated, coherent rocks form, the two groups reached very different conclusions. The disciples of each view feuded before the scientific gatherings and in the written communications of the time. Neither group turned out to be entirely correct. Today, however, looking back at their debate and how their views arose illustrates how observations, inference, and critical review come together in evolving scientific knowledge.

THE NEPTUNIST VIEW
Do Most Rocks Precipitate from Water?
A group of geologists later known as the Neptunists emphasized a cold Earth with nearly all rocks forming by chemical precipitation from water. The name *Neptunist* is derived from Neptune, the Roman god of the sea. Abraham Werner (1749–1817) developed the Neptunist view. He was a professor at a German mining academy whose stimulating instructional style attracted students from all over Europe to learn about the origin of rocks. Their studies also promoted understanding of the formation and location of economically important minerals utilized to fuel the industrial revolution. Werner went beyond having students examine rock collections in laboratories and giving lectures that conveyed second-hand information. He also took his students to study rocks in the mines near Freiberg. His real-world approach may have originated the tradition of field instruction that is required in most university geoscience curricula today.

Werner's concepts built upon prevailing views of eighteenth-century natural philosophers that Earth originated as a vast blob of material uniformly dispersed in water. He envisioned this primordial ocean as a thick, dense mixture of solids and water rather than a relatively thin, dilute, surficial fluid with dissolved salts like the modern seas. From this mucky beginning solids precipitated and the liquid somehow disappeared into space.

As illustrated in **Figure 3.10**, Werner held that the sequence of precipitation of different minerals was readily determined from looking at the rocks they form. Werner concluded that successive layers of rock must form in consecutive order much like a pile of newspaper may accumulate in your home—the oldest papers lie on the bottom, followed upward by successively more recent editions.

Werner noted that the most conspicuously layered rocks commonly rested above more massive rocks, where the latter were exposed in deep river canyons (e.g., Figure 3.1b) or heaved upward in mountains (e.g., Figure 3.7b). This observation implies that the more massive rocks formed first. He called these very thick, unlayered, and poorly layered rocks, whose bottoms were never exposed to view, the Primitive Rocks. These rocks formed, according to Werner, by precipitation and settling of the least soluble silicate minerals in the earliest time before dry land even appeared. The Primitive Rocks are overlain by the Stratified Formations, distinctively layered rocks composed of precipitates of more soluble minerals mixed in places with fragments worn from highlands of Primitive Rocks, which were gradually exposed as the enclosing ocean diminished in depth.

Of lesser significance and found only at the surface because of their recent formation, according to Werner, are rocks originating from volcanoes and unconsolidated gravel, sand, clay, and soil. The unconsolidated sediment composed the Washed Deposits, viewed more as residue left behind by the eroding waves of the huge shrinking sea than as a result of the ongoing modification of the planet surface by wind and moving water. Werner viewed volcanoes as minor aberrations formed only very recently, and only at places where underground coal beds had somehow burst into flames and melted overlying rock, which then extruded at volcanoes.

To get a better grasp of Werner's classification, try to equate his rock types to the classification presented in the previous section. To do this, visualize the

Primitive Rocks as including what the modern classification labels as plutonic-igneous and most metamorphic rocks. The Stratified Formations include sedimentary rocks as well as many volcanic rocks (e.g., layered lava flows like those illustrated in Figure 3.7a) and some conspicuously banded metamorphic rocks (e.g., Figure 3.9).

THE VULCANIST VIEW

Do Most Rocks Result from Earth's Internal Heat?

Opposing the Neptunists, the Vulcanists thought of an inherently hot planet with rocks forming from the fusing of particles by heat, or from solidification of magma rising from a molten interior. The name *Vulcanist* relates to Vulcan, the Roman god of fire. Principal among Werner's critical contemporaries was the man eventually labeled as the champion of the Vulcanists. James Hutton (1726–1797) was the Scottish gentleman farmer who helped to develop the principle of uniformitarianism (Section 1.4). Experienced and interested in scientific endeavors, he sought to make his farm as productive and profitable as possible by studying farming methods in his wide travels through Britain and the European continent. It is easy to understand how his agricultural experience led to a curiosity for the characteristics and origin of soil and, eventually, to the rocks found beneath the soil.

At several localities, Hutton made observations such as his illustration in **Figure 3.11** that were vital to understanding the origin and classification of rocks. Hutton observed Stratified Formations above Primitive Rock, as predicted by Werner's concept of successive precipitates from water. It was also quite clear to Hutton, however, that narrow veins of the Primitive Rock continued upward into the overlying Stratified Formations. This implied that the Primitive Rock could not always be the older of the two, because the Stratified Formations had to be present before the Primitive Rock could be injected into them. To Hutton, the fingerlike veins of the Primitive Rock suggested the injection of a liquid along cracks in the Stratified Formations. Furthermore, at the boundary of the two rock types the layered formations were modified as if baked at high temperature.

Hutton demonstrated circumstances where the Primitive Rocks, even though underlying the Stratified Formations, could not have come first, as prescribed by Werner. Furthermore, Hutton saw the action of heat (what we now call metamorphism) and the injection of a hot liquid from which veins of Primitive Rock solidified. This was contrary to the Neptunian view of a cold planet and the notion that rocks formed from the precipitation of minerals from water. Hutton cast further doubt on the precipitation of the silicate-rich Primitive Rocks from water by noting, from his experiments, that most silicate minerals are practically insoluble in water, so they could never precipitate from a watery solution.

Hutton also noted among the then-labeled Stratified Formations some rocks so closely resembling lava flows described at active volcanoes that they must surely also result by crystallization of molten material and not precipitation from water. Furthermore, the presence of these ancient lava flows among rocks of various ages demonstrated that volcanic activity was not a minor aberration of modern time but was a long-active geologic process. So impressed was

◄ **Figure 3.11 An example of Hutton's field observations.** Hutton made this sketch of geologic features exposed along a roadside in Edinburgh, Scotland. Thinly layered rocks (Stratified Formations, now called sedimentary rocks) are cut across by a massive rock type (Primitive Rocks, now called igneous rocks). Alteration at the contact between the two rock types implied that the processes forming the massive rock also heated the layered rocks. The massive rock appeared to Hutton to form by solidification of a molten liquid that injected through the layered rocks.

Hutton with the abundant evidence of subterranean heat in forming rocks that he went even further to suggest that the lithification of loose sediment into sedimentary rock was a consequence of heat causing sediment grains to fuse.

ANALYZE THE PROBLEM
How Did Geologists Resolve the Dispute?
The hot Earth, Vulcanist view clearly clashed with the Neptunist doctrine not only by inferring a different sequence of events in Earth's history but also by presenting contrasting explanations for the origins of rocks. Hutton's approach essentially followed the modern scientific method in that his ideas led from observations. In contrast, Werner's method was geared to explaining observations in terms of a presumed factual concept that lacked supporting scientific evidence. Hutton's conclusions were not all correct, however, because although he championed an internally hot Earth, as we do today, he took his Vulcanist doctrine too far by attributing a major role for heat in sedimentary rock formation. For example, today we know that chemical sedimentary rocks and the cementing minerals in the clastic sedimentary rocks owe their origin to precipitation from water and do not require heat in the process. This is one of the few components of the Neptunian doctrine that has held up.

The eventual acceptance by scientists of the Vulcanist explanation for what you now know as igneous rocks came through the gradual conversion of Neptunists, who made their own objective observations that led undeniably to supporting Hutton's conclusions. French geologists who had long suspected that volcanoes played a greater role in rock formation than that permitted by the Neptunists invited some of Werner's students to an area of recently extinct volcanoes in southern France. There, the Neptunists saw lava flows that had clearly issued from the volcanoes and were clearly composed of a rock called **basalt**. In Werner's view, basalt was assigned to the stratified formations and was not volcanic. In light of their field observations in France, Werner's students acknowledged that volcanic activity was more prevalent than the Neptunist view permitted, and as a process it must explain the formation of some rocks that the disciples of Werner mistakenly attributed to chemical precipitation from water. Another skeptic of Hutton's views melted basalt and allowed it to resolidify. The mineral content and crystal textures of the laboratory igneous rock so clearly matched that of the natural basalt that the skeptic shifted to the Vulcanist camp.

INSIGHT
How Do We Apply Knowledge Gained from This Historical Debate?
This recounting of the Neptunist–Vulcanist debate about the origin of rocks highlights the complexities of understanding the origin of rocks and, therefore, the difficulties in erecting a genetic classification. The significance of the debate goes beyond the implications for classifying rocks. The dispute between those allied with Werner and those convinced otherwise by Hutton's arguments is an important development in the history of geology as a scientific discipline. This is because Hutton, as well as others, demonstrated the application of the scientific method using observations of Earth materials and features. These observations, furthermore, appropriately tested hypotheses about the very origin of the planet.

Putting It Together–**How Do We Know** ... **How to Determine Rock Origins?**

- The threefold genetic classification of rocks largely arose from late eighteenth- to early nineteenth-century debates among geologists about the origin of rocks.
- Neptunists, led by Abraham Werner, asserted that nearly all rocks formed as chemical precipitates from water. This assertion was consistent with the prevailing, but not rigorously substantiated, views that the Earth had always been internally cold and originated as a predominantly watery orb.

■ Vulcanists, following the work of James Hutton, used careful field observations to refute the chemical-precipitate origin of many rocks and, by demonstrating the presence of igneous rocks, showed that the interior of the Earth was, and had probably always been, very hot.

3.4 How Are the Rock Classes Related to One Another?

The three rock classes—igneous, sedimentary, and metamorphic—are not isolated genetic groups but have relationships to one another in a changing Earth. Sedimentary rocks are derived from the breakdown of *preexisting* igneous, metamorphic, or other sedimentary rocks. Metamorphic rocks originate by changing the form and nature of minerals in a *preexisting* igneous, sedimentary, or other metamorphic rock. Igneous rocks crystallize from magmas that form by melting *preexisting* rocks, typically of metamorphic or igneous origin. It is difficult to melt a sedimentary rock under natural conditions, where temperatures generally increase very slowly, without first converting it to a metamorphic rock. Each rock type originates from preexisting rock, which implies a circular relationship among the classes.

The circular relationship of the three rock classes is portrayed as the **rock cycle** in **Figure 3.12**. The rock cycle is a useful concept to visualize the processes that form rocks and that relate each rock type to the others. Every process has a product, and some of those products are consolidated aggregates of minerals that meet the requirement for the definition of a rock.

The link between processes and products also emphasizes the genetic nature of the rock classification. Imagine a descriptive classification of rocks: If geologists classified rocks on the basis of color (all white rocks forming a single

▼ **Figure 3.12 Visualize the rock cycle.**
The rock cycle is a conceptual illustration of the relationships between the three rock types. A variety of process paths convert the components of one class of rocks into another rock class, with intermediate products forming along the way.

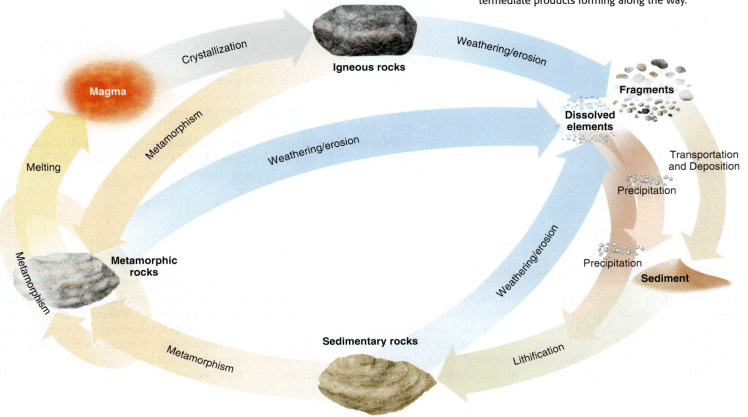

class, for example) or mineral content (with, perhaps, all rocks containing mostly quartz considered as one group), it would be difficult to know how these rocks formed because observations would not link to processes. This is because each such descriptive class would encompass rocks resulting from the wide array of processes that form igneous, sedimentary, and metamorphic rocks.

Another value of the rock-cycle concept is that it is a reminder of the dynamic, ever-changing nature of Earth and its constituent rocks—while not changing, in any significant way, the total components. Minerals or their constituent elements in an igneous rock may ultimately be recycled as components of sedimentary or metamorphic rocks. Mass transfers from one rock class to the others by the linking processes that continuously construct one type of rock from the others.

The process links in the rock cycle also relate, in large part, to plate tectonics. The limited distribution of active volcanoes (Figure 1.10) reveals that conditions within Earth required to produce magma correspond to processes nearly unique to divergent and convergent plate boundaries and hot spots. Sedimentary rocks are also not distributed everywhere across the surface of the planet. Clastic sediment erodes from high-standing areas and is transported primarily by water to be deposited in low areas, called sedimentary basins, where accumulations of rock several kilometers thick are found. Plate motions drive the uplifting of mountainous sediment sources and the downward sagging of basins where sediment deposits. The importance of convergent-plate-boundary processes to form many regional metamorphic rocks and hydrothermal metamorphic rocks at divergent boundaries has already been mentioned (Section 3.2).

Another important point to reflect on is that plutonic-igneous and metamorphic rocks form below the surface and yet are commonly exposed at the surface. In order for this to happen, the subterraneously formed rocks must rise toward the surface, and the material that originally resided above them must erode away. Tectonic forces drive this process of uplift and form steep slopes prone to erosion. Plutonic-igneous and metamorphic rocks, therefore, not only form in tectonically active regions but also become exposed at the surface as a result of tectonic activity.

*Putting It Together—**How Are the Rock Classes Related to One Another?***

■ The rock cycle links the three rock classes through a series of processes that convert the materials in one class of rock into new rocks of the same or a different class. Despite these changes, the chemical composition of Earth remains essentially constant.

■ Rock-forming processes and the relationships among rock classes are closely related to processes at plate boundaries. Plate tectonics is, therefore, a large-scale driving force in the rock cycle.

Where Are You and Where Are You Going?

A framework now exists for understanding the origins, properties, and uses of rocks. Geologists have used the threefold genetic classification of rocks efficiently for nearly two centuries. Igneous rocks form from the crystallization of molten magma. Volcanic, or extrusive, rocks crystallize at Earth's surface, whereas plutonic, or intrusive rocks crystallize at depth. Sedimentary rocks form from the deposition and precipitation of products produced by the breakdown of previously formed rocks. Minerals and rock fragments produced from weathering produce clastic sediment and when lithified form clastic sedimentary rocks. Chemical sedimentary rocks are aggregates of minerals that precipitate from water that is rich in elements previously dissolved from other rocks by weathering. Biologic materials, such as shell fragments, are important components of sedimentary rocks. Metamorphic rocks are derived from changes in preexisting rocks resulting from elevated temperatures and pressures and fluid-rock interactions that generate new minerals and rock textures. The rock cycle relates the rock types to one another. Plate tectonics, in turn, provides most of the driving force for the transformations in the rock cycle.

This chapter is just a glimpse into the understanding of the three rock groups. Many important questions remain to be answered. How is magma created and how does it reach the surface at volcanoes? How do rocks break down by weathering, and what constituents end up in sedimentary rocks? What temperatures and pressures are required for metamorphism? How do economically important resources, such as oil, coal, metal ores, and gems, form in rocks? These are but a few of the questions about rocks that the next three chapters address.

Confirm Your Knowledge

1. Where does the sand at the beach come from?

2. Why do some rocks consist of mineral grains with sharp edges and flat surfaces, while others consist of mineral grains with rounded edges and surfaces?

3. A rock exposure consists of layers of fine-grained rounded particles that are cemented together. What type of rock is it: igneous, metamorphic, or sedimentary?

4. A rock consists of small but well formed intergrown silicate-mineral crystals. What type of rock is it: igneous, metamorphic, or sedimentary?

5. A shiny rock contains minerals oriented preferentially in discontinuous layers. What type of rock is it: igneous, metamorphic, or sedimentary?

6. A rock contains fragments of shell and coral. What type of rock is it: igneous, metamorphic, or sedimentary?

7. Which part of the Neptunist doctrine on the formation of rocks was correct?

8. Which part of the Vulcanist doctrine on the formation of rocks was incorrect?

9. Define the three groups of rocks.

10. What are the conditions required for metamorphism?

11. What are the 3 major types of metamorphism? What condition dominates for each type? What plate tectonic conditions are associated with each type of metamorphism?

12. Compare the Neptunist and Vulcanist explanations for the origins of rocks.

13. Which scientists are the champions of Neptunism and Vulcanism?

14. Diagram and explain the rock cycle.

Confirm Your Understanding

1. Write out an answer for each question in the Chapter Outline for the chapter sections assigned by your instructor.

2. Distinguish the differences you would see between an intrusive and extrusive igneous rock formed from the same magma. Distinguish the differences you would see between two intrusive igneous rocks formed from different magmas.

3. If you were visiting Germany in 1800, what would you do to convince a geology student that not all rocks formed by precipitation from water?

4. Layering occurs in all type of rocks. How do we know they formed by different processes?

5. Based on what you know about the scientific method, did the Neptunists and Vulcanists follow the modern principles?

6. How could you test the rate of weathering in your area using gravestones?

7. The rock cycle describes recycling on a planetary scale. How would our 4.56-billion year-old Earth differ if there were no rock cycle or the recycling of materials? Give your answer in terms of topographic relief, depth of the oceans and variability of Earth materials.

8. The classification of rocks into igneous, metamorphic and sedimentary rocks is a genetic classification. Create a descriptive classification of rocks. Evaluate the descriptive classification system in comparison with the one in common use. Assess the strengths and weaknesses of each system. Which system would you recommend to use if your objective is to understand how Earth works?

9. Develop both genetic and descriptive classifications for the objects on your desk at home.

Key Terms

basalt (p. 60)
bedding (p. 55)
chemical sedimentary rocks (p. 55)
clastic sediment (p. 54)
contact metamorphic rocks (p. 56)

hydrothermal metamorphic rocks (p. 57)
igneous rocks (p. 54)
lithification (p. 55)
magma (p. 55)
massive (p. 56)
metamorphic rocks (p. 54)

metamorphism (p. 56)
plutonic rocks (or intrusive rocks) (p. 55)
regional metamorphic rocks (p. 56)
rock cycle (p. 61)
sandstone (p. 50)

sedimentary rocks (p. 54)
volcanic rocks (or extrusive rocks) (p. 55)
weathering (p. 54)

4 Formation of Magma and Igneous Rocks

Chapter Outline

Why Study Igneous Rocks?

IGNEOUS ROCKS RESULT FROM THE SOLIDIFICATION OF MOLTEN silicate mixtures called magma. Nowhere are igneous processes and products more dramatically demonstrated than at an erupting volcano. The melting of rock and the solidification of another rock from that melt are, however, processes that involve many geologic phenomena besides erupting volcanoes.

Why are igneous rocks and processes important? The thought of living at the base of an erupting volcano might motivate you to understand why, how, and when volcanoes erupt. If you understand the workings of volcanoes, then you can avoid volcanic hazards that are devastating to people and property. Understanding the formation of igneous rocks also provides clues to finding and recovering some of Earth's economic resources, including valuable metal ores and building materials. The formation of magma also offers critical insights into dynamic processes inside Earth and relates to plate tectonics.

There are three primary objectives in this chapter:

✔ To understand the conditions permitting rocks to melt into magma and how these conditions relate to plate tectonics

✔ To comprehend the processes that form different types of magma compositions and how these diverse processes relate to the many styles of volcanic eruptions and shapes of volcanoes

Lava flows from a volcanic cone on the flank of Mount Etna, Sicily, Italy, in May 2000. ▶

✔ To apply knowledge of igneous processes and products to understand how important economic resources form and where they are found

To accomplish these objectives, these ten questions are addressed:

4.1 What are igneous processes?

4.2 How are igneous rocks classified?

4.3 Where do igneous rocks appear in a landscape?

4.4 How and why do rocks melt?

4.5 *How do we know . . .* how magma is made?

4.6 How does magma generation connect to plate tectonics?

4.7 What makes igneous rock compositions so diverse?

4.8 Why are there different types of volcanoes and volcanic eruptions?

4.9 How are volcanoes hazardous?

4.10 Why don't all magmas erupt?

In the **FIELD**

Figure 4.1 illustrates three virtual fieldtrips to explore the nature of igneous rocks and the processes that form them.

On the slope of the Kilauea volcano in Hawaii (Figure 4.1a), a fountain of fiery, red-hot liquid rises straight up from a black cone that is about 100 meters high. Although it flows almost as quickly as water from a fountain in a city park, this liquid is not water; it is molten lava, with a temperature of 1000°C. Where the lava falls to the ground, some blobs congeal and stick together to build the cone even higher. The remainder of the molten liquid streams away as lava flows, which move downslope at speeds from 1 to 10 meters per second. Volcanologists carefully approach within a few meters of the flow and, dressed in heat-resistant clothing, insert tools into the lava to sample it for analysis back at the lab.

Mount Pinatubo, a volcano in the Philippines with no previous record of eruption in human history, stirs to life. The eruption reaches a climax with gigantic explosions that hurl porous pumice and small ash particles high into the air (Figure 4.1b). The ash clouds ultimately rise to a height of more than 30 kilometers and drift with the wind to encircle the globe. Avalanches of hot pumice and ash rush down the slopes of the volcano, obliterating everything in their path. Unlike in Hawaii, there are no spectacular lava fountains and captivating streams of flowing lava. No volcanologist will dare venture near this dangerous volcano until days after the explosions end.

Figure 4.1c shows a view over the inspirational landscape of Yosemite National Park in California. Not only are there deep valleys and high waterfalls but also vast expanses of rock laid bare by erosion. From a distance, the rock appears monotonous—uniformly gray, without contrasting colors or layering. Up close, though, the rocks are salt-and-pepper mosaics of mineral grains colored white, black, and light pink; some are glassy, some dull (Figure 4.1d). A geologist recognizes these rocks as products cooled from molten **magma**. What is the difference between magma and lava? Lava is molten material ejected from a volcano, whereas magma is molten material below Earth's surface. At Yosemite, the magma solidified into rock deep below ground. Uplift and erosion raised and wore away the rocks above the solidified magma. More than 100 million years later they are exposed for you to study.

Why are lava fountains and lava flows the most clearly visible products of the eruptions at Kilauea, while pumice and ash, flung high into the sky or advancing rapidly down slopes, are the main products of the Mount Pinatubo eruption? Why did the once-molten magma that solidified into the salt-and-pepper rocks at Yosemite never reach the surface at a volcano? How did the molten material form at these three locations (and many others) to begin with? As you read through this chapter, the answers to these and other questions will gradually come together to reveal a cohesive picture of how Earth forms igneous rocks, the most common rock type and host to vital natural resources.

Figure 4.1 What igneous rocks and processes look like. ▶

<u>*Field observations of igneous rocks*</u>

A. 1985 - Kilauea, Hawaii
Volcanologists make observations only a short distance from the Pu'u O'o cone on the flank of the Kilauea volcano. Red-hot lava fountains high above the cone and feeds lava flows with dark, solidified crust.

B. 1991 - Luzon, Philippines
Explosive energy propels pumice and ash more than 24 km above the Mt. Pinatubo volcano. The volcano is hidden behind ash clouds rising from incandescent avalanches of pumice and ash that rush at hurricane speed down all sides of the volcano.

C. 2005 - Yosemite National Park, California
Igneous rock formed when magma solidified deep below the ground. Erosion exposed these rocks to view at Earth's surface.

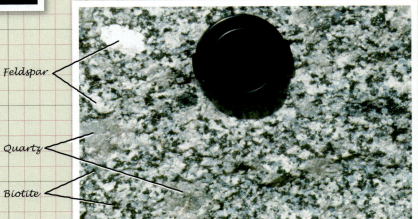

D. *A closer view at this igneous rock reveals the intergrown mosaic of minerals that crystallized from the molten magma as it cooled. Camera lens cap for scale.*

Feldspar

Quartz

Biotite

4.1 What Are Igneous Processes?

These three field trips show you different phenomena, but they all express the processes that form igneous rocks. In order to have magma, rock first has to melt at high temperature inside Earth. The liquid magma then moves upward, in some cases reaching the surface and in other instances remaining trapped underground. Wide varieties of hazardous phenomena result where molten material erupts on Earth's surface and into the atmosphere. The properties of magma determine how volcanoes erupt, either with fiery fountains and fluid lava, like at Kilauea, or with mind-boggling explosive force and global scattering of volcanic ash, as at Mount Pinatubo.

Geologists evaluate the nature of igneous rocks and processes that form them by combining knowledge gained from field observations, geochemical analyses, and laboratory experiments. All of these studies bring in essential data and complement each other to allow us to form inferences and draw conclusions.

Field studies of ancient volcanoes provide insights to volcanic processes that are not safely studied at active volcanoes. Studies of locations such as Yosemite Valley and other igneous-rock landscapes allow geologists to infer how magma moves and solidifies below Earth's surface. Geologists also conduct microscopic studies of rocks that they sample and bring to the laboratory. The microscope permits geologists to describe and identify the minerals found in rocks and the relationships between the minerals that determine how magmas crystallize to form rocks. Geochemists identify and measure the abundance of elements in rock samples. Volcanologists remotely or even directly analyze gases from volcanoes. Scientists combine the results of all of these analyses to infer the processes involved in the evolution of igneous rocks.

Geologists use experiments to simulate processes and conditions within the Earth that they cannot directly observe. Laboratory experiments include melting rocks and crystallizing the resulting liquid in high-temperature furnaces and high-pressure devices that reproduce specific temperature and pressure conditions inside Earth.

Results of observation, analysis, and experiment add up to answer your questions from the field trips about the nature of igneous processes and products. You will follow an approach of first describing igneous rocks and learning how they are encountered in the field; then you will seek to understand the origin of the observed features after you know what phenomena must be explained.

---

*Putting It Together–***What Are Igneous Processes?**

- Igneous processes involve the melting of rock to form magma and its subsequent solidification into new rock.
- Igneous products form where magma crystallizes below Earth's surface or erupts as lava and other materials onto Earth's surface.
- Geologists study processes forming igneous rocks through field observations and laboratory studies that include geochemical analyses and experiments.

4.2 How Are Igneous Rocks Classified?

Observations in Hawaii, the Philippines, and at Yosemite National Park provide glimpses into the variety of igneous rocks found on Earth. The characteristics of these rocks, in turn, are expressions of the processes that form them. This means that careful description and classification of the rocks are essential to understanding the phenomena that igneous processes must explain.

The field observations suggest dividing igneous rocks into two general categories based on where they formed. Rocks originating from the eruption of molten material at the surface are **volcanic** (or **extrusive**) rocks. Volcanic rocks form from flowing lava, as in Hawaii, or they form when explosions break apart the magma, shooting out sticky blobs that quickly quench into glass and fall to the ground. These fragmental **pyroclastic** materials (from the Greek *pyros*: fire, and *clastos*: broken) include the lava-fountain blobs building the cone at Kilauea and the pumice and ash forming the high eruption cloud and hot avalanches at Mount Pinatubo (Figures 4.1a, b). Igneous rocks that solidify below the surface comprise the **plutonic** (or **intrusive**) rocks. The word *plutonic* derives from *Pluto*, the Greek god of the underworld. The rocks exposed at Yosemite are plutonic rocks (Figure 4.1c, d).

Figure 4.2 shows a collection of igneous rocks that exhibit variations in appearance that form a basis for descriptive classification. Igneous rocks vary widely in color but most commonly range from black to very pale gray or nearly white with tinges of pink or red. These color variations correspond to different minerals composing the different rocks. All of the mineral grains are visible to the unaided eye in some rocks (Figure 4.2a), but in other samples you need a microscope to see the crystals (Figures 4.2b and c). In still other cases, the rock sample consists of both large, visible mineral grains and those that are nearly invisible (Figure 4.2d). Some igneous rocks appear to lack crystals of any size and instead resemble glass (Figure 4.2e).

Composition (minerals present) and texture (size of mineral grains) provide a framework for naming igneous rocks. Composition and texture in igneous products are a direct result of igneous processes. A wide range of magma compositions and cooling conditions assures diverse combinations of crystallizing minerals and rock textures. The result is a great variety of igneous rock types.

First Component in Classification: Composition

Magma cools when it loses heat to its cooler surroundings. The crystallization temperatures of specific minerals are reached as the magma cools, and the minerals appear and grow in a process similar to ice forming from liquid water. Ice is a single mineral, however, whereas many different minerals crystallize from any particular magma. The composition of the magma determines the minerals that form and the appearance of the resulting rock.

With only very rare exceptions, magma is rich in silica, the compound of silicon and oxygen expressed as SiO_2. Silica content in most magma ranges from 45 to almost 80 percent by weight, along with oxides of aluminum, magnesium, iron, calcium, sodium, potassium, and other elements in very minor concentrations. It is, therefore, not surprising that igneous rocks are overwhelmingly composed of silicate minerals.

Gases are also present in all magmas, with water vapor, carbon dioxide, and odorous sulfur compounds as the most abundant. When magma nears or reaches the surface, gases escape and account for the

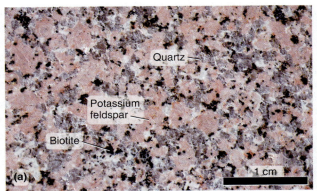

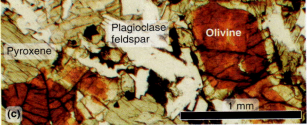

◀ **Figure 4.2** **Igneous rocks have different mineral contents and textures.**

This light-colored rock contains large, readily visible crystals of quartz, feldspar, and biotite. These minerals and coarse crystal sizes characterize the felsic, phaneritic (coarse-grained) rock called granite.

This rock from a lava flow is dark and contains no visible crystals. These are features of the mafic, aphanitic (fine-grained) rock called basalt. The vesicle holes represent former gas bubbles in the magma.

A microscopic view of basalt reveals the tiny crystals of olivine, pyroxene, and plagioclase feldspar formed by crystallization of the lava flow.

This rock is intermediate in color between granite and basalt, and contains visible plagioclase feldspar and hornblende in a gray background of microscopic crystals. This rock is porphyritic andesite.

Obsidian is volcanic glass that lacks crystals. Notice the curving fractures that are typical of broken glass.

holes caused by vacated bubbles, called **vesicles**, in some volcanic rocks (Figure 4.2b). A critical difference between magma and lava is entrapment of gas within magma at depth and its release during eruptions. Magma is a mixture of liquid, crystals, and dissolved gases below the surface, whereas lava is magma at the surface from which the gases have mostly escaped.

Figure 4.3 shows how both magma and igneous rock compositions are categorized as **felsic**, **intermediate**, **mafic**, or **ultramafic** according to the decreasing abundance of silica in the magma or in the rock-forming minerals. "Felsic" refers to *fel*dspar and *si*lica (expressed by the presence of quartz). The term "mafic" is derived from names of the elements *ma*gnesium and iron (*fer*ric), which are more abundant in lower silica magma. Natural rocks gradually vary

▼ **Figure 4.3** **Classification by composition and texture.**
The combination of composition, revealed by mineral identification or chemical analysis, and mineral grain size is the basis for classification and naming of igneous rocks. Each rock name is distinguished by predominantly aphanitic or phaneritic texture and the relative abundance of minerals pictured above the rock name. Light-colored felsic rocks are rich in silica (SiO_2) and contain mostly quartz and feldspar. Dark-colored mafic and ultramafic rocks have lower silica content and are also richer in iron and magnesium, permitting the formation of dark silicate minerals such as olivine and pyroxene.

from one compositional group to the next and are described by different abundances and combinations of minerals, as demonstrated in Figure 4.3. The key compositional characteristics are as follows:

- Mafic and ultramafic rocks mostly contain the magnesium-rich and iron-rich silicate minerals, olivine and pyroxene. Ultramafic rocks consist almost entirely of olivine and pyroxene, whereas mafic rocks also contain calcium-rich plagioclase feldspar (Figure 4.3). The dark magnesium- and iron-rich minerals dominate over the light-colored feldspar, producing dark gray to black rocks.
- Intermediate rocks commonly contain hornblende, biotite or pyroxene, and plagioclase feldspar with nearly equal amounts of calcium and sodium (Figure 4.3). Feldspar and magnesium- and iron-rich minerals are present in about equal proportions, so intermediate-composition rocks are lighter in color than ultramafic and mafic rocks; gray shades are most typical.
- Felsic rocks contain quartz, sodium-rich plagioclase feldspar, and potassium feldspar with very minor dark minerals, including hornblende, biotite, and less common pyroxene (Figure 4.3). These rocks are usually the lightest in color.

Second Component in Classification: Texture

The size of crystals in a rock is an important component of rock texture. Crystal size in igneous rocks depends on many factors, but the magma cooling rate is generally the most important one (Section 4.10 considers another factor). When magma cools down quickly, the resulting rock consists of very small crystals, which are generally invisible to the unaided eye (Figure 4.2b). This fine-grained texture is called **aphanitic** (Figures 4.2b, 4.3), which is derived from the Greek word *aphanes*, meaning "invisible." Very rapid crystallization occurs when volcanic rocks form because the temperature at the surface is many hundreds of degrees cooler than the solidification temperature of the melt, and heat quickly conducts and radiates to the surroundings. Intrusive rocks may also display aphanitic texture where small volumes of magma intrude into relatively cold rocks close to Earth's surface.

The melt solidifies into a noncrystalline glass if the cooling rate is so fast that few, if any, crystals have time to form. Glassy texture is common in rocks formed at the surfaces of lava flows (Figure 4.2e) and along the margins of some igneous intrusions. **Obsidian** is felsic volcanic glass with dark gray to black color that seems uncharacteristic of felsic rocks (Figure 4.2e). Extremely tiny grains of magnetite (and sometimes other minerals) are widely dispersed through the obsidian, making it opaque and dark despite the felsic composition. Pyroclastic fragments are commonly glassy because each small fragment quickly loses heat to the much colder air once ejected from a volcano.

In contrast, magma cools slowly where warm rock surrounds and insulates magma deep beneath the surface. Mineral crystals grow slowly in this deep intrusive environment to produce a coarse-grained texture known as **phaneritic** (Figures 4.2a, 4.3), which is from the Greek word *phaneros*, meaning "visible." Thin, aphanitic lava flows may cool within hours on Earth's surface, whereas deeply formed phaneritic intrusive rocks may require tens of thousands of years, or longer, to solidify completely.

In some instances, magma starts to crystallize slowly at depth, producing some large crystals, before moving into colder environments near or at the surface. Remaining melt crystallizes quickly in the colder environment, and the resulting igneous rock consists of some large crystals surrounded by smaller crystals. The resulting **porphyritic** texture consists of a mixture of fine and coarse crystals sizes (Figure 4.2d).

A special approach is needed to classify the pyroclastic materials erupted from volcanoes. Most pyroclastic materials are glassy, so the attribute that best

(a)

(b)

(c)

(d)

Basaltic bombs have aerodynamic ribbon and spindle shapes formed as the solidifying blobs of lava traveled through the air.

Dark basaltic and andesitic lapilli are called cinder or scoria.

Light dacitic and rhyolitic lapilli are called pumice.

Fine, light-colored volcanic ash collected 140 kilometers east of Mt. St. Helens, Washington during the May 1980 eruption.

▲ **Figure 4.4** **Pyroclastic deposits classified by fragment size.** All of the fragments pictured here fell to the ground after powerful volcanic explosions ejected them into the sky. Starting out as molten masses, they chilled into glassy fragments in the cool atmosphere. The largest fragments accumulate on the slope of the volcano, whereas wind carries smaller particles progressively farther distances. Table 4.1 shows how size determines the name applied to the individual fragments and the resulting deposits.

distinguishes them is the size of the fragments, as described in **Figure 4.4** and **Table 4.1**, rather than the size or types of minerals within the fragments. The largest fragments are called **bombs** (Figure 4.4a), **lapilli** are intermediate in size, and **volcanic ash** is the smallest. Contrary to more common meanings of bombs and ash, volcanic bombs do not explode and volcanic ash is not a product of combustion.

Naming Igneous Rocks: Composition + Texture

The classification scheme for igneous rocks combines composition and texture (Figure 4.3). Ultramafic and phaneritic **peridotite** composes Earth's mantle, and pieces of this rock are sometimes carried to the surface at erupting volcanoes. **Basalt** and **gabbro** form from mafic magma and contain the same minerals but have different textures (Figure 4.3); gabbro is the coarse-grained equivalent of basalt. The oceanic crust is composed mostly of these mafic rocks. **Andesite** and **diorite** solidify from intermediate-composition magmas. **Rhyolite** and **granite** form from the most felsic magmas. **Dacite** and its phaneritic equivalent **tonalite** have compositions that fall between andesite and rhyolite. Many recently active volcanoes, including Mount St. Helens (Washington, 1980–2004), Mount Pinatubo (Philippines, 1991), and Mount Unzen (Japan, 1991), erupted dacitic lava flows and pyroclastic fragments. A wide range of intermediate to felsic igneous rocks compose the continental crust, and the most accessible rocks of the upper crust average to the tonalite composition.

Pyroclastic deposits are named on the basis of fragment size and whether the fragments are loose or consolidated into rock (Table 4.1). Compositional terms are used as modifiers to complete a descriptive naming of pyroclastic deposits (Table 4.1). Lightweight, highly vesicular lapilli is called **pumice**, if dacitic or rhyolitic in composition, and **cinder** (or **scoria**) if basaltic or andesitic.

Igneous Rock Provide Essential Economic Resources

Igneous rocks are widely used in construction and industry. Many volcanic rocks are strong because the finely intergrown mineral crystals are difficult to

TABLE 4.1	**Classification of Pyroclastic Materials**		
Size of Fragments (mm)	0 ——— 2	——— 64	——→
Name of loose fragments	ash	lapilli*	bomb*
Name of rock composed of many fragments**	tuff / lapilli tuff	lapillistone	agglomerate

* Lapilli and bombs of mafic/intermediate composition are called *scoria* whereas those of felsic composition are *pumice*.
** Compositional terms can be used as modifiers (e.g., basaltic bomb, rhyolitic tuff).

fracture along grain boundaries. These rocks make good building stone and, when crushed, form excellent aggregate for highway and railroad-bed construction. Tuff is a strong yet lightweight rock used for building stone, as shown in **Figure 4.5**. Volcanic ash and pumice are mixed into superior lightweight concrete. Pumice is also mined for use as an abrasive in toothpaste and soap and is tumbled with denim for the manufacture of stonewashed jeans. Obsidian is easily chipped and flaked into incredibly sharp tools and weapons. Although mostly replaced by manufactured metal implements, the use of obsidian for weaponry and tools made it an essential resource in many ancient cultures and its availability determined migratory and trading routes. Obsidian knives are, however, still preferred by many plastic surgeons. Phaneritic igneous rocks are commonly quarried for decorative purposes, including use as building facades, monuments, and gravestones.

▲ **Figure 4.5 Building with volcanic rock.**
Hand-hewn blocks of light-colored tuff and dark scoria, about 15 cm across, form the walls of villas in the Roman city of Herculaneum, near modern Naples, Italy. Volcanic rocks are easily shaped for building purposes but are also strong and bear weight well. Proximity to volcanic building materials also places some habitations in peril. Pyroclastic flows from the famous volcano, Vesuvius, destroyed Herculaneum in August A.D. 79.

Putting It Together–*How Are Igneous Rocks Classified?*

■ Composition (minerals present) and texture (crystal size) are used to classify and name igneous rocks.

■ Pyroclastic deposits, igneous products formed by volcanic explosions into the atmosphere, are primarily classified on the basis of fragment size and degree of consolidation.

■ Ultramafic, mafic, intermediate, and felsic are compositional categories of magma and rocks. Light-colored silicate minerals dominate in silica-rich felsic rocks, whereas darker silicate minerals form mafic rocks.

■ Rapidly cooled volcanic and shallow plutonic rocks are aphanitic (fine-grained) whereas slowly cooling deep magma intrusions produces phaneritic (coarse-grained) rocks. Porphyritic texture describes rocks containing at least two distinct crystal sizes. Igneous rocks have properties that lend themselves to a variety of uses in construction and industrial processes.

4.3 Where Do Igneous Rocks Appear in a Landscape?

To understand the origin of igneous rocks requires knowing where they are found and how they appear in natural landscapes. Volcanoes consisting of lava flows and pyroclastic deposits form where magma reaches the surface (Figure 4.1). Plutonic rock bodies of varying shapes and sizes form where magma solidifies before reaching the surface. These plutonic rocks are only revealed at the surface after erosion of the rock that originally was above the place where the magma halted and crystallized (at Yosemite, for example; Figure 4.1c).

Modern volcanoes show where magmas form and rise to the surface today. Geologists see only the extruded volcanic materials, but they infer simultaneous magma intrusion in a subterranean plumbing beneath the volcanoes. The geographic coincidence of extrusive and intrusive processes is revealed by

- Earthquakes caused by magma movement beneath volcanoes
- Pieces of plutonic rocks ejected from volcanoes (Figure 3.4d, for example)
- Presence of plutonic rocks exposed by erosion of volcanic rocks

Most volcanoes are found at or near divergent and convergent plate boundaries and only a few are found within plates (see Figure 1.10). Volcanoes are especially prominent along continental margins and island chains above subduction zones in the Pacific Ocean, forming the Ring of Fire. Seafloor volcanoes compose the mid-ocean ridges (see Figures 1.10 and 1.11) and some rise to form islands, such as Hawaii, where magma rises below plate interiors at hot spots (Figure 1.12). Volcanoes appear within continents at points where plate divergence is beginning (e.g., East Africa) or at hot spot localities such as Yellowstone National Park, Wyoming. Igneous activity occurs today within active

▶ **Figure 4.6 Visualize intrusive features.**
Names assigned to igneous intrusions relate to their shape and size. Large magma chambers have very irregular shapes, because they form over long time periods by multiple injections of magma from deep in the crust or mantle. Smaller bodies of magma rise through vertical or near-vertical fractures to form dikes, which cut across sedimentary rock layers. Magma injected between layers forms sills. A volcanic neck is rock that solidifies in the volcano throat. Dikes, sills, and volcanic necks form close to the surface, so they are the first revealed at the surface by later erosion. Erosion to deeper levels reveals the larger, deeply rooted solidified magma chambers, called batholiths.

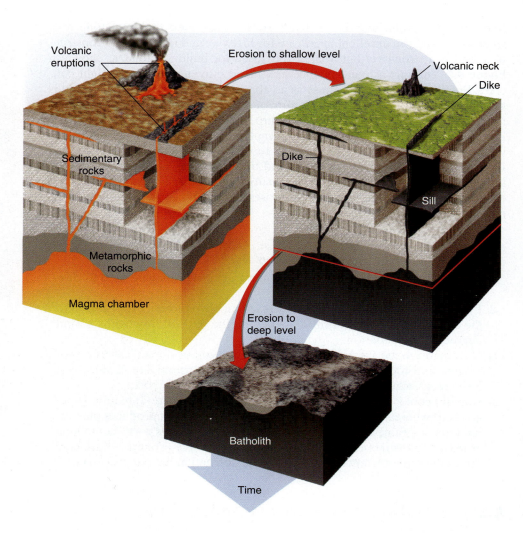

Forming Igneous Features and Landforms: *See how intrusions form and how they produce unique landforms when exposed by erosion.*

tectonic settings, so geologists infer that ancient volcanic rocks and their intrusive equivalents formed in tectonically active regions in the past.

Plutonic Rocks in the Landscape

Where erosion exposes plutonic igneous rocks, they are commonly more resistant to erosion than the surrounding rocks. This means that the plutonic rocks stand out on the landscape as hills, ridges, and even large mountains where softer enclosing rocks eroded away.

Figure 4.6 shows the wide variety of landforms that result from exposing igneous intrusions of various sizes and shapes. The sizes and shapes of plutonic-rock bodies relate to the volume of magma intruded and the mechanism of intrusion.

Figure 4.7 shows tabular intrusions, which are extensive in two dimensions but relatively thin in the third dimension. In some cases, magma rises to form long, steep, tabular **dikes** along fractures in the surrounding rocks (Figure 4.7a). Dikes cut across layers if the magma intrudes sedimentary rocks. In other cases, magma moves at low angles to the ground surface to form **sills**, which may be injected between sedimentary layers (Figure 4.7b).

Larger magma chambers form deep within the Earth's crust where the enclosing metamorphic rock is very hot and deforms in a plastic, ductile fashion rather than by brittle breakage. In these places, the magma squeezes aside the plastic surrounding rocks. Adjacent rocks melt into the magma, and large blocks detach from the roof of the magma chamber, which allows the melt to move farther upward. These large intrusions, which solidify at depths of several kilometers to more than 10 kilometers, are exposed only where substantial uplift and erosion of surrounding rocks occurred (Figure 4.6). The resulting large bodies of intrusive rocks are **batholiths** or **stocks**, depending on the area of exposure at the surface (batholiths are exposed over more than 100 square kilometers and stocks are smaller). This distinction is arbitrary, since stocks may simply be small parts of batholiths that have not yet been exposed by erosion.

When examining batholith rocks, geologists commonly notice that the composition of the rocks differs from place to place. From this observation, geologists infer that the large intrusion formed over a long period of time by injection of different batches of magma with different compositions. The intrusive rocks exposed at Yosemite National Park (Figure 4.1c) are merely a small part of the vast Sierra Nevada batholith. This batholith is exposed over nearly 100,000 square kilometers and contains countless, irregularly shaped intrusions, ranging in composition from gabbro to granite, and emplaced over a period of more than 50 million years.

Erosion in volcanic regions reveals the positioning of volcanoes above shallowly intruded dikes, sills, and small upward protrusions of large batholiths that anchored the igneous plumbing network at greater depth (Figure 4.6). In some cases, dikes extend to the surface to cause volcanic eruptions along elongate fissures. In other cases, nearly cylindrical pipes feed volcanoes, and when erosion later exposes the rocks filling the pipes, they form **volcanic necks**, as shown in **Figure 4.8**.

◀ **Figure 4.7 What dikes and sills look like.**

Dikes are steeply inclined intrusions that cut across sedimentary layers, if present. This basalt dike in the Grand Canyon, Arizona, cuts across red sedimentary layers. Notice the geologists (circled) standing on the dike for scale.

Sills are usually gently inclined or horizontal intrusions that are parallel to sedimentary layers, if present. This gabbro sill in Glacier National Park, Montana, is between horizontal sedimentary layers. The sill is more than 50 m thick.

Volcanic Rocks in the Landscape

Volcanic rocks take on a wide variety of landscape forms, both at the small scale as the products of individual eruptions and at the large scale as volcanoes formed over hundreds of thousands of years by many eruptions. Lava flows are

▼ **Figure 4.8 What volcanic necks look like.**
Necks mark the nearly cylindrical conduits of now-eroded volcanoes.

Shiprock, New Mexico is a volcanic neck, composed of tuff and intrusive rock that filled the conduit of an old volcano. Erosion of surrounding soft sedimentary rock exposed the neck.

Devil's Tower, Wyoming, resulted from erosion that exposed magma that solidified in the conduit of an eroded volcano. The long curving rock columns formed when the cooling rock contracted.

Actively flowing lava develops a glassy, smooth-to-wrinkled crust as it cools.

Some lava flows are very thin, attesting to the very fluid character of basaltic lava.

Smooth and ropy lava flow surfaces, called pahoehoe in Hawaii, form where the lava stops flowing while it is still very hot. If the flow continues to move after the crust hardens, then the rock crust breaks into pieces and forms rubbly lava flows, called a'a.

▲ **Figure 4.9 Flowing lava in Hawaii.**

the most common features. **Figure 4.9** shows basaltic lava flows in Hawaii that are only meters thick. **Figure 4.10** illustrates basaltic lava flows that cover very large areas. Intermediate and felsic lava flows, shown in **Figure 4.11**, are many tens to hundreds of meters thick but extend only a few kilometers from the vent where they erupt. **Lava domes** are an extreme case where lava seemingly does not flow at all but merely squeezes out as large mounds directly over the volcanic vent (Figure 4.11b). Some lava flows have smooth to slightly wrinkled surfaces, seeming to flow like cooling maple syrup (Figure 4.9). Other flows, like the one portrayed in **Figure 4.12**, are encased in broken blocks of rock that broke away from the cooler solidified outer margin of the lava as the partially molten, still fluid interior continued to flow. In Hawaii, the smooth or wrinkled flows are called pahoehoe and the rubbly lava is called a'a.

The characteristics of pyroclastic deposits depend on whether the explosively generated fragments fall from the sky or flow down the slopes of the volcanoes. **Pyroclastic-fall deposits** may extend more than 1000 kilometers from a volcano if the explosions carry the particles high into the atmosphere. Very violent explosions, such as the one that occurred at Mount Pinatubo in 1991 (Figure 4.1b), eject fine-ash particles so high into the atmosphere that they form a dusty ash veil that encircles the planet, traveling with prevailing winds. Wind disperses pyroclastic fragments away from the volcano, so the ash deposits are thickest in the downwind direction from the volcano. Larger and denser fragments settle out of the air first (Figure 4.4), so pyroclastic-fall deposits exhibit a uniform decrease in particle size with increasing distance from the volcano.

Pyroclastic-flow deposits, illustrated in **Figure 4.13**, form when avalanches of incandescent pumice and ash flow down the slope of a volcano. Moving pyroclastic flows at Mount Pinatubo are visible in Figure 4.1b. The deposits contain a wide range of ash- and lapilli-sized particles that form lapilli tuffs (Figure 4.13a, Table 4.1). Ash and lapilli fragments are sticky and plastic where pyroclastic-flow deposits retain considerable heat after coming to rest. Particles within the hot interior of the deposit squash and stick together under the weight of the overlying material, producing **welded tuff** (Figure 4.13b).

Types of Volcanoes

Volcanoes are hills, ridges, or mountains formed by the accumulation of lava flows and pyroclastic deposits around the conduit, or vent, from where they erupted. The size of a volcano relates to the volume of extruded volcanic materials. The shape of the conduit and the types of eruptions determine the shape of a volcano. Volcanoes with classic cone shapes form by eruptions through a single, centrally located crater. More irregular or elliptical shapes result from eruptions through many craters or elongation along fissures where dikes break through to the surface. Craters or deeply eroded river canyons

(a)

Basaltic lava flows poured across the Pacific Northwest about 15 million years ago. The Grande Ronde River canyon in Washington exposes hundreds of meters of basalt lava. Each layer in the photo is a separate lava flow. Some flows cover more than 100,000 square kilometers.

Lava flows

◄ **Figure 4.10 What mafic lava flows look like.**

(b)

Some basalt flows feature columnar joints formed when the rock contracts while cooling (also seen in Figure 4.8b). This thick lava flow is the highlight for visitors to Devil's Postpile, California.

(a)

Lava flow erupted from here

A thick rhyolite-obsidian lava erupted about 1000 years ago to form Glass Mountain in northern California. The lava completely filled and buried a crater near the center of the picture. The steep edge of the flow is higher than the pine trees visible at lower right.

(b)

Dacitic lava erupted at Mt. St. Helens, Washington, between 1980 and 1986 piled up over the volcano conduit to form a lava dome similar in size to a domed sports stadium.

(c)

1980-1986 dome

Lava extruding in October 2004

Dark volcanic ash in glacier ice

Steam and other volcanic gases partly obscure a fresh extrusion of solidified viscous lava at Mt. St. Helens in October 2004. Pressure exerted by the rising lava uplifted surrounding parts of the older 1980-1986 lava dome by more than 100 m.

▲ **Figure 4.11 What felsic lava flows look like.**
Some felsic lava flows are very thick but do not extend far from where they erupted.

▲ **Figure 4.12 Lava-flow rubble.**
Broken lava fragments completely surround and enclose many lava flows. A solid surface forms as the lava cools and insulates the flowing liquid below. This rock crust is too cold and brittle to flow like a liquid, so it breaks into fragments that travel on top of, and spill off the sides of, the flowing lava. This rubbly basaltic lava flow from Helgafell volcano is destroying a home in Iceland.

reveal the internal structure of volcanoes. **Figure 4.14** lays out a classification of volcanoes according to size, shape, and erupted volcanic materials.

Forming Calderas

Volcanic eruptions are usually thought of as processes that build up elevations through the addition of layers of lava and pyroclastic material, but some eruptions actually diminish elevation. The frequent ascent of magma warms up the surrounding shallow crust in regions experiencing volcanic activity for long periods of time. This heating permits ever-larger volumes of magma to rise closer to the surface without solidifying. In this fashion, large magma chambers form within a few kilometers of the surface and support the weight of the overlying rocks. If significant portions of magma erupt from the chamber without immediate replenishment from deeper levels, then the unsupported roof of the chamber collapses. The collapse produces a large, circular or elliptical pit at the surface, called a **caldera**.

Both of the volcanoes that you visited at the beginning of this chapter (Kilauea and Mount Pinatubo) have calderas. **Figure 4.15** shows how calderas form in shield volcanoes by withdrawal of magma from beneath the volcano summit to feed eruptions on the slopes. Since a geologic observation recorded in 1790, gradual subsidence at Kilauea formed a caldera 3.2 kilometers long, 2.6 kilometers wide, and hundreds of meters deep, but this depression partially filled with lava flows during later eruptions. In contrast, **Figure 4.16** illustrates how other calderas form dramatically within hours to days during explosive eruptions of pumice and ash that evacuate many cubic kilometers of magma from shallow magma chambers. The Mount Pinatubo caldera formed in 1991 during the eruption of about 10 cubic kilometers of magma during less than 24 hours of eruptions. The caldera is approximately 2.5 kilometers in diameter and lowered the elevation of the mountain by 250 meters. Crater Lake National Park

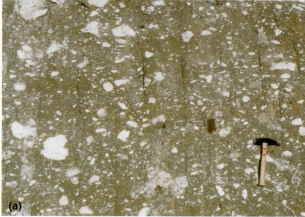

A nonwelded lapilli tuff, in New Zealand, consists of white pumice lapilli and bombs enclosed in a matrix of ash.

(a)

▶ **Figure 4.13 What pyroclastic-flow deposits look like.**
Pyroclastic-flow deposits are composed of an unsorted mixture of pumice and ash.

A welded tuff from Mexico formed when the pyroclastic-flow deposit remained sufficiently hot for the glassy particles to flatten under the weight of overlying pyroclastic debris. The black fragments, as much as 6 centimeters long, are pumice lapilli that compacted into lenses of black obsidian during welding.

(b)

Cinder cone. Small volcanoes, usually less than 600 m high, composed of basaltic to andesitic cinder (scoria). Slopes are near 30-35 degrees, the angle defined by the loose bombs and lapilli when they come to rest. Lava flows may issue from the base of the cone. Usually produced by single, prolonged eruptions lasting 1-20 years.

Example: SP Crater, near Flagstaff, Arizona.

Shield volcano. Small (300 m high) to giant (10,000 m high) volcanoes composed of many thin and widespread basaltic lava flows. Definitive shield shape is characterized by gentle slopes typically less than 15 degrees. Active for centuries to a few million years.

Example: Mauna Loa, Hawaii. The summit of Mauna Loa is nearly 4300 m above sea level; this volcano rises nearly 10,000 m above the sea floor. The summit of the Kilauea shield volcano is in the foreground.

Composite volcano. Modest (100 m high) to large (3000 m high) volcanoes composed of interlayered lava flows, lava-flow rubble, and pyroclastic deposits. Lava flows are typically thicker and shorter than those on shield volcanoes and may include lava domes. Volcanic rocks may range in composition from basalt to rhyolite at single volcanoes although most are dominated by basalt, andesite, and/or dacite. Slopes are generally greater than 25 degrees. Formed by rare to frequent eruptions over several hundred thousand years.

Example: Mount Fuji, Japan, rises to 3776 m above sea level and last erupted in 1707.

Dome complexes. Modest (500-2000 m high) volcanoes composed of multiple, overlapping volcanic domes. Usually of dacitic and rhyolitic composition but some are more mafic.

Example: Chaos Crags, northern California, is a series of dacitic domes that formed during eruptions 1100 years ago.

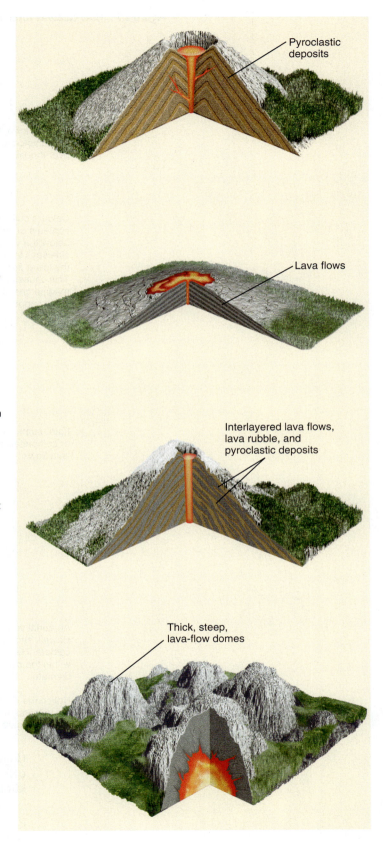

Pyroclastic deposits

Lava flows

Interlayered lava flows, lava rubble, and pyroclastic deposits

Thick, steep, lava-flow domes

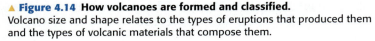

▲ **Figure 4.14 How volcanoes are formed and classified.**
Volcano size and shape relates to the types of eruptions that produced them and the types of volcanic materials that compose them.

Active Art

Forming Volcanoes: *See how different types of volcanoes form.*

▶ **Figure 4.15** **How calderas form in shield volcanoes.**

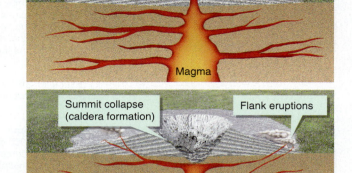

Caldera formation at Hawaiian shield volcanoes occurs gradually over thousands of years.

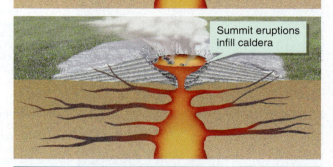

Caldera collapse occurs when the contents of the magma chamber below the volcano summit intrude sideways to supply eruptions on the lower flank of the caldera. This causes the roof of the magma chamber to subside as it loses support.

Later summit eruptions may partly or completely fill the caldera with lava flows.

An aerial view of Kilauea volcano, Hawaii, shows the summit caldera. Halemaumau Crater within the caldera is about 1 km in diameter.

Notice the eruption on the flank of the volcano, which is seen up close in Figure 4.1a.

(Figure 4.16) encloses a caldera lake, rather than a crater lake, produced by the collapse of Mount Mazama about 7600 years ago during eruption of 50 cubic kilometers of magma as pumice and ash.

Monstrous calderas, now filled and surrounded by extraordinarily thick pyroclastic-flow deposits, are found across the western United States and central and northern Mexico. Dozens of caldera-forming eruptions occurred in this region, beginning about 35 million years ago. The most recent such eruption extruded 1000 cubic kilometers of pumice and ash at Yellowstone,

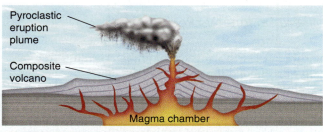

Pyroclastic eruption plume

Composite volcano

Magma chamber

◄ **Figure 4.16 How calderas form in composite volcanoes.**

Calderas form in composite volcanoes when large volumes of magma explosively erupt over periods of hours to a few days to form thick, widespread pyroclastic deposits.

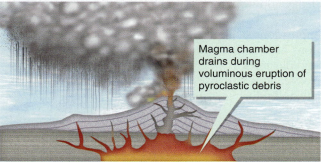

Magma chamber drains during voluminous eruption of pyroclastic debris

How Calderas Form: *See how explosive eruptions form calderas.*

The eruption of magma causes the roof of the magma chamber to subside as it loses support.

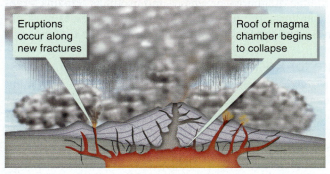

Eruptions occur along new fractures

Roof of magma chamber begins to collapse

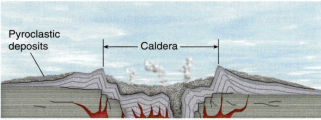

Pyroclastic deposits

Caldera

Crater Lake, Oregon, is a caldera formed at the top of ancient Mount Mazama volcano. Later eruptions created Wizard Island cinder cone, seen in the center. The caldera partly filled with water to form the deepest lake in North America.

Wyoming, about 600,000 years ago. The largest caldera in the United States is the 28-million-year-old La Garita caldera in the San Juan Mountains of Colorado. It is about 60 × 30 kilometers in size and formed during eruption of more than 4000 cubic kilometers of magma as pyroclastic flows. Very large calderas such as those at Yellowstone and La Garita encompass areas much larger than single volcanoes, such as Mount Mazama and Mount Pinatubo, and represent the collapsing roofs of unusually shallow batholiths.

Putting It Together—Where Do Igneous Rocks Appear in a Landscape?

■ Exposure of plutonic rocks at Earth's surface requires erosion to remove the rocks that originally surrounded the solidifying magma.

■ Once exposed to view, plutonic-rock landforms represent a variety of sizes and shapes that relate to the processes of intrusion and how erosion exposed the rocks.

■ Volcanic rocks include lava flows, pyroclastic-fall deposits, and pyroclastic-flow deposits.

■ Different magma compositions, conduit shapes, amounts of erupted material, and the proportions of lava flows and pyroclastic deposits determine the sizes and shapes of volcanoes.

■ Calderas form when volcanoes collapse downward into shallow magma chambers that partially drain during eruptions.

4.4 How and Why Do Rocks Melt?

At this point, your knowledge of igneous processes and products remains largely descriptive although you know the features that igneous processes must explain. The next problem is to understand how magma forms in the first place.

Melting Temperature for Rocks

If the temperature of a rock is raised high enough, the rock begins to melt. Every mineral has a measurable melting temperature that varies with pressure, and pressure equates to depth below the surface and the weight of overlying rock. Laboratory experiments reveal these melting temperatures at various pressures. Most silicate minerals melt at temperatures above 800°C. For example, olivine found in basalt melts at 1890°C at sea level. At a pressure equivalent to 100 kilometers below the surface, the melting temperature increases to 2050°C. Adding pressure, therefore, requires higher temperature to melt minerals.

Laboratory studies indicate that rock melting also depends on the minerals present in the rock and not only on temperature and pressure. Minerals react with each other at elevated temperatures, and these reactions affect the melting temperature. Specifically, mixtures of minerals melt at a lower temperature compared to experiments where the same minerals melt individually. At the surface, olivine present in a rock with pyroxene and plagioclase feldspar begins to melt at about 1200°C, almost 700°C lower than when olivine melts by itself.

Figure 4.17 illustrates laboratory rock-melting and magma-crystallization experiments. Although each mineral in a rock has a different melting temperature by itself, nearly all minerals begin to melt at a single, lower temperature when they are present together in a rock. Once one mineral completely melts, the temperature must increase to cause further melting of the remaining minerals, and each one completely transforms to liquid at a different temperature.

Water also influences the melting temperature. At the high pressure that equals the great depth where rocks melt, water does not boil off as steam. Instead, water remains in fluid form or exists as water molecules in minerals such as amphibole and mica. Experiments show that rocks melt at lower temperature when water is present than when the rock is dry.

How Magma Composition Relates to Rock Melting

Only when a rock completely melts does the magma have the same composition as the original rock. If the temperature does not exceed the final melting temperature for all of the constituent minerals, geologists say that the rock has only partially melted.

Stages in rock melting

1. Solid basalt at room temperature.

2. Rock remains solid until heated to 1100°C, where all three minerals begin to melt.

3. Pyroxene completely melts first and plagioclase and olivine continue to melt as temperature is increased.

4. At 1200°C the last of the plagioclase crystals melts, leaving only olivine crystals within the liquid melt.

5. Olivine continues to melt until about 1225°C, when the last traces of crystals disappear into the liquid melt.

6. Above 1225°C the sample is completely molten.

Lower temperature

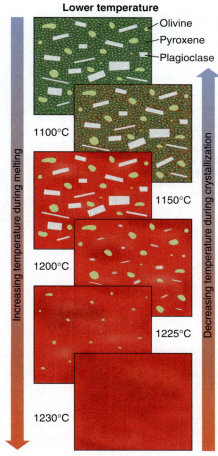

Olivine
Pyroxene
Plagioclase

1100°C

1150°C

1200°C

1225°C

1230°C

Increasing temperature during melting

Decreasing temperature during crystallization

Higher temperature

Stages in rock crystallization

6. Below 1100°C the sample is completely solid basalt.

5. At 1100°C pyroxene crystallizes along with remaining crystallization of olivine and plagioclase.

4. Olivine and plagioclase crystallize simultaneously as the temperature drops lower.

3. Plagioclase begins to crystallize in addition to olivine as the temperature reaches 1200°C.

2. Olivine begins to crystallize in the liquid melt at about 1225°C.

1. At a temperature above 1225°C the sample is completely molten.

▲ **Figure 4.17 How melting and crystallization work.**
These diagrams summarize results of experimentally melting and crystallizing a sample of basalt that erupted in Mexico in 1943. The rock does not completely melt at a single temperature but gradually transforms from solid rock to complete liquid over a range of more than 100°C. Similarly, the molten liquid gradually solidifies by sequential appearance of different minerals over the same temperature range during cooling experiments.

The composition of partial melts depends on the proportions of different minerals that melt and chemical reactions that occur between the magma and remaining crystals. In Figure 4.17, for example, the partial melt formed by melting basalt at 1150°C contains all of the chemical components of the melted pyroxene, but the components of olivine and plagioclase are mostly still in solid minerals. In most cases, partial melts contain a greater abundance of silica than the original rock. Experiments demonstrate, therefore, that partial melting of peridotite produces mafic magma, partial melts of mafic rocks generate intermediate magma, and so forth.

Why Rocks Melt

How does temperature rise to the point where rocks start to melt? Measurements in drilled wells and mine shafts confirm that temperature increases at greater depth. This relationship of increasing temperature as a function of depth is the **geothermal gradient**. Below continents the temperature increases about 25° to 30°C with each kilometer below the surface, and beneath oceans the average geothermal gradient averages 60°C per kilometer. Below 10 to 50 kilometers, the temperature does not increase so dramatically at increasing depths, for reasons explored in Chapter 8.

▶ **Figure 4.18** **How mantle rocks melt.**
Experiments provide insights into the processes that cause natural melting of rock to produce magma. These graphs illustrate conditions within Earth where magma forms.

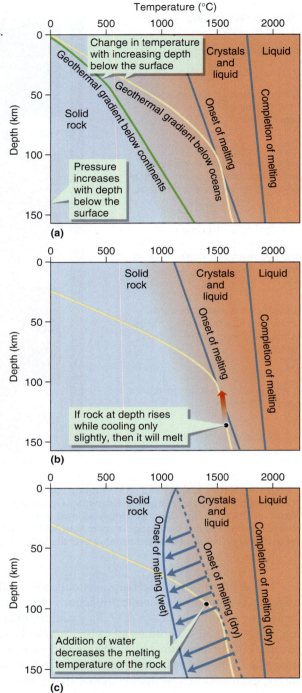

Melting along geothermal gradient:
The average geothermal gradient beneath continents does not intersect the onset-of-melting curve for peridotite. The gradient below oceans intersects the region of partial melting at about 100 km which accounts for the partly molten asthenosphere at that level in the mantle.

Melting by decreasing pressure:
Hot peridotite at depth partially melts and liquid rises upward but remains just as hot. The black circle denotes a location deep in the mantle. If that mantle material moves toward the surface (experiences a pressure decrease), then it crosses the onset-of-melting curve and partially melts.

Melting by adding water:
Rocks with water-bearing minerals melt at lower temperature than dry rocks. Water lowers the melting temperature only at elevated pressure, so the onset-of-melting curves under wet and dry conditions are the same at the surface. The black circle denotes a location deep in the mantle where dry peridotite is solid. If water is introduced, then this location will be hotter than the melting temperature.

Using Graphs to Understand Mantle Melting: *See how animated and annotated graphs explain how magma forms in the mantle.*

Figure 4.18 graphs the geothermal gradients in comparison to the melting temperature of mantle peridotite. Although temperature increases at depth, the melting temperature also increases because of increasing pressure. Figure 4.18a shows that average continental geothermal gradients are not hot enough to cause melting in the mantle, and only small amounts of partial melting at the depth of about 100 kilometers is likely beneath the oceans. No wonder volcanoes are not abundant on Earth's surface because magma seems hard to make with the relationships graphed in Figure 4.18a.

The graphs in Figure 4.18 imply three mechanisms for melting mantle peridotite.

• Add more heat: If hotter-than-average geothermal gradients were plotted in Figure 4.18a, then melting could occur in the mantle.

- Decrease the pressure: Figure 4.18b shows that rock does not cool off quickly if it rises toward the surface. Very hot, but solid, peridotite deep in the mantle melts as it rises to a shallow depth and lower pressure.
- Add water: Figure 4.18c illustrates the lower melting temperature of wet peridotite compared to dry peridotite. Hot, solid peridotite may melt if water is somehow added to the mantle.

Crystallization Resembles Melting in Reverse

Once rock melts, how does the magma crystallize? Cooling magma does not solidify all at once. As the temperature of the magma drops, individual minerals reach their crystallization temperatures and the minerals generally appear in the opposite order in which they melted (see Figure 4.17). Early-formed crystals react with the cooling magma as crystallization proceeds. These reactions cause changes in mineral composition or cause early-formed crystals to dissolve in the magma as new minerals crystallize. Reactions between minerals and the surrounding melt explain why quartz is not found in mafic rocks and why olivine is very rarely encountered in felsic ones. For those combinations of minerals and melt composition, the minerals react with the magma to form other minerals or completely dissolve.

EXTENSION MODULE 4.1
Bowen's Reaction Series.
Learn how early-formed crystals react with a cooling, crystallizing magma.

Putting It Together–*How and Why Do Rocks Melt?*

- Rocks melt gradually as the final melting temperatures of their constituent minerals are exceeded. Melting can happen in areas of unusually high temperature.
- Higher pressure requires a higher temperature for melting. This means that very hot rocks rising from great depth will melt when pressure decreases while temperature remains high.
- Increasing water content in rock decreases the melting temperature. Melting can occur, therefore, by adding water to hot rock at high pressure.

4.5 **How Do We Know** ... How Magma Is Made?

PICTURE THE PROBLEM

What Are the Necessary Conditions for Melting Rock?

The curves in Figure 4.18 illustrate whether peridotite is solid, partially melted, or completely melted for any combination of temperature, pressure, and water content. These graphs are essential for determining the conditions necessary for melting rock to make magma. Clearly, geologists cannot go deep inside Earth to watch natural magma production. To understand the conditions required to melt rock, geologists instead must rely on experiments that reproduce, in a simple way at least, the conditions inside Earth.

Geologists conduct experiments to relate temperature, pressure, and water content to magma production. Key laboratory investigations completed in the 1950s by O. F. Tuttle of Pennsylvania State University and N. L. Bowen of the Geophysical Laboratory at the Carnegie Institution of Washington considerably clarified geologists' understanding of the conditions required to make granitic magma. Although later research significantly extended the accomplishments of Tuttle and Bowen, their study stands as a classic illustration of how geologists use experiments to learn about igneous processes within Earth.

UNDERSTAND THE METHOD

How Is Rock Melting Studied in the Lab?

Tuttle and Bowen used the apparatus illustrated in **Figure 4.19** to conduct their experiments. The sample is placed in a furnace, heated to desired temperature, between rods that are squeezed together with a weight and lever to reproduce desired pressures. Water pumped into the chamber at very high pressure simulates the water present in felsic continental crust. In essence, Tuttle and Bowen

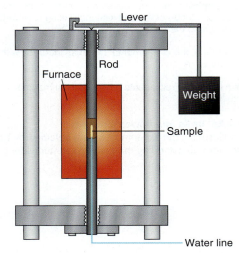

▲ **Figure 4.19 Tuttle and Bowen's experimental apparatus.**
This schematic diagram explains the device that Tuttle and Bowen used to produce experimental granitic magmas at different temperatures, pressures, and water contents. A small sample is heated by a furnace and compressed to high pressure by a weight pressed against a steel rod. Added water simulates conditions in the wet crust.

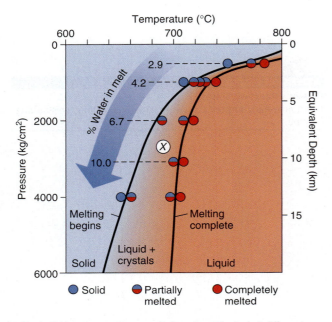

Each circle represents an experiment conducted at different combinations of temperature and pressure indicated by the values on the graph axes.

In each experiment, the sample either remained solid, partially melted, or completely melted. Curves representing the onset of melting and completion of melting with increasing temperature were drawn to fit to the data.

Water content of the experimental magmas increase with increasing pressure.

The point labeled X represents a hypothetical magma discussed in the text.

▲ **Figure 4.20** **Tuttle and Bowen's results.**
This graph illustrates the experimental data obtained by Tuttle and Bowen for understanding the melting conditions for granite.

Understanding Tuttle and Bowen's Data: *See how animated and annotated graphs explain Tuttle and Bowen's data.*

created a very elaborate pressure cooker, capable of much higher temperature and pressure than the familiar stovetop variety.

The starting materials for Tuttle and Bowen's experiments were synthetic mixtures of the same elements measured in granite by laboratory chemical analysis. Dozens of experiments were undertaken. Here we only follow the results of the experiments performed with elements found in quartz, potassium feldspar, and sodium-rich plagioclase feldspar. These elements—silica, aluminum, potassium, sodium, and oxygen—typically comprise more than 95 percent of typical granite, so these synthetic samples are a reasonable substitute for natural granite. Only about 1/100th of a gram of these powdered constituents were used for each experiment.

EVALUATE THE RESULTS
Under What Conditions Will Granite Melt?
Figure 4.20 graphs the results of one set of experiments. Each data point on the graph represents a single experiment. For each experiment, the sample was maintained at a selected temperature and pressure for several hours, after which the temperature was dropped abruptly so that any artificial magma quickly solidified into obsidian-like glass.

The quickly cooled sample was then examined with a microscope. If the sample was entirely crystalline, then the temperature was not high enough at that pressure to permit melting to begin. If the sample was a mixture of crystals and glass, then the experimental temperature was sufficient for partial melting. If the sample was entirely glass, then the temperature was sufficiently high to cause complete melting.

Tuttle and Bowen then drew curves through their data set to visualize the approximate conditions of temperature and pressure for the onset and completion of melting (Figure 4.20). They also analyzed the experimental glasses for water content to determine how much water was actually incorporated in the artificial magma at different pressures and temperatures.

Examination of these experimental results (Figure 4.20) reveals several critical features:

- The melting temperature of granite is much lower than that of peridotite (compare Figures 4.18 and 4.20b).
- The onset of partial melting in the presence of abundant water occurs at lower temperature as the pressure increases. This result is also seen for peridotite (Figure 4.18c) and is, indeed, a relationship seen for all magmas.
- At increasing pressure, more water dissolves into the magma.

INSIGHTS
How Do Laboratory Experiments Explain Field Geology?
Figure 4.20 shows the combinations of temperature, pressure, and water content that permit granite to melt. The melting temperature of wet granite decreases with increasing pressure. This result is the opposite from melting experiments with dry rock. Water-bearing minerals like amphibole and biotite are common in granite, so these experiments suggest that granite in the continental crust might melt at depths of 25 to 40 kilometers.

The experimental results also have important implications for what happens when wet granite magma moves toward the surface. Consider a granitic magma at conditions equivalent to "*x*" in Figure 4.20. If this magma rises toward the surface, then the point moves up on the graph. The magma crosses the melting curve and becomes solid granite at a depth of about 5 kilometers, even if it does not cool to a lower temperature. This means that wet magma "freezes" into rock without losing heat. Also, when at point "*x*," the magma contains a little less than 10 percent water, but just before it crystallizes at 5 kilometers it has 5 percent water. Somehow, the magma loses water as it rises. Where does the water go? Sections 4.8 and 4.9 consider igneous phenomena that are explained by behavior of water in magma formation and crystallization that Tuttle and Bowen's experiments reveal.

Putting It Together–How Do We Know … How Magma Is Made?

■ The ability to simulate the temperature, pressure, and compositional characteristics of the deep crust and mantle in the laboratory allows geologists to conduct experiments that portray the melting of rock and crystallization of magma.

■ Tuttle and Bowen's experiments show that magma under higher pressure can contain more water, and the presence of water lowers the melting temperature of magma.

4.6 How Does Magma Generation Connect to Plate Tectonics?

Variations in temperature, pressure, and water content determine melting conditions for different original rock compositions. Where do the melting conditions graphed in Figures 4.18 and 4.20 occur in Earth?

Magma is not generated everywhere on the planet but forms only where unique circumstances permit melting in the mantle or in the lower part of the crust. The association of igneous rocks with tectonically active areas, especially plate boundaries (Figure 1.10), suggests a link between plate tectonics and magma generation.

Decompression: Melting at Divergent Plate Boundaries and Hot Spots

Hot mantle rock can rise close to Earth's surface without losing much heat. **Figure 4.21**a shows that this process occurs at mid-ocean ridges (divergent plate boundaries) where the hot asthenosphere rises and fills the spaces where the oceanic lithosphere separates. The mantle peridotite experiences lower pressure as it rises. This decompression brings the peridotite to a combination of temperature and pressure that allows partial melting (Figure 4.18b). Experiments show that partial melting of the peridotite produces mafic magma, which is slightly more silica rich than the original peridotite. The mafic magma is also less dense than the surrounding mantle, and so it rises upward to erupt onto the seafloor as basalt.

Decompression melting also happens where unusually hot mantle rock rises at hot spots (Figure 4.21a), which are located below many volcanic islands such as Hawaii and Iceland (Figure 1.12). As the hot mantle rises and experiences progressively lower pressure, it partially melts to produce basaltic magma of the type erupted at Kilauea (Figure 4.1a).

Decompression melting of the mantle to form basalt also occurs where divergent plate boundaries or hot spots are located within continents. The hot magma introduces heat to the lower continental crust and, if temperatures are high enough, may melt parts of the continental crust to produce felsic magma. The newly generated felsic melt crystallizes to granite and rhyolite.

Addition of Water: Melting at Convergent Plate Boundaries

Figure 4.21b shows how magma forms at convergent plate boundaries where two plates collide, forcing a dense oceanic plate into the deeper mantle (Figure 1.11b). The upper part of the subducted plate is seafloor basalt that resided in

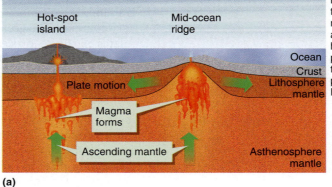

(a)

Mantle ascends toward the surface ridges and at mid-plate hot spots. If mantle ascent occurs with minimal heat loss, then the mantle peridotite begins to melt as the pressure decreases (the process illustrated in Figure 4.18b).

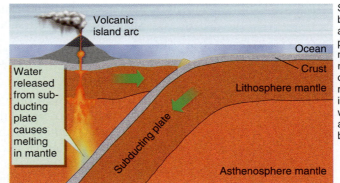

(b)

Subduction carries water-bearing minerals into the asthenosphere at convergent plate boundaries where metamorphic reactions release the water. The addition of water causes melting of the mantle peridotite, as illustrated in Figure 4.18c, and explains why volcanoes are present at convergent plate boundaries.

▲ **Figure 4.21** **How plate tectonics cause mantle melting.** The experimental insights illustrated in Figure 4.18 explain the relationships between plate tectonic processes and magma generation at divergent plate boundaries (mid-ocean ridges), hot spots (like Hawaii), and convergent plate boundaries (subduction zones).

Plate Tectonics and Magma Generation: *See how magmas form at plate boundaries and hot spots.*

seawater for tens to hundreds of millions of years. The water reacts with the basalt most intensely in areas close to the hot mid-ocean ridge where the water heats and circulates through fractures in the newly formed oceanic crust. By the time the seafloor basalt reaches the subduction zone it contains many water-bearing minerals. Water is also present in the sediment that accumulated above the seafloor basalt during its slow journey from the mid-ocean ridge to the subduction zone. Subduction, therefore, carries water deep into the mantle within basalt and sediment.

The subducting seafloor metamorphoses and releases water into the asthenosphere. At approximately 125 kilometers depth, the temperature and pressure causes metamorphic reactions where water-bearing minerals (amphibole, for example) turn into minerals lacking water (such as pyroxene). These reactions release water and a wide variety of dissolved elements that are not incorporated into the new metamorphic minerals. Recall that wet rocks melt at lower temperatures than dry rocks (Figure 4.18c). Laboratory experiments confirm that the expulsion of water from the subducting plate into the asthenosphere should cause partial melting of the asthenosphere peridotite.

Decompression melting of peridotite forms basaltic magma, so you may hypothesize that basaltic magma also forms near subduction zones. Indeed, basalt is found on volcanoes near subduction zones, but andesite and dacite are also very common. The greater diversity of magma composition at subduction zones relates partly to processes described in the next section. Compared to mid-ocean ridge and hot spot magmas, subduction-zone magma contains more water and also a bit more silica and other elements released from the subducting slab.

Does the subducted oceanic plate completely melt to form magma at convergent plate boundaries? This does not usually happen because the melting temperatures of the metamorphic minerals in the subducted plate are generally not exceeded within the mantle. Exceptions occur, however, where temperatures are unusually hot at the subduction zone or where the water content of the subducting plate is excessively high. In these cases, partial melting of a small amount of the subducted and metamorphosed basaltic crust produces dacitic magma. The dacite lava flows and pyroclastic debris erupted at Mount St. Helens in 1980 and 2004, and at Mount Pinatubo in 1991 may be derived from very slight melting of the subducted plate.

Putting It Together—*How Does Magma Generation Connect to Plate Tectonics?*

■ Magma forms only where unique circumstances take place that permit melting to occur in the mantle or lower crust.

■ Partial melting of the mantle happens where rising mantle remains hot but experiences decreasing pressure at divergent plate boundaries and hot spots.

■ Partial melting of the mantle also occurs by the addition of water by subduction at convergent plate boundaries.

■ The rising of very hot mafic magma generated in the mantle may cause partial melting of continental crust to form felsic magma.

4.7 What Makes Igneous Rock Compositions So Diverse?

The process of classifying igneous rocks (Section 4.2) demonstrated the remarkable diversity of igneous-rock compositions. The diversity of magmas solidifying to form these rocks is the result of one of four processes:

1. melting of equally diverse source rocks to different extents,
2. deriving one magma composition from another,

3. assimilating surrounding rock into the rising magma, and

4. blending magmas of different composition to produce a new hybrid composition.

Melting of Diverse Source Rocks

When examining plate-tectonic environments of magma formation, it was noted that different original materials produce different magmas (Section 4.6). Melting mantle peridotite generates basaltic magma. Melting subducted basalt produces dacitic magma. Melting continental crust leads to even more felsic magma. These results relate to the observation in Section 4.4 that partial melts are more silica rich than the original rock. A small degree of partial melting of a basalt, therefore, produces a relatively felsic magma, whereas complete melting generates a magma with a basaltic composition comparable to the original rock. So, not only different source-rock composition but also varying degrees of melting of the same source rock can generate a variety of magma compositions.

Deriving One Magma Composition from Another

If a basalt sample completely melts in the laboratory, then the result is a basaltic melt from which a basalt rock forms again if all of the liquid crystallizes (see Figure 4.17). What would happen if each mineral grain separated from the magma as soon as it formed while the melt cooled down? This scenario is the reverse of the situation explored in the previous paragraph where varying degrees of melting explained magma composition.

Figure 4.22 illustrates how melt composition changes during magma crystallization. The first mineral to crystallize while the magma cools does not have the same composition as the whole magma. An element preferentially incorporated within the mineral is less abundant in the remaining molten liquid than before crystallization began. In natural silicate magmas, the melt remaining after a fraction of magma crystallizes is almost always more felsic (more silica rich) than the original, completely molten liquid because of early crystallization of mafic minerals. If the first-formed mineral solids separate from the melt, then the remaining magma has a different composition from the original magma.

Figure 4.23 shows how early crystallized minerals separate from the parent magma to leave a melt with a different composition. In some cases the minerals are denser than the liquid and simply settle to the bottom of the magma chamber (Figure 4.23, left). These mineral accumulations can form significant natural economic resources. Metal ores of oxide minerals that settled in mafic magma chambers are mined at Stillwater, Montana, and most notably at the Bushveld Complex of intrusions in South Africa, which makes that country the world's principal supplier of chromium from the oxide mineral chromite.

Igneous intrusions that exhibit an inward zonation of progressively more felsic compositions suggest another process for separating crystals and melt. This arrangement likely happens as high-temperature minerals crystallize first along the cooler walls of the intrusion (Figure 4.23, right). As the intrusion

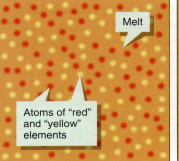
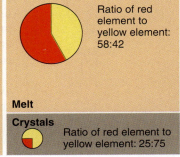
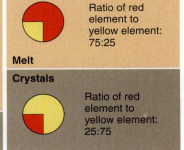

100% Melt, 0% Crystals **75% Melt, 25% Crystals** **50% Melt, 50% Crystals**

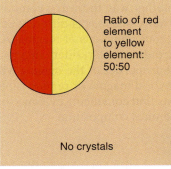

Ratio of red element to yellow element: 50:50

No crystals

Ratio of red element to yellow element: 58:42

Melt

Crystals Ratio of red element to yellow element: 25:75

Ratio of red element to yellow element: 75:25

Melt

Crystals

Ratio of red element to yellow element: 25:75

▲ **Figure 4.22 How crystallization changes melt composition.** This sample melt includes equal amounts of two elements represented by red and yellow dots. The first minerals to crystallize when the melt cools contain three times more of the "yellow" element than the "red" element. Crystallization preferentially removes the "yellow" element from the melt so that the remaining liquid becomes progressively richer in the "red" element. Melt composition changes, therefore, as crystallization takes place.

Fractional Crystallization: *See how magma composition can change during crystallization.*

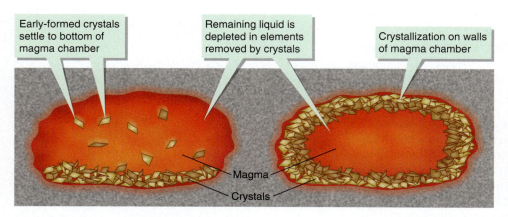

Early-formed crystals settle to bottom of magma chamber

Remaining liquid is depleted in elements removed by crystals

Crystallization on walls of magma chamber

Magma

Crystals

▲ **Figure 4.23 How crystals separate from magma.**
Crystals that form early during crystallization of the magma settle to the bottom of the chamber if they are denser than the melt (left). In some cases, crystallization takes place first on the wall of the magma chamber, where the temperature is coolest, and progresses inward (right). In both cases the crystals separate from the remaining molten liquid so that the remaining magma composition changes, as illustrated in Figure 4.22.

solidifies inward, the remaining magma is progressively enriched in silica and other elements that are less abundant in the early formed minerals.

In circumstances illustrated in Figure 4.23, the melt changes composition through time because minerals that form early during crystallization differ in composition from the magma and separate from it. This process is called **fractional crystallization**, because each fraction of the magma that crystallizes yields a melt of new composition (Figure 4.22). In this way it is possible to produce intermediate and felsic magmas from an original mafic melt. The compositional diversity of rocks at some volcanoes is convincingly shown to relate to this process. Fractional crystallization, for instance, likely accounts for many andesitic and dacitic volcanic products near subduction zones.

Magma Assimilation

When magma rises toward the surface, it comes in contact with, and may incorporate pieces of, the surrounding rock. These rocks may partially or completely melt and mix into the magma, which changes its composition. This process, called **assimilation**, is suggested by outcrops of some intrusive contacts and by chemical peculiarities in many igneous rocks. Large quantities of assimilated melt are not easily formed, however, because of the large amount of heat necessary for the magma to melt the surrounding rock. Such a loss of heat would, of course, lower the temperature of the magma and cause it to solidify before assimilation proceeds very far.

Magma Mixing

Another way to form a compositionally distinct magma is to mix together two or more magmas of different composition. Many large batholiths form from numerous injections of magma. As a new magma pulse arrives in the chamber, it encounters the resident magma, possibly of a different composition. The physical properties of the magma will determine if they can actually blend and form a hybrid magma or remain separated like oil and vinegar. In some cases, geologists find convincing evidence of magmas mixing to form a melt with a new geochemical composition.

Putting It Together–What Makes Igneous Rock Compositions So Diverse?

- Melting different types of rock produces magmas of different composition. The extent to which the rocks melt also determines the composition of the magma.
- Magma of one composition can form from magma of a different composition by fractional crystallization. No single mineral composition is exactly that of the whole magma, so crystallization of a fraction of the magma changes the composition of the remaining melt. A new magma composition forms if the remaining liquid melt separates from the crystals.
- Rocks surrounding a magma chamber may partly melt and assimilate to change the overall composition of the magma.
- Two or more magmas may mingle to produce a hybrid magma with a composition in between that of the original magmas.

4.8 Why Are There Different Types of Volcanoes and Volcanic Eruptions?

Many questions about the wide diversity of magma compositions and resulting igneous rocks, and the relationship of igneous rocks to tectonically active regions, are now answered. Go back, however, to the original observations of eruptions at Kilauea and Mount Pinatubo (Figure 4.1). Why do such different eruption styles occur? Why are there so many different types of volcanic landforms (Figure 4.14)? What determines why some volcanoes erupt mostly lavas whereas others explode pyroclastic materials? Why are some lava flows thin and widespread (Figures 4.9 and 4.10) whereas others are unable to move away from the vent and form lava domes (Figure 4.11)? Is there a link between the diversity of magma compositions and the variety of volcanic eruption styles and landforms?

The key to understanding volcanic phenomena rests in two properties of magma: gas content and viscosity. **Figure 4.24** shows how gas content and viscosity depend on magma composition and also relate to variations in volcanic eruptions and landforms.

First Clue: Gas Content

The most obvious difference in our observations of eruptions at Kilauea and Mount Pinatubo is the relatively quiet extrusion of lava at the former versus the explosiveness of the latter. What causes a liquid to explode into droplets that form pyroclastic fragments?

Figure 4.25 shows a familiar analogy to this phenomenon—opening a carbonated beverage. Some of the liquid converts into a bubbly foam or even ejects in a forceful spray, as happens when a champagne bottle pops open. Carbonated beverages contain carbon dioxide gas, formed by fermentation in alcoholic beverages and added to nonalcoholic ones. When looking at a carbonated beverage in an unopened, transparent container, you see few if any bubbles, and yet they appear in abundance as soon as you open the container and pour the contents. This is because the container was bottled under an applied pressure and, at this pressure, the carbon dioxide gas dissolves in the liquid. When the pressure is released, however, the gas comes out of solution in bubbles.

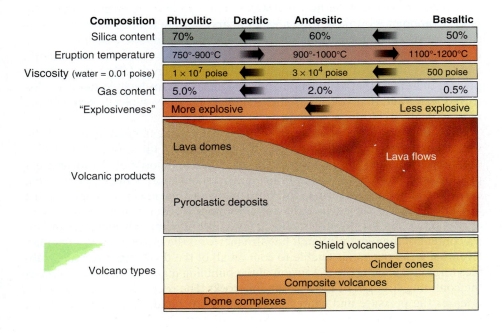

◄ **Figure 4.24 How magma composition determines volcano types and eruption style.**
Volcano types relate to the proportions of lava and pyroclastic materials composing them and the fluidity of lava flows, if present. Laboratory experiments demonstrate that gas content and viscosity vary according to the silica content of the magma. Higher gas content favors more explosive eruptions, which generate more pyroclastic deposits. Viscosity determines if lava forms fluid flows or accumulates in steep-sided domes. These measurable magma characteristics provide explanations for the commonly observed relationships among magma composition, explosiveness of eruptions, types of volcanic products, and types of volcanoes.

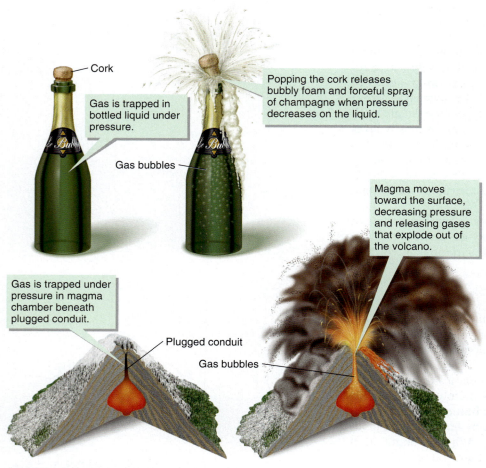

▲ Figure 4.25 Champagne as an eruption analogy.
Gas bubbles form abruptly when a champagne bottle is un-corked, which reduces pressure on the liquid contents. Gases also stream out of liquid magma when pressure decreases where the magma moves to the surface. In both champagne and magma, the bursting bubbles spray liquid into the air and form foam that flows down the side of the bottle, or volcano. In the case of magma, however, the spray and foam solidify into pyro-clastic fragments, whereas champagne remains liquid.

Labels in figure:
- Cork
- Gas is trapped in bottled liquid under pressure.
- Gas bubbles
- Popping the cork releases bubbly foam and forceful spray of champagne when pressure decreases on the liquid.
- Magma moves toward the surface, decreasing pressure and releasing gases that explode out of the volcano.
- Gas is trapped under pressure in magma chamber beneath plugged conduit.
- Plugged conduit
- Gas bubbles

Magmas also contain dissolved gases, as mentioned in Section 4.2. These gases remain dissolved only at elevated pressure, which equates to the weight of overlying rock. When magma moves toward the surface, the pressure decreases and the gases cannot remain dissolved in the magma. Tuttle and Bowen's experiment shows this result by the decreasing water content of the synthetic granitic magma at lower pressure (Figure 4.20). Bubbles form and stream toward the surface when water vapor and other gases come out of the solution (Figure 4.25). The bubbles have films of magma between them as they rise and explode out of the volcano like the foam spraying out of a champagne bottle.

There is, however, a major difference between magma and champagne; champagne spray remains liquid, but magma solidifies when it explodes out of the volcano because the temperature in the atmosphere is very cold compared to the solidification temperature of the melt. Glassy pyroclastic particles are the frozen fragments of bubbly magma. The fine-grained ash is the frozen liquid film between the bubbles, and the larger pieces of solidified foam form highly vesicular pumice and cinder lapilli and bombs (Figure 4.4).

The amount of gas that dissolves in magma strongly depends on the magma composition. Water vapor is the most abundant magmatic gas, and experiments show that it is generally four to five times more soluble in felsic magma than mafic magma at the same temperature and pressure (Figure 4.24). This relationship between gas dissolution and silica content of the melt explains why basaltic volcanoes rarely erupt violently but dacitic and rhyolitic volcanoes almost always do so. This also explains why basaltic volcanoes have a very high ratio of lava flows to pyroclastic deposits and why pyroclastic materials are progressively more abundant among andesitic, dacitic, and rhyolitic volcanoes (Figure 4.24).

Of course, you wonder how that water vapor got into the magma in the first place. Compounds prone to form a vapor at appropriate temperatures and pressures are present in small quantities throughout the mantle and incorporate into magma where melting happens. Subduction carries water into the mantle and triggers melting at convergent plate boundaries (Figure 4.21b). It is not surprising, therefore, that subduction-zone magmas contain high contents of dissolved water and that volcanoes above subduction zones typically produce the most violently explosive eruptions.

The buildup of gas pressure in shallow magma chambers may exceed the weight and strength of the overlying rock, triggering the onset of explosive eruptions. How does this buildup of pressure, akin to a bottle of champagne exploding on its own, take place? Fractional crystallization is the cause in some cases. The minerals that form at low pressure where magma crystallizes within a few kilometers of the surface do not accommodate water molecules in their crystal structures. As a result, fractional crystallization not only increases the silica content of the remaining melt but also increases the dissolved water content. Eventually, the magma is unable to contain all of the dissolved water and bubbles form in the melt. The expansion of the bubbling magma eventually exceeds the confining strength of the surrounding rocks. The rocks then fracture, which releases the pressure and causes more bubbles to rapidly form and the foam rushes upward at the onset of an explosive eruption.

Second Clue: Viscosity

Another critical feature of magma is its **viscosity**, or resistance to flow. Water, for example, flows readily on an inclined surface and has a very low viscosity. Other fluids, such as cake batter or tar, flow more slowly as thick, pasty masses, which indicates greater resistance to flow and high viscosity.

Viscosity partly relates to the molecular bonds within the liquid, so viscosity varies with composition. For silicate magmas, higher silica content favors greater viscosity. Shield volcanoes, therefore, comprise thin, laterally extensive lava flows of basaltic (low-silica) composition (Figure 4.10). On the other hand, silica-rich dacite and rhyolitic lava flows are thick and stubby or build high, steep-sided volcanic domes (Figure 4.11). Gas bubbles also separate more easily from low viscosity mafic magma than from high-viscosity felsic magma. Not only is the gas content high in a felsic magma, but the gas pressure rises dramatically when the bubbles cannot rise through the melt to the surface.

There are also some examples of basaltic lava domes and even more rare cases of rhyolitic lava flows that have traveled far. Experiments with molasses or tar reveal that the viscosity of both sticky liquids is substantially reduced when heated. Likewise, silicate melts are less viscous at higher temperatures than at low temperatures. Eruption temperature for basalt is typically higher than for andesites and dacites and much higher than for most rhyolites (Figure 4.24). Higher temperature, therefore, also favors lower viscosities for mafic melts compared to felsic ones.

Also, as the degree of magma crystallization increases, the mush of crystals and liquid becomes increasingly viscous. Most basaltic and andesitic lava domes are highly porphyritic, indicating that as much as 40 to 50 percent of the magma crystallized before eruption and providing an explanation for why relatively mafic lavas can be erupted in such a strongly viscous state.

Putting It Together—Why Are There Different Types of Volcanoes and Volcanic Eruptions?

- Volcanic phenomena and volcanic landforms can be understood from two properties of magma: gas content and viscosity.
- Gas content determines how explosive eruptions will be and the relative proportion of lava flows and pyroclastic deposits. More gas-rich magmas are more explosive.
- Viscosity determines if lava flows are thin and widespread, or thick and short. Lower viscosity produces more fluid flows.
- Viscosity and gas content are in large part determined by magma composition, and both properties increase with increasing silica content.

4.9 How Are Volcanoes Hazardous?

The serious hazard that volcanic eruptions present is one of the best reasons for geologists—and you—to understand igneous processes. More than 85,000 people died worldwide in the twentieth century as a result of volcanic activity.

Volcanoes are the sites of some of history's most dramatic and destructive natural disasters. From the destruction of the Italian cities of Pompeii and Herculaneum by the eruption of Vesuvius in A.D. 79 to the obliteration of Armero, Colombia, by erupting Nevada del Ruiz in 1985, history is full of accounts of destructive eruptions. Lava flows cover large areas of arable land and destroy buildings and crops, but fortunately they usually advance slowly enough for people to evacuate from harm's way. The lava flow depicted in **Figure 4.26**, for example, was one that traveled slowly enough so that people could avoid it. Other volcanic phenomena account for most eruption casualties.

▲ **Figure 4.26 River of lava.**
Basaltic lava streams from a crater on the flank of Mauna Loa volcano in Hawaii during a 1984 eruption. Although lava flows destroy buildings and bury agricultural land, they rarely flow fast enough to prevent people from escaping their path.

▲ **Figure 4.27** **Deadly pyroclastic flow.**
Ash billows above a fast moving pyroclastic flow unleashed from the obscured summit of Mount Unzen, Japan, in 1991. The combination of rapidly moving fragments, high temperature, and suffocating dust and gases makes pyroclastic flows one of the most destructive and deadly hazards of volcanic eruptions. This pyroclastic flow destroyed 50 homes and killed 28 people, including three volcanologists.

Hazards of Pyroclastic Flows

Fast-moving pyroclastic flows commonly cause horrific casualty tolls. **Figure 4.27** shows billowing ash from such a flow at Unzen volcano, Japan, in 1991. Typically moving at velocities in excess of 100 kilometers per hour, people in the path of these blasts have little chance of survival. That fact is driven home best by the destruction of St. Pierre, Martinique, French West Indies, in 1902 by a relatively small pyroclastic flow, erupted by nearby Mont Pelée. The city was flattened, and all but two of 18 boats that were anchored in the harbor sank. The photograph shown in **Figure 4.28** was taken after that disaster and still remains a striking record of this event, which killed nearly all of the 28,000 residents in the city.

Hazards of Lahars

The loss of nearly 25,000 people at Armero, illustrated in **Figure 4.29**, is the most tragic of many examples of destruction by rapid flow of water and loose debris down steep volcanic slopes. The Indonesian term **lahar** describes these events. Melting of snow and glacier ice by hot pyroclastic fragments caused the Armero catastrophe. The rapid erosion of thick, loose, pyroclastic deposits from steep hillsides also triggers lahars. Volcanoes in tropical areas experiencing heavy rainfall are most prone to this type of devastation. Lahars generated during the annual rainy seasons in the Philippines led to the burial of entire villages in the years following the 1991 eruption of Mount Pinatubo.

Hazards of Far-Flung Pumice and Ash

Ash and lapilli ejected high above volcanoes would, at first glance, seem more likely to be a nuisance than a hazard, but this is not always true. In tropical regions, volcanic ash weathers readily to produce highly fertile soils. This beneficial aspect of volcanism is sometimes offset, however, by the encroachment of agricultural villages too close to active volcanoes. Not only are these settlements at increased risk of destruction by lava flows, pyroclastic flows, and lahars, but pyroclastic-fall deposits at such close proximity to volcanoes may be thick enough to cause buildings to collapse. Nearly all of the 350 lives lost during the 1991 eruption of Mount Pinatubo were victims of falling roofs weighed down by ash saturated with rainfall from heavy typhoon rains that coincidentally occurred during the climactic eruption.

Volcanic ash is also strongly abrasive and can cause failure of machinery in areas hundreds of kilometers downwind of the volcano. During the 1980s and 1990s there was heightened awareness of the risks of high-flung ash clouds on jet aircraft. Unfiltered jet engines ingest ash, where it experiences temperatures above the ash melting point. This means that, in a sense, the ash converts to lava that coats the turbine blades and eventually causes engine failure because of the added weight stress. Following several near catastrophes with jumbo aircraft, new precautions were enacted to prevent an air catastrophe caused by a volcanic eruption.

Indirect Volcanic Hazards

In some cases, the destruction and loss of life is less directly related to the eruption itself. One example is the death of more than 36,000 Indonesian people as a result of the 1883 eruption of Krakatau. The volcano was a remote, uninhabited island where impact on humans was thought to be negligible. The eruption culminated, however, in caldera collapse, forming a deep depression on the ocean

◄ **Figure 4.28** **Pyroclastic-flow devastation.**
The deadly force of pyroclastic flows was made clear by the May 1902 eruption of Mont Pelée on the West Indies island of Martinique. A pyroclastic flow from the volcano, visible in the background, thoroughly devastated the city of St. Pierre, only 10 km away. Approximately 28,000 people died.

floor. The collapse disturbed the water surface, forming a huge wave more than 35 meters high that swept outward from the ruins of the volcanic island and devastated coastal villages on nearby Java and Sumatra. Another caldera-forming eruption in Indonesia, at Tambora in 1815, initially led to few casualties, but the devastation of farmland ultimately led to famine and the loss of more than 117,000 lives in an era before rapidly deployed international relief efforts.

Reducing Volcanic Hazards

Destruction by volcanic eruptions is substantially diminished when people practically apply knowledge of what causes different types of volcanic eruptions. The composition of erupted magma is the primary control on eruption style, as illustrated in Figure 4.24. It is possible, therefore, to loosely forecast the hazards of an eruption at its outset when the first magma is erupted, or based on the type of lava flows and pyroclastic deposits resulting from previous eruptions. Geologists then identify the geographic areas most susceptible to the expected eruption phenomena and evacuate the people most likely to be at risk. Increasing sophistication of volcanologic study in the latter part of the twentieth century has substantially reduced the casualties from volcanic activity.

Putting It Together–How Are Volcanoes Hazardous?
- Geologists relate the hazardous nature of eruptions (lava flows, pyroclastic flows, distant fallout of volcanic ash) to the composition of the erupted material.
- The most deadly and destructive volcanic phenomena are fast moving, far traveling pyroclastic flows and lahars.

4.10 Why Don't All Magmas Erupt?

A particularly nagging question remains. Earth's many extensive outcrops of plutonic rocks, such as those in the photo of Yosemite National Park (Figure 4.1c), reveal that a lot of magma never finds a way to the surface to erupt at a volcano. Why do some magmas reach the surface to be erupted at volcanoes whereas others solidify below ground?

How to Make a Pluton

Magma forms at high temperatures within Earth and rises into rocks of progressively lower temperature at shallow levels. As it rises, the hot magma conducts heat to the surrounding, cooler rocks. The magma temperature decreases to the value where it solidifies if it is not ascending rapidly enough. Indeed, this is why no ultramafic volcanic rocks form on Earth today. The high temperatures (>1600°C) required to keep ultramafic magmas in a largely molten state cannot be maintained at sufficiently shallow depths (note the geothermal gradient in Figure 4.18) to permit the magma to reach the surface without completely solidifying on the way. The presence of ultramafic volcanic rocks older than 2.5 billion years old is, however, powerful evidence of a formerly hotter planet, with higher temperatures at shallow depths.

Why does magma rise to begin with? Once magma forms, it is less dense than the surrounding rock and rises upward, as shown in **Figure 4.30**. If you try to draw a direct parallel to ice cubes (solid H_2O) floating in water (liquid H_2O), which clearly requires the liquid to be denser than the solid, this will not be perfectly clear. For most compounds, including silicates, the liquid phase is always less dense than the solid form of the same composition.

Mafic magma generated by partial melting in the mantle rises through, and shoulders aside, the surrounding denser peridotite and solid basalt in the oceanic crust but experiences greater difficulty rising in continental regions. This is because the intermediate and felsic rocks of the continental crust, by virtue of

▲ **Figure 4.29** **Lahar devastation.**
A rapidly flowing mixture of water from melted snow, volcanic ash, and other debris from Nevada del Ruiz volcano buried the city of Armero, Colombia, in 1985 and killed 25,000 people. The visible buildings are in what was a hilly part of the city, most of which is completely buried beneath gray volcanic mud in the background.

EXTENSION MODULE 4.2

Mitigating and Forecasting Volcanic Hazards.
Learn more about the types of volcanic hazards.

▶ **Figure 4.30 How magma rises.**
Mafic magma formed in the mantle rises upward because the magma is less dense than solid mantle peridotite. The rising magma stalls at the base of continental crust, however, because mafic magma is denser than the crust. The stalled magma may continue to rise along deep fractures because the weight (pressure) of the crust on the magma forces it upward through the fractures. Alternatively, the heat of the stalled mafic magma melts the continental crust to form felsic magma. The felsic magma rises toward the surface because it is less dense than the solid crust.

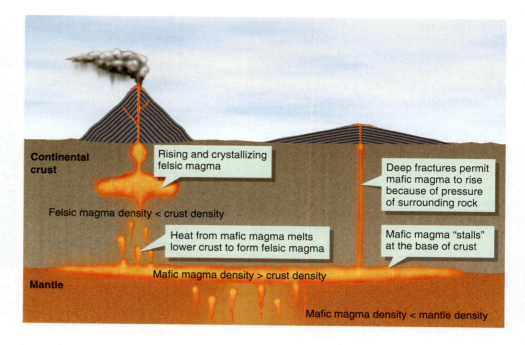

Continental crust

Rising and crystallizing felsic magma

Deep fractures permit mafic magma to rise because of pressure of surrounding rock

Felsic magma density < crust density

Heat from mafic magma melts lower crust to form felsic magma

Mafic magma "stalls" at the base of crust

Mafic magma density > crust density

Mantle

Mafic magma density < mantle density

Density and Magma Movement: *See how magma rises or stalls because of the density contrast between magma and rock.*

their different composition, are actually less dense than the mafic magma. The mafic magma, therefore, rises through the mantle but stalls at the base of the continental crust (Figure 4.30). If tectonic processes deeply fracture the crust, the melt may make it through to the surface. Otherwise the magma is stuck near the base of the crust and solidifies to form a gabbro sill. The mafic magma may, however, conduct enough heat into the crust to melt some relatively low-melting-temperature minerals to form a felsic magma. This felsic magma then rises farther upward because it is less dense than the surrounding crustal rocks (Figure 4.30).

Even most felsic magmas will not make it to the surface because of the behavior of water in the melt. Subduction-zone magmas are very water rich, as are most felsic magmas caused by melting of continental crust. The implications of water content for magma ascent were discussed in Section 4.5, and it is worthwhile to revisit Figure 4.20.

The hypothetical water-rich magma at point "*x*" in Figure 4.20 is a partially melted granite. Tracing a vertical line upward from "*x*" is equivalent to visualizing the granitic magma moving toward the surface. As the pressure decreases, the water content of the magma decreases, and the temperature for the onset of melting, which is the same as the temperature for the completion of crystallization of the rising magma, increases. There is no realistic way to increase the temperature of magma as it rises through progressively cooler rocks toward the surface, so the path of the magma must, at best, remain a vertical line on the graph. Now notice the key point—this magma crosses the completion-of-crystallization (also the onset of melting) curve before reaching the surface. The hypothetical felsic magma is destined to become a granitic intrusion, not a rhyolitic volcanic deposit.

Indeed, it is a wonder that magmas reach the surface to be erupted at all. The hurdles of decreasing temperature near the surface, density barriers, and crystallization as gases are released assure that all magma crystallizes to at least some extent before reaching the surface at a volcano. By far, the largest volume of Earth's magma ultimately solidifies as plutonic rocks. For every cubic kilometer of lava and pyroclastic debris erupted at the surface there is likely at least 50 cubic kilometers of igneous rock that solidified below the surface.

Pluton Crystallization Produces Economic Resources

The overwhelming majority of gold, silver, copper, and other metal resources are exploited from rocks within or close by igneous intrusions. Combining understanding of fractional crystallization and how gases (water and other vapors) release from crystallizing magma provides key insights to how a wide variety of useful ore deposits form.

Many economically important metallic elements (e.g., gold, silver, copper, lead, zinc, molybdenum, nickel) are present as minuscule constituents in magma. These elements readily concentrate in sufficient abundance for profitable mining because they are not included within the crystal structure of the dominant silicate minerals when the magma solidifies. Although they are a minor part of the original voluminous magma, these elements increasingly concentrate by fractional crystallization in the last dregs remaining after most of the silicate minerals form. It is also at this stage that gas-vapor pressure is highest (Section 4.8) so that fractures form within the solidified intrusion and surrounding rock. The remaining metal- and vapor-rich magma intrudes into these fractures, producing ore veins. Many metallic elements have an affinity to bond with sulfur, which is commonly present as a gaseous phase in the magma. As the sulfur-rich fluids move away from the magma during crystallization of the silicate minerals, the metals move away as well. Both the sulfur and metal dissolve in hot ground water that encircles and envelops the already solidified outer part of the intrusion. Changes in temperature, pressure, or water composition cause the metals to precipitate as sulfide minerals.

Especially notable in this regard are the huge copper mines of southern Arizona, Bingham, Utah, and Butte, Montana, which produced much of the world's copper from copper sulfides like chalcopyrite in the late nineteenth century through the middle of the twentieth century. These mining districts are located in stocks and batholiths of diorite to granite composition. The Bingham mine, shown in **Figure 4.31**, is the world's largest open pit, about 3.5 kilometers across and excavated 1 kilometer deep. At peak productivity, more than 400,000 metric tons of rock was moved *each day* to recover 1200 tons of copper.

Heated ground water near shallow intrusions produces hot springs and geysers at the surface. Although most of the hot water and steam are simply the local ground water, some of the water vapor and other gases originate from the magma, accounting for the sulfurous odors around many thermal springs. These are the same fluids that produce the voluminous metal ores at deeper levels below the ground. Where these hot fluids are located close to the surface, they are used directly for heating purposes or as steam to turn turbines to generate electricity. **Figure 4.32** shows that geothermal energy is locally important in volcanic regions throughout the world, such as in New Zealand, Iceland, Indonesia, and the Philippines.

Pegmatite, shown in **Figure 4.33**, is another economically important igneous rock that forms very late in the crystallization of intrusions. Very large crystals, approaching several meters in size, define these extraordinary phaneritic rocks. The large crystals form not because of slow cooling but because the high fluid content of the magma inhibits the onset of crystal formation but favors rapid growth once it is initiated. The size of the crystals, while spectacular in itself, is not the principal economic value of pegmatites. Pegmatites crystallize last, typically from vapor-rich felsic magma, so they are also enriched in elements that were excluded from typical rock-forming silicates. In pegmatites, these are uncommon elements that form silicate and oxide minerals only when concentrations are hugely enriched in the melt because of fractional crystallization. These pegmatites include ore minerals of rather exotic elements such as beryllium, lithium, and tantalum that have important uses in metallurgy and the production of ceramics. Some pegmatites are also the source of spectacular gemstones such as tourmaline, kunzite, and aquamarine.

▲ **Figure 4.31 World's largest open-pit mine.**
The Bingham Canyon Mine, Utah, is roughly 3.5 kilometers by 2.5 kilometers across and more than 1 kilometer deep. Sulfide ore minerals of copper and other metals are extracted from a felsic stock and surrounding rocks that were mineralized by fluids released from the magma as it crystallized. Note the circled buildings for scale. Waste rock and ore are removed from the terraced walls of the mine by power shovels that scoop 20 cubic meters of rock at a time.

Putting It Together—Why Don't All Magmas Erupt?

■ When rocks partially melt, the resulting liquid magma is less dense than the remaining solids and rises upward. If, however, less dense rock is encountered on the way up toward the surface, then the magma stalls and crystallizes.

■ Rock temperature decreases upward toward the surface. If magma does not rise quickly, then sufficient heat is conducted to surrounding rocks to cause crystallization of the melt before it can be erupted on the surface.

■ Most water-rich magma crystallizes below ground because the magma solidifies as it releases dissolved water at shallow depth.

■ Economically important elements present in minor amounts in the magma are concentrated into pegmatites and sulfide-mineral deposits by fractional crystallization.

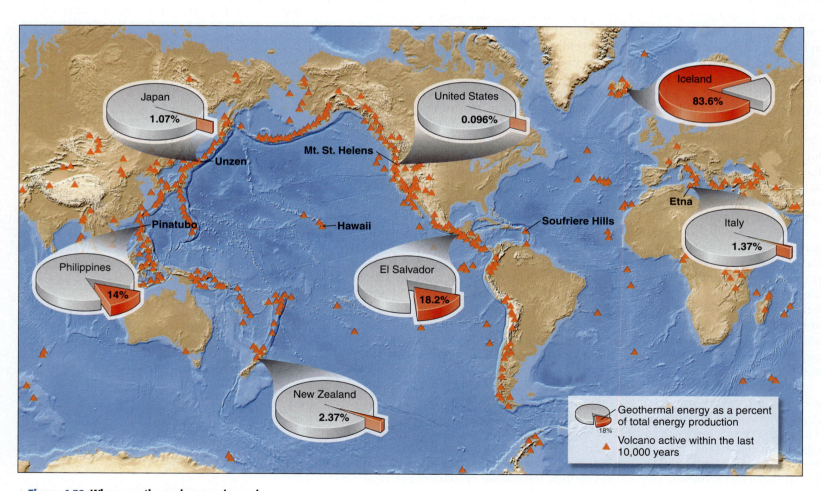

▲ **Figure 4.32** **Where geothermal energy is used.**
These pie diagrams show the varied significance of geothermal energy in various countries in volcanically active areas around the world. The United States produces the most geothermal energy, but this source provides very little of the total energy produced in the country. Geothermal energy is, however, a very significant energy source in the Philippines and El Salvador and is the dominant energy source in Iceland.

Spodumene crystals

Where Are You and Where Are You Going?

You observed the rocks and landscapes formed at three very different geographic localities, Hawaii, Mount Pinatubo, and Yosemite. These landscapes are partly a result of melting to produce magma and its subsequent solidification into igneous rock. Igneous rock types are classified using mineral composition and texture. Pyroclastic deposits, such as those observed at Mount Pinatubo, are classified by using fragment size and degree of consolidation.

Magma crystallizes above the surface, to produce volcanic rocks as observed in Hawaii and Mount Pinatubo, or it solidifies below the surface to form plutonic rocks that are possibly exposed by later erosion, as illustrated at Yosemite. Plutonic igneous rocks form intrusions with a variety of shapes and sizes depending on how much magma intrudes and the manner in which it

intrudes into surrounding rocks. Volcanoes vary in shape and size depending on the amount of erupted material and different eruption styles, which are the result of different compositions of magma. Gas content and viscosity of magma play a big role in determining eruption style and volcanic products. These variations in eruption style, in turn, determine the nature and extent of volcanic hazards at each volcano.

The different compositions of magma relate to how rocks melt and how the magma crystallizes. Temperature, pressure, and water content are the key controls. Experiments show that rocks melt (a) in areas of unusually high temperature, (b) where pressure decreases while maintaining high temperature, (c) or where water is added to hot rock at high pressure. Partial melting of mantle peridotite to produce basaltic

magma at divergent plate boundaries and hot spots results from decompression of the mantle as it rises upward. Magma forms at convergent boundaries where subduction carries water into the mantle and lowers the melting temperature of peridotite. A variety of igneous rocks forms by (a) melting diverse source rocks, (b) deriving one magma composition from another by fractional crystallization, (c) assimilating rock adjacent to magma, and (d) blending melts of different compositions.

Although magma forms deep in the crust or mantle, only a small fraction reaches the surface through volcanoes—most magma solidifies within Earth's crust. This process occurs where the magma runs into less dense rock on its way to the surface and is blocked, causing it to solidify in place, or the magma conducts away heat to its cooler surroundings on its way up and crystallizes before reaching the surface. Water dissolved in magma lowers its melting temperature, and the amount of water magma can hold decreases with decreasing pressure. Most water-rich magma crystallizes below ground because the water releases from the magma at shallow depth and the temperature is too low to maintain a dry melt. Some unusual elements are concentrated into mineral deposits during the crystallization of intrusions, creating valuable economic resources, including pegmatites (a source for gemstones and rare elements) and large-volume metal ores.

Now you understand the formation of magma and igneous rocks. How do other rock types form? How are sedimentary and metamorphic rocks related to igneous rocks? In the next chapter, you learn that most minerals composing igneous rocks disintegrate or dissolve at Earth's surface. Igneous and other rocks are subjected to a series of surface and near-surface processes that recasts their constituents as sediment and sedimentary rocks.

 ## Active Art

Forming Igneous Features and Landforms. See how intrusions form and how they produce unique landforms when exposed by erosion.

Forming Volcanoes. See how different types of volcanoes form.

How Calderas Form. See how explosive eruptions form calderas.

Using Graphs to Understand Mantle Melting. See how animated and annotated graphs explain how magma forms in the mantle.

Understanding Tuttle and Bowen's Data. See how animated and annotated graphs explain Tuttle and Bowen's data.

Plate Tectonics and Magma Generation. See how magmas form at plate boundaries and hot spots.

Fractional Crystallization. See how magma composition changes during crystallization.

Density and Magma Movement. See how magma rises or stalls because of density contrast between magma and rock.

Extension Modules

Extension Module 4.1: Bowen's Reaction Series. Learn how early-formed crystals react with a cooling, crystallizing magma.

Extension Module 4.2: Mitigating and Forecasting Volcanic Hazards. Learn about the types of volcanic hazards.

Confirm Your Knowledge

1. How does magma differ from lava? How are magma and lava classified?

2. Geologists gain understanding of the formation of magma and igneous rocks by making field observations, geochemical analyses and laboratory experiments. Give an example of each.

3. What are the basic types of magma compositions, and how do they differ in abundance of silica, iron and magnesium?

4. What property of igneous rock formation is most reflected by crystal grain-size? How?

5. How are pyroclastic deposits classified?

6. Which of the nine igneous rocks shown in Figure 4.3 were formed during last century's eruptions of Mount St. Helens and Mount Pinatubo?

7. How do we know that there is a magma chamber and the formation of plutonic rocks below an active volcano? List three reasons.

8. What determines the size and shape of a volcano?

9. Explain how peridotite can melt at a divergent plate boundary without an increase in temperature.

10. What is the origin of the magmas that form at a convergent plate boundary?

11. Explain how calderas form and at what rate.

12. Explain what factors control mantle melting.

13. Define partial melting and explain how it affects the magma composition.

14. What are the processes that lead to the generation of intermediate and felsic magmas, from a melt that is originally mafic?

15. What is the relationship between gas content and viscosity of the magma, and the type and location of volcanoes that form?

16. Obsidian is a felsic rock, yet has a black color. Explain.

17. Pahoehoe and a'a are types of lava flows found in Hawaii. Define and explain why they differ.

18. There is a relationship between areal extent of a lava flow, and magma type. Explain why some lava flows cover large areas and others do not.

19. What properties of magma determine how volcanoes erupt?

20. Identify the following igneous rocks based on their texture and mineral content

Fine-grained rock with 0% Quartz, 0% Potassium Feldspar, 61% Plagioclase Feldspar, 13% Biotite, 0% Muscovite, 17% Amphibole, 9% Pyroxene, 0% Olivine

Fine-grained rock with 0% Quartz, 0% Potassium Feldspar, 67% Calcium-rich Plagioclase Feldspar, 0% Biotite, 0% Muscovite, 0% Amphibole, 25% Pyroxene, 8% Olivine

Coarse-grained rock with 0% Quartz, 0% Potassium Feldspar, 52% Calcium-rich Plagioclase Feldspar, 0% Biotite, 0% Muscovite, 0% Amphibole, 32% Pyroxene, 16% Olivine

Coarse-grained rock with 0% Quartz, 0% Potassium Feldspar, 0% Plagioclase Feldspar, 0% Biotite, 0% Muscovite, 0% Amphibole, 56% Pyroxene, 44% Olivine

Fine-grained rock with 32% Quartz, 26% Potassium Feldspar, 24% Sodium-rich Plagioclase Feldspar, 3% Biotite, 5% Muscovite, 11% Amphibole, 0% Pyroxene, 0% Olivine

Confirm Your Understanding

1. Write out an answer for each question in the Chapter Outline for the chapter sections assigned by your instructor.

2. What is the meaning of geothermal gradient? Using an average geothermal gradient for continents and a surface temperature of 15°C, what is the temperature at a depth of 10 kilometers?

3. Given that the Sierra Nevada batholith is a series of plutons emplaced over 50 million years, how are the older plutons preserved when new, hot magma comes into the region?

4. What evidence is necessary to prove that batholiths form over a long period of time by multiple intrusions of differing composition?

5. What factors affect the melting temperature of a mineral? How do they affect it? Give three factors.

6. What type of volcanic products (lava flows, lava domes or pyroclastic material) and types of volcanoes would you expect from an eruption of basalt? … Of andesite? … Of dacite? … Of rhyolite? If there is more than one type of volcanic product, rank each from most abundant to least abundant

7. What processes lead to the variety of igneous rock compositions seen on Earth?

8. You are planning to move to a volcanic island. You have to decide between an island with a composite volcano composed of dacite and andesite or an island with a shield volcano consisting of basaltic lava flows. You have to decide which island is safer. Which island do you choose? Why?

Key Terms

andesite (p. 72)
aphanitic (p. 71)
assimilation (p. 90)
basalt (p. 72)
batholiths (p. 75)
bombs (p. 72)
caldera (p. 78)
cinder (or scoria) (p. 72)
dacite (p. 72)
dikes (p. 74)
diorite (p. 72)
felsic (p. 70)

fractional crystallization (p. 90)
gabbro (p. 72)
geothermal gradient (p. 83)
granite (p. 72)
intermediate (p. 70)
lahar (p. 94)
lapilli (p. 72)
lava domes (p. 76)
mafic (p. 70)
magma (p. 66)
obsidian (p. 71)
pegmatite (p. 97)

peridotite (p. 72)
phaneritic (p. 71)
plutonic rocks (or intrusive rocks) (p. 69)
porphyritic (p. 71)
pumice (p. 72)
pyroclastic (p. 69)
pyroclastic-fall deposits (p. 76)
pyroclastic-flow deposits (p. 76)
rhyolite (p. 72)
sills (p. 74)
stocks (p. 75)

tonalite (p. 72)
ultramafic (p. 70)
vesicles (p. 70)
viscosity (p. 93)
volcanic ash (p. 72)
volcanic necks (p. 75)
volcanic rocks (or extrusive rocks) (p. 68)
volcanoes (p. 76)
welded tuff (p. 76)

5 Formation of Sediment and Sedimentary Rocks

Chapter Outline

Why Study Sedimentary Rocks?

SEDIMENTARY ROCKS ARE PRODUCTS OF THE WEATHERING OF older rocks that are subsequently deposited at Earth's surface by water, wind, and glaciers. Sedimentary rocks are the most common rocks on Earth's surface. They contain an archive of the ancient history of the planet—a record of shifting seas and desert dunes, of changing climate, the raising and erosion of mountains, and life. Sedimentary rocks preserve remains of once-living organisms and record the emergence and evolution of life on Earth. Over time, buried biologic remains chemically transform into our most abundant energy resources—oil, natural gas, and coal. The open spaces between sediment particles are microscopic storage bins for ground water, oil, and natural gas. Clearly, because sedimentary rocks contain such critical resources, they are not only scientifically important, they are also essential to society.

This chapter challenges you with four primary objectives:

✔ To understand how rocks break down by physical and chemical weathering to produce sediment, the ingredients of sedimentary rocks

✔ To understand how loose sediment converts into hard sedimentary rock

✔ To learn how to examine sedimentary rocks in order to interpret the materials from which the sediment was derived and the environment where the sediment was deposited

✔ To appreciate the current economic importance of sedimentary rocks, particularly for energy and industrial resources

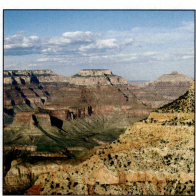

Scenic sandstone in Paria Canyon, Arizona, is the record of desert dunes that moved through the area more than 150 million years ago. ▶

Along the way to achieving these objectives, you will answer the following questions:

5.1 How and why do rocks disintegrate to form sediment?

5.2 What is the link between weathering and sediment?

5.3 Why are fossils found in sedimentary rocks?

5.4 How does loose sediment become sedimentary rock?

5.5 How are sedimentary rocks classified?

5.6 How do sedimentary rocks reveal ancient environments?

5.7 *How do we know . . .* how to interpret unseen turbidity currents?

5.8 How do plate tectonics and sedimentary rocks connect?

In the
FIELD

You hike along a dry stream channel, shown in **Figure 5.1**. With each step, your feet sink slightly into the pea-size gravel and dry sand, and you occasionally step over larger cobbles. Grass and shrubs grow sparsely along the banks, rooted in layers of sand and mud. Pausing to rest, and examining a fist-sized stone from the streambed, you quickly recognize the coarse mosaic of quartz, pink and white feldspars, and less abundant flakes of black biotite—minerals that together form granite. Flowing water transported the cobble at some time in the past from another location. You head up the nearby slope, looking for granite outcrops similar to the cobble.

A short distance above the stream, you discover a large granite outcrop. Crumbly and brown, the rock is unlike the solid, pinkish-gray granites you have seen elsewhere in the field and lab. Chipping with your rock hammer, you notice mineral grains readily falling off to join the similar loose fragments surrounding the base of the outcrop and extending downhill toward the stream. More forceful swings knock loose a large fragment from the stronger rock that resides below the crumbly surface. You look carefully at this larger chunk of granite (Figure 5.1c) and see the hard, pinkish-gray rock with the mosaic of quartz, feldspar, and biotite crystals.

So what is the story with the brownish outer part of the rocky outcrop and the pile of rock crumbs at your feet? The fragile brown rock is also granite; an unsurprising observation because there does not seem to be a sharp boundary between crumbly and hard material to suggest that two different rock types compose the outcrop. The brown rock fragments (Figure 5.1d) clearly contain quartz and feldspar. Biotite is also present, but not as abundantly as in the hard granite. Yellow-brown stains surround the biotite crystals, and the cleavage surfaces, instead of being shiny black, flash brassy yellow reflections in the sunlight.

Grains of all three minerals break loose to the ground as you handle the crumbly rock. Stooping down, you pick up a large handful of the loose mineral and rock particles littering the slope below the outcrop. These particles, visible in Figure 5.1e, range in size from brown-granite pebbles a few centimeters across to dust. Most grains are single crystals of quartz and feldspar, a millimeter or two across. Rec-

ognizing quartz and feldspar as the dominant components of the granite, you look through several handfuls in search of biotite. You pick out a few biotite flakes, but they are not as easy to find as they are in the granite. Your hands become increasingly stained with a brown dusty residue, which turns to sticky mud when you spare some drinking water to wash it off.

Returning to the streambed, you now see the sediment deposited there during past floods in a new light—it originated from nearby outcrops of granite. This is clear not only for the cobbles of recognizable granite but also for the quartz and feldspar composing most of the sand grains in the wash. The sediment on the stream bank includes the same brownish, muddy dirt that is mixed with the deteriorated granite on the hillside above, along with scattered flakes of brassy biotite. You smile in satisfaction with your reasoned discovery that the sediment carried by the stream originated through the disintegration of hard, granite rock.

Now, all sorts of other questions arise. How does the granite disintegrate? Why does the quartz and feldspar seemingly survive this process more readily than the biotite? What is the brown dust that turns to mud when moistened? Does its appearance in crumbly brown granite as the biotite become less abundant indicate a relationship between formation of some minerals at the expense of others? How does loose sediment like that seen in the dry wash today become hard sedimentary rock in the future? Will some of the loose sediment eventually wash into larger rivers and ultimately travel to the sea? Will the sediment carried by larger rivers, found on the beach, or lying on the ocean floor differ in any significant way from the deposits in the dry streambed? What processes at Earth's surface contribute to the formation and transportation of sediment? Clearly, you have only scratched the surface of your curiosity about sediment and sedimentary rocks.

In this chapter, you will learn how geologists seek answers to these and other questions. The methods they employ are broadly similar to those used to study igneous rocks in Chapter 4: field and microscopic examination of sediment and sedimentary rocks, analyses of minerals composing the rocks, and experiments designed to duplicate processes that form different sedimentary deposits.

Figure 5.1 Stages in the transformation of granite to sediment. ▶
These field photos illustrate the transformation of hard granite to crumbly granite to loose mineral fragments. The sand and gravel along the stream consist of loose mineral fragments similar to those found at the base of the granite outcrop.

Field observations on hike along stream bed

A. Sand and gravel are deposited on this dry stream bed during rare floods. Similar sediment, along with some mud, forms the eroded stream bank.

B. This crumbly granite forms outcrops uphill from the stream.

C. The interior of the granite outcrop consists of hard igneous rock, with light-colored feldspar and quartz, and dark biotite.

D. The outer part of the outcrop is crumbly granite, notably browner than the interior sample and with less biotite.

E. Fragments that accumulated at the base of the outcrop are almost entirely quartz and feldspar, coated in fine brown dust.

5.1 How and Why Do Rocks Disintegrate to Form Sediment?

By studying the disintegrating granite outcrop and the stream sediment, you observed the products of **weathering**, the processes that break down preexisting rocks at Earth's surface. Weathering reduces rocks into loose particles, dissolves some minerals, and produces new minerals.

Observations show that rocks weather through physical and chemical processes. The physical processes, **physical weathering**, naturally break large rocks into smaller fragments. The chemical processes, **chemical weathering**, dissolve some minerals and produce new ones at the expense of original minerals in the rock while generating ions carried away in watery solution. Organisms play an important role in chemical weathering and, to a lesser extent, in physical weathering.

Physical Weathering

Natural rock outcrops are not smooth, continuous rock faces. Cracks, like those seen in **Figure 5.2**, always break the rock surface. These fractures form in many ways:

- In some igneous rocks, the fractures and blocks of rock relate to the processes that formed the rock, such as cracks formed by cooling (Figures 4.8b and 4.10b) and breakage associated with flowing lava (Figure 4.12).
- Stresses in the crust resulting from plate tectonics cause many fractures, oriented in many directions. You will learn more about these features in Chapter 11.
- Cracks open up parallel to the ground surface when rock expands slightly outward where overlying materials erode away (Figure 5.2b). This process of forming rock sheets parallel to the ground surface is **exfoliation** (from the Latin *exfoliatus*, which describes stripping leaves or bark from trees).
- Bedding (layering) in sedimentary rocks causes planar breaks in rock (Figure 3.1a).

No rock, therefore, is a continuous, unbroken mass of crystals. Fractures break rocks into pieces, but most of the initial pieces remain as very large blocks. Weathering processes at or near Earth's surface exploit the earlier formed breaks in rock and create more fractures while dislodging smaller and smaller fragments.

Figure 5.3a shows how water, when frozen in the cracks of rocks, becomes an effective agent of physical weathering. Liquid water expands when it freezes into ice. The expansion in volume exerts sufficient force to break rock. Water enters rock along existing fractures and then moves along boundaries between mineral grains or into open spaces, called **pores**, between minerals. When the water freezes, the expanding ice forms new cracks that provide new openings for more water to enter after the ice thaws. In regions where the climate causes many freeze-thaw cycles during a single year, repeated freezing and thawing effectively disaggregate rocks. Figure 5.3a shows a freeze-thaw weathered outcrop resulting from this process.

Other weathering processes also exploit existing fractures to further disaggregate rock. One example visible in some places every spring is the common nuisance of crumbling pavement that results when communities apply salt to melt ice in winter months. Freezing and thawing of water can cause this problem, but salt spreading greatly speeds up the process. Salt dissolves in water, and the salty water percolates into cracks. When the water evaporates, salt crystals form in the cracks and break the rock apart as they grow. This same salt weathering process occurs along shorelines of oceans and salty lakes and by evaporation of rainwater in deserts. Salty water seeps into crevices and pore spaces and evaporates to permit growth of salt crystals that break the rock further apart, as illustrated in Figure 5.3b.

(a)

Cracks are pathways for water to enter into rock to cause physical weathering, by processes like freeze and thaw of water and ice, and chemical weathering by reactions between water and minerals. Trees commonly root in fractures and the growing roots expand to pry the rock apart. Fractured granite dominates this view of the Sandia Mountains in New Mexico.

(b)

Fractures can open parallel to the ground surface, as seen here in granite at Yosemite National Park, California. The fractures form when erosion removes the weight of overlying rock and permits the once buried rock to expand upward and outward.

▲ **Figure 5.2 Natural exposures of rocks are always fractured.**

Water enters rock along fractures and mineral-grain boundaries.

Water freezes forming ice, which forces the rock apart.

Repeated freezing of water to ice and thawing pries the rock apart.

(a)   **Time**

Salty water enters rock along fractures and mineral-grain boundaries.

Water evaporates causing crystallization of soluble minerals, which forces the rock apart.

Repeated wetting and drying loosens mineral grains that separate from the original rock.

(b)   **Time**

▲ **Figure 5.3 How physical weathering works.**
(a) The most important physical weathering process is the freezing and thawing of water and ice to pry rocks apart along existing fractures and mineral-grain boundaries. Repeated freeze and thaw breaks rock outcrops into angular fragments of various sizes, as seen here at Beartooth Pass, Montana. (b) Growth of salt crystals as water dries in cracks disintegrates rock along the shorelines of the ocean and salty lakes and in arid deserts. The pitted sandstone surface along the California coast results from salt weathering.

A closer look at Figure 5.2a reveals the trees growing in the rock; this is another process that exploits cracks. Root growth is inadequate to make new cracks in most rocks, but as roots grow to a larger diameter within an existing crack, they wedge the rock apart.

Expansion and contraction of minerals in rocks also cause disintegration. Some rocks contain clay minerals, a mineral group with crystal structure similar to mica (see Chapter 2). These minerals generally form at or near Earth's surface, mostly by chemical weathering processes, which you will explore very soon. Clay minerals incorporate water within their crystal structures. During cycles of wetting and drying, water is added and then drawn away from the edge of the clay grains, causing the mineral crystals to expand and contract. Repeated wetting and drying causes grains to separate from one another and substantially decreases the strength of the rock to the point where it falls apart.

The volume change of minerals caused by the heating and cooling of air in contact with rocks also aids disintegration. In deserts, where daily temperature fluctuations are greatest, air temperature changes cause minutely small volume changes in minerals. The amount of expansion and contraction varies for different minerals and forms microscopic cracks along mineral-grain boundaries. Water and wind-blown clay and salt particles enter these cracks and lead to further disintegration by freeze-thaw and wetting-drying cycles.

Chemical Weathering

The fact that rocks, such as the streamside granite, formed in the past under one set of conditions and deteriorate later under another set of conditions sets up a

Active Art

Physical Weathering: *See how the freezing and thawing of water and the evaporation of salty water pry rocks apart.*

EXTENSION MODULE 5.1

Chemical Reactions and Chemical Equations.
Learn about chemical reactions and how to write chemical equations.

question: What is so unusual about conditions at Earth's surface that causes rocks to weather? Most igneous and metamorphic rocks (Chapter 3) form within the Earth where oxygen gas (O_2) is nonexistent and water (H_2O) is a very minor constituent. Even lava flows erupted onto the surface solidify without incorporating oxygen and water, because atmospheric gases and water cannot incorporate into lava at low surface pressure.

The key to understanding chemical weathering, therefore, is to understand how rock-forming minerals react to the surface environment that is rich in oxygen and water. Nearly all minerals are unstable to some degree in the presence of water and oxygen gas. The reactions that take place with oxygen (O_2) and water (H_2O) are not only important in rock weathering at the surface but also in the shallow subsurface where ground water interacts with minerals.

Dissolution reactions break apart mineral molecules, and the constituent molecules disperse in water. Many minerals readily dissolve in water, especially those with ionic bonds (Section 2.4).

Halite (NaCl) undergoes this type of reaction, where halite crystals completely disappear while Na^+ and Cl^- ions appear between the H_2O molecules (Figure 2.16). It is easy to determine this process by noting the salty taste of the originally flavorless water. A simple chemical equation to represent this reaction is:

$$NaCl(s) + H_2O \rightleftharpoons Na^+(aq) + Cl^-(aq) + H_2O$$

halite   water   sodium   chloride   water
                          ion   ion

(5.1)

The (s) indicates compounds that are solid, in this case the halite, whereas the (aq) indicates components present in watery (aqueous) solution. So the equation states that halite reacts with water to produce sodium and chloride ions in solution.

For minerals such as halite, dissolution happens as soon as it becomes wet, as seen in **Figure 5.4**a. For others, dissolution is not so rapid as to be instantly noticeable but it is detectable over short periods of time, as illustrated in Figure 5.4b. Dissolution of other minerals is slower, but given enough time and the proper conditions of temperature and water chemistry, all of the principal rock-forming minerals will dissolve.

More complicated **hydrolysis** reactions consume both the mineral and some of the water molecules while reorganizing the elements into new compounds and dissolved ions. The newly formed solid compounds in these reactions are minerals. Weathering of potassium feldspar to make kaolinite, a clay mineral, is an example of hydrolysis (Figure 5.4c). Here is a chemical equation to represent this reaction:

$$2\,KAlSi_3O_8(s) + 11\,H_2O \rightleftharpoons 2\,K^+(aq) + 2\,OH^-(aq) + 4\,H_4SiO_4(aq) + Al_2Si_2O_5(OH)_4(s)$$

potassium feldspar   water   potassium ion   hydroxyl ion   silicic acid   kaolinite

(5.2)

This equation states that feldspar reacts with water to produce kaolinite plus a solution of ions and compounds that include the potassium and some of the silica originally present in the feldspar. The hydrogen and oxygen atoms in the reacting water molecules redistribute among the solid kaolinite and the dissolved compounds.

The weathering reactions represented by Equation (5.2) have two important implications:

1. Chemical weathering does not just destroy minerals (e.g., potassium feldspar), it also forms new minerals (e.g., kaolinite) that are more chemically stable in the surface environment than were the original compounds. **Table 5.1** lists the weathering products of some common minerals. Dust among the fragments of weathered granite (Figure 5.1e) is mostly clay minerals. This clay formed not only from weathering of feldspar but also from the weathering of the biotite (Table 5.1), a conclusion supported by the diminished amount of biotite in the weathered granite (Figure 5.1d).

EXTENSION MODULE 5.2

Why Is Seawater Salty?
Learn why some ions are concentrated in seawater to give its salty taste.

(a)

Dissolution of halite (rock salt) happens quickly enough to watch in the laboratory or kitchen.

(b)

Calcite dissolves very slowly in mildly acidic water, but the dissolution is noticeable over long time periods. The letters on this gravestone were clearly defined when first engraved, but weathering dissolved the outer surface of the calcite-rich rock (limestone) so that no letters remain legible.

0.02 mm

Kaolinite

Potassium feldspar

(c)

Microscopic examination shows the results of hydrolysis reactions between feldspar and water. New crystals of the clay mineral kaolinite formed where the corroded potassium feldspar reacted with water.

▲ **Figure 5.4** **Minerals undergo chemical reactions with water.**

2. Although the reaction began with pure water, the result is an aqueous solution containing other dissolved components [e.g., K^+, OH^-, H_4SiO_4 in Equation (5.2)].

No natural water is pure H_2O. Water always reacts with minerals where rainwater and snow melt percolate through soil and rock fractures, and ground water moves slowly between mineral grains. Salty seawater contains the dissolved products of weathering on the continents as well as from reactions of water with seafloor rocks and gases rising from submerged volcanoes. Dissolved constituents in water ultimately bond to form new minerals that precipitate from the water. The presence of dissolved ions may also enhance reactions with other minerals. This is especially true where reactions cause the water to become acidic [e.g., the formation of silicic acid in the Equation (5.2) reaction] because acids enhance weathering reactions of most minerals with aqueous solutions.

The most abundant acid in chemical weathering forms from mixing carbon dioxide (CO_2) with water.

TABLE 5.1	**Weathering Products of Some Common Rock-Forming Minerals**	
Original Mineral	Mineral Weathering Products	Ions in Solution
Halite (NaCl)		Na^+, Cl^+
Gypsum ($CaSO_4$ plus water)		Ca^{2+}, SO_4^{2-}
Calcite ($CaCO_3$)		Ca^{2+}, HCO_3^-
Quartz (SiO_2)		SiO_4^{4-}
Plagioclase feldspar (Ca, Na, Al silicate)	Clay	Ca^{2+}, Na^+, SiO_4^{4-}
Potassium feldspar (K, Al silicate)	Clay	K^+, SiO_4^{4-}
Olivine (Mg, Fe silicate)	Limonite, clay	Mg^{2+}, SiO_4^{4-}
Pyroxene (Ca, Mg, Fe silicate)	Limonite, clay	Ca^{2+}, Mg^{2+}, SiO_4^{4-}
Amphibole (Ca, Mg, Fe silicate)	Limonite, clay	Ca^{2+}, Mg^{2+}, SiO_4^{4-}
Biotite (Fe, Mg, K, Al silicate)	Limonite, clay	K^+, Mg^{2+}, SiO_4^{4-}
Muscovite (K, Al silicate)	Clay	K^+, SiO_4^{4-}

Pyrite reacts with oxygen in water to form iron hydroxide minerals called limonite. Notice that the typical cube shape of pyrite crystals is preserved in the limonite.

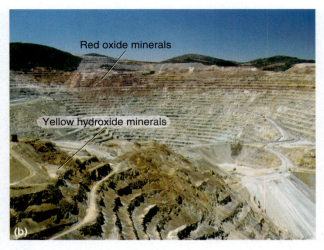

Iron-bearing sulfide minerals present in rocks near the top of this copper mine in New Mexico are weathered to oxide and hydroxide minerals by oxidizing ground water. The gray igneous rocks exposed deeper in the open pit are not oxidized. Geologists use the presence of oxidation minerals to locate metal ores for mining.

Many sedimentary rocks, such as these sandstones at Arches National Park, Utah, have red and yellow coloration from small quantities of hematite, an iron oxide mineral, and limonite, an iron hydroxide.

▲ **Figure 5.5 What the results of oxidation reactions look like.**

Water and CO_2 readily mix because carbon dioxide is present in the atmosphere and organisms respire it into soil. Reaction of water and CO_2 produces weak carbonic acid (H_2CO_3). The mixing reaction is written as:

$$\underset{\text{water}}{H_2O} + \underset{\text{carbon dioxide}}{CO_{2(g)}} \rightleftharpoons \underset{\text{carbonic acid}}{H_2CO_{3(aq)}} \quad\quad (5.3)$$

The (g) indicates that the carbon dioxide is a gas. The increased acidity of water that reacts with carbon dioxide is especially effective in dissolving carbonate minerals, such as calcite (Figure 5.4b). Other acids released from plant tissues (e.g., citric and ascorbic acids in fruit), and a variety of organic acids produced by bacteria flourishing on dead, fermenting tissue (e.g., acetic acid, commonly known as vinegar) also contribute to chemical weathering of minerals.

Reactions between minerals and atmospheric oxygen (O_2) are also important. **Oxidation** is the process by which substances react with oxygen to form new substances by exchanging electrons. You observe the result of oxidation whenever you see rusted metal. Iron reacts with oxygen and oxidizes to a new, much weaker compound, represented by the rust. Iron is also one of the most abundant elements in Earth's crust (Figure 2.24) and is well represented in rock-forming minerals, such as biotite in granite. Plutonic and metamorphic rocks form below Earth's surface in the absence of oxygen gas, so they are especially prone to oxidation reactions in the surface weathering environment.

Oxidation reactions involve the transfer of electrons from a substance to the O_2 molecule, which transforms into O^{2-} ions that bond into new compounds. The most common iron ion in igneous and metamorphic minerals is Fe^{2+}. In the presence of O_2, iron readily donates an additional electron to oxygen and becomes the Fe^{3+} ion. Fe^{3+} is smaller than Fe^{2+} and, clearly, has a different charge. These size and charge differences affect mineral structures where Fe^{2+} oxidizes to Fe^{3+}. As a result, minerals containing abundant Fe^{2+} (such as olivine, pyroxene, amphibole, biotite, and pyrite) readily weather, not only because of reactions with water but also because oxidation of Fe^{2+} breaks down the original crystal structure.

Oxidation weathering produces oxide and hydroxide minerals, of which those containing iron are the most common. The iron hydroxides are a complex group, and not all of them have a sufficiently well-organized internal structure to qualify as minerals. They tend to form yellowish to brownish grains or stains loosely referred to as "limonite" (Table 5.1) and illustrated in **Figure 5.5**.

EXTENSION MODULE 5.3

Geochemistry of Calcite.
Learn the factors that determine whether calcite dissolves or precipitates in water, which explains many features of rocks and landscapes.

Hematite is the most common iron oxide mineral. Hematite usually forms from rock weathering but commonly also forms from reactions of oxygen-rich ground water with sediment containing iron-bearing minerals, including iron hydroxides and iron silicates. Hematite colors rocks red, even when present in very small amounts, as shown in Figure 5.5b and c. The brown hue of weathered granite (Figure 5.1d, e) and soil reveals limonite produced by oxidation of iron within the biotite and hornblende. The brassy color of the biotite flakes in the weathered granite reveals partial conversion to the mineral vermiculite (also used in potting soil), which contains oxidized iron.

Putting It Together—*How and Why Do Rocks Disintegrate to Form Sediment?*

- Weathering is the interaction of the geosphere with the atmosphere, hydrosphere, and biosphere.
- Physical weathering disaggregates rocks by mechanical means. The most effective agent of physical weathering is the volume change associated with the freezing and thawing of water. Other processes include salt weathering and cracks formed by expansion and contraction of rocks when they heat up and cool down.
- Chemical weathering involves chemical reactions of minerals with water and oxygen. The reactions cause dissolution of some minerals, formation of new minerals by hydrolysis and oxidation, and the contribution of dissolved ions to natural water.

5.2 What Is the Link Between Weathering and Sediment?

Within the rock cycle (Figure 3.12), weathering represents the processes that form sediment, the raw material for sedimentary rocks. Two types of sediment are derived from the solid minerals and dissolved ions that are the two products of weathering.

Making Clastic Sediment

Clastic sediment is the residue of particles that remain after rocks weather. These particles are the physically weathered parts of the original rock along with minerals, such as clays, that form by chemical weathering. Weathered rock residue that remains more or less where it formed is a principal component of soil, along with decaying organic matter and windblown dust (you will learn more about soil in the context of landscape evolution in Chapter 14).

Minerals are not equally affected by chemical weathering processes. It is challenging to rank common minerals according to resistance to weathering because of large natural variations in abundance of moisture and water chemistry that drive weathering reactions. Nonetheless, some generalizations can be made. Minerals dominated by ionic bonds (e.g., halite, gypsum, calcite) most readily dissolve in water, especially if the water is acidic. Silicate minerals are generally more resistant to chemical weathering, and those with a greater abundance of strong Si–O bonds (e.g., quartz and feldspar) weather more slowly than minerals with a greater abundance of ionic bonds, especially when those ions include easily oxidized Fe^{2+} (e.g., olivine, pyroxene, biotite, amphibole).

Among common minerals, quartz is the most resistant to weathering because of its strong, mostly covalent bonds and lack of iron. Recall that the biotite in your granite sample (Section 5.1) was the most readily weathered. With more time and water for chemical reaction, all of the biotite will be destroyed along with most or all of the feldspar. Only quartz, clay, and limonite will remain, with a large mass of ions carried away in solution (Table 5.1). Quartz and clay are, in fact, the most abundant constituents of sediment and sedimentary rock.

Moving currents of wind or water, and in some places the slow movement of glaciers, erode (pick up) clastic particles and transport them away from where they formed by weathering. Water and wind currents exert sufficient force on the particles to roll them or even pick them up and suspend them in the current. Thus begins an odyssey where sediment grains may ultimately be transported thousands of kilometers, mixed with sediment from countless other weathered rock outcrops, and ultimately deposited to form thick layers of sedimentary rock. The quartz, feldspar, and clay grains and larger fragments of granite you observed are the beginning of that process along the streambed below the granite outcrop (Figure 5.1).

Making Chemical Sediment

Not all of a weathered rock is represented by clastic particles, however. Chemical weathering forms dissolved ions that move away in solution. What is the fate of those constituents of the original rock? Most ions dissolved in water eventually precipitate as solid ionic compounds called **chemical sediment**. Precipitation is the opposite of dissolution. As you will see in later sections, ions in solution bond to form mineral grains, such as halite and calcite, when water chemistry or temperature change. Lakes and oceans are the ultimate destination for the dissolved weathering products of continents. Most precipitation of chemical sediment at Earth's surface takes place in these bodies of water. Precipitation of minerals from water within open pores between sediment grains also occurs; this process ultimately cements the grains together into coherent sedimentary rock (more on this in Section 5.4).

*Putting It Together—***What Is the Link Between Weathering and Sediment?**

- Weathering breaks down rock into the components that form sediment.
- Clastic sediment consists of mineral and rock fragments remaining from physical weathering and newly formed mineral grains produced by chemical weathering.
- Chemical sediment consists of minerals precipitated from water. The ions composing these minerals are mostly generated by chemical weathering reactions.

5.3 Why Are Fossils Found in Sedimentary Rocks?

Sediment forms and accumulates at the surface of the planet where it serves as an archive of the interaction of the geosphere with the hydrosphere, atmosphere, and biosphere. Organisms live and die in the same environments where sediment accumulates, so an archive of the biosphere is sometimes preserved within the resulting sedimentary rocks.

Fossilized Organisms

You have long been aware of fossils, and it is likely that dinosaur remains were your childhood favorites. **Fossils** are the remains of organisms preserved in rock, and there are basically three types.

1. Fossil remains of shelly organisms and vertebrate teeth and bone, shown in **Figure 5.6**a and b, are the original mineral matter secreted by the organism.
2. Ions dissolved in pore water replace the original cellular compounds of organic matter buried in sediment with minerals. The most common minerals formed in this process are calcite and quartz that petrify the organic remains, as illustrated by petrified wood in Figure 5.6c.
3. In some cases only an impression remains where most or all of the organic material was completely destroyed by organic decay or mineral dissolution, as is seen in Figure 5.6d.

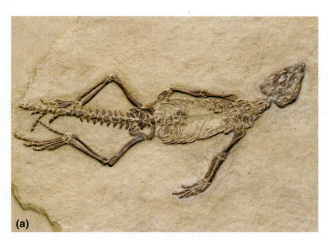

(a)

Bones, such as this fossil reptile skeleton, commonly preserve as fossils because they are hard mineral matter. Minerals that precipitate from ground water also replace the bone minerals to form some fossils.

(b)

Shells, such as these fossil clams from California, consist of hard minerals secreted by invertebrate animals.

(c)

This petrified wood in eastern Arizona formed when quartz and hematite replaced the original woody tissue.

(d)

Soft-bodied animals and plant leaves preserve as coaly organic films and impressions between layers of sedimentary rock, as in the case of this fossil fern.

Extremely few organisms become fossils upon death. Shell, bone, and teeth, in addition to soft tissue, are subject to destruction by weathering. Nonetheless, sufficient numbers of organisms become fossils to provide insights into the history of life on Earth and to help geologists define the environment of deposition for the enclosing sediment. Some volcanic rocks, especially those formed from rapidly accumulating ash layers, also contain fossils, but sedimentary rocks are the primary hosts for fossils because they are the most abundant rocks formed at Earth's surface.

Fossil Fuels

Organic matter buried with sediment provides combustible energy sources called **fossil fuels**. Coal, oil, and natural gas originate from organic matter deposited with the sediment. The energy in fossil fuels is a type of stored energy. Primarily, this energy comes from the Sun and is utilized during photosynthesis to create organic molecules in plants and some microscopic organisms. Organisms (like us) that consume photosynthesizers can then use this converted solar energy. Where fossil organic molecules are found in sedimentary rock, the stored energy can be liberated as heat when the fuel burns (burning is a rapid and energetic oxidation reaction). The heat, in turn can be converted to motion energy that turns turbines to generate electricity or drives pistons inside vehicle engines. Although the consumption of fossil fuels in all aspects of modern living seems like an immense amount of energy, it is really a small amount of Earth's energy budget. All of the known fossil-fuel reserves on Earth add up to about 10 days worth of energy delivered to the planet from the Sun.

Burial of organic matter in sediment requires special circumstances. Carbon-rich organic tissue usually is consumed by decomposing organisms or is oxidized to form carbon dioxide and other gases. In some deep lakes and ocean basins, swamps, and nearshore lagoons, however, the oxygen in the water is completely consumed by oxidation of organic matter and Fe^{2+}-bearing minerals. In these situations where oxygen is unavailable, additional organic matter cannot be oxidized and instead is buried along with inorganic sediment. As more sediment accumulates above, the pressure and temperature rise in the organic-rich deposit, which promotes chemical reactions that change the organic molecules into new compounds. **Coal** is a sedimentary rock composed almost entirely of the compacted remains of fossil plants, such as that shown in **Figure 5.7**a. Oil and natural gas are organic compounds produced by the high-temperature alteration of organic molecules that are mostly the remains of aquatic protozoa and plants. These important fluid energy resources accumulate in the pore spaces within sediment and sedimentary rock (Figure 5.7b).

To locate oil and gas, geologists must use knowledge of sedimentary rocks. They need to know where organic matter is likely to be buried with the sediment rather than destroyed by oxidation. They also need to know where to find the rocks with abundant, interconnected pore spaces through which oil and gas move and accumulate in sufficient volume to constitute an economic resource.

(a)

Coal consists of compressed plant remains. Microscopic examination (inset) reveals the original cell structure of the plant matter

(b)

Oil and natural gas are organic compounds that originate in plankton, algae, and plants. The gray stain in this sedimentary rock in northern Alaska is caused by crude oil within the pore spaces between the sediment grains.

▲ **Figure 5.7 Fossil fuels come from sedimentary rocks.**

Putting It Together—*Why Are Fossils Found in Sedimentary Rocks?*

- Many sedimentary rocks also contain biologic remains called fossils.
- Living organisms abound in the same environments where sediment accumulates, and their remains are buried with the sediment. Geologists interpret ancient environments with the help of biologic remains preserved as fossils.
- Organic matter buried with sediment and sedimentary rock changes at elevated temperature and pressure to form combustible organic compounds that yield large amounts of energy when burned. These organic materials are the fossil fuels: coal, oil, and natural gas.

5.4 How Does Loose Sediment Become Sedimentary Rock?

Sediment is a mixture of residue from weathering, precipitated ionic compounds, and organic materials, but how does a loose assortment of grains become a hard, coherent rock? For most igneous rocks, the intergrowth of crystals forms a consolidated rock at the outset. The precipitation of minerals from water can produce a similar rock, as observed by the crystallization of calcite around a spring in Section 3.1. In other cases, however, individual mineral grains precipitate in the water and settle to the bottom in much the same way as would a clastic particle. How does loose clastic or chemical sediment lithify—transform into rock?

First Step in Making a Sedimentary Rock: Compaction

Lithification involves the processes that convert sediment into rock. The most common first step in lithification is **compaction**, illustrated in **Figure 5.8**. When you walked along the dry streambed full of weathered granitic sediment (Section 5.1), your feet sank into the sand leaving footprints. The sand grains are inefficiently packed, leaving large open spaces between them. Your weight causes the grains to slide past one another and reorganize into a smaller space (Figure 5.8). Some fragile grains (thin flakes of biotite or weathered feldspar, for example) break. Natural sediment compaction takes place while layers of sediment accumulate and lower particles experience the weight of overlying sediment. The total volume of solid material remains the same during compaction, but the air- or water-filled pore space between grains decreases.

Does compaction of sediment grains really cause them to stick together to make a rock? Mostly, compaction just decreases the volume of pore space and simply packs the grains closer together. While not making a hard rock, compacted sediment may, however, be somewhat consolidated. Have you squeezed slightly moist sand in your hand and noticed the grains clumping into fragile clods? Add more water (or finer clay particles) and you can make more coherent clods, like the ones you possibly threw at siblings or friends when you were younger. This consolidation results from weak electrical attractions between the surfaces of the mineral grains and between mineral grains and pore water.

Natural broken surfaces on minerals commonly expose atoms with unbalanced charges that tend to attract oppositely charged atoms in adjacent grains. Clay minerals contain many stray charges on their outer surfaces, and this enhances clumping of clay particles. The asymmetry of the water molecule (Figure 2.15) also helps hold grains together. The greater the compaction, the greater the mineral-surface contacts within the sediment and the greater the amount of mild electrical attraction between atoms in adjacent particles. Although the compacted sediment may not strictly be loose, it is still far from being a hammer-ringing-hard rock.

Loosely packed sediment

Compacted sediment

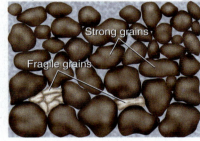

Strong grains

Fragile grains

Time

◀ **Figure 5.8 How sediment compacts.**

During compaction, sediment grains rotate and repack closely together, which reduces the pore space between the grains.

Some fragile grains may break or squash between stronger grains.

Weak electrical forces at grain boundaries are more effective at holding compacted, rather than uncompacted, sediment together because compacted grains touch along larger surfaces.

The weight of accumulating sediment provides the pressure to compact sedimentary layers.

Second Step in Making a Sedimentary Rock: Cementation

To make hard rock from sediment, the pore spaces fill partly or completely with precipitated minerals that glue the grains together. This is the **cementation** process illustrated in **Figure 5.9**. Mineral grains not only precipitate from water and accumulate as chemical sediment but minerals also precipitate from water in the pore spaces between sediment particles and cement them together into rock. The most common cementing agents are calcite, quartz, clay minerals, and hematite. The elements composing most of these minerals are available in solution as a result of chemical weathering (Table 5.1). Additional dissolved ions come from continued reaction of pore water with mineral grains in the sediment. Examples are reaction of pore water with feldspar grains in sand to form clay or with iron-silicates to form hematite.

There is, however, an upper limit to how much of each ion can dissolve in water. As the concentration of ions increases, it is increasingly likely that some will bond together to make minerals.

Precipitation is not only caused by the increasing concentration of ions in the water, it is also strongly affected by temperature. Calcite, for example, dissolves more readily in cold water and precipitates in warmer water. So pore water with abundant calcium and carbonate ions could precipitate calcite as binding cement when experiencing higher temperature at depth as it is buried with sediment. Quartz, on the other hand, dissolves slightly in warmer water but precipitates in cooler water. Consider the scenario of warm pore water at depth that becomes increasingly enriched in dissolved SiO_4^{4-} from reactions with feldspar and other silicate grains in the sediment. If the water moves upward as the sediment compacts, then it cools and precipitates quartz as a cement.

Time

▲ **Figure 5.9 How cementation works.**
Loose sediment becomes a sedimentary rock because cementing minerals precipitate in the pore spaces between the sediment grains.

Compaction and cementation lithify unconsolidated sediment into sedimentary rock at the expense of pore spaces that squeeze shut or fill with cement. It follows that this process progressively diminishes the ability of fluid to move through the rock. Ground water, oil, and natural gas move easily through loose sediment and slightly consolidated sedimentary rocks but do not readily move through rocks where most or all of the pore spaces are either closed off from one another or completely filled with cement. Geologists determining the extent of resources must, therefore, have knowledge of how lithification varies from place to place and at different depths within a sedimentary deposit.

> *Putting It Together*–*How Does Loose Sediment Become Sedimentary Rock?*
> ■ As sediment accumulates, older sediment compacts under the weight of overlying younger sediment. Compaction packs sediment grains closer together and permits a minor amount of cohesion because of weak electrical attractions between mineral-grain surfaces.
> ■ Strong sedimentary rocks form by precipitation of cementing minerals in the spaces between sediment grains. Calcite, quartz, clay minerals, and hematite are the most important mineral cements.

5.5 How Are Sedimentary Rocks Classified?

You know how sediment forms and how it lithifies into rock. It seems reasonable, therefore, to develop a classification scheme that incorporates these concepts. You could start with the three types of sediment grains—clastic, chemical, and biogenic (or organic). Some geologists adopt such a threefold classification scheme. This system is awkward, however, because most biologic sediment particles are biochemically precipitated inorganic compounds (like calcite shells), making for a fuzzy distinction between chemical and biogenic sediment in many cases. Lumping chemical and biogenic sediment into a single class avoids this problem. The two primary categories then become:

1. Clastic sedimentary rocks composed primarily of mineral grains produced by the weathering of preexisting rocks and cemented by minerals precipitated from pore water or which formed from the interaction of the grains with pore water
2. Chemical and biogenic sedimentary rocks composed of minerals precipitated from water by biologic or inorganic processes or simply the remains of organisms

Texture (grain size) and composition (minerals present) are easily observed features of the rocks that are used for classification and naming. These two characteristics are just as suitable for naming sedimentary rocks as for igneous rocks. Geologists emphasize texture to name clastic sedimentary rocks and composition to name chemical sedimentary rocks.

Naming Clastic Sedimentary Rocks

The most important aspect of clastic-sediment texture, explained in **Figure 5.10**, is grain size. Although weathering produces a wide range of sediment grain sizes, the vast majority of sedimentary rocks consist of particles transported by water or wind to an ultimate site of deposition. These fluid currents exert forces on the sediment grains to transport them. Larger forces, such as fast-flowing water or strong wind, are necessary to move large grains in contrast to tiny dust particles. A fast-moving flood in a stream may, for example, move grains ranging in size from dust to boulders more than 1 meter across. As the flood current slows down, the large boulders come to rest first, while sufficient force remains to move the smaller pebbles, sand, and mud farther downstream. With further decrease in the current energy, successively smaller particles come to rest at greater distances downstream. The result is that many sedimentary deposits consist of a relatively narrow range of grain sizes. **Sorting** describes the range in grain size, with well-sorted specimens containing mostly one grain size and poorly sorted sediment containing a wide range of grain sizes (Figure 5.10).

The classification of clastic sediment and sedimentary rocks, illustrated in **Figure 5.11**, is built around three grain-size categories—gravel, sand, and mud. You probably already use these words to denote the relative sizes of sedimentary particles, with gravel being largest, or coarsest, and mud being smallest, or finest. Geologists attribute specific particle sizes to each of the categories and to subcategories within them (Figure 5.11). For rocks composed primarily of gravel, sand, or mud (silt + clay) size particles, the terms **conglomerate** (or **breccia**), **sandstone**, and **mudstone** (or **shale**), respectively, apply. If the rock is poorly sorted, containing more than one grain-size class, then it is a conglomeratic sandstone, sandy conglomerate, muddy sandstone, and so on.

Another aspect of clastic-rock texture is the **rounding** of the grains. Most clastic grains have sharp edges and corners when initially produced by weathering and are described as angular. The edges and corners abrade during transport by wind and water as the particles collide with each other, and the grains progressively get more rounded, as depicted in Figure 5.10. Conglomerate and breccia are distinguished by rounded or angular fragments, respectively. So you can now refer to clastic sedimentary rocks (Figure 5.11) in terms of the three ingredients of sedimentary-rock texture:

1. The dominant grain size
2. The range of grain size (degree of sorting)
3. The extent to which the grains are rounded

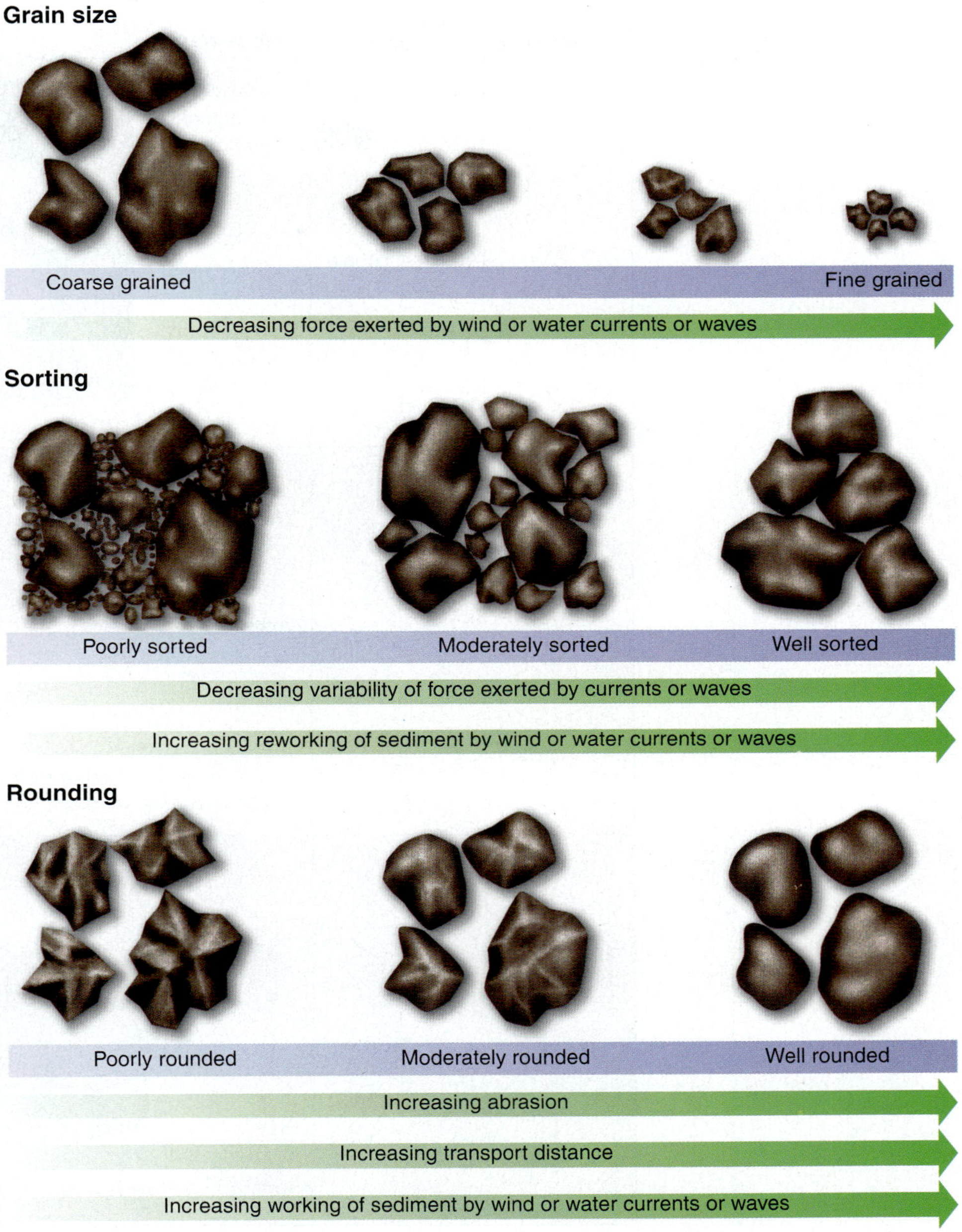

Perhaps confusingly, "clay" refers both to a group of minerals and to a grain size (Figure 5.11) regardless of the composition of the particles. It may help to know that clay-mineral grains in sediment are also almost always clay size. However, not all clay-size grains are clay minerals. So it is always important to understand whether the term "clay" is being used to describe the composition or texture of the sediment.

Composition plays a secondary role in classification of clastic sedimentary rocks, and it is most commonly applied to sandstones (Figure 5.11). **Quartz sandstone** consists almost entirely of quartz. Quartz is abundant in sedimentary rocks because it is the abundant rock-forming mineral that is most resistant to chemical weathering. Quartz is also sufficiently hard and difficult to break, since it lacks cleavage, to survive long-distance transport. **Arkose** is a sandstone containing at least 25 percent feldspar. The remainder of an arkose is mostly quartz, usually with minor amounts of mica and other silicate minerals. **Lithic sandstone** consists of sand-size rock fragments. Lithic sands form by weathering of

Grain size

Coarse grained | Fine grained

Decreasing force exerted by wind or water currents or waves

Sorting

Poorly sorted | Moderately sorted | Well sorted

Decreasing variability of force exerted by currents or waves

Increasing reworking of sediment by wind or water currents or waves

Rounding

Poorly rounded | Moderately rounded | Well rounded

Increasing abrasion

Increasing transport distance

Increasing working of sediment by wind or water currents or waves

▲ **Figure 5.10 Texture of clastic sediment.**
Grain size, sorting (the range of grain sizes present), and the rounding of the grains describe the texture of clastic sediment. Larger forces are required to move larger grains. Sorting improves with persistent moving of particles and with application of transporting forces that do not vary much in strength. Abrasion with greater distance of transport and persistence of grain movement shapes well-rounded grains. For example, sediment deposited by flash floods of a steep mountain stream will likely be coarse grained, poorly sorted, and poorly rounded. In contrast, sediment deposited on a beach will be finer grained, better sorted, and better rounded.

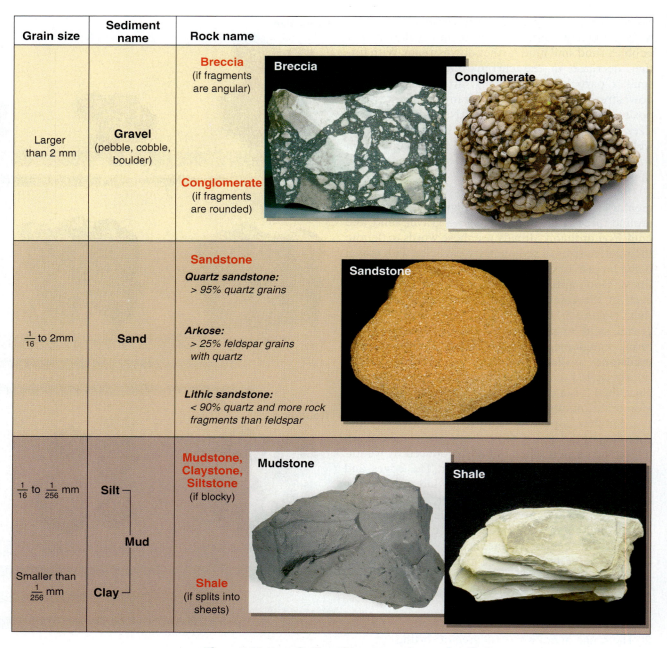

Grain size	Sediment name	Rock name
Larger than 2 mm	**Gravel** (pebble, cobble, boulder)	**Breccia** (if fragments are angular) **Conglomerate** (if fragments are rounded)
$\frac{1}{16}$ to 2mm	**Sand**	**Sandstone** *Quartz sandstone:* > 95% quartz grains *Arkose:* > 25% feldspar grains with quartz *Lithic sandstone:* < 90% quartz and more rock fragments than feldspar
$\frac{1}{16}$ to $\frac{1}{256}$ mm — **Silt** ⌍ **Mud** Smaller than $\frac{1}{256}$ mm — **Clay** ⌎		**Mudstone, Claystone, Siltstone** (if blocky) **Shale** (if splits into sheets)

▲ **Figure 5.11 How clastic sedimentary rocks are classified.**
The first step in naming a clastic sedimentary rock is to determine the dominant grain size. For gravel-size sediment, the next step is to determine if the grains are rounded or angular. Sandstones are further subdivided on the basis of grain composition. Fine-grained sedimentary rocks can simply be called mudstone (or also siltstone or claystone if dominated by one grain size or the other) or can be called shale if the rock shows a tendency to split into thin sheets.

very fine-grained rocks, such as basalt, where individual sand grains are rocks because they contain many mineral crystals.

The composition of clastic sedimentary rocks reveals the types of rocks that weathered to produce the sediment and the amount of weathering that occurred. For example, both quartz sandstone and arkose could represent weathering and erosion of granite, which also contains quartz and feldspar. A greater amount of weathering, or breakage of cleaveable feldspar during transport, might account for the greater quartz content of the quartz sandstone compared to the arkose. Conglomerate and breccia fragments are large chunks of rock eroded from the sediment source and allow interpretation of the rocks exposed at the sediment source.

Naming Chemical and Biogenic Sedimentary Rocks

Composition is the most useful characteristic for classifying and naming the chemical and biogenic sedimentary rocks. It is very unusual for the conditions to be just right for two or more minerals to precipitate from water simultaneously, so most chemical sedimentary rocks consist overwhelmingly of only one mineral, as depicted in **Figure 5.12**.

Limestone, composed of mainly calcite, and **chert**, composed of mostly quartz, are the most abundant rocks formed by chemical precipitation. These common rocks form by both inorganic and biologic processes, which highlights the difficulty of separating biogenic and chemical sedimentary rocks. Calcite and quartz dissolve to a limited extent in water and readily precipitate from water when chemistry or temperature slightly change. Many invertebrate animals, protozoans, and algae secrete shells and other hard tissues composed of calcium carbonate as calcite or as aragonite that slowly transforms to calcite over time. Other protozoans and algae, such as diatoms, and some invertebrates, including sponges, secrete hard tissues composed of silica. Most limestone and chert form primarily by these biologic precipitation processes, with additional inorganic precipitation of calcite or quartz as cement after the sediment is buried.

Dolostone is a chemical sedimentary rock composed of the calcium-magnesium carbonate mineral dolomite. No organisms secrete shells of dolomite, and dolomite precipitates from seawater only under conditions of unusual water chemistry in some nearshore lagoons. Nonetheless, dolostone is abundant in many ancient sedimentary-rock successions although notably sparse in younger ones. How does it form? Most dolostone forms by reactions of ground water with calcite in limestone to produce dolomite long after deposition of the original sediment. The Mg^{2+} ion is smaller than the Ca^{2+} ion, resulting in tighter packing of atoms in the dolomite crystal structure compared to calcite. This means that when calcite reacts with water to produce dolomite, the resulting mass of dolostone occupies a smaller volume, leaving behind open pores that may fill with oil or natural gas.

Other minerals, such as halite and gypsum, are very soluble in water. The water must undergo considerable evaporation so that the appropriate ions can bond together to make these minerals. Rock salt and rock gypsum are, therefore, commonly referred to as **evaporites** to emphasize their formation in arid lagoons or drying, desert lakes where evaporation allows precipitation of halite and gypsum.

The one remaining rock in this category is coal, which consists entirely of plant organic matter (Section 5.3). Coal geologists use sophisticated classifications of coal based on the degree of transformation of the original plant tissue into new organic compounds. The greater the extent of transformation under elevated heat and pressure during burial, the greater the combustibility or heat production from the coal when it burns. **Peat** describes plant material before it transforms into the low-density, black rock that you normally think of as coal

Composition	Rock name	
Calcite [CaCO₃]	Limestone	Limestone
Dolomite [CaMg(CO₃)₂]	Dolostone	
Quartz (microscopic) [SiO₂]	Chert	Chert
Halite [NaCl]	Rock Salt — Evaporite —	Rock salt / Rock gypsum
Gypsum [CaSO₄•H₂O]	Rock Gypsum	
Plant organic matter	Coal	Coal

▲ **Figure 5.12 How chemical and biogenic sedimentary rocks are classified.**
To name a chemical or biogenic sedimentary rock, it is necessary to know the composition of the sediment because a single, dominant component forms each rock.

but that is more specifically referred to as **bituminous coal**. If metamorphosed, the coal transforms to **anthracite**, which is more accurately classified as a metamorphic rock although it is still commonly called coal. Purity also determines the quality of coal as a fuel. If the plant material accumulated in the nearly complete absence of other sediment, then nearly the whole rock burns and leaves behind little waste material (called fly ash). If substantial clay and silt washed into the coal swamp, then the quality of the coal is substantially less. The high abundance of organic matter in coal swamps also exhausts all of the oxidizing capacity of the water. Microbes in this oxygen-deficient environment reduce the S^{6+} in pore-water sulfate (SO_4^{2-}) to S^-, which then bonds with Fe^{2+} ions to produce pyrite (FeS_2). When coal burns in furnaces or power plants, the pyrite oxidizes and produces sulfur dioxide gas, which releases to the atmosphere and mixes with water vapor to produce acid rainfall composed of sulfuric acid.

Putting It Together—How Are Sedimentary Rocks Classified?

■ Origin, texture, and composition of the sediment grains form the basis for naming sedimentary rocks.

■ Clastic rock names rely mostly on sediment grain size and secondarily on grain rounding and composition.

■ Clastic-sediment texture relates to the strength and steadiness of currents transporting the sediment and the distance of transport.

■ Clastic sediment composition reveals the type of rock that weathered to create the sediment and the extent to which this source rock weathered.

■ Chemical and biogenic sedimentary rocks consist of sediment produced by inorganic chemical precipitation, biologic processes, or both. Some rocks, such as limestone, form by both inorganic and biologic chemical processes. Dominant mineral composition determines the names of these rocks.

5.6 How Do Sedimentary Rocks Reveal Ancient Environments?

An important objective while studying sedimentary rocks is comprehending how the rocks reveal the conditions and processes acting on Earth's surface where and when the sediments accumulated. The depositional environment might be a streambed, a beach, a desert sand dune, the seafloor, or any of a number of Earth-surface features where sediment comes to rest or precipitates from water. If rocks contain some sort of memory of where deposition occurred, then it is possible to reconstruct ancient geography by mapping out the extent of different recorded environments. Reconstructing ancient geography is fascinating because it tells geologists where mountains, shorelines, coral reefs, and many other surface features existed in the past, often in locations that have very different geography today.

This information is more useful, however, than just for imagining how places looked in the distant geologic past. Some environments, for example, favor the accumulation of organic material that becomes fossil fuels. Other environments feature deposition of the types of sediment that eventually form rocks with pore spaces that fill with ground water, oil, or natural gas resources.

What Fossils Reveal

Fossils (Figure 5.6) provide the most easily interpreted record of past environments. Organisms adapt to living in particular conditions: land versus water, shallow lagoons versus deep ocean, dry deserts versus tropical rainforests, and so forth. By identifying the environmental requirements for organisms preserved as fossils, paleontologists simultaneously provide key information about the ancient depositional environment.

What Rock Types Reveal

When geologists decipher details about past depositional environments from ancient rocks, they start by examining what type of sediment accumulates in those environments today. **Figure 5.13** illustrates examples of how sedimentary rock composition and texture permit environmental interpretation. Coarse-grained clastic rocks, like conglomerate, require environments with strong transporting currents, whereas fine-grained mudstone requires quiet water where fine silt and clay settle slowly from suspension. Limestone reveals submerged lake bottoms or seafloor. Evaporites indicate arid lakes and shoreline lagoons where water evaporates and concentrates ions that precipitate as minerals.

What Sedimentary Structures Reveal

If some physical characteristics of modern sediment relate to the processes that deposit it, then these same characteristics in ancient rocks imply that the same processes occurred in the past to form the rock. Sedimentologists commonly look to these physical features in the rock, called **sedimentary structures**, which form during sediment deposition or shortly after deposition and before the sediment becomes sedimentary rock. Sedimentologists watch sedimentary structures form in modern environments or during laboratory experiments. In Chapter 1 (Figure 1.8), you saw how application of the principle of uniformitarianism

▼ **Figure 5.13 Using modern environments to interpret ancient environments.**
These illustration pairs are examples of how geologists interpret the ancient depositional environments recorded in sedimentary rocks by comparison to where sediments of similar texture and composition are observed to accumulate today.

The photo on the left shows very coarse, angular, poorly sorted gravel that accumulates during flash floods at the base of steep mountains in Death Valley, California. The photo on the right shows a coarse, angular, poorly sorted conglomerate that geologists interpret to have formed by flash floods near steep mountains in Death Valley, millions of years ago.

The satellite image on the left shows light-colored areas where limey sediment accumulates in shallow, tropically warm water near Florida, the Bahamas, and Cuba. Geologists interpret the limestone layers on the New Mexico mountainside in the right photo to have formed in a similar warm, shallow sea hundreds of millions of years ago.

▶ **Figure 5.14** **What bedding looks like.**

The horizontal rock layers in this view of the Grand Canyon, Arizona, are large-scale sedimentary beds. Changes in depositional environment through time caused deposition of different sediment types in successive beds, one above the other. Hard sandstone and limestone form cliffs because these rocks are more resistant to weathering than the slope-forming mudstone and shale.

Small-scale bedding, as seen here in old lake beds in Washington (with a pocket knife for scale), commonly forms by changes in depositional processes or the type of sediment deposited. Each pair of these dark and light beds represents a single year of deposits in a glacier-fed lake. Light-colored silt settled during spring and summer snow-melt runoff. Dark organic matter settled to the bottom during the winter when the lake surface froze. Can you determine how many years were required to deposit the sediment visible in the photo?

permits interpretation of the processes that formed the same structures when they are seen preserved in rocks.

Figure 5.14 illustrates a fundamental sedimentary structure, called **bedding** or **stratification**, which simply refers to the layering that always exists in sedimentary rocks (also see Figure 3.1a). Bedding reflects changes in depositional processes. Some bedding planes represent pauses in deposition, or even erosion, whereas others represent changes in the sediment-depositing processes. Bedding shows that accumulation of sediment in any environment is not a monotonously continuous process. The energy of moving currents and waves varies with time, eroding sediment on some occasions and depositing it at other times. Fluctuations in sediment-transport ability also cause variations in the grain size and sorting of the deposited sediment. Some depositional environments may only receive clastic sediment during river floods or strong oceanic storms, or the volume and composition of sediment may vary with the seasons. All of these variations produce separate beds in the resulting rock.

Figure 5.15 illustrates a simple sedimentary structure called **mud cracks**. You probably have seen mud cracks, because they form wherever wet mud dries up. Muddy sediment usually contains abundant clay minerals that absorb water and swell when wet but then lose water, shrink, and crack while drying. If sediment washes or blows into open cracks, then the polygon shapes of the cracked ground are preserved in rock. Presence of mud cracks in sedimentary rocks implies a depositional environment exposed to the air where sediment dries, like a river floodplain, rather than an environment always submerged in water, like the seafloor.

▼ **Figure 5.15** **What mud cracks reveal about past environments.**
This illustration shows how mud shrinks and cracks when it dries out. Mud cracks, commonly filled in with sandstone, are preserved in some sedimentary rocks. The appearance of mud cracks reveals an environment that was alternately wet and dry.

Fine-grained mud settles from suspension in quiet water.

Water evaporates and mud shrinks, causing deep cracks to form in sediment.

Water again covers the area, and sand grains fill in the mud cracks.

Later exposure of the top of the mudstone layer reveals mud cracks filled with sand.

Time

Figure 5.16 illustrates **cross-beds**, which are inclined layers in sediment or sedimentary rock that reveal current or wave transport of sediment. Cross-beds, as the name implies, are crosswise to the bedding that forms parallel to nearly horizontal land and sea floor surfaces.

Cross-beds form by the movement of sediment **dunes** and **ripples**. Dunes and ripples are curving ridges of loose sediment that move along with water or wind currents or move back and forth beneath oscillating water waves. The detailed physics of dune and ripple formation are different, but in a general way you can think of ripples as small features, less than 3 centimeters high, and dunes as similar but larger features ranging in height from several centimeters to many meters. Although the term "dune" usually conjures up images of

Forming Cross-beds: *See how cross-beds form and how sand dunes migrate through time.*

▼ **Figure 5.16** **How cross-bedding forms.**

Cross-bedding forms by movement of dunes and ripples, which are ridges of loose sediment shaped by wind or water currents.

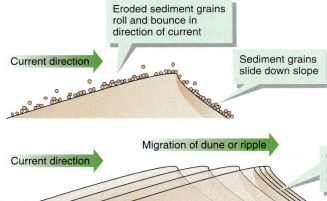

Eroded sediment grains roll and bounce in direction of current

Current direction

Sediment grains slide down slope

Sediment grains, usually sand, move with the current and slide down the slope on the down-current side of the dune or ripple. Successive positions of this down-current inclined slope form cross-beds as the dune or ripple migrates in the direction of the current.

Migration of dune or ripple

Current direction

Cross-beds mark shifting position of down-current side of dune

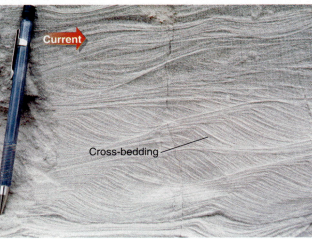

Large-scale dune (left) and small-scale ripple (right) cross-bedding appear in ancient sedimentary deposits. The cross-beds incline downward in the direction of the transporting current.

Current

Sediment grains of many sizes suspend in a rapidly moving current.

Grains settle as current slows down; larger grains settle faster than small grains.

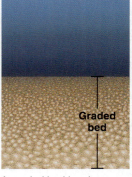

Graded bed

A graded bed has larger grains at the bottom and smaller grains at the top.

It is easy to make graded beds in the laboratory by stirring a sand and mud mixture in water and letting the sediment grains settle.

◄ **Figure 5.17 How graded bedding forms.**
Graded beds are recognized by coarser grains concentrated at the base of the bed whereas finer grains are mostly at the top. This sorting of grains by size occurs when currents slow down, which allows larger grains to settle out to the resulting deposit before smaller grains accumulate.

Graded bed of gravel and sand deposited by a flood.

windblown sand dunes, flowing water also forms submerged dunes, and both wind and water generate ripples (Figure 5.16).

Figure 5.16 illustrates how cross-beds form when sediment accumulates on the steep, down-current side of a moving dune or ripple ridge. The recognition of cross-beds in sedimentary rock not only indicates an environment where water or wind move across the surface but also reveals the direction the currents were traveling.

Figure 5.17 illustrates the formation of **graded beds**, where sediment grain size uniformly changes from coarser at the base of a bed to finer at the top. Graded beds reveal intermittent currents that transport a wide mixture of sediment grain sizes and then slow down so that large particles deposit first and smaller particles progressively accumulate on top of the larger ones. Graded beds commonly result from rapid sediment deposition during river floods, settling of sediment stirred up by big storm waves on the seafloor, and by episodic currents of sediment-laden water, called **turbidity currents**, that sweep across lake bottoms and the seafloor (more coming up in Section 5.7).

Evidence of Changing Environments

A vertical transition from one sedimentary rock type to another represents a change in the depositional environment. **Figure 5.18**, for example, illustrates an upward progression from sandstone, to shale, to limestone. All of these layers contain fossils of marine organisms. How did the marine environment change through time in order to cause deposition of sandstone, followed by deposition of shale, and then precipitation of limestone?

Figure 5.19 shows how geologists interpret this succession by thinking about where these rock types likely form relative to a shoreline. Clastic sediment is produced by weathering on continents and delivered to the coastline by rivers. Coarser-grained sediment deposits closest to shore as the river currents slow down when entering the ocean, and waves wash the sand alongshore to form beaches. Wave agitation suspends the silt and clay, and these fine grains settle to the seafloor farther offshore. Calcite-secreting organisms live throughout the area of sand and mud deposition, where their mineral hard parts are

Limestone

Shale

Sandstone

◄ **Figure 5.18 Rocks record changing environment.**
This upward succession of sedimentary rock types in the Grand Canyon, Arizona, records changing environments 500 million years ago. The transitions between rock types are not sharp. The lower part of the shale is slightly sandy, and there are some thin limestone layers in the upper part of the shale before it gives way to mostly limestone with a few shale layers. These observations suggest that the changes in the depositional environment causing the upward transition from sandstone to shale to limestone were gradual and not abrupt.

buried in the sediment washed from the continent. Far from shore, however, there is less continent-derived clastic debris settling to the seafloor, so abundant calcite remains and precipitated crystals accumulate on the seafloor in the near absence of clastic sediment. In this scenario, you can see how sediments that will someday become sandstone, shale, and limestone simultaneously accumulate in different environments defined by proximity to a shoreline (Figure 5.19).

This information allows geologists to derive one explanation for the vertical arrangement of sandstone, shale, and limestone at a single location. Figure 5.19 shows that if sea level rises and the shoreline position changes, then the area where sand was once deposited is now buried in mud, and the previous area of mud deposition is covered with an accumulation of limey sediment. The vertical change in rock types, therefore, records changing environmental conditions—in this case, a rise in sea level.

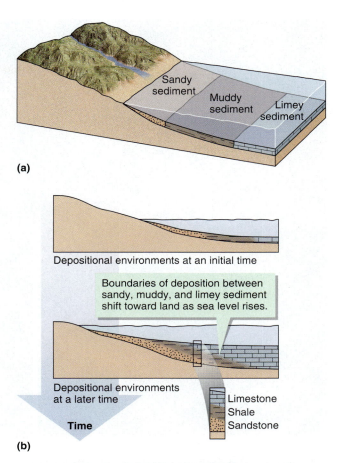

(a)

Depositional environments at an initial time

Boundaries of deposition between sandy, muddy, and limey sediment shift toward land as sea level rises.

Depositional environments at a later time

Limestone
Shale
Sandstone

Time

(b)

▲ **Figure 5.19** **How shifting environments form different rocks.**
Each successive sedimentary layer accumulates on the layer below, so vertical changes in rock type represent environmental changes during the time of sediment deposition. (a) Bands of different sediment type accumulate parallel to shore, with coarser clastic sediment deposited closest to shore and limey sediment farther from shore. (b) If sea level rises, then the environments represented by each rock type shift toward land. The resulting vertical succession of rock layers records upward transitions to sediment types that accumulated progressively farther from shore. Compare this diagrammatic sequence of sedimentary rocks with real rocks illustrated in Figure 5.18.

Putting It Together—How Do Sedimentary Rocks Reveal Ancient Environments?

■ Different sedimentary rock types indicate the physical and chemical processes active in the depositional environment.

■ Sedimentary structures reveal details about physical processes in the depositional environment. Cross-beds also reveal the direction of moving water and air currents that transport sediment.

■ Changes in environments or processes produce boundaries between beds of sedimentary rock. Observations of vertical successions of rocks with different features allow geologists to interpret environmental change over time during sediment deposition.

5.7 **How Do We Know** … How to Interpret Unseen Turbidity Currents?

PICTURE THE PROBLEM

How Is Coarse Sediment Deposited in the Deep Sea?

Figure 5.20 illustrates a commonly seen sedimentary sequence of alternating sandstone and shale beds interpreted (on the basis of their fossil content) to have been deposited on the seafloor. Even before 1900, sedimentologists made these observations and interpretations, highlighted in Figure 5.20:

• The most conspicuous sedimentary structure is the presence of graded beds.

• The graded beds imply sediment settling out of suspension in the water, with larger particles reaching the seafloor first, followed by successively smaller grains.

• The next most notable structures within the graded bed are ripple-scale cross-beds.

• The base of each graded bed is very sharp and typically not flat, indicating erosive scour of the underlying mud before sand deposition.

• The scoured base of the graded bed and the current-formed cross-beds indicate that the sediment did not simply settle quietly to the seafloor but was also swept along by a current. The current eroded the muddy seafloor before the sand began to accumulate.

• The intervening shale layers reveal that the depositional environment was characterized by slow accumulation of fine clay particles in between the times when the sandy graded beds were deposited.

Paleontologists contributed further observations and interpretations during the early twentieth century:

• The shale layers between the graded beds contain fossil plankton that secrete microscopic shells of calcite and silica. These fossils accumulate only in

Changing Shorelines and Sedimentation: *See how sedimentation near shorelines differs with changing sea levels.*

Interpreted sedimentary structures

Current

Small ripple cross-beds

Graded bed

▲ **Figure 5.20** **What deep-sea sandstone and shale look like.** Tilted layers of light-colored sandstone and dark shale crop out along the coast near San Francisco, California. The field photo shows a close view of a sandstone layer sandwiched in shale. The enhanced view of the sandstone bed labels the features that a geologist sees:

• The layer is a graded bed, with the sizes of the grains decreasing upward until the distinction between sandstone and shale is unclear.

• The upper part of the bed contains a thin interval of small-scale cross-bedding produced by ripples moving with a current from right to left.

sediment deposited in deep water far from shore, and there are no fossils of organisms that prefer shallow, nearshore conditions.

• The sandstone graded beds contain fossil remains of nearshore animals and shreds of land-plant debris.

• The fossils reveal that the sediment composing the sandy graded beds was swept from near a shoreline into deep water, where it was deposited with fine mud and settling plankton carcasses on the deep ocean floor, far from shore. Many geologists treated this interpretation with disbelief, for how could sandy, and sometimes even gravelly, sediment, which typically accumulates close to shorelines, be deposited in deep water? These relationships contradict the simple pattern illustrated in Figure 5.19.

CLUES FROM THE FIELD

How Does Sediment Travel on Lake Bottoms?

Geologists working on problems far removed from the origin of deep-sea graded beds provided observations that supported the existence of deep-water currents. The geologists noted that muddy river water disappears from the surface of the sparkling clear lakes very close to river mouths. This includes observations of muddy river water seemingly disappearing at the upstream end of an artificial clear-water reservoir and the reappearance of that muddy water from the outlet at the base of the dam 20 kilometers downstream. Geologists suggested that sediment-laden river water is denser than the clear lake water and sinks downward to flow along the lake bottom. The turbid-water current flows along the sloping bottom of the lake, below still water, for the entire length of the reservoir. The name "turbidity current" was applied to this unseen, hypothesized process. Could turbidity currents, caused by the higher density of sediment-laden water moving below less dense clear water, also form in the ocean? If possible, what would be the erosional and depositional effects of these currents? Could these currents produce graded beds?

CLUES FROM EXPERIMENTS

How Does Sediment Move on Submerged Slopes?

Geologic oceanographer Philip Kuenen conducted laboratory experiments in the late 1940s to understand the erosional properties of turbidity currents. He constructed a large, aquarium-like tank with gently sloping bottoms, shown in **Figure 5.21**. He filled the tank with seawater and mixed fine sand, mud, and seawater in a separate container. The sediment-water mixture was vigorously stirred to suspend the sediment and was then quickly dumped through a trap door into the upslope end of the large tank. Figure 5.21 shows the result of such an experiment. The turbid sediment-water mixture does not simply mix with the clear seawater but produces a turbulent, swirling current that sweeps rapidly down the sloping bottom of the tank to the opposite end. Kuenen made laboratory-scale turbidity currents.

INTEGRATING EXPERIMENTS AND FIELD DESCRIPTIONS INTO A HYPOTHESIS

Do Turbidity Currents Deposit Deep-Sea Graded Beds?
Many field geologists read the results of Kuenen's experiments with great interest. Up until this point, there was no link between hypothesized turbidity currents in lakes and deep-marine graded beds of sandstone and conglomerate. Visualizing laboratory turbidity currents suggested a link. The hypothesis goes like this:

- River floods, storm waves, or submarine landslides triggered by earthquakes stir up nearshore sediment.
- The sediment and water mixture sweeps downslope into deep water as a turbidity current, carrying along nearshore organic remains.
- The energetic current erodes some of the seabed clay, and then, as the current energy wanes, the sediment grains settle to form the graded beds.
- Continued motion of the slowing current moves the settled sediment to produce ripple cross-beds.
- Peaceful accumulation of clay and plankton remains resumes between these punctuated events.

Geologist C. I. Migliorini saw this potential link between field observation and experiment and invited Kuenen to visit his Italian field-study sites with graded beds. Kuenen saw the potential to link his experiments to understanding sedimentary deposits rather than simply to evaluate potential erosion by turbidity currents. He returned to his laboratory tanks and adjusted his experiments and measurements to focus on the deposits left by artificial turbidity currents. Indeed, these deposits, shown in Figure 5.21c, closely resembled Migliorini's graded beds, and the two scientists teamed up to publish their results. The deposits of turbidity currents became known as **turbidites**.

COMPLETING A UNIFORMITARIAN ANALYSIS

Do Ocean Turbidity Currents Happen in Nature?
You learned in Section 5.6 (Figures 5.13–5.17) that formation of ancient sedimentary rocks is explained by watching sediment accumulate in modern environments and by experiments. Clearly, it is a tall order to watch a real, natural, rare, submarine turbidity current and then look at the resulting deposit for comparison to ancient rocks with graded beds. Watching waves on a beach to interpret beach sandstone or diving on a modern coral reef to interpret an ancient limestone is not the same as trying to be in the right place at the right time to witness (and survive!) a catastrophic turbidity current on the deep-sea floor.

This does not mean, however, that the interpretations by Migliorini and Kuenen lack application of uniformitarianism. The field observations and laboratory results are facts, which logically link to provide a reasonable explanation for rocks like those illustrated in Figure 5.20. Indeed, the interpretation of such beds as turbidites is the simplest explanation, and the logical progression to the simplest explanation is at the heart of uniformitarianism and all scientific pursuits. Nonetheless, sedimentologists and oceanographers did eventually link turbidite beds with actual historical turbidity current events.

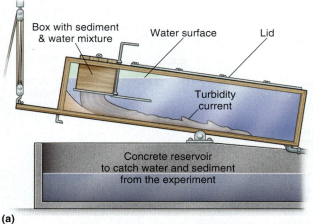

(a)

This sketch shows Kuenen's experimental apparatus for producing small turbidity currents. A tilted glass-walled tank simulates the sloping seafloor. Sediment and water are mixed in a box suspended above the upslope end of the tank. A trap door opens in the bottom of the box and releases the sediment and water mixture into the tank, where it flows along the bottom as a turbidity current.

(b)

This photograph was taken through the glass sidewall of the 2-m-long tank during one of Kuenen's 1950 experiments. The milky white, sediment-laden turbidity current moves along the bottom of the tank below less dense, clear water.

(c)

Each experimental current produced a graded bed. The white bars mark the base of three graded beds. The black marks on the left are 1 cm apart.

▲ **Figure 5.21** **How to make graded beds in the laboratory.**

Geologists and oceanographers used remotely operated cameras and sophisticated sonar to map the outline of the 1929 Grand Banks, Canada, turbidity current deposit.

The positions of telegraph cables broken following the earthquake, and the time elapsed between the earthquake and the break, are also depicted on the map.

Many cables broke instantaneously at the time of the earthquake along the steep continental-shelf margin. The earthquake triggered a submarine landslide along this steep submarine escarpment, and the landslide broke these cables. The turbidity current formed when landslide sediment mixed into seawater. The turbidity current broke more cables as it moved across the seafloor.

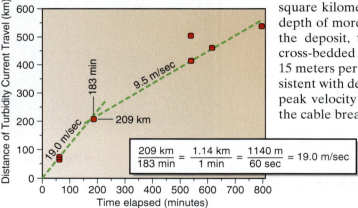

A graph of time elapsed for each cable break after the earthquake plotted against the distance of the cable break from the earthquake location reveals how fast the turbidity current moved. Velocity is a rate, obtained by dividing distance by time, as shown in the example calculation. The dashed lines are possible interpretations of the velocity of the current, beginning at 19 m/sec (about 65 km/hr) and then slowing to half that fast.

▲ **Figure 5.22 Documenting a real turbidity current.**

A 1929 earthquake off the Grand Banks of southeastern Canada generated the best-documented historic turbidity current, as illustrated in **Figure 5.22**. This part of the North Atlantic Ocean was crisscrossed with telegraph cables. Several cables broke almost instantaneously at nearly the same time as the earthquake. Another cable broke about an hour later at some distance from Grand Banks, and then cables at successively greater distances broke in succession over the next 17 hours. Oceanographers reexamined the cable-break times in the 1950s when Kuenen and others developed the turbidity-current concept. Was it not likely that a southerly moving turbidity current generated by the earthquake also caused the successive cable breaks? They used the known times of the earthquake and the cable breaks to estimate the velocity of the current (Figure 5.22). The calculated velocities are quite high and indicate that even gravel clasts a few centimeters across could be suspended in the current.

More recent technological advances permit retrieval of photographs of the seafloor ravaged by the Grand Banks turbidity current. Carefully retrieved samples of the deposit show a graded bed. The current deposited 175 cubic kilometers of sediment (enough to bury all of New York City to a depth of 225 meters) while covering more than 150,000 square kilometers (a little larger than Indiana) to a depth of more than 1 meter. In the northern part of the deposit, the surface undulates with dunes of cross-bedded gravel, indicating a velocity of about 15 meters per second based on experiment, and consistent with deposition as the current slowed from its peak velocity of 19 meters per second suggested by the cable breaks.

No person has observed an actively moving underwater turbidity current of the size required to produce turbidite-sandstones that extend across thousands of square kilometers. Documenting the Grand Banks turbidity current by the cable breaks and later study of the deposits, however, is still a powerful illustration of the role of modern observation for uniformitarian interpretation of ancient sedimentary rocks.

INSIGHTS

How Do Geologists Use Interdisciplinary Studies?
Similar examples of the interpretation of sedimentary structures and depositional environments could be recounted for all of the processes and settings that sedimentologists interpret from the study of sedimentary rocks. All of these interpretations build from the combination of observations of ancient rocks, processes in modern environments, and laboratory experimentation.

This story of the discovery of turbidity currents and turbidites also reveals that scientific progress to resolve questions is not always a systematic plan. Although field sedimentologists and paleontologists hypothesized turbidity currents, Kuenen's initial experiments were not designed to test that hypothesis.

Like many geologic problems, this one was solved by successfully integrating different observations by workers in different disciplines. Geologists with broad perspectives to appreciate the interrelationships of different research results and those who strike up collaborative projects make some of the greatest scientific contributions.

Putting It Together–How Do We Know … How to Interpret Unseen Turbidity Currents?

■ Combining field observation of rocks and modern processes with laboratory experiments reveals that certain graded sandstone beds interlayered with deep-water marine shale originate from rapidly moving currents.

■ Turbidity currents are mixtures of sediment and water that are denser than the clear water above and hug the seafloor or lake bottom as they move.

■ Solving some geologic problems requires integrating field observation and laboratory experimentation with study of ancient rocks and modern processes.

5.8 How Do Plate Tectonics and Sedimentary Rocks Connect?

Modern global geography corresponds closely to plate tectonic processes (see Section 1.5). High mountain ranges form where plates converge, whereas wide, deep oceans flank divergent boundaries along mid-ocean ridges. The patterns of ancient environments preserved in sedimentary rocks permit interpretation of ancient geography, which in turn provides a history of plate tectonics.

Sediment Accumulates in Tectonic Basins

Sedimentary rocks on continents are thickest where depressions, called **basins**, form on Earth's surface. Tectonic processes raise mountains and also cause basins to subside. The presence of thick sedimentary-rock accumulations, therefore, tells geologists where tectonic processes actively formed basins at various times during geologic history. Where basins formed, the depositional environments revealed in the sedimentary rocks provide insights into the size of the basin and the direction toward, and proximity to, mountainous highlands. The composition of clastic-sediment grains reveals the nature of the rocks in the mountains. These observations integrate to reveal the tectonic processes affecting a region in the distant geologic past.

Example of Sedimentary Rocks and Plate Tectonics

Figure 5.23 illustrates an ancient geography interpreted from sedimentary rocks. The map shows distributions of sedimentary rocks that accumulated in the eastern United States about 425 million years ago. These rocks outline a sedimentary basin that subsided far enough for seawater to extend through the middle of the continent. The distribution of rock types shows the location of a shoreline along the eastern edge of the basin. The absence of sedimentary rocks of this age farther east implies an area where erosion provided the clastic sedimentary materials. The presence of conglomerate suggests steep, mountainous slopes where fast-moving streams transported coarse gravel. Some conglomerate fragments include volcanic and plutonic rocks, which suggest erosion of igneous rocks

▼ **Figure 5.23 Making a paleogeographic reconstruction.** This map depicts the presence of different sedimentary rock types deposited in the eastern United States about 425 million years ago. The sandstone and conglomerate mostly contain fossils of land organisms whereas some of the sandstone and all of the other rocks contain marine fossils. The sedimentary rocks permit this interpretation of the ancient geography:

• The westward change from sandstone through shale to limestone is the pattern expected along a westward-deepening, shallow sea (compare to Figure 5.19). This suggests dry land to the east and rivers flowing westward to a warm, shallow sea, which is consistent with current directions from cross-bedding shown by the arrows. So the area of missing rock of this age near the present Atlantic coast likely represents a relatively high area undergoing erosion to provide the sediment found farther west.

• The presence of conglomerate in the far northeast suggests that streams flowed on steeper slopes to transport the coarser fragments, implying more mountainous topography in that area.

• The geographic picture reconstructed from the sedimentary rocks (on the right) is of a mountainous landmass near where the Atlantic Ocean is today and a shallow sea in the midcontinent, which is presently high and dry.

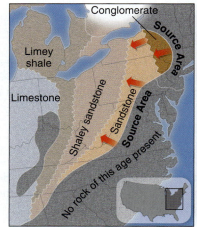
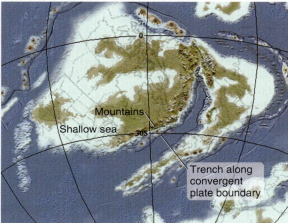

near a subduction zone. The picture, therefore, is of a mountainous, volcanically active landmass along a convergent plate boundary, near where the Atlantic Ocean is today, with tectonic subsidence of a submerged sedimentary basin in the midcontinent, which is presently high and dry.

> ### Putting It Together—*How Do Plate Tectonics and Sedimentary Rocks Connect?*
>
> - Plate tectonic forces cause depression of basins where sediment eroded from mountains accumulates.
> - The composition of the sediment reveals the nature of the rocks in the sediment source region, possibly uplifted by tectonic activity.
> - The depositional environments of the sedimentary rock reveal the landscapes and seascapes resulting from tectonic processes.

Where Are You and Where Are You Going?

You have completed two-thirds of a circuit around the rock cycle.

Processes forming sediment and sedimentary rocks are geologic recyclers. Older rocks physically and chemically break down into components that are shuffled, redistributed, and formed into new rocks. Clastic sedimentary rocks contain fragments created from rocks during physical weathering and new minerals produced by chemical-weathering reactions. Chemical sedimentary rocks form when minerals precipitate from ionic solutions that result in large part from dissolving other rocks during chemical-weathering. Chemical precipitation in the pore spaces between sediment grains cements both clastic and chemical sediment into rock.

Observation of modern processes provides a useful basis for understanding the origin of sedimentary rocks. You can witness the weathering processes that disaggregate rocks and analyze water to track the ions carried away in solution. You can witness sediment transport and deposition by a variety of processes and in countless environments, and use key features observed in the modern sediment to interpret ancient depositional environments. These features include sedimentary structures such as cross-beds, which also reveal direction that currents moved while transporting sediment.

Sedimentary rocks are the archive of processes at Earth's surface and of the organisms that live there. Vertical changes in the rock type record changing processes of sediment deposition and shifting environments, like those caused by fluctuating sea level. Horizontal changes in rocks of the same age reveal the elements of ancient landscapes and seascapes from which tectonic processes are interpreted. Reconstructing past views of our planet does not require a fanciful imagination once you understand how to translate the rich archives of sedimentary rocks.

Now you are ready to complete a trip around the rock cycle. You appreciate that rocks formed well below the surface of the Earth are not stable at the Earth's surface. Likewise, rocks formed at or near the surface at relatively low temperature and pressure are unstable at the higher temperature and pressure found at greater depth. This is especially true where tectonic forces also exert strong horizontal pressure and where water is available to assist chemical reactions that transform one group of minerals to another. You are now ready to investigate metamorphic rocks.

Active Art

Physical Weathering. See how the freezing and thawing of water and the evaporation of salty water pry rocks apart.

Forming Cross-beds. See how cross-beds form and how sand dunes migrate through time.

Changing Shorelines and Sedimentation. See how sedimentation near shorelines differs with changing sea levels.

Extension Modules

Extension Module 5.1: Chemical Reactions and Chemical Equations. Learn about chemical reactions and how to write chemical equations.

Extension Module 5.2: Why Is Seawater Salty? Learn why some ions are concentrated in seawater to give its salty taste.

Extension Module 5.3: Geochemistry of Calcite. Learn the factors that determine whether calcite dissolves or precipitates in water, which explains many features of rocks and landscapes.

Confirm Your Knowledge

1. Define weathering and provide examples of the two types of weathering processes.

2. Why do rocks weather?

3. What is the role of (CO_2) in weathering? Explain Equation (5.3).

4. Oxidation is a common weathering process. Which group of minerals does it affect most, and what minerals does it commonly form? How can you recognize the newly formed minerals?

5. Define clastic sediment. How does it differ from soil?

6. What three parameters do we use to describe the textures of clastic sedimentary rocks? Describe how one of the parameters can be used to provide information about the environment of deposition.

7. What are fossil fuels and how are they formed?

8. Define a fossil and list the three ways that fossils are formed. Explain why fossils are usually found in sedimentary rocks.

9. Explain the processes involved in lithification.

10. How are the two classes of sedimentary rocks named and what is that classification based on?

11. What are the most common biogenic and chemical rock types, and what types of environments favor their deposition?

12. What tools do geologists use to reconstruct an ancient environment?

13. List some sedimentary structures and what they indicate.

14. How are plate tectonics and sediment deposition linked?

Confirm Your Understanding

1. Write out an answer for each question in the Chapter Outline for the chapter sections assigned by your instructor.

2. Consider a modern lagoon system. It is dominated by mud. What clues would you look for to identify an ancient lagoon in the geologic record? How would you differentiate it from a floodplain, where mud collects during floods?

3. Rocks chemically precipitate from ions in solution. Why is this precipitation commonly associated with lakes and oceans, but not rivers? Where else might you expect to find significant accumulations of chemically precipitated rocks?

4. Fossilization includes replacement of organic material. How does this occur? Are there any environments of deposition that would lead to better preservation?

5. A quartz sandstone and an arkose differ significantly. Identify a likely history for each rock.

6. Out hiking with two of your classmates, you observe an outcrop with sandstone rich in quartz and feldspars. Friend A says that the sandstones are derived from a basaltic volcano nearby. Friend B says that the sandstones are derived from a granite outcrop nearby. Which one of your friends is right and why?

7. Explain why the sediment in wind-blown sand dunes is better sorted than that in flood deposits. In other words, explain why sand dunes are well sorted and why flood deposits are poorly sorted.

8. Shales are the most abundant sedimentary rocks on the Earth's surface. What is the relationship of the abundance of shales to the chemical weathering of feldspars?

19. At a field outcrop, you notice a sequence of sedimentary layers. From the bottom up, sandstone is overlain by mudstone and limestone followed by another sequence, only this time, it's limestone overlain by mudstone and sandstone. What can you interpret about the environment during the time of deposition of these two sequences of rock?

Key Terms

anthracite (p. 120)
arkose (p. 117)
basins (p. 129)
bedding (or stratification) (p. 122)
bituminous coal (p. 120)
breccia (p. 117)
cementation (p. 115)
chemical weathering (p. 106)
chemical sediment (p. 112)
chert (p. 119)

clastic sediment (p. 111)
coal (p. 114)
compaction (p. 114)
conglomerate (p. 117)
cross-beds (p. 123)
dissolution (p. 108)
dolostone (p. 119)
dunes (p. 123)
evaporites (p. 119)
exfoliation (p. 106)
fossil fuels (p. 113)

fossils (p. 112)
graded beds (p. 124)
hydrolysis (p. 108)
limestone (p. 119)
lithic sandstone (p. 117)
lithification (p. 114)
mud cracks (p. 122)
mudstone (p. 117)
oxidation (p. 110)
peat (p. 119)
physical weathering (p. 106)

pores (p. 106)
quartz sandstone (p. 117)
ripples (p. 123)
rounding (p. 117)
sandstone (p. 117)
sedimentary structures (p. 121)
shale (p. 117)
sorting (p. 117)
turbidites (p. 127)
turbidity currents (p. 124)
weathering (p. 106)

6 Formation of Metamorphic Rocks

Chapter Outline

Why Study Metamorphic Rocks?

SOME ROCKS ARE NOT THE RESULT OF IGNEOUS OR SEDIMENTA-ry processes. Features in these rocks show that they underwent considerable changes in mineral content, texture, or both after they originally formed. The rocks have been so transformed that the original minerals, and the story of how they first formed, can be hard for geologists to determine. These transformations result from changes in the physical and chemical conditions compared to those that formed the original rock. Rocks transformed from previously existing rocks make up the third major group of rocks found on Earth—metamorphic rocks.

As complex as it may seem, if you understand how temperature and pressure vary within Earth and the processes that deform rocks below Earth's surface, you can understand metamorphism. Metamorphic processes cannot be seen because almost no metamorphic rocks form on Earth's surface. Instead, geologists use theoretical and experimental approaches to identify and explain the processes that form metamorphic rocks.

Metamorphic rocks are widely exposed in actively forming mountain ranges. They are always found in eroded ancient mountain belts in the interior of continents. The oldest

Swirling mineral bands characterize this metamorphic-rock outcrop in the Shining Rock Wilderness Area of the Appalachian Mountains of western North Carolina. ▶

rocks on Earth are metamorphic rocks, as you will learn more about in Chapter 7. Metamorphic rocks are key, therefore, to reconstructing the history of Earth, especially the history of mountain-building events.

Many metamorphic rocks or their constituent minerals are used as economic resources. These include talc for talcum powder, graphite for pencils and lubricants, marble for building material, garnet and corundum as industrial abrasives, and metamorphosed coal for fuel. Metamorphic processes also form some metal ores. Understanding the metamorphic history of a region aids the discovery and extraction of economic resources.

So how does an igneous or sedimentary rock transform into a metamorphic rock? How does changing the temperature and pressure cause this transformation? What are the roles of deformation and fluid during metamorphism? Your objectives in this chapter are as follows:

✔ To learn how to identify metamorphic rocks

✔ To understand how metamorphic rocks form

✔ To examine where metamorphic rocks form

✔ To understand how geologists use metamorphic rocks to interpret the geologic history of a region

To do this, you need to answer the following questions:

6.1 What is metamorphism?

6.2 What is the role of temperature in metamorphism?

6.3 What is the role of pressure in metamorphism?

6.4 What is the role of fluid in metamorphism?

6.5 Why do metamorphic rocks exist at the surface?

6.6 *How do we know* ... how to determine the stability of minerals?

6.7 What were the conditions of metamorphism?

6.8 How are metamorphic rocks classified?

6.9 What was the rock before it was metamorphosed?

6.10 Where does metamorphism occur?

In the FIELD

You are on a road trip when you notice an interesting road cut, seen in **Figure 6.1**, and pull over to take a look. This may be a chance to identify some of the rocks you are learning about in your geology class. What catches your eye from the car is that the outcrop is very shiny, with a very subtle layering that runs parallel to the hillside but without obvious bedding of the sort common in sedimentary rocks.

A closer examination reveals that the rock consists mostly of muscovite mica, along with scattered crystals of two darker minerals (see Figure 6.1). The thin, flaky mica crystals are several millimeters across and arranged like sheets of paper strewn across the floor. The cleavage faces all point in the same direction so that light reflects off the mineral surfaces to produce the shiny characteristic that was obvious from the car. The darker minerals do not resemble the minerals you learned as common constituents of igneous or sedimentary rocks. One mineral type, forming equidimensional crystals, with a deep, red-wine color and well-formed crystal faces, does remind you of a mineral illustrated in Chapter 2; this is garnet (see Figure 2.27). The brown crystals stump you, so you return to your car and retrieve your field-worn, dog-eared field-guide-to-minerals book. You find a descriptive match to a silicate mineral called staurolite. A key identifying feature is the tendency for staurolite crystals to grow through one another to form crosses (the mineral name derives from the Greek word *stauros*, which means "cross").

What kind of rock is this? You have sometimes seen muscovite as a minor constituent in igneous rocks along with much more abundant quartz and feldspar, whereas muscovite forms most of this rock. Also, you do not recognize garnet or staurolite as typical igneous-rock minerals. Although the outcrop has a subtle layering formed by the parallel orientation of the thin, flat muscovite crystals, it bears no other resemblance to a sedimentary rock. The garnet and staurolite crystals have well-defined crystal faces with sharp edges and corners, so they cannot be clastic sediment grains, which should be rounded by transport in flowing water. A rock composed of silicate minerals is not likely to be a chemical sedimentary rock because silicate minerals are barely soluble in water.

The texture and mineral content of this rock are very different from what you expect for sedimentary and igneous rocks. You have encountered an example of a third group of rocks—this is an outcrop of metamorphic rocks.

Metamorphic rocks result from changes in the appearance of preexisting rocks (Section 3.2 provides some background). To change the appearance of a rock, the minerals that compose it must rearrange in some fashion, must change to new ones, or both. How do metamorphic rocks form? What was the rock before it was metamorphosed? Where do metamorphic rocks form? How do you name this rock you have picked up?

Figure 6.1 Field observations of metamorphic rock. ▶

Field observations of metamorphic rocks

Roadside outcrop of shiny rock

Subtle layering parallel
to the hillside is caused
by parallel orientation
of flaky mica crystals.

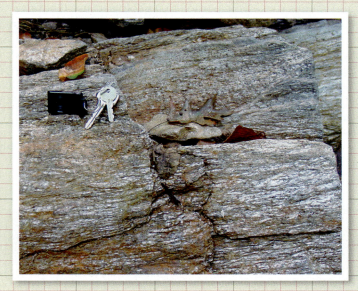

Closer view of rock, with car keys to show the scale.
The rock is shiny because light reflects off of large
crystals of muscovite mica, all of which are oriented
the same way so that light reflects off of the flat
cleavage surfaces.

Close examination shows 3 minerals in the rock:
- shiny <u>muscovite mica</u>
- red, equidimensional crystals
  with shiny faces: <u>garnet</u>
- brown, elongate crystals; some cross over
  each other: <u>staurolite</u>

6.1 What Is Metamorphism?

Metamorphism (from the Greek terms *meta* for change and *morphe* for form) describes the mineralogical, chemical, and textural changes to preexisting rocks that occur in an essentially solid state—that is, without substantial melting. Metamorphism occurs in Earth's crust and mantle at conditions different from those where the original rock formed. Increasing temperature and pressure are very important in metamorphism. Chemically reactive fluids and rock deformation within the dynamic planet also contribute to metamorphic transformations. If substantial melting takes place, you know that these melts crystallize to form igneous rocks. You also know that relatively low temperatures and pressures may produce reactions that lithify sedimentary rocks, but not enough to change the minerals and textures to the extent seen in metamorphic rocks. This means that conditions for metamorphism can be thought of as ranging from those lithifying sediment into sedimentary rocks, through to the conditions of temperature and pressure just before rock melts to make magma, as illustrated in **Figure 6.2**. Metamorphic petrologists describe the intensity of temperature and pressure during metamorphism as ranging from **low grade** to **medium grade** to **high grade**. Figure 6.2 illustrates the approximate temperature and pressure conditions for the different metamorphic grades.

What Happens during Metamorphism?

Nearly all metamorphic rocks form at elevated temperature and pressure deep below the surface; therefore, the processes that produce them are not observable. So how do geologists know about metamorphic processes if they cannot be observed?

Laboratory experiments are necessary to decipher metamorphic processes. While studying igneous rocks, you learned that geologists use laboratory experiments to recreate the conditions deep inside Earth by controlling magma composition, temperature, pressure, and abundance of water (see Chapter 4, Section 4.5). Similarly, to reproduce the minerals found in metamorphic rocks, geologists in the laboratory vary (1) the original rock composition, (2) temperature, (3) pressure, and (4) abundance and composition of fluid. Experiments indicate that changes in any of these four parameters determine the mineral composition and texture of the resulting metamorphic rocks. Rocks exhibit two types of metamorphic changes: (1) New minerals form at the expense of the original ones because of chemical reactions, or (2) the rock texture is altered by changes in size, shape, and orientation of constituent minerals.

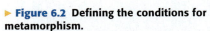

▶ **Figure 6.2 Defining the conditions for metamorphism.**
The range of temperatures and pressures where metamorphism occurs exceeds that for lithification of sedimentary rocks but falls short of the conditions where rock melts to form magma. Grades of metamorphism approximately correspond to the labeled areas on the graph. The melting temperature and pressure partly depends on the water content of the rock, so potential melting curves are shown for both wet and dry crust. The conditions for high-grade metamorphism of dry rocks overlap with melting of water-rich rocks.

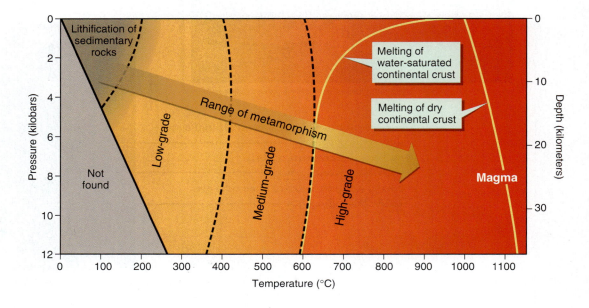

Significance of the Original Rock

Although changes of temperature, pressure, and fluid composition cause metamorphism, the original composition of the rock—the parent composition—is essential to determining what reactions will actually take place. The overall chemical composition of the metamorphosed rock can be quite similar to the starting material, with only minor modification. Some chemical and mineralogical changes, such as those caused by removal or addition of water, are minor. In other words, the composition you start with, with slight modification, is the composition you end up with, except that the minerals, texture, or both will be different. In some cases, features in the original rock, such as igneous crystals and sedimentary bedding, may still be recognizable.

The process of making a cake offers an analogy. After mixing the ingredients and baking it, the cake has a very different appearance and texture than the original bowl of ingredients. Yet the bulk chemical composition of the cake is nearly the same as the mixture of ingredients in the batter. The metamorphic rock you picked up along the road has the same bulk chemical composition as shale. So geologists infer that if shale metamorphoses, then it can end up looking like your field sample (Figure 6.1).

In other cases, significant changes to the composition of the rock occur because large amounts of chemically active fluids participate in the metamorphism. This happens, for example, when hot, water-rich solutions from magma move into adjacent rocks during metamorphism. The metamorphic minerals incorporate elements from the fluid to produce a rock with different composition from the parent rock.

The conditions of metamorphism determine the minerals formed in the rock (composition), or the shapes and arrangement of minerals in the rock (texture), or both. Metamorphic changes in rocks are the response to changing factors: changes in temperature, pressure, and fluid availability and composition. What role does each factor play in forming the metamorphic rock? Addressing each factor independently is quite tricky since there is rarely only one variable in any metamorphic reaction. Still, it can be useful to examine the role of each separately to the extent possible.

Putting It Together–**What Is Metamorphism?**

■ Metamorphism describes the mineralogical, chemical, and textural changes to preexisting rocks in an essentially solid state by changing environmental conditions. These conditions include changing temperature, changing pressure, and the addition or subtraction of fluid.

■ Metamorphic rocks may retain the bulk chemical composition of the original parent rock or the composition may significantly change. Whether or not the composition changes depends on the mobility of fluid to introduce new components and remove some constituents during metamorphic reactions.

6.2 What Is the Role of Temperature in Metamorphism?

The cake-baking analogy in the last section also offers insights into the role of temperature in metamorphism. No matter how long the batter sits in the bowl, it will never become a cake unless it is put into the oven and baked. Temperature is a measure of the intensity of heat that is present. Most metamorphic reactions require heat in order to happen. A key question is, How can temperature increase for metamorphism to take place?

How Rock Temperature Increases

Temperature increases with depth below Earth's surface (Section 4.4). Some metamorphic reactions occur when minerals that formed at low temperature

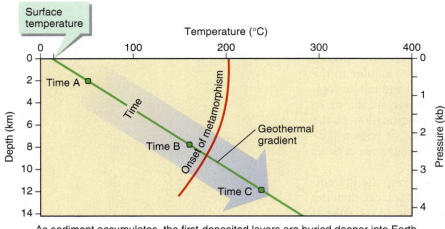

As sediment accumulates, the first-deposited layers are buried deeper into Earth, raising the temperature of the rocks.

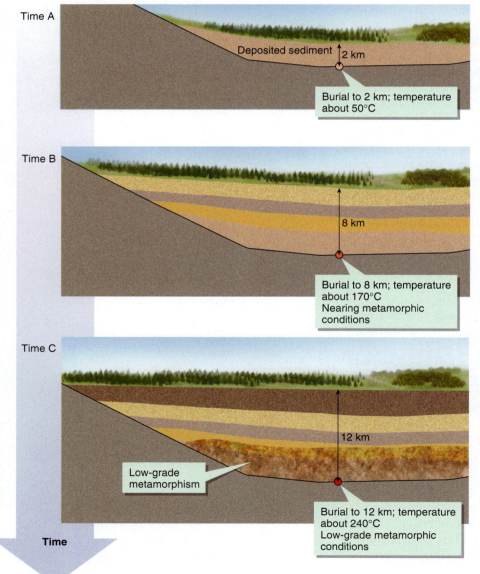

▲ **Figure 6.3** **Raising the temperature of rocks through sediment burial.**
Sedimentary rocks experience increasing temperature and pressure while sediment accumulates. If the rocks are buried sufficiently deeply, then low-grade metamorphism occurs.

later transform into different minerals as temperature increases. The rates of metamorphic reactions also increase at higher temperature.

Three processes most commonly cause heat transfer to rocks so that they experience increasing temperature.

- **Figure 6.3** depicts the effects of sediment burial. Temperature increases with depth along the geothermal gradient. Where sedimentary rocks accumulate in sinking basins, the early-deposited sediment experiences progressively higher temperature as more sediment accumulates on top. Eventually, the sedimentary rocks are buried sufficiently deeply for there to be enough heat to cause metamorphism.
- **Figure 6.4** depicts the similar effects of tectonic burial. Where one block of crust is forced over the top of another block, the lower block heats to the higher temperature associated with a greater depth below Earth's surface.
- **Figure 6.5** depicts the effect of magma intrusion. The injection of hot magma into rocks changes the local geothermal gradient by increasing the temperature. The transfer of heat from magma metamorphoses the surrounding rocks.

Notice that increasing temperature by burial occurs along with increasing pressure exerted by the weight of overlying rock, whereas increasing temperature near an intrusion need not require a change in pressure.

Heat Drives Away Water and Gases

High-temperature metamorphism causes minerals containing water or gas molecules to break down and release these components. These are dehydration (loss of water) and degassing (loss of gas) reactions. An example of a high-temperature dehydration reaction is the formation of sillimanite and potassium feldspar from muscovite and quartz. Water releases by the breakdown of muscovite, and two new minerals form that lack water. The reaction is written like this:

$$\text{Add heat}$$
$$\downarrow$$
$$\underset{\text{muscovite}}{KAl_3Si_3O_{10}(OH)_2} + \underset{\text{quartz}}{SiO_2} \rightarrow$$

$$\underset{\text{sillimanite}}{Al_2SiO_5} + \underset{\substack{\text{potassium-}\\\text{feldspar}}}{KAlSi_3O_8} + \underset{\text{water}}{H_2O}$$

Sillimanite is a common mineral in high-temperature metamorphic rocks that you will learn more about in Section 6.6.

Dehydration reactions release water or water vapor into the spaces between the mineral grains and may speed up further metamorphic reactions that re-

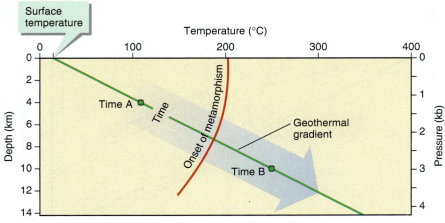

◄ **Figure 6.4 Raising the temperature of rocks through tectonic burial.** Tectonic processes can cause rocks to be moved deeper into Earth, which causes the rocks to experience higher temperatures and pressures where metamorphism occurs.

At time A, the green location is 4 km below the surface and has a temperature of 100°C. Following tectonic processes that displace the blocks of crust, the green location at time B is now at 10 km depth, with a temperature of 260°C.

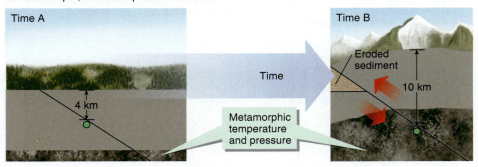

quire the presence of water. The presence of water can also decrease the melting temperature of silicate minerals and enhance formation of magma (see Figures 4.18c and 4.21b). Dehydration reactions during metamorphism of subducted oceanic crust lead to magma generation at convergent plate boundaries (see Section 4.6).

The most common degassing reactions release carbon dioxide from carbonate minerals. One such reaction is applied artificially to make commercial cement when carbon dioxide degasses from calcite ($CaCO_3$) in limestone:

$$\begin{array}{c} \text{Add heat} \\ \downarrow \\ CaCO_3 \rightarrow CaO + CO_{2(gas)} \\ \text{\small calcite} \qquad \text{\small lime} \quad \text{\small carbon dioxide} \end{array}$$

Roasting the limestone in a kiln produces lime (CaO), which is the primary ingredient in cement. A natural example is the "baking" of dolostone near a hot igneous intrusion. The dolomite breaks down at high temperature to calcite and periclase (a magnesium-oxide mineral) and releases carbon dioxide gas:

$$\begin{array}{c} \text{Add heat} \\ \downarrow \\ CaMg(CO_3)_2 \rightarrow CaCO_3 + MgO + CO_{2(gas)} \\ \text{\small dolomite} \qquad \text{\small calcite} \quad \text{\small periclase} \quad \text{\small carbon dioxide} \end{array}$$

The Concept of Mineral Stability

The dehydration and degassing reactions also illustrate the concept of mineral stability. Stability simply is the tendency to remain the same rather than to change. Most minerals exist—are stable—only at particular ranges of temperature and pressure and react to form different minerals if conditions change.

▶ **Figure 6.5 Raising the temperature of rocks through magma intrusion.**
Magma intrusions conduct heat into cooler surrounding rock, which raises the temperature and causes metamorphism.

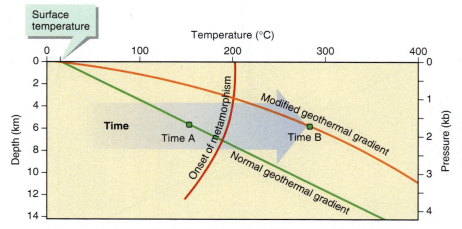

Prior to the magma intrusion, the temperature of the rock at the green location, 6 km below the surface, is 160°C, based on a normal geothermal gradient. Heat conducted from intruding magma modifies the geothermal gradient, and the rock at the green location heats to 285°C and metamorphoses. Rocks closer to the magma experience even higher temperature, higher-grade, metamorphic effects.

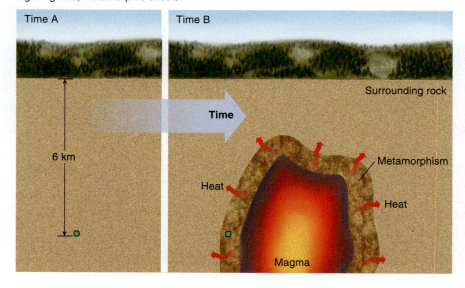

Metamorphism, therefore, may involve the crystallization of new minerals at the expense of preexisting minerals because mineral stability changes. The production of a new mineral requires the breaking of the chemical bonds in the original mineral and rearrangement of the freed-up atoms into a new mineral structure. Using the dolomite example, dolomite is stable at low temperature but is unstable and breaks down to other minerals and gas at higher temperature.

Water as liquid, solid, or steam provides a familiar example of stability. **Figure 6.6** shows the temperature and pressure conditions where liquid water, water vapor (steam), and solid ice are each stable. At atmospheric pressure (one bar), water boils at 100°C to become steam and freezes at 0°C to become ice. If pressure decreases, the boiling temperature of water drops. This is why heating water boils sooner when you are camping in the mountains than at home at a lower elevation (Figure 6.6). The diagram shows that at distinct combinations of temperature and pressure, more than one phase can coexist: liquid water and steam, or ice and liquid water, or even liquid, ice, and steam. So when you put an ice-cube tray of water into the freezer, the temperature conditions are dropped significantly to convert the water into ice. Once the temperature drops to the freezing point of water, the atoms reorganize into the more orderly arrangement of atoms found in ice because this configuration is more stable

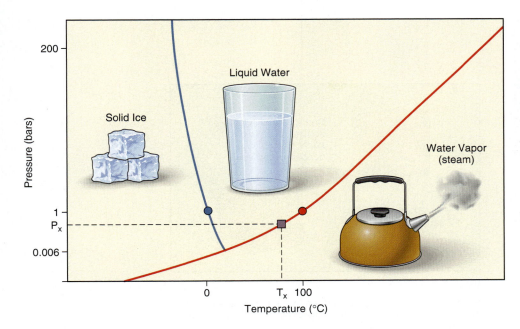

◀ **Figure 6.6** **Visualizing the pressure and temperature stability conditions of water.**
This diagram illustrates the temperatures and pressures where solid, liquid, and vapor water are stable. At the appropriate temperature and pressure, liquid and ice coexist along the blue line. The blue dot indicates the temperature where water freezes on Earth's surface at one bar, which is atmospheric pressure at sea level. The red dot indicates the temperature where water boils at one bar. When pressure drops below atmospheric pressure (P_x), such as in mountains above sea level, the temperature at which water converts to steam also drops (T_x).

at lower temperature. Since the conditions are different inside the freezer than they are outside, the water changes, by freezing, to achieve stability in its new environment. Similarly, rocks under different environmental conditions change to achieve the most stable conditions.

Putting It Together–*What Is the Role of Temperature in Metamorphism?*

■ Temperature increases along the geothermal gradient as rocks move deeper into Earth's crust through tectonic processes or are buried deeper as rocks are deposited on top. Rocks also experience higher temperatures near igneous intrusions.

■ Increasing temperature is an expression of the transfer of heat energy that is necessary to drive chemical reactions during metamorphism. These reactions include dehydration and degassing of minerals that contain water or gas molecules in their structure.

■ Minerals are stable within a defined range of temperature and pressure. Outside of this stability range the minerals break down and form new minerals.

6.3 What Is the Role of Pressure in Metamorphism?

Pressure also affects mineral stability and influences crystal size and orientation, which describe rock texture. Pressure is related to the concept of **stress**. Both stress and pressure are defined as the magnitude of a force divided by the area of the surface on which the force is applied. **Figure 6.7** illustrates two types of stress. **Normal stress** is force applied perpendicular to a surface (Figure 6.7a), and **shear stress** describes force applied parallel to a surface (Figure 6.7b). The increasing force exerted by the weight of overlying rock causes pressure to increase at greater depth. Pressure is a special type of normal stress where the forces are equal in all three dimensions (Figure 6.7c).

Strain is the deformation of rock as a result of an applied stress. Normal stress causes change in volume and commonly also the shape of material (Figure 6.7a). Shear stress always causes changes in shape but not in volume (Figure 6.7b)

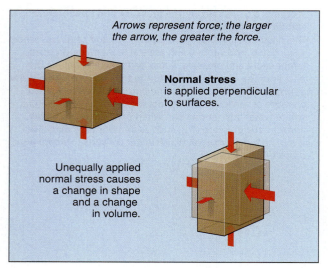

Arrows represent force; the larger the arrow, the greater the force.

Normal stress is applied perpendicular to surfaces.

Unequally applied normal stress causes a change in shape and a change in volume.

(a)

Normal stresses applied perpendicular to the surface but not equally cause a change in shape and in volume of the cube.

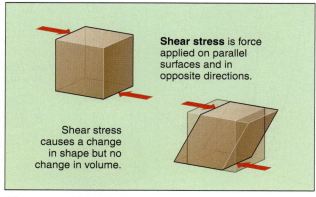

Shear stress is force applied on parallel surfaces and in opposite directions.

Shear stress causes a change in shape but no change in volume.

(b)

Shear stress acts parallel to surfaces and in opposite directions. This stress changes the shape of the cube but does not affect volume.

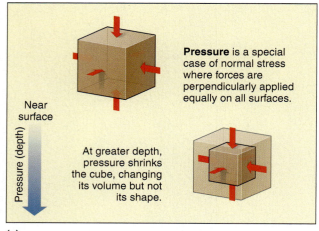

Pressure is a special case of normal stress where forces are perpendicularly applied equally on all surfaces.

Near surface

Pressure (depth)

At greater depth, pressure shrinks the cube, changing its volume but not its shape.

(c)

Normal stress applied in all directions equally is pressure. At greater depths, increasing pressure causes a change in volume without a changing shape.

▲ **Figure 6.7 How normal and shear stresses deform rock.**

Changes in Mineral Stability

Mineral stability relates not only to temperature but also to applied stress, especially equal-dimensional normal stress, or pressure (Figure 6.7c). Atoms rearrange into denser, more closely packed crystal structures at high pressure, such as illustrated by the conversion of graphite to diamond (Figure 2.23). Laboratory experiments show that this metamorphic reaction occurs when pressure exceeds 25 kilobars, or 25,000 times the pressure exerted on Earth's surface by the atmosphere.

Very few natural metamorphic reactions occur only as a consequence of increasing pressure alone, because both pressure and temperature increase at depth. Laboratory experiments do show, however, that increasing pressure drives some reactions more than increasing temperature. Calcite and aragonite are examples of polymorphs whose stabilities are mostly defined by pressure (see Figure 2.21). Calcite is more stable than aragonite at lower pressure. Calcite becomes unstable at high pressure. In response to increasing pressure, the atoms in calcite rearrange into the more closely packed crystal structure of aragonite (see Figure 2.20). The new mineral aragonite crystallizes as calcite breaks down.

Recrystallization of Minerals into New Sizes and Shapes

Rock texture may change during metamorphism without changing chemical composition or minerals. **Figure 6.8** provides a simple example of this process by examining snowflakes and glacial ice as a readily examined metamorphic change. Snowflakes precipitate at atmospheric conditions where each snowflake grows as a delicate crystal (Figure 6.8a). Snow that is deeply buried in a glacier (Figure 6.8b) no longer consists of delicate six-sided crystals. Rather it consists of polygonal ice grains that commonly join at three-sided junctions. First, the snowflakes pack closely together and eliminate the spaces between grains under the weight of the accumulating ice. Second, the smaller grains shrink as the larger grains grow by **recrystallization**, the transfer of atoms from one part of a crystal to another part of the crystal or to an adjacent crystal. The atoms within the crystals rearrange as a result of the increasing pressure; they fit tighter together, making the minerals more compact and denser. As a result of

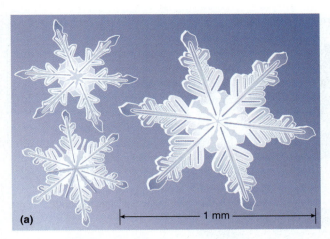

Each snowflake is unique and initially crystallizes into six-pronged crystal when formed.

▲ **Figure 6.8** **Metamorphic change from snowflake to ice.**

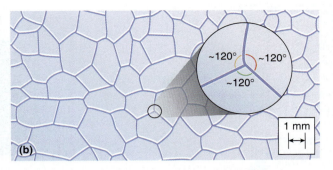

The pressure exerted by the weight of accumulating snow causes buried snowflakes deep within a glacier to recrystallize. Compared to the original snowflakes, the ice crystals are more equidimensional in shape and size with nearly straight edges that can join at three-sided junctions, forming approximately 120° angles.

recrystallization, the crystals form a mosaic of large grains. Recrystallization, then, is a metamorphic process that changes the size and shape of existing minerals rather than making new minerals.

Another example of recrystallization occurs when quartz sandstone metamorphoses as shown in **Figure 6.9**. You can see the individual, round quartz grains within the sedimentary rock (Figure 6.9a). Once metamorphosed, however, the quartz grains have a texture of straight-sided crystals that form three-sided junctions (Figure 6.9b).

How Rock Textures Record Strain

Strain during metamorphism changes the rock texture. **Figures 6.10**, **6.11**, and **6.12** illustrate how crystals rearrange into planes, called **foliation**, depending on the type and orientation of the stress. The term foliation comes from the Latin word *folium*, meaning *"leaf."* Foliation planes are recognized by

- preferred orientation of minerals (Figure 6.10)
- alternating bands of different minerals (Figure 6.11)
- the flattening and stretching of minerals (Figure 6.12)

This is a microscope photo of sandstone, exhibiting rounded, sand-size quartz grains surrounded by deep-red hematite cement.

When metamorphosed, the quartz grains recrystallize into straighter-sided grains that seem to penetrate into one another, and locally intersect at three-sided junctions (example junctions are circled).

◀ **Figure 6.9** **From sandstone to metamorphic rock.**

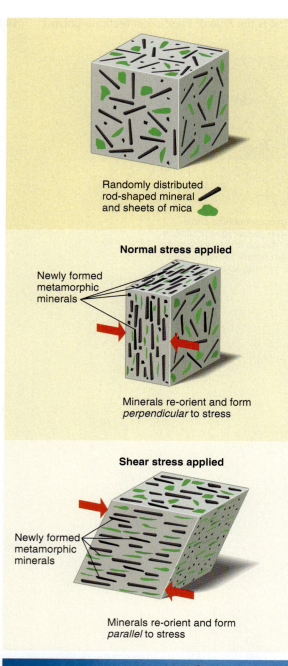

Randomly distributed rod-shaped mineral and sheets of mica

Normal stress applied

Newly formed metamorphic minerals

Minerals re-orient and form *perpendicular* to stress

Shear stress applied

Newly formed metamorphic minerals

Minerals re-orient and form *parallel* to stress

Metamorphic rock with foliation defined by oriented mica crystals

▲ **Figure 6.10 How minerals reorient and crystallize to form foliation.**

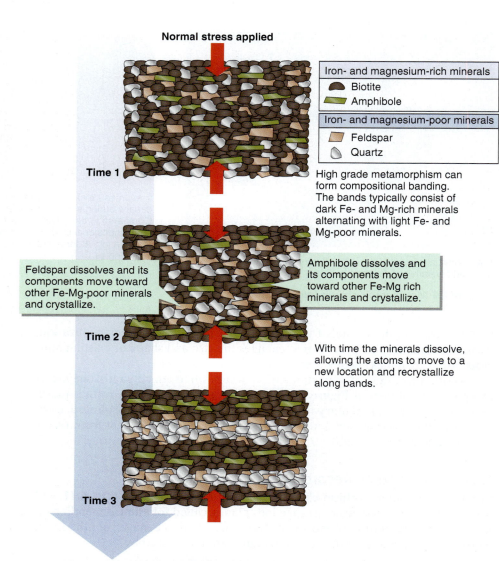

Normal stress applied

Iron- and magnesium-rich minerals	
⬤	Biotite
▬	Amphibole

Iron- and magnesium-poor minerals	
◿	Feldspar
◻	Quartz

Time 1

High grade metamorphism can form compositional banding. The bands typically consist of dark Fe- and Mg-rich minerals alternating with light Fe- and Mg-poor minerals.

Feldspar dissolves and its components move toward other Fe-Mg-poor minerals and crystallize.

Amphibole dissolves and its components move toward other Fe-Mg rich minerals and crystallize.

Time 2

With time the minerals dissolve, allowing the atoms to move to a new location and recrystallize along bands.

Time 3

This metamorphic rock illustrates compositional banding foliation formed during metamorphism.

▲ **Figure 6.11 How compositional banding forms foliation.**

1 cm

Foliation forms by mechanical rotation of preexisting minerals or dissolution and new mineral growth along preferred orientations. Mechanical rotation or growth of crystals parallel to the direction of least stress typically forms foliation in rocks with platy minerals such as clay and mica (Figure 6.10). Minerals also dissolve and recrystallize, or their atoms rearrange into new minerals that form parallel to the direction of the smallest normal stress (Figure 6.12). In other cases, the minerals segregate into compositional layers that are oriented according to the stress orientation (Figure 6.11). All types of foliation are planes in the rock that are perpendicular to the greatest normal stress or parallel to shear stress.

Putting It Together–*What Is the Role of Pressure in Metamorphism?*

■ Pressure increases with increasing depth into Earth. Increasing pressure causes many minerals to metamorphose to denser minerals that are more stable at high pressure.
■ Recrystallization at high pressure changes rock texture without changing chemical composition or mineral content.
■ Foliation describes planes of minerals formed in response to stress. Foliation is recognized by preferred orientation of minerals, reshaping of minerals, or alternating bands of different minerals.

6.4 What Is the Role of Fluid in Metamorphism?

Fluids participate in metamorphism in two ways:

1. Fluid reacts with minerals to form new minerals that contain components of the fluid molecules. Primarily, these reactions create minerals containing constituents of water or carbon dioxide molecules.
2. Fluids usually change the metamorphic rock composition compared to the parent rock by delivering and removing dissolved ions. Fluid makes metamorphic reactions occur more easily and faster.

Adding Fluid and Gas during Metamorphism

In Section 6.2, you learned that water and gases, like carbon dioxide, are driven away from minerals at high temperature. In some cases, however, these components end up being added to minerals. This mostly happens during low-grade metamorphism where the temperature is relatively low. But it can happen during high-grade metamorphism if the high pressure keeps water and gas from escaping the environment even when the temperature is high.

You already know about processes of chemical weathering where water is added to minerals (Section 5.1). An example is the weathering of feldspar, which lacks water, to clay, which contains water molecules in its crystal structure. Metamorphic reactions that produce different minerals by adding water or carbon dioxide occur at higher temperature and pressure than chemical weathering.

◄ **Figure 6.12** How minerals flatten to form foliation.

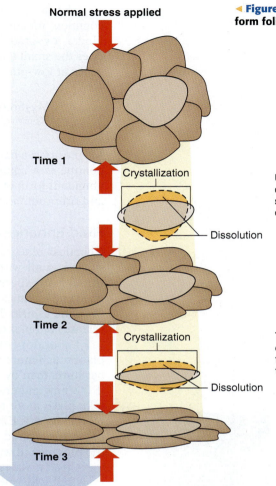

Normal stress applied

Time 1

Crystallization

Dissolution

Under normal stress, quartz grains dissolve along the axis of maximum stress and crystallize along the axis of minimum normal stress.

Time 2

Crystallization

Dissolution

This process of recrystallization causes the grains to elongate or flatten along the axis perpendicular to maximum stress.

Time 3

The pebbles in a conglomerate were flattened to form foliation in this metamorphic rock.

Forming Foliation: *See how the three types of foliation form.*

The metamorphism of olivine in mafic and ultramafic igneous rocks is a common metamorphic reaction involving both water and carbon dioxide. Under low-grade metamorphic conditions water reacts readily with olivine, as does the small amount of carbon dioxide that is almost always dissolved in the water at low-grade temperatures and pressures. This reaction is

$$2Mg_2SiO_4 + 2H_2O + CO_{2(gas)} \rightarrow Mg_3Si_2O_5(OH)_4 + MgCO_3$$

Mg-olivine   water   carbon dioxide   serpentine   magnesite

Magnesite is a carbonate mineral with a crystal structure similar to calcite. Serpentine is a complex silicate mineral (see Table 2.2) that includes the most abundant form of asbestos, which was once abundantly used to make insulation and fireproof fabric.

Fluid Enhances Metamorphism

Chemical reactions between solids take place extremely slowly or never get started at all without adding water. **Figure 6.13** illustrates this role of water through an experiment that you can do in your kitchen. If you mix powdered aspirin and baking soda in a shallow dish, nothing happens. If you add water, however, the aspirin dissolves to form an acid that reacts with the baking soda and releases bubbles of carbon dioxide gas.

Laboratory experiments repeatedly show that metamorphic reactions take place much more quickly in the presence of water, and commonly at lower temperature, than by merely placing the dry ingredients together. This is true even if the new minerals do not contain water. In fact, even dehydration reactions take place more readily when the minerals are immersed in water rather than remaining dry. Experimental mineral reactions that require many days at temperatures around 1000°C may take place in minutes at half the temperature in the presence of water.

Clearly, water plays an impressive role in enhancing metamorphic reactions, but why is this the case? The most important factor is the role of water as a solvent (see Section 5.1). Water dissolves minerals, and the ions then move through the water to react with other ions. Experiments also show that it is much easier for crystals to precipitate out of aqueous solution than to form from a combination of solids. Where, however, do the fluids that drive metamorphism come from?

The Origin of Fluids

Fluid may either be part of the original parent rock or be introduced into the metamorphic environment. All rocks that form at or near the surface contain water in open pore spaces or fractures, or along boundaries between mineral grains. When these rocks are buried to low-grade metamorphic conditions, the fluids are already there to participate in the reactions. The high-grade dehydration and degassing reactions described in Section 6.2 generate fluids that rise upward toward regions of lower temperature and pressure. These fluids participate in and help drive metamorphic reactions at these less extreme temperatures and pressures.

Igneous intrusions are another important source of migrating fluids. Hot, aqueous solutions commonly form during the late stages of magma crystallization (see Section 4.10). These fluids, rich in reactive ions, move into the surrounding rock as it metamorphoses in response to heating from the intrusion. This fluid from the magma not only delivers reaction-enhancing water but also heat (which speeds up reactions) and ions from the magma (which participate in reactions). These solutions may introduce economically valuable metal ions that form ores of metal sulfide and oxide minerals in the metamorphic rocks. Even if the magma lacks abundant water to be forced into the surrounding rocks, the high temperature of the magma causes water in the

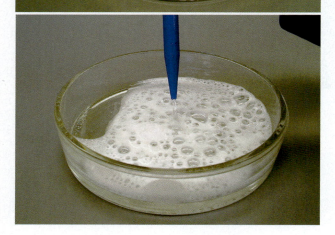

▲ **Figure 6.13 Adding water to make the reaction go.**
A dry mixture of crushed aspirin and baking soda does not react. Adding water to the mixture, however, causes a visible, bubbling reaction between the two compounds.

surrounding rock to heat up and move away from the intrusion. This process moves warm, reactive water into rocks where all of the ingredients for metamorphism are already present except for a bit more heat and water to initiate the reactions.

Fluid Changes the Composition of the Rocks

The ability of water to transport chemical components means that metamorphism in the presence of abundant fluid always produces a metamorphic rock with a bulk composition very different from the parent rock. Ions in the water deliver new components, and elements liberated by mineral breakdown are transported away in solution.

Putting It Together—What Is the Role of Fluid in Metamorphism?
- Fluids, present in the original rock or introduced from magma or high-temperature dehydration and degassing reactions, create metamorphic minerals with components from the water and gas molecules.
- Metamorphic reactions take place faster and at lower temperature in the presence of water than under dry conditions.
- Movement of fluid during metamorphism delivers some ions and takes others away in solution so that the metamorphic rock has a different bulk composition than the parent rock.

6.5 Why Do Metamorphic Rocks Exist at the Surface?

If temperature and pressure are critical factors of metamorphism and can convert graphite to diamond, then why is diamond not converted back to graphite at Earth's surface temperature and pressure? Will the diamond ring you inherited become a piece of graphite? Will any metamorphic mineral formed at great depth convert back to its original starting material when later exposed at Earth's surface? How do metamorphic rocks even end up being exposed at the surface?

Most metamorphic rocks form a few to many kilometers underground (see Figures 6.3, 6.4 and 6.5) and are exposed to view for study only by removing overlying rock. The primary reason that metamorphic rocks are most commonly associated with active or ancient mountain belts is that mountain building involves uplift of rock, a process to be explored more thoroughly in Part III of this book. **Figure 6.14** illustrates that as mountains rise, erosion removes the uplifted rock, eventually exposing metamorphic rocks that formed at great depth.

The unlikely transformation of a metamorphic mineral back to the original starting mineral is explained by the same factors that cause metamorphism. Heat is commonly needed for chemical reactions to take place, but the temperature decreases as the rocks are uplifted toward the surface, so there is not enough heat to drive the reactions. The loss of fluid during metamorphic dehydration also inhibits metamorphic minerals from converting back to their original minerals. The reactions cannot reverse if fluids moved away from the rock and are no longer available to participate in the reactions. In some cases, the time it took for uplift and exposure is short relative to the time it took for the metamorphism of the rocks. Reaction rates are typically very slow, and if the rocks rise toward the surface rapidly, there is not enough time for the reactions to reverse while enhancing fluid and heat are still present. For these reasons it is possible to study high-grade metamorphic rocks at the surface.

If high-grade metamorphic rocks rise to depths associated with low-grade metamorphism and reside there for a long time before later uplift and exposure, then they may be re-metamorphosed, especially if fluids are abundant. It

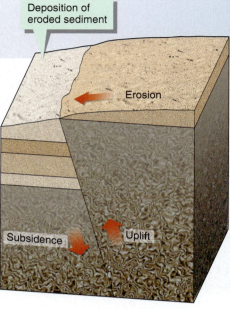
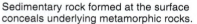
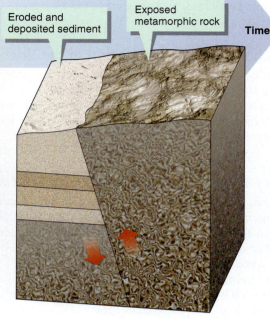
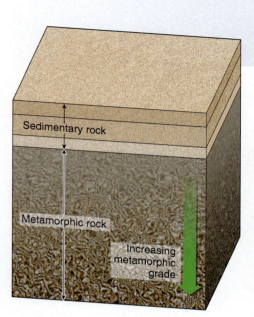

Deposition of eroded sediment

Eroded and deposited sediment

Exposed metamorphic rock

Time

Sedimentary rock

Metamorphic rock

Increasing metamorphic grade

Erosion

Subsidence

Uplift

Sedimentary rock formed at the surface conceals underlying metamorphic rocks.

Tectonic stress causes a break in the rock and displacement of blocks of crust.

During uplift, the sedimentary rocks erode away to gradually expose the underlying metamorphic rocks.

▲ **Figure 6.14** **How metamorphic rocks are exposed at Earth's surface.**
A combination of tectonic uplift, weathering, and erosion exposes metamorphic rocks at Earth's surface.

Exposing Metamorphic Rocks: *See how metamorphic rocks end up at the surface after forming at great depth.*

is important to remember that metamorphic rocks are not simply transformed sedimentary and igneous rocks. Metamorphic rocks can be metamorphosed, too.

As for your diamond, it will not convert to graphite on Earth's surface because the transformation of diamond to graphite requires high temperature to initiate the reaction. The temperature at the Earth's surface is not sufficiently high for the reaction to take place. As a result, not only are most metamorphic rocks not converted back to their original form at Earth's surface, but your family heirloom will still be a diamond for many, many, many generations to come.

Putting It Together—Why Do Metamorphic Rocks Exist at the Surface?
- Mountain-building processes uplift and expose metamorphic rocks at the surface as erosion removes overlying rocks.
- Once formed, metamorphic rocks do not revert back to the original minerals at Earth's surface. There is not sufficient temperature, pressure, fluid, or time to promote the necessary reverse reactions.

6.6 How Do We Know . . . How to Determine the Stability of Minerals?

PICTURE THE PROBLEM
Under What Conditions Are Metamorphic Minerals Stable?
How do geologists know what minerals are stable under specific metamorphic conditions? How high does the temperature have to be for muscovite mica to dehydrate? At what pressure does calcite transform to aragonite? Under what conditions of temperature, pressure, and fluid composition does olivine metamorphose to hydrous serpentine and the carbonate mineral magnesite? These mineral reactions served as examples of metamorphic changes in the previous three sections, but how are these reactions known to occur?

Geologists infer the reactions from observations, such as the microscopic view of a metamorphic rock shown in **Figure 6.15**. One mineral, called sillimanite, has grown within another mineral grain of andalusite. This observation suggests that the andalusite grew first in the metamorphic rock. Then the andalusite stopped growing and sillimanite formed, at least partly by metamorphic reactions that destroyed andalusite. What caused andalusite to become unstable while sillimanite became stable? Furthermore, if the necessary variations in temperature, pressure, or fluid composition can be determined, then it is possible to reconstruct the history of changing metamorphic conditions that produced this rock.

EXPERIMENTAL SETUP

How Are Metamorphic Conditions Reproduced in the Lab?

Geologists explore metamorphic mineral stability by conducting laboratory experiments. You readily understand the stability conditions of liquid water at Earth's surface (Figure 6.6). On winter days when temperatures descend below 0°C, liquid water freezes to ice. Heating water on a stove to temperatures above about 100°C causes transformation of liquid to water-vapor gas. Full reproduction of the stability relationships represented in Figure 6.6 requires laboratory facilities that produce pressures higher or lower than those observed on Earth's surface. For each chosen pressure, the temperature is varied to see when liquid water freezes or boils. Geologists similarly investigate mineral stability in laboratory experiments by progressively changing temperature, pressure, and fluid content to simulate metamorphic conditions within Earth. The approach resembles Tuttle and Bowen's experiments to determine conditions required to melt granite (Section 4.5).

The changing mineral stability captured in the rock shown in Figure 6.15 was explained by laboratory experiments conducted in the 1970s by Michael Holdaway at Southern Methodist University, in Dallas, Texas. Andalusite, sillimanite, and a third mineral, kyanite, are polymorph minerals. All three minerals share the same chemical composition, Al_2SiO_5, but have different atomic structures. These aluminum-silicate polymorphs are common in Al-rich metamorphic rocks, so determining their stability conditions provides tremendous insights into the temperature and pressure conditions that rocks experience during metamorphism.

It is also important for the experiments that these minerals have the identical chemical composition and do not contain water in their crystal structures. These features make their reaction relationships simple compared to chemical reactions involving multiple minerals of different composition and some including water. Temperature and pressure are the only laboratory conditions that need measurement to determine the stability of the three minerals.

Figure 6.16 illustrates an experimental setup for determining the stability of minerals at different pressures and temperatures. A heated pressure vessel encloses a sample capsule containing the reaction components, including water to speed up the reaction. Temperature and pressure are changed in a controlled manner and measured.

To explore the stability of andalusite compared to sillimanite, Holdaway placed a single andalusite crystal of known mass into the sample capsule and surrounded it with ground-up sillimanite. After heating the capsule at a measured temperature and pressure for several weeks, he turned off the heater. The removal of heat energy stops whatever reactions may

▲ **Figure 6.15 Microscopic evidence of changing metamorphic conditions.**
In this metamorphic rock, the mineral sillimanite crystallized while the mineral andalusite dissolved, so that the sillimanite replaced the andalusite. This observation reveals that the metamorphic conditions changed from circumstances where andalusite is stable to conditions where sillimanite is stable. Polarized lighting produces unnatural colors in this microscope photo. Both minerals are actually tan to brown in color.

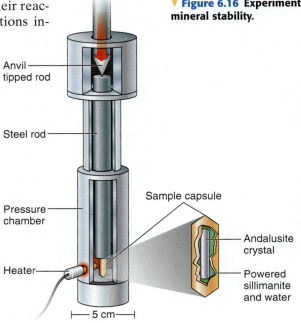

▼ **Figure 6.16 Experimental apparatus for determining mineral stability.**

Mineral-stability studies use a cylindrical pressure chamber.

A capsule containing the mineral samples is placed at the bottom of the chamber. A steel rod fills the remainder of the chamber above the sample capsule.

The top of the chamber closes by turning a nut that lowers an anvil tip onto the steel rod to apply a measured pressure against the sample capsule.

A heater heats the chamber to the temperature required for the experiment.

Anvil tipped rod

Steel rod

Pressure chamber

Sample capsule

Andalusite crystal

Heater

Powered sillimanite and water

⊢ 5 cm ⊣

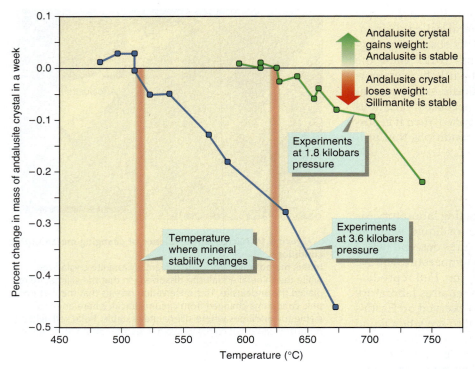

▲ **Figure 6.17 Visualize the experimental results.**
Andalusite is stable at temperatures and pressures where the andalusite crystal grew during the experiment. Conditions where the andalusite crystal lost weight reveal where andalusite is unstable and sillimanite is stable instead. The temperature where mineral stability changes is different at different pressures.

occur in the capsule. Then Holdaway removed and weighed the andalusite crystal.

The mass of the andalusite reveals mineral stability. If andalusite mass increased during the reaction, then andalusite was stable and sillimanite was not. The instability of sillimanite is inferred because the only source of aluminum and silica for andalusite growth is the breakdown of unstable sillimanite. On the other hand, if the mass of the andalusite crystal decreased, then andalusite was not stable and sillimanite was the stable polymorph of Al_2SiO_5.

VISUALIZE THE RESULTS
At What Temperatures and Pressures Are Aluminum Silicate Minerals Stable?

Figure 6.17 graphically illustrates the results of several of Holdaway's experiments. The changing mass of the starting andalusite crystal reveals whether andalusite or sillimanite are the stable mineral at the measured temperature and pressure. At the relatively low pressure of 1.8 kilobars (equivalent to a depth of about 6 kilometers below Earth's surface), andalusite is stable to a temperature of about 625°C. Double the pressure to 3.6 kilobars, however, and andalusite is stable only to about 520°C. Different experiments yielded slightly different results, so there is about 10–20° uncertainty in the temperatures that separate the stability of sillimanite and andalusite at different pressures.

Figure 6.18 summarizes the results of many similar experiments, including experiments with kyanite. Holdaway's results, along with those of other geologists, outline the stability conditions of the three aluminum-silicate minerals in terms of the experimentally varied temperature and pressure. The experiments reveal that kyanite is more stable at higher pressure and lower temperature than andalusite or sillimanite; that sillimanite is more stable at higher temperature than either kyanite or andalusite; and that andalusite is more stable at lower pressure with a variable range in temperature.

INSIGHTS
How Do Lab Results Increase Understanding of Metamorphism?
What do the laboratory results mean for interpreting real metamorphic rocks? Take another look at Figure 6.15 and examine the graph in Figure 6.18. Two conclusions emerge:

1. The metamorphic reaction observed in Figure 6.15 requires an increase in temperature, or an increase in pressure, or both, in order for andalusite to become unstable and sillimanite to be stable. The rock, therefore, experienced an increase in metamorphic grade.
2. The metamorphic temperature had to exceed 500°C, which is the lowest experimental temperature where sillimanite is stable.

Nearly all metamorphic rocks contain more than one mineral. A variety of experimental results defines the stability conditions for most of the minerals that are commonly seen. By combining the stability conditions for all coexisting minerals in a rock, it is possible to interpret the temperature of metamorphism within 50°C and the pressure to within 0.5 kilobar. Knowledge of mineral stability, therefore, is critical to reconstructing metamorphic conditions.

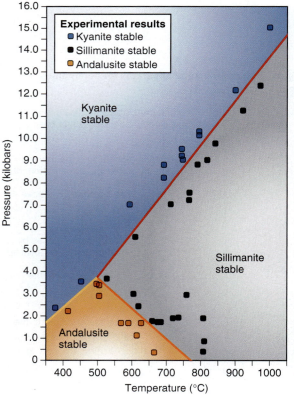

Each dot on this graph represents an experimental result that indicates which of the three aluminum-silicate minerals is stable at the indicated temperature and pressure.

Lines drawn between the data points separate the graph into three areas representing the ranges of temperatures and pressures where kyanite, sillimanite, and andalusite are each stable.

◄ **Figure 6.18 Diagram of aluminum-silicate mineral stability.**

Putting It Together–How Do We Know . . . How to Determine the Stability of Minerals?

- Many minerals are stable over limited ranges in temperature and pressure. Changes in these two variables cause a mineral to react to form a new mineral or minerals.
- Laboratory experiments conducted with different minerals and at measured temperatures and pressures reveal the conditions where metamorphic minerals are stable.

6.7 What Were the Conditions of Metamorphism?

Once geologists recognize from recrystallization, foliation, or presence of metamorphic minerals that a rock is metamorphic, it is natural to wonder how strongly it has been metamorphosed. Geologists want to know what temperature and pressure the rock experienced so that they can reconstruct how deep it once was in Earth and how it formed. If the rock contains one or more of andalusite, kyanite, and sillimanite, then the data graphed in Figure 6.18 answer that question. Similar stability-field graphs documenting metamorphic reactions have been made for most of the commonly observed metamorphic minerals.

Using Index Minerals

Minerals used to estimate the pressure and temperature conditions of metamorphic rock formation are known as **index minerals**. Geologists use many field observations in conjunction with experiments on mineral stability fields to develop a generalized guide to minerals that represent increasing grades of metamorphism.

   Figure 6.19 summarizes the relationship between the minerals present in a rock and metamorphic grade. Index minerals reveal metamorphic grade. Only minerals with limited ranges of stability serve as the key index minerals. Other minerals, such as quartz and feldspar, are stable over large temperature and pressure ranges and do not indicate metamorphic grade (Figure 6.19).

▼ **Figure 6.19** Using index minerals.
Metamorphic grade corresponds to the minerals observed together in metamorphic rocks and the temperatures and pressures for mineral stability determined by experiments. For example, chlorite, a green iron- and magnesium-rich mica, indicates low-grade metamorphism, whereas sillimanite indicates high-grade conditions. Each of these minerals, therefore, is an index of metamorphic grade. Some minerals, such as quartz, feldspar, and muscovite, are not good indicators of grade because they are stable over a wide range of metamorphic grades. This diagram emphasizes minerals whose stability is affected by both pressure and temperature. Other stability relationships are determined for minerals whose stability is mostly a function of temperature or pressure alone.

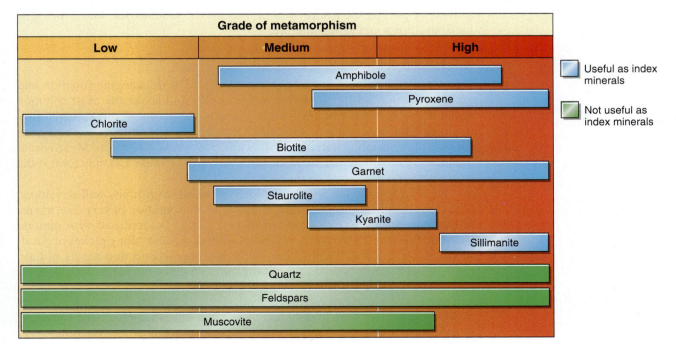

Consider again the rock you picked up on your road trip and refer to Figure 6.19. The presence of muscovite is not very diagnostic of the metamorphic conditions. Garnet reveals that the rock could be medium or high grade. Staurolite is the most useful index mineral in this rock because its presence indicates metamorphism toward the upper limits of medium-grade conditions. The stability conditions of staurolite determined in experiments suggest that metamorphism reached a minimum temperature of about 550°C and pressure of 2 to 7 kilobars. Keep in mind that pressure also correlates with depth, so if the pressure range is known, you also know the depth in the crust or mantle where these processes were occurring. For your rock metamorphosed at a pressure of 2 to 7 kilobars, the equivalent depth is approximately 6.5 to 23 kilometers.

> ### *Putting It Together—What Were the Conditions of Metamorphism?*
> - Index minerals reveal metamorphic temperature and pressure because they are stable over limited ranges of temperature, pressure, or both.

6.8 How Are Metamorphic Rocks Classified?

Observed variations in the mineral content and texture of metamorphic rocks reveal the temperature and pressure of metamorphism and the composition of parent rocks and reactive fluids. Composition and texture, therefore, are useful criteria for classifying metamorphic rocks. In metamorphic rocks the primary textural attributes are presence or absence of foliation and mineral grain size.

Using Composition and Texture to Classify Metamorphic Rocks

Figure 6.20 illustrates how to use composition and texture to classify and name metamorphic rocks. The first distinction is whether the rock contains a well-developed foliation. If it does, then the rock is named primarily on the basis of the type of foliation and whether the minerals are too fine to distinguish with the naked eye or are coarse and readily visible. Notice that each foliated rock type potentially contains many different minerals, so mineral content is not distinctive. For rocks with no foliation or an indistinct foliation, the mineral content is the primary basis for naming the rock. Some low- to medium-grade metamorphic rocks retain many of their original features and are simply described by putting the prefix *meta* in front of the parent rock name, such as a metabasalt or a metaconglomerate.

Rocks with Foliation

Figure 6.21 illustrates the rock names associated with changing foliation and grain size produced by increasing metamorphism of shale, the most common sedimentary rock. At progressively higher temperature and pressure, the predominant clay minerals present in shale at relatively low temperature and pressure become unstable. Initially the clay minerals convert to micas. Muscovite is the most abundant low-grade mica, but chlorite (see Table 2.2) may form if the original clays contain iron or magnesium.

The first-formed metamorphic rock is **slate**, which exhibits well-developed rock cleavage planes produced from the parallel orientation of very fine, microscopic mica grains. Rock-cleavage foliation, not to be confused with mineral cleavage (Section 2.1), is the preferential splitting of rock along planes of parallel microscopic micas. The property of slate to break into thin sheets of hard rock makes it ideally suited as a roofing material and even for making pool tables (the slate tabletop is covered with felt).

Increasing metamorphic grade leads next to **phyllite**. The mica grains in phyllite are coarser than in slate and generate a silky sheen from the reflection of light from the parallel mica cleavage surfaces that define the foliation.

Texture	Typical minerals	Metamorphic rock	Parent rock
Foliated rocks			
Fine-grained with minerals aligned along planes (rock cleavage)	Clay, muscovite, biotite, chlorite, quartz	Slate	Shale, tuff
Fine-grained with minerals aligned along planes; distinctive sheen	Muscovite, biotite, chlorite, quartz	Phyllite	Shale, tuff
Coarse-grained, minerals have a distinct layering; shiny	Muscovite, biotite, chlorite, amphibole, quartz, garnet, talc, staurolite, kyanite, andalusite, sillimanite, graphite	Schist	Shale, igneous rocks
Compositional banding of minerals along parallel, commonly contorted, bands	Muscovite, biotite, amphibole, quartz, feldspar, garnet, talc, staurolite, kyanite, andalusite, sillimanite	Gneiss	Shale, igneous rocks

▲ **Figure 6.20 Classification of metamorphic rocks.**
Texture and minerals content form the basis of this classification scheme of metamorphic rocks. Foliated rocks are divided primarily on texture whereas the nonfoliated rocks are named primarily on composition (minerals present).

(continued on next page)

Schist forms at still higher temperature and pressure. This rock has even larger mica grains that are strongly parallel, easily seen with the naked eye, and reflect light to make for a shiny rock. Your road-trip sample is schist (Figure 6.1). The mica grains commonly surround scattered, well-formed crystals of minerals such as garnet, staurolite, and kyanite. If chlorite was present in the slate and phyllite, then it reacts to form biotite at the grade where schist forms. If the parent rock was a black, organic-rich shale, then the schist may contain graphite produced by the metamorphism of organic matter.

Gneiss (pronounced like "nice") forms when temperature and pressure rise to high grade. Gneiss is defined by its characteristic foliation of parallel

Figure 6.20 (*continued*)

Texture	Typical minerals	Metamorphic rock	Parent rock

Non-foliated rocks or weakly foliated rocks

Texture	Typical minerals	Metamorphic rock	Parent rock
Mosaic of coarse grains	Calcite, dolomite	Marble	Limestone, dolostone
Mosaic of coarse grains	Quartz	Quartzite	Quartz, sandstone
Mosaic of fine, microscopic grains	Quartz, feldspar, muscovite, biotite, garnet, andalusite	Hornfels	Any fine-grained rock
Fine-grained, commonly fibrous or greasy	Serpentine	Serpentinite	Peridotite
Shiny, curving broken surfaces	Carbon (not a mineral)	Anthracite coal	Coal
Fine-grained	Chlorite, amphibole, feldspar, quartz	Greenstone	Mafic igneous rocks
Coarse-grained; minerals commonly aligned	Amphibole, feldspar, garnet, quartz	Amphibolite	Mafic and intermediate igneous rocks
Mosaic of coarse grains	Pyroxene, garnet	Eclogite	Mafic igneous rocks

▶ Figure 6.21 **Progressive metamorphism of shale.**
Visible changes in rock texture and mineral content take place as
the grade of metamorphism increases. These are some of the
possible changes observed in the progressive metamorphism of
shale to gneiss.

compositional layers of light-colored (e.g., quartz, feldspar) and dark-colored (e.g., biotite, amphibole, pyroxene, garnet) minerals. At the highest metamorphic temperatures gneisses lack mica or amphibole because these water-bearing minerals break down by dehydration reactions.

If the high-grade temperature, pressure, and fluid conditions are appropriate for melting to begin, then the resulting rock contains the textures of both metamorphic and igneous rock. An example of such a rock, called **migmatite**, is shown in **Figure 6.22**. The migmatite resembles gneiss except that the light-colored bands have the igneous-crystallization texture of granite, whereas the dark layers reveal metamorphic crystal growth and recrystallization.

The dominant coarse-grained minerals in foliated rocks are commonly used as modifiers when naming the rock. These adjectives are particularly useful if referring to index minerals, because the name of the rock then conveys information about grade independent of the type of foliation. For example, your schist has staurolite and garnet surrounded by muscovite. You name this rock staurolite-garnet-muscovite schist, indicating the dominant component is muscovite, followed by garnet and staurolite.

Rocks without Foliation

If the rock is nonfoliated or only weakly foliated, then the mineral content is more important for naming the rock. For instance, the rock composed mostly of metamorphically recrystallized calcite is **marble**, whereas **quartzite** consists of metamorphically recrystallized quartz. These nonfoliated rocks are distinct from limestone and sandstone because they have metamorphic recrystallization textures, such as large grain size, crystals meeting along straight edges (Figure 6.9), or both. Other metamorphic rocks composed primarily of a single mineral include **amphibolite** (amphibole minerals with plagioclase feldspar) and **serpentinite** (typically composed almost entirely of serpentine).

Organic compounds experience metamorphic reactions. Coal contains few minerals, but as temperature increases, hydrogen- and nitrogen-rich organic compounds break down and escape as gas from the coal, converting the organic matter to 90 percent or more pure carbon. Pyrite, commonly found in coal (Section 5.6), also breaks down, and the sulfur is likewise released as gaseous compounds. Metamorphosed coal, called **anthracite**, is the highest-quality

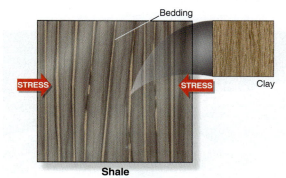

Bedding

Shale

Shale, composed of compacted clay-size, clay minerals. A sedimentary rock.

Clay

Low grade

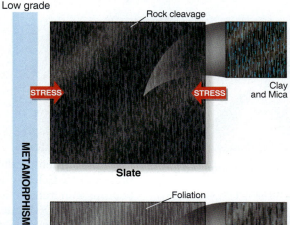

Rock cleavage

Slate

Clay and Mica

New minerals such as muscovite and chlorite start to form. This produces fine banding or cleavage in the rock that is perpendicular to the maximum direction of normal stress.

Foliation

Phyllite

Mica

Clay completely converted to muscovite and chlorite. Feldspar and biotite starts to form. The micas also start to grow in size and are barely visible with the naked eye, giving the rock a shiny appearance. Foliation is perpendicular to the principal direction of stress.

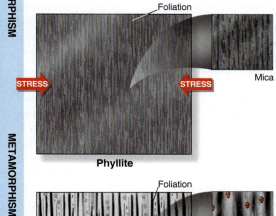

Foliation

Schist

Mica and Garnet

Micas recrystallize to large, easily seen crystals. New minerals such as garnet and staurolite form. Foliation is perpendicular to the principal direction of stress.

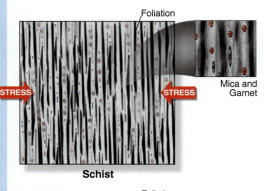

Foliation

Gneiss

Quartz and Feldspar

Biotite and Amphibole

Minerals recrystallize and segregate into bands or layers. The foliation bands typically are feldspar and quartz, alternating with bands of darker biotite or amphibole. Minerals are coarse-grained and show preferred orientations perpendicular to the principal direction of stress.

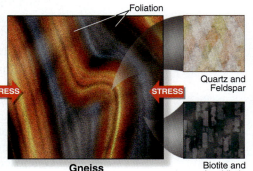

High grade

METAMORPHISM

STRESS STRESS

▲ **Figure 6.22** **A rock at the transition from metamorphism to melting.**
Migmatites possess the texture of both a metamorphic and igneous rock, as seen in this outcrop near Tucson, Arizona. When this gneiss started to melt to form granitic magma, the melt accumulated in bands parallel to the foliation. Both the melt bands and foliated minerals were further contorted by metamorphic stress, resulting in light-colored granite bands within the gneiss.

coal because it burns with the most heat, and it contains almost none of the sulfur compounds that pollute the atmosphere when sedimentary coal burns. At both high pressure and temperature, the organic carbon converts entirely to graphite, which, being a platy mineral like mica, forms a dark, greasy schist.

Hornfels is the name applied to any very hard, nonfoliated, metamorphic rock composed mostly or entirely of microscopically small crystals, regardless of composition. Typically, the newly formed metamorphic minerals are those whose origin more closely relates to changes in temperature rather than changes in pressure. The insignificance of pressure in the formation of hornfels is also suggested by the lack of foliation.

Other nonfoliated or weakly foliated metamorphic rocks containing many minerals include **greenstone** and **eclogite** formed by metamorphism of mafic igneous rocks. Greenstone is a very descriptive name for a metamorphic rock that contains abundant green minerals. Usually the iron-magnesium mica, chlorite, is the primary constituent, although green amphibole, feldspar, and quartz are usually present, too. Eclogite is a very high-grade metamorphic rock that lacks water-bearing minerals. It is dominated by garnet and pyroxene, sometimes with minor quartz. Eclogite is denser than peridotite. When basaltic oceanic crust subducts at convergent plate boundaries, it metamorphoses to dense eclogite, which drags the subducting plate deeper into the mantle.

Putting It Together–*How Are Metamorphic Rocks Classified?*

■ The classification of metamorphic rocks is based on texture (foliation and grain size) and mineral content. The names of foliated rocks are based on texture with the addition of modifying adjectives that describe the index minerals that are present. Most nonfoliated rocks are named on the basis of mineral content.

6.9 What Was the Rock before It Was Metamorphosed?

You now know how to name a metamorphic rock and how to estimate the grade of metamorphism that occurred. You also know that all metamorphic rocks start out as something else, so it would be useful to know how to study a metamorphic rock, to work back and determine what the parent rock was. This can be very challenging. If a lot of fluid moves through the rock during metamorphism and transports ions to and away from the reaction sites, then the end result is a metamorphic rock containing minerals that are inconsistent with the starting bulk composition of the parent rock.

Metamorphosed Sedimentary Rocks

Figure 6.23 illustrates some possible metamorphic pathways for different parent sedimentary and igneous rocks. The illustrated rock transformations assume that water participates primarily to make water-bearing minerals and does not substantially change the bulk composition of the rock. The metamorphic changes for shale are illustrated in Figure 6.21.

Quartz sandstone and limestone generally metamorphose to nonfoliated rocks composed of a single mineral because a single, nonplaty mineral dominates the rock. Arkose or lithic sandstone containing significant nonquartz grains may form micas at the expense of feldspar and other reactive minerals

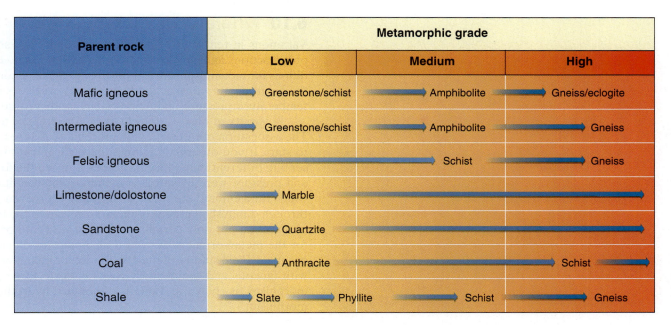

Parent rock	Metamorphic grade		
	Low	**Medium**	**High**
Mafic igneous	Greenstone/schist	Amphibolite	Gneiss/eclogite
Intermediate igneous	Greenstone/schist	Amphibolite	Gneiss
Felsic igneous		Schist	Gneiss
Limestone/dolostone	Marble		
Sandstone	Quartzite		
Coal	Anthracite		Schist
Shale	Slate Phyllite	Schist	Gneiss

▲ **Figure 6.23 Possible metamorphic pathways for parent rocks.**
This diagram outlines some of the possible metamorphic rocks that form as various parent rocks metamorphose to different grades. Some parent rocks, such as limestone, produce the same metamorphic product, such as marble, regardless of the grade of metamorphism. Other parent rocks, such as the mafic igneous rock basalt, produce a variety of metamorphic rocks depending on the minerals that are stable at different metamorphic grades.

during low-grade metamorphism in the presence of water. The resulting rocks may be weakly foliated mica-rich quartzite or even schist if the abundance of quartz in the sandstone is minor.

Metamorphosed Igneous Rocks

Low- to medium-grade metamorphism of mafic and intermediate igneous rocks in the presence of water generates dark-colored metamorphic rocks with abundant iron- and magnesium-bearing mica, like chlorite or biotite. If foliated, these rocks are chlorite or biotite schists, and if not foliated they are greenstones. At higher metamorphic grade hornblende becomes increasingly stable and forms amphibolites. At highest grade the metamorphic product is gneiss or eclogite (Figure 6.23).

Felsic igneous rocks metamorphose to schist if water is available to convert feldspar to muscovite. Otherwise, there may not be substantial textural or mineralogical changes, especially for coarse-grained granite, until high-grade conditions cause segregation of minerals into bands to form gneiss.

Gneiss formed from metamorphism of igneous rock may be difficult to distinguish from a highly metamorphosed shale. Igneous gneisses are less likely, however, to contain the very aluminum-rich minerals such as muscovite, staurolite, corundum, kyanite, and sillimanite formed by metamorphism of clay.

Putting It Together—What Was the Rock before It Was Metamorphosed?

■ The parent rock is interpreted from the mineral composition of the metamorphic rock, so long as fluids have not extensively removed or delivered chemical components during metamorphism.

■ The major elements found in the constituent metamorphic minerals, such as Fe and Mg, are similar to those found in the minerals from the parent rock. Felsic igneous rocks, therefore, metamorphose to rocks containing abundant quartz, feldspar, and muscovite, whereas mafic igneous rocks contain abundant amphibole, biotite, and pyroxene.

■ If the parent rock contains mostly one mineral, such as in limestone and quartz sandstone, the metamorphic equivalent may also contain mostly the same mineral, just recrystallized into a metamorphic texture, as is seen in marble and quartzite.

6.10 Where Does Metamorphism Occur?

Variations in temperature, pressure, and fluid availability determine metamorphic reactions, and increased heat and pressure inside Earth provide suitable conditions for metamorphism. Is it possible to make a more specific interpretation of the environment of metamorphism? What processes account for the changes in temperature, pressure, and fluid availability that cause the metamorphism?

A variety of observations help us interpret the metamorphic environment. The association of different metamorphic rock types in the field is a good place to start. If you observe horizontal or vertical transitions between rocks representing different metamorphic grades, you see the location where temperature and pressure were elevated to metamorphic conditions or the direction from which fluid entered the rock. In many cases it is even possible to follow the rock through diminished grades of metamorphism to the unaltered parent rock. This observation not only permits straightforward interpretation of the parent rock for the metamorphic rocks but it also helps to determine how metamorphism occurred. Field studies of the distribution of metamorphic rocks and association with igneous and sedimentary rocks indicate three general settings for metamorphism, as shown in **Table 6.1**—contact, hydrothermal, and regional metamorphism.

Contact Metamorphism

Contact metamorphism occurs near the contacts of igneous intrusions, as depicted by **Figure 6.24**a, or less extensively beneath erupted lava flows. Heat causes the metamorphism, which is restricted to the region adjacent to the magma or lava. The amount of heat introduced by the magma or lava and the amount of fluid movement determines the volume of rock that is metamorphosed. The heat associated with lava flows is trivial compared to that associated with large subsurface intrusions, so it is more useful to focus on metamorphism adjacent to intrusions.

Heat is the dominant factor in contact metamorphism, so the resulting rocks are generally nonfoliated. Hornfels is the most common contact-metamorphic rock and forms adjacent to magma bodies that intruded within a few kilometers of the surface into sedimentary and volcanic rocks. Where limestone and dolostone are interbedded with chert or shale, the heat from the intrusion causes reactions between the carbonate and silicate minerals and forms

TABLE 6.1	The General Types and Settings of Metamorphism		
Type of Metamorphism	How Metamorphism Occurs	Where Metamorphism Occurs	Characteristics of Metamorphic Rocks
Contact	Increase in temperature and, in some cases, migration of hot fluids	In tectonically active areas adjacent to igneous intrusions and, to a lesser extent, below lava flows. Metamorphism extends over distances of less than a meter to as much as 20 kilometers from the igneous body, depending on size, temperature, and composition of the intrusion (or lava flow) and the composition of the surrounding rocks.	Nonfoliated rocks, of which hornfels is typical, form around near-surface intrusions. Contact metamorphic zones around deeper intrusions are sometimes recognized by unusually high temperature minerals within regional-metamorphic rocks.
Hydrothermal	Large-volume interaction of hot fluid with rocks	Where water is abundant and temperature is high; especially below mid-ocean ridges and near some igneous intrusions. Metamorphism may affect areas as small as hundreds of square meters to more than hundreds of square kilometers.	Nonfoliated rocks containing water-rich minerals, including micas and amphiboles. Metamorphic rocks commonly occur along fractures that cut across other rocks. Commonly contain economically valuable sulfide minerals.
Regional	Rock transformation over large regions affected by high tectonic stress and geothermal gradients typically higher or lower than average	In rocks near convergent plate boundaries where magma is formed and mountain building takes place. Regional metamorphic rocks typically cover tens of thousands of square kilometers, or more.	Foliated rocks, except where parent rock types do not metamorphose to platy minerals.

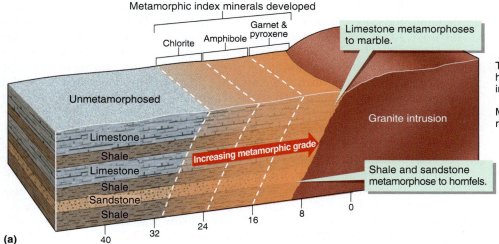

This diagram illustrates the change in index minerals within hornfels and marble that occurs because of temperature increase in close proximity to an igneous intrusion.

Metamorphism occurs when the magma was intruded and the rocks are later exposed by erosion.

(a)

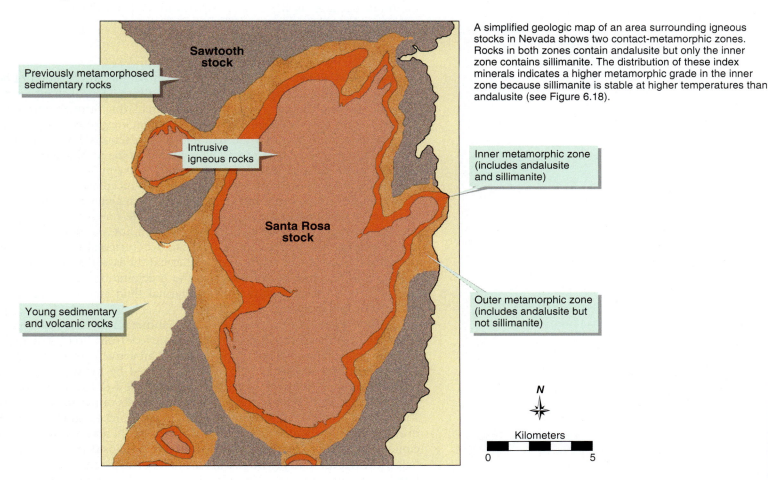

A simplified geologic map of an area surrounding igneous stocks in Nevada shows two contact-metamorphic zones. Rocks in both zones contain andalusite but only the inner zone contains sillimanite. The distribution of these index minerals indicates a higher metamorphic grade in the inner zone because sillimanite is stable at higher temperatures than andalusite (see Figure 6.18).

(b)

▲ **Figure 6.24** **Contact metamorphism.**

coarse-grained rocks composed of calcium and magnesium silicates, such as garnet, talc, and pyroxene. At greater depths the surrounding rocks are already metamorphosed and the contact metamorphism simply causes higher-temperature metamorphic minerals to form near the intrusion than are found farther from where the magma solidified.

Metamorphic grade is highest close to the contact with the igneous rock and decreases away from the intrusion. An observation of the decreasing grade of metamorphism away from an intrusion indicates that heat from the intruding magma body caused the metamorphism. This is the most diagnostic field criterion for recognizing contact metamorphism.

A field example of contact metamorphism superimposed on previously metamorphosed rocks is illustrated in Figure 6.24b. This geologic map shows intrusive stocks in Nevada surrounded by concentric zones of contact metamorphic rocks. The close association of these zones with the outlines of the intrusions indicates a relationship between the igneous intrusion and the metamorphism. When the magma intruded, it cooled by conveying heat to the surrounding rock. The heat caused the temperature in the surrounding rock to rise considerably. The rocks now show a progressive series of changes in the minerals as they are examined outward from the igneous contact. The new mineral constituents closest to the igneous contact are the highest-temperature minerals, and the outer zone contains lower temperature minerals. Two minerals found in the inner zone of the Santa Rosa stock include andalusite and sillimanite, whereas the outer zone has andalusite but no sillimanite. This indicates that the temperature and pressure conditions in the inner zone were stable for both andalusite and sillimanite or that andalusite was reacting to form sillimanite (Figure 6.18). At even greater distance the rocks do not appear to be modified because as the heat dissipates away from the stock, the grade of metamorphism drops.

Hydrothermal Metamorphism

Hydrothermal metamorphism involves the migration and reaction of hot, ion-rich fluids. This metamorphism, therefore, involves chemical change in the rock because of the substantial role of hot water circulating through pore spaces and cracks. The term hydrothermal conveys the equal importance of water and elevated temperature to drive the metamorphic reactions. Hydrothermal metamorphism is commonly associated with contact metamorphism on a local scale but also occurs on a large scale at volcanically active mid-ocean ridges.

Figure 6.25 shows how hydrothermal metamorphism occurs along mid-ocean ridges. Metamorphism occurs where hot magma rises to create ocean floor and circulating seawater is readily available for metamorphic reactions. This causes metamorphism of the newly formed hot seafloor crust. These sites are commonly associated with ore genesis, where metal-rich sulfide minerals precipitate from hot, circulating hydrothermal fluids within and above fissures in the seafloor. **Figure 6.26** illustrates the precipitation of sulfide minerals in seawater where hot metal-rich, hydrothermal fluid emerges from depth and mixes with cold seawater.

A field example of active hydrothermal metamorphism has been studied below the Mid-Atlantic Ridge. Rocks retrieved from holes drilled into the seafloor provide insights into the anatomy of an actively forming seafloor mineral deposit and the rocks below it that formed by hydrothermal metamorphism. **Figure 6.27** illustrates the cross-sectional view of the ore deposit and the seafloor below. The bulk of the ore deposit consists of pyrite and quartz along with sulfide minerals containing valuable copper, nickel, and zinc. The hot hydrothermal fluids, nearing 360°C, mixed with the cold seawater to cause the precipitation of the ore minerals. The hydrothermal fluids originated from magma below the deposit and provided the ions that participated in the chemical reactions to produce the ore minerals. The rocks below the ore are chlorite-rich basalts metamorphosed by hydrothermal fluids flowing through the cracks and fissures of the seafloor. Further hydrothermal metamorphism converted some of the chlorite-rich basalt into a sodium-rich mica. The sodium is derived from reactions of minerals with the seawater. Below the hydrothermally metamorphosed basalt is original, unaltered basalt.

Regional Metamorphism

Regional metamorphism refers to metamorphism over large areas and to rock volumes not related to specific igneous intrusions or sources of hydrothermal fluid. This metamorphism typically relates to the formation of mountain belts

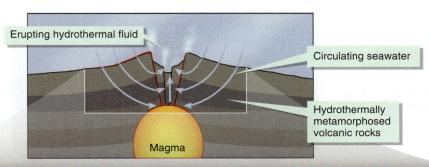

Hydrothermal metamorphism occurs where
magma heats circulating seawater. Magmatic gases also dissolve in the hot fluid. The
fluid provides sources of ions and pathways
for ions to travel and the magma provides
heat to drive the metamorphic reactions.

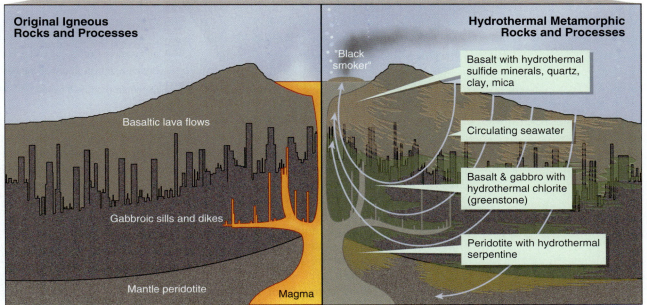

along subduction zones. Regional metamorphism involves progressively increasing temperature- and pressure-driven mineralogical and textural changes
to rock. It encompasses large volumes of continental crust, and in some cases
oceanic crust and mantle. This kind of metamorphism occurs under the influence of tectonic stresses in addition to the pressure exerted by overlying rock.
These stresses include near-horizontal normal and shear stresses associated
with the horizontal convergence of plates at subduction zones. As a result,
regional-metamorphic rocks are almost always foliated, but the foliation may
be vertical or at some intermediate angle between horizontal and vertical depending on the orientation of stresses. The heat driving the metamorphic reactions is enhanced in many cases by the rise of magma above subduction zones.

The tectonic setting for regional metamorphism along a convergent plate
boundary is illustrated in **Figure 6.28**. This diagram includes a schematic view of
the thermal structure of a subduction zone between two lithospheric plates and
highlights regions of various temperature and pressure conditions of metamorphism. The lithospheric plate that subducts into the mantle stays relatively cool,

▶ **Figure 6.26** Ore deposits in the making.
This photograph taken from a research submarine shows hydrothermal fluid emerging on the
seafloor. Geologists call this feature a "black smoker," but the black cloud is not smoke. The
"cloud" is fine particles of pyrite, galena, and sphalerite (iron, lead, and zinc sulfides). The mineral components dissolve in the high-temperature (>300°C) hydrothermal fluid below the
seafloor but precipitate in contact with cold seawater. Metal sulfides fill fractures and pore
spaces in the seafloor basalt and form the chimney through which the hydrothermal fluid
erupts. Similar hydrothermal deposits are important metal ores where ancient seafloor is uplifted onto the margins of continents.

▶ **Figure 6.27 Hydrothermal metamorphism of the seafloor with an ore deposit.**
This is a cross section of the rocks encountered by drilling through a black smoker along the Mid-Atlantic Ridge. Hydrothermal solutions mixed with seawater to form deposits of sulfide ore minerals and quartz. Below these deposits are hydrothermally metamorphosed basalts rich in sodium-rich mica and chlorite, which form as hydrothermal solutions and seawater invade fractures through the rock.

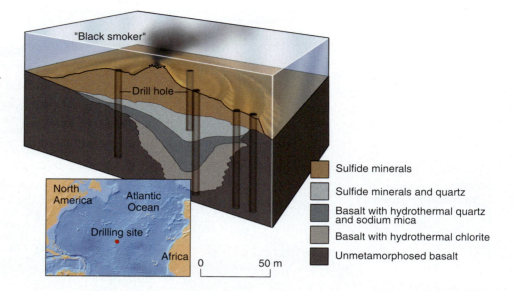

even at great depth. This produces a region of relatively low-temperature and high-pressure metamorphism. Magma forms deeper in the subduction zone and results from melting the mantle above the subducting plate. Here, water is a product of metamorphic dehydration reactions in the subducted plate and induces melting of the hot mantle (see Section 4.6). The rising magma close to the surface produces high-temperature, low-pressure metamorphic conditions in the crust above the subducted slab. At greater depth below the volcanic arc, the pressure and temperature are high enough for high-temperature and high-pressure metamorphism.

Japan offers an example of a subduction-zone setting with contrasting types of regional metamorphism. **Figure 6.29** shows the pairing of belts of low-

▶ **Figure 6.28 Tectonic setting for regional metamorphism at a convergent plate boundary.**
This cross section illustrates the different temperature and pressure conditions that exist near a subduction zone. The 300°C and 600°C temperature curves indicate cooler conditions near the cold subducting slab and warmer conditions under the volcanic arc where hot magma rises toward the surface. Relatively low temperature but high pressure metamorphism occurs near the subducted lithosphere. As the subducted slab releases water by metamorphic dehydration reactions, magma forms and rises into the overriding plate. The heat from the magma causes high-temperature, low-pressure metamorphism in the crust. At greater depths and higher pressures, below the volcanic arc, conditions are such that high-temperature and high-pressure metamorphism takes place. Erosion of uplifted rocks near convergent plate boundaries exposes these different types of metamorphic rocks.

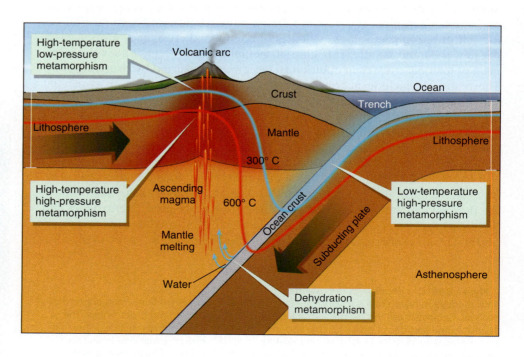

temperature, high-pressure and high-temperature, low-pressure metamorphism. Magma rising in the overriding plate provides high temperatures with low-pressure, near-surface conditions. Closer to the cold subducting plate, the metamorphic rocks contain minerals stable at high pressure applied by the plate-boundary stresses but also at relatively low temperature.

Figure 6.30 summarizes the different metamorphic environment conditions. Different types of metamorphism relate to the pathways of changing temperature and pressure that rocks experience.

Putting It Together–*Where Does Metamorphism Occur?*

■ There are three primary types of metamorphism—contact, hydrothermal, and regional.

■ Contact metamorphism occurs along the contacts of intruding igneous bodies. It is local and driven primarily by heat. Metamorphism is most intense closest to the contact with the magma and decreases away from the intrusion.

■ Hydrothermal metamorphism on a local scale involves hot fluids derived from intruding igneous bodies or infiltrating ground water that circulate through the rock. This metamorphism occurs on a large scale at divergent plate boundaries, where large volumes of circulating hot seawater promote metamorphism and ore production.

■ Regional metamorphism involves increasing temperature and pressure over very large volumes of crust and produces extensive tracts of foliated rocks. It typically occurs at convergent plate boundaries.

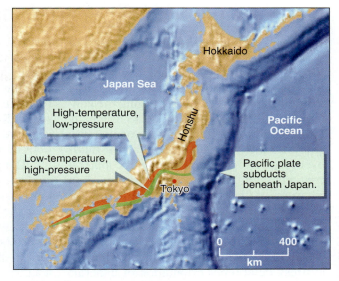

▲ **Figure 6.29** **Paired metamorphic belt in Japan.** This map shows the location of paired metamorphic belts near the subduction zone beneath Japan. These belts formed at a subduction zone with temperature and pressure conditions similar to those illustrated in Figure 6.28 and were later exposed by erosion. The rocks indicate a condition combining both low-temperature and high-pressure conditions near the subduction zone and high-temperature, low-pressure conditions farther from the subduction zone.

EXTENSION MODULE 6.1

Metamorphic Isograds, Zones, and Facies.
Learn how geologists use metamorphic minerals and the chemical reactions that form them to determine the metamorphic history of a region.

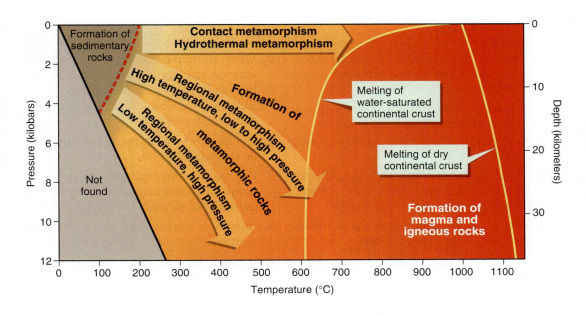

◀ **Figure 6.30** **Summarizing metamorphic environments.**
The metamorphic temperature and pressure environments are between the conditions where sedimentary rocks and magma form. The paths of changing temperature and pressure that rocks experience during progressive metamorphism further distinguish the metamorphic environments. Most observed examples of contact and hydrothermal metamorphism occur near the surface under conditions of low pressure and increasing temperature. Regional metamorphism follows paths of both increasing temperature and pressure.

Where Are You and Where Are You Going?

You have just been introduced to the last of the three major rock groups found on Earth. Metamorphic rocks originate from any of the major rock groups, including previously metamorphosed rocks, by changing the conditions from those where the rocks originally formed. Increasing temperature or pressure and introduction of chemically active fluids over long periods of time modify the mineral contents and textures. Temperature is a measure of the heat energy required for metamorphic reactions. Fluids enhance metamorphic reactions by transporting ions in solution. Pressure, which is simply the weight of overlying rocks or directed tectonic stress, determines orientation of minerals in metamorphic rocks and the packing of atoms that define different mineral structures. Texture (foliation and grain size) and composition (minerals present) are used to classify metamorphic rocks. Experiments define the temperature and pressure conditions at which metamorphic minerals are stable. Key metamorphic index minerals reveal the pressure and temperature conditions of metamorphism.

The general types of metamorphism are contact, hydrothermal, and regional metamorphism. Contact metamorphism occurs near the contact of intruding magma and is primarily driven by elevated temperature. Hydrothermal metamorphism involves hot fluids that are derived from intruding magma or from infiltrating ground water and seawater heated by the magma. Regional metamorphism occurs where variations in temperature and pressure affect very large volumes of crust to produce large regions of foliated rocks, commonly associated with mountain building near convergent plate boundaries. Uplift and erosion bring the metamorphic rocks to the surface.

You have examined the three major rock types. The processes forming these rocks provide the *what, why,* and *how* for interpreting Earth history. The remaining question is to determine *when* the processes were active. The next chapter wraps up your study of Earth materials by revealing how rocks and minerals are used to reconstruct the sequence of geologic events and determine when the events took place.

 # Active Art

Forming Foliation. See how the three types of foliation form.

Exposing Metamorphic Rocks. See how metamorphic rocks end up at the surface after forming at great depth.

Extension Modules

Extension Module 6.1: Metamorphic Isograds, Zones, and Facies. Learn how geologists use metamorphic minerals and the chemical reactions that form them to determine the metamorphic history of a region.

Confirm Your Knowledge

1. What is metamorphism?

2. Why are the processes that produce metamorphic rocks not observable in nature?

3. What are the four factors that determine the mineral content and texture of a metamorphic rock?

4. List and describe the three common processes that cause rocks to experience increasing temperature.

5. What are the three ways foliation can form in a metamorphic rock?

6. Describe the role fluids play in the formation of a metamorphic rock.

7. For each of the mineral reactions listed below, determine if the change results in hydration or dehydration.
   - Muscovite and quartz reacts to form sillimanite and potassium feldspar
   - Olivine reacts to form serpentine and magnesite

8. Since diamonds form from graphite at high pressure, why don't they convert back to graphite at Earth's surface?

9. Why are some minerals useful as metamorphic index minerals and some are not?

10. What is mineral stability? How does mineral stability change with temperature and pressure? Give an example of metamorphic alteration at low pressures and at high pressures.

11. Define pressure, stress, and strain. How do they all relate?

12. Explain how metamorphic rocks, which require high temperatures and pressures to form, can be abundant at the surface.

13. Identify a parent rock for each metamorphic rock: slate, marble, gneiss.

14. Explain the type of rocks produced by contact metamorphism and regional metamorphism.

Confirm Your Understanding

1. Write out an answer for each question in the Chapter Outline for the chapter sections assigned by your instructor.

2. For the following pressures and temperatures determine if a rock formed at these conditions would typically be sedimentary, metamorphic or igneous. If metamorphic, determine if the rock would be considered low, medium, or high-grade.

   - Pressure = 1 kb, Temperature = 700 °C
   - Pressure = 2 kb, Temperature = 100 °C
   - Pressure = 3 kb, Temperature = 300 °C
   - Pressure = 4 kb, Temperature = 500 °C
   - Pressure = 6 kb, Temperature = 1100 °C
   - Pressure = 7 kb, Temperature = 1000 °C

3. Rocks exhibit two types of metamorphic changes, a change in mineral content or a change in texture or both. For each of the metamorphic changes listed below determine the type of metamorphic change.

   - Shale metamorphoses into a schist
   - Dolostone metamorphoses near an intrusion to a rock containing calcite and periclase
   - Graphite metamorphoses into diamond
   - Quartz sandstone metamorphoses into quartzite
   - Limestone metamorphoses into marble

4. Based on Figure 6.18 determine if andalusite, kyanite or sillimanite is stable at the following sets of conditions.

   - P = 1 kb, T = 500 °C
   - P = 2 kb, T = 650 °C
   - P = 4 kb, T = 450 °C
   - P = 5 kb, T = 650 °C

5. Classify the following metamorphic rocks into low, medium, or high-grade based on the minerals present in the rock.

   - Quartz, biotite, chlorite
   - Quartz, feldspar, biotite, garnet, sillimanite
   - Quartz, feldspar, biotite, garnet, staurolite
   - Quartz, feldspar, muscovite, biotite, chlorite
   - Quartz, feldspar, muscovite, biotite, garnet
   - Quartz, feldspar, pyroxene

6. What would you call the following metamorphic rocks?

   - Foliated rock with alternating compositional bands rich in quartz and feldspar or biotite, garnet and sillimanite
   - Shiny foliated rock rich in muscovite and biotite, with numerous crystals of coarse-grained garnet
   - Weakly foliated rock consisting entirely of coarse grained recrystallized calcite
   - Weakly foliated rock consisting entirely of fine grained recrystallized quartz
   - Non-foliated green rock consisting of quartz, feldspar, chlorite and amphibole

7. Define recrystallization. Describe the changes a crystal undergoes during recrystallization. How does recrystallization differ from cementation of sedimentary particles? How might you distinguish between them?

8. Would you expect the chemical composition of a metamorphic rock to be dramatically different from the composition of its parent rock? Why?

Key Terms

amphibolite (p. 155)
anthracite (p. 155)
contact metamorphism (p. 158)
eclogite (p. 156)
foliation (p. 143)
gneiss (p. 153)
greenstone (p. 156)
high-grade metamorphism (p. 136)

hornfels (p. 156)
hydrothermal metamorphism (p. 160)
index minerals (p. 151)
low-grade metamorphism (p. 136)
marble (p. 155)
medium-grade metamorphism (p. 136)

metamorphism (p. 136)
migmatite (p. 155)
normal stress (p. 141)
phyllite (p. 152)
quartzite (p. 155)
recrystallization (p. 142)
regional metamorphism (p. 160)
schist (p. 153)
serpentinite (p. 155)

shear stress (p. 141)
slate (p. 152)
strain (p. 141)
stress (p. 141)

7 Earth Materials as Time Keepers

Chapter Outline

Why Study the Ages of Rocks?

KEYS FOR EXPLAINING ANY PROCESS ARE THE SEQUENCE OF steps involved and the time requirements for each step. For example, how useful is a recipe that lists ingredients but does not describe preparation and cooking time, or product-assembly instructions that do not give the order of each step in the process? The previous chapters addressed the geologic processes that form igneous, sedimentary, and metamorphic rocks and the minerals that compose them. Rock-forming processes also have order and duration, which are explored in this chapter. How can you look at a complicated rock outcrop, such as the chapter opener photo, and determine the sequence of events that formed it? How do geologists determine how long it takes to deposit and lithify thousands of meters of sedimentary rock? How do they determine how long it takes to uplift a mountain range?

Geologists use rocks to tell time in two ways. In the field, it is possible to look at a complicated geologic landscape and decipher the order of events that produced it. However, if geologists want to know how old the rocks are or how long it took for the events to occur, they must employ complicated laboratory procedures. Therefore, geologists use various methods to unlock the history of Earth that rocks archive.

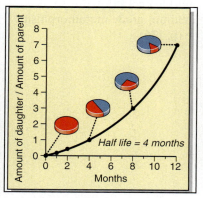

A geologist's rock hammer rests on steeply tilted sedimentary layers that are overlain by less steeply tilted layers. This outcrop, at Siccar Point on the east coast of Britain, was central to James Hutton's interpretation of an extraordinarily long Earth history. ▶

Many generations of humans before us had an avid curiosity about the age of Earth. To figure out the age of the planet, you first need to understand how the age of a rock is measured and why scientific experiments permit a high degree of confidence in the results.

Earth formed about 4.5 billion years ago. How do geologists know that? First, can you even begin to fathom how long 4.5 billion years is? Who can? Geologists have to put their amazement aside, for they always work within the context of the vast length of Earth history that makes the average human lifespan, the rise and fall of nations, even the evolution and extinction of individual species seem insignificantly short. Popular nonfiction writer John McPhee aptly referred to this geological perspective on history as "deep time."

The objectives for this chapter are as follows:

✔ To learn and apply the principles for placing geologic events in order from oldest to youngest

✔ To examine different approaches to estimating the age of Earth

✔ To understand how geologists measure ages of rocks

To understand the duration of Earth history and the ordering of events preserved in rock archives, you need to answer these questions:

7.1 How do you determine the order of events?

7.2 How are geologic events placed in relative order?

7.3 How do geologists determine the relative ages of rocks in widely separated places?

7.4 How was the geologic time scale constructed?

7.5 How do you recognize gaps in the rock record?

7.6 How have scientists determined the age of Earth?

7.7 How is the absolute age of a rock determined?

7.8 *How do we know . . .* how to determine half-lives and decay rates?

7.9 How do you reconstruct geologic history with rocks?

In the **FIELD**

Enjoying a stroll along the seashore, you take a few moments to contemplate the rocks forming the steep cliff that rises above the sandy beach. The rocks exhibit a variety of colors and an eye-catching pattern of horizontal, vertical, and inclined transitions from one rock type to another. You examine these rocks a bit more closely and begin, as field geologists typically do, by sketching the scene in your notebook.

Your sketch, reproduced as **Figure 7.1**, shows red sandstone and shale beds, inclined at an angle downward to the right, overlain by horizontal layers of sandstone, limestone, and shale. A nearly vertical band of basalt cuts across the sedimentary layers forming a dike. Discolored, hardened sedimentary rock borders the basalt and implies contact metamorphism adjacent to the dike.

You have a handle on the origin of each rock type, but an explanation for the origin of the whole outcrop remains elusive—how was it all put together? The sedimentary rocks indicate deposition of clastic and chemical sediment in sedimentary basins. In some layers, cross-bedding reveals the direction of currents that deposited the sediment. Different fossils reveal that the reddish strata formed in a continental depositional environment, whereas the overlying layers were deposited beneath the sea. The dike implies a tectonically active area where magma formed and moved upward, perhaps even reaching the surface at volcanoes that have eroded away. The heat of the rising magma metamorphosed the adjacent sedimentary rocks. Each rock type by itself is identifiable, measurable, and readily explained since you are familiar with rock-forming processes. There is another dimension at play, however, and that is time. The sea-cliff outcrop reveals a history that requires sorting out the order of geologic events and an explanation for why the rocks are oriented at different angles.

Figure 7.1 How do geologists interpret Earth's history? ▶
A field sketch illustrates rocks exposed along a sea cliff. History unfolds as each rock reveals its own record of the processes that formed it. With careful study you can understand the order in which these processes took place and even when they happened.

7.1 How Do You Determine the Order of Events?

Ancient rocks reveal the history of our planet in a way similar to how documents reveal recorded human history. Historians track the sequence of events by studying written papers, art, photographs, and oral recordings that serve as the historical record. For geologists deciphering the long history of Earth, the record is in the rocks. Sedimentary and volcanic rocks record processes that occur on the surface of Earth, whereas plutonic and metamorphic rocks reveal processes that take place in the interior.

To establish the order of events that compose the historical record, historians use specific clues. For example, a historian might put a pile of undated pho-

Field observations of rocks exposed along the beach

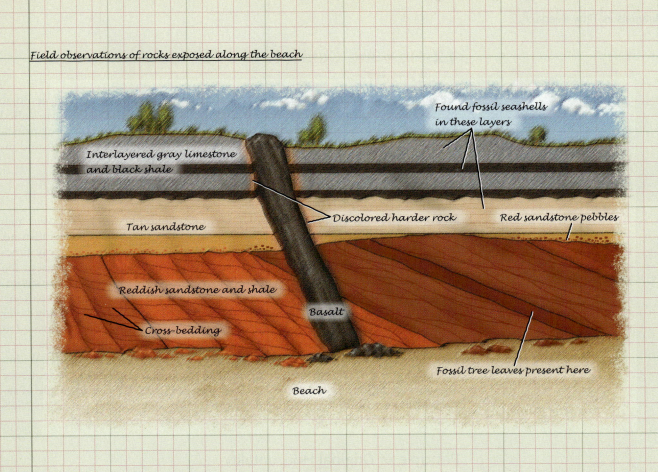

tographs of a city skyline in sequence from oldest to most recent by noticing the appearance of new buildings and the disappearance of older buildings in the pictures. Ordering of objects or features from oldest to youngest determines the **relative age** of each one; this process establishes the age of one thing as older or younger than another.

Historians also work to establish exactly when an event took place; they may establish the date when a document was written, a photograph taken, or an art object completed. Establishing the date of an event provides its **absolute age**. In history, this means the calendar date and possibly even the hour of the event.

So you need to describe the geologic history of the events preserved in the rock record of the sea cliff in two ways. In what order did the events occur, and when did they occur? You need to determine either the relative or absolute age of each rock type exposed in the cliff.

Putting It Together–*How Do You Determine the Order of Events?*

- Relative ages establish the sequence of events without establishing exactly when they occurred. Items or events are placed in order of what happened first, what happened next, and what happened last.
- Absolute ages establish when an event took place or when a feature formed. Absolute ages assign a specific age in years to an object or states that an event occurred a particular number of years in the past.

7.2 How Are Geologic Events Placed in Relative Order?

Geologists establish a set of rules to determine the relative ages of geologic events preserved in the rock record of Earth history. These rules, also called principles, are mostly self-evident yet very powerful. Together, the four principles of superposition, original horizontality, cross-cutting relationships, and inclusions provide a system for ordering geologic events.

Principle of Superposition

In Figure 7.1, start by looking at the horizontal sedimentary layers exposed high on the sea cliff. From bottom to top these consist of two layers of tan sandstone, black shale, gray limestone, another black shale layer, and a final gray limestone. What are the relative ages of these rock layers? The layers represent the accumulation of sediment on an underlying surface. There are marine fossils entombed in the rock, so in this case you can tell that the surface was the seafloor. The mud of the lowest shale layer was deposited on top of the sandy layer below. This means that the sand making up the tan sandstone was deposited at some earlier time and indicates that the sandstone is older than the shale.

You have established a simple principle for determining relative ages. Within a sequence of rock layers formed at Earth's surface, those lower in the sequence are older than those found above, as illustrated in **Figure 7.2**. This **principle of superposition** was formally written down in 1669 by the Danish physician Niels Steensen, who is better known by his Latinized name, Nicolaus Steno.

Principle of Original Horizontality

Steno also noticed that the surfaces where sediment usually accumulates—the seafloor, a riverbed, the bottom of a lake—are nearly flat. Steno's **principle of original horizontality** states that sediment tends to be deposited in horizontal layers. As a result, you expect the bedding planes in sedimentary rock to also be horizontal or nearly so, as seen in Figure 7.2. Sedimentary layers are usually not *exactly* horizontal because the seafloor, riverbeds, and other depositional surfaces are not perfectly flat. Nonetheless, these surfaces rarely slope at angles of more than 0.5 degree, which appears horizontal to the human eye.

▶ **Figure 7.2 Applying the principle of superposition.** The principle of superposition requires that the oldest visible sedimentary rocks along the Colorado River, in Utah, are exposed at the bottom of the canyon while the youngest rocks form the top rim of the canyon.

Nonhorizontal sedimentary layers, therefore, require an explanation. Consider, for example, the sedimentary rocks illustrated in **Figure 7.3**. These nonhorizontal rocks were somehow tilted *after* the deposited sediment lithified to rock. This observation establishes the relative ages of sediment deposition, lithification, and rock deformation. Having this information, you can now fine-tune the relative-age relationships shown by the horizontal sedimentary rocks in the upper part of the sea-cliff outcrop and the inclined red layers at the bottom of the outcrop in Figure 7.1. The principle of superposition requires that the lower red layers are older than the tan sandstone, gray limestone, and the black shale forming the upper part of the cliff. You can also safely assert that the red layers were originally horizontal and then tilted at a later time. The sequence of events is (1) deposition of the red sand and mud; (2) lithification; (3) tilting and erosion of the red sedimentary rock layers; and (4) deposition of horizontal sediment layers that subsequently become tan sandstone, black shale, and gray limestone.

Principle of Cross-Cutting Relationships

The basaltic feature sketched in Figure 7.1 is a dike, because it is a tabular mass of igneous rock that cuts across sedimentary layers (see Section 4.3); and the magmatic heat caused contact metamorphism of the adjacent sedimentary rocks. The sedimentary rocks existed before the dike-forming magma intruded; otherwise there would not have been surrounding rock for the magma to intrude.

The **principle of cross-cutting relationships**, applied by James Hutton in Scotland (Figure 3.11), states that geologic features that cut across rocks must form after the rocks that they cut through. **Figure 7.4** illustrates applications of this principle. Besides igneous intrusions, other common cross-cutting features include **faults**—fractures where rocks are displaced as a result of tectonic forces. Faults clearly must develop after the formation of the rocks they displace.

Principle of Inclusions

As simple as they seem, care must be taken in applying the principles of superposition and cross-cutting relationships. Igneous intrusions form by magma

▲ **Figure 7.3 Applying the principle of original horizontality.** The inclined layering in these sedimentary layers at Glacier National Park, Montana, indicates that the layers were tilted from an initial horizontal orientation after they were deposited.

▼ **Figure 7.4 Applying the principle of cross-cutting relationships.** In the photo on the left, the geologist examines where a relatively young basalt dike cuts across part of an older granite batholith in northern Michigan. In the photo on the right, a fault offsets sedimentary layers. The dashed lines show how much one layer was displaced by fault movement. The sedimentary deposits formed prior to movement along the faults.

▶ **Figure 7.5 Determining which rock is older.**
Granite underlies sedimentary rocks in both of these outcrop sketches, but the age relationships of granite and sedimentary rocks are different.

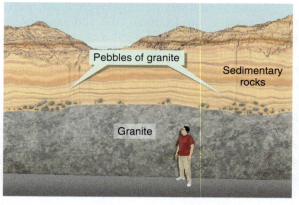

In this diagram, geologists can tell that the sediment was deposited on top of the granite because the lowest sedimentary bed contains pebbles eroded from the granite. Superposition reveals that the sedimentary rocks are younger than the granite.

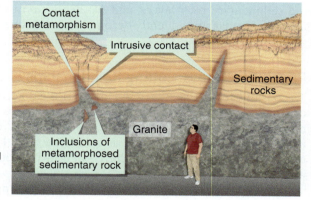

In this diagram, the granite cuts across the sedimentary strata along an intrusive contact and the sedimentary rock is metamorphosed near the contact with the granite. Therefore, the granite must be younger than the sedimentary rock.

rising from great depth; therefore, plutonic igneous rocks commonly appear below sedimentary rocks into which the magma was rising when it crystallized. A study of **Figure 7.5** shows that if a geologist does not look carefully at the contact between the plutonic-igneous rock and the overlying rock, he or she might mistakenly apply the principle of superposition and conclude that the plutonic igneous rock is older. However, the plutonic rock might actually be younger, a relationship that is demonstrated by the plutonic rock cutting across layers in the overlying rock and by causing contact metamorphism of the overlying rock.

Another concept originating with Steno applies to determine relative-age relationships in Figure 7.5. The **principle of inclusions** states that objects enclosed in rock must be older than the time of rock formation. Granite pebbles in sedimentary rock reveal that the granite is older than the sedimentary rock. Metamorphosed inclusions of sedimentary rock within granite indicate the contrasting interpretation that the granite is younger than the sedimentary rock.

Putting It Together—How Are Geologic Events Placed in Relative Order?

■ The principle of superposition states that many rocks form at Earth's surface in layers, one above the other, so that the lowest rock formed first and each successively higher layer is younger than the one below.

■ The principle of original horizontality states that sedimentary layers are horizontal, or nearly so, when they are deposited. Nonhorizontal layering indicates disruption of the beds at some time following deposition.

■ The principle of cross-cutting relationships states that geologic features, such as dikes and faults, that cut across otherwise continuous rocks must form after the rocks that they cut across.

■ The principle of inclusions requires that objects enclosed by rock must have formed prior to inclusion in the rock.

Relative Dating Principles: *See how the relative dating principles are used to decipher the sequence of geologic events.*

7.3 How Do Geologists Determine the Relative Ages of Rocks in Widely Separated Places?

The principles of superposition, original horizontality, cross-cutting relationships, and inclusions allow geologists to order events archived in the rock record in one location, such as the sea cliff shown in Figure 7.1. These principles do not, however, reveal the relative ages of these rocks compared to rocks observed elsewhere. To construct the geologic history of a wide region, or even across the entire Earth, geologists must determine relative ages without relying on features seen only at single outcrops. Two additional principles, lateral continuity and faunal succession, are applied to the problem.

Principle of Lateral Continuity

In some regions, sedimentary rock layers continue in outcrops for long distances. This possibility is especially common where the climate is dry and dense vegetation does not obscure the rocks, and in places where deep canyons lay bare thick successions of rocks for many kilometers. Perhaps the most dramatic example of such continuous exposures are those in the Grand Canyon of the Colorado River in northern Arizona. **Figure 7.6** shows parts of the canyon where intervals of sandstone, shale, and limestone can be traced continuously for 200 kilometers, establishing the chronological sequence of events for a large area.

Steno anticipated that sedimentary rock layers would be continuous because he assumed that sediment beds accumulate in a continuous pattern until encountering some obstruction. This concept is the **principle of lateral continuity** of beds.

In some places sedimentary layers may erode away and in other localities they may be buried beneath younger rock. Just because ancient sediment was continuously deposited over large areas does not necessarily mean that the resulting rocks are now exposed everywhere in that region. The principle of lateral continuity, however, encourages geologists to relate rocks in isolated outcrops to one another in some fashion (Figure 7.6). Combining principles of lateral continuity and superposition extends relative age relationships over larger areas.

Principle of Faunal Succession

How can geologists determine the relative ages of rocks in widely separated regions where the rocks bear little resemblance to one another? **Figure 7.7** illustrates this problem. Englishman William Smith solved this puzzle with careful observations in the late 1700s and early 1800s.

Smith made key observations while surveying coal mines and canal excavations during the blossoming of the Industrial Revolution in Britain. He recognized that the coal seams in different underground mines were found in predictable positions between other sedimentary layers. He reached this conclusion not only by using superposition to put the different rock layers in relative order within each mine but also by noting that some layers contained unique fossils. For example, two different coal layers in one mine are each overlain by limestone layers that contain distinctly different fossils. When

▼ **Figure 7.6** **Lateral continuity of sedimentary rocks.**

In this photograph of the Grand Canyon, sedimentary-rock layers are continuous as far as the eye can see.

The diagram below shows how geologists use the lateral continuity of sedimentary layers to interpret the extent of layers concealed beneath the surface, and the original continuity of layers interrupted by erosional irregularities of the surface.

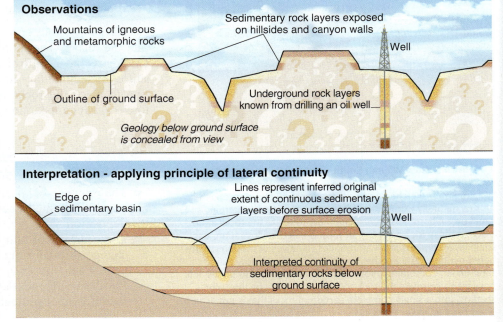

Observations

Mountains of igneous and metamorphic rocks

Sedimentary rock layers exposed on hillsides and canyon walls

Well

Outline of ground surface

Underground rock layers known from drilling an oil well

Geology below ground surface is concealed from view

Interpretation - applying principle of lateral continuity

Edge of sedimentary basin

Lines represent inferred original extent of continuous sedimentary layers before surface erosion

Well

Interpreted continuity of sedimentary rocks below ground surface

▶ **Figure 7.7 The problem of determining relative ages between distant outcrops.** These two outcrops, 1000 kilometers apart, are similar but are not exactly the same. Which rocks are older and which are younger?

• Are the rock sequences of different ages? If so, which set of rocks is older and which is younger?

• Are the rock sequences the same age but appear as different rock types because they formed in different depositional environments? If so, how can their similar age be determined?

• How do geologists determine the relative-age relationships between distant rock outcrops?

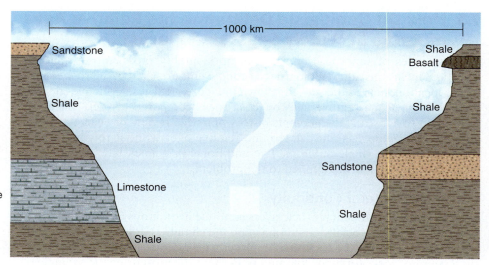

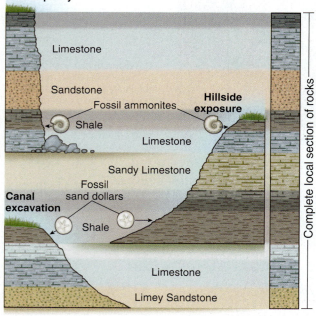

▲ **Figure 7.8 How William Smith used fossils to correlate rocks.**
William Smith sketched and described vertical rock sections that he saw naturally exposed on hillsides and in mines, quarries, and canal excavations. Fossils revealed how to relate the rocks in one place to those found at another location. In this illustration, one shale layer contains fossil ammonites (relatives of modern chambered nautilus) and another contains a fossilized, now extinct, sand dollar. Combining information on rock types and fossils, it is possible to correlate the rock sections, as shown by the color shading, into a single complete section of rocks that is not actually exposed at any one place.

encountering coal and limestone in another mine, Smith confidently related layers between the two mines by comparing their fossils. **Correlation** is the process of matching up rocks found in different places.

Smith saw long, high exposures of rock while he surveyed canal excavations dug for transporting coal from the mines into cities where it was used to heat homes and fuel factories. He carefully noted the succession of rock types and the fossils found within each layer at each artificial and natural exposure. **Figure 7.8** illustrates how Smith used fossils to correlate the rocks at various localities to establish a complete vertical sequence of the beds, ordered chronologically. Smith became so knowledgeable at placing fossil-bearing rocks in the appropriate relative order that amateur fossil collectors could show him specimens and he would tell them exactly which rock layers each had come from and where those rock layers are exposed.

Using this knowledge, William Smith established the **principle of faunal succession**, which states that fossil plants and animals appear in the rock record according to definite chronological patterns:

• Fossils of different organisms first appear at different times.
• Fossils of related organisms change in the same fashion in progressively younger rocks every place they occur.
• Fossil species disappear from the rock record everywhere when they become extinct and do not reappear in younger rocks.

Each type of fossil represents a definite interval of geologic time, which means that the relative ages of fossil-bearing rocks are defined by the included fossils. Geologists have confirmed this principle by examining thick successions of fossiliferous sedimentary rock in many different places and, after using superposition to place the rocks in relative order, discover that the different types of fossils always appear and disappear in the same order.

Putting It Together–How Do Geologists Determine the Relative Ages of Rocks in Widely Separated Places?

■ The principle of lateral continuity states that sedimentary beds are continuously deposited over large areas until some sort of barrier limits their deposition.

■ The principle of faunal succession states that fossil assemblages in rocks change through time as some species become extinct and new ones appear. Each species of organism has a limited time interval of existence. Fossil-bearing rocks are placed in relative-age progression by determining the interval of geologic time represented by the fossils that they contain.

7.4 How Was the Geologic Time Scale Constructed?

If unique assemblages of fossil organisms characterize each interval of geologic history, as revealed by William Smith's work, then geologists can determine the relative ages of rocks in distant places by comparing their fossils. For ease of reference, the time intervals represented by different fossil associations are named **periods**. **Figure 7.9** illustrates the **geologic time scale**, which orders, groups, and subdivides the periods.

The names of many periods derive from places where the rocks of that age were first described. For example, the Cambrian Period is named for Cambria, the Latin name for Wales; the Devonian Period is named for rocks in Devonshire, England; and the Jurassic Period, well known to all dinosaur lovers, is named for the Jura Mountains along the border between France and Switzerland. Other periods took their names from dominant rock types, such as the Carboniferous Period in Europe (divided into Mississippian and Pennsylvanian in the United States), which was a prominent time interval for coal (carbon)

▼ **Figure 7.9 The geologic time scale.**
The geologic time scale is a chronological listing of time intervals of varying duration. The time intervals are arranged within a hierarchy. **Periods** are the fundamental time interval. Periods are grouped into **eras**, which further group into **eons**. Periods consist of shorter intervals called **epochs**. Fossils contained in sedimentary rocks define each Phanerozoic time interval. Age boundaries between Precambrian time intervals are adopted by international convention. Quaternary and Tertiary are traditionally defined periods of the Cenozoic Era, although Neogene and Paleogene are alternative names. The Pennsylvanian and Mississippian Periods in the United States are recognized elsewhere in the world to compose the Carboniferous.

Eon	Era	Period		Epoch	Age (millions of years)
Phanerozoic	Cenozoic	Quaternary	Neogene	Holocene (Recent)	0.01
				Pleistocene	1.8
		Tertiary		Pliocene	5
				Miocene	23
			Paleogene	Oligocene	34
				Eocene	56
				Paleocene	65
	Mesozoic	Cretaceous			145
		Jurassic			200
		Triassic			251
	Paleozoic	Permian		Epochs are defined for each period although only those of the Cenozoic era are commonly referred to by specific names. Epoch names in other periods are indicated by the adjectives "Early", "Middle", and "Late" with the period name; e.g., Late Devonian Epoch.	300
		Carboniferous	Pennsylvanian		318
			Mississippian		359
		Devonian			416
		Silurian			444
		Ordovician			488
		Cambrian			542
Precambrian	Proterozoic	Neoproterozoic			1000
		Mesoproterozoic			1600
		Paleoproterozoic			2500
	Archean	Neoarchean			2800
		Mesoarchean			3200
		Paleoarchean			3600
		Eoarchean			4500

▶ **Figure 7.10** Assigning depositional ages to rocks.

Fossils are collected from the sedimentary rock layers along the sea cliff.

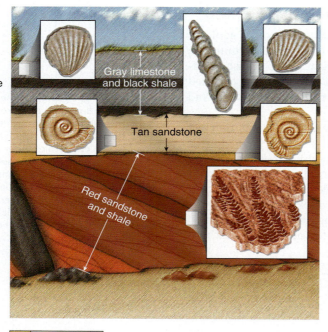

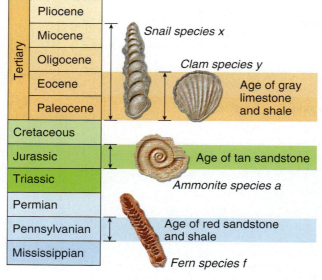

Each fossil is diagnostic of a particular geologic time interval. The fossil fern from the tilted red strata lived during the Pennsylvanian Period. Ammonite species *a* indicates deposition of the tan sandstone during the Jurassic Period. Snail species *x* lived during many epochs of the Tertiary Period. Clam species *y*, however, restricts deposition of the gray limestone and black shale to one or both of the Eocene and Paleocene Epochs.

formation; and the Cretaceous Period, which was named for the limestone variety chalk (*creta* in Latin) found in rocks of this age in northwestern Europe.

Geologists gradually constructed the time scale during the nineteenth century largely on the basis of fossil-bearing sedimentary rocks. Sedimentary rocks anywhere in the world can be assigned to a particular interval on the time scale based on the fossils in the rock. Fossils define the periods so ages of igneous and metamorphic rocks were inferred by relative-age relationships to datable sedimentary rocks. The time scale in Figure 7.9 also lists absolute ages, in millions of years, for the boundaries between periods. These ages were added to the time scale starting about 1960, through use of techniques to be explained in Section 7.6.

Sedimentary rocks containing fossils can be placed in relative-age position by comparing the fossils with those defining the various periods of geologic history. The geologic puzzle illustrated in Figure 7.7 is solved by establishing the period of deposition for each sequence of rocks through the study of the fossils that they contain.

Now you can assign the sedimentary rocks at the seaside cliff to the different periods on the geologic time scale. The rocks contain fossils, so you can study those fossils and match them up to the appropriate geological period. The assistance of a professional paleontologist, a geologist who studies fossils and the history of life on Earth, will help.

Figure 7.10 shows the result of the paleontological investigation. Only the sedimentary rocks are assigned to geologic periods at this stage because the basalt, an igneous rock, does not contain fossils. The periods of deposition of the sedimentary rocks are consistent with your earlier application of the principle of superposition. The red layers at the bottom of the cliff are oldest, the limestone and black shale are the youngest, and the tan sandstone is of intermediate age.

Putting It Together–How Was the Geologic Time Scale Constructed?

■ The geologic time scale is an established chronological order of time intervals. The fossils that are found in rocks deposited during that time define each named interval.

7.5 How Do You Recognize Gaps in the Rock Record?

Take another look at the ages established for sedimentary rocks at the sea cliff (Figure 7.10). Although these ages are consistent with the principle of superposition, the three intervals of deposition are not adjacent to each other on the time scale (Figure 7.9). This incomplete geologic record is analogous to a biography constructed from a diary that has many missing pages.

Why is the rock record incomplete? The discovery that the tan sandstone is substantially younger than the red sandstone and shale (Figure 7.10) is not surprising. The once-horizontal red sediment was tilted and eroded to produce a flat surface before deposition of the tan sand began. Pebbles of red sandstone and shale in the lowest bed of the tan sandstone (Figure 7.1) indicate that the red sediment lithified to rock before it was tilted and before deposition of the tan sand. Lithification, tilting, and erosion of the red sediment require some amount of time that is not represented by rocks at the sea cliff.

The time gap above the tan sandstone was not expected. Going back to your sketch, you realize that the contact between the tan sandstone and the overlying black shale is irregular rather than flat, which suggests erosion of the tan sandstone. A time interval of erosion separated two intervals of sediment deposition. Using fossils to determine the age of sedimentary rocks commonly reveals gaps in deposition so that a single cross section of rock rarely reveals a complete geologic history.

Unconformities

Gaps in the rock record, when erosion rather than deposition happened, are **unconformities**. There are two unconformities in the sea-cliff section, one above and one below the tan sandstone. Unconformities are significant for two reasons. First, to construct the geologic history of an area, it is important to know what part of the rock record does not exist, just like a biographer must know what years are missing from an incomplete diary. Second, geologists seek to identify any event or succession of events that caused the break in the rock record, because these events also are part of the geologic history. Three types of unconformities are generally recognized.

An **angular unconformity**, illustrated in **Figure 7.11**, is present between intervals of layered rocks (sedimentary beds or lava flows) that are inclined at different angles. There must be a time interval when the lower layers of rock were tilted and eroded at Earth's surface prior to deposition of the overlying rocks. An angular unconformity separates the red sandstone and shale from the overlying tan sandstone at the seashore.

▼ **Figure 7.11 What an angular unconformity looks like.**

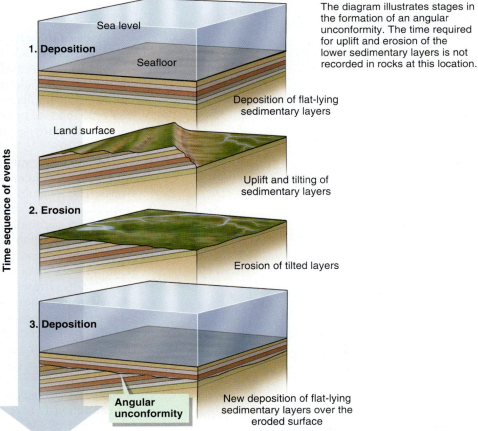

The diagram illustrates stages in the formation of an angular unconformity. The time required for uplift and erosion of the lower sedimentary layers is not recorded in rocks at this location.

1. **Deposition**
Sea level
Seafloor

Deposition of flat-lying sedimentary layers

Land surface

Uplift and tilting of sedimentary layers

2. **Erosion**

Erosion of tilted layers

3. **Deposition**

Time sequence of events

Angular unconformity

New deposition of flat-lying sedimentary layers over the eroded surface

The black line traces part of an angular unconformity in the Grand Canyon. Beds above the unconformity are horizontal, whereas beds below the unconformity incline down to the right. Approximately 500 million years of geologic history are missing along the unconformity.

▼ **Figure 7.12** **What a disconformity looks like.**

The black line traces part of a disconformity in the Grand Canyon. Fossils indicate that several million years are missing from the rock record along this sharp, eroded boundary between sedimentary rock layers.

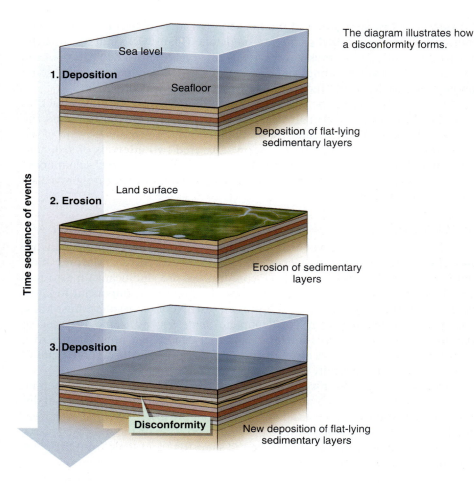

The diagram illustrates how a disconformity forms.

Time sequence of events

1. **Deposition**

Sea level

Seafloor

Deposition of flat-lying sedimentary layers

2. **Erosion**

Land surface

Erosion of sedimentary layers

3. **Deposition**

Disconformity

New deposition of flat-lying sedimentary layers

Unconformities: *See how the three types of unconformities form.*

A **disconformity**, illustrated in **Figure 7.12**, is also present between intervals of layered rock, but where all layers are either horizontal or inclined at the same angle. A surface of erosion, which might be a deeply eroded channel or a subtle and almost planar feature, marks a disconformity. The geologic record is incomplete across the disconformity because no rock exists to show the time interval of erosion or nondeposition separating the deposited layers. A disconformity separates the tan sandstone and black shale on the sea cliff (Figure 7.10).

A **nonconformity**, illustrated in **Figure 7.13**, is present where sedimentary or volcanic rocks accumulate on top of eroded plutonic-igneous or metamorphic rocks. This contact must be an unconformity because plutonic-igneous and metamorphic rocks form beneath Earth's surface, whereas sedimentary and volcanic rocks accumulate on the surface. When metamorphic and plutonic rocks form, there is other rock above that extends to the surface. All of that other rock, possibly many kilometers thick, must erode to expose the metamorphic and plutonic rocks at the surface before the metamorphic and plutonic rocks then can be covered by sediment or volcanic deposits. The existing rocks, then, do not record the time required to erode away the rock that was originally above the metamorphic and plutonic materials.

The rock record at a single locality is usually incomplete and may be riddled with unconformities, but no one unconformity extends completely around the planet. Erosion in one locality produces sediment that is deposited at another place during the same time. The complete geologic time scale was carefully constructed by tracing unconformities toward locations where sediment accumulated continuously over the missing intervals of time and the unconformities did not form. This approach permits the boundaries between periods to be defined by changes in the fossil record of life where the rock record is most complete.

The black line traces part of a nonconformity in the Grand Canyon. Metamorphic and plutonic-igneous rocks in the bottom of the canyon are overlain by sedimentary rocks. Approximately one billion years of Earth history is missing along the unconformity.

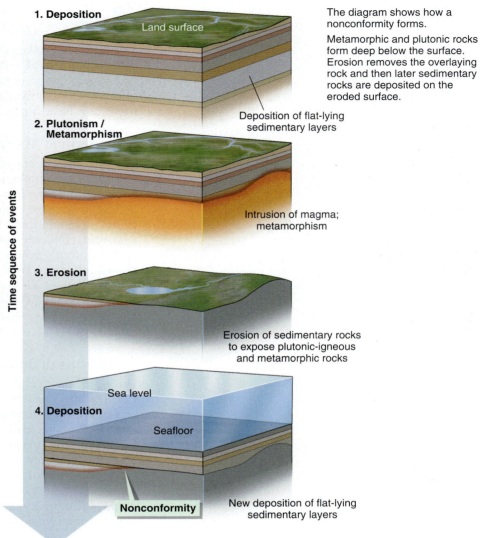

1. Deposition

Land surface

The diagram shows how a nonconformity forms.

Metamorphic and plutonic rocks form deep below the surface. Erosion removes the overlaying rock and then later sedimentary rocks are deposited on the eroded surface.

Deposition of flat-lying sedimentary layers

2. Plutonism / Metamorphism

Intrusion of magma; metamorphism

3. Erosion

Erosion of sedimentary rocks to expose plutonic-igneous and metamorphic rocks

Sea level

4. Deposition

Seafloor

Nonconformity

New deposition of flat-lying sedimentary layers

Time sequence of events

▲ **Figure 7.13 What a nonconformity looks like.**

Putting It Together—How Do You Recognize Gaps in the Rock Record?

- Unconformities are former land or seafloor surfaces between rock layers that represent time intervals *not* recorded in the local rock record.
- Angular unconformities separate sedimentary or volcanic rocks that are inclined at different angles. The time required for tilting the lower rocks and eroding them to produce a new surface of deposition for the upper rocks is *not* represented in the geologic record.
- Disconformities separate sedimentary or volcanic rocks that are inclined at the same angle but are separated by an irregular erosion surface. This surface indicates a break in deposition or erosion of the older rocks.
- Nonconformities separate sedimentary or volcanic rocks from underlying plutonic-igneous or metamorphic rocks. The time required to erode the materials that overlie the plutonic and metamorphic rocks until the onset of sediment accumulation on the eroded surface is *not* represented in the geologic record.

7.6 How Have Scientists Determined the Age of Earth?

How long has it taken for all of the geologic events recorded in the rocks to take place—how old is Earth? Over hundreds of years, both scientific and nonscientific approaches were taken to answer this question. Considering some of these efforts shows the varied ways scientists try to reach quantitative answers. Geologists now have strong evidence that Earth is about 4.5 billion years old.

Nonscientific Assessments of the Age of Earth

The age of Earth engaged people long before scientists developed methods for addressing the problem. Ancient Greek philosophers envisioned the universe as continually cycling in a circular fashion without a beginning or an end. Hindus believe in cycles of cosmic destruction and renewal that do not close in a circle. Based on Hindu traditional accounts of these cycles, Earth is about two billion years old. Numerous efforts were made between the second and seventeenth centuries to determine the age of Earth from biblical chronologies. The best-known effort was by the Irish Bishop James Ussher during the 1650s. Bishop Ussher adopted a completely literal interpretation of time on the basis of the Bible's first book, Genesis, asserting that God created the planet and all that is on it in six days. Ussher used biblical genealogies to conclude that creation occurred in 4004 B.C., or an age for the Universe and Earth of 6009 years in 2005.

Indications of an Older Earth

During the eighteenth century several people attempted to determine the age of Earth by applying logical scientific reasoning to observations of natural processes. The French diplomat Benoit de Maillet, for example, found marine fossils in rocks high above sea level. He used historical records of measured small variations in sea level to conclude that a period of 2.4 million years would account for these fossils if they were left behind by the fall of a once all-encompassing sea. Geologists now know that sea level rose and fell countless times during Earth history, and that either falling sea level or uplift of the land explain marine rocks found above present sea level. Nonetheless, de Maillet's estimate was based on realistic observation and stated assumptions that were widely accepted at the time.

James Hutton also challenged the notion of an Earth as young as thousands of years. He was particularly struck by the significance of unconformities as indicators of the antiquity of Earth. Hutton illustrated his conclusion of an old Earth with a field trip to Siccar Point, on the east coast of Britain, which is pictured in the chapter opener photo. He pointed out that huge stretches of time are required for the accumulation of sediment layers, based on the slow rates of erosion and sedimentation recorded during human history. The tilting of the older sequence of rocks at Siccar Point, below the angular unconformity, also requires an immense period of time in human terms, given that there is no indication of single upheavals on this scale during recorded history. Next is the time that went by while these older rocks eroded to a low surface. Then renewed, slow deposition of the upper sequence of sedimentary beds occurred. Additional time is required to account for the tilting of the rocks again and the erosion of the modern landscape to expose these geologic relationships. A participant on Hutton's field trip later recalled listening to his explanation of the sequence of events recorded at Siccar Point:

> The mind seemed to grow giddy by looking so far into the abyss of time; and while we listened with earnestness and admiration to [Hutton] who was now unfolding to us the order and series of these wonderful events, we became sensible how much farther reason may sometimes go than imagination can venture to follow. (*The Works of John Playfair*, Archibald Constable and Co., 1822, p. 81)

Hutton knew that if interpretation of geologic phenomena followed the logic later called uniformitarianism, then Earth must be far older than Bishop Ussher's pronouncement. Although Hutton did not estimate the absolute age of the planet, he was so impressed by the need for an antiquity nearly unfathomable to the human perspective of time that he concluded his 1788 volume, *Theory of the Earth*, with the following:

> The result, therefore, of our present enquiry is, that we find no vestige of a beginning—no prospect of an end. (*Transactions of the Royal Society of Edinburgh*, vol. 1, p. 304)

How Long Would It Take for Earth to Cool?

British physicist William Thomson, better known as Lord Kelvin, made a rigorous attempt to calculate the duration of Earth history in the late nineteenth century. Kelvin reasoned that if Earth was originally hotter than at present, then its age can be calculated by knowing (a) the original temperature, (b) the current distribution of temperatures within the planet, and (c) the rate at which heat dissipates from Earth's interior into space. He developed what modern scientists would call a conceptual model. **Figure 7.14** illustrates Kelvin's concept.

How did Kelvin obtain the values that he used for his calculation? There was abundant evidence from temperature measurements in deep coal and metal-ore mines that temperature increases with depth into the planet. This is the geothermal gradient described in Section 4.4. The rate of heat conduction through rock is readily measured in the laboratory. An estimate of the original temperature of Earth is a matter of greater uncertainty. Kelvin followed his contemporaries, and consistent with modern theory, that Earth coalesced from the amalgamation of smaller objects in the early history of our solar system (to be explored further in Chapter 9). When fast-moving objects collide, the energy of motion converts to heat when one or both objects instantaneously decelerate to zero velocity (slap your hands together to experience this effect on a small scale). The challenge becomes to determine how hot early Earth was as a result of this heating from collisions. Kelvin and others assumed initial surface temperatures varying from 1200°C to 3870°C.

Lord Kelvin admitted the uncertainty of the values used in the calculations and the nature of his assumptions. The biggest and most important assumption was that the original heat of planetary amalgamation is the only source of Earth's internal heat. Kelvin arrived at a range of values because of the many uncertainties, but argued that the most likely age is between 20 million and 40 million years.

Most geologists at the time felt that the complex rock record required an even longer Earth history, in light of the slowness of observed rates of Earth processes. Biologist and paleontologist Thomas Huxley was one of those who considered Kelvin's age of Earth, 20–40 million years, to be too young. Huxley wrote words that are still cited today as a caution to overreliance on conceptual mathematical models:

> Mathematics may be compared to a mill of exquisite workmanship, which grinds you stuff of any degree of fineness; but nevertheless, what you get out depends on what you put in ... so pages of formulae will not get a definite result out of loose data. (*Quarterly Journal of the Geological Society of London*, 1869, vol. 25, p. xxxviii)

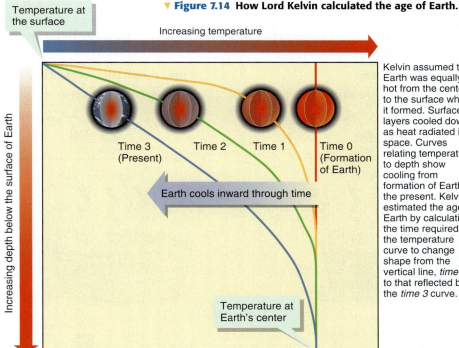

▼ **Figure 7.14 How Lord Kelvin calculated the age of Earth.**

Temperature at the surface

Increasing temperature

Increasing depth below the surface of Earth

Time 3 (Present)   Time 2   Time 1   Time 0 (Formation of Earth)

Earth cools inward through time

Temperature at Earth's center

Kelvin assumed that Earth was equally hot from the center to the surface when it formed. Surface layers cooled down as heat radiated into space. Curves relating temperature to depth show cooling from formation of Earth to the present. Kelvin estimated the age of Earth by calculating the time required for the temperature curve to change shape from the vertical line, *time 0*, to that reflected by the *time 3* curve.

Kelvin's Calculation of Earth Age: *See how Kelvin determined the age of Earth.*

In the modern age of elaborate computer-calculated models of natural phenomena, Huxley's admonition is commonly paraphrased as "garbage in, garbage out," a phrase that reminds you how a final calculated answer is only as good as the certainty of the values used in the calculation and the assumptions used in constructing the formula. This does not mean that mathematical solutions are not worth seeking, but that these solutions, like any scientific result, must be carefully scrutinized. In this case, as you shall soon see, there is no error in the math but rather that the most important of Kelvin's assumptions was later proven false.

Using the Salty Ocean as a Clock

Irish geology professor John Joly developed a different conceptual model in about 1900 to determine the age of Earth. His idea was that if the saltiness of the ocean results from salt carried to the ocean by rivers, then the age of the oceans, as an approximation of the age of Earth, can be calculated if (a) the amount of salt in the ocean and (b) the quantity of salt carried by rivers are known. **Figure 7.15** illustrates Joly's logic.

What are the key assumptions and level of certainty in Joly's calculation? In Joly's time there were many measurements of the amount of salt (NaCl) dissolved in seawater, and this value did not vary much from place to place. There were also measurements of the salt content of river water flowing to the ocean, but the values ranged considerably so estimating the total salt delivery to oceans from weathering on continents was highly uncertain. The area of Earth's oceans was well known at the time, but there were very few measurements of ocean depth, so any selection of a value for seawater volume had a substantial uncertainty. This value is critical because the total amount of ocean salt is calculated by multiplying the average salt concentration in seawater by the total volume of seawater. Joly knew that some salt in seawater chemically precipitated during evaporation as rock salt (halite), so the volume of ancient rock salt also was estimated with great uncertainty. Joly assumed that his value for salt input from rivers is constant through geologic time and also estimated an initial saltiness for the primeval ocean instead of assuming that seawater started out completely fresh.

Acknowledging his assumptions and the uncertainty of the values he used, Joly estimated the antiquity of the oceans to be on the order of 80–100 million years.

Although Joly calculated a greater age than that settled on by Lord Kelvin, many geologists still believed Earth to be older. Today, geologists see many problems with Joly's calculation. Oceanographers later explored the ocean depths and now know that oceans hold a much larger volume of seawater than estimated by Joly. Subsurface explorations for oil and gas and field excursions in remote lands reveal that rock-salt deposits are also much larger than known by Joly. These newer data result in a substantially older age for Earth using Joly's logic.

The Significance of Radioactivity to Understanding Earth's Age

The discovery of radioactivity in the 1890s was a major scientific breakthrough with two immediate implications for the age of Earth.

1. Radioactivity provides a persistent heat source inside Earth that negates Lord Kelvin's assumption that all of the heat remains from Earth formation. Kelvin's estimated age of Earth, 20–40 million years, is too young.

▼ **Figure 7.15 How Joly calculated the age of Earth.** Evaporation removes water from the ocean but the salt stays behind and accumulates over time. Joly mistakenly assumed that the amount of salt in the ocean today, salt$_{today}$, was simply that amount that was originally present, salt$_{original}$, plus the annual input of salt by rivers, salt$_{added}$, multiplied by the number of years elapsed since Earth formed.

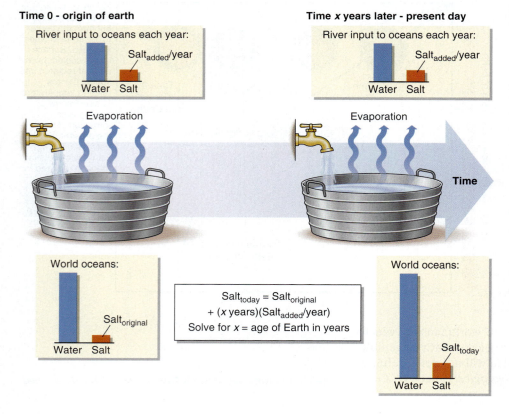

$$Salt_{today} = Salt_{original}$$
$$+ (x \text{ years})(Salt_{added}/\text{year})$$
Solve for x = age of Earth in years

2. Radioactivity provides a method for measuring the absolute ages of rocks and reveals that Earth is about 4.5 billion years old.

Radioactivity is the energy and subatomic particles released when atoms of one element transform into atoms of another element by processes that change the number of protons and neutrons in the nucleus. A characteristic number of protons defines each element. **Figure 7.16** illustrates that the number of neutrons can fluctuate between atoms of the element carbon.

Isotopes are atoms of the same element that have the same number of protons but a different number of neutrons. This means that all isotopes of the same element have the same atomic number (the number of protons) but different atomic mass numbers (the number of protons plus the number of neutrons). Scientists have identified 339 isotopes among the 84 naturally occurring elements. **Appendix B** lists isotopes that are important in geologic studies.

Only certain combinations of the number of protons and neutrons can be present in an atomic nucleus without causing instability between forces in the nucleus. **Radioactive decay** is a change in the number of protons, neutrons, or both that transforms an unstable isotope to a stable one. Only 70 of the 339 natural isotopes are unstable, or radioactive, isotopes.

Radioactive decay permits geologists to date the age of minerals because the abundances of some isotopes change through time. Radioactive-isotope abundance decreases, whereas the abundance of isotopes created by decay increases. If these abundances are measured in a mineral, and the rate of decay is known, then the elapsed time since the mineral formed is calculated. The next section explores this application of natural radioactivity to determine absolute ages of rocks and the 4.5-billion-year antiquity of Earth.

Carbon 12 (^{12}C)
6 protons + 6 neutrons

98.9% of all carbon atoms

Carbon 13 (^{13}C)
6 protons + 7 neutrons

1.1% of all carbon atoms

Stable, unchanging carbon isotopes

Carbon 14 (^{14}C)
6 protons + 8 neutrons

0.0000000001% of all carbon atoms

Nitrogen 14 (^{14}N)
7 protons + 7 neutrons

Radioactive decay → Radiation

Unstable, radioactive carbon isotope

○ Protons
◉ Neutrons

▲ **Figure 7.16 Visualizing isotope examples.**
Carbon exists as three isotopes. All carbon atoms contain six protons, and most of these atoms also contain six neutrons so that the total atomic mass number is 12. One carbon isotope has seven neutrons, for a total mass number of 13. Another very rare isotope, ^{14}C, has eight neutrons. ^{12}C and ^{13}C are stable isotopes whose abundances do not change over time. ^{14}C, however, is an unstable isotope that undergoes radioactive decay to the most common isotope of nitrogen. During this radioactive decay one carbon neutron converts to a proton.

EXTENSION MODULE 7.1
Radioactivity and Radioactive Decay.
Learn the different ways that radioactive isotopes decay and the resulting levels of natural radioactivity.

Putting It Together–How Have Scientists Determined the Age of Earth?

■ Eighteenth-century geologists, including James Hutton, applied scientific observations and principles to conclude that the geologic rock record requires an extraordinarily ancient age for Earth.

■ Lord Kelvin calculated the age of Earth based on a conceptual model of how much the planet has cooled since its origin. His calculations suggested an age in the range of 20–40 million years.

■ John Joly calculated the age of the oceans based on a conceptual model of how long it would take Earth's rivers to transport enough salt to account for the current saltiness of the ocean. His calculations suggested an age of 80–100 million years.

■ The discovery of radioactivity was critical in efforts to determine an accurate age of Earth. First, radioactivity is a source of internal heat not considered by Kelvin, meaning that his calculated age must be too young. Second, an understanding of rate of radioactive decay of one isotope to another provides a basis for measuring the absolute age of rocks.

7.7 How Is the Absolute Age of a Rock Determined?

Your absolute age, in years, is calculated by subtracting the year of your birth from the current year. To establish the age of a rock, you need to establish the year of its "birth"—the date when it formed. Geologists use the natural radioactive decay of elements commonly found in rock-forming minerals to determine the "birth date" of many minerals. This date is the basis for establishing the absolute age of rocks.

Measure the Isotope Abundances

Figure 7.17 illustrates how to determine the absolute age of a geologic sample by measuring the changing abundance of selected isotopes. The sample contains atoms of many elements, but it is simpler to focus on just the abundance of a particular radioactive **parent isotope** that decays through time to produce a **daughter isotope** of another element. In this hypothetical example, no daughter isotope is originally present, but over a period of time, radioactive decay causes the number of daughter isotope atoms to increase while the abundance of the parent isotope decreases. Each daughter atom originates from the decay of a parent atom, so the sum of daughter and parent isotopes remains the same. The ratio of daughter isotope to parent isotope relates directly to how much time elapsed since radioactive decay began in the sample.

Apply the Half-Life

Notice in Figure 7.17 that the number of parent-isotope atoms transformed to daughter-isotope atoms is not the same during each one-month interval. There are 10 million transformations in the first month but only 9 million in the second month. The rate of transformation of parent to daughter atoms depends on the number of parent atoms present. This relationship indicates that each atom decays independently of all the others, so that over any particular interval of time there is a particular probability for a given atom that a decay will take place. The greater the abundance of parent atoms, the more decays that occur. As the number of parent atoms decreases, progressively fewer decays occur over the same time interval.

Notice that 32 million parent atoms, or half of the original parent-isotope atoms, remain after four months. Another four months later, after a total of eight months, the 32 million parent atoms are halved again to 16 million. After twelve months this number once again decreased by half to 8 million atoms. The **half-life** of the radioactive decay process is the time interval during which the number of parent-isotope atoms decreases by half. The half-life is one way to express the rate of radioactive decay from parent to daughter isotope. The half-life of the hypothetical decay in Figure 7.17 is four months.

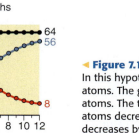

◄ **Figure 7.17 How radioactive-isotope abundances change with time.**
In this hypothetical example, red atoms of a parent isotope decay to blue daughter-isotope atoms. The graphs show the number of atoms of each isotope within a total sample of 64 million atoms. The total number of atoms does not change through time, but the number of parent atoms decreases while the number of daughter atoms increases. The number of parent atoms decreases by half during every four-month interval, indicating a half-life of four months. The inset yellow graph shows how the ratio of daughter to parent atoms changes during radioactive decay. If this ratio is measured in a sample, then the sample age can be read from the graph.

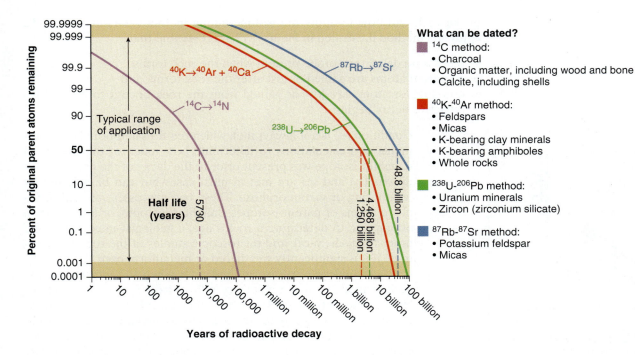

What can be dated?

■ ^{14}C method:
• Charcoal
• Organic matter, including wood and bone
• Calcite, including shells

■ ^{40}K-^{40}Ar method:
• Feldspars
• Micas
• K-bearing clay minerals
• K-bearing amphiboles
• Whole rocks

■ ^{238}U-^{206}Pb method:
• Uranium minerals
• Zircon (zirconium silicate)

■ ^{87}Rb-^{87}Sr method:
• Potassium feldspar
• Micas

◀ **Figure 7.18 Radioactive-decay schemes used by geologists.** Geologists use more than a dozen radioactive-isotope-decay schemes to date rocks. The chosen method depends on the composition of the dated material and its estimated age. The right-hand column shows what materials are dated by each method. The graph shows the changing abundance of parent isotopes through time for four commonly used dating methods with a wide range in half-life. Notice that the axes are not linear scales. The shaded areas near the top and bottom of the graph mark the practical limits of applying each method. The upper shaded bar covers that part of the graph where so little decay has taken place that the measuring instruments may not detect a decrease in the amount of parent isotope or that any daughter isotope has been produced. The lower shaded area represents conditions after nearly all of the parent isotope has decayed and may not be detected.

Radioactive Decay: *See how the abundance of parent and daughter isotopes changes over time because of radioactive decay.*

The two keys to determining the absolute age of a rock are (1) measuring the abundances of parent and daughter isotopes, and (2) knowing the half-life value for the rate of decay of parent to daughter. Geologists routinely measure these abundances and values with high precision and accuracy in specially equipped geochemical laboratories. The decay rates of many radioactive isotopes commonly found in rock-forming minerals have also been determined by laboratory experiments. You will learn more about this in Section 7.8.

Figure 7.18 summarizes four radioactive-decay schemes that geologists commonly use to determine the age of Earth materials and illustrates how the parent-isotope abundance decreases through time. This figure also shows the practical limits of dates for each method. When the half-life is long and the decay has only recently begun, the abundance of parent atoms has barely changed. In this case the abundance of daughter isotopes is very small, and the precision of the geochemical instruments is insufficient to determine that any decay has occurred. Likewise, when nearly the entire parent isotope has decayed, there is too little remaining to be accurately detected by the instruments.

Radioactive-Isotope Decay Provides Absolute Ages

Each isotope system has its own time range of usefulness depending on the magnitude of the half-life. The radioactive decay of one particular element, ^{14}C, is useful for dating relatively young samples. It is especially useful for determining the age of historical and archaeological materials dating back to around 50,000 years. An early test of the validity of radioactive-isotope dating used the ^{14}C method to calculate ages for materials of known age is shown in **Figure 7.19a**. For dating older materials and most rocks, geologists use isotope-decay systems with longer half-lives, on the order of billions of years (Figure 7.18). In some cases, the half-lives of two or more methods are appropriate for dating the same sample, and multiple age determinations provide a further test of the approach (Figure 7.19b). Consistency of relative and absolute ages also validates the radioactive-isotope method.

Understanding the Geologic Conditions

Determining when a rock formed requires more than analyzing its constituent minerals for parent and daughter isotopes. Understanding the geologic origins of the rock and how to apply the isotope-dating methods is fundamental. To apply the method illustrated in Figure 7.17, it is essential that

a. there is no daughter isotope present when the rock forms,
b. no parent atoms are gained or lost after the rock forms, and
c. no daughter atoms are lost or gained from the rock once it forms and radioactive decay begins.

Condition (a) is not always met, which requires additional measurements and calculations to determine how much of the measured daughter isotope was present prior to the decay of the parent. Conditions (b) and (c) are met in most cases but can be affected by weathering or metamorphism. These two processes can add or remove atoms of parent isotopes, daughter isotopes, or both. Before attempting to determine a rock's age, a geologist must undertake other field, chemical, and microscopic studies of the rock to determine if it is significantly weathered or metamorphosed.

A radioactive-isotope age indicates when the *minerals* in the rock formed. When a mineral crystallizes from magma, precipitates from a watery fluid, or forms from metamorphic reactions, the radioactive "clock" starts to run as atoms of daughter isotopes accumulate from decay of parent atoms in the crystal structure. For example, the age of a feldspar crystal collected from granite usually records when the granitic magma crystallized. If a feldspar crystal from that granite erodes as a sand grain that ends up in sandstone many millions of years later, the age of the crystal remains that *of when it formed as part of the granite*, not the age of deposition of the sandstone. If the granite metamorphoses to gneiss, the potassium feldspar crystals in the granite heat up and may exchange atoms with adjacent crystals and fluids and recrystallize into new feldspar crystals. Metamorphism usually resets the radioactive-isotope clock. The age determined by a geochemist represents the *age of metamorphism* rather than the age of crystallization of the original granite.

An Example—Potassium-Argon Dating

The potassium-argon (^{40}K-^{40}Ar) dating method may be the easiest method to understand and is conceptually illustrated in **Figure 7.20**. Several common minerals contain abundant potassium, such as potassium feldspar, muscovite, and biotite. Others, including amphibole and plagioclase feldspar, contain readily measured trace amounts of potassium. These potassium-bearing minerals commonly form by crystallization of magma or during metamorphism, so the ^{40}K-^{40}Ar method applies to dating igneous and metamorphic rocks. Unlike potassium, argon is a pure gas that does not readily bond to other elements to form compounds. For this reason, the crystal structures of growing minerals do not incorporate argon.

The ^{40}K-^{40}Ar system is, therefore, similar to the hypothetical example in Figure 7.17, which specifies that the starting material contains parent isotope but no daughter isotope (Figure 7.20). A difference from the hypothetical example in Figure 7.17 is that ^{40}K simultaneously decays to ^{40}Ar *and* ^{40}Ca. The dating method uses only the decay of ^{40}K to ^{40}Ar because ^{40}Ca is the most common isotope of calcium and is initially present in almost all potassium-bearing minerals.

Argon is a nonreactive element, and therefore it is not always retained inside a mineral once it is formed by radioactive decay. If the mineral heats up during metamorphism, expansion of the

(a)

The ^{14}C dating method is tested by dating archaeological materials of known age, and tree wood whose age is known by counting the annual growth rings. The graph shows the results. For a perfect match between known and isotope ages, all data points should lie on the purple line, as nearly all of them do within the known uncertainties. This consistency demonstrates the validity of the method.

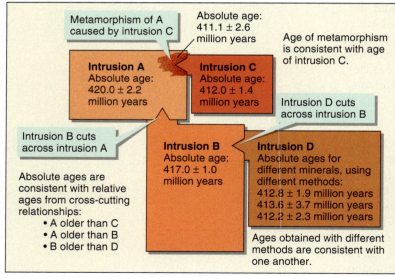

(b)

This diagram schematically shows the intrusive relationships between igneous intrusions, and isotope ages determined for rocks from each intrusion. Within the uncertainty of measurement all of the ages agree with relative ages determined from cross-cutting relationships. Ages obtained by three different methods for three different minerals in intrusion D are also identical within measurement uncertainty.

▲ **Figure 7.19 How radioactive-isotope ages are confirmed.**

▼ Figure 7.20
How potassium-argon dating works.

Igneous Rock

The "clock" starts running when the rock cools down to < 400°C, when daughter argon atoms begin accumulating in the rock or mineral. If no ^{40}K or ^{40}Ar are added or removed, the ratio of the two isotopes will indicate the age of the rock.

1

Magma contains both ^{40}K and ^{40}Ar

Gases, including argon, escape into atmosphere.

Potassium is incorporated into crystallizing minerals.

"Isotope clock" set

"Isotope clock" starts

Weathering can cause loss of potassium, argon or both from some minerals although the clock usually keeps running.

Melting

Weathering and Erosion

3
Metamorphic Rocks

After the metamorphic rock cools and retains argon formed by decay of ^{40}K, the clock starts running again, so that the age of metamorphism can be determined.

"Isotope clock" reset

Clock will start when rock cools and retains ^{40}Ar

During intense metamorphism, the argon that previously formed by radioactive decay is released as gas, and the potassium may be reshuffled into new minerals that initially lack argon.

Metamorphism

2
Sedimentary Rocks

When weathered minerals are incorporated into sediment, the age of the minerals is still their crystallization age and not the depositional age of the sediment. Cement minerals, like clay, typically incorporate potassium but no argon at the time of formation; therefore, the clock starts running when clay forms and argon accumulates from decay of ^{40}Ar.

minerals from older rocks

newly-formed cement minerals

"Isotope clock" keeps running

"Isotope clock" starts when minerals form

The upper diagram shows how the isotope clock works in relationship to the processes and products of the rock cycle. Potassium-argon dating works best for dating the crystallization of igneous rocks and metamorphic minerals, and is less useful for dating the deposition of sedimentary rocks.

^{40}K simultaneously decays to ^{40}Ar and ^{40}Ca. There is so much ^{40}Ca found in minerals that it is impossible to distinguish the atoms that were there originally from those formed by decay of ^{40}K. Only the measured ratio of $^{40}Ar/^{40}K$, therefore, is used to determine rock ages.

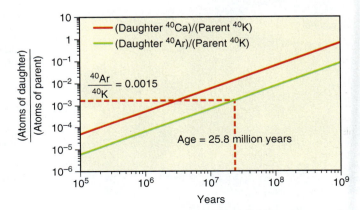

— (Daughter ^{40}Ca)/(Parent ^{40}K)
— (Daughter ^{40}Ar)/(Parent ^{40}K)

$\frac{^{40}Ar}{^{40}K} = 0.0015$

Age = 25.8 million years

$\frac{\text{(Atoms of daughter)}}{\text{(Atoms of parent)}}$

Years

The basalt dike at the seaside outcrop in Figure 7.1 has a measured $^{40}Ar/^{40}K$ ratio of 0.0015, indicating an age of 25.8 million years.

EXTENSION MODULE 7.2
The Mathematics of Radioactive-Isotope Decay.
Learn how to use mathematical equations, instead of graphs, to calculate mineral ages.

crystal structure may permit the loosely caged argon atoms to escape. In this case, the isotope clock resets and the calculated age from measurements of ^{40}K and ^{40}Ar reflects the time of reheating rather than the original crystallization age.

The ^{40}K-^{40}Ar method is used to determine when the basaltic dike exposed at the beach crystallized (Figure 7.1). Careful chemical analyses of a 100-gram sample of the basalt detect 7.54×10^{21} atoms of ^{40}K and 1.13×10^{19} atoms of ^{40}Ar produced by decay of ^{40}K. This means that the $^{40}Ar/^{40}K$ ratio is 0.0015. On the chart given in Figure 7.20, you can see that a ratio of 0.0015 corresponds to an age of 25.8 million years.

Putting Absolute Ages on the Geologic Time Scale

The distribution of fossils within sedimentary rocks defines the geologic time scale (Figure 7.9). The scale was established decades before the discovery of radioactivity and about a century prior to geologists' routine determination of absolute ages for rocks. Obtaining absolute ages for the boundaries between periods and eras on the time scale is not simple because radioactive-isotope methods are most readily used to determine the ages of igneous and metamorphic rocks, but the time scale is based on fossils found in sedimentary rocks.

Unfortunately, direct isotope dating of sedimentary rocks is very difficult. Clastic grains eroded from older rocks yield the absolute age of the original source rock and not the age the sediment was deposited. Cementing minerals provide the age of cementation, which occurs after deposition. Calcite in limestone commonly contains some uranium, which allows application of uranium-lead dating methods (Figure 7.18). Calcite, however, is prone to recrystallization and reaction with pore water, which changes abundances of both the parent and daughter isotopes so that calculated ages may be meaningless.

Figure 7.21 demonstrates how to combine absolute and relative ages to decipher the age of sedimentary rocks. The principles of superposition and cross-cutting relationships relate the relative age of sedimentary rocks to igneous rocks, and igneous rocks provide absolute-age measurements. The approximate absolute age of sedimentary rocks deposited between volcanic ash or lava layers, which can be dated, is determined with this approach. **Figure 7.22** applies this approach to obtain ages for boundaries on the time scale. It is extraordinarily rare to find dateable igneous rocks right at the boundary between two periods defined by fossils, so the age of the boundary is estimated. Geologists revise the ages on the time-scale boundaries to new values with less uncertainty as they obtain more absolute dates that are relevant to establishing boundary ages. For this reason you may find the ages on the time scale in this book (Figure 7.9) to differ from those in older texts, and the ages listed here will doubtless be revised in the future. The names of the time intervals on the time scale do not change, and the fossils that define each time interval are also agreed upon by international convention, but the boundary ages continually fluctuate by small amounts as new absolute-age data become available.

The Oldest Rocks and Age of Earth

Finding the oldest rocks on Earth is a daunting task because the planet is very dynamic. Metamorphism resets most isotope-dating clocks in rocks. Weathering and erosion recycle material from old rocks into younger ones. The accumulation of sedimentary and volcanic rocks at the surface progressively buries older rock, which is reexposed only when tectonic forces and erosion conspire to uplift and remove the

▼ **Figure 7.21** **How to estimate absolute ages of sedimentary rocks.**
This diagram shows an example of how geologists combine relative-dating principles and absolute-dating methods to determine the depositional age of sedimentary rocks.

Relative ages:

Sandstone is younger than basalt lava by principle of superposition.

Basalt dike is younger than sandstone by principle of cross-cutting relationships.

Radioactive isotope absolute ages:

Basalt lava is 25 million years old.

Basalt dike is 20 million years old.

Approximate absolute age of sandstone:

Sandstone deposited between 25 million and 20 million years ago.

covering strata. Earth's oldest rock may lay buried deep below the surface and out of view.

The oldest dated materials on Earth are more than 4 billion years old. The oldest rock found so far is gneiss resulting from metamorphism of a tonalite intrusion in northwestern Canada. The tonalitic gneiss contains the zirconium-silicate mineral zircon, which incorporates radioactive uranium when it crystallizes. Uranium-lead isotope measurements reveal a 4.030-billion-year age for the centers of zircon crystals in the gneiss and are interpreted to represent the original crystallization age of the tonalite before metamorphism. Sandstone in Australia contains zircon sand grains as old as 4.4 billion years, and although no rock this old has been found, the source of these zircon grains may be concealed beneath younger rocks. These sand grains are the oldest dated minerals on Earth.

Earth likely formed at about the same time as other objects in the solar system, as you will learn more about in Chapter 9. The crust of the Moon is laid bare for observation; there is no tectonic activity to cause metamorphism or erosion and no sedimentary processes to recycle and bury old rocks. Samples returned from the Moon by the NASA Apollo missions yield radioactive-isotope ages of 3.05–4.3 billion years for the dark-colored areas visible from Earth, and 3.65–4.56 billion years for rocks collected in the light-colored areas. Meteorites found on Earth are mostly fragments of asteroids that orbit between Mars and Jupiter. A majority of these meteorites exceed 4.4 billion years old, and the oldest is 4.568 billion years old. It is possible that the fully formed Earth is slightly younger than the oldest meteorites. Various estimates of Earth's age range from 4.4 to 4.56 billion years, so 4.50 ± 0.06 billion years old is a reasonable expression of the planet's age as currently interpreted (Figure 7.9).

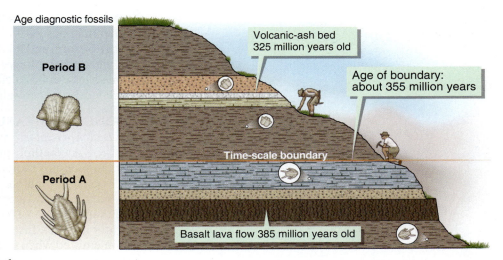

▲ **Figure 7.22 How to determine absolute ages of geologic-time-scale boundaries.**
Fossils found in sedimentary deposits define the geologic time scale but radioactive-isotope dating methods rarely give reliable measurements of the depositional age of sediment. This diagram illustrates how to use absolute ages of igneous rocks to estimate the age of a hypothetical time-scale boundary.

EXTENSION MODULE 7.3
Using Geologic Clocks.
Learn the different radioactive-isotope dating methods and how they are used to determine the age of geologic materials.

*Putting It Together–**How Is the Absolute Age of a Rock Determined?***
- Radioactive dating methods provide absolute ages of minerals. Naturally occurring radioactive parent isotopes decay to stable daughter isotopes at known measured rates. Measurement of parent and daughter isotope abundances permits calculation of rock age.
- Absolute ages are determined for the geologic time scale by combining radioactive-dating and relative-dating methods. Age-diagnostic fossils present in sedimentary rocks define the time-scale boundaries. Time-scale boundary ages are estimated from absolute ages of igneous rocks whose relative age relationships to fossiliferous sedimentary layers are known.
- The oldest rock measured thus far on Earth is 4.030 billion years old, and the oldest mineral is 4.4 billion years old. Even older rocks are found on the Moon and in collections of meteorites that have landed on Earth. The current estimated of the age of Earth is 4.50 ± 0.06 billion years.

7.8 How Do We Know ... How to Determine Half-Lives and Decay Rates?

UNDERSTAND THE PROBLEM
Why Is It Important to Measure Radioactive-Decay Rates?
Measuring the abundances of parent and daughter isotopes in a mineral is insufficient for determining the mineral's age unless the rate is known for the decay of the parent to the daughter isotope. Chemical analyses of the abundance of elements, and individual isotopes of elements, have been routine for

several decades. The abundances of most isotopes are measured down to minute fractions of 1 percent. Measuring isotope abundances is reasonably straightforward, and it is factual data. Different values for decay rates will, however, yield different ages based on these same factual analyses. No radioactive-isotope ages can be calculated without confident knowledge of the decay rates. Geochemists use carefully designed experiments to measure these rates.

MAKING THE MEASUREMENTS
What Data Are Required to Calculate Decay Rate?
The ^{40}K-^{40}Ar method provides an example of how geologists design experiments to measure radioactive decay. The half-life for the decay of ^{40}K to ^{40}Ar plus ^{40}Ca is 1.25 billion years, as Figure 7.18 shows. The very long half-life implies that the decay rate is very slow. Clearly, no one conducts an experiment for more than 1 billion years to see that half of the original ^{40}K atoms converted to daughter isotopes. Instead, the rate of decay is determined by measuring the radiation emitted when each parent atom decays. The number of decays during a particular time interval depends on how many parent atoms exists in the sample, as noted from the example in Figure 7.17. The decay-rate calculation, then, requires two types of data:

1. Measure the amount of parent isotope in the sample, which is possible to far better than 1 percent accuracy.
2. Measure the amounts and types of radiation emitted during decay, which requires further consideration.

Two radiation-producing processes take place during decay of ^{40}K:

1. Conversion of ^{40}K to ^{40}Ar occurs when an electron orbiting near the nucleus enters into the nucleus and combines with a proton to produce a neutron. This causes rearrangement of electrons circling the nucleus, which produces x-rays.
2. Conversion of ^{40}K to ^{40}Ca happens when a neutron in the nucleus changes to a proton with simultaneous production of a negatively charged electron that exits the nucleus.

Geochemists use detectors to measure the release of x-rays and electrons from highly purified samples containing a measured amount of potassium. Each detection of an x-ray or an electron represents the decay of a ^{40}K atom.

VISUALIZE THE RESULTS
How Fast Does ^{40}K Decay?
Figure 7.23 summarizes many laboratory measurements of the radiation released during decay of ^{40}K. As measurements continue to improve, more recent experiments provide more reliable values. Many different scientific research groups, in different laboratories using different starting samples and different types of radiation-counting instruments, obtained very similar values. Reproducibility is an essential part of establishing the merits of scientific results.

In 1976 an international commission of geochemists adopted values for the two decay paths of ^{40}K to its daughter isotopes. Figure 7.23 depicts these selected values. The geochemists did not simply average all of the available numbers but gave greater importance to the more recently obtained values as these were found to reflect smaller uncertainty.

These data confirm a very slow radioactive decay. Adding the two decay paths together reveals that between 31 and 32 ^{40}K atoms decay each second in every gram of potassium. This may seem like a lot of decays, but one gram of potassium contains 1.8×10^{18} atoms of ^{40}K atoms, so the ^{40}K abundance is decreasing very, very slowly, so that 1.25 billion years must go by before half of the ^{40}K decays to its daughter isotopes, ^{40}Ar and ^{40}Ca.

Is the half-life of 1.250 billion years *exactly* correct? Not in the strictest sense, because the experimental results yield small differences. The differences

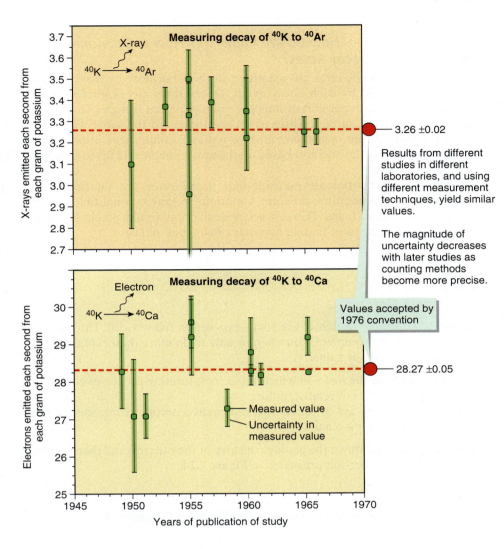

Measurements of radioactivity emitted from potassium reveal the rate of decay of ^{40}K to its two daughter isotopes. X-ray measurements reveal decay of ^{40}K to ^{40}Ar and the number of electrons released reveals decay to ^{40}Ca. These graphs show different measurement results of both decay rates over a 20-year period.

in decay rates are, however, so slight that you can have a high degree of confidence that the half-life is uncertain by no more than one-half of 1 percent (or about six million years).

INSIGHTS

How Reliable Are Decay Rates?

The consistent reproduction of experimental determinations of decay rates leave geologists very confident of their conclusion that Earth is about 4.5 billion years old. Measurements of the abundances of the parent and daughter isotopes in minerals and the rate of decay of parent to daughter isotopes are confirmed by repeated laboratory measurements. Another key question, however, is whether these decay rates are always the same or if they might change under different conditions. If decay rates are not constant, then the laboratory measurements are not valid for calculating mineral ages.

Many experiments have been conducted to test the constancy of decay rates, typically for radioactive isotopes that disintegrate faster than ^{40}K. These experiments checked for changes in decay rates by varying temperature from $-250°C$ to $1550°C$, by applying pressure as great as 2000 times that experienced at Earth's surface, by varying the force of gravity, and by employing magnetic fields more than 160,000 times as strong as the natural field at Earth's surface. No results showed variations in the rate of decay that would have a noticeable effect on the calculation of a mineral's age.

Putting It Together–How Do We Know . . . How to Determine Half-Lives and Decay Rates?

■ Radioactive-decay rates are measured in the laboratory by detecting the energy released by each decay event. Decay rates are expressed as the number of decays per unit of time per gram of parent isotope.

■ Decay rates measured in different laboratories are the same within the uncertainty of the measurement. International committees establish the universally used decay rates based on the most reliable and up-to-date laboratory values.

■ Laboratory experiments indicate that decay rates are unaffected by changes in temperature, pressure, variations in gravitational field, or different magnetic fields. There is no evidence that would preclude use of measured decay rates to date minerals of all types and ages.

7.9 How Do You Reconstruct Geologic History with Rocks?

Return to your field sketch of the rocks exposed in the sea cliff. The picture is now literally more complete. You are now able to do more than explain the origin of the rocks, one at a time.

• You can explain the *order* in which these rock-forming processes took place, using the principles of relative dating.

• You can, with the aid of fossils and radioactive-isotope dates, specify *when* these processes took place.

Stop here and write down the geologic history of the outcrop and then compare your history to the version presented in **Figure 7.24**.

Putting It Together–How Do You Reconstruct Geologic History with Rocks?

■ Both relative and absolute dating methods, which reveal the order and the age of events, allow geologists to determine the geologic history of an area.

■ Combined with knowledge of rock-forming processes and the origins of other geologic features, relative and absolute dating methods permit narrative descriptions of geologic history.

▶ **Figure 7.24 The geologic history recorded in the sea cliff.**
Here is a reasonable historical narrative for the sea-cliff outcrop illustrated in Figure 7.1. This history is consistent with principles for establishing the relative order of events recorded in rocks, the knowledge of how to use fossils to establish the geologic-time-scale age of sedimentary strata, and the use of radioactive-isotopes to establish the absolute age of rock-forming minerals. Only events occurring during time represented by rocks preserved in the outcrop are described with great confidence. The processes occurring during the intervening times are inferred only by the presence of unconformities.

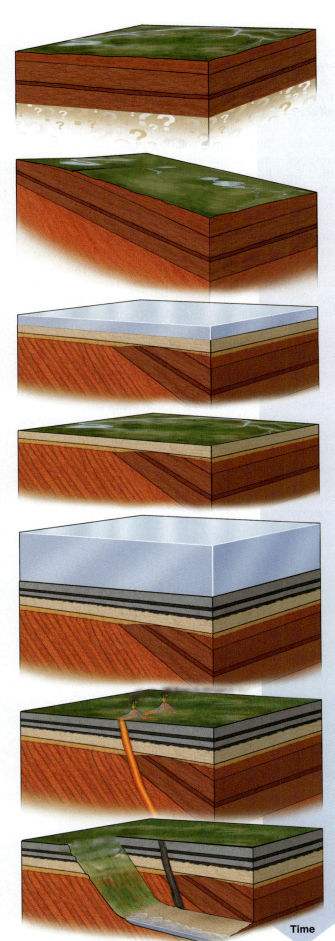

Rivers deposit sediment during the Pennsylvanian Period

Application of the principle of superposition requires that the lowest exposed layers reveal the earliest known history. The red sandstone and shale layers contain fossils of nonmarine organisms characteristic of the Pennsylvanian Period (see Figure 7.10). The cross-bedded sandstone records deposition by rivers with fine silt and clay accumulating on floodplains now represented by the shale. The geologic time scale (Figure 7.9) indicates that the Pennsylvanian sedimentary strata formed during some part of the interval between 300 and 318 million years ago.

Late Paleozoic or early Mesozoic Tilting

Tectonic forces tilted the rocks during the time following Pennsylvanian deposition but before accumulation of overlying Jurassic tan sandstone. You know this because the inclined red Pennsylvanian strata were originally nearly horizontal. Tilting of the Pennsylvanian rocks and erosion to produce a new surface for later deposition took place during some part of the late Paleozoic or early Mesozoic Eras, between about 300 and 200 million years ago (see Figures 7.9 and 7.10).

Sedimentation in the Jurassic Sea

Submergence of the tilted and eroded red sandstone and shale caused deposition of tan beach sandstone containing Jurassic fossils. Fragments of Pennsylvanian sedimentary rocks were included in the Jurassic beach sand. Based on the time scale (Figure 7.9) this sea existed during part or all of the Jurassic Period between 146 and 200 million years ago.

A pause in deposition

A disconformity forms the top of the Jurassic rocks (see Figure 7.10), so there was a pause in deposition and, quite likely, some erosion of rock. You do not know how long this period of nondeposition and erosion persisted.

Deposition in the early Tertiary Sea

The area submerged again during the early Tertiary, to account for deposition of gray limestone and black shale. The fossils in these rocks suggest that deposition took place during some part of the time interval between 36 and 65 million years ago (see Figure 7.10).

Oligocene igneous activity

During the late part of the Oligocene Epoch, at about 25.8 million years ago based a ^{40}K-^{40}Ar age (Figure 7.20), tectonic processes formed basaltic magma that intruded as a dike. You can speculate that the dikes fed volcanoes at the surface, but if this is true, those volcanic rocks eroded away.

Erosion to form the present landscape

Erosional processes returned sometime after 25.8 million years ago. You know that rock eroded away because the basaltic dike is exposed at the surface. Dikes are intrusive igneous rocks that are exposed only by erosion of overlying rock.

Time

Where Are You and Where Are You Going?

Rocks record the antiquity of Earth's history and the multitude of events, slow and fast processes, and the ever-changing environments that characterize this history. Not only do features in a rock tell geologists about the processes that formed it, but it is also possible to determine the relative and absolute ages of these rock-forming processes.

Relative ages describe the sequence of events. The principles of superposition, original horizontality, inclusions, and cross-cutting relationships allow you to place the events recorded in rocks into a sequential order from oldest to youngest. The principle of lateral continuity of beds permits you to extend knowledge from one location to another where more rocks may be exposed to gain a more complete picture of geologic history. Fossils play an important role in determining the relative ages of rocks over large areas and even globally. Within a relative-age framework defined by superposition of fossiliferous sedimentary layers, it is clear that most organisms existed for only short intervals of the vast history of the planet. As a result, the types of fossils in rock reveal that the rock formed during a discrete interval of Earth history. The geologic time scale formally defines and names these intervals.

Absolute ages of rocks refer to how much time passed since the rocks formed. Natural radioactive decay of elements commonly found in rock-forming minerals serves as the foundation for dating rocks. Parent isotopes decay to daughter isotopes at rates that are measured and verified by laboratory experiments. These values for the rocks continue to be improved and refined. Half-life is a convenient way to express the rate of decay and is the elapsed time required for half of the parent to convert into an equal amount of daughter products. The abundances of the parent to daughter isotopes are routinely measured with high precision. Along with careful field and microscopic study of the sample, the measurements of isotope abundances are combined with knowledge of the half-life in order to determine when component minerals formed in the rock.

Radioactive-isotope dating allows estimation of boundary ages on the geologic time scale. This work is required because isotope-dating techniques are best suited for providing crystallization ages for igneous and metamorphic rocks, but fossils found in sedimentary rocks define the geologic time scale. Applying principles of superposition and cross-cutting relationships, it is possible to estimate the age of the time-scale boundaries by obtaining isotope ages on interlayered and cross-cutting igneous rocks. Both relative-age relationships and absolute-age measurements are key to deciphering geologic history.

Radioactive-isotope ages allow estimates of the age of Earth. Nineteenth-century estimates of Earth's age based on the geothermal gradient and the saltiness of the ocean produced estimates between about 20 and 100 million years. The assumptions involved in making these calculations have since been shown to be invalid. The oldest absolute-age determination for minerals formed on Earth is 4.4 billion years, but older rocks have been returned from the Moon and found among meteorites that have fallen to Earth. These factors have been combined to suggest that Earth originated about 4.5 billion years ago.

Your knowledge of Earth materials, extensive as it now is, still pertains only to what is known about the outermost skin of the planet. In the next three chapters you will learn how geologists determine the materials and processes within Earth's interior.

 ## Active Art

Relative Dating Principles. See how the relative dating principles are used to decipher the sequence of geologic events.
Unconformities. See how the three types of unconformities form.

Kelvin's Calculation of Earth Age. See how Kelvin determined the age of Earth.

Radioactive Decay. See how the abundance of parent and daughter isotopes changes over time because of radioactive decay.

Extension Modules

Extension Module 7.1: Radioactivity and Radioactive Decay. Learn the different ways that radioactive isotopes decay and the resulting levels of natural radioactivity.

Extension Module 7.2: The Mathematics of Radioactive-Isotope Decay. Learn how to use mathematical equations, instead of graphs, to calculate mineral ages.

Extension Module 7.3: Using Geologic Clocks. Learn the different radioactive-isotope dating methods and how they are used to determine the age of geologic materials.

Confirm Your Knowledge

1. Distinguish between relative age and absolute age. Give an example of each.

2. Which principle(s) would you use to determine the relative ages of a beds in sequence of flat-lying sedimentary rocks? Why?

3. Which principle(s) would you use to determine the relative sequence of events that accounts for an exposure of tilted beds of sedimentary rocks? Why?

4. Which principle(s) would you use to determine the relative ages of an intrusive igneous rock, such as a dike or a stock, and the rock adjacent to the dike? Why?

5. What principle(s) would you use to determine the relative ages of a granitic intrusion when there are granite pebbles in the sedimentary rock overlying the granite? Why?

6. What is an isotope?

7. What percentage of the known naturally occurring isotopes are radioactive?

8. What is the half-life of a radioactive-decay process?

9. In order to apply the isotope-dating method, what must be known or measured, and what conditions must be met?

10. What is correlation? How can you be sure you are correlating the same layers?

11. What are the three types of unconformities? How does each form?

Confirm Your Understanding

1. Write out an answer for each question in the Chapter Outline for the chapter sections assigned by your instructor.

2. Explain a hypothetical example where the oldest rock exposed in a deep canyon is not at the bottom of the canyon.

3. Approximately what percent of geologic time is represented by the Precambrian? Which Phanerozoic period represents the longest period of time? How long is it? Which Phanerozoic era represents the longest period of time? How long is it?

4. Early scientific estimates of the age of Earth used observations of exposed marine fossils requiring a drop in sea-level (Benoit de Maillet), cooling of a hotter Earth (Lord Kelvin), and the salinity of the ocean (John Joly). What age of Earth did each of these estimates come up with? What assumptions were made for each of these estimates that we now know to be incorrect?

5. How do we know that the ^{14}C method is valid?

6. You need to know the age of a rock collected during a field research project. How do you select the particular radioactive dating method to use?

7. Age determinations from radioactive-isotope decay methods are usually performed on a mineral crystal. Assume that a crystal of feldspar is plucked from a sandstone, analyzed and determined to be 200 million year old. What can you say about the age of the sandstone? What other methods can be used to determine the age of the sandstone?

8. Why do the absolute ages of the boundaries in the geologic time scale change?

9. The K-Ar dating method is applied to dating potassium feldspar in rocks. But it is important to keep in mind just exactly what the resulting numbers mean. For each example, indicate what geologic process or event is being dated:

- A K-Ar date of 2.3 million years on potassium feldspar in a volcanic rock
- A K-Ar date of 535 million years on potassium feldspar in a metamorphosed volcanic rock
- A K-Ar date of 164 million years on potassium feldspar in a clastic sedimentary rock

10. Assume that a particular radioactive-isotope dating scheme has a half-life of 150 million years. Also assume that the parent isotope is commonly incorporated in igneous-rock minerals but that the daughter isotope does not occur in such minerals by the crystallization of magma. Measurements show that one crystal in this rock contains 500 atoms of the parent isotope and 7500 atoms of the daughter isotope. How old is this crystal?

Key Terms

absolute age (p. 169)
angular unconformity (p. 177)
correlation (p. 174)
daughter isotope (p. 184)
disconformity (p. 178)
eon (p. 175)
epoch (p. 175)
era (p. 175)
fault (p. 171)

geologic time scale (p. 175)
half-life (p. 184)
isotopes (p. 183)
nonconformity (p. 178)
parent isotope (p. 184)
period (p. 175)
principle of cross-cutting
   relationships (p. 171)

principle of faunal succession
   (p. 174)
principle of inclusions (p. 172)
principle of lateral continuity
   (p. 173)
principle of original
   horizontality (p. 170)

principle of superposition
   (p. 170)
radioactive decay (p. 183)
radioactivity (p. 183)
relative age (p. 169)
unconformities (p. 177)

8 Journey to the Center of Earth

Chapter Outline

8.1 How do geologists know about rocks in Earth's interior?

8.2 How do earthquakes help make images of Earth's interior?

8.3 How do we know … how to determine velocities of seismic waves in rocks?

8.4 What composes the interior of Earth?

8.5 How hot is the interior of Earth?

Why Study Earth's Interior?

HUMANS HAVE ALWAYS BEEN CURIOUS ABOUT THE INTERIOR OF Earth. Myths tell of gods, demons, or whole civilizations that reside in an inner world. R. D. Oldham, an early twentieth–century geologist, appropriately captured the imaginations of scientists and nonscientists, alike: "Of all the regions of the earth none invites speculation more than that which lies beneath our feet" (R. D. Oldham, 1906, *Quarterly Journal of the Geological Society*, v. 62, p. 456). Perhaps no one captured that speculative imagination better than nineteenth-century author Jules Verne in his epic *Journey to the Center of the Earth*, in which explorers descend a volcano in Iceland and undertake a perilous journey through a hypothetical underworld before emerging from another crater in Italy.

Humans have always known that inner Earth is dynamic. Magma rises from below and erupts at volcanoes. Earthquakes, apparently driven by unseen forces from below, heave and split the land surface. Therefore, it is not surprising that earthquakes and volcanic eruptions are central to both myth and scientific investigation about Earth's interior. These events reveal an active interior where energy drives many processes, some visible on the surface that profoundly affect human lives. Along with the age-old natural curiosity about the most inaccessible part of the planet, understanding earthquakes, volcanoes, and the persistent motion of continents provides strong motivation for studying the interior of Earth.

Even if you visited every country on every continent and sailed the seven seas, you would see very little of Earth. Most of the planet is beneath you. The deepest diamond mine in South Africa is 3.6 kilometers deep. The deepest well, in northeastern Russia, penetrates to 12 kilometers. The entire radius of Earth is 6371 kilometers, making these human explorations into the interior of the planet insignificant in comparison.

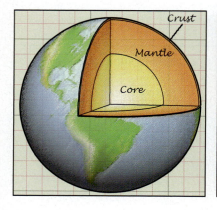

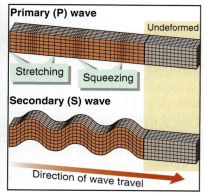

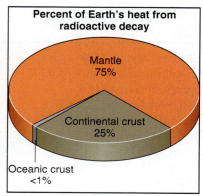

The diamonds in these specimens from Africa and Siberia formed 150 kilometers below Earth's surface. ▶

You have two related objectives in this chapter:

✔ To learn what geologists know about Earth's interior: layers arranged concentrically, in nearly spherical shells of different composition and physical properties

✔ To understand how geoscientists constructed this view of the interior

In this chapter you will learn that earthquakes are key to our knowledge of inner Earth. Energy released by earthquakes provides a way to make images of the interior.

To achieve the two objectives for this chapter, you will address the following questions:

8.1 How do geologists know about rocks in Earth's interior?

8.2 How do earthquakes help make images of Earth's interior?

8.3 *How do we know ...* how to determine velocities of seismic waves in rocks?

8.4 What composes the interior of Earth?

8.5 How hot is the interior of Earth?

In the
LAB

Determining the internal structure and composition of Earth is similar to trying to figure out what is inside a baseball. Imagine picking up a baseball in your geology laboratory and being told to determine what it consists of—without cutting it in half. A rock saw is available but off limits right now. You must restrict your observations to the ball as a whole, and you are not allowed to cut deeper than through the leather covering. This exercise in "scratching the surface" is analogous to studying Earth. Geologists' direct knowledge of the interior arises from studying little more than the outer skin of the planet.

It makes sense to start making observations as deeply as you can. When you take a knife to the baseball, you see that the stitched leather cover is only a few scant millimeters thick, with tightly wound wool yarn beneath it, as seen in **Figure 8.1**. You have determined that the ball is inhomogeneous—that is, it consists of more than one type of material. Does the yarn continue to the center of the ball? The rules forbid you to dig deeper, so how can you tell?

One option is to compare some property of the entire ball to the properties of the observed leather and yarn. The easiest property to measure is density. If the density of the whole ball is similar to the density of the yarn, then perhaps the remainder of the ball is made of yarn. It takes just a minute to weigh the ball and measure its outer dimensions. From these measurements, written out in Figure 8.1, you determine that the density of the ball is more than the stated densities of yarn and leather. There must be something in the interior that is denser than the materials you found in your shallow excavation at the surface.

What next? Is there a way to look inside and learn about the invisible interior? You might think of x-rays. Doctors figure out much about the interior processes of a living person by using x-rays and other imaging technology. Security officers rapidly determine the contents of airline luggage with scanning devices. A way to interpret the deeper interior of the baseball simply would be to use an x-ray machine, but none is available in your geology lab.

Your instructor ends the suspense and allows you to survey more deeply into the ball. As shown in Figure 8.1, you cut open the ball with the rock saw. Sure enough, the center is higher-density rubber, which fits your conclusion that something else composes the ball besides cowhide and yarn.

How does this exercise provide insights into how to determine the nature of the inaccessible Earth interior? Extending the idea of calculating density, can you determine the density of Earth by comparing it to that of some rocks found near the surface? This would be similar to comparing the density of your baseball to the density of its near-surface components. You will discover that this process provides some clues but not a complete picture. Extending the idea of imaging to study Earth's interior, can you zap the planet with x-rays? No, aside from the fact that no x-ray machine is large enough, these electromagnetic waves are not likely to be helpful for imaging Earth because rock, like bone, is mineral matter and nearly opaque to x-rays. How then have geologists developed cross sectional views of Earth like the one sketched in the lab book, at the bottom of Figure 8.1? The answer is that geologists have found another way to get a three-dimensional image of the planet.

Figure 8.1 What the insides of a baseball and Earth look like. ▶
The concentric layers of different materials composing the interior of a baseball are analogous to the interior of Earth. You can cut open a baseball to reveal the insides, but how do geologists draw a picture of Earth's interior? Your primary objective in this chapter is to see how scientists learn about the interior of the planet.

Lab notes: Baseball analogy for Earth's interior

Cut into baseball with a knife.

Cowhide leather
makes up the
outer layer.

Wool yarn
makes up next
lowest layer.

Calculation: density of baseball

$Density = \dfrac{mass}{volume}$

Mass measurement = 146 grams

Sphere volume $= \dfrac{4}{3}$ (radius)^3

Radius measurement = 3.81 cm

Volume of ball $= \dfrac{4}{3}$ (3.81 cm)^3 = 231 cm^3

$Density = \dfrac{146 \ grams}{231 \ cm^3} = 0.632 \ g/cm^3$

Is the rest of the baseball made of yarn?
Prof says that the overall density of the tightly wrapped yarn and leather cover is 0.611 g/cm^3. Calculated density of ball is greater than this value. This implies that there is something else in the interior of the ball that is denser than yarn and leather.

Looking farther inside the baseball
We cut the baseball open and found a black rubber ball at the center.

Rubber ball has a radius of 1.61 cm and a mass of 22.4 grams. The density of the rubber ball, therefore, is 0.775 g/cm^3. The rubber-ball center is denser than the yarn and leather, as predicted.

Both the baseball and Earth have concentric layers:

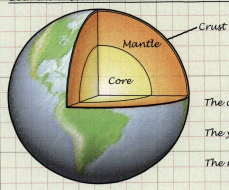

Crust

Mantle

Core

The cowhide cover on the baseball is analogous to Earth's crust.

The yarn composing most of the interior of the baseball is analogous to Earth's mantle.

The rubber ball at the center of the baseball is analogous to Earth's core.

8.1 How Do Geologists Know about Rocks in Earth's Interior?

Geologists gain knowledge about Earth's interior in a manner similar to your analysis of the baseball. They make real observations as far as they can—by looking as deeply as possible in drilled boreholes and mines, and by studying rocks that formed deep below the surface and were later brought to the surface by dynamic Earth processes. Beyond these observations, geologists use two approaches similar to the study of the baseball:

- Geologists infer the interior composition at deep levels from calculations of Earth's density.
- Geologists use energy from earthquakes to construct images of Earth's interior.

How Far Can We See into Earth?

You learned in Chapter 6 about outcrops of metamorphic rocks that originated far below the surface. The metamorphic minerals in most of these rocks formed at temperatures and pressures that are present at depths of 10–50 kilometers within Earth. Later uplift and erosion exposed these metamorphic rocks, providing insights into deep continental crust.

Surface outcrops and drill-hole data reveal that continental crust consists primarily of quartz-rich igneous and metamorphic rocks, depicted in **Figure 8.2**a. Sedimentary strata are also common at the surface, but these rocks are rarely more than 5 kilometers thick. Felsic igneous rocks, such as tonalite and granite, are present at deeper levels along with gneiss containing high abundances of quartz, feldspar, and mica. Metamorphic rocks from deeper than about 15–25 kilometers are typically amphibolite or eclogite, which contain less quartz and more amphibole, pyroxene, and garnet. These deeper rocks are probably metamorphosed mafic igneous rocks.

Volcanoes provide additional clues about the interior because some eruptions eject pieces of rock torn loose by magma rising from deep below the surface. **Figure 8.3** shows two examples of these rocks. More exotic samples like those pictured at the beginning of this chapter contain diamonds, which form at depths of about 150 kilometers. Far more common are pieces of granite, gneiss, and other materials similar to those found in outcrops of plutonic and metamorphic rocks that are exposed at the surface by uplift and erosion. Also found are chunks of peridotite that contain garnet in addition to olivine and pyroxene. Laboratory experiments indicate that garnet peridotite forms at depths greater than 50 kilometers. Taken together, these pieces of evidence imply that below the typical quartz-bearing rocks of continental crust there must be rocks of strongly contrasting ultramafic composition. This is the boundary between crust and mantle pictured in Figure 8.2a.

Geologists also have substantial knowledge of rocks found beneath the oceans. In the nineteenth century, ocean-going scientists explored the composition of the seafloor by dragging buckets across the bottom of the ocean at the end of long chains and hauling them back to the ship's deck. These buckets contained muddy oozes composed primarily of the remains of microscopic plankton and chunks of basalt.

▼ **Figure 8.2 Visualizing the thickness and composition of crust.**
Continental and oceanic crusts, with different thickness and composition, overlie peridotite mantle. Notice that the vertical scale is different for the two illustrations—continental crust is thicker than oceanic crust.

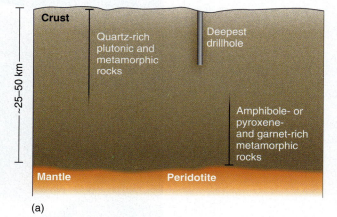

Continental crust:
- Typically ranges from 25–50 km thick.
- Mostly a mix of plutonic and metamorphic rocks.
- Rock compositions are more felsic in the upper crust, and more mafic in the lower crust.

(a)

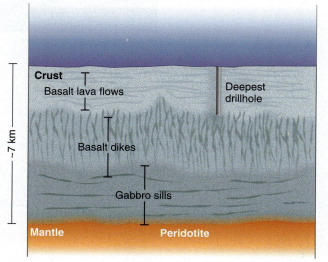

Oceanic crust:
- Usually about 7 km thick.
- Composed of mafic igneous rocks.
- Gabbro sills pass upward into closely spaced basaltic dikes, which were the volcanic feeders to basalt lava flows found on the sea floor.

(b)

More sophisticated exploration beneath the seafloor began in the 1960s, when ship-based drill rigs began operation. This international scientific undertaking has drilled holes in the sea bottom at more than 2000 locations worldwide. Many of these wells penetrate deeply into the seafloor basalt, to as much as 2.1 kilometers. Rocks recovered from the wells reveal crust that is nearly devoid of quartz and overwhelmingly dominated by basaltic lava flows and intrusions as shown in Figure 8.2b. These mafic oceanic rocks contrast sharply with the felsic composition of most continental rocks at similar depths.

Oceanic crust is also present within mountain belts on continents, in places where subduction-zone processes shove slivers of seafloor onto continental crust. **Figure 8.4** illustrates a landscape made of this kind of uplifted seafloor. Sedimentary layers of chalk and chert embedded with deep-sea fossils resting on basaltic lava flows reveal the oceanic origin of these rocks. Stacks of lava flows, like those encountered in the deep-ocean drill holes, are commonly more than 2 kilometers thick, and they are increasingly interrupted at greater depths by dikes, which rise from a yet lower zone of gabbro sills. In other words, this rock is strikingly similar to the oceanic crust illustrated in Figure 8.2b. In the thickest uplifted seafloor slices the mafic igneous rocks are underlain by peridotite or its metamorphosed equivalent, serpentinite. The total thickness of the uplifted mafic oceanic-crust rocks is about 7 kilometers, resting on peridotite mantle.

The crusts of oceans and continents, therefore, differ in composition and thickness (Figure 8.2). Continental crust contains mostly quartz-bearing igneous and metamorphic rocks, with an average composition of diorite to tonalite, and it is commonly greater than 25 kilometers thick. Oceanic crust consists of mafic igneous rocks and is typically about 7 kilometers thick. Both types of crust rest on top of peridotite mantle.

Remarkable as it is that we have samples of rocks that formed tens of kilometers below the surface, geologists overall have a rather poor sampling of the whole Earth. Samples available for study represent less than the outermost one

▲ **Figure 8.3 What volcanoes bring up from inside Earth.**
Volcanoes sometimes eject pieces of rock from deep inside Earth that are torn loose as magma moves toward the surface. The rock on the left is a fragment of green peridotite enclosed in basalt. The peridotite contains an assemblage of minerals not seen in rocks known from Earth's crust and is interpreted to be a piece of the mantle. The fragment of metamorphic rock on the right contains minerals that indicate pressures of formation more than 15 km below the surface and represents a piece of deep continental crust.

▲ **Figure 8.4 What uplifted mantle and oceanic crust looks like.**
The photo on the left shows a sliver of mantle peridotite and basaltic oceanic crust that was shoved up and over shallow-water limestone and uplifted to form this mountainous landscape in Turkey. A closeup of a similar uplifted mantle sliver in Oman, on the Arabian Peninsula, shows peridotite sliced through with closely spaced gabbro dikes.

percent of the radius of the planet. They tell you no more about the whole Earth than the leather tells you about the entire baseball. As with the baseball, however, Earth's visible rocks do reveal inhomogeneity, with a variety of rock types in the crust resting on peridotite. Does this peridotite continue on to the center of Earth? If not, what else may be there?

What Is the Density of Earth?

You deduced that the baseball's center could not be composed entirely of yarn by comparing the density of the entire baseball to the materials you could see. Is it possible to conduct a similar test to help understand the composition of the interior of Earth? To calculate the density of Earth you need to know its shape, its size, and its mass.

In the fifth century B.C., Greek observers established that Earth was nearly spherical in shape. They reached this conclusion by observing how ships seem to disappear as they sail beyond the curving horizon and the curved outline of Earth's shadow on the Moon during lunar eclipses.

The astronomer Eratosthenes used some simple surveying and measurements of shadows in sunlight to estimate Earth's size, placing the planet's radius at about 6350 kilometers. In the seventeenth century Sir Isaac Newton calculated that a rotating planet would not be perfectly spherical but must bulge at the equator and be somewhat flatter at the poles. Refinement of surveying methods led to the current calculation that Earth has a radius of 6378 kilometers at the equator, 6357 kilometers through the poles, and an average radius of 6371 kilometers. Using these values, one can easily calculate the volume of Earth.

Determining the mass of Earth is much more challenging than placing a baseball on a scale, and this mass-measurement process has played out over centuries. Newton formulated the laws of gravity that explain the mutual attraction of objects and the orbit of planets around the Sun. Other scientists combined Newton's mathematical expression of gravitational force, along with ingenious experiments during the late eighteenth and early nineteenth centuries to calculate Earth's mass at 5.976×10^{24} kilograms. More precise measurements during the twentieth century, which calculate the effects of Earth's mass on satellite orbits, arrive at a value of 6.001×10^{24} kilograms.

Combining the measurements of shape, size, and mass, the density of Earth turns out to be about 5.5 g/cm³. Typical mantle peridotite has a density of only about 3.2 g/cm³. As with the baseball, Earth's near-surface material has a different density than that of the whole sphere. The deeper interior must be much denser than the rocks that geologists know to form the outermost layers.

Is it possible that peridotite is denser at greater depth due to the weight of overlying rock compressing mineral structures more tightly? On the other hand, would not minerals expand as temperature increases deeper and deeper into Earth? Laboratory experiments show that the density increase caused by increasing pressure is greater than the expansion due to higher temperatures. These experiments suggest, however, that the density of hypothetical peridotite at the center of Earth would still be insufficient to explain the density of the whole planet.

The notion of a very dense core surrounded by a less dense mantle and a thin covering of crust has been a part of scientific knowledge since the late nineteenth century. Measurements of Earth's wobbly rotation on its axis allow astronomers and geophysicists to conclude that the planet has a distinct

▼ **Figure 8.5** Acceptable dimensions of the mantle and core.

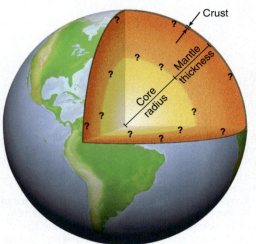

The average Earth density is 5.5 g/cm³, and calculations based on the planet's wobbly rotation require that most of the mass is concentrated toward the center. If Earth is mostly composed of mantle and a denser core, how thick are each of these layers?

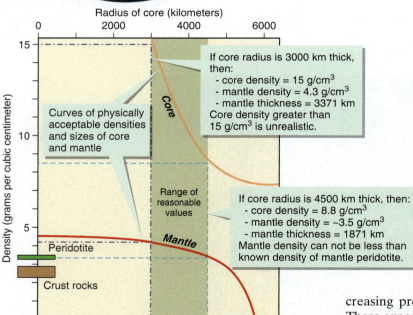

If core radius is 3000 km thick, then:
- core density = 15 g/cm³
- mantle density = 4.3 g/cm³
- mantle thickness = 3371 km

Core density greater than 15 g/cm³ is unrealistic.

Curves of physically acceptable densities and sizes of core and mantle

Range of reasonable values

If core radius is 4500 km thick, then:
- core density = 8.8 g/cm³
- mantle density = ~3.5 g/cm³
- mantle thickness = 1871 km

Mantle density can not be less than known density of mantle peridotite.

Active Art

Dimensions of the Mantle and Core: *See how to determine physically possible thicknesses of Earth's mantle and core.*

concentration of high-density material near the center rather than a gradual inward transition from low-density to high-density rock.

Combining knowledge of Earth's density and the mechanics of rotation places limits on the density of mantle and core, as shown in **Figure 8.5**. The exact dimensions and densities of these concentric layers remained speculative, however, until the early twentieth century. What was missing was a way to create an image of the interior in much the same way you thought of x-raying your baseball to determine its internal structure.

Putting It Together—*How Do Geologists Know about Rocks in Earth's Interior?*

■ Some rocks seen at Earth's surface originated 10–50 kilometers, or rarely as far as 150 kilometers, below the surface. These rocks reveal mostly felsic continental crust and mafic oceanic crust, both of which overlie peridotite mantle.

■ The rocks found near the surface cannot explain the density of the whole Earth. Along with wobbles in Earth's rotation, the high overall density of the planet implies a central core of material that is much denser than mantle peridotite.

8.2 How Do Earthquakes Help Make Images of Earth's Interior?

Earthquakes occur around the globe, with particular concentrations near plate boundaries (see Figure 1.10). News headlines focus on large earthquakes that cause tremendous destruction and loss of life, but thousands of smaller quakes happen everyday. Chapter 11 explores the causes and effects of earthquakes and at this point you will consider a scientifically beneficial aspect of earthquakes—the use of earthquake waves to make images of Earth's interior.

Visualizing Waves

Consider what happens when you throw a pebble into a pool of water. **Figure 8.6** shows ripples forming on the surface of the water and, moving outward in ever-widening circles. These ripples are waves. **Waves** are disruptions that move through a medium, in this case water, without any overall transport of the medium in the direction of wave movement. The amount of energy released at the water surface by the pebble as it hits determines the height of the water waves. Movement of water by the waves is greatest at the surface and decreases downward in the pool (Figure 8.6). If you are underwater when a pebble plunks into the pool, you may not feel the surface waves pass over you, but you may hear a sound. Sound waves move radially outward in three dimensions away from the splash point and not just along the surface (Figure 8.6).

Water molecules move when the waves pass by but then return to their original position. For this reason the waves are **elastic**—they do not permanently deform the water. Elastic deformation also

EXTENSION MODULE 8.1
Sizing Up Earth.
Learn how Eratosthenes determined the radius of Earth and how eighteenth-century scientists calculated the mass of Earth.

▼ **Figure 8.6** Waves in water.
Circular ripples expand outward from where a pebble impacts the water surface. When the pebble hits the water, waves move on the water surface, and sound waves move outward and downward through the water. The inset shows that the water motion caused by the passing surface waves diminishes downward and is not noticed by the swimmer. Sound waves, however, move as a pulse of changing water pressure through the water so the swimmer hears the sound of the pebble hitting the surface.

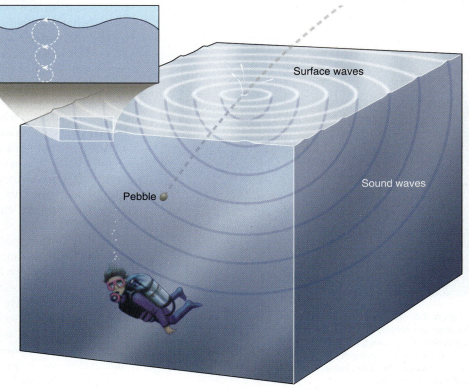

Circular particle motion caused by passing surface wave decreases downward

Surface waves

Sound waves

Pebble

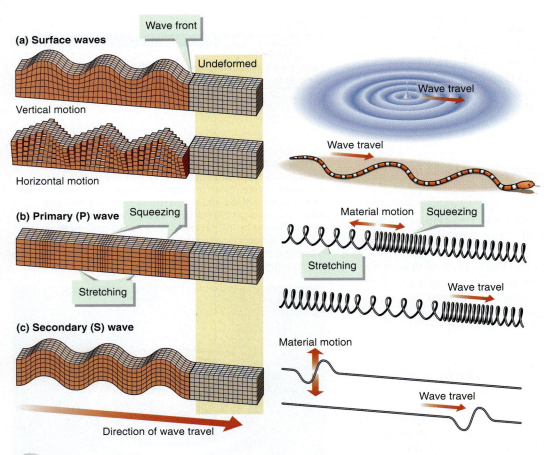

(a) Surface waves

Wave front

Undeformed

Vertical motion

Horizontal motion

(b) Primary (P) wave   Squeezing

Stretching

(c) Secondary (S) wave

Direction of wave travel

Wave travel

Wave travel

Material motion   Squeezing

Stretching

Wave travel

Material motion

Wave travel

Active Art

Seismic Wave Motion: *See how the different seismic waves move.*

▲ **Figure 8.8 How seismometers record earthquakes.**
These seismometers consist of pens suspended above rotating drums that are covered with a sheet of paper. Earthquake waves move the drums, causing the pens to trace out the movement on the paper while the drum turns. This paper record of the earthquake is a seismogram.

◄ **Figure 8.7 How earthquake waves move through rock.**
Each diagram portrays how earthquake waves deform originally cubic blocks of rock, along with an analogy.
a. Surface waves distort the rock vertically, like ripples in water, or horizontally, like a slithering snake. The deformation decreases downward into Earth.
b. The P body wave moves through rock as a pressure pulse, alternately squeezing and stretching the blocks. Pushing on a spring produces a similar kind of wave.
c. The S body wave moves through rock by vertical twisting distortions of the blocks. Shaking a string produces a similar kind of wave.

happens when you stretch a rubber band and then release it, because the rubber band returns to its original shape after being deformed. In contrast, if you stretch a piece of clay it does not return to its original shape when you let go of it; this type of deformation is **plastic**, and it is permanent.

Earthquake Waves

Earthquakes are nearly instantaneous releases of stored energy resulting from the breaking and sudden movement of rock under stress. The release of energy generates elastic waves of three types illustrated in **Figure 8.7**: surface waves, primary (P) waves, and secondary (S) waves.

The **surface waves** are similar to the ripples on a pond surface, except they not only disrupt water surfaces but also cause motion in rock and soil when they pass. Some surface waves cause Earth's surface to roll up and down like ocean waves whereas others cause sideways motion like a slithering snake. Like water waves, the surface wave disturbance decreases downward into Earth (Figure 8.7a). Most destruction by earthquakes results from the passage of surface waves, which cause the ground to shake up, down, and sideways by as much as a meter or more.

The other two types of waves are **body waves**, which move through Earth below the surface. Body waves consist of primary (P) and secondary (S) waves. **P waves** displace material in the same direction that the wave is moving, which causes alternating squeezing and stretching of the material as the wave passes (Figure 8.7b). **S waves** twist material at right angles to the direction of wave motion (Figure 8.7c). Unlike surface waves, which are not confined by the pressure of overlying rock, body waves move rock a few millimeters or centimeters. The magnitude of this motion abruptly decreases as the wave spreads out through Earth.

The different motions of P and S waves determine what these body waves can pass through. Gases, liquids, and solids can all deform by the squeezing and stretching caused by P waves; therefore, P waves can travel through all media. Only solids can twist, so S waves do not travel through gases or liquids.

Seismometers are instruments designed to detect, amplify, and record surface and body wave motion, some of which may not be felt by humans. The instrument name derives from *seismos*, a Greek word that describes ground shaking. This word root also appears in **seismogram**, which is the record from a seismometer like that shown in **Figure 8.8**, and in seismologist, a scientist who studies earthquake records. Similarly, earthquake waves are also called **seismic waves**.

What do earthquakes have to do with deciphering the internal structure of Earth? Records of earthquake body waves allow us to construct an image of Earth's interior, not unlike x-rays reveal an image of the interior of your body. The key is to understand the factors that determine how fast seismic waves travel, especially body waves that penetrate deeply into the planet. The velocity of seismic waves depends on the properties of the material through which the waves move. The velocity also determines the elapsed time between an earthquake and the detection of seismic waves by a seismometer. Seismologists measure these elapsed times at different locations on Earth's surface in order to learn about the properties of materials within the planet through which the waves passed. This knowledge permits imaging of Earth's interior.

A Simple Seismic Experiment

Seismologists learn about seismic-wave properties not only from earthquakes but also from energy released by artificial explosions. Worldwide networks of seismic stations were established during the 1950s and 1960s not to measure earthquakes but instead to monitor underground nuclear-bomb tests during the Cold War. Explosions in rock quarries and mines, and even disasters such as the collapse of the World Trade Center towers in September 2001, generate seismic energy detected by seismometers.

Figure 8.9 illustrates how a seismometer records a hypothetical dynamite explosion at a gold mine. The seismogram records the arrival of three waves. The primary (P) and secondary (S) body waves arrive first and second; now you can see that these names relate to the order of wave arrival at a seismic station. Later arriving surface waves cause the remaining disturbances on the seismogram. Figure 8.9 shows how you can calculate the speeds of the seismic waves because the mine geologist recorded the precise time of the dynamite detonation.

Figure 8.10 illustrates different ways of describing the distance between a seismometer and an earthquake. The **focus** is the location of an earthquake within Earth, and the point on the surface directly above the focus is the **epicenter**. The distance along the surface between the epicenter and a distant seismometer is much longer than the straight-line distance through Earth. Geologists usually describe this distance as an angle rather than as a length.

Assume for a moment that the P- and S-wave velocities measured during the mine-explosion experiment (Figure 8.9) also apply to P- and S-wave motion everywhere within Earth. This assumption may or may not be true, but it provides a starting point for investigating Earth's interior. *If this assumption is true, then you can predict the time when the P and S waves from an earthquake will be recorded anywhere around the world. The travel-time curves graphed in*

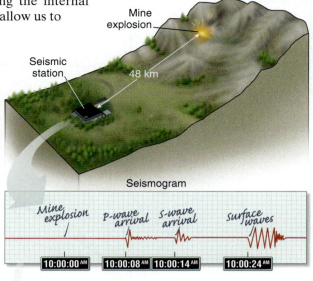

A seismic station records body and surface waves produced by a precisely timed mine explosion.

The seismogram shows the recorded arrival times of P, S, and surface waves from the mine explosion.

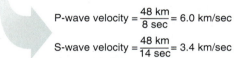

P-wave velocity = $\frac{48 \text{ km}}{8 \text{ sec}}$ = 6.0 km/sec

S-wave velocity = $\frac{48 \text{ km}}{14 \text{ sec}}$ = 3.4 km/sec

The velocities of the P and S waves are calculated by dividing the travel distance by the travel time.

▲ **Figure 8.9** Determining the velocity of earthquake waves.

▼ **Figure 8.10** Describing the distance from an earthquake.

The distance between an earthquake epicenter and a seismometer that records it can be defined by:
• The straight-line distance through Earth,
• The circumference distance along the surface, or
• The angle between lines drawn from the center of Earth to the epicenter and the seismometer.

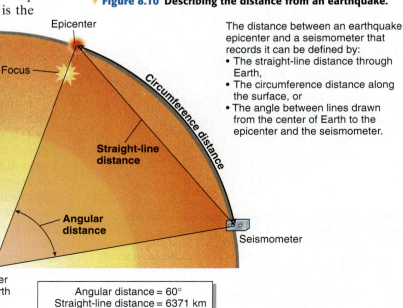

Angular distance = 60°
Straight-line distance = 6371 km
Circumference distance = 6668 km

▶ **Figure 8.11 What earthquake-wave travel-time curves look like.**
The P- and S-wave velocities calculated from the mine-explosion data (from Figure 8.9) predict the travel time of the waves over any distance on Earth; these predictions are shown by the dotted lines. *Actual* arrival times of the waves agree with predicted values only when the travel distances are short. At greater distances, the waves arrive at seismic stations increasingly earlier than predicted. There are also "shadow zones" where the waves are not detected at all, or where P waves are weak and arrive at unexpected times.

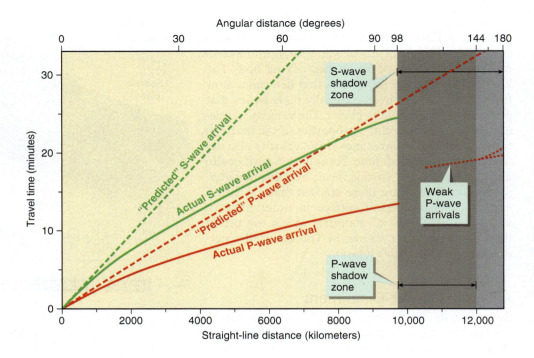

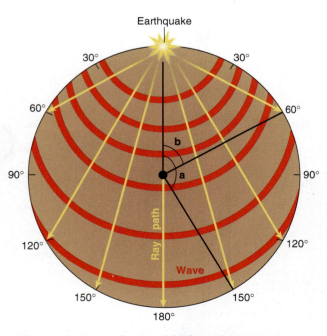

▲ **Figure 8.12 Waves that travel farthest also penetrate deepest into Earth.**
This cutaway view of Earth illustrates how a seismic wave moves through the planet from a hypothetical earthquake focus at the North Pole. It is sometimes easier to visualize the ray path, which are arrows drawn perpendicular to the wave. Ray paths to seismometers at large angular distance from the epicenter (angle *a*, for example) travel through deeper parts of the planet than those that travel small angular distances (such as *b*).

Figure 8.11 show this prediction of P- and S-wave travel times between the focus and seismometers. The predicted curves compare poorly to the real travel times determined from decades of earthquake records from all over the globe. The predicted travel times match up with the actual data only where the travel distance is small. At greater distances the real waves arrive much earlier than predicted. The assumption that P- and S-wave velocities are the same throughout Earth does not hold true.

Why Seismic Waves Speed Up

The arrivals of far-traveled earthquake P and S waves at earlier times than predicted from the mine-blast data (Figure 8.9) indicate that seismic waves speed up as they travel to more distant locations. **Figure 8.12** illustrates the paths of P and S waves through Earth and shows that waves detected at progressively more distant locations from the focus pass through progressively deeper parts of the planet. Geologists conclude that seismic-wave velocities are faster at greater depth into Earth because the average wave velocity increases with increasing distance between the focus and the seismic station.

Careful measurements show that seismic waves move at different velocities through rocks with different properties. Faster wave velocities in Earth's interior, compared to near-surface rocks, imply that rocks in the interior differ from those seen at the surface.

The seismic-wave travel times also help us understand how waves move through rocks with different properties. **Reflection** and **refraction**, illustrated in **Figure 8.13**, describe important processes that occur where waves encounter boundaries between materials with different properties. When a wave encounters a boundary, such as that between layers of different rock types, some of the energy reflects, which means that it bounces back from the boundary as a new wave. The wave energy that crosses the boundary travels at a different velocity consistent with the change in rock properties across the boundary. The velocity change also causes the wave to refract, which means that the wave bends and moves off in a new direction. **Figure 8.14** demonstrates that as seismic velocity increases with depth, refraction causes a wave path to curve upward and eventually emerge at the surface. So the shape of the travel-time curve (Figure 8.11) is consistent with changes in seismic velocity.

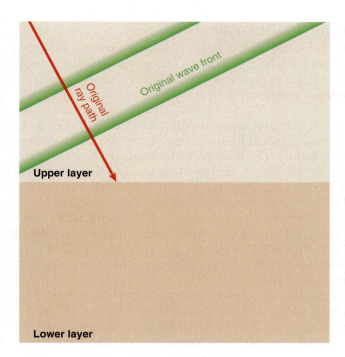

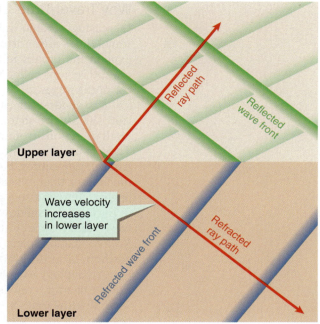

▲ **Figure 8.13** **How waves reflect and refract.**
Two things happen when waves encounter a boundary between layers with different physical properties:
(1) Some of the wave energy reflects off of the boundary like a ball bouncing off of the floor, and (2) some
of the wave energy continues into the lower layer but speeds up or slows down depending on the properties of the two layers. In this example, the wave speeds up in the lower layer causing a change in the direction of the ray path. This change in the direction of wave motion is refraction.

Wave Reflection and Refraction: *See how waves
reflect and refract through rock.*

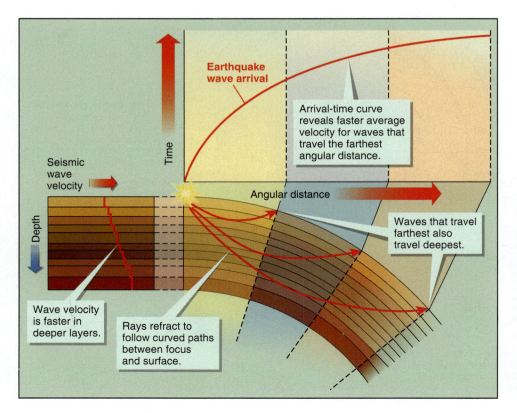

◄ **Figure 8.14** **Earthquake-wave arrival times show that
velocity increases at depth.**
This diagram schematically shows how waves move through a
planet where the wave velocity increases downward in each
deeper layer. Refraction bends the ray paths back to the surface. The farthest traveled ray paths also travel through the
deeper layers where velocity is fastest. This means that ray
paths through deeper layers have a higher average velocity
than waves traveling through shallower layers. As a result,
earthquake-wave arrival times at greater distances are progressively earlier than predicted using constant wave velocities
measured over short distances in shallow layers (compare this
graph to Figure 8.11).

Why Travel-Time Curves Are Discontinuous

Increasing seismic-wave velocities within Earth's interior explains why the travel times are shorter than initially expected, but increasing velocities do not explain the breaks in the curves depicted in Figure 8.11. Particularly notable are the regions called **shadow zones** where seismometers do not record P or S (or both) waves.

The shadow zones are very important to the interpretation of Earth's interior. The S-wave shadow zone begins at locations that are 98° around the globe in all directions from an epicenter. The P-wave shadow zone is more complex. There is an abrupt break in the P-wave arrivals at an angular distance of 98° from the epicenter, but then the waves reappear at 144° although they arrive much later than expected by simple extrapolation of the arrival times at lower angles. There is also a second set of P waves that show up as weak arrivals within parts of the shadow zone, and these waves overlap with the more typical P waves that reappear at 144°. **Figure 8.15** illustrates the distribution of the single S-wave shadow zone and the two P-wave shadow zones for a hypothetical earthquake at the North Pole.

An analogy to light and shadows helps to illustrate these complexities. The S-wave shadow zone is completely "dark" and receives no S waves. The two P-wave shadow zones are "dimly lit" by P-wave arrivals that are inconspicuous except on the most sensitive seismometers. The region between the two P-wave shadow zones is unusually "bright" because seismic stations in this region receive two sets of P waves.

Something dramatic must happen inside Earth at a depth that corresponds to an angle of 98° between epicenter and seismic station in order to account for

▶ **Figure 8.15 How seismic waves move through Earth.**
The curving lines show ray paths through Earth for P waves (on the right) and S waves (on the left) for an earthquake at the North Pole. Earthquake-wave behavior defines the boundaries of the mantle, outer core, and inner core.

• S- and P-wave shadow zones reveal the presence of a liquid outer core at 2900 km depth. S waves do not pass through the outer core, whereas P waves refract strongly at the boundary between solid mantle and liquid core.

• Weak, delayed P-wave arrivals (red ray paths in illustration) reveal wave reflection and refraction from a solid inner core.

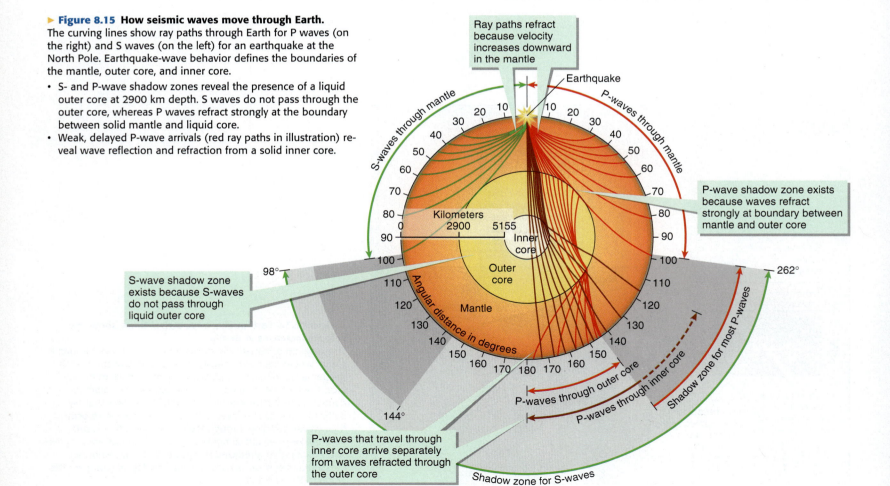

shadow zones. This depth is about 2900 kilometers, and it is within the range of reasonable depths for the boundary between mantle and core (Figure 8.5). What happens here also affects P and S waves differently.

Figure 8.15 explains the shadow zones. The absence of S waves on the opposite side of Earth from the epicenter indicates that these waves do not pass through the core. S waves do not pass through liquid, so this observation indicates that at least the outer part of the core is liquid. Inward refraction of P-wave paths at the core-mantle boundary (Figure 8.15) accounts for the P-wave shadow zones. The pattern of refraction requires a large decrease in wave velocity where P waves pass into the outer core, and this is consistent with a liquid outer core because P waves move more slowly through liquids than solids. The weak P waves recorded in the P-wave shadow zone reflect from a boundary within the core at a depth of 5155 kilometers below the surface. Wave reflection requires distinct inner and outer parts to the core with different properties. Furthermore, travel-time curves indicate that P waves travel faster in the inner core than in the outer core, which implies that the inner core is solid.

The behavior of seismic waves provides a remarkable record of changing physical properties within Earth. Therefore, geologists use earthquake waves to identify confidently the compositionally different interior parts of Earth—crust, mantle, outer core, and inner core. What materials make up these different concentric layers? To answer this question, you need to know the relationship between seismic velocities and rock types so that the velocities can be translated into an interpretation of the materials making up the interior of the planet.

EXTENSION MODULE 8.2
How to Locate an Earthquake.
Learn how to use seismograms and travel-time curves to determine where an earthquake happened.

Putting It Together–*How Do Earthquakes Help Make Images of Earth's Interior?*

- Earthquakes produce surface waves and two distinct body waves, called primary (P) and secondary (S) waves, which travel through the planet. P and S waves reflect and refract when encountering boundaries between materials.
- Seismograms are records of earthquake waves detected by seismometers located around the globe. The time required for different earthquake waves to reach these instruments reveals the velocity at which the waves move through Earth.
- Travel-time curves reveal that, in general, seismic-wave velocity increases at deeper levels in the mantle. Shadow zones where S, P, or both waves are not recorded reveal that Earth has a liquid outer core, beginning at 2900 kilometers below the surface, surrounding a solid inner core at 5155–6371 kilometers depth.

8.3 How Do We Know … How to Determine Velocities of Seismic Waves in Rocks?

UNDERSTAND THE PROBLEM

How Fast Do Seismic Waves Move in Different Rock and in Different Conditions?
The records of seismic-wave arrivals at various distances from epicenters clearly reveal different wave velocities in different parts of Earth. Geologists want to know what is different about rocks in different layers to account for the changing velocities.

Most of the rocks that earthquake body waves move through are concealed from view beneath Earth's surface. In addition, temperature and pressure vary dramatically inside Earth, and these conditions may cause variations in seismic-wave velocities compared to measurements at Earth's surface temperature and pressure. To permit inferences of what rock types relate to different seismic velocities and how velocities vary with increasing temperature and pressure inside Earth, geophysicists design laboratory experiments to measure P- and S-wave velocities in different Earth materials under varying conditions.

EXTENSION MODULE 8.3
Velocity of Seismic Waves.
Learn about the physical properties of rocks that determine how fast P and S waves move.

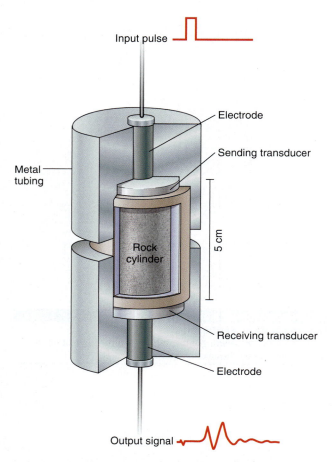

Input pulse

Electrode

Sending transducer

Metal tubing

5 cm

Rock cylinder

Receiving transducer

Electrode

Output signal

▲ **Figure 8.16 How seismic-wave velocities are measured.**
This diagram illustrates an experimental setup used to measure seismic velocities in the laboratory. Metal tubing confines a cylinder of rock, 2.5 centimeters in diameter and 5 centimeters long. A transducer converts an electrical impulse to a body wave. When the wave exits the lower edge of the rock, another transducer converts the body wave into another electrical signal. The time elapsed between the input and output signals reveals the body-wave velocity through the rock.

PICTURE THE EXPERIMENT
How Are Seismic-Wave Velocities Measured in the Lab?
Figure 8.16 schematically illustrates how a typical experiment works. A mechanical signal enters at one end of a cut cylinder of mineral or rock. The time that elapses before the signal arrives at the opposite end is measured. This allows the experimenters to calculate the velocity of wave travel. The setup shown in Figure 8.16 is modified for experiments at different temperatures and pressures to permit measurements at conditions like those inside Earth.

Seismic waves travel at velocities of several kilometers per second, so they pass through a 5-centimeter-long rock sample in mere hundredths of thousandths of a second. Electrical devices are much more accurate than mechanical devices for measuring such short time intervals. For this reason, the original input pulse of energy is electrical. A transducer converts the electrical pulse to a mechanical thump on the rock.

A transducer is simply a device that converts one form of energy to another. You use transducers every day when you make a telephone call. The vibrating sound waves of your voice enter the phone as mechanical energy. A transducer converts the sound to electrical energy that transmits by wire or microwave signal to a receiver. Another transducer in the receiver converts the transmitted energy back to mechanical sound waves.

The transducer at the bottom of the rock sample works in reverse to convert the received mechanical vibration into an electrical signal that is recorded by an instrument. The time between the input pulse and the recorded output pulse reveals the seismic velocity of the rock.

VISUALIZE THE RESULTS
How Does Seismic-Wave Velocity Relate to Rock Properties and Conditions?
Figure 8.17 depicts typical ranges of P-wave velocities in a variety of materials. Notice that unconsolidated sediment has lower seismic velocity than sedimentary rock. This is because the waves move very slowly through the air or water that fills the pore spaces between the solid sediment grains. Seismic waves in igneous rocks are generally fast, and they move faster in mafic gabbro than felsic granite and even faster in ultramafic rocks.

Seismic velocity also varies with changes in temperature and pressure, as the data graphed in **Figure 8.18** show. The experiments show that velocity increases with increasing pressure (Figure 8.18a) and decreases with increasing temperature (Figure 8.18b).

INSIGHTS
How Do Rock Properties Determine Seismic Velocities within Earth?
Velocity varies considerably for several of the materials illustrated in Figure 8.17. Some of these variations result from differences in the abundance of open fractures or other pore spaces where the seismic waves move very slowly through air or water. Variation in mineral abundances within different samples of the same rock also causes variations in seismic velocity. A related observation from Figure 8.17 is that different materials can have the same seismic velocity. A measured P-wave velocity of 4.75 kilometers per second, for example, is compatible with granite, salt, limestone,

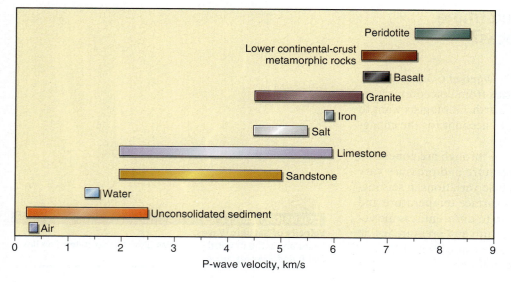

Peridotite

Lower continental-crust metamorphic rocks

Basalt

Granite

Iron

Salt

Limestone

Sandstone

Water

Unconsolidated sediment

Air

0   1   2   3   4   5   6   7   8   9
P-wave velocity, km/s

◀ **Figure 8.17 How fast P waves move in Earth materials.**
This bar graph summarizes laboratory and field measurements of P-wave velocities in a variety of materials at surface temperature and pressure. Velocities will be different for the temperature and pressure conditions within Earth. In all rocks there are natural variations in mineral abundance and open pore space that affect wave velocity, and the lengths of the bars reveal typical velocity ranges.

or sandstone at surface temperature and pressure (see Figure 8.17). Seismic velocity, alone, therefore does not allow geologists to identify rock type.

Figure 8.18 illustrates that increasing pressure and temperature have significant, but opposite, effects on seismic velocity. Both properties increase in value at greater depth within Earth, so experiments at various combinations of pressure and temperature are critical for determining how seismic velocity changes for different materials at different depths below Earth's surface.

How about the effect of density? You might be tempted to speculate that seismic velocity increases when density increases, since high pressure causes minerals to compact more closely together in rocks. Experiments show, however, that although density affects velocities, the velocities are even more strongly influenced by how the minerals in the rock elastically deform when the waves pass through. These elastic properties are more accurately measured in the lab than are the rapid seismic velocities, so the velocities can be calculated from these other results rather than measured directly.

> **Putting It Together–How Do We Know ... How to Determine Velocities of Seismic Waves in Rocks?**
>
> ■ Laboratory experiments relate seismic-wave velocities to different types of rocks, over ranges of temperature and pressure comparable to those in the deep interior of Earth.
> ■ Seismic waves move more slowly in unconsolidated sediment than through sedimentary rocks, and they move even faster in igneous and metamorphic rocks. Within igneous rocks, felsic rocks transmit waves more slowly than mafic rocks, and ultramafic rocks display the most rapid velocities.

8.4 What Composes the Interior of Earth?

A hypothesis for the composition of Earth's interior must take into account all available data. These include seismic data that image the interior, experimental data on seismic velocities of minerals and rocks, the geochemistry of magma resulting from mantle melting, and even the composition of meteorites that arguably represent pieces of small planets that initially formed at the same time as Earth. There is more than one way to combine these observations, so geologists continually test prevailing ideas against new results and modify the hypothesis when necessary. **Table 8.1** summarizes the generally accepted characteristics of

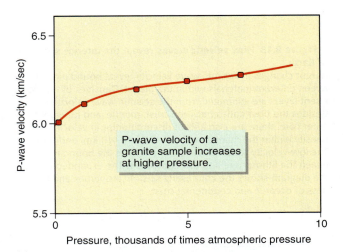

(a)

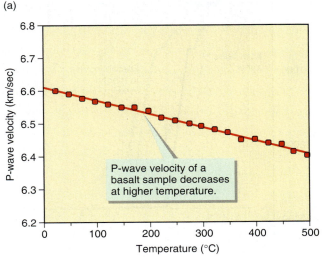

(b)

▲ **Figure 8.18 How pressure and temperature affect seismic velocity.**
Experimental measurements illustrate how P-wave velocity changes at different pressures and temperatures.

TABLE 8.1	Summary of Characteristics of Earth's Inner Layers		
Layer	Thickness	Composition	Solid or Liquid?
Crust			
Continental	Range: 25–85 km; mostly 30–40 km	Mostly igneous and metamorphic rocks with average composition similar to diorite; upper crust has tonalite composition, and lower crust is probably gabbroic	Solid, except for local accumulations of magma
Oceanic	Range: 5–25 km; mostly 5–10 km.	Mafic igneous rocks (basalt and gabbro)	Solid, except for local accumulations of magma
Mantle	~2900 km	Ultramafic igneous composition. Upper mantle is olivine-rich peridotite with pyroxene and garnet. Transition zone (410–660 km depth) and lower mantle (below 660 km) probably have peridotite chemical composition but with silicate minerals that are stable at high pressure and not present in the upper mantle	Almost entirely solid. Small amounts of partially molten rock are likely present at base of the lithosphere (around 100 km depth, on average), and possibly in discontinuous zones just above the core-mantle boundary
Core			
Outer	2255 km	Iron and nickel with as much as 10 percent lighter elements	Liquid melt
Inner	1215 km	Very iron rich, iron-nickel alloy	Solid

▼ **Figure 8.19** **How seismic waves reveal the interior structure of Earth.**
Abrupt changes in seismic-wave velocity reveal boundaries between different materials within the planet. Densities of the different layers are estimated from the seismic-wave velocities. Besides the boundaries between crust, mantle, and inner and outer core, there are also abrupt discontinuities in velocity and density within the mantle at about 100 km, 410 km, and 660 km. The low-velocity zone below 100 km defines the boundary between the lithosphere and asthenosphere. The expanded part of the diagram shows the contrasting seismic properties and thickness of oceanic and continental crust.

each Earth layer. The thickness and physical properties of each layer draw completely from seismic imaging of the interior. The objective of this section is to understand how geologists determined the composition of the hidden mantle and core.

The Velocity and Density Structure of Earth

Figure 8.19 depicts Earth's interior using seismic-wave velocity data. This mapping of the interior using earthquake and explosion data is the starting point for identifying the rocks and other materials present in the interior.

It is not possible to get a clear picture of what rocks are present at what depth with just a quick comparison of Figures 8.17 and 8.19, because the seismic velocities within Earth are much faster than those predicted from laboratory experiments. It is important to remember that the velocities summarized in

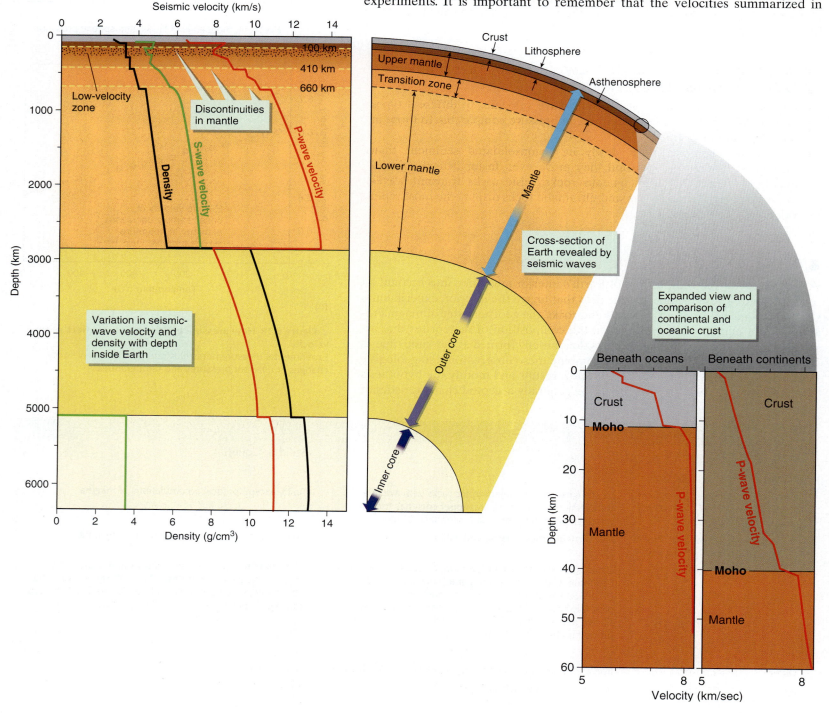

Figure 8.17 were measured at surface pressure and temperature and need to be adjusted for conditions within Earth.

Figure 8.19 also depicts the variation in density with depth. Density is estimated from seismic-wave velocity. These estimates show that density increases as depth increases. Furthermore, the density is, on average, three times higher in the core than in the mantle, which is consistent with arguments presented in Figure 8.5.

You now have the basic knowledge for translating seismic velocity into a prediction of geologic materials in order to complete an image of the interior. The gradual changes in density and velocity within the mantle and core might simply relate to increasing pressure. The more critical features to explain are the abrupt steplike changes in velocity, density, or both that show up in Figure 8.19. These changes occur not only at the base and top of the mantle but also within mantle and core. There is also considerable velocity variation within the crust, which is readily explained by the wide variety of rock types composing the crust. Therefore, we can focus on the deeper layers of Earth.

Crust-Mantle Boundary

The P-wave velocities in continental and oceanic crust differ in an expected way because of the different rock types composing each type of crust (Figure 8.19; also see Figure 8.2). Continental-crust velocity increases downward, consistent with data from outcrops, deep drill holes, and fragments from volcanoes that suggest a downward transition from felsic to more mafic rocks (Section 8.1). Mafic rocks have higher seismic velocities than felsic rocks (Figure 8.17). Relatively high oceanic-crust velocity is consistent with the abundance of basalt and gabbro found in drill holes in the sea floor and slivers of oceanic crust found on continents.

Velocity increases abruptly at the crust-mantle boundary (Figure 8.19). This change corresponds to the abrupt downward change to peridotite known from geologic data collected at the surface (Section 8.1, Figure 8.2). Yugoslavian seismologist Andrija Mohorovičić discovered this sharp change in seismic velocity in 1909 and, in his honor, the crust-mantle boundary is known as the **Mohorovičić discontinuity**, or more simply as the Moho. Estimates of depth to the base of the crust based on field-collected geologic data are available only in very few places. Seismic data, however, allow geophysicists to map the Moho around the globe. The boundary is typically 5–20 kilometers deep beneath ocean basins, where it is thinnest near mid-ocean ridges and thickest near continental margins. Continental crust varies from 25 to 85 kilometers thick although values of 30–40 kilometers are most typical. The thickest crust is present beneath the highest mountains.

Zones and Boundaries in the Mantle

Three discontinuities in the seismic velocity curves, labeled in Figure 8.19, appear globally within the mantle. Remember that each change in velocity requires a change in the properties of the mantle. The three discontinuities seem abrupt in Figure 8.19, but none of them are boundaries that are sharp enough to reflect seismic waves. This observation suggests that the changes in mantle properties are spread out over thicknesses on the order of 5–15 kilometers. Moving downward from the surface, the first discontinuity coincides with a decrease in velocity at an average of about 100 kilometers below the surface (Figure 8.19). Farther down are velocity increases at about 410 and 660 kilometers (Figure 8.19).

The Mantle Low-Velocity Zone

The **low-velocity zone** (Figure 8.19) is the only part of the mantle where velocity decreases systematically with depth. The top of this zone is at 10–75 kilometers below oceans and 100–300 kilometers below continents. At depths between

▼ **Figure 8.20 Why there are abrupt increases in seismic velocity within the mantle.**
High-pressure repacking of the crystal structures of olivine and pyroxene explain sharp discontinuities in seismic-wave velocity at about 410 and 660 km below the surface. These metamorphic transformations change the mineral structure but not the mantle chemical composition. The observed seismic-wave velocities are consistent with laboratory measurements of velocity in olivine and pyroxene, and in the minerals that they transform to at greater depth.

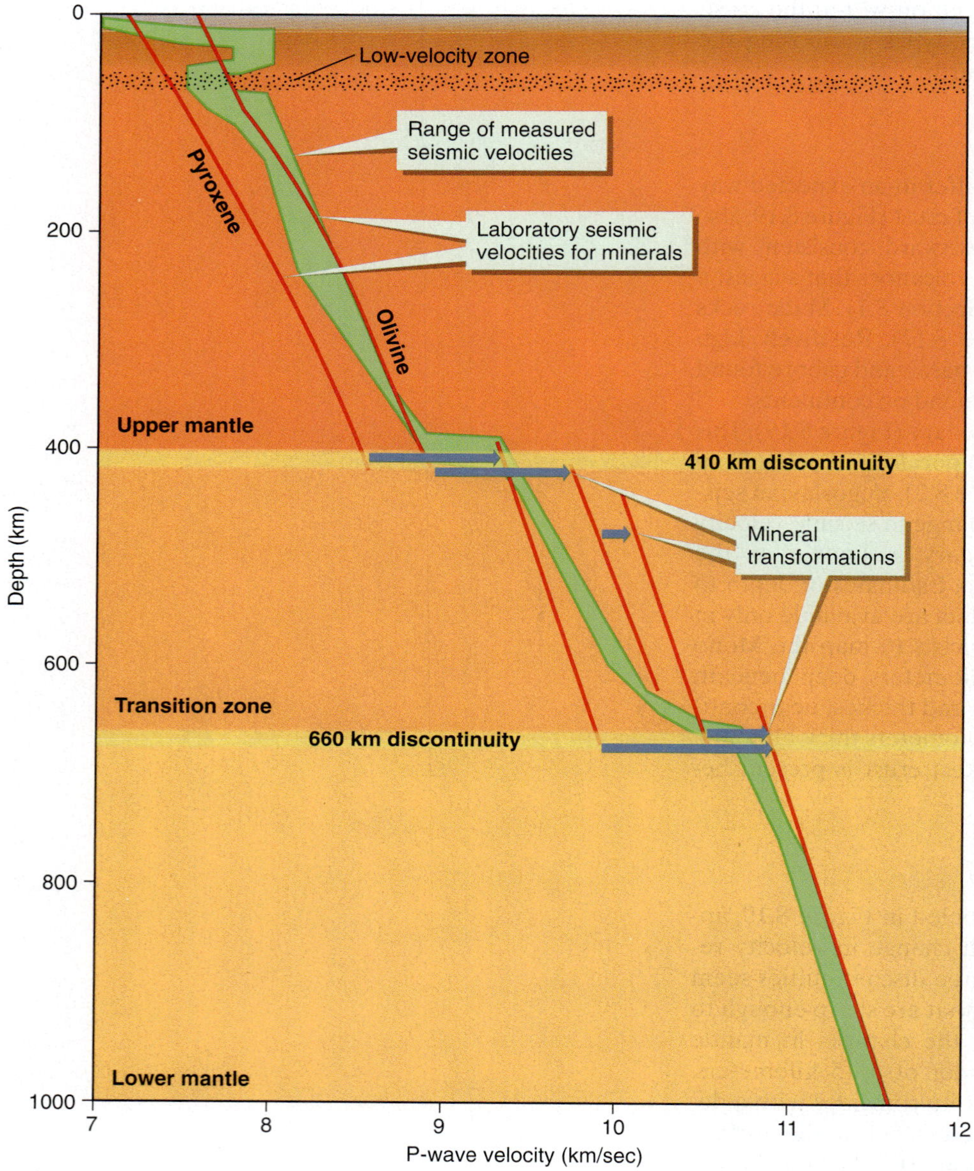

220 and 400 kilometers the P- and S-wave velocities typically return to values comparable to those just above the low-velocity zone.

Laboratory studies show that decreasing velocity can be caused by the presence of rocks that are less rigid, meaning that they yield more readily to stress. Rocks becomes less rigid when heated close to the melting point. Magma has no rigidity. The low-velocity zone, therefore, represents a somewhat squishy, less rigid layer in the mantle. S waves do not travel through liquids but do pass through the low-velocity zone, so this zone cannot be entirely molten. It is likely, however, that there is a small amount of melt in the low-velocity zone, which along with rocks close to their melting point accounts for the slowing of seismic waves.

The abrupt change at the top of the low-velocity zone marks the boundary between the strong **lithosphere** (rocky sphere) consisting of crust and uppermost mantle and the less rigid **asthenosphere** (weak sphere). These layers play an important role in plate tectonics, as mentioned in Section 1.6 and explored more thoroughly in Chapter 12.

The Mantle Transition Zone

Sharp discontinuities in P- and S-wave velocities appear globally at about 410 and about 660 kilometers (Figure 8.19). In detail, the depths vary as much as 50 kilometers above and below these locations at different places around the globe. Given the similar changes in seismic velocities at the crust-mantle boundary, you may suspect that peridotite in the upper mantle is underlain by some other rock type at greater depth. This is not a necessary conclusion.

Minor changes in chemical composition within the mantle almost certainly exist, but an effect of increasing pressure can also account for these discontinuities in seismic-wave velocity. Geologists explain the increases in seismic wave velocity at 410 and 660 kilometers by changes in the crystal structure of peridotite minerals at pressure that is more than 100,000 times greater than at Earth's surface. At this extremely high pressure, atoms within minerals reconfigure with a tighter fit to produce new crystal structures. Nearly all of the upper-mantle peridotite recovered at Earth's surface consists of olivine, pyroxene, and garnet. Lab experiments show that olivine and pyroxene transform to new minerals at progressively higher pressures and temperatures that coincide with the depths of 410 and 660 kilometers below Earth's surface. Experimental measurements of seismic velocity within these high-pressure olivine and pyroxene equivalents indicate substantial velocity increases that account for the discontinuities illustrated in Figure 8.19.

The region of the mantle between 410 and 660 kilometers is a **transition zone** where minerals undergo changes between peridotite upper mantle and a remarkably uniform lower mantle of high-pressure minerals. **Figure 8.20** illustrates that reconfiguration of atoms within peridotite minerals, rather than substantial changes in rock composition, readily explains the observed changes in seismic velocity near this transition zone.

EXTENSION MODULE 8.4
Mantle Minerals.
Learn about the changes in mantle minerals that occur at high pressures and temperatures deep within Earth.

Changes in the Lowermost Mantle

In the lowest part of the mantle, just above the core-mantle boundary, there is a zone as much as 300 kilometers thick where seismic velocities increase little, if any, with depth and in some places even decrease. Figure 8.19 does not label this zone because it varies considerably in thickness from place to place around the core and is not even present in some locations.

The origin of this poorly defined zone is uncertain and the subject of considerable ongoing research. These spots in the lowermost mantle may be partly molten or be a compositional mixture of mantle and core components. The presence of this unusual lowermost mantle plays a role in understanding dynamic motion within Earth, and you will revisit its significance in Chapters 10 and 12.

Core Composition

P- and S-wave behaviors imply a solid inner core surrounded by a liquid outer core and a substantial density increase compared to the mantle (Figure 8.19). Geophysicists have three reasons to infer that the core consists of iron:

1. Iron is the only common element that accounts for the high density of the core.
2. Liquid iron in the core can explain Earth's magnetic field (to be examined more thoroughly in Chapter 10).
3. Some meteorites consist mostly of iron and appear to be likely analogs for planet cores.

The density of the inner core matches that of nearly pure iron with perhaps some nickel at very high pressure, but the outer-core density is a little bit lower than expected for liquid iron. The density inferred from the seismic-wave velocities for the outer core requires the presence of about 10 percent elements that are less dense than iron and nickel. Different geologists argue for oxygen, sulfur, silicon, potassium, or hydrogen as this minor, lighter component. Without actual samples from Earth's core, it is difficult to know which of these elements is actually present.

EXTENSION MODULE 8.5

Meteorites as Guides to Earth's Interior.
Learn about meteorites and why some meteorites likely represent analogs of Earth's interior.

Putting It Together—*What Composes the Interior of Earth?*

- The internal composition of Earth is inferred by relating calculated seismic-wave velocities to properties of rocks and minerals at temperatures and pressures equivalent to Earth's interior.
- The crust-mantle boundary, called the Moho, displays a sharp discontinuity in seismic velocity where mafic to felsic igneous and metamorphosed igneous rocks of the crust are underlain by mantle peridotite.
- The low-velocity zone that exists, on average, about 100 kilometers below the surface is evidence that part of the upper mantle is less rigid and may be partly molten. This less rigid zone is at the top of the asthenosphere and is below the more rigid lithosphere consisting of the crust and uppermost mantle.
- Crystal-structure transformations of peridotite minerals at high pressure and temperature explain the abrupt increases in seismic velocity at about 410 and 660 kilometers below the surface. The interval between 410 and 660 kilometers depth is the transition zone between upper mantle (shallower than 410 kilometers) and lower mantle (deeper than 660 kilometers).
- The core is probably composed mostly of iron or iron-nickel alloy. The outer core is molten, and the inner core is solid. About 10 percent lighter elements—likely oxygen, sulfur, silicon, potassium, or hydrogen—must be present along with iron in the outer core to match the calculated seismic velocities.

8.5 How Hot Is the Interior of Earth?

Seismic-wave data also provide insights into the internal temperatures within Earth. Seismic-wave properties distinguish solids and liquids and therefore indicate whether material is above or below its melting temperature. You have frequently encountered statements about the temperature inside Earth, beginning in Chapter 3, but have not learned (a) how geologists determine the planet's internal temperature, or (b) why the temperature is so high inside the planet. You are now equipped with enough information to address both of these questions.

Determining the Geothermal Gradient

The geothermal gradient is simply the increase in temperature with increasing depth, and at shallow depths it is calculated from temperature measurements in deep wells and mines (see Section 4.4). These measured geothermal gradients vary considerably from place to place but typically range between 25 and 30°C per kilometer on continents and average about 60°C per kilometer beneath the oceans.

Is it reasonable to use geothermal gradients, which are based on near-surface measurements, to determine temperatures deep within Earth? A realistic geothermal gradient of 30°C per kilometer near the surface under continents would, if extended deeper, reach the melting temperature of mantle peridotite at 50 kilometers and an outrageous value approaching 200,000°C at Earth's center. The travel of S waves throughout the mantle indicates, however, that the peridotite is mostly solid, so the geothermal gradient must decrease at greater depth in the mantle.

Integration of different geologic data sets provides more realistic estimates of the geothermal gradient deep within the mantle.

- If, as seems likely, the low-velocity zone in the mantle relates to small degrees of partial melting at the top of the asthenosphere, then temperatures of about 1300°C must be reached at depths of 75–100 kilometers beneath oceans, and somewhat deeper beneath continents.
- Laboratory experiments with diamond-bearing mantle rocks brought to the surface through continental volcanoes indicate that these rocks formed at depths of 100–150 kilometers below the surface but only at temperatures of ~1000–1200°C. So the overall gradient must decrease to less than 10°C per kilometer below the continents.
- Laboratory experiments on the transformation of olivine to a denser mineral at 410 kilometers suggest a temperature of only 1450 ± 150°C. This result suggests a further decrease in the geothermal gradient to about 3.5°C per kilometer beneath both continents and oceans.

Figure 8.21 summarizes these data to sketch the geothermal gradient in the upper mantle.

Continuing the geothermal gradient deeper into the mantle and core requires considerable speculation, because the precise melting temperatures of lower mantle and core rocks are unknown. Figure 8.21 portrays one *possible* geothermal gradient that is consistent with what geologists do know.

The geothermal gradient depicted in Figure 8.21 shows temperature rising very gradually through most of the lower mantle and then more abruptly to a value of 3500°C just above the core-mantle boundary. We know the temperature in the outer core exceeds the melting temperature of the iron alloy present there because S waves do not pass through it. In the uppermost outer core, current best estimates suggest a melting temperature of 3500 ± 500°C. A temperature of about 3500°C is also close to estimated melting temperature for high-pressure mantle rock. A small degree of melting in the lowermost mantle is one explanation for the curiously low seismic velocities observed in some places just above the core-mantle boundary.

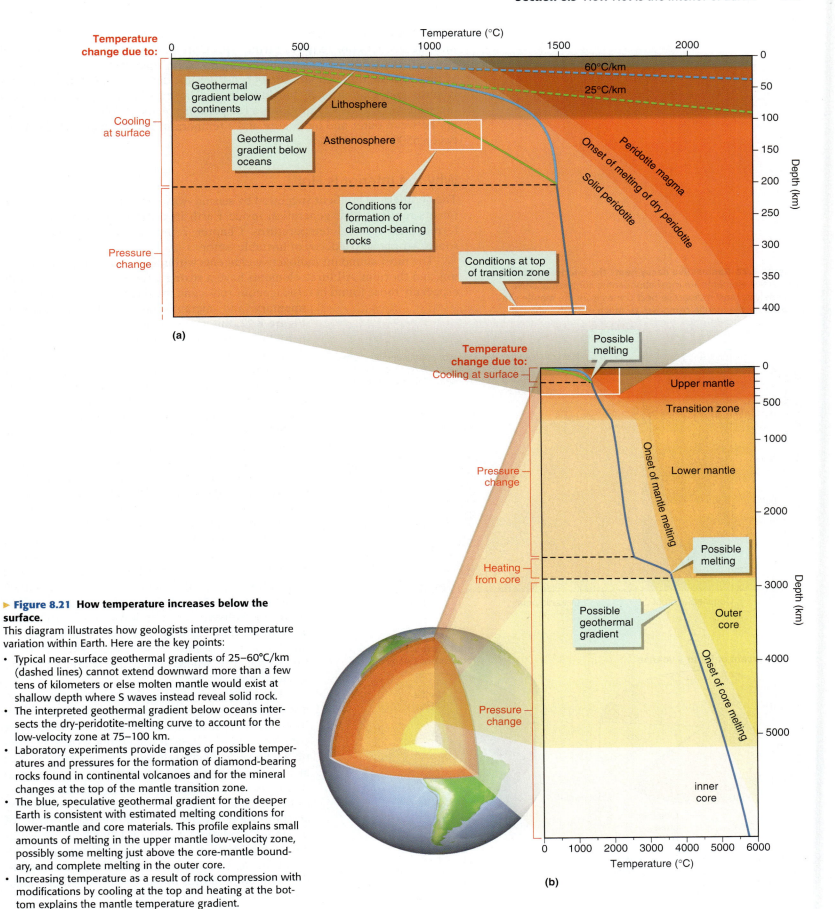

▶ **Figure 8.21 How temperature increases below the surface.**
This diagram illustrates how geologists interpret temperature variation within Earth. Here are the key points:

- Typical near-surface geothermal gradients of 25–60°C/km (dashed lines) cannot extend downward more than a few tens of kilometers or else molten mantle would exist at shallow depth where S waves instead reveal solid rock.
- The interpreted geothermal gradient below oceans intersects the dry-peridotite-melting curve to account for the low-velocity zone at 75–100 km.
- Laboratory experiments provide ranges of possible temperatures and pressures for the formation of diamond-bearing rocks found in continental volcanoes and for the mineral changes at the top of the mantle transition zone.
- The blue, speculative geothermal gradient for the deeper Earth is consistent with estimated melting conditions for lower-mantle and core materials. This profile explains small amounts of melting in the upper mantle low-velocity zone, possibly some melting just above the core-mantle boundary, and complete melting in the outer core.
- Increasing temperature as a result of rock compression with modifications by cooling at the top and heating at the bottom explains the mantle temperature gradient.

Extending the meager information on the mantle temperature gradient into the deeper core requires further speculation. The melting temperature of iron in the core increases with increasing pressure. The inner core must be cooler than this melting temperature because seismic data are consistent with nearly pure and solid iron in the inner core. The possible geothermal gradient illustrated in Figure 8.21 proposes a temperature of about 5000°C at the boundary between outer and inner core and a temperature at the center of Earth between 5500 and 6000°C.

Why the Interior Is Hot

The geothermal gradient suggested by seismic-wave data and laboratory experiments requires heat generation inside Earth in order to account for the inferred high internal temperatures. Chapter 7 introduced the primary heat source—radioactive decay of unstable isotopes.

The most abundant radioactive elements within Earth are potassium, uranium, and thorium. All three elements have a strong affinity for silicate minerals, especially those found in felsic igneous rocks and in mineralogically similar sedimentary and metamorphic rocks. As shown in **Figure 8.22**, this means that a cubic meter of continental-crust rock generates far more heat from radioactive decay than does an equal volume of basaltic oceanic crust or peridotitic mantle. Although some geologists speculate on the presence of significant potassium in the core, most scientists think this is unlikely. This lack of potassium means that there is probably little or no radioactive-decay heat generated in the core.

How do geologists explain the high temperature of the mantle when radioactive elements concentrate into the continental crust? The total heat production from the mantle is huge because the mantle composes 83 percent of the volume of Earth. Figure 8.22 shows that even though a small sample of mantle rock contains miniscule concentrations of radioactive elements, 75 percent of Earth's radioactive heat originates in the mantle simply because the mantle occupies such a huge volume. Even this enormous source of radioactive heat is, however, insufficient to account for temperatures near 3500°C at the base of mantle, and it does not explain the higher temperature in the core, where radioactive isotopes are rare or lacking altogether.

The additional heat required to keep the core so hot probably originates from crystallization of iron from the molten outer core to form the solid inner core. You add heat to melt something, so heat is given off whenever melt crystallizes. This may seem strange at first because you know that decreasing temperature causes crystallization. Yet crystallization releases heat that increases temperature. Heat released by crystallization greatly slows down the cooling of magma. Calculations based on laboratory measurements suggest that crystallization of iron might account for the high temperature of the core as well as the high temperatures in the lowermost mantle adjacent to the core.

▼ **Figure 8.22** **Radioactive decay heats the interior of Earth.** Radioactive elements are most abundant in continental crust. Nonetheless, most radioactive heat comes from the mantle simply because the mantle composes so much of Earth's volume.

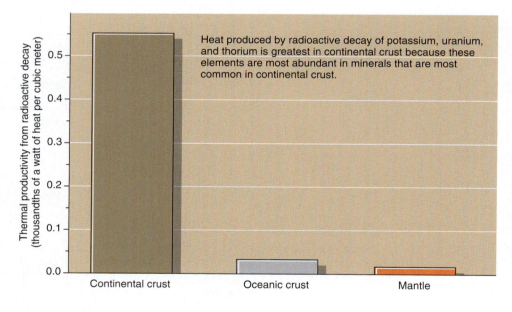

Heat produced by radioactive decay of potassium, uranium, and thorium is greatest in continental crust because these elements are most abundant in minerals that are most common in continental crust.

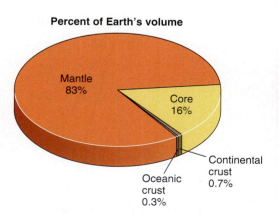

The mantle overwhelmingly composes the majority of the volume of Earth, even though it contains low abundances of radioactive, heat-generating elements.

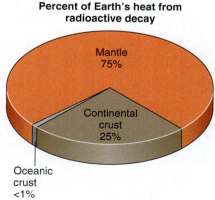

Multiplying heat production by the volume of rocks reveals that about 75% of all of Earth's radioactive heat originates in the mantle.

Why Temperature Increases with Depth

Three factors play a role in why temperature increases with increasing depth inside Earth.

1. The outer surface of the planet cools by radiating heat to the atmosphere and space.
2. Heat conducts outward from the core into the lower mantle.
3. Rocks must heat up when pressure increases.

This last factor may not be obvious and requires elaboration. Laboratory experiments show that whenever a solid, liquid, or gas is compressed, its temperature rises while its volume decreases. The total amount of heat energy does not change, but it is distributed over a smaller volume, causing the temperature to rise. Likewise, temperature decreases whenever a solid, liquid, or gas expands. You have experienced this phenomenon if you ever noticed an aerosol can become cold during use. The gas in the can cools when it expands as some of the gas is released.

Rock and magma are far less compressible than gases, so the increase in temperature with increasing pressure and depth within Earth is much less than that observed in aerosol cans. Laboratory experiments and calculations indicate that increasing pressure should cause temperature to rise about 0.5 °C per kilometer in the lithosphere, 0.3°C per kilometer through the transition zone, and about 0.1–0.2°C per kilometer in the lower mantle. This temperature increase, caused only by increasing pressure, explains the possible geothermal gradient through most of the mantle and core depicted in Figure 8.20.

Putting It Together–How Hot Is the Interior of Earth?

- Geothermal gradients measured near the surface are higher than in the interior. Temperature is on the order of 1000–1200°C at 100–150 kilometers depth, perhaps 3000–4000°C at the core-mantle boundary, and greater than 5000°C in the inner core.
- Heat sources within Earth include (a) radioactive decay of elements found in the mantle and crust; and (b) crystallization of iron in the core.
- The shape of the geothermal gradient relates to (a) loss of heat at the surface to the hydrosphere, atmosphere, and space; (b) heating of the lower mantle by the hotter core; and (c) increasing temperature with compression of rock at high pressure.

Where Are You and Where Are You Going?

Thanks to earthquake waves, you now have a descriptive image of the interior of the planet. Rocks found at the surface or retrieved from deep mines and drill holes provide insights into the outermost 50 kilometers or so of Earth, but earthquake data permit an understanding of the deeper, layered inner structure. To quote again from R. D. Oldham,

> The seismogram, recording the unfelt motion of earthquakes, enables us to see into the earth and determine its nature with as great a certainty, up to a certain point, as if we could drive a tunnel through it and take samples of the matter passed through.
> (R. D. Oldham, 1906, *Quarterly Journal of the Geological Society*, v. 62, p. 456)

The deduction of the internal composition and temperature of Earth follows from a series of related studies. Seismic stations around the world record earthquake-wave arrival times that allow calculation of earthquake-wave velocities. These data demonstrate that wave velocity, for the most part, increases with depth. A shadow zone where S waves do not reach Earth's surface requires a liquid outer core. The otherwise persistent S waves, which do not travel through liquid, indicate that the remainder of the inner planet is solid with only minor amounts of melting within it. Laboratory measurements reveal relationships between seismic velocity and rock type. These relationships allow geologists to use seismic-velocity data to predict what physical materials occur at specific depths within the planet.

Earth's crust is usually 25–50 kilometers thick beneath continents and averages 7 kilometers thick below the oceans. Continental crust is largely quartz-bearing igneous and metamorphosed igneous rocks that become progressively more mafic at greater depth. Oceanic crust is composed almost entirely of mafic lava flows and intrusive rocks. The base of the crust is a boundary called the Moho, where a sharp increase in seismic velocity indicates the existence of peridotite in the upper mantle.

The mantle continues downward to about 2900 kilometers below the surface. Variations in seismic velocity reveal a number of internal boundaries within the mantle.

Seismic velocity increases downward within the crust and uppermost mantle that together compose the lithosphere, but then it decreases in the low-velocity zone at the top of the asthenosphere. The mantle is less rigid and probably contains a small volume of magma in the low-velocity zone. Sharp increases in velocity near 410 and 660 kilometers depth imply repacking of atoms in mantle-peridotite minerals into more compact and seismically faster crystal structures. Patches of unusually slow velocities near the core mantle boundary remain to be explained with a high degree of confidence.

The dense core consists of two parts, a liquid outer core and a solid inner core. The core's high density is most consistent with a dominance of iron or iron-nickel alloy. The outer core may contain small amounts of lighter elements.

Now you know about Earth's layered internal structure along with how geologists discovered this structure. The next chapter investigates how this zonation formed, and Chapter 10 shows why geologists infer that Earth's interior is in persistent motion.

Active Art

Dimensions of the Mantle and Core. See how to determine physically possible thicknesses of Earth's mantle and core.

Seismic Wave Motion. See how the different seismic waves move.

Wave Reflection and Refraction. See how waves reflect and refract through rock.

Extension Modules

Extension Module 8.1: Sizing Up Earth. Learn how Eratosthenes determined the radius of Earth and how eighteenth-century scientists calculated the mass of Earth.

Extension Module 8.2: How to Locate an Earthquake. Learn how to use seismographs and travel-time curves to determine where an earthquake happened.

Extension Module 8.3: Velocity of Seismic Waves. Learn about the physical properties of rocks that determine how fast P and S waves move.

Extension Module 8.4: Mantle Minerals. Learn about the changes in mantle minerals that occur at high pressures and temperatures deep within Earth.

Extension Module 8.5: Meteorites as Guides to Earth's Interior. Learn about meteorites and why some meteorites likely represent analogs of Earth's interior.

Confirm Your Knowledge

1. How was the radius of Earth determined? Is Earth a perfect sphere?

2. How do geologists determine the composition of Earth's interior? Why not access it directly?

3. How do we get samples of rocks from the continental crust that formed below 5 kilometers?

4. How thick is continental crust and what non-sedimentary rocks comprise the continental crust?

5. How thick is oceanic crust and what non-sedimentary rocks comprise the oceanic crust?

6. How do oceanic and continental crusts differ? How do scientists determine this?

7. What is the difference between elastic and plastic deformation? What type of deformation does a rock experience when an earthquake wave passes through it?

8. What is a wave? How does a wave move through rock?

9. Describe the three types of earthquake waves.

10. What instrument records seismic waves?

11. Which type of wave arrives first at an earthquake recording station? Why do the waves not arrive at the same time?

12. What is the difference between an earthquake's focus and its epicenter?

13. What causes the mantle low-velocity zone?

14. What is the most likely cause of the mantle seismic velocity transition zones at 410 and 660 kilometers?

15. What causes the S wave shadow zone? What causes the P wave shadow zone?

16. What do shadow zones reveal about the interior of Earth?

17. What is the evidence that the outer core consists mostly of liquid iron? What is the evidence that the outer core contains elements other than iron?

18. What is the source of heat responsible for the high temperatures in Earth's core? What is the source of heat responsible for the high temperatures in Earth's mantle?

Confirm Your Understanding

1. Write out an answer for each question in the Chapter Outline for the chapter sections assigned by your instructor.

2. How do we know that Earth is inhomogeneous?

3. What evidence tells us that Earth has a distinct concentration of high-density material near the center and not a gradual transition from low-density to high-density rock?

4. How do we know that P and S waves do not travel through Earth at constant velocities?

5. How could you use seismic-wave travel velocities to determine if magma is present below a restless volcano?

6. You have collected data on seismic-wave velocities in rocks that exist several kilometers below the surface. Use information in Figure 8.17 to determine what kind of rock(s) are suggested by these P wave velocities: 2.25 km/s; 4 km/s; 6.25 km/s; 6.75 km/s; 8 km/s.

7. Based on Figure 8.19 compare the P-wave velocities of oceanic and continental crust at a depth of 10 km. Why are these velocities different?

Key Terms

asthenosphere (p. 214)
body waves (p. 204)
earthquakes (p. 204)
elastic (p. 203)
epicenter (p. 205)
focus (p. 205)

lithosphere (p. 214)
low-velocity zone (p. 213)
Mohorovičić discontinuity (Moho) (p. 213)
plastic (p. 204)
P waves (p. 204)

reflection (p. 206)
refraction (p. 206)
seismic waves (p. 204)
seismogram (p. 204)
seismometers (p. 204)
shadow zones (p. 208)

S waves (p. 204)
surface waves (p. 204)
transition zone (p. 214)
waves (p. 203)

9 Making Earth

Chapter Outline

Why Study the Origin of Earth?

WHEN ASTRONAUTS VIEW EARTH FROM SPACE, THEY SEE ROCKY continents and wide oceans blanketed by an atmosphere. Have you ever wondered how the geosphere, hydrosphere, and atmosphere originated?

Sometimes people call Earth the "Goldilocks planet" because it is "just right." In the fairytale of Goldilocks and the three bears, Goldilocks ate the baby bear's porridge because it was neither too hot nor too cold. Likewise, our planet is the perfect distance from the Sun—Earth is neither too hot nor too cold to sustain liquid water, an atmosphere, and life. Some other planets in the solar system have atmospheres, but none of them currently have liquid water. Quartz-rich rock in the crust is also apparently unique to Earth. In addition, no other planet is as dynamic as Earth with its moving, deforming lithosphere, and an ever-changing landscape crafted by weathering and erosion.

The last chapter ventured deep into Earth to learn how seismic waves and geophysical principles reveal the planetary interior as a series of concentric layers distinguished by different compositions and physical properties. Curiosity about Earth now leads to the next question—*how* did Earth become layered?

How the layers formed leads to other questions: How did Earth form? How did Earth end up with distinct oceanic and continental crusts? How did the atmosphere and hydrosphere form? A complete look at the chemistry and physics behind the hypotheses for the formation of the planet, its layers, and the origin of its water are beyond the scope of this text. A look at the fundamental ideas for how Earth formed and evolved in its youth can be informative, however, and this knowledge forms a good basis for going further in your study of geology.

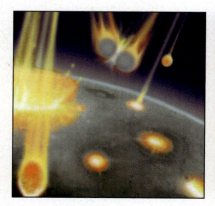
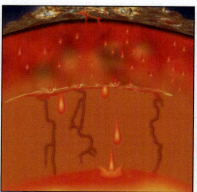

NASA satellite *Galileo* recorded parting images of Earth and Moon, combined here into one photo, as it headed off to explore the solar system. ▶

Your objectives in this chapter are as follows:

✔ To examine the prevailing theories for the construction of our solar system

✔ To understand how Earth's layered internal structure relates to theories and hypotheses that explain how the planet formed

✔ To understand data that place constraints on the early development of the lithosphere and the formation of the two distinct types of crust

✔ To learn how the origins of the atmosphere and the hydrosphere relate to the formation of the planet and to the history of the lithosphere

To achieve these objectives, the following questions are addressed:

9.1 How did Earth form?

9.2 How did the core and mantle form?

9.3 How does the crust form?

9.4 How did the hydrosphere and atmosphere form?

9.5 *How do we know* ... the hydrosphere came from the geosphere?

In the **FIELD**

You are outside on a clear night brightened by the light of a full Moon. A backyard telescope enlarges your view of the Moon and you take a picture for your notebook, as seen in **Figure 9.1**.

As you look at the Moon, you easily observe the contrasting whitish gray and black colors of its rocky surface, and perhaps you recall how old these rocks are (see Section 7.6). Among the rocks returned from the NASA Apollo visits to the Moon, the youngest are 2.9 billion years old and the oldest are about 4.56 billion years old. The lack of rocks younger than 2.9 billion years old implies that the Moon has not changed much since very early in its history. Unlike Earth, the Moon is geologically "dead." It has no oceans, no atmosphere, no active volcanoes, and only rare moonquakes. Earth is dynamic; it has oceans, an atmosphere, volcanoes, and abundant earthquakes. These changing features and processes cause Earth to look very different today from the way it did when it first formed. The Earth and the Moon are two planetary bodies in complete contrast, yet they are closely linked.

Did Earth and Moon form at the same time, from the same material? In Chapter 7 you learned how scientists infer that the Earth and Moon probably formed at roughly the same time as the other planetary bodies of the solar system. Moon likely originated from the same material as Earth. To understand how Earth and Moon formed, it makes sense to broaden your view to contemplate the origin of the Sun and our planetary neighbors.

Pursuit of knowledge about the origin of Earth must be consistent with seismic data that reveal the three distinct layers: core, mantle, and crust. Did the Earth's layers form concurrently with planetary formation, or did the layers somehow form at a later time? Why is the iron core separate from the silicate materials? How did the thick silicate part of Earth divide into crust and mantle? Why are there distinct continental and oceanic types of crust?

Earth is unique because it has the right temperature and pressure conditions to contain water as solid, liquid, and gas. Water is particularly critical for geologic processes because it explains many metamorphic reactions, aids in forming magma near convergent plate boundaries, causes explosive volcanic eruptions, weathers rocks to make sediment, and erodes landscapes. How did the hydrosphere originate? How did Earth form an atmosphere?

On any clear night, as you gaze at the Moon and think about Earth and the solar system, you might imagine the variety of suggested hypotheses for the formation of Earth, its layers, and water. How Earth formed is one of the fundamental questions in geology. Geologists combine their knowledge with that of astronomers and astrophysicists to investigate and explain how Earth formed.

Figure 9.1 Examining Earth's Moon. ▶
Moon is the closest planetary object to Earth, yet it looks very different from Earth. How do planets and moons form and why are they different?

9.1 How Did Earth Form?

One of the fundamental challenges in geology is grasping the physical and chemical events and processes that formed Earth. Understanding the origin of planets involves consideration of a wide variety of physical and chemical processes. Various theories establish scientific understanding of these processes. Rather than examine each elaborate theory separately, you will consider the condensed and integrated highlights, which we will simply call the planetary-origin theories. Many of the fundamental ideas originated in the seventeenth and eighteenth centuries and remained hypothetical until space-age observations were made to test them.

Telescope observation of Earth's Moon

Moon surface consists of light-colored rock and dark rock. Most of the dark rock occupies more or less circular areas, which are big craters formed by meteor impacts.

Finding the Evidence

To understand the origin of Earth, geologists first examine the geologic record. The earliest rocks on Earth provide clues, but even these oldest rocks are not as old as the planet. Earth is about 4.5 billion years old, but the oldest dated minerals on Earth are slightly younger, at about 4.4 billion years old (see Section 7.6). The rocks that originated during or immediately after planetary formation are not preserved for study. They were recycled in the rock cycle or consumed during resurfacing of the planet by erosion and plate tectonics. The geologic record is not a complete dead end, however, because knowledge of Earth's composition and internal layered structure is composed of observations that any theory of planet formation must explain.

Data from elsewhere in the cosmos round out the evidence for planet formation. These data include meteorite samples and rocks from the Moon that are older than Earth's oldest rocks. Additional information comes from studies of the other planets in our solar system, along with their moons and other small objects such as rocky asteroids and icy comets. Information on our planetary neighbors partly comes from Earth-based telescope observations but mostly from data-collecting instruments on spacecraft that pass close to these other bodies. Also offering key pieces of evidence are telescopic observations of other solar systems, including those where planets are probably forming as we watch.

Whatever processes account for how Earth formed must explain the similarities and differences among all of the objects that formed in our solar system at the same time. **Table 9.1** summarizes some important statistics about the planets and emphasizes their differences. The inner planets (Mercury, Venus, Earth, and Mars) are relatively small and dense, which implies that they are mostly solid rock. The outer planets, with the exception of Pluto, are huge in size but with low densities, which implies that they mostly consist of compressed gases with

TABLE 9.1	**Basic Planet Properties**			
Planet	Distance from the Sun (millions of km)	Diameter (km)	Density (g/cm^3)	Composition
Inner Planets				
Mercury	58	4,878	5.4	Silicate crusts and mantles with iron-nickel cores.
Venus	108	12,104	5.2	
Earth	150	12,756	5.5	Atmospheres are thin or lacking altogether.
Mars	228	6,794	3.9	
Outer Planets				
Gas giants				Mostly hydrogen and helium gases, liquids, and ices with small rock and iron cores.
Jupiter	778	143,884	1.3	
Saturn	1,427	120,536	0.7	
Uranus	2,870	51,118	1.2	
Neptune	4,497	50,530	1.7	
Pluto	5,900	2,300	1.8	Water ice and rocky core with nitrogen-rich atmosphere.

▶ **Figure 9.2 Views of other worlds.**
The inner, rocky planets contrast with the gas giants in the outer solar system. NASA's Mars Exploration Rover *Spirit* took the image on the left in June 2004 while traveling the rocky desert plains of Mars. NASA spacecraft *Cassini* took the image of Saturn, on the right, as it approached the planet for closer study in March 2004. Ice and rock fragments encircle the great planetary ball consisting mostly of gas and ice.

EXTENSION MODULE 9.1
Geologic Tour of the Solar System.
Learn the geologic basics of the planets, moons, asteroids, and comets of the solar system.

relatively small centers of rock or ice. **Figure 9.2** provides recent images of the surface of Mars, a rocky world that resembles an earthly desert, in contrast with Saturn, a gas giant. Pluto is a small ice-and-rock oddball in comparison to its large, gas-rich neighbors, and you will soon return to this notable exception to an otherwise neat division of inner and outer planets. The distinction between inner rocky planets and the gas giants must be explained by any theory of planet formation.

Starting with a Cloud of Gas and Dust

The planetary-origin theories explain the formation of planets by the gradual accumulation of small dust particles and gas molecules to form increasingly larger objects that eventually achieve the status of planets. **Figure 9.3** illustrates the key stages of this process.

Astronomical observations show that stars and planets originate from **nebulae**, clouds of gas and dust in space. Gas molecules, mostly hydrogen, strongly outnumber dust particles, which compose less than 1 percent of a nebula's mass. Uneven distribution of matter within a nebula causes uneven gravitational forces that attract particles and molecules toward one another according to their masses. Gravity causes the particles and gas molecules to contract inward toward the center of the cloud. The gas and dust rotate around the center at increasing speed as mass concentrates more and more toward the center. This speed-up effect is similar to the accelerating spin of an ice skater who pulls his or her arms in close to the body. Forces of motion associated with the increasingly faster spin also flatten the cloud into a pancake-shaped disk (see Figure 9.3), much the same way that a pizza chef makes a wider and thinner crust by spinning the dough in the air.

Forming the Protosun

The disk-shaped cloud is called the **solar nebula** because the concentrated mass near the center eventually forms into the Sun. **Figure 9.4** illustrates recent

▶ **Figure 9.3 How the solar system probably formed.**
Planetary-origin theories explain the origin of the planets by accretion of progressively larger masses of dust and gas into planetesimals, planetary embryos, and finally into planets. Smaller objects left over from planet formation form the asteroid and Kuiper belts. The outer, gas-giant planets formed before the inner, rocky planets.

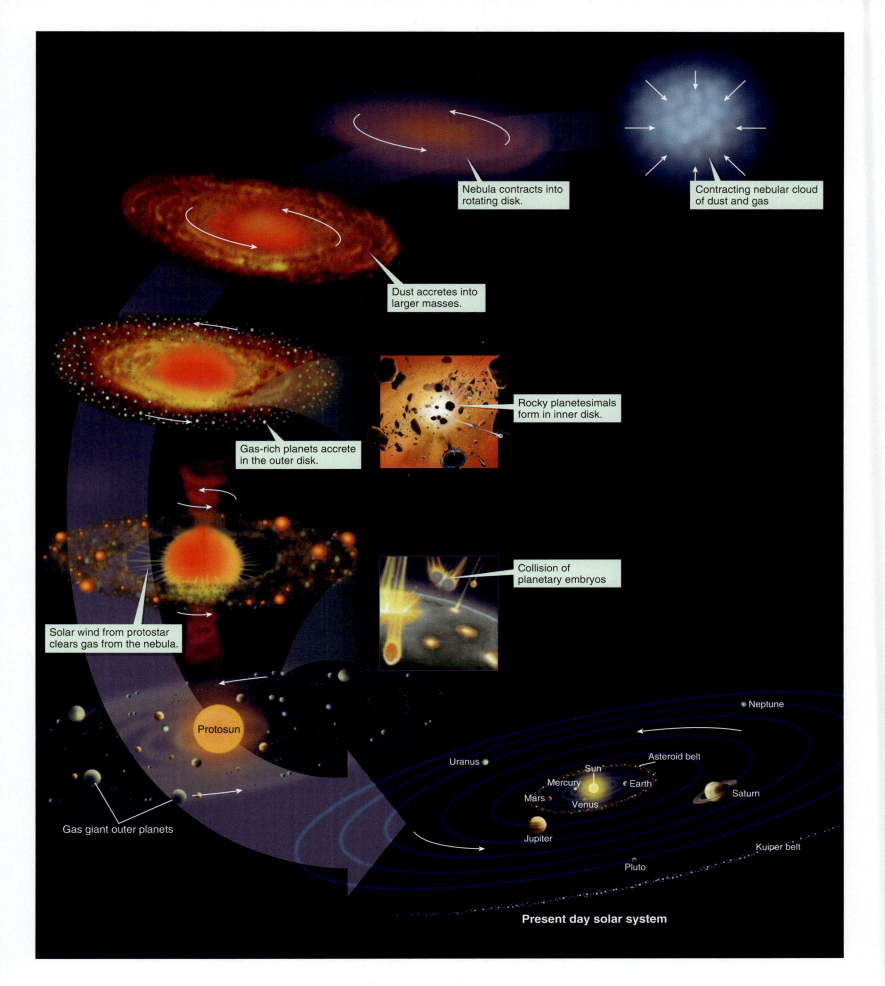

Contracting nebular cloud of dust and gas

Nebula contracts into rotating disk.

Dust accretes into larger masses.

Rocky planetesimals form in inner disk.

Gas-rich planets accrete in the outer disk.

Collision of planetary embryos

Solar wind from protostar clears gas from the nebula.

Protosun

Gas giant outer planets

Present day solar system

Neptune

Uranus

Asteroid belt

Sun

Mercury

Earth

Mars

Venus

Saturn

Jupiter

Kuiper belt

Pluto

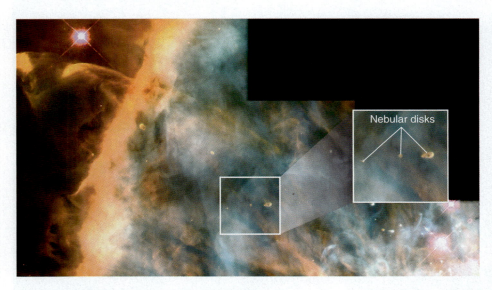

Nebular disks

▲ **Figure 9.4 Witnessing star and solar system formation.** This Hubble Space Telescope image of the Orion Nebula shows a giant cloud of dust and gas illuminated by countless bright, young hot stars. The closer view of a small part of Orion shows faint stars surrounded by dust and gas. These solar nebulae are the beginnings of planetary systems.

observations of nebular clouds and disks that are the forerunners of stars and planets. The density and temperature within the spinning disk are highest at the center. The density is high because gravity pulls an increasing mass of dust and gas into a small volume of space. Heat energy can be thought of as the motion energy of atoms and their internal particles, such as electrons. When a large number of moving atoms occupy a small volume, the heat energy is also concentrated in that small volume and radiates into space only along the edges. Heat energy trapped in the interior of the solar nebula causes the temperature to increase, too.

About 100,000 years after the beginning of nebular contraction, the dense, hot center of the disk has an internal temperature greater than 1 million degrees and a surface temperature of about 3000°C to 4000°C. While not yet having the internally fueled furnace of a true star, this central mass is bright and constitutes a **protostar**, or for our solar system, the protosun. The heat radiating from the protosun is sufficient to split the dust particles in the inner part of the spinning nebular disk into their constituent atoms. Where spinning forces carry atoms farther from the protosun, they cool and condense into new dust particles with compositions determined by the temperature at each condensation location.

The original nebula is peppered with dust grains of various compositions. When the dust reforms, however, after being atomized near the protosun, the particle composition is determined by location and temperature within the disk. **Figure 9.5** shows how ice particles of various frozen gas molecules formed in the relatively cool, outer part of the solar system, whereas only silicate, oxide, and metal dust particles condensed in the hotter inner part of the solar system and were surrounded by the always more voluminous gas.

Making Planetesimals and Planetary Embryos

The concentration of matter along the center plane of the spinning nebular disk caused collisions between dust particles. Colliding particles stuck to one another to make larger clumps. Unless you frequently clean under your bed, you are well aware of how effectively tiny dust particles clump into larger aggregates. This process in space, called **planetary accretion**, gradually accumulates small particles into large objects that we recognize as the planets.

The initial small objects, called **planetesimals** (see Figure 9.3), were only the sizes of football stadiums up to large cities. Planetesimal growth resembles the construction of a snowman. When you roll a small snowball along the ground more small snowflakes stick to it and the snowball grows larger and heavier. Planetesimals did not roll on a surface, however, so the better analogy might be throwing a snowball through a heavy blizzard of snowflakes, with the snowball enlarging with every accreted flake. The attraction of water molecules binds a snowball together whereas in planetary accretion gravity is the force that pulls masses together and holds them together in a growing planetary ball.

Once planetesimals were a few kilometers across, their gravity forces perturbed the motion of their neighbors and caused the planetesimals to collide. Some of these collisions undoubtedly pulverized planetesimals into smaller chunks. Computer modeling of the physics of these collisions suggest, however, that most collisions would produce overall accretion of the planetesimals into larger bodies. This is sort of like piling up your large snowballs to construct the snowman. These enlarged objects are **planetary embryos** (see Figure 9.3), with sizes as large as Moon or Mars, or approximately one-hundredth to one-tenth the mass of Earth.

Each embryo staked out a swath of space near its orbital path in the nebula from which it swept up and gravitationally pulled in nearby planetesimals. This

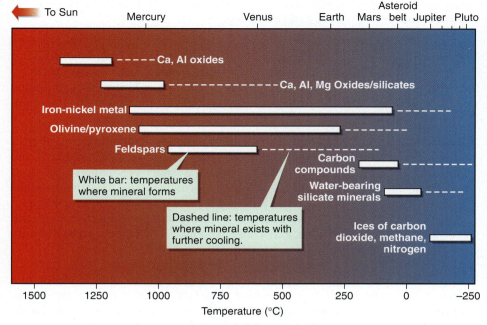

◀ **Figure 9.5 Where minerals crystallized in the solar nebula.**
The chart illustrates where minerals crystallized in the solar nebula depending on temperature and distance from the Sun. The planets are plotted at the temperature values coinciding with the distance of each planet from Sun within the early nebula. Condensed dust particles that accreted into planetesimals in the vicinity of Earth contained iron-nickel metal and silicate minerals such as olivine, pyroxene, and feldspar. Water-bearing silicate minerals, carbon compounds, and ices of gas-rich elements accumulated farther out in the solar system where nebular temperatures were much cooler.

zone that fed mass into the growing embryos contained materials of different compositions in different parts of the disk, because of the different condensation temperatures of particles at various distances from the hot protosun.

The first planetesimals formed in the outer disk where cooler temperatures permitted condensation of ice particles while the inner solar nebula remained too hot for significant formation of any solids. The icy planetesimals and embryos became sufficiently large for their gravitational forces to pull in surrounding envelopes of gas molecules from the nebula. In this fashion the gas giants were born first in the outer solar system (see Figure 9.3).

Planetesimals were probably just starting to accrete into embryos in the hotter inner solar system when the protosun underwent a violent stage of its own history that altered the composition of the solar nebula. A few million years after the nebular disk formed, the internal temperatures in the continually contracting protosun exceeded five million degrees. The energetic motion of atoms tore them apart into the constituent protons and electrons. These charged particles exploded out into space in spectacular displays that are sometimes also witnessed in other nebulae by today's astronomers. The streams of charged particles produce the **solar wind**, which is not moving air but is, instead, fast-moving protons and electrons blasting outward from a star. The solar wind remains a feature of the present Sun but not with the intensity as in the early history of our local star.

The early solar wind had little effect on the condensed dust particles and planetesimals but effectively blew the lightweight gas molecules out of the nebula except where they were strongly bound by gravitational forces around the gas giants in the outer solar system. This clearing out of the nebula (see Figure 9.3) had a profound effect on the continued growth of the inner planets, where high temperatures had not permitted formation of ices or water-bearing minerals and the solar wind removed all of the more abundant gaseous molecules. The distinction between the inner, rocky planets and the outer, icy and gas-rich planets was inherited from this early event in the history of planet and Sun formation.

Planets and Sun

The final stages in the formation of the inner planets were complicated by the increasingly strong gravitational interactions between the enlarging planetary embryos and the gravitational pull of the early formed gas giants in the outer part of emerging solar system. Gravitational push and pull directed embryos into different orbits, which had two effects.

1. The embryos began to accrete particles from outside of their original feeding zones. This blurred the original compositional gradients from inner to outer parts of the solar system imparted by varying condensation temperatures.
2. The embryo orbits crossed one another, setting up collisions that continued the planetary accretion process to the next stage—the making of the inner, rocky planets (Figure 9.3).

There were perhaps a few dozen embryos and an uncountable number of remaining planetesimals in the inner solar system as Mercury, Venus, Earth, and Mars started to form. Over a period of perhaps as short as 5 million years or as long as 100 million years, embryos collided to form the planets. Earth and Venus probably formed from a dozen or more embryos, along with continued sweeping up of smaller planetesimals pulled in by the gravitational attraction of each growing planet. Smaller Mars may consist of only a few embryos and Mercury may be a single embryo.

The final accretion of Earth included giant embryo collisions. The scientifically popular giant-impactor hypothesis states that a catastrophic collision of a large embryo with Earth near the end of planetary accretion formed the Moon. Vaporized debris from the impacting embryo, along with chunks from Earth, spun into a ring around the planet, that accreted to form the Moon.

The protosun matured into a star early in the formation of the inner planets. The key final step occurred when continued contraction of the dense, gas-rich ball drove the temperature at the interior above 10 million degrees. The rapid motion of atomic particles to account for so much heat is sufficient to fuse them together when they collide. **Nuclear fusion** describes the process where protons fuse together into larger atoms with the simultaneous release of huge amounts of energy. The nuclear furnace ignited and our Sun became a true star.

The Leftovers

Not all of the planetesimals and embryos accreted to form planets. Many in the inner solar system fell into the Sun. Hundreds of other objects were flung out to the outer edges of the solar system to form the Kuiper belt (see Figure 9.3). Most astronomers argue that Pluto is one such castoff embryo rather than qualifying as a major planet. Several dozen fragments escaped further collisions to become Moons trapped by the gravitational field of planets to follow orbits around the larger bodies. Some of the leftover debris spins as rings around the outer planets, most notably Saturn (see Figure 9.2).

A larger cluster of irregularly shaped, rocky planetesimals called **asteroids** form an orbiting belt between Mars and Jupiter (see Figure 9.3). Some small asteroids and pieces flung off by collisions are pulled into highly stretched orbits by the gravitational power of gigantic Jupiter. These odd orbits bring the objects on potential collision courses with Earth. These objects are **meteors** as they streak through the atmosphere and **meteorites** when they land on the surface.

Comets are mixtures of ice and rock that astronomers compare to dirty snowballs. Most comets originate beyond the Kuiper belt but some have orbits that cross into the inner solar system.

The arrival of meteorites and smaller dust particles provide critical evidence for the planetary-origin theories. Some meteorites have mineral textures suggesting that they are pieces left over from very early stages of dust accumulation and never formed into larger embryos. These meteorites are artifacts preserving the materials that formed the planets without the chemical and physical modifications that accompanied accretion. Others seem to be chunks of embryos that were smashed to bits during collisions and provide pieces of planetary interiors for geologists to study.

EXTENSION MODULE 9.2
Origin of the Moon.
Learn about the different hypotheses for the origin of the Moon.

The accretion process is not entirely over. Small meteors and dust particles continue to accrete about 100,000 kilograms to Earth each day. Most of these colliding objects are too small to notice, other than the occasional bright fireballs and meteor streaks (shooting stars) visible in the night sky. Most of the incoming mass is fine space dust.

Craters that are only slightly modified by erosion, such as the one illustrated in **Figure 9.6**, mark large collisions in the recent geologic past. Geologists hypothesize that the impact of an asteroid about 20 kilometers in diameter caused catastrophic environmental change on Earth around 65 million years ago, leading to the extinction of the last of the dinosaurs and about 75 percent of species then present on the planet. More catastrophic collisions will happen in the future, as the orbits of some remaining asteroids cross Earth's orbit. Every few years, one of these objects passes by Earth at a distance closer than the distance from Earth to the Moon. These events show that planetary accretion is still occurring, although nearly all of the mass in the original solar nebula now resides within the Sun and planets, and only very little additional growth of these bodies is possible. Earth accreted to more or less its current volume and mass by about 4.5 billion years ago (see Section 7.6 for discussion of the age of Earth).

Planetary Differentiation

The collision of planetesimals and embryos generates a tremendous amount of heat. When fast-moving objects with large masses collide, their motion energy instantaneously converts to enormous heat energy. Clap your bare hands together repeatedly with some force and then press them to your face to feel the warmth. Now, imagine the heat generated when objects the size of Moon collide with one another.

Collisional heating is sufficient to melt planetary embryos. Molten metallic compounds do not mix with molten silicates. Gravity pulls the molten metal toward the center of the embryo, leaving less dense molten silicates near the surface. **Figure 9.7** illustrates the somewhat similar process of unmixing and separation observed in a bottle of oil and vinegar salad dressing. The denser molten metallic compounds analogous to the vinegar in the bottle are drawn by gravity toward the center of Earth.

Planetary differentiation is the name applied to the planetary-scale process of separation of matter based on density. Differentiation within embryos could produce denser metallic cores and less-dense silicate mantles. Many meteorites collected on Earth consist only of silicate minerals or only of iron-nickel metal, which suggests that the meteorites are fragments of ancient differentiated planetesimals or planetary embryos.

▲ **Figure 9.6 What a meteorite-impact crater looks like.**
Meteor Crater in northern Arizona formed about 25,000 years ago when a meteor crashed into Earth. The crater is 1.2 kilometers wide and 180 meters deep. The meteorite fragments are iron metal, which suggests that the meteor was a piece of the core from a shattered planetary embryo in the asteroid belt.

Oil and vinegar are well mixed in vigorously shaken salad dressing.

Over time, oil and vinegar separate from one another. Oil is less dense than the vinegar, so oil occupies the top of the bottle while vinegar segregates on the bottom.

▲ **Figure 9.7 Oil and vinegar as an analogy for planetary differentiation.**
Chemical characteristics of oil and vinegar cause them to separate into two different fluids. Vinegar is denser than oil, so the vinegar moves to the bottom of the bottle, which leaves oil to occupy the top of the bottle. In a similar fashion, molten iron metal separated from molten silicates in planetary embryos. Gravity pulled the denser iron toward the embryo centers to make metallic cores surrounded by silicate minerals.

Putting It Together—*How Did Earth Form?*

■ The planetary-origin theories explain the formation of the Sun and planetary bodies in the solar system. Earth, as well as other solar system bodies, gradually accreted from collisions of progressively larger objects formed from dust and gases composing a nebular disk surrounding the central Sun.

■ Earth formed in a part of the solar system where silicate minerals and iron-metal alloys crystallized. Gaseous compounds formed icy planets in the colder, outer solar system.

■ Moon-sized embryos collided to form Earth. The embryos may have differentiated, with metal-rich cores and silicate mantles. One such collision may have formed the Moon from a mixture of materials from the early Earth and a colliding embryo.

9.2 **How Did the Core and Mantle Form?**

The major chemical and physical boundaries within Earth are between the mantle and core and between the mantle and crust. The contrast between the metallic core and the mostly silicate mantle is the more distinctive of the two boundaries, and the core and mantle compose almost all of Earth's mass. Therefore, it makes sense to consider first the origin of the distinct mantle and core layers.

Core Formation from a Magma Ocean

The possibility that Earth was once partly or largely molten was once thought unlikely, but scientists now view this idea as highly probable. **Figure 9.8** illustrates the current thinking—called "the magma-ocean hypothesis"—which calls for separation of dense, iron-rich metal from extensively melted silicate minerals to form Earth's core and mantle, respectively.

Differentiation of materials by density occurs readily in a liquid state, as described in the previous section. High-density metallic elements settle inward to form a planetary core, while remaining low-density silicate compounds compose the enveloping mantle. This is a *physical* process and, when thought of in terms of the oil and vinegar analogy, is fairly intuitive. Scientists who study core formation and the magma-ocean hypothesis, however, concentrate on the *chemical* process of differentiation, in particular the behavior of iron-seeking elements.

Just as the name implies, iron-seeking elements are those that prefer to bond with metallic iron (Fe) rather than silicon. Do not confuse the metallic, uncharged version of iron with its common ions (Fe^{2+} and Fe^{3+}), which readily fit into silicate mineral structures. There are many iron-seeking elements and their affinity for bonding with metallic iron ranges from "strong" to "moderate" to "slight." Analyses of a class of meteorites not modified by accretion into planetesimals provide estimates of the starting abundances of elements in the solar system and are a window into the chemical composition of early Earth. In these meteorites, geochemists find an undifferentiated soup of elements with varying affinities to metallic iron or to silicate minerals.

If the magma-ocean hypothesis is correct, then two things must have happened as Earth differentiated.

1. Under the physical influence of gravity, the heavy, dense iron sank to form the core, leaving the less dense silicates to form the mantle.

2. As the iron sank, one can reasonably expect that it took with it most of the "strongly" iron-seeking elements and some of the "moderately" iron-seeking elements and left behind some of the "moderate" and most or all of the "slightly" iron-seeking elements. Unfortunately, we cannot test the core to see if it is full of strongly iron-seeking elements. On the other hand, we can test the other side of the assertion—if the strongly iron-seeking elements all went to the core with the iron, there should be little, if any, left behind in the mantle.

To test the magma-ocean hypothesis, geochemists compare Earth rocks with the unmodified meteorites. Differences between elemental abundances in mantle rocks compared to the meteorites assumed to represent the original, overall Earth composition imply what elements moved into the core.

Figure 9.9 illustrates what the geochemists found. The measured abundances of the iron-seeking elements in mantle peridotite form a remarkable "stair step" pattern: Slightly iron-seeking ele-

▼ **Figure 9.8** **Visualizing core formation in a magma ocean.** This diagram summarizes how the magma-ocean hypothesis accounts for core formation on Earth. Very dense, molten, metallic iron sinks through the silicate magma ocean in the upper mantle and temporarily accumulates at a depth of about 400 km where the lower mantle is only partly melted. The molten-iron ponds enlarge until they are heavy enough to sink downward through the partially melted lower mantle to form the central core of Earth. Smaller volumes of molten iron and other metals also percolate downward between crystals in the mostly solid lower mantle.

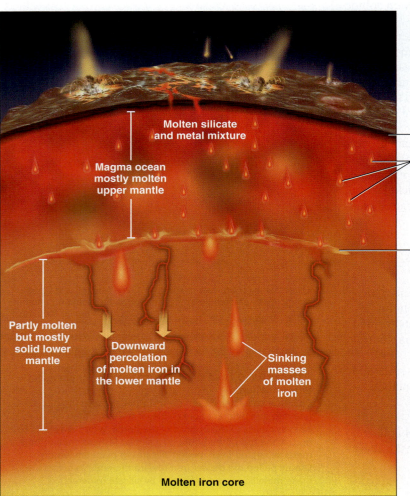

Molten silicate and metal mixture

Magma ocean mostly molten upper mantle

Chilled crust

Iron droplets separate from silicate melt

Partly molten but mostly solid lower mantle

Downward percolation of molten iron in the lower mantle

Sinking masses of molten iron

Temporarily ponded iron

Molten iron core

ments are most abundant, moderately iron-seeking elements are less abundant, and strongly iron-seeking elements are present at less than 1 percent of expected solar-system abundances found in the meteorites.

Many geochemists argue that the data plot in Figure 9.9 supports the magma-ocean hypothesis. Slightly and moderately iron-seeking elements only partly sank with metallic iron into the core, leaving behind residual abundances in the mantle. Elements with stronger affinity to iron metal have very low mantle abundances because they more efficiently separated from the mantle when the core formed. Experiments successfully reproduce the steplike pattern of elemental abundances in the mantle, as shown in Figure 9.9, and suggest that the metallic core components first separated from a silicate-rich magma ocean that was at least 400 kilometers deep. The dense core material sank downward to the center of Earth through the mushy, and partly melted, silicate lower mantle.

As Earth cooled, iron metal crystallized and settled under the force of gravity toward the center, toward the initial core, which was entirely molten. Crystallization of the core from the inside outward gradually led to the distinction between solid inner core and liquid outer core. The inner core continues to enlarge very slowly at the expense of the outer core because of this crystallization. Cooling also led to almost complete solidification of the enveloping mantle.

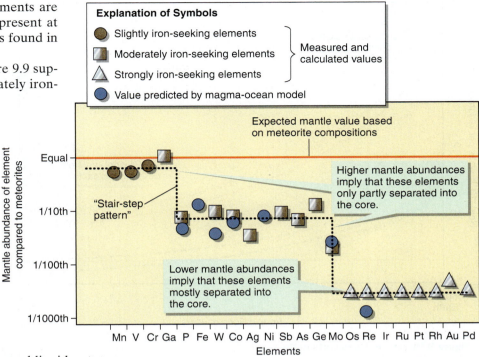

▲ **Figure 9.9 How element abundances support the magma ocean hypothesis.**
Graphed abundances of the iron-seeking elements in Earth's upper mantle exhibit a stair-step pattern. The pattern is consistent with the hypothesis that strongly iron-seeking elements more completely separated from the mantle into the core compared to slightly iron-seeking elements. The actual mantle abundances for all of these elements are close to those predicted by separation of molten iron and silicate compounds in a magma ocean.

Heat Sources for Magma Ocean and Core Formation

Geologists are confident that early Earth was much hotter than at present. Lord Kelvin also reached this important conclusion when he tried to calculate the age of Earth (see Section 7.6). Apart from data supporting the magma-ocean hypothesis, geologists have other evidence to indicate that early Earth was much hotter than today.

1. The present-day heat flow from within Earth is approximately twice as high as can be accounted for by present-day radioactive decay in the crust and mantle (Figure 8.22 summarizes this heat production). This suggests excess heat that remains from processes active in the past when Earth was hotter.
2. Ultramafic lava flows are found on Earth only among rocks older than 3 billion years. Experiments by petrologists reveal that these odd lava flows, the only known extrusive equivalents of peridotite, could form only at mantle temperatures several hundred degrees hotter than those estimated today.

So where did the heat come from to make the magma ocean so that the core formed? Collisional heating and radioactive decay are the likely heat sources.

Radioactive-isotope decay is one potential source of heat for a hotter early Earth. The internal temperature of the young Earth was higher than today partly because of higher abundances of radioactive elements. Decay of potassium, uranium, and thorium isotopes provide the source for much of Earth's current heat energy (see Section 8.5). Over the last 4.5 billion years, large amounts of these radioactive isotopes have decayed to daughter isotopes, which mean the radioactive parents were much more abundant early in Earth history than they are today. **Figure 9.10** illustrates how geologists use the decay rates of these three isotopes to calculate that radioactive heat production in the first two billion years on Earth was four times higher than it is today.

The total heat production by radioactive decay must, however, have been even greater than that produced by decay of potassium, uranium, and thorium.

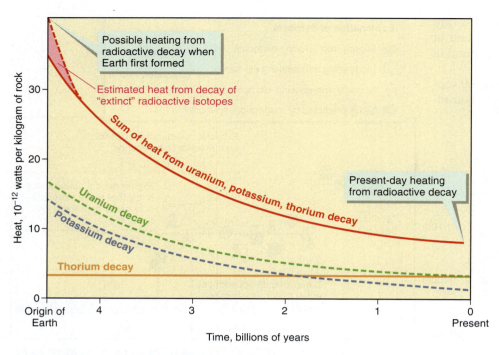

Figure 9.10 **How hot was Earth in the past?**
Geologists estimate the amount of heat energy released by radioactive decay of uranium, thorium, and potassium in Earth by combining measurements of the present-day abundances of the parent and daughter isotopes with calculations of the half-lives of their decays. When Earth formed, the total heat produced by decay should have been about four times greater than it is today. Many "extinct" isotopes with short half-lives have completely decayed away to their daughter isotopes but contributed heat very early in Earth history.

Experiments produce other radioactive isotopes in the laboratory that are not found on Earth. These isotopes are missing in nature because they have very short half-lives compared to the old age of Earth and have long since completely decayed to their daughter isotopes. The daughter products *are present* in Earth and Moon rocks and in meteorites. This means that the "extinct" radioactive isotopes were present during the early history of Earth and must have contributed additional heat as they decayed.

Finally, heating by collision during accretion, added to that from radioactive decay, may have been sufficient to melt the embryonic Earth to produce a magma ocean. The giant-impactor hypothesis for the origin of the Moon provides heat energy to account for the simultaneous formation of magma oceans on both bodies at a very early stage of planet formation. Wholesale melting by the impact was also possible by the already hot state of the planet resulting from radioactive decay.

Putting It Together–How Did the Core and Mantle Form?

■ Earth's metallic core separated from the mantle by planetary differentiation in a global magma ocean. Very dense metallic compounds segregated from lower density silicate compounds to produce a metallic core and a silicate-mineral mantle.

■ As Earth cooled, crystallization and inward settling of iron metal caused further distinction of a solid inner core and a remaining molten outer core. The mantle solidified into silicate minerals with a chemical composition similar to peridotite.

■ Very high temperatures within early Earth resulted from collisions with accreting embryos and higher heat production by radioactive decay than at present.

9.3 How Does the Crust Form?

With a reasonable hypothesis established for explaining the separation of the core from the mantle, we can move on to the question of how Earth's crust formed. There is no direct geologic information about the earliest crust, simply because none has survived through the active tectonic processes that reshape the surface of the planet. Geochemical characteristics of silicate-mineral crust, which is composed mostly of igneous rock, suggest that it forms from melting in the mantle. The resulting magmas are less dense than the mantle peridotite, so they migrate upward and solidify to form the crust.

Oceanic Crust Compared to Continental Crust

Sampling in the field, experimental studies, and seismic data provide the basic information about Earth's crust (Table 8.1). Oceanic crust, of basaltic composition, covers roughly 70 percent of the Earth's surface, and the more silica-rich continental crust forms the remainder. Oceanic crust reaches a maximum age of about 180 million years, under the western Pacific Ocean, and the average age of all the current oceanic crust is about 100 million years. Continental crust, on the other hand, has a maximum recorded age of approximately 4.4 billion years and an average age of about 2.2 billion years.

Plate tectonics explains the greater antiquity and permanence of continental crust compared to oceanic crust. Subducting basaltic oceanic crust metamorphoses in the mantle to very dense eclogite (Section 6.8). This allows the dense oceanic lithosphere to sink into the asthenosphere at subduction zones, whereas the less dense continents do not sink (you will learn more about this concept in Chapter 12). The density comparison explains why continental crust is much older than oceanic crust and why the oldest history of Earth is found in continental crust.

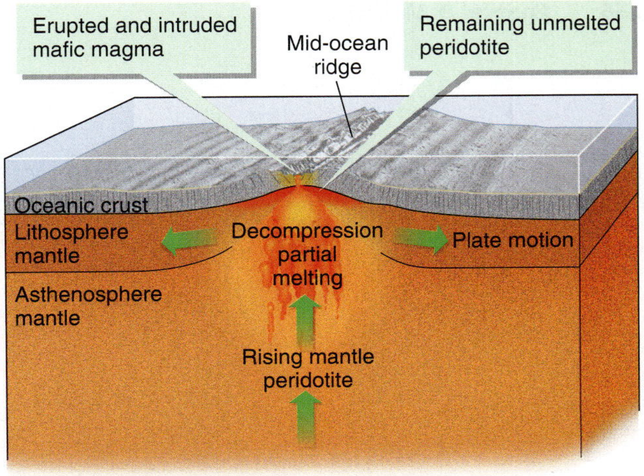

Origin of Oceanic Crust

It is perhaps a bit surprising that the formation of the basaltic oceanic crust, despite its relative inaccessibility, is better understood than that of continental crust. Section 4.6 explained the recipe for making basalt—partial melting of mantle peridotite produces basaltic magma, as illustrated in **Figure 9.11**. The nonmelted peridotite minerals remain behind and make up an important constituent of the mantle part of the lithosphere. Oceanic crust, therefore, is simply extracted from the mantle by partial melting and crystallization.

▲ **Figure 9.11 How oceanic crust forms.**
Mantle peridotite rises below mid-ocean ridges and partially melts when pressure decreases closer to the surface. The resulting mafic magma crystallizes at or close to the surface to produce basaltic oceanic crust. The unmelted peridotite forms part of the lithospheric mantle below the crust.

Origin of Continental Crust

Deciphering the origin of continental crust is much more challenging. In comparison to oceanic crust, which currently forms at mid-ocean ridges, the great antiquity of most continental crust suggests processes that are partly or wholly unique to conditions on early Earth. Experiments show that the melting of the mantle produces basaltic magma, whereas the average composition of the continental crust is more similar to andesite, and many continental igneous rocks are still more felsic. These observations imply different processes of magma generation than are witnessed at mid-ocean ridges.

Subduction zones, where intermediate and felsic composition magmas currently form and erupt at volcanoes, are a good place to start investigating the recipe for continental crust. Looking to subduction zones also makes sense because the oldest Archean rocks include metamorphosed volcanic and sedimentary rocks that are consistent with formation near a convergent plate margin. Mantle melting near convergent plate boundaries happens because silica-rich fluids expelled by dehydration metamorphism of the subducted plate cause adjacent peridotite to melt (Section 4.6 describes this melting process). **Figure 9.12** illustrates how continental crust may form near convergent boundaries.

The initial subduction-zone magma can contain a bit more silica than basalt forming at mid-ocean ridges, but it is still less felsic than the average andesitic composition of continental crust. Fractional crystallization (explained in Section 4.7) can produce intermediate and felsic magma from the original mafic melt, but it leaves thick accumulations of olivine and pyroxene crystals at the bottom of the

Modern subduction zones:
- Water from metamorphic dehydration causes partial melting in the asthenosphere.
- Fractional crystallization of mafic magma forms intermediate and felsic upper crust.
- Crystals separating from melt during fractional crystallization accumulate to form mafic and ultramafic lower crust.
- The average composition of upper and lower crust remains close to basalt, rather than the andesitic composition observed for continents.

Crust Delamination:
- Dense metamorphosed lower crust and lithospheric mantle detaches and sinks into the asthenosphere.
- Intermediate and felsic rocks remain as continental crust.

Early Precambrian subduction zones:
- Melting of subducted mafic crust in the very hot, early Earth produces tonalitic magma.

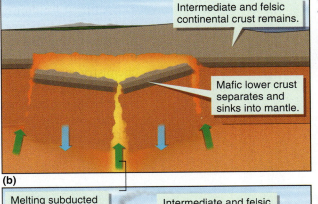

▶ **Figure 9.12 How continental crust forms.**

235

magma chambers. So, although some intermediate and felsic rocks form, Figure 9.12a shows that the bulk composition of the total igneous crust remains close to basalt, which is the average of intermediate and felsic upper crust and the mafic and ultramafic mineral accumulations in the lower crust. The conclusion, therefore, is that modern subduction-zone processes do not easily produce continental crust, because although felsic magmas form near convergent margins, a thick, mafic lower crust also forms and the overall crust composition does not match with average andesitic continental crust.

Geologists have tackled this problem with two alternative hypotheses. One option, illustrated in Figure 9.12b, is to get rid of the mafic rocks forming the initial lower crust above subduction zones. Detachment of the mafic rocks leaves behind the intermediate and felsic rocks that are more typical of continental crust. The detachment is physically possible because at depths of 50 kilometers or more, the mafic rocks should metamorphose to eclogite. Eclogite is denser than mantle peridotite, whereas the intermediate and felsic upper-crust rocks are less dense than peridotite. It is possible, therefore, that the ecologitic lower crust pulls away and sinks into the mantle while the intermediate and felsic rocks remain as typical continental crust.

The second option is that early Precambrian magma-forming processes were different than those observed today. Section 4.6 explained that melting of the water-bearing minerals in subducting mafic crust yields tonalitic magma rather than basalt. This process for making large volumes of felsic magma is not very important today because geologists see evidence for plate melting at only a few modern subduction zones. Figure 9.12c shows, however, that melting of basaltic crust could have been more important in early Earth history when the mantle was much hotter. In this recipe, the first melting of the mantle at divergent plate boundaries formed basaltic oceanic crust, which later melted at convergent plate boundaries to form felsic magma.

The generation of abundant granite in continental crust requires an additional melting step and the presence of water. Early steps in the recipe form dioritic and tonalitic rocks at convergent plate boundaries (Figure 9.12c) but not much granite. In the next step, these early-formed igneous rocks partially melt when heated by additional rising magma. Granite magma results from partial melting of diorite or tonalite that contains water-bearing minerals.

A tiny, 4.4-billion-year-old sand grain composed mostly of zircon is the oldest known geologic record of crust (see Section 7.6). The zircon crystal encloses tiny quartz and feldspar crystals. This mineral association strongly suggests an origin from crystallization of a granitic magma. Other details of the mineral chemistry suggest the presence of dissolved water in the magma. From this evidence, geologists know that less than 150 million years after Earth formed, rock resembling the continental crust seen today was forming and water was probably present. The next problem to contemplate is how water originated on Earth.

Putting It Together–**How Does the Crust Form?**
- Basaltic oceanic crust is the direct result of the partial melting of peridotite in the mantle.
- Intermediate and felsic continental crust cannot form by simple mantle melting. Physical separation of metamorphosed mafic igneous rocks may leave behind typical, more felsic crust. Partial melting of basaltic oceanic crust in hot early Earth may have generated magmas with continental-crust composition that form rarely today.
- All processes for producing intermediate and felsic magma require melting in the presence of water. Water was present in minerals melted to form granitic magma 4.4 billion years ago.

9.4 How Did the Hydrosphere and Atmosphere Form?

According to the planet-origin theories, Earth (along with Mercury, Venus, and Mars) formed in a part of the solar system where water and gaseous compounds did not exist (see Figure 9.5). Water and gaseous compounds had not yet condensed in the hotter, inner solar system when solar winds from the protosun cleared the light elements out of the solar system. Gaseous compounds like water vapor condensed and ices formed only in the cold temperatures in the vicinity of the outer planets prior to this violent phase of solar evolution.

Two contrasting hypotheses endeavor to explain the paradox of Earth having water even though it formed in a part of the solar system where water probably did not exist. Each hypothesis has merit and both have strong supporters, so it is only fair to examine each one.

Late Delivery of Water to an Initially Dry Earth

One hypothesis for the origin of Earth's water assumes that most of Earth formed without water and that the planet acquired water late in the accretion process. Planetesimals growing in the orbits of the inner rocky planets accreted in a dry state because they were too close to the hot early Sun for water molecules to exist. This hypothesis requires later delivery of water to Earth from comets and meteors that originated farther out in the cooler, gaseous part of the solar system. Earth-based and spacecraft analyses indicate that comets, such as those seen in **Figure 9.13**, are mainly composed of water ice. Some varieties of meteorites also have water-bearing minerals. The hypothesis states that these types of water-bearing objects accreted to Earth during the very latest stages of planet formation, mixed water into the upper mantle, and produced the water-bearing minerals found in Earth rocks today.

The heavily cratered lunar surface (Figure 9.1) attests to a large number of encounters with stray comets and asteroids. Based on studies of Moon rocks, most of these impacts occurred between 3.8 and 4.4 billion years ago. Given the proximity of Earth and Moon and Earth's larger size, it seems likely that millions of planetesimals left over from planetary accretion likewise pummeled Earth at this same time. Some of these objects hypothetically include icy comets and meteorites with water-bearing minerals, either of which could deliver water to Earth.

Planetary Embryos Form an Initially Wet Earth

In a more recently proposed and evolving hypothesis, researchers suggest that planetary embryos rich in water-ice or water-bearing minerals accreted to form the planet—that is, the planet initially accreted wet, rather than having water added near the end of planet formation.

Perhaps some planetesimals and embryos contributing to Earth's growth did not originate in the nearby orbital feeding zone for the accreting planet. Instead, some of these accreting objects condensed farther out from the Sun, near the present-day asteroid belt, where water-bearing minerals did form (see Figure 9.5). Erratic orbits influenced by the proximity of giant Jupiter sent one or more these water-bearing embryos on a collision course with Earth. Late-arriving comets are of minor significance in this hypothesis, perhaps providing only 10 percent of Earth's total water.

What is the evidence for accretion of a wet planet? The predicted abundances of iron-seeking elements depicted in Figure 9.9 are based on an assumption of water in the mantle. The predicted abundances are so close to actual measured elemental abundances that they imply the presence of a hydrous magma ocean during core formation. Supporters of an initially hydrous Earth also point to the Moon as evidence that Earth's water did not arrive as late as 3.8–4.4 billion years ago. If Earth received most of its water late in accretion

▲ **Figure 9.13 Watery comets impact Jupiter.**
An artistic reconstruction, based on spacecraft images, shows fragments of Comet P/Shoemaker-Levy 9 slamming into Jupiter. This first-ever observation of two colliding solar system bodies occurred in July 1994 when 20 comet fragments, as large as 2 kilometers across, collided with Jupiter. The icy comets delivered at least 2 million metric tons of water to Jupiter. Similar impacts of icy comets with early Earth are one hypothesized source of water for the planet.

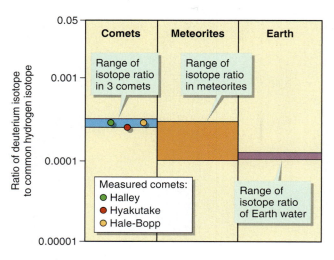

▲ Figure 9.14 **Where did Earth's water come from?**
This chart compares the deuterium/hydrogen isotope ratios from three comets, various meteorites, and Earth. The deuterium/hydrogen ratio of Earth does not resemble that of measured comets; so comets may not be the primary source of water on Earth. Some meteorites have a similar deuterium/hydrogen ratio to water on Earth, but meteorites contain very little water.

▼ Figure 9.15 **Artistic rendition of early Earth.**
Scientific illustrators use geological data and hypotheses to portray the early Earth surface. The initial atmosphere consisted of gases emitted from volcanoes that must have been much more numerous on the hot early Earth than at present. Meteors and comets plunged through the atmosphere to the surface.

from watery bodies that impacted it, then the Moon should have also received a supply of water. Moon rocks, however, do not have water-bearing minerals. In addition, the 4.4-billion-year-old Australian zircon eroded from granite indicates the presence of water before the late heavy bombardment had hardly gotten underway.

Insights on Water Delivery from Comets and Meteorites

One way to test whether comets and meteorites originating far from Earth could deliver water to Earth is to look more carefully at the chemical data. Geochemists commonly compare the abundances of two nonradioactive isotopes of hydrogen that compose water molecules on Earth, as well as in other planetary objects. Most hydrogen has no neutrons in its nucleus, and this hydrogen is almost 10,000 times more abundant on Earth than deuterium, a variety of the hydrogen atom with one neutron. If the deuterium/hydrogen ratio of water on Earth matches that measured in comets, then there would be support for the hypothesis that comets delivered Earth's water after planetary accretion was nearly completed.

Figure 9.14 compares the deuterium/hydrogen ratios from comets and meteorites to that of water measured on Earth. The existing data, which includes measurements from only three comets, do not support the hypothesis of comets delivering much water to Earth because comets have a different isotopic composition from Earth water. Some meteorites have similar deuterium/hydrogen ratios as Earth water (Figure 9.14), but meteorites contain much less water than comets. This means that if meteorites, rather than comets, delivered the water, then a very large mass of meteorite material has to be added to Earth; this mass is greater than can reasonably be accounted for by the late-delivery hypothesis.

Taken together, the data graphed in Figure 9.14 suggest that Earth's water mostly arrived during the primary accretion process within embryos with compositions similar to those represented by meteorite analyses. In this case, the later heavy bombardment of objects whose scars remain on the Moon delivered very little, if any, water. Supporters of the late-delivery hypothesis point out that data from only three comets may not be sufficient to represent most potential comet contributions of water to early Earth. As a result, both hypotheses remain viable to varying degrees. Intensive research continues to investigate when and how Earth obtained the water for its hydrosphere.

Making the Early Atmosphere

Regardless of when water arrived on Earth, it largely mixed into the interior and was incorporated into water-bearing minerals, rather than accumulating as liquid on the surface. Experiments show that water immediately releases into magma right at the onset of partial melting of rocks containing these water-bearing minerals. Therefore, both hypotheses for the origin of Earth's water propose that magma formation exported steam and other gases from the mantle through volcanoes to produce the atmosphere and oceans. **Figure 9.15** portrays the volcanically active, hot early Earth spewing out gases through innumerable volcanoes. These gases accumulated as the first atmosphere and steam condensed to form the hydrosphere.

Gases from active volcanoes are potential examples of the gases that formed the early atmosphere. **Figure 9.16** compares the composition of the modern atmosphere to the composition of volcanic gas.

Water vapor is the most important constituent in volcanic gas. This observation is consistent with the

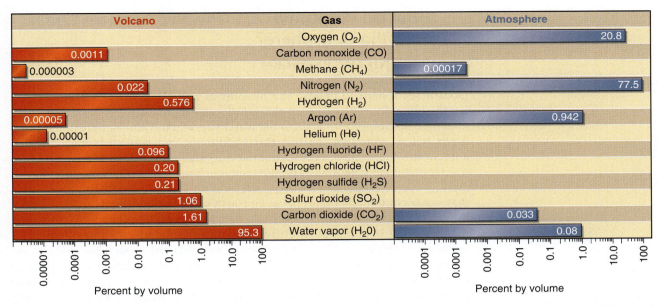

Volcano	Gas	Atmosphere
	Oxygen (O₂)	20.8
0.0011	Carbon monoxide (CO)	
0.000003	Methane (CH₄)	0.00017
0.022	Nitrogen (N₂)	77.5
0.576	Hydrogen (H₂)	
0.00005	Argon (Ar)	0.942
0.00001	Helium (He)	
0.096	Hydrogen fluoride (HF)	
0.20	Hydrogen chloride (HCl)	
0.21	Hydrogen sulfide (H₂S)	
1.06	Sulfur dioxide (SO₂)	
1.61	Carbon dioxide (CO₂)	0.033
95.3	Water vapor (H₂O)	0.08

Percent by volume   Percent by volume

hypothesis that water degassed from the mantle to form the atmosphere and hydrosphere.

Otherwise, the analyzed abundances compared in Figure 9.16 are very different. Perhaps the initial atmosphere composition more closely resembled the composition of the volcanic gases, but the abundances of different compounds in the atmosphere change through time. Weathering reactions between atmospheric gases and surface minerals, as well as biological processes after life first appeared on the planet, would cause these changes.

Notably, Figure 9.16 shows that oxygen gas (O_2), an obviously critical ingredient for most life on present-day Earth, does not erupt from volcanoes. In agreement with the hypothesis that volcano degassing produced the atmosphere, there is abundant geological evidence for a lack of oxygen in the early atmosphere. The vast iron-ore resources of the Great Lakes region, South America, and Australia indicate the lack of atmospheric oxygen during early Earth history. These iron ores are peculiar chemical sedimentary rocks that cannot form from water that contains oxygen, as in the modern oceans. Almost all of these iron-rich sedimentary rocks formed prior to 2.3 billion years ago and indicate a lack of oxygen on the early Earth. Oxygen most likely accumulated in the atmosphere very gradually after the appearance of photosynthetic organisms, which consume carbon dioxide (CO_2), and manufacture oxygen.

▲ **Figure 9.16** **Gas content from a volcano compared to the modern atmosphere.**
These charts compare the gas content of an active volcano in eastern Russia with the composition of the atmosphere. The most abundant gas erupted from volcanoes is water vapor. Oxygen is an important atmospheric gas not erupted by volcanoes.

Putting It Together—How Did the Hydrosphere and Atmosphere Form?

■ Two contrasting hypotheses explain the origin of Earth's atmosphere and hydrosphere. One hypothesis suggests delivery of water-rich materials from meteorites and comets that accreted to the initially dry Earth very late in the formation of the planet. The other hypothesis suggests that Earth accreted wet, from planetesimals or embryos containing water-bearing minerals.

■ Water vapor is the most abundant gas erupted from volcanoes. This observation is consistent with the hypothesis that the hydrosphere derived from melting water-bearing mantle minerals and erupting water vapor through volcanoes.

■ Atmosphere composition changed through Earth history as a result of weathering reactions and photosynthetic organisms, which consume carbon dioxide and create oxygen.

9.5 How Do We Know . . . The Hydrosphere Came from the Geosphere?

UNDERSTAND THE PROBLEM
Did Early Earth Rocks Contain Oceans of Water?
Does it seem hard to believe the hypothesis that all of the water you see on Earth originally escaped from the interior as vapor through erupting volcanoes? Do not forget, either, all of the water vapor in the atmosphere and ground water below your feet. To test this hypothesis, can geologists calculate the amount of water that was present in the early mantle?

Unfortunately, geologists do not gain much insight into this question by analyzing water contents of mantle peridotite or volcanic gases. The problem is that water continually cycles between Earth's interior and exterior. Water exits the interior through volcanoes, but subduction of crust and sediment containing water-bearing minerals also returns water to the mantle at convergent plate boundaries (Figure 9.12a). Dehydration metamorphic reactions release the water from the subducted plate into the mantle. Some of this water resides in the mantle for a long time before joining partial melts of peridotite and returning to the surface through volcanoes. These observations mean that although water vapor is the most abundant gas emanating from erupting volcanoes (Figure 9.16), most of this water is recycled surface water, so it does not explain the origin of the first surface water.

STATING THE HYPOTHESIS
Could the Hydrosphere Originate from the Geosphere?
Consider a hypothetical time period near the beginning of Earth's history before volcano degassing substantially exported mantle water to the surface. In that primeval time nearly all of the hypothesized water supply was in the geosphere, whereas today nearly all of it exists in the hydrosphere and atmosphere. Geologic processes, according to proponents of the hypothesis, redistributed the original water reservoir in the geosphere so that most of the water now exists in the atmosphere and hydrosphere. The hypothesis claims that most of the primordial water has left the mantle, so how can geologists test the hypothesis that all of the water started out in the mantle?

A concise statement of the hypothesis is a good starting point for designing the test: The original mantle contained enough water to generate the modern hydrosphere by gas release through volcanoes. The stated hypothesis implies working with numbers because it is necessary to quantify how much water is enough water. The test, therefore, requires some simple calculations. To determine if the volume of the hydrosphere can come from the mantle, you need to estimate, and then compare, two values:

1. How much water is *currently* on Earth?
2. How much water was in the *original* mantle?

COLLECTING THE DATA
How Much Water Is on Earth, and How Much Was in the Original Mantle?
Figure 9.17 shows the estimated volumes of the major water reservoirs on Earth. The largest volume of water is in the ocean, and this is a reliable value because the depth and area of Earth's oceans are well known. The sizes of most of the other reservoirs are almost as well known but they are clearly much smaller than the ocean, so any uncertainties in these values are unimportant for the calculations. The amount of water retained by minerals in crust and

▼ **Figure 9.17** **Reservoirs of water on Earth.**
This diagram schematically illustrates the volumes of water in various reservoirs at or near Earth's surface, in the atmosphere, and bound within rocks in the mantle and crust.

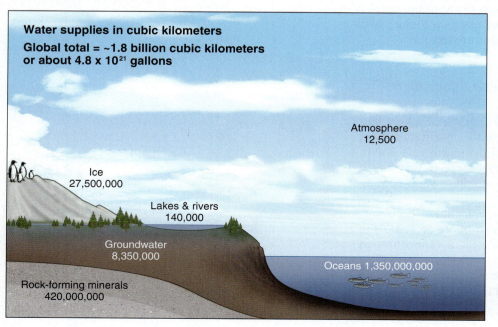

Water supplies in cubic kilometers

Global total = ~1.8 billion cubic kilometers or about 4.8×10^{21} gallons

Atmosphere
12,500

Ice
27,500,000

Lakes & rivers
140,000

Groundwater
8,350,000

Oceans 1,350,000,000

Rock-forming minerals
420,000,000

mantle rocks is the least well known value, is fairly large, and is subject to speculation. Geochemical arguments suggest that the value shown in Figure 9.17 is reasonable and that the water in rocks cannot be more than twice the given value. The total estimated volume of the water on Earth is, therefore, 1.8 billion cubic kilometers. This is a very large volume, equivalent to 4.8×10^{21} gallons. To put this number in perspective, if everyone on Earth consumed 100 gallons of water each day, then it would take 22 million years to use all of this water.

How do geologists determine the amount of water in the original mantle? It is not helpful to look at the composition of *modern* mantle rocks for this value, because the hypothesis assumes that the mantle has already lost much of its water to create the present hydrosphere. You need, instead, to examine the water content of rock that might represent the *original* mantle. Geologists commonly turn to some classes of meteorites to play this role. This is because some meteorites are parts of small asteroids that never accreted in embryos, so they likely represent the pristine starting materials for planet building. The measured water content of these meteorites, enclosed in mineral crystal structures, is highly variable and ranges from 0.5 percent to 9.0 percent, by mass.

TESTING THE HYPOTHESIS

Do the Numbers Add Up?

Figure 9.18 illustrates the calculations required to test the hypothesis that the original mantle contained enough water to produce the present-day hydrosphere. The calculated volume of water in the early mantle is 10 times larger than the volume currently present on, below, and above Earth's surface. The result successfully tests the hypothesis so long as the assumption that certain meteorites resemble the composition of the early mantle is valid. The validity of the assumption is, itself, difficult to assess. This is why it is considered an assumption and not a separate hypothesis. Nonetheless, the calculation provides a relatively small number, rather than a suspiciously large value, for two reasons:

1. The calculations assume a conservative estimate of 0.5 percent water by mass from meteorites. Although some meteorites contain as much as 10 percent water, these types are not as abundant in meteorite collections as those with less water, so the calculation uses a smaller percentage.
2. The calculations ignore any water delivered from comets, which are as much as 80 percent water. If the calculations included cometary water, then the volume of original water in the mantle would be even larger. Comets were left out of the calculation because there currently is no compelling evidence that comets were a major source of Earth's water (see Figure 9.14), but they may have been a minor player.

INSIGHTS

Did Earth Lose Some of Its Water?

Assumptions, measurements, estimates, and calculations compose the test for the hypothesis that Earth's water originated in the geosphere. Some values used

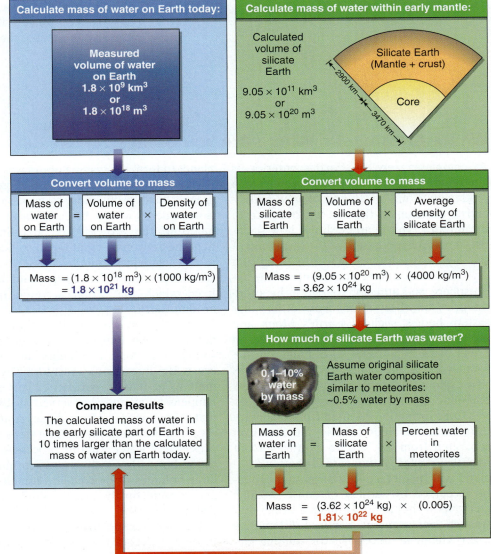

▼ **Figure 9.18** **Calculations to test the hypothesis: The hydrosphere came from the geosphere.**
This diagram illustrates the calculations that test the hypothesis that the original mantle, with a composition similar to meteorites, contained enough water to account for the water on and within Earth today. The masses of water present on Earth today and in the early silicate Earth are calculated separately and then compared.

Calculate mass of water on Earth today:

Measured volume of water on Earth
1.8×10^9 km³
or
1.8×10^{18} m³

Convert volume to mass

Mass of water on Earth = Volume of water on Earth × Density of water on Earth

Mass = $(1.8 \times 10^{18} \text{ m}^3) \times (1000 \text{ kg/m}^3)$ = $\mathbf{1.8 \times 10^{21} \text{ kg}}$

Compare Results
The calculated mass of water in the early silicate part of Earth is 10 times larger than the calculated mass of water on Earth today.

Calculate mass of water within early mantle:

Calculated volume of silicate Earth
9.05×10^{11} km³
or
9.05×10^{20} m³

Silicate Earth (Mantle + crust)
Core
2900 km
3470 km

Convert volume to mass

Mass of silicate Earth = Volume of silicate Earth × Average density of silicate Earth

Mass = $(9.05 \times 10^{20} \text{ m}^3) \times (4000 \text{ kg/m}^3)$ = 3.62×10^{24} kg

How much of silicate Earth was water?

0.1–10% water by mass

Assume original silicate Earth water composition similar to meteorites: ~0.5% water by mass

Mass of water in Earth = Mass of silicate Earth × Percent water in meteorites

Mass = $(3.62 \times 10^{24} \text{ kg}) \times (0.005)$ = $\mathbf{1.81 \times 10^{22} \text{ kg}}$

are well known, others poorly known and some are fairly speculative, but the result is reasonable for two reasons:

1. The largest water reservoir on Earth, the oceans, is the best known, and uncertainties in the other reservoir volumes will not significantly affect the calculated volume of Earth's water.
2. Although the estimated volume of water in the early mantle rests on assumptions, it is also probably a minimum value, and it is still ten times greater than the amount required to successfully test the hypothesis.

Indeed, there may be a problem of missing water. The assumptions and calculations permit the interpretation that Earth only has 10 percent, or less, of the water that it started with. Geologists think that it is very likely that Earth lost lots of water early in its history. Recall that Earth probably experienced countless giant impacts between 3.8 and 4.4 billion years ago when Moon was heavily bombarded. The largest impacts may have vaporized the first-formed oceans and blown the early atmosphere into space. These losses of the early hydrosphere and atmosphere likely occurred several times.

Putting It Together–*How Do We Know . . . The Hydrosphere Came from the Geosphere?*

■ Calculations of the amount of water available in the early mantle, using conservative estimates of water in meteorites, suggest that early Earth contained many more times the water than is currently found on or near the surface.

Where Are You and Where Are You Going?

A complete view of Earth's interior is emerging. Chapter 8 provided you with a fundamental understanding of how geoscientists determined the layered structure of the interior. This chapter shows how scientists derived hypotheses for the formation of the layers by differentiation, melting and crystallization, and the appearance of the hydrosphere and atmosphere. These hypotheses draw on observations from other solar system bodies, such as Earth's Moon and meteorites, in addition to data collected on Earth. The hypotheses also build on the framework of the planetary-origin theories that explain the origin of the diverse planets and minor bodies in our solar system.

Core formation resulted from planetary differentiation where dense metallic iron separated from the silicate mantle. Recent hypotheses suggest that core formation coincided with planetary formation and probably occurred by settling of dense metallic elements within a global magma ocean that was at least several hundred kilometers deep. The heat necessary for melting the

planet to form the magma ocean was provided by giant impacts, as well as from much greater heat production from radioactive-isotope decay within the early Earth as compared to the present. The giant-impactor hypothesis proposes that Earth collided with a large planetary embryo late in its formation history to produce a vapor and dust ring from which the Moon accreted in orbit around Earth.

Melting in the mantle produced less dense magmas that rose to the surface and solidified to form continental and oceanic crusts. Basaltic oceanic crust forms by partial melting of mantle peridotite. This process happens today and is observed at divergent plate boundaries. More felsic continental crust requires multiple melting steps and the presence of water. Melting processes unique to the hot early Earth favored formation of large volumes of dioritic and tonalitic crust and account for the great antiquity of most continental crust.

Gases escape from Earth's interior through volcanoes to provide components

of the atmosphere and hydrosphere. Partial melts of the mantle produce water-rich magmas and the water vapor escapes to the atmosphere through erupting volcanoes and then condenses to liquid water. With the exception of oxygen, volcanic gases can produce Earth's atmosphere composition after accounting for reactions among the gases, between gases and rocks, and organic processes that consume and create atmospheric gases. Photosynthesis, not geologic processes, best explains atmospheric oxygen content. Calculations successfully test the hypothesis that Earth's early mantle minerals contained more than enough water to account for Earth's current hydrosphere.

Two hypotheses exist to explain how water got into the mantle during planetary accretion. Long-held interpretations favor a late-stage delivery of water brought in by comets and meteorites that formed much farther from the Sun than did Earth. New data support the contrasting hypothesis that water-rich planetesimals and embryos were part of the early accretion history of the planet. Water is critical not only to the formation of the hydrosphere but is an essential ingredient in the recipes for making continental crust. Without water there would be no continents on Earth.

Understanding the internal make-up of the planet provides the knowledge you need to continue your quest for understanding how processes in the interior of the planet drive the external processes observed at the surface. The next step, then, is to understand the active processes within the mantle and core rather than viewing the mantle and core as static layers within the planet.

Extension Modules

Extension Module 9.1: Geologic Tour of the Solar System. Learn the geologic basics of the planets, moons, asteroids, and comets of the solar system.

Extension Module 9.2: Origin of the Moon. Learn about the different hypotheses for the origin of the Moon.

Confirm Your Knowledge

1. With respect to water, why is Earth unique compared to the other planets in our solar system?
2. What are the main differences between the inner and outer planets?
3. What event caused the region of our solar system that contains the inner planets to lose its lightweight gas molecules?
4. How can we estimate the abundances of elements in the early solar system?
5. How did the formation of the inner planets lead to the formation of Earth's Moon?
6. Has the innermost part of Earth's core always been solid and constant in size?
7. What is the process of planetary differentiation? How did this process determine the composition of Earth?
8. What do meteorites tell us about Earth?
9. What was the source of the heat that made the early Earth much hotter than it is today?
10. What evidence tells us that early Earth contained radioactive isotopes that are no longer present?
11. Why is the oldest continental crust (4.4 billion years) so much older than the oldest oceanic crust (180 million years)?
12. How was the first continental crust formed?
13. Summarize the process of formation of oceanic crust.
14. What rocks support the hypothesis that Earth's early atmosphere (before 2.3 billion years ago) did not contain oxygen?

Confirm Your Understanding

1. Write out an answer for each question in the Chapter Outline for the chapter sections assigned by your instructor.
2. Water is crucial to many geologic processes. Using the rock cycle as a guide, describe what would not happen if Earth was a waterless planet.
3. What is the evidence that the outer planets started to form before the inner planets.
4. How do a bottle of salad dressing and clapping your hands offer analogs for planetary differentiation?
5. How is the composition of Earth's present day mantle used to support the magma-ocean hypothesis?
6. What evidence from volcanic rocks indicates that Earth was much hotter in the past?
7. What evidence helps us determine when Earth received its water?
8. Compare and contrast the compositions of volcanic and atmospheric gases.

Key Terms

asteroids (p. 230)
comets (p. 230)
meteorites (p. 230)
meteors (p. 230)
nebulae (p. 226)
nuclear fusion (p. 230)
planetary accretion (p. 228)
planetary differentiation (p. 231)
planetary embryos (p. 228)
planetesimals (p. 228)
protostar (p. 228)
solar nebula (p. 226)
solar wind (p. 229)

10 Motion Inside Earth

Chapter Outline

Why Study Earth's Internal Motion?

A MOTIVATION TO LEARN ABOUT EARTH'S INTERIOR IN CHAPTER 8 was to gain a better understanding of the internal forces that cause earthquakes and volcanoes. To complete this goal you need to know more about the internal workings of the planet.

This chapter focuses on the dynamic inner Earth processes taking place throughout the Earth's concentric layers rather than viewing the interior as static. Earth's mantle is in constant, albeit slow, motion. You will find out how heat and gravity drive that motion and will begin to relate this motion to plate tectonics. You will learn about vigorous currents in the outer core and how these currents generate the planet's magnetic field. This concept alone is worth understanding further because so much about Earth relates to the magnetic field. The field permits a compass to work for navigation, protects Earth from harmful cosmic radiation, and even allows migrating animals to find their way.

Motion inside Earth is fundamental to understanding processes at Earth's surface. Your objectives in this chapter are as follows:

✔ To understand how motion is possible within Earth

✔ To learn where and how fast the motion occurs

✔ To understand how motion in the outer core generates the magnetic field

✔ To consider why the magnetic field changes in orientation and intensity with time

Following this chapter, Part III of the text more fully develops the consequences of internal motion in the mantle to explain phenomena at the surface. In order to achieve the objectives for this chapter, you will answer the following questions:

10.1 How does convection work?

10.2 What does mantle convection look like?

10.3 How does outer-core convection generate the magnetic field?

10.4 *How do we know* . . . Earth's core is a dynamo?

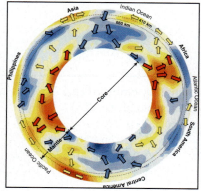
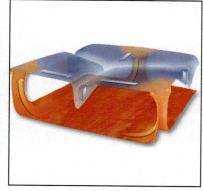
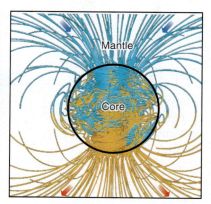

Molten lava rises as fountains to the surface of a lava lake in Africa and then cools to a dark crust that sinks back to the depths along the hot, glowing cracks. Convection stirs this lava lake and also stirs Earth's interior. ▶

In the
LAB

A geophysicist visits your lab class and demonstrates a lava lamp, a popular mood setter in college dorm rooms in the 1960s, 1970s, and even today. When the lamp turns on, blobs of orange liquid move up to the top, pause, and sink back down, as **Figure 10.1** shows. What causes this motion?

Heat and density are key ingredients to motion in the lava lamp. Heat must be essential to the movement of the orange liquid, because there is no motion without supplying heat at the base of the lamp. Density differences between the orange and blue liquids might cause the motion, just as ice cubes rise through denser water. Can density explain both up and down motion, however? Perhaps the density contrast between the liquids varies with temperature. This suggestion makes sense—most compounds expand when heated, which means density of a substance is greater when it is cold and less when it is hot. The geophysicist confirms that the orange liquid is denser than the blue liquid at room temperature, but it expands more than the blue liquid when heated. This means that the orange liquid is less dense than the blue liquid when it heats up and so it rises to the top of the lamp. As the orange blob cools at the top of the lamp, it once again becomes denser than the blue liquid and it sinks back to the bottom (see Figure 10.1).

The motion of liquids in the lava lamp not only represents movement of mass but also movement of heat. Heat moves with the liquid from the hot light at the bottom of the lamp upward to the top, where it dissipates into the surrounding cooler environment. Motion that simultaneously moves matter and heat is convection, a process introduced in Section 1.6.

Watching the lava lamp is interesting, but what does this have to do with geology? The geophysicist indicates that the lava lamp is, in a simple way, an analog for motion inside Earth. It is this internal motion that causes volcanoes to erupt, earthquakes to happen, mountains to rise, basins to sink, and your magnetic compass to function in the field. Motion inside Earth is fundamental to understanding geology at the surface.

Chapters 8 and 9 present arguments that most of Earth's interior is solid and that density increases downward within Earth (see Figure 8.19), so how is convective motion like that seen in the lava lamp possible inside the planet? To appreciate how convection occurs within Earth, you need to understand more about how heat and gravity cause convection.

Figure 10.1 Convection explains how a lava lamp works. ▶
Liquids move up and down inside a lava lamp because they have different densities at different temperatures. The less dense, warmer liquid rises while the denser, cooler liquid sinks.

10.1 How Does Convection Work?

You can visualize how heat moves by a quick analysis of an everyday kitchen experience. **Figure 10.2** depicts an all-metal pot on a stove filled with heating soup. Initially, both the pot and its contents are at room temperature (Figure 10.2a). Shortly after turning on the stove, the end of the pot handle is too hot to touch without a hot pad, and the air temperature around the pot is slightly warmer than before. Heat conducts from the stove burner through the pot, out to the end of the handle, and radiates into the surrounding air (Figure 10.2b). Conduction and radiation transport heat without moving matter.

After a few more minutes of heating there is motion in the soup. Soup wells up to the surface at the edge of the pot and then seems to sink near the center. This movement is convection—hot, expanded, less dense soup moves from the bottom to the top, while an equal volume of cooler, denser soup moves downward (Figure 10.2c).

Your first thought might be that this makes sense for the kitchen but not for Earth. In the kitchen, you see heat transfer by conduction in solids and by convection in liquids. Earth is mostly solid, so how is convection possible?

Cause of Convection Instead of Conduction

Conduction is only the transfer of heat from hot regions to cold regions, whereas **convection** is the simultaneous movement of matter and heat. The rate of heat conduction varies for different materials because heat transfer relates to properties of atoms and mineral crystal structure. Convection is commonly paraphrased as "warm material, like air or water, rises while cold material sinks." While temperature is certainly part of convection, it is gravity rather than heat that directly causes convection.

To understand the role of gravity in convection, it is useful to consider the analogy of two objects of unequal density—such as water and ice cubes. The force of gravity exerted by the mass of Earth pulls downward more strongly on the denser water than on the less dense ice cubes. This is why ice cubes rise to the top of a glass of water. Most compounds expand when they heat up (become less dense) and contract when cooled (become denser). Convective motion occurs in heated soup because gravity pulls cooler, denser soup downward relative to warmer, less dense liquid, which then rises upward to replace the sinking cool liquid. Strictly speaking, it is the

Lab notes: Demonstration of convection in a lava lamp

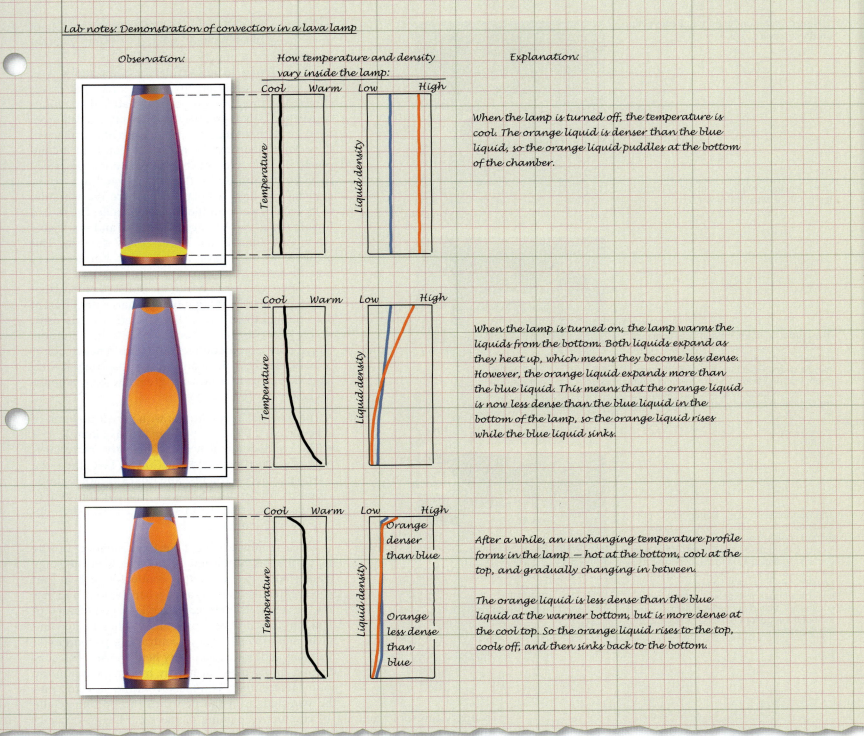

| Observation: | How temperature and density vary inside the lamp: | Explanation: |

When the lamp is turned off, the temperature is cool. The orange liquid is denser than the blue liquid, so the orange liquid puddles at the bottom of the chamber.

When the lamp is turned on, the lamp warms the liquids from the bottom. Both liquids expand as they heat up, which means they become less dense. However, the orange liquid expands more than the blue liquid. This means that the orange liquid is now less dense than the blue liquid in the bottom of the lamp, so the orange liquid rises while the blue liquid sinks.

After a while, an unchanging temperature profile forms in the lamp — hot at the bottom, cool at the top, and gradually changing in between.

The orange liquid is less dense than the blue liquid at the warmer bottom, but is more dense at the cool top. So the orange liquid rises to the top, cools off, and then sinks back to the bottom.

density contrast rather than the temperature difference that causes convective motion.

Heat transfer other than convection is also important in the soup pot. The cooled-off sinking soup warms up again by conduction from the burner through the bottom of the pot, while the rising hot soup cools by conducting and radiating heat to the surrounding air. The continual heat transfer by conduction and radiation at the boundaries of the liquid maintains the necessary density contrast between top and bottom for convection to happen.

A couple of questions regarding the convection analogies remain unanswered. Why does convection not begin immediately when the base of the lava

Active Art

Convection in a Lava Lamp: *See how convection works inside a lava lamp.*

▶ **Figure 10.2 Conduction and convection on the kitchen stove.**

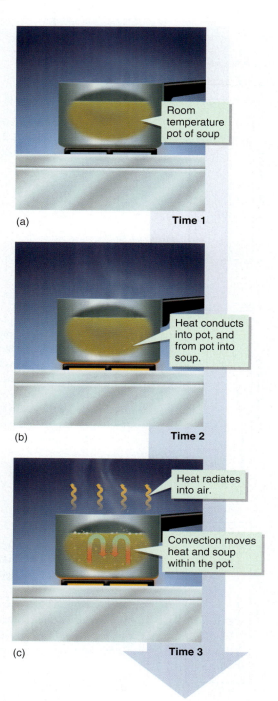

Soup and the metal pot start out at room temperature when the stove is off.

(a)   **Time 1**

Room temperature pot of soup

The burner is turned on and heat conducts from the stovetop into the metal pot. Heat conducts through the metal making the handle warm. Heat also conducts into the soup, but the liquid does not conduct heat as readily as the pot does, so the soup in the bottom of the pot is much warmer than at the top.

Heat conducts into pot, and from pot into soup.

(b)   **Time 2**

The warmer soup at the bottom of the pot expands to the point where it is less dense than the cooler soup above. The warmer, less dense soup rises to the top of the pan, cools by radiating heat to the overlying air, becomes denser, and sinks back to the bottom of the pot. Most of the soup has a uniform warm temperature because of the convective mixing of heat and soup, although the bottom remains hotter and the top is cooler.

Heat radiates into air.

Convection moves heat and soup within the pot.

(c)   **Time 3**

lamp and soup pot are heated? Why does a metal pot not convect when heated? What determines whether convection or conduction takes place?

A key step to answering these questions is to consider that while gravity is the force that drives convection, there is a property of matter that inhibits convection—that property is viscosity. Recall from Section 4.8 that viscosity is a measure of resistance to flow determined by the nature of bonds in a compound. Liquids are not the only materials described by viscosity. Solids also flow if subjected to a stress, especially at high temperatures where the vibration of molecules weakens the chemical bonds. The rate of flow in solids can be very, very slow (and not visible on human time frames), but solids do flow. This means that solids have a measurable, but very very high, viscosity.

For convection to occur, the effect of gravity that causes the vertical motion of materials with different densities has to be greater than the effect of viscosity, which slows down or even prevents that motion. Gravity has a greater effect if the difference in density between the colder and hotter materials is very large. The density contrast is great and convection is likely if

- The material expands a lot with very little input of heat (like the orange blobs in the lava lamp).
- The conduction of heat through the material is very slow. If heat conduction is fast, then the whole object expands, and there is not enough density difference between top and bottom to drive convection. In this case, heat transfers entirely by conduction.

We can now complete our understanding of why heat conducts in the pot and convects in the soup. Watery soup is a very slow conductor of heat, but water expands considerably when heated and has low viscosity. Metals are excellent conductors of heat, but they do not expand much when heated and have high viscosity. These contrasting properties of the soup and pot explain why heat transfers by conduction in the pan, but mostly by convection in the soup. Heat initially conducts from the stovetop into the pan and from the pan into the bottom of the soup. A threshold amount of conductive heating of the soup has to occur in order to cause enough expansion to start convection. When the warm soup convects to the top of the pot, it cools by radiating heat into the air. The cooling increases the liquid density so the soup sinks. An important part of convection is that over any time interval, an equal mass of liquid moves from top to bottom as moves from bottom to top.

How Solids Can Convect

So now you know what causes convection rather than conduction of heat, but perhaps you still have trouble picturing how convection can ever occur in a highly viscous solid. Most silicate rocks (that dominate the crust and mantle) exhibit very low heat conductivity and significant expansion when heated. These

After one hour   **After one day**

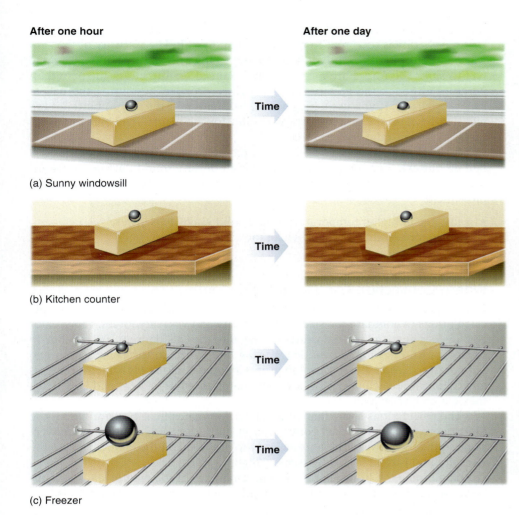

(a) Sunny windowsill

A steel ball is denser than a stick of butter, but that does not necessarily mean that the ball sinks through the butter.

(b) Kitchen counter

The viscosity of the butter resists motion caused by the force of gravity. When viscosity changes, so does the ball's movement. The warmer the butter, the less viscous it is and the faster the ball sinks through the butter.

(c) Freezer

The viscosity of frozen butter resists gravity, and a small steel ball does not sink noticeably. However, a larger ball weighs more, so the gravity force overrides the viscosity resistance, and the ball sinks into the butter.

characteristics favor the transfer of heat by convection *if* the large viscosity force is overcome. The key variables to evaluating potential convection in a solid are

- The mass of material that gravity tries to move because of density contrast; large masses favor convection.
- The time permitted for gravity-driven motion to occur; long time periods favor convection.
- The viscosity of the solid, which typically decreases with heating; low viscosity favors convection.

Another kitchen analogy depicted in **Figure 10.3** evaluates these three factors. Picture three sticks of butter: one in the freezer, one at room temperature, and one heated on a sunny windowsill to become soft but not melted. Imagine placing steel balls of equal mass on top of each butter stick. The density of the steel ball is more than five times greater than that of butter. Gravity, therefore, pulls down on the ball with more force than it pulls on the butter. From gravity considerations alone, you might predict that the steel ball will move down into the butter (for the same reason that ice cubes move up through water). The ball on the frozen butter, however, just sits there—no movement. Initially, there also does not seem to be any movement of the ball on the room-temperature butter stick, but after a day, you return and find a small dimple under the ball—the ball is slowly moving downward. The ball immediately settles downward slightly in the warmed butter on the windowsill and sinks even farther by the next day (Figure 10.3). The gravity force, which relates to the mass of the steel ball, is the same for all three cases, but the viscosity resistance is different. For the frozen butter, viscosity is too great to permit the ball to sink. The viscosity of the sun-warmed butter is sufficiently low compared to the gravity force and allows the ball to sink. The forces are more closely balanced for the butter at room temperature although the gravity force is slightly stronger, permitting the ball to sink very slowly. If you use a bigger steel ball on top of the frozen butter, it too will sink because the gravity force is greater for the bigger steel ball and eventually overcomes the viscous force.

If a high-density solid is placed on top of a low-density solid (like the steel ball on butter), then it is possible for the high-density solid to move downward while the low-density solid moves upward. The key observations are that movement

- is favored at higher temperature where solid viscosity is lower,
- occurs more readily if the masses of the solids are large,
- takes place very slowly.

Where Convection Can Occur Inside Earth

Determining whether or not convection occurs in solids involves using simple calculations to compare the sizes of the gravity and viscosity forces. The greater challenge is determining the values to plug into the calculations. It is possible to proceed with reasonable confidence using inferences on the composition and temperature inside Earth developed in Chapter 8, along with measurements and theoretical calculations of how well materials conduct heat and how much they expand when heated.

The calculations show that conduction dominates heat transfer within the lithosphere, except for local convection in molten magma chambers. Convection is unlikely between the bottom and top of the solid lithosphere because the viscosity in this rigid layer is extremely high.

Combining the lithosphere with the remainder of the mantle leads to a different result. Calculations show that mantle convection is expected because

- Viscosity beneath the lithosphere is lower because the rock is close to its melting temperature (Figure 8.21).
- Some of the lithosphere is denser than the asthenosphere.
- The thickness and mass of the mantle are huge (analogous to moving a big steel ball through butter).

Convection is slow because viscosity is high. In the mantle, the average viscosity is 10^{20} times greater than cold molasses; as a result, the calculated velocity of mantle convection is very slow—about 10 centimeters per year. In the asthenosphere, estimated viscosity is 1 million times lower than the lithosphere viscosity, which explains why convection occurs deeper within the mantle but not within the lithosphere. Even in the lower mantle, the highly compacted minerals here have a calculated viscosity 100 times lower than in the lithosphere, because the lower mantle is much hotter than the lithosphere.

What about convection in Earth's core? Geophysicists apply the same concepts to evaluate this question. The liquid outer core has an extremely low viscosity, about equal to that of water at room temperature and pressure. The liquid iron of the core, however, is an extremely efficient heat conductor. This conduction efficiency means that heat readily transfers by conduction before convection driven by heating and cooling can start. The combination of high heat conductivity of solid iron and high viscosity of solid iron (similar to the soup pot) makes convection even less probable in the inner core.

These observations do not mean that the liquid outer core is motionless. Keep in mind that imbalance between gravity and viscous force drives the movement of mass and that the gravity force depends on density contrast. Viscosity is very low in the outer core, but could compositional differences, rather than temperature differences, provide the necessary density contrasts for con-

EXTENSION MODULE 10.1
Is Mantle Convection Physically Possible?
Learn the simple mathematical calculations that determine the likelihood of convection in Earth's mantle.

vection? Seismic data suggest that the inner core is nearly pure iron, whereas the outer core has more nickel and light elements in addition to iron. Chapters 8 and 9 noted that the inner core forms from crystallization of an initially all-liquid core, so the inner core is getting larger as iron crystallizes from the outer core. The crystallization of iron leaves the nearby remaining liquid with less iron than the average outer core. **Figure 10.4** illustrates how this compositional change causes convection. The iron-depleted liquid adjacent to the inner core is less dense than the average outer core. This means that the newly formed parcels of iron-depleted liquid move up while denser, more iron-rich liquid moves down. The upward-moving fluid also transports heat from the hotter inner-outer core boundary region.

Changing composition, instead of changing temperature, causes density variations in the outer core. Currents formed in this fashion describe *chemical convection* (density differences caused by composition) rather than the *thermal convection* (density difference caused by temperature) that you considered for the lava lamp, soup pot, and the mantle. As a consequence of the low viscosity of the outer core, convection velocities are on the order of 10 kilometers per year—in other words, about 10,000 times faster than in the mantle.

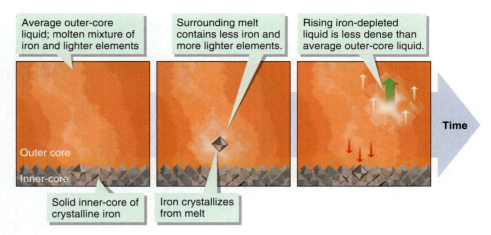

▲ **Figure 10.4 How chemical convection works in the core.** Compositional differences, rather than temperature differences, produce density variations that cause fluid motion in the outer core. When iron crystallizes from the liquid of the outer core, the surrounding liquid contains less iron than it did before crystallization. This iron-depleted liquid is less dense and rises while denser, iron-rich fluid sinks. The rising liquid also carries heat upward.

Putting It Together–*How Does Convection Work?*

■ Convection is the simultaneous movement of matter and heat, whereas conduction transfers only heat from hot to cold regions.

■ Convection occurs when gravity forces that cause motion are greater than viscosity forces that resist motion; otherwise, heat transfers by conduction.

■ The gravity force driving thermal convection originates from density contrasts between hot and cold materials. Higher density contrast, which favors convection, is present in materials that have low heat conductivity and large expansion when heated.

■ Calculations show that convection should occur in the whole mantle and in the outer core, but not within the lithosphere or inner core.

■ Convection in the outer core results from density variations caused by compositional differences, rather than by temperature differences that cause mantle convection.

10.2 What Does Mantle Convection Look Like?

Admittedly, the calculations summarized in the previous section include some speculation and inference about the properties of inaccessible materials deep within Earth, so you can rightly ask if there is any evidence for convection in the mantle to support the results. You may also ask how convection can even occur when the picture of Earth's interior painted by earthquake data in Chapter 8 clearly suggests that density increases with depth (Figure 8.19). How is it possible to get material to move up if lower-density rock overlies higher-density rock? Even if convection occurs in the mantle, does it really resemble what happens in the lava lamp and soup pot?

What Heat-Flow Data Show

You know that temperature increases downward into Earth. Temperature measurements in deeply drilled wells permit calculation of the total amount of heat escaping the surface of Earth—a quantity called **heat flow**. The source of that heat is radioactive decay.

▶ **Figure 10.5 Heat flow from Earth's surface.**
Temperatures measured in deep drill holes in both continental and oceanic crust permit calculation of the amount of heat energy emerging from within Earth. Geologists measure the quantity of heat in thousandths of a watt of energy per square meter of Earth's surface. The values range greatly from place to place but, on average, more heat moves through oceanic crust than through continental crust.

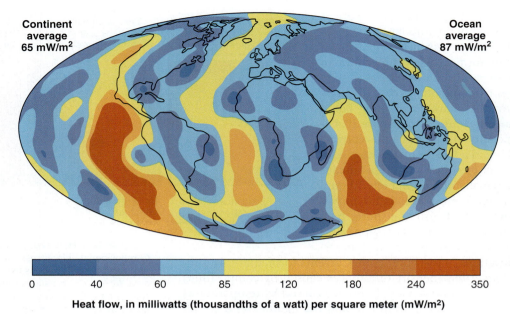

Continent average 65 mW/m²

Ocean average 87 mW/m²

100 watt bulb

2 million mW/m²

Earth heat flow seems tiny compared to the output of a 100-watt light bulb, but when multiplied by the surface area of Earth, this heat energy is more than 3 times greater than annual human energy consumption on the entire planet.

| 0 | 40 | 60 | 85 | 120 | 180 | 240 | 350 |

Heat flow, in milliwatts (thousandths of a watt) per square meter (mW/m²)

Shortly after the discovery of radioactivity in the early 1900s, geologists hypothesized that heat flow through continents should be greater than through ocean floors. This conclusion arose from these two observations:

1. Granite, which is common in continental crust, contains a greater abundance of radioactive-heat-producing elements than basalt, which composes oceanic crust (see Figure 8.22).
2. Continental crust is much thicker than oceanic crust (see Figure 8.19)

Nearly all heat-flow measurements at that time were from continents, so there were insufficient data to test the hypothesis.

By the mid-1950s, enough heat-flow data existed from beneath ocean floors to disprove the hypothesis, as demonstrated in **Figure 10.5**. Heat-flow values vary considerably across both oceans and continents but are, on average, higher in oceans, not lower as hypothesized. This implies that most of the heat must be coming from below the crust because radioactive-heat production in thin, basaltic oceanic crust should be small. The only way to account for so much heat flow from the mantle would be to multiply the huge volume of the mantle by its miniscule radioactive heat production (see Figure 8.22). While this adds up to much more heat than is generated in the thin crust, the vast majority of the heat is way too far below the crust to conduct effectively to the surface because of the extraordinarily low heat conductivity of silicate rock. The best way to account for the high heat flow from the mantle is to conclude that it convects to the base of the lithosphere from deep in the mantle. From the base of the lithosphere, which is generally thinner beneath oceans than continents, the heat conducts to the surface, where it is measured as heat flow.

How Dense Mantle Rock Can Move by Convection

Figure 10.6 shows how it is possible for the mantle to convect. Section 8.5 presents estimates of the temperature variation within Earth. Most of the temperature increase with depth results from the work done when rock compresses under the increasing weight of overlying material (see Figure 8.21). The key to mantle convection is that the temperature profiles at the top and bottom of the mantle are different from the profile explained only by changes in pressure throughout most of the mantle thickness. Conduction from the outer core causes the high temperature gradient at the base of the mantle, whereas conductive and radiative cooling at Earth's surface causes the abrupt temperature decrease at the top of the mantle. If upward motion starts in the lower mantle, perhaps because of rock expansion during heating from the core, the moving parcel of mantle expands further as it rises simply because it decompresses at lower pressure (Figure 10.6). Expansion causes the temperature of the parcel of rock to decrease as it decompresses, and decreasing temperature causes the rock to contract.

So, is this parcel of rising mantle now less dense, denser, or of equal density to its surroundings? Density, *on average*, increases downward and decreases upward through the mantle. Calculations show that rising mantle rock expands more because of decompression than it contracts because of cooling. Sinking mantle rock contracts more because of compression than it expands because of heating. In other words, changing pressure influences additional density change more than changing temperature. If the parcel of rock at a particular horizon is sufficiently less dense than its surroundings so that it overcomes viscous resistance, then it rises, and if denser it sinks.

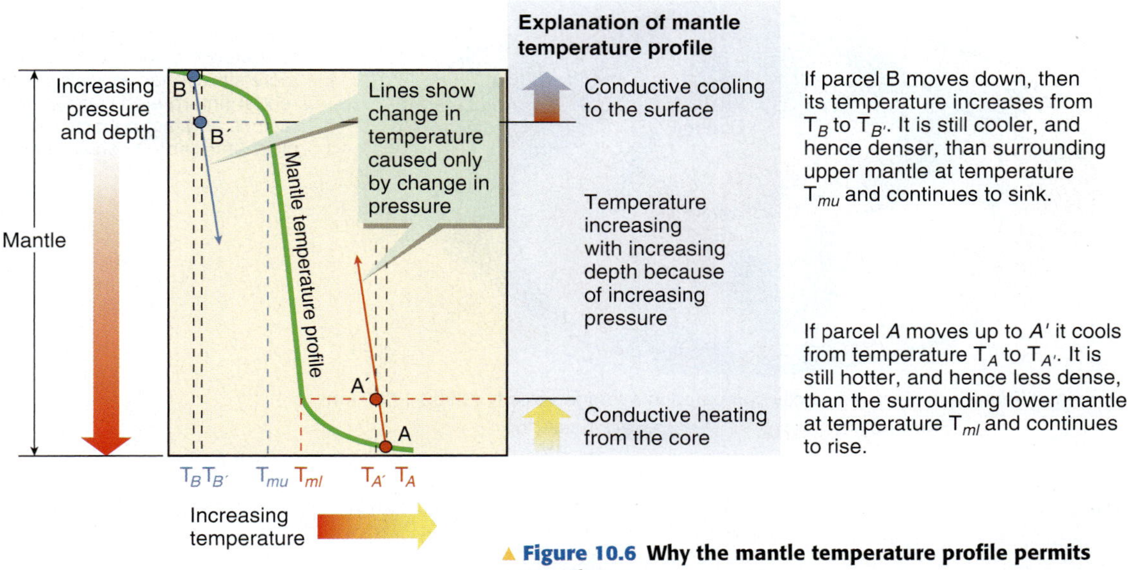

How Solid Convection Generates Magma

Without convection, there would be no volcanoes on Earth. Convective rise of mantle rock accounts for one of the most important mechanisms for producing magma explored in Chapter 4. Decompression melting takes place where hot rock rises from depth while cooling only slightly (Figures 4.18 and 10.6). Eventually, the hot rock rises to where the pressure is sufficiently low for the rock to melt while surrounding, cooler rock remains completely solid. Petrologists were proponents of convection in the mantle before other evidence for this process was known, basically because they knew how difficult it is to come up with a mechanism to partly melt peridotite to make basaltic magma without convection.

What Seismic Data Show

The preceding arguments suggest that if convection occurs in the mantle, then at any level in the mantle there should be some variation in density. There should be regions of lower density representing upwardly mobile mantle and denser regions where mantle is sinking. If the density is different in adjacent regions, then you should expect seismic velocity also to differ because velocity partly depends on density. Calculations show that if a parcel of mantle is both less dense and hotter than its surroundings, then the P and S waves should move through that parcel more slowly.

Chapter 8 did not provide insights into any such variations in seismic-wave velocity. The curves in Figure 8.19 depicting seismic-wave velocities at different depths in Earth portray the interior as radially symmetric. A radially symmetric model means that the only thing determining seismic velocity is depth below the surface (or distance from the center). For example, the curves in Figure 8.19 assume that the wave velocities at 500 kilometers below Los Angeles, California, are the same as at 500 kilometers below Beijing, China.

Actually, seismologists know from earthquake data that there are local heterogeneities in Earth that cause wave velocities to vary as much as 2 percent at any particular level. Two percent seems like a very small difference, and ignoring it in Chapter 8 made comparing internal properties of Earth by looking at

Explanation of mantle temperature profile

Conductive cooling to the surface

Temperature increasing with increasing depth because of increasing pressure

Conductive heating from the core

If parcel B moves down, then its temperature increases from T_B to $T_{B'}$. It is still cooler, and hence denser, than surrounding upper mantle at temperature T_{mu} and continues to sink.

If parcel *A* moves up to A' it cools from temperature T_A to $T_{A'}$. It is still hotter, and hence less dense, than the surrounding lower mantle at temperature T_{ml} and continues to rise.

▲ **Figure 10.6 Why the mantle temperature profile permits convection.**
Through most of the mantle, temperature increases downward because of the increase in pressure at greater depth (see Section 8.5). Vertically moving parcels of mantle, such as those schematically represented on the graph by circles A and B, also warm or cool according to this temperature dependence on pressure. The mantle temperature profile, however, curves into a backward-S shape because of conductive cooling at the top, and conductive heating from the core, at the bottom. Therefore, parcels of rising lower mantle are always warmer than adjacent rocks and parcels of sinking upper mantle are always cooler than adjacent rocks, which permits convection to happen.

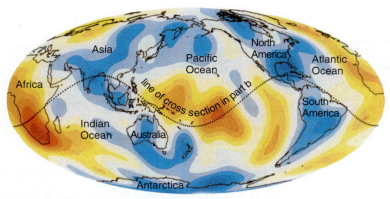

This map shows the global velocity variation in S-wave velocity between 900 km and 2880 km below the surface. Seismic waves move more slowly than expected through the redder (interpreted warmer, less dense) regions and faster than expected through the bluer (interpreted cooler, denser) regions.

Percent difference in S-wave velocity compared to average velocity at 900–2880 km

Slower by 1% — Hotter   Average   Faster by 1% — Colder

(a)

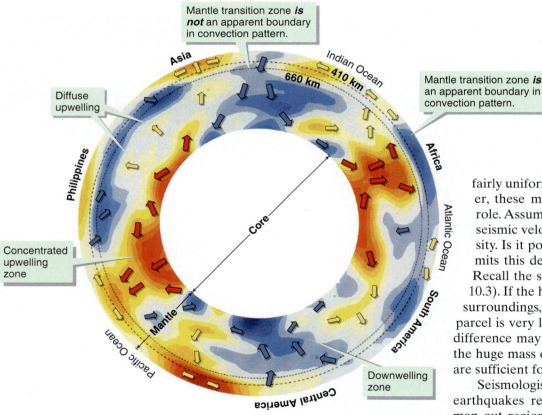

Mantle transition zone **is not** an apparent boundary in convection pattern.

This cross-section shows the variation in S-wave velocity within the mantle along the dotted line shown in the map (a).

- Red and yellow arrows show interpreted upwelling of slower, hotter mantle depicted in red shades.

- Blue arrows show interpreted downwelling of faster, cooler mantle depicted in blue shades.

Mantle transition zone **is** an apparent boundary in convection pattern.

Diffuse upwelling

Concentrated upwelling zone

Downwelling zone

Asia · Indian Ocean · 410 km · 660 km · Africa · Atlantic Ocean · South America · Central America · Pacific Ocean · Philippines · Core · Mantle

(b)

▲ **Figure 10.7** **Seismic-velocity variations map convection in the mantle.**
The seismic tomographic map (a) and cross section (b) of the mantle show variations in S-wave velocity that most geophysicists equate to variations in temperature and density. Arrows show the interpreted pattern of convective upwelling and downwelling implied by the temperature variations. Deep upwelling zones, like those beneath Africa and the Pacific Ocean, are round in map view and relatively narrow in cross section—their three-dimensional shape is like a column. The blue downwelling zones are irregularly shaped and elongate on the map and narrow in the cross section—their three-dimensional shape is like a rumpled sheet.

fairly uniform radial slices a simpler process. Now, however, these minor variations may play a more important role. Assume for a moment that a 0.5 percent increase in seismic velocity equates to a 0.5 percent increase in density. Is it possible that such a small density increase permits this denser parcel to sink in viscous, solid mantle? Recall the steel balls and frozen butter example (Figure 10.3). If the high-density parcel is very small relative to its surroundings, then it may not move. If, on the contrary, the parcel is very large relative to its surroundings, this density difference may be adequate for convection to occur. Given the huge mass of the mantle, these small density differences are sufficient for convection.

Seismologists use seismic records from thousands of earthquakes recorded at thousands of seismic stations to map out regions of the mantle where seismic waves travel slightly faster or slightly slower than in adjacent rock at the same depth. This method of determining the internal structure of Earth closely resembles the diagnostic medical tomography methods for imaging the human body—so the method applied to Earth is called **seismic tomography**.

Figure 10.7 shows a seismic-tomographic map and cross section of Earth based on variations in S-wave velocities. The illustrations reveal complicated patterns. There are areas of seismically slow mantle, presumed to be warmer, less dense, and rising compared to adjacent regions. There are also complementary regions of seismically fast mantle, interpreted to be colder, denser, and sinking. The cross section (Figure 10.7b) shows that some of the areas of interpreted upward or downward flow transect the entire thickness of the mantle. In other places, the pattern seems interrupted by the transition zone between 410 and

660 kilometers depth (see Section 8.4). The map shows that the cold areas of downward flow tend to be very elongate, so they look like rumpled sheets in three dimensions. Some warm areas have diffuse boundaries, whereas others are sharply defined columns rising through the mantle with circular shapes in map view. In order to maintain a reasonably smooth surface at the top of the mantle, nearly horizontal flows must also connect the upwelling and downwelling regions. This horizontal flow near Earth's surface relates to the motion of plates, as mentioned in the first chapter (Section 1.5). Therefore, a connection between convection in the mantle and deformation at Earth's surface is emerging, and Chapter 12 fully develops this connection.

What Mantle Convection Should Look Like

If the seismic-tomographic images in Figure 10.7 truly represent upward and downward movement in the mantle caused by convection, then the pattern is very complicated and not very similar to the lava lamp and soup pot. It is fair to ask, then, if the pattern of mantle convection interpreted from the seismic data is physically plausible. To answer this question requires consideration of a vast amount of physics that is beyond the scope of this book. The results of certain experiments and computer simulations are pertinent to the problem, however.

Figure 10.8 portrays a simple view of convection. In this case, heat enters at the bottom of a flat plate, like the light in the lava lamp and the stovetop burner under the soup pot. The heated material has uniform viscosity from top to bottom. Convection produces very symmetrical zones of hot upwelling and cold downwelling. Notice that the convective motion is not simply up and down. There is also horizontal flow at the top and bottom, which connects each upwelling zone with adjacent downwelling regions.

Does this convection resemble the motion implied by the seismic-tomographic images in Figure 10.7? One feature does seem similar—the upwelling zones are narrow columns whereas the downwelling regions are elongate sheets (Figure 10.8). The simulated convection pattern, however, is far more regular than the pattern implied for Earth. In addition, the distance between an upwelling region and the closest downwelling region in the simulations is almost exactly the same as the thickness from the top to the bottom of the convecting layer. The convection pattern implied by the seismic data does not show a consistent pattern. In most cases, however, the distance between seismically slow upwelling zones and the closest downwelling zone is much greater than the thickness of the convecting mass, especially where most of the interpreted motion occurs only in the upper mantle (Figure 10.7).

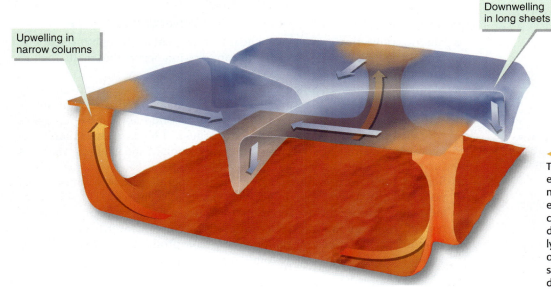

Upwelling in narrow columns

Downwelling in long sheets

◄ **Figure 10.8** **What simple convection looks like.** This artistic rendition of convection is based on computer simulations. The simulated convection describes the motion of material that is of uniform viscosity and heated only at the base. Hot material rises in narrow columns, spreads out and cools near the surface, and descends in long, thin sheets. This pattern agrees closely with the photograph of the lava lake at the beginning of the chapter, where lava rises along central columns, spreads out and cools, and then sinks in sheets of dense liquid.

Is this simple form of convection really applicable to the mantle? No, three complexities need to be added:

1. The mantle is heated from within and not simply from the core below. Radioactive decay produces heat *throughout* the mantle (Section 8.5). Current measurements and estimates suggest that only about 10 percent of the heat rising to the base of the crust originates in the core; the bulk of the heat is generated in the mantle. Computer simulations must, therefore, include heat distributed throughout the mantle rather than just heating at the base.
2. Mantle rocks almost certainly do not have the same viscosity from top to bottom. Geophysicists estimate, for example, that the lower mantle viscosity is about 1000 times greater than the asthenosphere viscosity. For most materials, viscosity decreases at higher temperature. The hotter lower mantle, however, is more viscous than the colder upper mantle. This is because pressure causes dense packing of atoms in the lower-mantle minerals. The top of the asthenosphere has the lowest viscosity because the peridotite is close to its melting temperature, or even partly molten, so that it flows more readily.
3. Earth is a sphere, not a plate, so heat moves out through an ever-increasing volume of rock as it moves upward from depth.

Plugging these complexities into the computer simulations produces a much more complicated convection pattern, as shown in **Figure 10.9**. The regular spacing of upwelling and downwelling zones depicted in Figure 10.8 is missing, and the distances between them tend to be greater than the thickness of the convecting layer. In some places, these upwelling zones pass through the entire simulated mantle. In other places, motion in the upper mantle seems disconnected from motion in the lower mantle, thereby emphasizing the importance of the viscosity change where mineral transformations occur in the transition zone. The downwelling zones seem sharply defined in narrow zones. In contrast, most upwelling zones have irregular shapes with diffuse outlines that reflect internal heating by radioactive elements in the mantle. Much of this diffuse upwelling occurs simply because sinking regions of dense, colder mantle displace adjacent regions of less dense, warmer mantle toward the surface. There are a few more intense, localized upwelling zones that seem to originate at the core-mantle boundary. The pattern in Figure 10.9 more closely resembles the pattern implied by the seismic-tomographic data illustrated in Figure 10.7, and adds confidence to the conclusion that the mantle does convect.

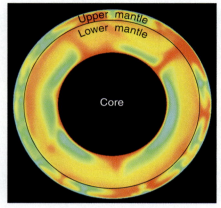

(a)

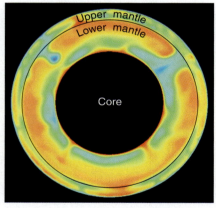

(b)

Red shades represent warmer, upwelling regions.

Blue shades represent cooler, downwelling regions.

Boundary between upper and lower mantle drawn at 660 km.

▲ **Figure 10.9 What complex convection looks like.**
These two diagrams are from a computer simulation of convection under more Earth-like conditions than shown in Figure 10.8. This model does not capture all of Earth's complex properties, but it does include three important real-Earth variables:
1. Internal sources of heat in the mantle
2. Viscosity that increases below the boundary between upper and lower mantle (660 km)
3. Spherical shape

The two illustrations are drawn from different times within the simulation and show that the convection pattern changes with time.

Putting It Together–**What Does Mantle Convection Look Like?**
- Mantle convection explains why heat flow is greater through the seafloor than through continents and why decompression occurs that melts rock to form magma.
- Mantle convection occurs because upward-moving rock expands more by decompression than it contracts by cooling, allowing it to remain hotter and less dense than its surroundings so that it continues to move up. Likewise, sinking mantle compresses more because of high pressure than it expands by warming, so that it remains cooler and denser than its surroundings, and it continues to move down.
- Seismic tomography shows regions of the mantle where seismic waves travel slightly slower or slightly faster than in adjacent rock. The slower regions are interpreted to represent warmer, upwelling mantle, and the faster regions are interpreted to represent cooler, downward-flowing rock.
- Computer simulations of mantle convection produce patterns similar to those inferred by the seismic-tomographic data.

10.3 How Does Outer-Core Convection Generate the Magnetic Field?

Motion in the outer core is so distant from the surface that it may, at first, seem too esoteric to worry about. At the beginning of this chapter you read that convection in this deeply remote part of Earth produces the magnetic field. The magnetic field allows humans to navigate with a compass, permits migrating animals to navigate using magnetic cell structures in their brains, and deflects hazardous cosmic radiation away from Earth. It is, therefore, very important to understand why Earth has a magnetic field and how that field relates to convection.

What the Magnetic Field Looks Like

Before you can understand how Earth creates magnetic force, you first need to identify the phenomena that require explanation. For starters, consider what a compass reveals about Earth's magnetic field. A typical compass, shown in **Figure 10.10**, consists of an arrow-shaped magnetized needle. A magnetized object is one that consists of material, like iron or the mineral magnetite, in which atoms align with the direction of magnetic force. Earth's magnetic force is oriented almost parallel with its rotation axis, so the compass needle points, more or less, to the north. In the most sophisticated compasses, the end of the needle opposite the point has a weight on it. This is because the magnetic force not only causes the compass needle to pivot toward a northerly point but also deflects the needle up or down (Figure 10.10). The weight keeps the compass needle freely spinning in a horizontal plane. Columbus and Magellan used compasses when they sailed to explore the Americas in the fifteenth and sixteenth centuries, so the knowledge that Earth has a magnetic field has been utilized for a long time, even though the origin of the field has remained a mystery until recently.

You can map out the orientation of the magnetic force by taking careful measurements with a freely suspended magnetized needle at many places on Earth and noting the direction that the needle points and the extent to which it deflects up or down. Lines are drawn to show the direction of the magnetic force at each location. These lines define the magnetic field, which is illustrated in **Figure 10.11**a. Notice that the field extends into space as well, where it deflects electrically charged particles and cosmic rays away from the planet.

Two measurements define the orientation of the field at any location on Earth's surface (Figure 10.11a). **Declination** is the horizontal angle made by lines connecting the measurement location to the magnetic and geographic north poles. **Inclination** is the vertical angle between the field-force line and Earth's surface.

Two key features emerge when we examine Earth's magnetic field:

1. The pattern of Earth's magnetic field very closely, but not exactly, resembles the field of a bar magnet (Figure 10.11b). The analogy works only if the hypothetical magnet producing Earth's field is deep in the interior. This observation, made in the nineteenth century, led to the conclusion that Earth's magnetism originates deep within Earth and not at the surface or somewhere in space.
2. The north magnetic pole, where the force lines go into Earth (Figure 10.11a), and south magnetic pole, where the force lines emerge, do not coincide exactly with the geographic north and south poles. Currently, the angular distance between the magnetic and geographic poles is about 18 degrees.

The Magnetic Field Changes with Time

Careful measurements show that the magnetic field is not a static feature of Earth. Curiously, the location of the magnetic poles constantly changes, as

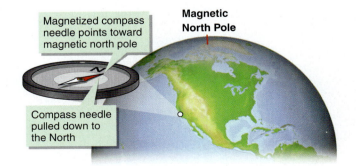

▲ **Figure 10.10 How a compass reveals the magnetic field.** The magnetized arrow in a compass is pulled toward the magnetic North Pole and, in the northern hemisphere, also pulled downward.

► **Figure 10.11 What Earth's magnetic field looks like.**

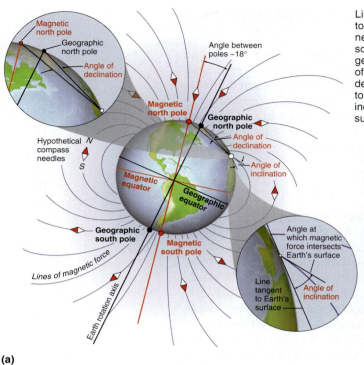

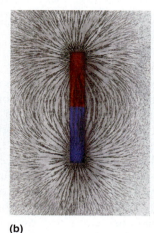

Lines of magnetic force pass through and encircle the planet to define Earth's magnetic field. Hypothetical compass needles show the orientation of the field. The north and south magnetic poles do not coincide exactly with the geographic poles along Earth's rotation axis. The orientation of the field at any point on the surface is described by the declination (the angle between the directions from the point to the geographic and magnetic north poles) and the inclination (the angle between the magnetic force and the surface at the point).

This photograph illustrates the magnetic field of a bar magnet. The red and blue magnet is below a transparent sheet of plastic partly covered in iron filings. The magnetic iron fragments align themselves with the field of the magnet. The circular pattern resembles Earth's magnetic field depicted in (a) and suggests that most of the magnetic field originates deep within Earth and has a geometry similar to that of a bar magnet.

(a)   (b)

illustrated in **Figure 10.12**. This means that the field shifts position, which also affects declination and inclination angles everywhere on Earth. Accurate navigation instruments must be adjusted for this shift. The strength of the field also changes by as much as 0.01 percent per year in some parts of the world.

Magnetic minerals in rocks preserve a record of the orientation of the magnetic field at the time the rock formed. Measurements of this orientation show great variations in inclination and declination, which is not surprising since such changes also occur in historical time (Figure 10.12). The measurements of ancient magnetic-field orientations also reveal **magnetic reversals**, when the magnetic poles swap positions. During periods of magnetic reversals your compass would point south rather than north! Times when the field is oriented as it is now are **normal polarity intervals,** and those times when the field is oriented in the opposite direction are called **reversed polarity intervals**. The field reversed several hundred times during the last 160 million years. Over the last 15 million years the average polarity interval has been about 200,000 years, although the intervals range considerably in duration. The last reversal was 780,000 years ago, so polarity is long overdue for a change when this duration is compared with the average polarity interval.

The record of the ancient magnetic field in rocks indicates that reversals occur fairly abruptly, over intervals of one thousand to six thousand years, compared to the intervals of fixed polarity. The intensity of the field decreases significantly during reversals. This may affect organisms on the surface that are subjected to dangerous levels of cosmic radiation due to the fact that the radiation is less effectively repelled when the magnetic field is weak. Regardless of how the magnetic field forms, the process is quite capable of causing rapid and substantial changes in the orientation and strength of the field.

Why the Magnetic Field Must Form in the Core

The pattern of the magnetic field implies that the field originates within Earth (Figure 10.11). So, where does it form within the planet?

Although the magnetic fields of Earth and a bar magnet appear very similar, there cannot be a big mass of magnetized material deep within Earth that is

Understanding Inclination and Declination: *See how inclination and declination change with location on Earth.*

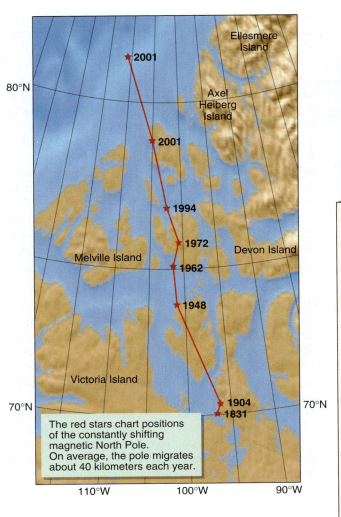

The red stars chart positions of the constantly shifting magnetic North Pole. On average, the pole migrates about 40 kilometers each year.

2004 declination angles across the United States

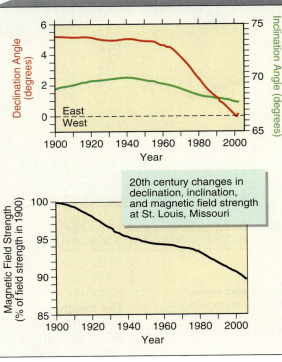

20th century changes in declination, inclination, and magnetic field strength at St. Louis, Missouri

◄ **Figure 10.12 The magnetic poles are on the move.** The magnetic North Pole shifts about 40 kilometers each year and, if it stays on its current trajectory, will cross the Arctic into Siberia by 2050. Currently, there is almost no magnetic declination in St. Louis, but there is a 13° westerly declination at New York City and a 15° easterly declination at San Francisco. The declination angle at St. Louis changed by about 5 degrees during the twentieth century because of the moving magnetic pole. This means that an uncorrected compass that points north today in St. Louis would have pointed 5 degrees east of north in 1900. The strength of the magnetic field at St. Louis also diminished by more than 10 percent since 1900.

a close analog to a bar magnet. Geologists know this because magnets function only at relatively low temperatures. The mineral magnetite, for example, loses its magnetism when heated above 580°C, although it will reacquire magnetization parallel to Earth's field when it cools back down below that temperature. Even pure iron loses its magnetization above 800°C. Given a reasonable near-surface geothermal gradient of 25–30°C/km, this means that no permanent magnetized minerals should exist below roughly 25 or 30 kilometers depth. Since the field originates deeper within Earth, this observation also means that the field does not emanate from permanently magnetized material but must constantly regenerate as it fades at high temperature.

A fundamental observation from basic physics is that magnetism and electricity are related forces. **Figure 10.13** shows that an electrical current generates a magnetic field. Likewise, a changing magnetic field generates an electric current. A **dynamo**, which transforms motion energy to electrical energy, exploits this principle to generate electricity. As shown in **Figure 10.14**, the rapid

▶ **Figure 10.13 Electrical current generates a magnetic field.**
This picture shows a vertical electrical wire passing through a horizontal board that is partly covered with iron filings. The electrical current produces a magnetic field that encircles the wire. The magnetic iron filings outline the resulting circular magnetic field.

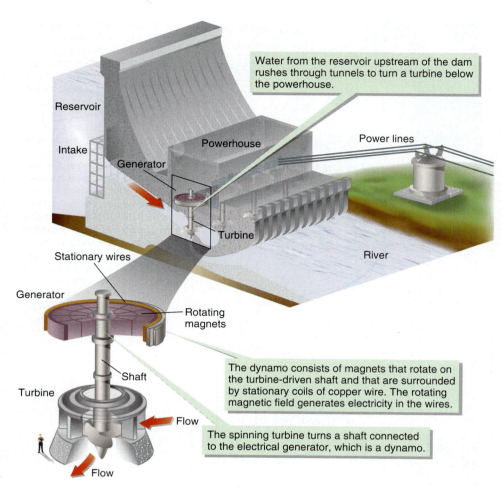

Water from the reservoir upstream of the dam rushes through tunnels to turn a turbine below the powerhouse.

Reservoir

Intake

Powerhouse

Power lines

Generator

Turbine

River

Stationary wires

Generator

Rotating magnets

Shaft

The dynamo consists of magnets that rotate on the turbine-driven shaft and that are surrounded by stationary coils of copper wire. The rotating magnetic field generates electricity in the wires.

Turbine

Flow

The spinning turbine turns a shaft connected to the electrical generator, which is a dynamo.

Flow

▲ **Figure 10.14** **How an electromagnetic dynamo works.**
Dams generate electricity when waterpower spins magnets surrounded by electrically conductive copper wire. The changing magnetic field produced by the spinning magnets generates electricity in the wires.

rotation of a magnet inside coils of electrically conductive wire produces an electrical current in the wire.

Scientists hypothesize that a natural dynamo somehow exists deep within Earth. The natural dynamo requires materials that are both electrically conductive and capable of being magnetized. Rocks dominated by silicate minerals are extraordinarily poor electrical conductors. Iron, on the other hand, is about 10^{16} times more conductive than granite or basalt but still about a million times less conductive than water, which can conduct sufficient current to cause electrocution. Iron can also be magnetized. The conditions for generating a magnetic field by a natural dynamo are, therefore, more consistent with the iron-rich composition of the core than with the silicate composition of the mantle.

How the Magnetic Field Forms

The next piece of the puzzle is to determine just how this hypothesized dynamo works in the core. Geologists cannot directly examine the core, so the process for generating the magnetic field is only a hypothesis, but it is one that remains consistent with observed data and computer simulations.

A self-sustaining natural dynamo, sometimes called the geodynamo, is illustrated in **Figure 10.15**. It requires an electrical conductor, such as iron, and motion caused by convection and Earth's rotation. Movement of the conductor within a magnetic field generates an electrical current. The current, in turn, produces a magnetic field. The magnetic field generates an electrical current in the convecting molten iron, which continually contributes to the magnetic field and so on, perpetually. The continual generation of the magnetic field avoids the impossible requirement for permanently magnetized objects at high temperature deep within Earth. In essence, the motion of magnetic material produces an electrical current and the current generates a magnetic field.

The geodynamo hypothesis, as illustrated in Figure 10.15, also offers an explanation for why the magnetic field closely parallels Earth's rotation axis. Rotation may cause spiraling convection currents closely parallel to the rotation axis. Earth's rotation more likely affects convection in the outer core than in the mantle because outer-core viscosity is very low, close to that of water. In addition, convection in the extremely low viscosity outer core is much faster than mantle convection, so outer-core motion more easily explains the measured rapid changes in the magnetic field than does sluggish movement of the mantle.

The one remaining stumbling block is that there needs to be an initial magnetic field from somewhere else to get the geodynamo started. The initial field does not have to be very strong, however, and might have originated within the Sun at the time Earth formed. That hypothesis begs the question because we don't know how the magnetic field first formed in the Sun, so you can understand why this concept is still the subject of considerable research.

Why the Magnetic Field Changes

Scientists have a reasonable hypothesis for explaining the presence of Earth's magnetic field. How can geologists explain why the field changes orientations through time, varies from the ideal bar-magnet analogy, or reverses polarity?

A schematic diagram for the geodynamo

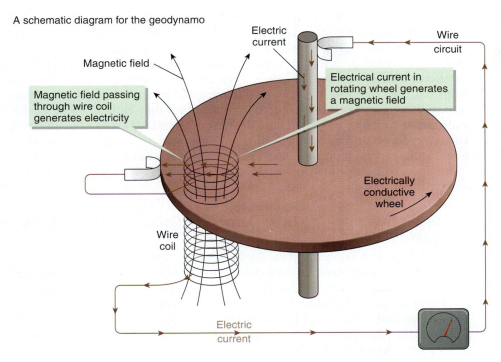

◀ **Figure 10.15 Visualizing the geodynamo.**
This diagram schematically illustrates how the self-sustaining geodynamo might work in the outer core. The magnetic field results from a changing electrical current which, in turn, is produced by the magnetic field. The convectively circulating, electrically conductive molten iron in the outer core plays the roles of the wire coil and the spinning wheel in the schematic diagram. Magnetization fades away at high core temperatures but is continually regenerated by convection. Earth's rotation likely influences the convection pattern and, therefore, the orientation of the continually regenerated magnetic field. All that is needed is a small magnetic field to get the geodynamo started.

Simple view of the geodynamo

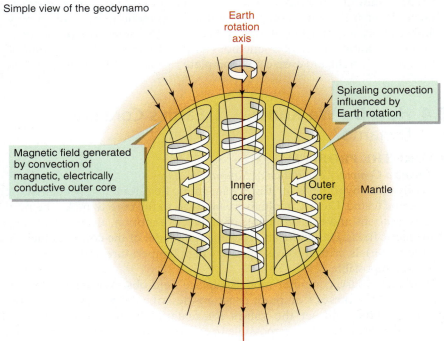

Geophysicists still wrestle with answers to these questions. What is clear is that there are many natural complications in Earth that likely cause these phenomena.

For starters, convection in the outer core is probably much more complicated than depicted in Figure 10.15. Once the magnetic field forms in the outer core, it becomes a force that interacts with gravity and viscosity to affect the pattern of convection. Although the inner core is too hot to retain a magnetic field for very long, the iron is very electrically conductive, so it should conduct the electrical currents generated by the geodynamo in the outer core. This electrical energy in the inner core affects the orientation of the resulting magnetic field that forms in the outer core.

Scientists have yet to understand how convection currents in the outer core likely interact with the mantle. Recall from Section 8.4 that there are scattered low-seismic-velocity patches in the lowermost mantle, which suggest either partial melting, mixing of core and mantle ingredients, or both at those locations. If these regions are hotter, then they will affect how heat conducts from the core into the mantle, thus affecting the outer-core convection pattern. If these mantle regions contain iron, then they may be both more electrically and thermally conductive than the normal mantle, thereby affecting both core convection and magnetic field generation. Any or all of these hypothesized complexities could reasonably account for variations in the strength and orientation of Earth's magnetic field over time as well as explain the polarity reversals.

Putting It Together—How Does Outer-Core Convection Generate the Magnetic Field?

■ The shape of Earth's magnetic field resembles that produced by a laboratory bar magnet. Geologists know from the lines of magnetic force that the field originates deep within Earth, where temperatures are too high for the existence of permanently magnetized material.

■ The magnetic field must be continually generated by the geodynamo—a term that describes rapid convective motion of electrically conductive, iron-rich liquid in the outer core.

■ Complexities in rapid outer-core convection caused by electrical conductance through the inner core, interaction of magnetic forces with gravity and viscosity, and transfer of mass or heat from the outer core to the mantle may account for rapid fluctuations in the strength and orientation of the magnetic field and even cause complete reversals of the field.

10.4 How Do We Know . . . Earth's Core Is a Dynamo?

PICTURE THE PROBLEM
Why Create a Computer Simulation of Natural Processes?
The geodynamo hypothesis for Earth's magnetic field remains speculative unless it can be tested. Clearly, geologists cannot go to the core to make the pertinent observations and measurements of convective motion, electrical currents, and magnetic fields. Designing an experimental model of the core that includes all of the materials, critical forces, and very high temperature and pressure also borders on impossible.

When necessary, computer models replace field observations and experiments to test hypotheses. Computer programs describe the processes by mathematical equations and set the values for material properties and rates of processes that are difficult or impossible to control in the laboratory. Even computer simulations have limitations not only because of uncertainties in values for deep-Earth materials and processes, but also because it takes immense computing power to simulate what happens within the huge volume of Earth during long periods of geologic time.

BUILDING THE SIMULATION
What Goes Into a Computer Model of the Core?
A well-known geodynamo simulation was constructed by Gary Glatzmaier, formerly of Los Alamos National Laboratory and now at the University of California at Santa Cruz, and Paul Roberts, from the University of California at Los Angeles. The model includes equations to represent the fundamental physics of magnetism and electricity, Earth's rotation, and convective motion. Values are assigned to various Earth properties.

Some aspects of the real world are not incorporated in the simulation. The majority of the assigned values in the simulation, such as the size, mass, rotation rate, and most of the material properties in the core and mantle, are very Earth-like. To have the computer program reproduce the compositionally driven convection thought to occur in the outer core (Figure 10.4) would add considerably to the computational time. Rather than sacrifice other elements of the simulation, Glatzmaier and Roberts substituted thermal convection in place of chemical convection. They accomplished this substitution by increasing the heat input across the boundary from inner core to outer core and increasing the viscosity of the liquid outer core. Such computational short cuts are commonly necessary when building simulations of very complex natural processes. Even with this simplification, the initial simulation of 45,000 years of core processes required more than a year to run on a huge supercomputer. Subsequent computer runs simulated as much as 300,000 years.

VISUALIZE THE RESULTS

Do Simulations Support the Geodynamo Hypothesis?

Figure 10.16 shows that the simulation reasonably reproduces Earth's magnetic field and supports the geodynamo hypothesis. The simulated magnetic-field lines resemble a tangled ball of yarn in the core but are organized into the expected shape of a bar magnet's magnetic field outside of the core (compare Figure 10.16a with Figure 10.11b). The simulated variation in the strength of the field at

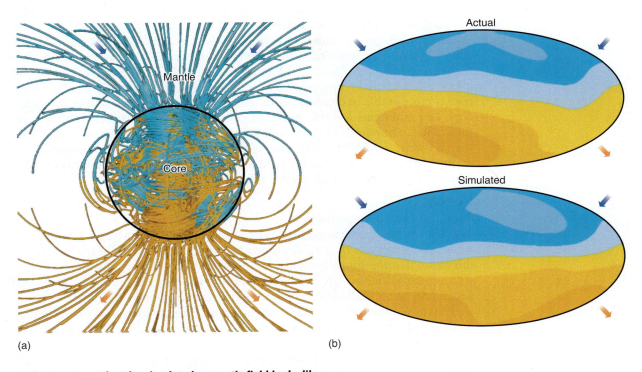

(a)   (b)

▲ **Figure 10.16 What the simulated magnetic field looks like.**
In these two diagrams summarizing results of the Glatzmaier-Roberts simulations, blue shows the inward-directed part of the magnetic field and yellow shows the outward-directed part of the field.

a. Simulated magnetic field lines resemble a tangled ball of yarn in the core, but farther out in the mantle the field orientation resembles that of a bar magnet, as is actually measured (compare to Figure 10.11b). Earth's surface is approximately located at the outer edge of the illustration.

b. This diagram compares a computer simulation of the field on Earth's surface with the measured field. Denser shading represents higher field intensity than lighter shading. The simulation reproduces the general patterns of the actual field intensity and direction. Although the simulated and actual patterns are not exactly the same, both change rapidly through time, so only the general appearance is expected to be the same.

the surface also resembles the actual field (see Figure 10.16b). Glatzmaier and Roberts numerically introduced an initial magnetic field into the simulation to represent the hypothesized field that seeded Earth's magnetism and then allowed this initial field to fade away. As predicted by the geodynamo hypothesis, the interaction of electrical currents and magnetic fields in the convecting, electrically conductive outer core produced a field to replace the initial fading one. The computer-simulated core generated a self-sustaining magnetic field that showed no indication of decaying away during the simulations.

UNEXPECTED RESULTS

Why Does Earth's Magnetic Field Reverse?

The first Glatzmaier-Roberts simulation produced an unexpected result even more noteworthy than successfully testing the physics behind the geodynamo hypothesis—the simulated magnetic field reversed on its own. **Figure 10.17** shows snapshot views of the simulated magnetic field reversing. In the simulation, about 1000 years elapsed in the transition from normal to reverse polarity. Combining the results from several simulations, reversals occur about every 200,000 simulated years, which is about the average period between real reversals. The computer model calculates that the strength of the magnetic field weakens during reversals (see Figure 10.17), which is consistent with measurements of magnetic intensity in rocks that record reversals during Earth history. None of these results were obvious outcomes of the way the computer simulation was designed, so magnetic-field reversal relates to fundamental physical processes that were not fully appreciated before the simulations were made.

Analysis of the computer-model output suggests that interactions between magnetic fields in the inner and outer cores cause polarity reversals. Rapid, chaotic convection currents cause a very complicated magnetic field in the outer core (Figure 10.16a) that attempts to change its overall orientation about every 100 years. The temporary magnetization induced in the hot, solid, inner core is more stable in direction, counteracts the attempted reversal in the outer core, and maintains a stable orientation of the overall Earth field for tens of thousands to hundreds of thousands of years. Only once in many attempts does the convectively induced magnetic field in the outer core remain flipped long enough to induce a reversal in the inner-core part of the field, and this process allows full polarity reversal to take place.

INSIGHTS

Are Computer Models Sufficient to Test Hypotheses?

The Glatzmaier-Roberts simulations reproduce observed natural phenomena while admittedly not realistically representing all of the natural variables. Does this mean that the simulations are insufficient tests of the geodynamo hypothesis? Similar criticisms arise whenever scientists apply computer models, with necessarily simplified input data and equations, to offer explanations of complex natural phenomena. With no hope of directly observing slow-acting processes deep within Earth, computer models are the best approaches to hypothesis testing.

Although the Glatzmaier-Roberts computer simulations do not include all of the complex processes likely to occur in the core, they add confidence to the adoption of the geodynamo hypothesis to explain Earth's magnetic field. Not only do the computer models reproduce a self-sustaining, Earthlike magnetic field, but the physical processes captured by the mathematical formulas pro-

▶ **Figure 10.17 A computer-simulated magnetic-field reversal.** These illustrations show the magnetic field at Earth's surface at three time steps in the computer model while the magnetic field reversed. The shading represents the intensity of the field and blue areas show where the field is directed into the planet and yellow areas show where the field is directed outward. The paler colors during the transitional period show that the magnetic field strength decreases during a reversal.

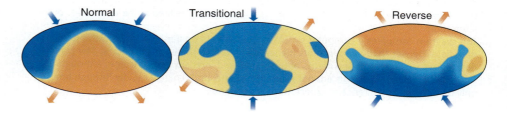

duce magnetic reversals and variations in the strength and orientation of the field just as geologists actually observe.

If, on the other hand, the simulations revealed something altogether different, then geologists would not know if the hypothesis was wrong or whether the model was too unrealistic to simulate nature. One could argue, therefore, that simplified models cannot falsify a hypothesis because the failure might be attributed to the model rather than the hypothesis.

Glatzmaier recognizes and states this problem clearly: "We still have a long way to go before we can be confident that these results are robust—they may not be." Glatzmaier and his colleagues know what is necessary to make more realistic models but "the computational costs will be staggering by today's standards.... Significant breakthroughs that will fundamentally improve our understanding of the geodynamo will come from young people who are inspired to work on what others proclaim is impossible." Those are the breakthroughs that continually advance science. For now, Glatzmaier and Roberts provided the breakthrough that strongly suggests that the geodynamo hypothesis is more correct than it is incorrect.

Putting It Together—How Do We Know … Earth's Core Is a Dynamo?

- Computer-simulation results, which include most but not all of the interpreted complexity of core processes, are consistent with the geodynamo hypothesis.
- The Glatzmaier-Roberts simulations not only reproduce the key features of the observed magnetic field at the surface but also generate simulated polarity reversals.
- The computer output suggests that magnetic-field-polarity reversals are caused by complex interactions between magnetism in the inner and outer cores.

Where Are You and Where Are You Going?

You have now completed a study of Earth's interior. You know that physical properties of Earth measured at the surface, such as seismic-wave velocities and the magnetic field, are consistent with dynamic motion inside the planet.

Convective motion transports matter and heat within Earth. Heat transfers by convection, rather than conduction, when the gravity force acting to move materials of different density is sufficiently greater than the viscosity force that resists this movement. Calculations show that conduction, rather than convection, is likely in the lithosphere and inner core, but that convection should occur in the mantle as a whole as well as in the outer core.

Mantle convection occurs because the silicate minerals composing mantle rock are very poor conductors of heat but expand when heated. Mathematical calculations indicate that when large masses of rock are heated at depth and expand, they rise upward as cooler, denser mantle sinks. Seismic-

tomographic maps, depicting small variations in seismic velocity at different levels within Earth, reveal subtle density differences that outline cylindrical upwelling zones and sheetlike downwelling bands in the mantle. This pattern is consistent with computer simulations of convection where heat is provided within and at the base of the mantle and where mantle viscosity increases at greater depth.

Outer-core convection depends on density differences more likely caused by different liquid compositions than by temperature changes. Crystallization of iron at the margin of the inner core removes iron from the adjacent outer-core liquid. The relatively iron-poor liquid is less dense than average outer-core melt and rises as the denser liquid sinks.

A self-sustaining geodynamo in the outer core generates Earth's magnetic field. Rapid convective motion of low-viscosity and electrically conductive outer-core liquid causes electrical currents that, in turn, produce mag-

netic fields. Computer simulations indicate that changes in the intensity and orientation of the field measured at the surface likely result from two factors: (1) the complex convection in the outer core and (2) interactions between the magnetic field in the outer core and that temporarily induced in the inner core. These interactions also cause the magnetic field to reverse polarity at very irregular intervals that average about 200,000 years.

In the next group of chapters, you will embark on further studies of the consequences of dynamic motion within Earth. Convection provides an important mechanism for generating magma by decompression of hot, rising mantle rock and provides a sought-after connection between volcanoes and internal processes. Convective motion describes the motion of lithospheric plates, which in turn explains the formation of mountains and the occurrence of devastating earthquakes. The evidence of this deformation at the surface is laid out in the next part of the book.

Active Art

Convection in a Lava Lamp. See how convection works inside a lava lamp.

Understanding Inclination and Declination. See how inclination and declination change with location on Earth.

Extension Modules

Extension Module 10.1: Is Mantle Convection Physically Possible? Learn the simple mathematical calculations that determine the likelihood of convection in Earth's mantle.

Confirm Your Knowledge

1. What is convection? What causes convection? How does convection differ from conduction?

2. Explain what chemical convection is and where it may be occurring within Earth.

3. Explain the factors that determine if convection occurs.

4. Why is conduction the dominant form of heat transfer within the lithosphere?

5. Which is more important during mantle convection, density changes due to pressure or density changes due to temperature?

6. How does evidence from seismic tomography support the hypothesis of mantle convection?

7. What is the difference between magnetic declination and magnetic inclination?

8. List the different ways Earth's magnetic field changes over time.

9. Why is it thought that Earth is overdue for a magnetic reversal?

10. What causes Earth to have a magnetic field?

11. What are polarity intervals?

Confirm Your Understanding

1. Write out an answer for each question in the Chapter Outline for the chapter sections assigned by your instructor.

2. Describe the effects of thermal expansion, conduction of heat, and viscosity to determine if heat transfers through material by conduction or convection.

3. Explain why there would be no volcanoes on Earth without convection.

4. Using what we know about thermal expansion, conduction of heat, and viscosity, should we expect thermal convection within Earth's core?

5. How do we know that most of the heat leaving Earth's interior is coming from below the crust?

6. How do heat flow measurements help support the hypothesis of mantle convection?

7. How do we know that Earth's magnetic field originates deep within Earth?

8. How do we know that Earth's magnetic field does not originate from permanently magnetized material?

9. Explain what will eventually happen to the magnetic field as Earth continues cooling down in the future.

Key Terms

conduction (p. 246)
convection (p. 246)
declination (p. 257)
dynamo (p. 259)

heat flow (p. 251)
inclination (p. 257)
magnetic reversals (p. 258)
normal polarity intervals (p. 258)

reversed polarity intervals (p. 258)
seismic tomography (p. 254)

11 Deformation of Rocks

Chapter Outline

Why Study Rock Deformation?

SOME ROCKS SEEM SHATTERED LIKE GLASS AND OTHERS CONtorted to resemble swirling taffy. Geologists describe broken and folded rocks as *deformed*.

Earthquakes result from rapid and extremely hazardous rock deformation. During a strong earthquake, rocks shift by many meters, accompanied by violent shaking that destroys buildings and possibly kills thousands of people, all within a matter of seconds. With the goal of anticipating and predicting earthquakes, geologists strive to understand how and why rocks deform.

On the other hand, crustal deformation plays a role in forming valuable economic resources, such as accumulations of natural gas, petroleum, and some important metal ores. Understanding how rocks deform helps us to understand how resources form and how to extract them. Mineral deposits rich in valuable metals (e.g., gold, copper) commonly form in areas of bent and contorted rocks, where metal-rich fluids move through the fractures resulting from deformation. Folded and broken rocks also provide conditions for concentrating and trapping oil and natural gas.

Strike-slip fault, Nevada

Low-porosity rock

Oil well

Gas

Oil Water

High-porosity rock

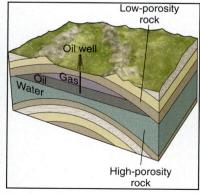

Building collapse. Yulin, Taiwan, 1999. Moment magnitude 7.8.

Uplifted and folded marine sedimentary rocks form the scenic coastline of Dorset, England. ▶

Geologists collect outcrop information about deformed rocks over large regions to locate economic resources, to interpret the geologic history of an area, and to assess risks from earthquakes. Geologists study deformed rocks at the surface, but the concealed convection of Earth's interior is what causes the deformation. Chapter 12 pulls together the exact connection between convection and deformation, but first you must understand how to recognize rock deformation and how it is possible to shatter, fold, and swirl rocks.

Your goals in this chapter are as follows:

✔ To observe and describe the types of deformation visible at Earth's surface

✔ To understand the relationship between deformation and the forces that drive it

✔ To learn the causes and effects of earthquakes

✔ To understand the relationship between Earth resources and rock deformation

To accomplish these goals, you must address the following questions:

11.1 What do deformed rocks look like?

11.2 How are resources related to geologic structures?

11.3 Why do rocks deform?

11.4 *How do we know* … why some rocks break and others flow?

11.5 How do geological structures relate to stress, strain, and strength?

11.6 How does strength vary in the lithosphere?

11.7 How do earthquakes relate to rock deformation?

11.8 How are earthquakes measured?

11.9 Why are earthquakes destructive?

In the FIELD

From a high vantage point, looking out across the rough, rocky landscape illustrated in **Figure 11.1**, you notice that the multicolored sedimentary rocks are tilted and folded, and not lying flat. In the broad valley in front of you is an oil well. How did the oil company know that there was oil below ground at that exact location? You decide to take a short hike around the area. Here and there you stop to take photos of what you see; some of these photographs appear in Figure 11.1 along with your field notes. Climbing across the outcrops, you identify layers of sedimentary rocks: pink mudstone, white limestone, interlayered red sandstone and mudstone, tan sandstone, and gray shale.

One key observation you make is that the rock layers are inclined at angles (Figure 11.1b); you also notice displaced layers, as if the rock broke and moved along the break (Figure 11.1d). You clamber across a valley eroded in soft shale and encounter a ridge of the same tan sandstone that you just left on the opposite side of the valley. Except, here the sandstone bed is inclined in the opposite direction. As you hike back to your starting point and look back at where you walked, you notice that the rocks bend to form a pattern like the letter "S" on its side. How can you describe these rocks to others? A sketch of the area, shown in Figure 11.1e, seems like a good decision.

As you complete your sketch, you wonder how the rocks deformed. What forces are necessary to break or bend rocks? How long does it take to deform rock? Images on television of rocks broken during earthquakes attest to the fact that it can happen quickly. Do all rocks respond similarly to deforming forces, or do some rocks fold or break more easily? Does the location of the oil well (Figure 11.1b) have something to do with the deformed rock?

11.1 What Do Deformed Rocks Look Like?

Before you can understand the forces that deform rocks, you first need to understand the characteristics of deformed rocks that the forces must explain. Field observations, like those documented in Figure 11.1, record the shapes of folded rocks, the orientations of tilted rock layers, and the directions that rocks have moved relative to one another across breaks. Geologists combine these observations into systematic types of measurements and terms that describe the appearances of deformed rocks.

Describing the Orientation of Rocks

What is the best way to describe the orientation of the tilted and broken rocks in the field area? Recall the principles of original horizontality and superposition (Section 7.2), which state that sediment accumulates in nearly flat layers, with older layers located below younger layers. The sedimentary rocks in your field area are not horizontal, so they were clearly deformed after deposition.

Field observations of deformed rocks

A. Inclined sedimentary rock layers seen while hiking

B. Orientation of inclined rock layers

C. Pumping oil well located in valley just to left of the top picture (A)

These are the same tan sandstone layers, but they do not line up.

D. All of the tilted rock layers are broken and displaced along this line.

E. Sketch map of tilted and broken rock layers

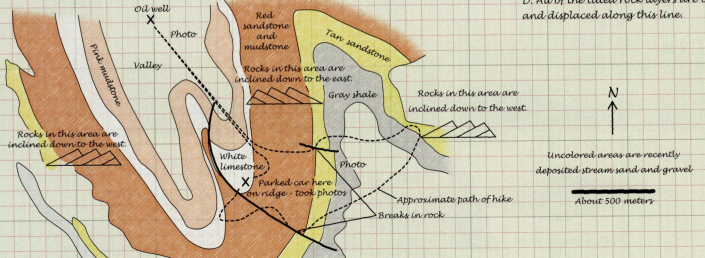

Oil well

Photo

Valley

Pink mudstone

Red sandstone and mudstone

Tan sandstone

Rocks in this area are inclined down to the east.

Gray shale

Rocks in this area are inclined down to the west.

Rocks in this area are inclined down to the west.

White limestone

Parked car here on ridge - took photos

Photo

Approximate path of hike

Breaks in rock

N

Uncolored areas are recently deposited stream sand and gravel

About 500 meters

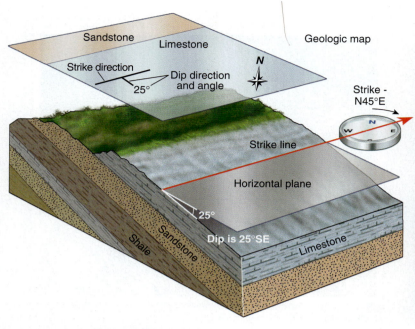

▲ **Figure 11.2 Visualizing strike and dip.**
The diagram illustrates the geometry of strike and dip. Imagine a horizontal plane that intersects an inclined layer of rock. The strike line is the intersection of the inclined layer with the horizontal plane. The strike direction is the angle measured from the strike line to true north. The dip is the angle formed between the horizontal plane and the inclined bed. The strike-and-dip symbol is sketched on the simple geologic map above the block. Wave erosion of sedimentary rocks along the California coast, seen on the right, exposes cliffs that parallel the dip of the sedimentary layers while the cliff trends in the direction of bedding strike.

The tilted bedding planes in the rocks provide a useful benchmark for describing the orientation of the deformed layers. Geometric elements—lines, planes, and angles—provide measurement guidelines to describe the orientation of the rock layers.

- The **strike** of a nonhorizontal bed, as illustrated in **Figure 11.2,** is the compass orientation of a line formed by the intersection of an imaginary horizontal plane with the inclined bedding plane.
- The **dip** of the inclined rock layer is the angle between the imaginary horizontal plane and the inclined rock layer.

Geologists describe the orientation of any planar geologic feature, not just bedding planes, by measuring strike and dip and recording them on a map with a standard geologic symbol (Figure 11.2).

Describing Folded Rocks

The folded rocks at the field site dip in different directions and these dip directions are key to describing a fold, as shown in **Figure 11.3**. Rocks commonly dip in opposite directions on either side of a fold. These oppositely dipping sides of

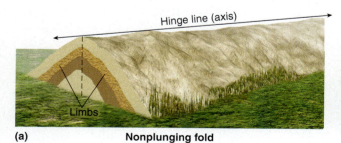

(a)   **Nonplunging fold**

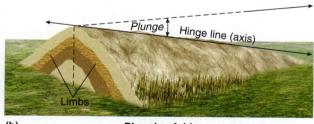

(b)   **Plunging fold**

Folding hinged wood

◄ **Figure 11.3 Visualizing parts of a fold.**
Simple folds in layered rocks are defined by limbs that dip in opposite directions from a hinge line, which is also known as the fold axis. Two sheets of hinged wood provide an analogy for the hinge line and limbs of a fold. Plunging folds have an inclined hinge line.

the fold are its **limbs**. The limbs join along a **hinge line** (also called the axis), which you can draw on a deformed layer through the series of points where the dip direction changes. Figure 11.3 draws an analogy between the limbs and hinge line of a fold and wood sheets connected by door hinges. In Figure 11.3a, the hinge line is horizontal, but in many cases the hinge line plunges downward into the ground and upward into the air, which defines a **plunging fold** like that illustrated in Figure 11.3b. You can think of a plunging fold as the combination of folding and tilting the rock layers in the direction of plunge.

Folded rocks exhibit a variety of very complex shapes, but for the simpler and most common folds it is easiest to focus on whether the rock layers in the limbs dip down toward the hinge line or dip down away from the hinge line. **Figure 11.4** shows how this simple distinction describes two basic types of folds, or four types by including the possibility of both horizontal and plunging hinge lines. **Anticlines** are arched folds where limbs dip *away* from the hinge line, and **synclines** are trough-shaped folds where limbs dip *toward* the hinge line. Figure 11.4 also illustrates the surface patterns of eroded and folded rocks. Notice that the oldest rocks are exposed along the axis of an anticline, whereas the youngest rocks are found along the axis of a syncline.

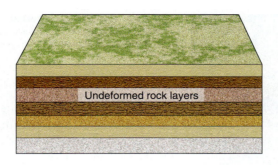
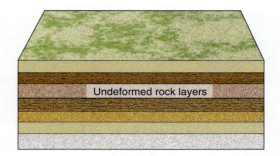
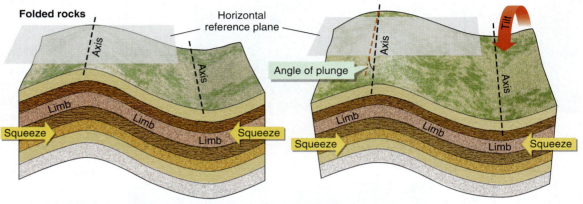
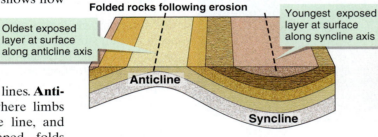
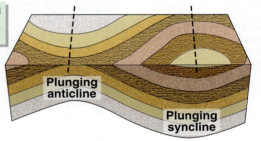
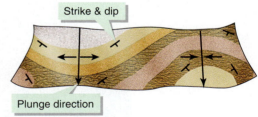
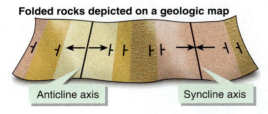

► **Figure 11.4 Visualizing folds.**
Folds form by forces that squeeze rocks horizontally. Plunging folds form where rocks both fold and tilt. The diagrams show the shapes of folded layers and how the layers look when folds erode, which exposes older layers along the axis of anticlines and younger layers along the axis of synclines. Geologists use special symbols to show the trace of fold axes on maps.

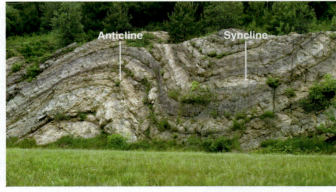
Anticline and syncline exposed in roadcut, New Jersey

Aerial view of plunging anticline, Sheep Mountain, Wyoming

Folding Rock: *See how rocks fold.*

◀ **Figure 11.5** **Highly contorted folds.**
Foliation bands in this gneiss exhibit an extremely contorted folding pattern caused by rock flow at high temperature and pressure.

Figure 11.5 illustrates a rock that is so highly deformed it does not seem reasonable to identify anticlines and synclines. The folds illustrated in Figure 11.4 resemble a rumpled rug, but the outcrop in Figure 11.5 more closely resembles the swirling of fudge sauce and melted ice cream. These rocks have flowed rather than simply folded along a hinge line.

Describing Broken Rocks

There are also terms for describing broken rocks. Fractures of various lengths and shapes are common in almost every rock outcrop, just as you always see cracks in building walls and ceilings and in pieces of lumber. Geologists make an important distinction between situations where rocks simply crack and where rocks slide past one another along a fracture.

- **Joints** are cracks where very little or no displacement of the rock occurred, as pictured in **Figure 11.6**.
- **Faults** are fracture planes along which rocks moved.

Miners and geologists assigned names to the rocks adjacent to fault surfaces while excavating tunnels to extract minerals. The link between mining and faults is that economically important minerals precipitate from water within open spaces in the crushed and fractured rock found along faults. Faults are rarely vertical. **Figure 11.7** shows why you can describe the rocks below the fault as the **footwall** in contrast to the **hanging wall** composed of rocks above the fault.

The next step in describing a fault is to determine how the footwall and the hanging-wall blocks moved past one another. As an analogy, place your palms together in front of you. Now, move your hands up and down past one another. This simulates movement parallel to the dip of the fault plane. Next, move one hand away from your body while moving the other one toward you. This motion compares the movement along the strike of the fault plane. A third possibility is to simultaneously move your hands up and down while also moving them toward and away from you. **Figure 11.8** applies these types of motion to classify faults.

Movement of rock along the dip of the fault plane defines **dip-slip faults** (Figure 11.8). There are three kinds of dip-slip faults. A **normal fault** forms when the hanging wall rock moves downward compared to the footwall. On the other hand, if the hanging wall moves up along a dip-slip fault compared to the footwall, then the fault is called a **reverse fault**, if the fault dips at an angle steeper than 45 degrees, or a **thrust fault**, if the angle is less.

Checkerboard Mesa at Zion National Park, Utah, shows a geometric pattern of vertical joints that break across inclined cross-beds in the sandstone. Notice that the joints do not displace the cross-beds.

Parallel joints in sandstone appear to be etched into the landscape in this aerial view in central Utah. Water preferentially accumulates in and flows along the joints, weathering and eroding the sandstone to accentuate the joints.

◀ **Figure 11.6** **What joints look like.**
Joints are fractures along which the rocks are not displaced.

Strike-slip faults form by horizontal movement along the strike direction of the fault plane (Figure 11.8). Comparing the relative movement of the rocks on either side of the fault distinguishes two types of strike-slip faults. If features appear shifted to the left from one side of the fault to the other, then a **left-lateral strike-slip fault** exists. For a **right-lateral strike-slip fault**, features shift to the right across the fault. You might think that right- and left-lateral motion depends on your perspective in looking at features displaced across the fault, but it does not matter. To check this, look at the strike-slip fault diagrams in Figure 11.8 by turning the book to the left and then to the right and you will see that displacement type is the same regardless of which side of the fault you are on.

Faults can show a combination of dip-slip and strike-slip movements. These are **oblique-slip faults**, where diagonal motion occurs along the fault plane, both along the strike and dip as shown in Figure 11.8.

Making a Geologic Map and Cross Section

Armed with these various devices for describing the orientation of deformed layers and the types of folds and faults, you can reexamine the sketch of the rocks at the field site (Figure 11.1e). Your sketch is a roughly drawn **geologic map**, which shows the locations of the different rock types, their strike and dip, and locations of fold hinge lines and faults. **Figure 11.9** shows the actual, scaled geologic map for the field site. Color bands trace the distribution of different rock types. Strike-and-dip symbols show the orientations of the inclined sedimentary layers. The backward "S" pattern of deformed beds that is conspicuous in the field sketch (Figure 11.1e) reveals a side-by-side plunging anticline and syncline on the geologic map.

Geologic cross sections interpret subsurface geology based on measurements of the orientations of rock layers, folds, and faults at the surface. The schematic diagrams of folds and faults in Figures 11.4 and 11.8 include both map views and cross-section views of each structure. Figure 11.9 includes a cross section of the field site and shows how the rock layers are folded into an anticline and a syncline and offset by faults.

The cross section is made by drawing the subsurface continuation of the rock layers seen at the surface. The measured dip angles at the surface provide the necessary data for drawing the subsurface rock layers in the proper orientation. Faults mapped at the surface continue downward and offset layers in the subsurface. The cross section matches with the geologic map to complete a three-dimensional illustration of the deformed rocks.

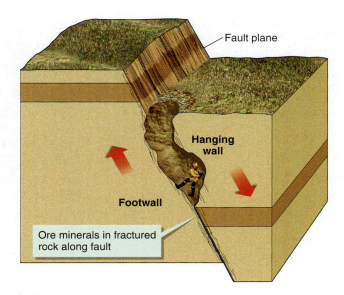

▲ **Figure 11.7 Describing faulted rocks.**
Fault planes separate footwall and hanging-wall blocks of displaced rock. The terms for rocks on either side of a fault originated with miners who tunneled to extract ore minerals that precipitate from water in fractured rock along the fault. Imagine walking along an open tunnel following the sloping fault plane. You would walk on the footwall, and the hanging wall would hang over your head.

Putting It Together–What Do Deformed Rocks Look Like?

- Strike and dip describe the orientation of planar geologic features, such as inclined sedimentary layers and faults.
- Anticlines and synclines are distinguished by noting the dip directions of rocks forming the limbs of the fold.
- Breaks in rock include (1) joints, where little or no displacement has taken place along fractures; and (2) faults, where offset of rock occurred.
- Dip-slip faults include normal, reverse, and thrust faults. Strike-slip faults include left-lateral and right-lateral faults. A combination of movements along both the dip and strike of the fault plane produces oblique-slip faults.
- Geologic maps contain colors, patterns, and standard symbols that depict rock types, their location and orientation, and the nature and location of folds and faults that disrupt the rocks.
- Geologic cross sections are interpretations of rocks and structures below the ground surface, mostly based on the measured orientations of rocks and structures observed on the surface.

▶ **Figure 11.8 Visualizing faults.**
These diagrams and photographs illustrate the principal types of faults. You distinguish the different types by recognizing whether displacement parallels the dip, strike, or both strike and dip of the fault plane and the relative displacement of the hanging wall compared to the footwall.

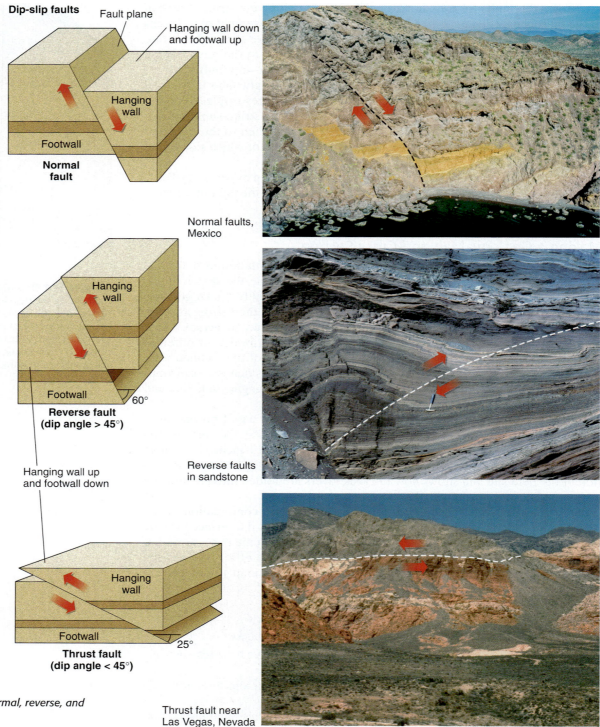

Dip-slip faults

Fault plane

Hanging wall down and footwall up

Hanging wall

Footwall

Normal fault

Normal faults, Mexico

Hanging wall

Footwall 60°

Reverse fault (dip angle > 45°)

Reverse faults in sandstone

Hanging wall up and footwall down

Hanging wall

Footwall 25°

Thrust fault (dip angle < 45°)

Thrust fault near Las Vegas, Nevada

Fault Motions: *See how motion along normal, reverse, and strike-slip faults displaces rocks.*

11.2 How Are Resources Related to Geologic Structures?

Now you can consider why there is an oil well in your field area (Figure 11.1c). Geologists purposely located the oil well on the axis of the anticline.

Oil and Gas Traps

In Chapter 5 (Section 5.3), you learned that oil and natural gas are fossil fuels, which form from organic materials deposited with sediment. Oil and gas form

Figure 11.8 (*continued*)

Strike-slip faults

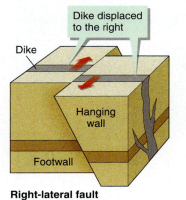

Left-lateral fault

Right-lateral fault

Strike-slip fault, Nevada

Oblique-slip faults

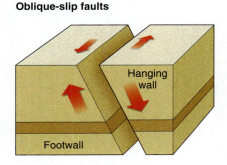

from algae, bacteria, and tiny plants and animals (plankton) that live in the sea or in lakes. When the sediment and organic material are buried to depths where the temperature rises above 100°C, the organic compounds undergo changes analogous to mineral metamorphism. These changes convert the organic material into oil and gas.

Oil, gas, and ground water move through pore spaces in rock. All three fluids have different densities, as shown in **Figure 11.10**. Oil is less dense than water, and natural gas is less dense than both oil and water. This means that oil and gas migrate upward through ground water in porous rock and eventually leak out to the surface unless they are trapped below less porous rock. For example, poorly cemented sandstone typically has large pores between sand grains whereas shale has virtually no pore space. If a shale layer overlies a sandstone layer, then the shale forms a roof rock that traps oil and gas in the sandstone.

Anticlines commonly form oil and gas traps. **Figure 11.11**a shows the trapping process. If a low-porosity rock layer overlies a high-porosity layer, then the oil and gas migrate upward to the highest part of the porous layer and are trapped at the top of the anticline.

Thrust and reverse faults also trap oil and gas (Figure 11.11b). These traps form where movement along a fault displaces low-porosity rock on top of high-porosity rock. This type of trap is common throughout the Rocky Mountain region and in northern Alaska.

Metallic Mineral Resources

Economically important mineral deposits commonly coincide with faults. Most mines extracting metal ores occur in highly fractured rock. Underground shafts and tunnels in mines sometimes follow faults (Figure 11.7). **Figure 11.12** shows an example of a mineral vein along a fault. The minerals typically precipitate

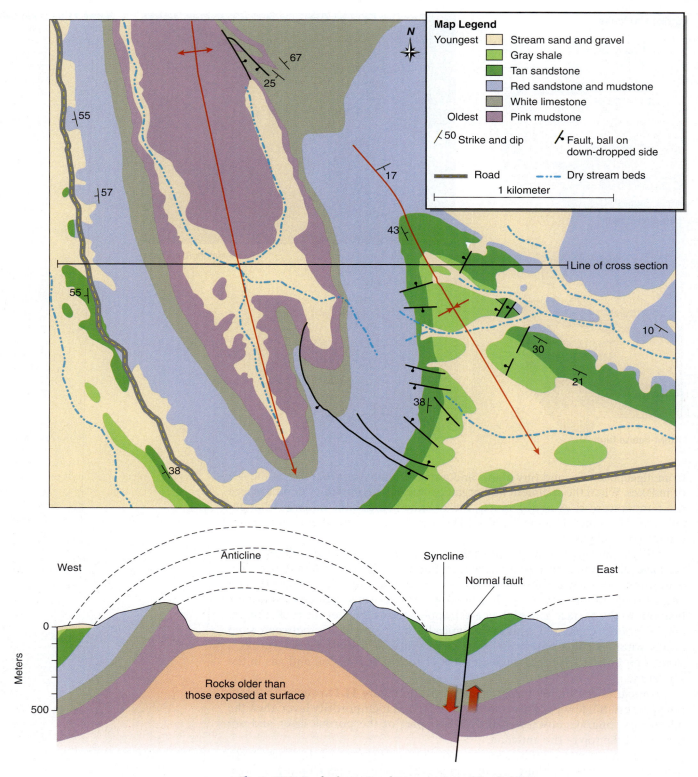

▲ **Figure 11.9 Geologic map and cross section of the field site.**
The geologic map looks similar to the field sketch (Figure 11.1) but this map has more detailed geologic information, such as strike-and-dip symbols, fold axes, and displacement directions along faults. The cross section illustrates an interpretation of the rock layers below ground level and how the rocks projected above the current surface before eroding. The cross section is based on the strikes and dips that are measured in the field and labeled on the map.

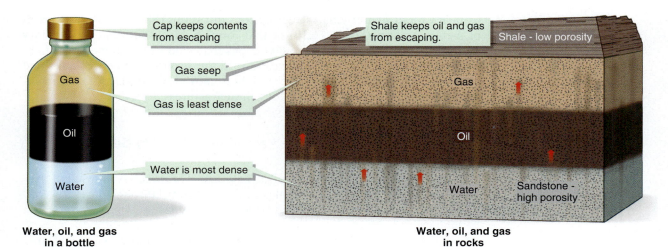

Cap keeps contents from escaping

Gas seep

Gas is least dense

Water is most dense

Gas

Oil

Water

Water, oil, and gas in a bottle

Shale keeps oil and gas from escaping.

Shale - low porosity

Gas

Oil

Water

Sandstone - high porosity

Water, oil, and gas in rocks

◀ **Figure 11.10 How density determines where oil and gas are found.**
Gas, oil, and water separate according to their densities, with least-dense gas filling the top of a bottle, most-dense water at the bottom, and intermediate-density oil in the middle. The lower densities of oil and gas compared to water allow the fossil fuels to rise upward through water-filled pores in rocks. This upward movement stops only where low-porosity rock traps the oil and gas below the surface.

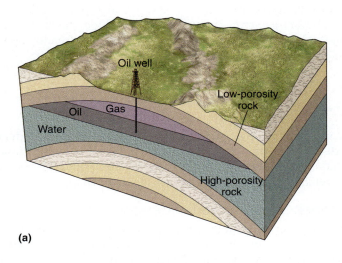

Oil well

Low-porosity rock

Oil   Gas

Water

High-porosity rock

(a)

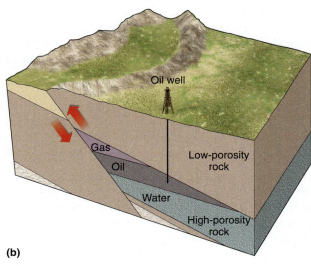

Oil well

Gas

Oil

Water

Low-porosity rock

High-porosity rock

(b)

▲ **Figure 11.11 How geologic structures trap oil and gas.**
(a) Anticlines with layers of low-porosity rocks overlying high-porosity rocks commonly provide traps for oil and gas. The oil and gas rise up through water-saturated pores in the porous rock and are trapped beneath the low-porosity rock along the axis of the anticline. (b) Fault movements trap oil and gas by displacing high-porosity rocks next to low-porosity rocks.

▲ **Figure 11.12 Minerals accumulate along faults and joints.**
Many economically important mineral deposits form along fractures in rock. Shiny sulfide ore minerals form a vein that follows a fault cutting through volcanic rocks in this mine in Peru. The sulfide minerals include ores of copper, silver, lead, and zinc.

from hot water that originates near igneous intrusions (Section 4.10) as a result of metamorphism (Section 6.10) or that forms by deep burial in sedimentary basins. Hot water dissolves more metal ions than cold water, so metal ore minerals precipitate as the water moves toward the surface and cools. Faults and joints provide avenues for metal-rich fluid to move toward the surface so mineral deposits preferentially form along these fractures.

Putting It Together–*How Are Resources Related to Geologic Structures?*

■ Gas is less dense than oil, and oil is less dense than water. As a result of the density contrast, oil and natural gas migrate upward through pores toward Earth's surface unless impeded by low-porosity layers, which trap the oil and gas.

■ Anticlines and faults form common traps for oil and natural gas. The oil and gas migrate upward to the highest point below a nonporous layer along the hinge line of an anticline and are trapped beneath reverse and thrust faults where low-porosity layers are displaced on top of high-porosity layers.

■ Metal-rich mineral resources commonly form along joints and faults where warm fluids with high concentrations of metal ions readily flow and then cool to precipitate metallic minerals.

11.3 Why Do Rocks Deform?

Describing and mapping geologic structures are essential steps to understanding *how* rocks deform, but it is at least equally important to understand *why* they deform. Put simply, rocks deform in response to applied forces, such as squeezing and stretching, that result in changes in size or shape.

Rocks Strain When Stress Exceeds Strength

Introduction of a few terms permits easier discussion of why rocks deform. **Stress** is the force applied over a given area (also see Section 6.3) and **strain** describes the measurable rock deformation resulting from stress. Small stresses do not typically cause obvious strain because the chemical bonds within minerals and contacts between mineral grains resist deformation. Strain occurs only when the applied stress exceeds the rock **strength**, which is a measure of the amount of stress that a material can endure before it fails, either by breaking or flowing.

Three Types of Stress and Strain

Figure 11.13 demonstrates basic types of stresses, and resulting strains, by the analogy of deforming a block of modeling clay. Imagine a vertical plane passing through the middle of the block with pins inserted in the clay on either side of this plane. Three scenarios for stress and strain are easily demonstrated:

• If you push the block from opposite sides *toward* the imaginary plane, then the block gets *shorter* in the direction of the applied stress and also gets *thicker*. The pins end up closer to one another and closer to the imaginary plane than they were prior to deformation. This pushing-together stress is **compression** and the resulting strain is **shortening** (also called contraction).

• If you pull the block from opposite sides *away* from the imaginary plane, then the block gets *longer* in the direction of the applied stress and also gets *thinner*. The pins end up farther from one another and farther from the imaginary plane than they were prior to deformation. This pulling-apart stress is **tension** and the resulting strain is **elongation** (also called extension).

• If you pull the block on one side and push it on the other parallel to the imaginary plane, then the block warps but the pins remain the same distance from the imaginary plane that they were prior to deformation. The term **shear** de-

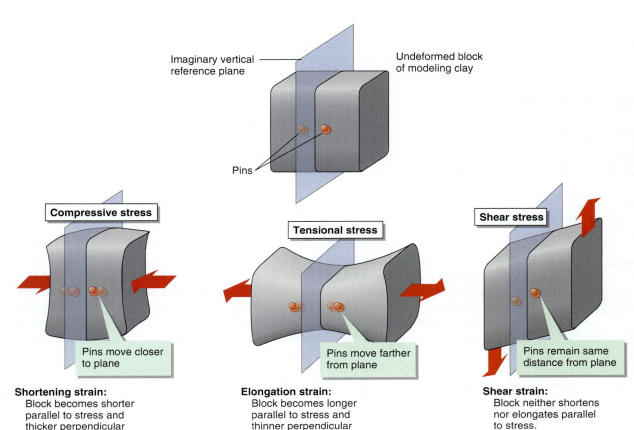

◄ **Figure 11.13** **Visualizing stress and strain.**
Pushing, pulling, and sliding a clay block demonstrates the stresses and strains that deform rocks. Movement of pins compared to an imaginary reference plane helps to visualize the strain. Compression is a pushing or squeezing stress, which shortens the block and moves the pins closer to the plane. Tension is a pulling or stretching stress, which elongates the block and moves the pins farther from the plane. Shear stress is a sliding stress, which neither shortens nor elongates the block parallel to the stress.

Imaginary vertical reference plane

Undeformed block of modeling clay

Pins

Compressive stress

Pins move closer to plane

Shortening strain:
Block becomes shorter parallel to stress and thicker perpendicular to stress.

Tensional stress

Pins move farther from plane

Elongation strain:
Block becomes longer parallel to stress and thinner perpendicular to stress.

Shear stress

Pins remain same distance from plane

Shear strain:
Block neither shortens nor elongates parallel to stress.

scribes both the stress and the strain of sliding rocks in different directions without overall shortening or elongation.

Faulting, Folding, and Flowing Are Strain

Geologists design laboratory experiments to reproduce and explain the structures of deformed rocks. Some informative experiments use materials that are weaker than real rocks so that structures form more easily with small stresses.

The relationship between stress, strain, and structures is illustrated by the sandbox experiment shown in **Figure 11.14**. Tension in the left part of the sandbox causes the layers of loose sand and clay powder to elongate. Compression in the right part of the sandbox causes the layers to shorten. The strains in each part of the sandbox occur mostly along faults. If you look at the elongating side of the box, you will notice the normal faults that formed where the hanging-wall

▼ **Figure 11.14** **Making faults and folds in a sandbox.**
This sandbox experiment illustrates how different stresses form folds and faults.

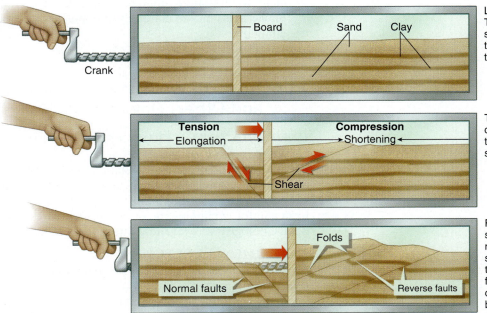

Board   Sand   Clay

Crank

Tension
Elongation

Compression
Shortening

Shear

Folds

Normal faults

Reverse faults

Layers of sand and clay fill the box. Turning the hand crank on the left side causes the vertical board in the center of the sandbox to shift to the right.

The board slowly moves to the right, causing elongation of the layers on the left side of the box and shortening on the right side.

Folds and reverse faults form in the shortening side of the sandbox and normal faults form in the elongated side. Normal faults form in response to tensional stress and reverse faults form by compression, but shear stress causes motion along the faults because the fault planes are not perpendicular to the stresses.

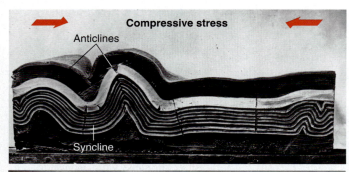

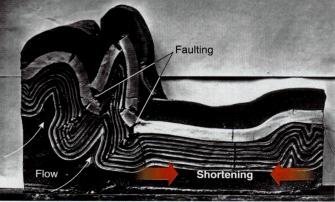

▲ **Figure 11.15 Visualizing folds and faults with wax.**
These photos show cross sections through a progressively deformed wax model. Anticlines and synclines form as the wax layers fold during initial shortening. With increased shortening, the
layers break along faults, and some wax flows into the growing
anticlines.

block slid off of the footwall block. If you look on the shortening side of the box,
you will see that reverse faults formed where the hanging wall block shoved
over the footwall block. Small folds also account for some of the shortening.

It is important to notice that the faults are not perpendicular to the stresses
created when the vertical board is cranked through the sandbox, because this
reveals something consistent with what geologists know about faults. Sand and
clay particles on opposite sides of a fault move past one another without getting
closer to or farther from the fault plane. The way layers move past each other
along the fault is, therefore, shear strain. At the same time, however, the movement of hanging wall blocks relative to footwall blocks shows that normal faults
cause elongation where tension occurs and reverse faults cause shortening
where compression occurs. Stated another way, in either compartment of the
sandbox, the larger-scale deformation is elongation or shortening, and locally, or
at the smaller scale, shear occurs along faults.

Figure 11.15 illustrates a classic nineteenth-century experiment where weak
wax layers were stand-ins for the strained sedimentary beds. Colored wax layers
shortened when squeezed from opposite sides. The layers initially folded into
anticlines and synclines. More stress caused the folds to grow higher and narrower and faults appeared. The bottom wax layer flowed into the centers of the
arching anticlines. The wax-model experiment shows that folds are a form of
shortening caused by compression. Also notice how the total thickness of wax
increased as the wax model shortened.

The stress required to deform real rocks is much greater than the stresses
applied in the sandbox or wax experiments. So how do geologists determine the
strength of real rocks and whether the rocks deform by faulting, folding, or
flowing? In order to determine how rocks deform, you need to delve further
into relationships between stress, strain, and strength.

---

*Putting It Together—***Why Do Rocks Deform?**

■ Stress is the magnitude of a force applied over an indicated area, and
strain is the measure of the amount of deformation resulting from the
stress.

■ Rock strength is the amount of stress that the material can endure before
it strains.

■ When applied stress exceeds the strength of a rock, the rock deforms by
folding, faulting, or flowing. Compressive stress shortens rocks perpendicular to the stress direction, whereas tensional stress elongates rocks.
Shear stress moves materials in opposite directions without shortening or
elongation.

11.4 How Do We Know . . . Why Some Rocks Break and Others Flow?

UNDERSTAND THE PROBLEM
How Strong Are Rocks?
Are all rocks equally strong? What factors determine if rocks break or flow? Do
the higher temperature and pressure below Earth's surface affect rock
strength? These are very important questions for understanding how rocks deform and for explaining the origins of structures observed and measured in the
field. Geologists answer these questions with laboratory experiments.

EXPERIMENTAL SETUP
How Do Geologists Deform Rocks in the Lab?
Every rock has a stress limit where it breaks or flows if stress increases beyond
the limit. This critical stress that exceeds rock strength is determined for different rock types in a laboratory apparatus.

Figure 11.16 depicts a typical experimental setup. Small cylindrical rock samples are ground flat on the top and bottom, and the length and diameter of the cylinder are measured. The rock cylinder is placed within a thin-walled metal cylinder and set between an anvil and a moveable piston. Hydraulic fluid is poured into a chamber encircling the rock and metal jacket and maintained at chosen confining pressures to represent surrounding rocks at various depths below Earth's surface. For example, by applying 250 bars of confining pressure, which is about 250 times greater than air pressure at sea level, a geologist can measure rock properties in conditions equivalent to one kilometer below the surface. The whole apparatus can also be heated to reproduce temperatures found below the surface. Pushing the piston down on the sample produces a compressive stress that is calculated from the amount of force exerted on the piston. Measured vertical shortening of the rock specimen records the resulting strain.

FIRST RESULTS

How Do Rocks Strain?

Figure 11.17 illustrates results of rock-deformation experiments with cylinders of limestone subjected to compressive stress. At low confining pressure and small stress, the cylinder does not change. With increasing stress the cylinder shortens by breaking and sliding along fractures that are analogous to faults seen in the field. At higher confining pressure, equivalent to greater depth within Earth, the cylinders shorten by flow without visible fractures. The observations are that small stress produces no deformation and that rocks strain by either breaking or flowing when subjected to large stresses. A clearer understanding of what actually happens during deformation is revealed if we compare the amount of strain that results from applying different amounts of stress.

Figure 11.18 illustrates the relationship between stress and strain for three different rock-deformation experiments. The rock cylinders exhibit different shortening strains at the same stress, and one deforms by breaking, whereas the other two flow (Figure 11.18a). These results translate into graph curves (Figure 11.18b).

The plots for all three rocks start with a line where strain increases as stress increases (Figure 11.18b). It turns out that if the stress is reduced, the strain also diminishes within this straight-line part of each plot. Stated another way, the rock returns to its original dimensions after the stress is removed, just like an elastic rubber band returns to its original shape after you stretch it and then let go. Deformation along this line where stress and strain are both low is, therefore, **elastic** (also see Section 8.2). When stress increases to the **yield strength** of the rock, the deformation is permanent and cannot be reversed by decreasing the

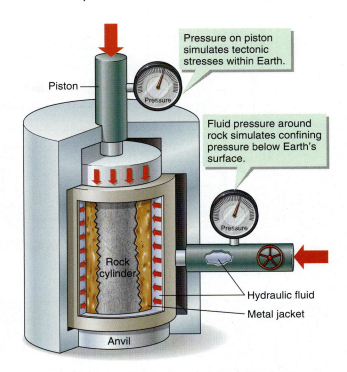

▲ **Figure 11.16** **Experimental setup for measuring rock strength.**
This schematic diagram shows an experimental apparatus for subjecting rocks to compressive stresses and then measuring and describing the resulting strains. A small, cylindrical rock specimen is placed inside the metal jacket in the center of the sample chamber. Compressional stress is applied from the top by pressing down on a piston, which forces the sample against a fixed anvil at the bottom of the chamber. Confining pressure is controlled by pumping in hydraulic fluid to surround the jacket and keep it from buckling. The chamber can also be heated to investigate deformation at different temperatures.

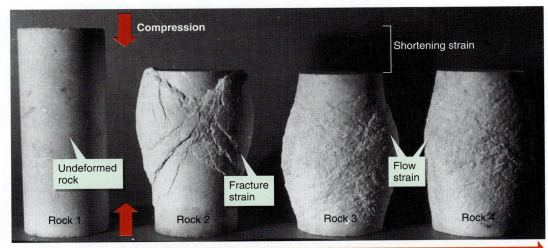

◄ **Figure 11.17** **How rocks deform in the lab.**
These are examples of rock cylinders used to determine the strength and type of strain exhibited by limestone. Flow strain occurred at higher confining pressure than fracture strain. Notice that rocks 2, 3, and 4 all exhibit the same vertical shortening compared to the undeformed sample (rock 1).

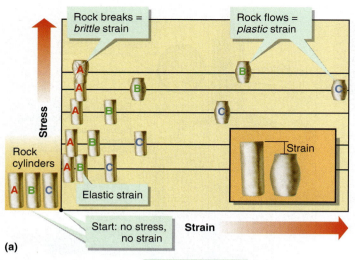

When rock cylinders initially compress, they shorten slightly in an elastic fashion. At higher stress, rock A breaks by fractures (like rock 2 in Figure 11.17). Rocks B and C deform by flow at yet higher stresses.

(a)

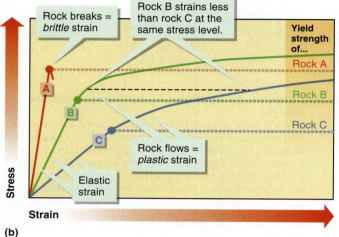

This graph summarizes the stress and strain relationships for the three rocks. Initial elastic strain makes straight lines on the graph up to the stress value that equates to the yield strength of the rock. Brittle fracture occurs at the end of the elastic strain line for rock A, and curves track the plastic flow of rocks B and C at higher stress.

(b)

stress. When stress exceeds the yield strength, the rock either breaks or flows.

Rock A in Figure 11.18 breaks when stress exceeds the yield strength. The term **brittle** describes the deformation of rock A because it broke at a relatively small strain.

Rocks B and C do not break when stress exceeds the yield strength. These rocks permanently deform by flow, which results from microscopic shifts within the crystal structures of the minerals composing the rock. Geologists use the term **plastic** (or ductile) to describe rocks that flow without breaking.

How do you compare the strengths of the three rocks in Figure 11.18? Rock A is the strongest because it has the highest yield strength, which is the same as saying that it sustains more stress before exhibiting permanent strain. Rock B is stronger than rock C not only because rock B has the higher yield strength but also because rock C undergoes much more strain than rock B at the same stress.

Table 11.1 summarizes rock strengths of common rocks at temperature and pressure conditions that represent Earth's surface. These surface conditions clearly do not represent all of the conditions where rocks deform so it is essential to understand how temperature and pressure affect rock strength and deformation.

SECOND RESULTS
What Determines Rock Strength?
The observation that different rocks possess different strengths at surface conditions indicates that the types of minerals found in rocks and how the mineral grains interconnect with one another must be an important factor that determines rock strength. Temperature, pressure, and water content are other factors that should come to mind for consideration, based on the persistent reappearance of these three variables in previous explanations of

Graphing Stress and Strain: *See how to interpret graphs of stress and strain in rocks.*

TABLE 11.1	Relative Rock Strength at Surface Temperature and Pressure	
Strength and deformation		**Rock type**
High strength, brittle deformation		
		Quartzite
		Granite
		Quartz-cemented sandstone
		Basalt
		Limestone
		Calcite-cemented sandstone
		Schist
		Marble
		Shale and mudstone
		Rock salt
Low strength, plastic deformation		

rock origins and rock properties. Temperature, pressure, and water content are easily changed in the experimental setup.

Figure 11.19 compares the stress and strain of limestone at different confining pressures that reproduce different depths below the surface. At confining pressure equivalent to about 1 kilometer, the limestone experiences brittle fracture at a higher stress than at surface pressure. High pressure compresses the chemical bonds within minerals and compresses the mineral grains against one another, which increases rock strength and makes it increasingly difficult to break the rocks. At higher confining pressures, representing depths of 6 and 12 kilometers, the limestone exhibits plastic, rather than brittle, deformation. Two conclusions emerge from these experiments:

1. The strength of the rock increases with increasing confining pressure, which implies that rock becomes stronger where it is deeper below the surface.
2. When the stress exceeds rock strength, brittle fracture occurs close to the surface but plastic flow occurs at greater depth.

This experiment with limestone does not account for the possible effect of increasing temperature below the surface. In addition to confining pressure and temperature, the presence or absence of water also affects rock strength. Now let us look at results of experiments carried out at various temperatures and with differing water contents but at a constant confining pressure.

Figure 11.20 shows how increasing temperature and water content in marble lowers the strength of the rock. The graph of stress and strain shows that the same stress causes much more plastic flow at 150°C than at room temperature. Molecules vibrate more rapidly at high temperature than at low temperature, and the vibration weakens the bonds that resist strain. Adding water at this elevated temperature decreases the strength even more. Water lubricates grain boundaries, breaks bonds, and enhances microscopic slips that add up to produce plastic flow of the rock.

INSIGHTS

How Does Deformation Vary from Place to Place?
The lab results have important implications for how deformation should vary within Earth. Some rocks are strong and others are weak; some are more prone to brittle fracture and others to plastic flow. At increasing depth within Earth, the confining pressure and temperature both tend to decrease brittle fracture and increase plastic flow. Faults and joints, which are expressions of brittle deformation, should occur only near Earth's surface, whereas plastic flow is expected at depth. A few geologic materials are so weak that they actually flow even close to the surface. The most common examples are rock salt, which deforms plastically at depths of only 300 meters, and glacial ice, which flows when it is only 50–60 meters thick. The distribution of water in rocks will also affect deformation because water lowers rock strength. The experimental results that describe and explain brittle and plastic deformation can now be combined with stress types so that we can fully understand rock deformation.

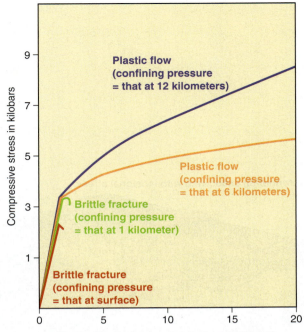

▲ **Figure 11.19** **Experimental results—stress and strain at different confining pressures.**
This graph summarizes the results of four compressive-stress experiments on limestone at different confining pressures, which correspond to different depths within Earth. At pressures between the surface and 1 kilometer, the limestone fails in a brittle manner, with greater strength at higher pressure. At confining pressures corresponding to depths of 6 and 12 kilometers, the limestone deforms by plastic flow. Less plastic strain occurs at a particular stress at a pressure equivalent to 12 kilometers than at 6 kilometers, which indicates that rocks are stronger at greater depth.

▶ **Figure 11.20** **Experimental results—stress and strain at different temperature and water content.**
The experimental stress and strain results graphed here show that marble is weaker at higher temperature than at low temperature and is also weaker when wet than when dry. The same confining pressure, equivalent to about 30 km deep in the crust, was applied in all three experiments. Marble deforms plastically at this pressure. Increasing temperature from room temperature to 150°C lowers the strength of the rock as demonstrated by the greater strain at the same compressive stress. At the same elevated temperature, adding water to the marble lowers the rock strength even further.

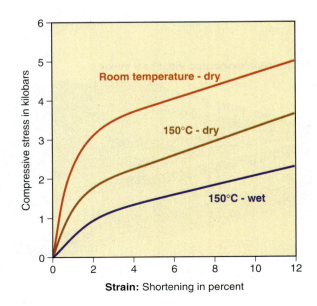

Putting It Together–How Do We Know ... Why Some Rocks Break and Others Flow?

■ At different combinations of stress and strain rocks exhibit elastic, brittle, or plastic deformation.

■ Increasing confining pressure increases rock strength and leads to plastic flow rather than brittle fracture.

■ Increasing temperature decreases rock strength and enhances plastic flow.

■ Wet rocks are weaker than dry rocks at the same temperature and confining pressure.

Deformation associated with tensional stress

Strike-slip fault
Horst
Tensional stress
Graben
Crust
Normal faults
Graben
Mantle
Plastic flow

Crust stretches and thins by brittle normal faulting in upper crust and plastic flow in lower crust.

Deformation associated with compressional stress

Strike-slip fault
Compressional stress
Folds
Reverse fault
Thrust fault
Crust
Mantle
Plastic flow

Crust shortens and thickens by brittle faulting and folding in upper crust and plastic flow in lower crust.

Deformation associated with shear stress

Shear stress
Strike-slip fault
Crust
Mantle
Plastic flow

11.5 How Do Geologic Structures Relate to Stress, Strain, and Strength?

Sandboxes, wax models, and compression of rock cylinders all provide key pieces of understanding about how and why rocks deform. These insights combine with field observations of structures to develop a comprehensive picture of the deformed crust summarized in **Figure 11.21**. **Figure 11.22** illustrates examples of regional landscapes in the United States that relate to rock deformation.

It is also important to keep in mind the origin of stress. Up to this point you have seen the results of squeezing wax, cranking a board through a sandbox, and compressing rock cylinders between pistons and anvils. In nature, plate motions are the origin of most of the stresses that deform rocks. You will learn more about plate motions in the next chapter. Here it is simpler to think of these plate-tectonic compressive, tensional, and shear stresses as caused by motions within the interior of Earth and expressed at the surface as deformation.

Deformation by Tension

Figure 11.21 illustrates the deformation that occurs in areas subjected to tensional stress. Tension is typical of divergent plate boundaries and locations where upwelling mantle stretches the lithosphere. The overall deformation pattern is one of elongation and thinning of the crust. Plastic flow in the lower

◄ **Figure 11.21 What rock deformation looks like on a large scale.**
Stresses deform rocks by brittle fracture at shallow depths along with folding of sedimentary layers by bending and slipping along bedding planes. Plastic flow occurs in the lower crust and mantle where pressure and temperature are much higher than in the upper crust. Tensional stress forms horst and graben blocks bounded by normal faults. Compressional stress forms thrust and reverse faults along with folds. Strike-slip faults form where faults are parallel to extensional and compressional stresses and are also the dominant structure where shear stress is the principal deforming stress.

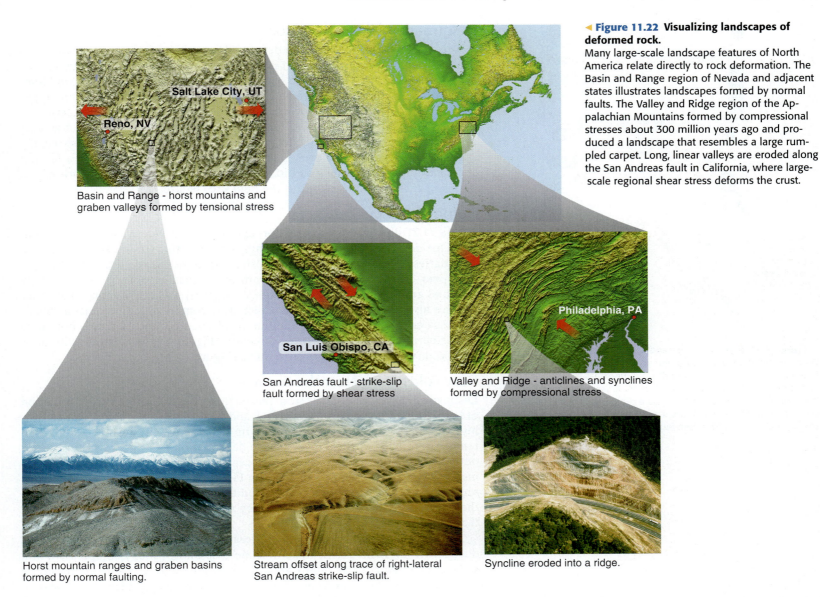

◄ **Figure 11.22 Visualizing landscapes of deformed rock.**
Many large-scale landscape features of North America relate directly to rock deformation. The Basin and Range region of Nevada and adjacent states illustrates landscapes formed by normal faults. The Valley and Ridge region of the Appalachian Mountains formed by compressional stresses about 300 million years ago and produced a landscape that resembles a large rumpled carpet. Long, linear valleys are eroded along the San Andreas fault in California, where large-scale regional shear stress deforms the crust.

Basin and Range - horst mountains and graben valleys formed by tensional stress

San Andreas fault - strike-slip fault formed by shear stress

Valley and Ridge - anticlines and synclines formed by compressional stress

Horst mountain ranges and graben basins formed by normal faulting.

Stream offset along trace of right-lateral San Andreas strike-slip fault.

Syncline eroded into a ridge.

crust resembles stretching taffy, whereas normal faults break the brittle upper crust and thin the crust as hanging-wall blocks slide down and away from foot-wall blocks. Where fractures closely parallel the stress direction, rocks slide past one another on strike-slip faults.

Figure 11.22 illustrates the typical jumbled landscape resulting from tensional tectonic stress. Blocks of crust jostled upward along normal faults are **horsts**, while those jostled downward are **grabens**. "Graben" originates from the German term *grabe*, meaning "ditch," and "horst" describes the nests built by large birds of prey, typically on a high cliff. Many parallel horsts and grabens form the Basin and Range region of Nevada and parts of adjacent states; the valley basins are grabens and the mountain ranges are horsts (Figure 11.22).

Tectonic stresses form most joints and faults but some joints also form independently of tectonic stresses. One example of nontectonic joints is where rocks expand toward the surface as overlying rock erodes away, which reduces the confining pressure. The decreasing rock weight causes tensional stress perpendicular to the ground surface so that rocks spall off in sheets by the exfoliation processes illustrated in Figure 5.2b. Shrinkage of igneous rocks during cooling produces tensional stresses that crack the rocks apart along cooling joints, such as those illustrated in Figures 4.8b and 4.10b.

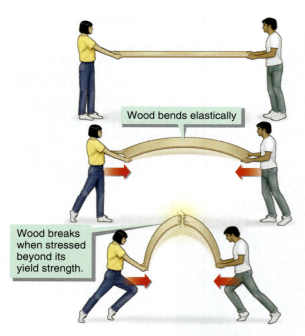

▲ **Figure 11.23 Plywood as an analogy for rock bending.**
When two people walk toward one another while grasping a sheet of plywood, the wood bends until the stress reaches the yield strength of the wood, at which point it breaks. Thin, wide sheets of plywood have similar relative dimensions to thin sedimentary beds that are continuous over large areas. In a similar fashion, rock layers fold before faulting when compressed.

Deformation by Compression

Figure 11.21 also summarizes the deformation that occurs in areas subjected to compressional stress, usually near convergent plate boundaries. Shortening occurs by plastic flow in the lower crust, where relatively high pressure and temperature inhibit brittle fracture and by faulting and folding in the brittle upper crust. The faults are mostly reverse and thrust faults that shorten the crust by sliding the hanging-wall block up and over the footwall block. Where fractures closely parallel the stress direction, rocks slide past one another on strike-slip faults. Shortening also thickens crust and raises surface elevations in a fashion similar to that seen in the wax model (Figure 11.15). Compression formed the plunging anticline and syncline at the field site (Figure 11.9).

Deformation by Shear

Figure 11.21 illustrates structures formed in a region where shear stress dominates the deformation without shortening or elongating strains. Transform plate boundaries are an example of such a stress scenario. Strike-slip faults form in the upper crust and continue downward into a plastic zone where rocks flow past one another in opposite directions. Nonetheless, keep in mind that shear stresses also act locally in regions affected by compression and tension. Shear stresses occur wherever brittle faulting or plastic flow occurs that is not along planes perpendicular to the stress direction (Figure 11.14), so strike-slip faults also occur with normal and reverse faults (Figure 11.21).

How Folds Form Near the Surface

Distinguishing brittle fracture from plastic flow does not fully explain formation of folds. Some folds record plastic flow and these flow folds are dramatically illustrated in metamorphic rocks that form deep below the surface, where high temperature and pressure favor plastic deformation (see Figure 11.5). Clearly not all folding occurs by plastic flow because folds are most common in sedimentary rocks, such as those pictured in Figure 11.4 and at your field site. Most sedimentary rocks could not have folded at temperatures and pressures where they deform by plastic flow, or else they would be metamorphic rocks rather than sedimentary rocks.

Folding at low temperature and pressure, especially in sedimentary rocks, does not result from plastic flow. In some cases, horizontally compressed, thin sedimentary layers bend significantly before they break. An analogy would be to bend a sheet of plywood considerably before it breaks, as **Figure 11.23** shows. Similarly, the top layer of wax in the model illustrated in Figure 11.15 folded to some extent before it started to break. The sedimentary layers commonly crack into joints and small faults as they bend, which further demonstrates that folding takes place under conditions of brittle strain. These small fractures are visible in Figure 11.4 and in the wax model in Figure 11.15.

Sedimentary rocks also fold because weak bedding planes allow rock layers to slip past one another without breaking. **Figure 11.24** draws an analogy to folding a stack of papers—the sheets bend and slip past one another when a compressive stress is applied. Distortion of writing on the edge of the stack of sheets demonstrates that the sheets move past one another.

Some folds form near the surface in lay-

▼ **Figure 11.24 Folding a stack of paper is like folding sedimentary rocks.**
When a stack of paper folds the sheets slide past one another, as indicated by the changing shape of the letters, written on the edge of the stack. Similarly, sedimentary layers slide along bedding planes when compressive stress folds the rocks.

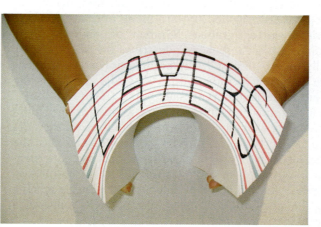

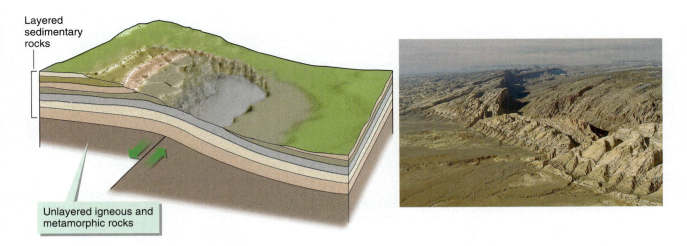

Layered sedimentary rocks

Unlayered igneous and metamorphic rocks

◄ **Figure 11.25 Folds may form above faults.** The diagram shows how movement along a fault causes overlying sedimentary rocks to fold over, or drape, the uplifted fault block. Although the fault displacement may later break to the surface, at this stage the folded sedimentary layers are the only hint of the presence of the buried fault. The photo shows a locality in central Utah where sedimentary rocks are folded above a concealed fault.

ered rocks that overlie faulted rocks. **Figure 11.25** illustrates this scenario where a fault in unlayered igneous rock passes upward into folded sedimentary rocks. The deforming sedimentary rocks slip along their bedding planes to produce a fold rather than breaking along the fault. In many cases, the movement along a fault at depth eventually folds the sedimentary rocks beyond their ability to bend and then the fault ruptures up through the sedimentary rocks, too. These types of folds and related underlying faults are common features in the Rocky Mountains of the western United States.

Putting It Together—How Do Geologic Structures Relate to Stress, Strain, and Strength?

■ Brittle faults and joints form in rocks close to the surface whereas plastic flow occurs at deeper levels where high pressure inhibits fracturing and temperature lowers rock strength.

■ Compression shortens and thickens crust by producing folds, reverse faults, and thrust faults in shallow rock, and plastic flow in deeper rock.

■ Tension elongates and thins crust by movement on normal faults in shallow rocks and taffy-like stretching of plastic rocks at deeper levels.

■ Folds form by plastic flow deep in the crust but also develop at and near the surface in sedimentary rocks where bending of thin layers and slippage along layers take place.

11.6 How Does Strength Vary in the Lithosphere?

Before leaving the topic of rock strength and deformation style, it is worthwhile to contemplate large-scale strength variations within the lithosphere that will ultimately be important for understanding plate tectonics. While brittle deformation occurs near the surface and plastic deformation occurs at greater depth, these deformation styles do not reveal the strength of the deforming rocks. In a general way, strength increases with increasing pressure (Figure 11.19) but decreases with increasing temperature (Figure 11.20), so does strength increase or decrease downward in the lithosphere? How does strength vary in the different rock types that compose the continental crust, oceanic crust, and the mantle?

Contrasting Strength of Oceanic and Continental Lithosphere

Figure 11.26 illustrates a reasonable pair of interpretations of how strength might vary along a vertical profile downward through average oceanic and continental lithosphere. It is important to emphasize that these represent *average*

▶ **Figure 11.26 Strength of the lithosphere.** The strengths of oceanic and continental lithosphere are compared by using (1) laboratory rock-strength data, (2) the contrasting compositions and water contents of the two lithosphere types, and (3) the different thicknesses of oceanic and continental crust. Strength initially increases downward because of increasing pressure. Eventually, strength then decreases at greater depth because of increasing temperature. The temperature-induced strength decrease occurs within the mantle part of oceanic lithosphere, but happens within the lower-crust part of continental lithosphere. The strength profile of continental lithosphere resembles a jelly sandwich—weakest in the middle and stronger above and below.

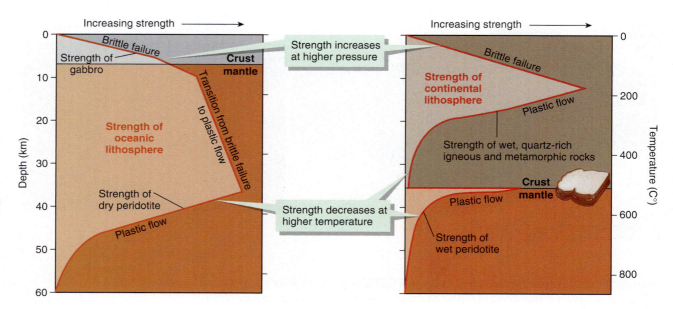

lithosphere because strength differs from place to place depending on geothermal gradient and rock types. The continental lithosphere has a strength profile analogous to a jelly sandwich. Two relatively strong layers, upper continental crust and upper mantle, sandwich the weaker lower crust. This profile is very different from the oceanic lithosphere, which is relatively strong down to about 40 kilometers.

The contrasts between these hypothetical strength curves are geologically significant. Overall, oceanic lithosphere is stronger than continental lithosphere, which suggests that continental lithosphere should be more easily deformed by plate-tectonics stresses than oceanic lithosphere. Brittle failure to form faults usually occurs down to about 12–15 kilometers for average crust, and plastic flow typically occurs at greater depths.

The Significance of Composition and Crust Thickness

The difference in the hypothetical strengths of average oceanic and continental lithosphere relates to compositional differences and the greater thickness of continental crust. These are the key observations:

- Knowledge of mantle rocks from peridotite fragments ejected from volcanoes and details of igneous rocks formed by magmas extracted as partial melts from the mantle show that oceanic mantle in the lithosphere is probably dry whereas continental lithospheric mantle contains water and water-bearing minerals. Wetter mantle beneath continental crust is weaker than the drier mantle beneath oceanic crust because water lowers rock strength (see Figure 11.20).

- Continental crust is thicker than oceanic crust and experiences higher temperatures that reduce rock strength and enhance the likelihood of plastic flow. Lower continental crust is weak because of the high temperatures at depth whereas thin oceanic crust rocks rarely experience such high temperatures.

- Wet continental crust is weaker than dry oceanic lithosphere mantle. Continental crust contains water and water-bearing minerals. The wet crust is very weak below about 15 kilometers in contrast to the dry mantle rocks composing oceanic lithosphere at the same depth. Continental lithosphere becomes progressively weaker below about 12 to 15 kilometers, whereas oceanic lithosphere becomes stronger at depth until plastic flow strength decreases sharply below 40 kilometers.

11.7 How Do Earthquakes Relate to Rock Deformation?

Although rock, sand, and wax deformation take place rapidly in laboratory experiments, most natural rock deformation occurs slowly. Some motion along faults, however, is very rapid and causes damaging, and often deadly, earthquakes. This hazardous aspect of rock deformation represents a major area of geoscience research.

Earthquakes resemble the laboratory fracture of rock illustrated in Figure 11.17 (rock 2). Strong rocks absorb large stresses during elastic strain. That stored stress is a form of potential energy that instantly releases, mostly as motion energy, when the stress finally exceeds the rock yield strength and the rock breaks. An analogy is the compression and release of a spring; the stress applied to compress the spring is stored energy that becomes rapid motion energy when the spring is released. Earthquakes occur when the stress exceeds the rock strength, causing the rock to fracture and releasing the stored-up energy as seismic waves (also see Section 8.2).

How Earthquakes Relate to Faults

Historic observations demonstrate that earthquakes result from movement along faults. **Figure 11.27**, for example, shows how earthquake foci and epicenters line up along recognizable faults and **Figure 11.28** illustrates examples of observed displacement along faults during historic earthquakes. Movement along a dip-slip fault causes an abrupt vertical displacement of the ground surface called a **fault scarp**.

▼ **Figure 11.27 Earthquakes occur by movement along faults.** The map and cross section show locations of earthquakes near the central California coast during October 1989. The larger circle marks the location of the Loma Prieta earthquake, which caused substantial damage in Santa Cruz and the San Francisco Bay area. Notice that most of the earthquakes coincide closely with the location of the San Andreas fault.

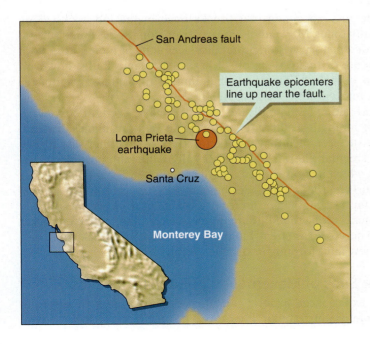

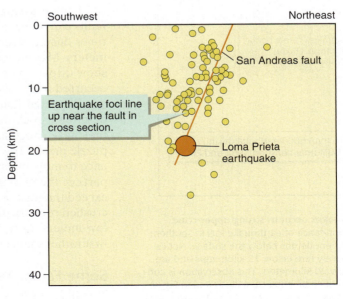

▶ **Figure 11.28 Faults move during earthquakes.**
These photographs, taken right after powerful earthquakes, document that blocks of crust move along faults during earthquakes.

1-meter offset of highway pavement in southern California occurred during movement on strike-slip fault in 1992.

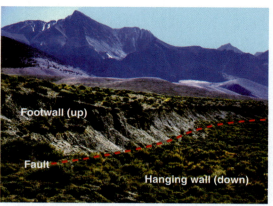

3-meter high scarp in central Idaho occurred during movement on normal dip-slip fault in 1983.

Millions of faults criss-cross Earth's crust and mark planes of brittle rock failure. Some faults, such as the famous San Andreas fault in California, coincide with tectonic plate boundaries. Most faults are relatively insignificant producers of modern earthquakes—they are only a few kilometers long, or have experienced very small amounts of slip, or do not show evidence of displacement in the recent geologic past. Not all faults are exposed at Earth's surface. Instead, recently deposited sediment or folded rock layers conceal many deeper faults until earthquakes reveal their locations.

Earthquakes result from brittle failure along faults, so you can predict that earthquake foci should occur at relatively shallow depths, where temperature and pressure are low. An example graph in **Figure 11.29** shows that earthquakes in continents overwhelmingly occur at depths less than 15 kilometers. The shallow occurrence of earthquakes is consistent with the lithosphere-strength curve in Figure 11.26, which predicts a dramatic decrease in continental crust strength below approximately 15 kilometers. Deformation at greater depth should be plastic flow rather than brittle fracturing that produces faults.

This is not to say that earthquakes do not occur at greater depths. Rare earthquake foci in continental crust are as deep as 70 to 90 kilometers under Tibet, where old, cold crust remains strong. Earthquakes deeper than 50 kilometers, and as deep as 640 kilometers, are restricted to narrow zones near convergent plate boundaries, and Chapter 12 examines clues for the origin of these very deep earthquakes.

Unlike a single rock specimen that breaks once in the lab, rocks along faults can move many times. Movement has occurred along some faults several times during recorded history yielding total displacements of meters to tens of meters. Measurements of rock layers offset across the same faults, however, show displacements of many kilometers, which implies that many earthquakes occurred over long intervals of geologic time.

Repeated failure along faults indicates that they have strength, just as we describe a rock has strength. Faults are not perfectly smooth planes. Instead, they have irregular surfaces such that the fractured rocks on either side fit like puzzle pieces. The irregular interlocking surfaces along the fault produce resistance to movement, which increases the strength of the fault. Even on smoother parts of the fault surface, friction impedes movement. The tectonic stress must exceed the rock or fault strength in order for displacement to occur. Interlocked crushed rock fragments along the fault are much weaker than unbroken rock a few meters away. Therefore, as stress builds up, the relatively weak fault fails first rather than a new fault forming in stronger, unbroken rock.

Some Faults Are Sites of Movement without Earthquakes

Not all faults have experienced historic earthquakes, and yet historic displacement is readily visible, as illustrated in **Figure 11.30**. Some faults are too weak to

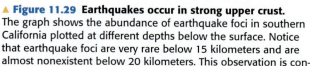

▲ **Figure 11.29 Earthquakes occur in strong upper crust.**
The graph shows the abundance of earthquake foci in southern California plotted at different depths below the surface. Notice that earthquake foci are very rare below 15 kilometers and are almost nonexistent below 20 kilometers. This observation is consistent with the strength of continental crust illustrated in Figure 11.26; below 20 kilometers average continental crust is too weak to fail by brittle fracture along faults.

store up the energy of applied stress. As a result, rocks creep past one another on either side of the fault without generating earthquakes.

Elastic Rebound Theory Explains Building Stress along Faults

The **elastic rebound theory**, illustrated in **Figure 11.31**, explains earthquakes as sudden brittle failure following the buildup of elastic strain. The concept compares well with the representation of rock A on the stress-strain graph in Figure 11.18, where some degree of elastic strain occurs before sudden brittle fracture. Rocks on either side of a locked-up fault slowly bend as stress builds up. When the stress eventually exceeds the strength of the fault, the bent rocks suddenly snap back to their prestressed orientation, with an abrupt displacement along the fault. The bending of the rock is nonpermanent elastic deformation, which is recovered, or rebounds, after the fault breaks; hence the name "elastic rebound." The ability of the rock to bend and store energy is much like the illustrated bending of plywood prior to breaking (Figure 11.23), or the stretching of a rubber band until it breaks.

Careful surveying of locations on either side of faults before and after earthquakes provides evidence supporting the elastic rebound theory. Example data are mapped in Figure 11.31 and show elastic strain along the Hayward fault in California. Almost no displacement occurs near the locked strike-slip fault. Instead, the shear stress bends the rock on either side of the fault. The bending is elastic strain in response to stress along the fault that will eventually reach the yield strength of the fault and result in an earthquake. Unfortunately, it is difficult to determine the threshold of stress accumulation that is required for the fault to rupture, or else specific earthquake predictions might be possible.

▲ **Figure 11.30 Creeping deformation along a fault.** Earthquakes do not accompany all movements along faults. This photo shows right-lateral strike-slip offset of a sidewalk in Hollister, California. This displacement did not occur during earthquakes but, instead, takes place by persistent, slow, creeping movement along the fault.

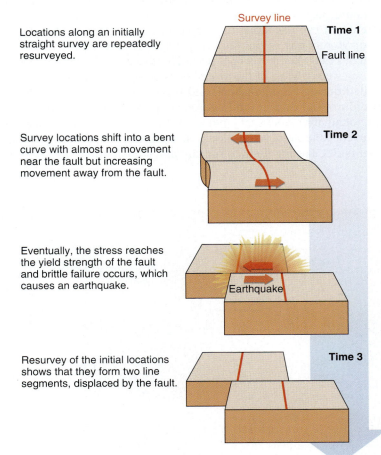

Locations along an initially straight survey are repeatedly resurveyed.

Survey locations shift into a bent curve with almost no movement near the fault but increasing movement away from the fault.

Eventually, the stress reaches the yield strength of the fault and brittle failure occurs, which causes an earthquake.

Resurvey of the initial locations shows that they form two line segments, displaced by the fault.

Survey line
Time 1
Fault line
Time 2
Earthquake
Time 3

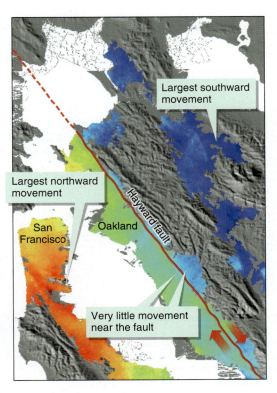

Largest southward movement

Largest northward movement

San Francisco

Oakland

Hayward fault

Very little movement near the fault

Detected motion, (mm/yr)

−6   −4   −2   0   2   4
South   North

◄ **Figure 11.31 Visualizing elastic rebound theory.** The blocks schematically illustrate surveyed locations that shift because of elastic deformation in blocks of rock on either side of the fault. The bending continues until the stress exceeds the yield strength of the fault, at which time the fault breaks during an earthquake. Rebound of elastic strain returns the survey locations to a straight line that is offset by the fault. The map uses colors to show surface motion detected by satellite radar on either side of the Hayward fault in California. The pattern of displacement indicates elastic strain across the fault.

Active Art

Elastic Rebound: *See how elastic strain deforms rocks along a fault before an earthquake and what happens during an earthquake.*

Foreshocks, Mainshock, Aftershocks

All the stored energy in the bent rocks rarely releases in a single, large earthquake. *Most* of the energy releases during a single earthquake, called the **mainshock**, but there are smaller **foreshocks** and **aftershocks** that precede and follow the mainshock. Foreshocks represent small displacements of the weaker parts of the fault surface as stress rises, and aftershocks are breaks along small areas of the strongest part of the fault surface, whereas the mainshock records breakage of most of the fault surface. Returning to the stressed plywood analogy (Figure 11.23), cracking sounds heard before the board breaks are comparable to foreshocks. After the board breaks, representing the mainshock, there may be a few splinters that still connect the pieces until you bend the board further. Breaking these last connecting splinters is analogous to aftershocks that complete the slip along the fault plane.

Small earthquakes are usually recognized as foreshocks only after the mainshock occurs, so foreshocks are rarely useful in forecasting a major earthquake. After all, it is difficult to know if a small earthquake is a foreshock preceding a major earthquake or simply a minor solo event.

Hundreds of aftershocks commonly follow large earthquakes and may persist for weeks to months. Most of the aftershocks are relatively small, but some are large enough to cause collapse of buildings damaged by the mainshock. Aftershocks, therefore, greatly hamper relief and rescue efforts following earthquake disasters.

Figure 11.32 illustrates foreshocks and aftershocks related to a great earthquake in central Alaska in 2002. The modest Nenana Mountain earthquake and related aftershocks resulted from brittle failure along part of the Denali strike-slip fault. These earthquakes were apparently foreshocks to the larger Denali earthquake, which happened about a week after the Nenana Mountain events. The Denali fault earthquake started with failure at the focus along the Susitna Glacier thrust fault, and displacement continued eastward and southeastward to cause 200 kilometers of rupture along two strike-slip faults. The earthquake mainshock lasted for 90 seconds as the fault displacement cracked across the landscape at a speed of 11,000 kilometers per hour. Hundreds of aftershocks occurred in the following weeks. Most of the aftershocks occurred along the same three faults that failed during the mainshock and represent small areas of the fault surfaces that did not fail during the larger earthquake. It is important to note, however, that not all aftershocks take place on the same fault plane where the mainshock occurred. Release of stress along the mainshock fault may transfer stress to adjacent faults, causing them to rupture and contribute to the aftershock pattern.

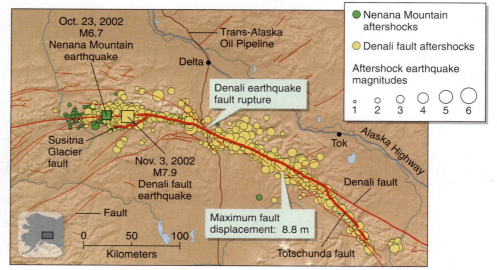

▶ **Figure 11.32 Foreshocks and aftershocks show rupture along faults.**
This map plots earthquake epicenters in central Alaska in fall 2002. The Nenana Mountain earthquake and its aftershocks were possibly foreshocks for the larger Denali fault earthquake. Movement occurred on all three labeled faults during the 90-second Denali fault earthquake. Small areas of these fault planes that did not break during the Denali fault earthquake ruptured over the next few weeks, producing the plotted aftershocks. Some aftershocks also occurred on nearby faults where stress transferred after the big earthquake. The Denali fault rupture passed beneath the Trans-Alaska Oil Pipeline but did not break the pipeline.

11.8 How Are Earthquakes Measured?

News reports of earthquakes always include the earthquake magnitude. The magnitude is a way of communicating the size of an earthquake and the amount of strain energy released. Early efforts to measure earthquake size relied on describing earthquake effects. Geologists now measure earthquake size based on measurement of earthquake waves recorded on seismometers (go to Figure 8.8 to see a seismometer).

Measuring Intensity

Intensity measures the violence of an earthquake in terms of the extent to which people felt the earthquake, the damage to structures, and secondary effects such as landslides. Giuseppi Mercalli developed the first intensity scale in 1902, and the updated version is the Modified Mercalli Intensity Scale. **Table 11.2** describes the scale, which assigns the typical observed effects of earthquakes Roman numerals from I to XII. The effects of an earthquake vary from place to place, so the intensity of the earthquake also differs at different locations and the earthquake may have more than one intensity value.

Figure 11.33 illustrates Modified Mercalli Intensity maps for several large U.S. earthquakes. To make these maps, geologists examine the damage done by earthquakes in different locations and interview people to determine how severe the earthquake shaking was at these locations. Historic accounts in newspapers and diaries provide the information to determine the effects for earthquakes that happened prior to the development of the intensity scale. The appropriate Roman numeral is placed on a map at each location to indicate the intensity of the earthquake at that point. Then, lines are drawn to show areas of equal intensity similar to contour lines on topographic maps, which show lines of equal elevation. The maximum intensities, indicated by the

TABLE 11.2	Modified Mercalli Intensity Scale
Intensity	Abbreviated Description of Effects
I	Not felt except by a very few under especially favorable conditions.
II	Felt only by a few persons at rest, especially on upper floors of buildings.
III	Felt quite noticeably by persons indoors, especially on upper floors of buildings. Many people do not recognize it as an earthquake. Standing motor cars may rock slightly. Vibrations similar to the passing of a truck. Duration estimated.
IV	Felt indoors by many, outdoors by few during the day. At night, some awakened. Dishes, windows, doors disturbed; walls make cracking sound. Sensation like heavy truck striking building. Standing motor cars rocked noticeably.
V	Felt by nearly everyone; many awakened. Some dishes, windows broken. Unstable objects overturned. Pendulum clocks may stop.
VI	Felt by all, many frightened. Some heavy furniture moved; a few instances of fallen plaster. Damage slight.
VII	Damage negligible in buildings of good design and construction; slight to moderate in well-built ordinary structures; considerable damage in poorly built or badly designed structures; some chimneys broken.
VIII	Damage slight in specially designed structures; considerable damage in ordinary substantial buildings with partial collapse. Damage great in poorly built structures. Fall of chimneys, factory stacks, columns, monuments, walls. Heavy furniture overturned.
IX	Damage considerable in specially designed structures; well-designed frame structures thrown out of plumb. Damage great in substantial buildings, with partial collapse. Buildings shifted off foundations.
X	Some well-built wooden structures destroyed; most masonry and frame structures destroyed with foundations. Rails bent.
XI	Few, if any (masonry) structures remain standing. Bridges destroyed. Rails bent greatly.
XII	Damage total. Lines of sight and level are distorted. Objects thrown into the air.

(*Source*: USGS–http://neic.usgs.gov/neis/general/mercalli.html)

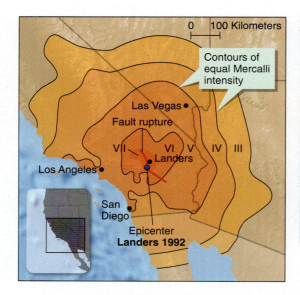

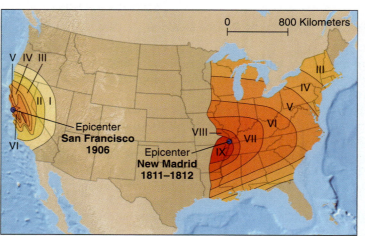

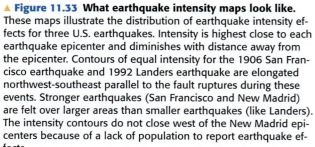

▲ **Figure 11.33 What earthquake intensity maps look like.** These maps illustrate the distribution of earthquake intensity effects for three U.S. earthquakes. Intensity is highest close to each earthquake epicenter and diminishes with distance away from the epicenter. Contours of equal intensity for the 1906 San Francisco earthquake and 1992 Landers earthquake are elongated northwest-southeast parallel to the fault ruptures during these events. Stronger earthquakes (San Francisco and New Madrid) are felt over larger areas than smaller earthquakes (like Landers). The intensity contours do not close west of the New Madrid epicenters because of a lack of population to report earthquake effects.

highest Roman numerals are located near the epicenters of the earthquakes depicted in Figure 11.33. Notice that the numbers generally decrease away from the epicenter. Before the development of seismographs, intensity maps were commonly used to approximately locate earthquake epicenters. The Modified Mercalli Intensity map of the New Madrid, Missouri, earthquakes shows how much damage can result at great distances from the epicenter. The largest New Madrid earthquake was felt 2000 kilometers away, and the maximum intensity reached Roman numeral XII. Damage was reported over a 500,000-square-kilometer area, which is larger than the area of California.

The Modified Mercalli Intensity Scale contains numerical values but is still really a qualitative measure of an earthquake. It is challenging to apply this scale when earthquakes take place in unpopulated areas with few reports and a lack of buildings to record the intensity of the shaking. Nonetheless, maps like these shown in Figure 11.33 are useful to emergency management agencies and relief organizations because they provide an indication of the areas potentially affected by large earthquakes in the future.

Measuring Magnitude

Magnitude is a measure of earthquake size that corresponds to the energy released during the earthquake. It is not a directly measured physical quantity, such as temperature, but rather a value calculated from measurements made after an earthquake. Some magnitude calculations rely on measurement of the height of seismic waves (also known as the amplitude) on seismograms. Different magnitude scales use the amplitude of primary (P), secondary (S), or surface waves (see Section 8.2 for a refresher on types of earthquake waves).

Charles Richter developed the first magnitude scale in 1935, and **Figure 11.34** shows how to calculate the Richter magnitude. To calculate the Richter magnitude, seismologists measure the maximum amplitude recorded on a specific type of seismometer. The wave amplitude is greater on seismometers closer to the earthquake and smaller at longer distances between the epicenter and the seismometer. To adjust for the different wave amplitudes at different distances, the magnitude calculation also uses the time elapsed between the arrival of P and S waves at the seismic station. If necessary, look back at Section 8.2 and Figure 8.9 to refresh your memory of how this time elapsed on the seismogram relates to the distance between the earthquake focus and the seismic station. For every whole-number increase on the magnitude scale, the measured seismic-wave amplitude increases tenfold (Figure 11.34); therefore, a magnitude 5 (M5) earthquake has a wave amplitude ten times larger than a M4, and 100 times larger than a M3 earthquake.

Unlike intensity, which varies from place to place for a single earthquake, there is only a single magnitude assigned to an earthquake, based on any particular magnitude scale. The news media commonly report magnitude of a newsworthy earthquake and may refer to this value as the "Richter magnitude," even though seismologists more frequently apply different, more recently developed methods to determine magnitude.

Magnitude scales based on seismograms assume that an earthquake occurs at one point and releases all of its energy instantaneously. This approach is contrary to observations such as those recorded for the Denali fault earthquake in Figure 11.32. Single earthquakes can rupture long faults over periods of many seconds to a few minutes.

The problem of large earthquakes not fitting the assumption of the Richter scale is avoided by calculating the "moment magnitude," which is a better measure of the actual energy released. Moment magnitudes do not depend on measuring the response of a seismometer to an earthquake but instead rely on calculation of a physical parameter known as seismic moment. The **seismic moment** is calculated from measurements of the area of the fault plane that ruptured, how far the rocks moved on either side of the fault, and the rock strength. In three dimensions, the earthquake results from a rupture of a fault plane, not a line or a point. Field measurements reveal the length of the fault and the amount of displacement, whereas the mainshock and aftershock foci reveal how deeply the fault ruptured. The moment magnitude measures

Earthquake A

Largest surface wave

Amplitude of largest wave is 8 mm

Time between arrival of first P and first S wave is 6 seconds.

Plot time between P and S arrivals on this line.

Plot amplitude of largest surface wave on this line.

M = 3.5

M=3.5

Connect points and read magnitude off of this line.

Magnitude (M)

Amplitude (millimeters)

Distance (kilometers)   S-P (seconds)

Earthquake B

Largest surface wave

Amplitude of largest wave is 80 mm

Time between arrival of first P and first S wave is 6 seconds.

M = 4.5

Magnitude (M)

Amplitude (millimeters)

Distance (kilometers)   S-P (seconds)

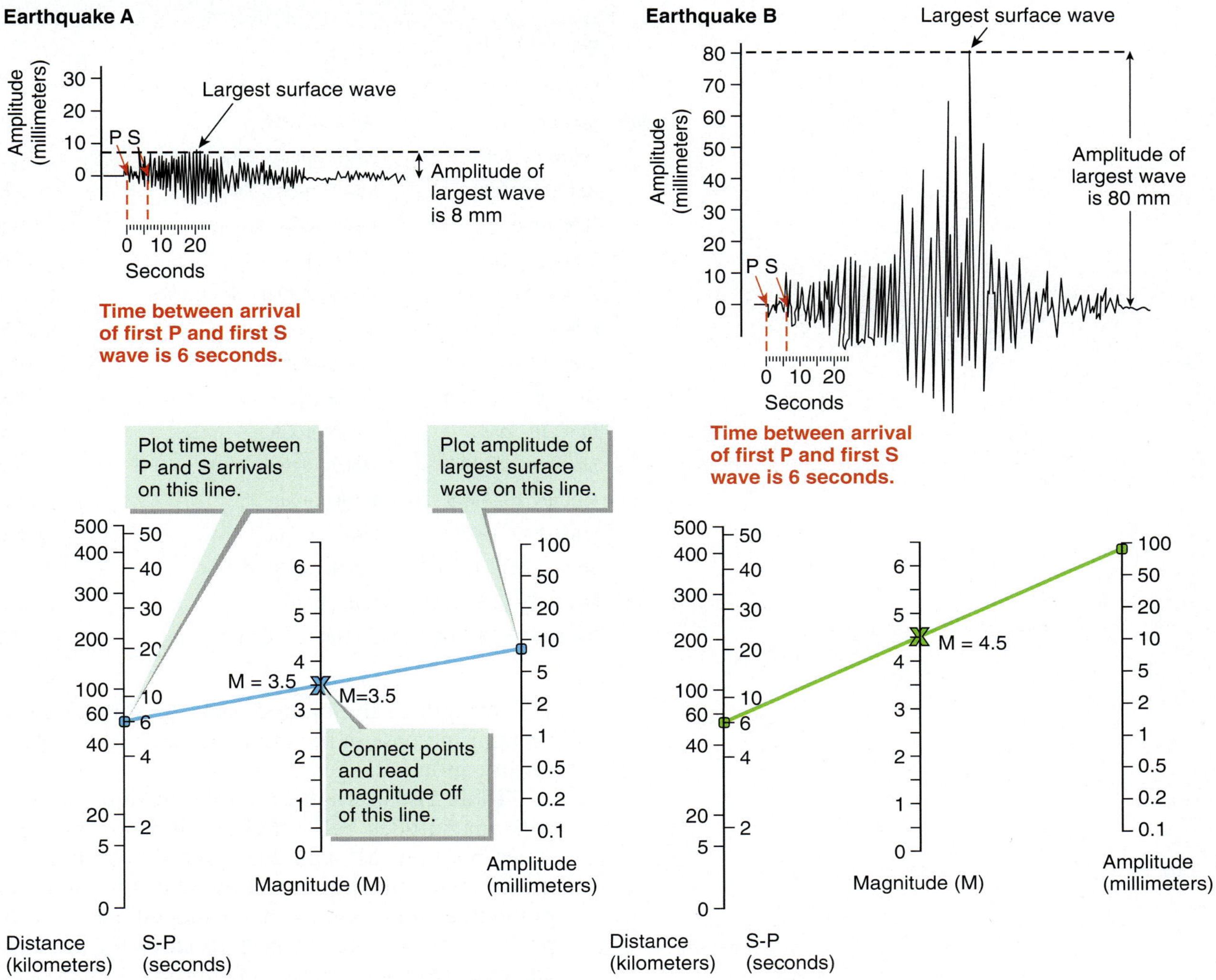

▲ **Figure 11.34 How to determine Richter magnitude.**
This diagram shows the graphical calculation of Richter magnitudes for two earthquakes. The magnitude is determined by making two measurements on the seismogram: (1) the time that passes between arrival of P and S waves (which relates to the distance of the seismic station from focus), and (2) the amplitude, or height, of the largest recorded surface wave. Drawing a line between plots of these two measurement points on the special graph intersects the magnitude scale. Notice that the distance from the seismic station to the focus is the same for the two earthquakes. Earthquake B produced surface waves that were ten times higher than those produced in earthquake A, but the earthquake magnitudes differ by only one unit (3.5 and 4.5).

the energy radiated from the entire plane on which there was movement over the time required for all of the rupture to take place. By contrast, magnitudes based on seismograms underestimate the energy release because the energy is spread out over all of the waves generated during the duration of the earthquake so that no single wave height really represents the energy of the whole event. The seismic moment calculation, schematically shown in **Figure 11.35**, involves multiplying together the area of the fault surface, the average displacement that took place, and

▶ **Figure 11.35 Determining seismic moment.**
The moment-magnitude scale uses the seismic moment to describe the size of an earthquake. This diagram shows how the seismic moment is calculated from measuring (1) the strength of the rocks offset along the fault, (2) the area of the fault plane along which movement occurred during the earthquake, and (3) the average amount of slip that occurred on the fault.

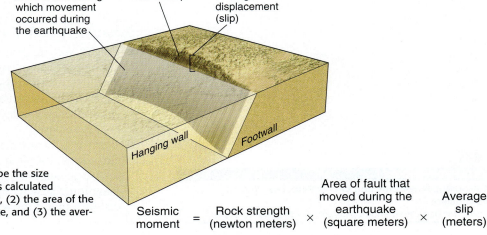

Area of fault along which movement occurred during the earthquake

Fault scarp

Average displacement (slip)

Hanging wall

Footwall

$$\text{Seismic moment} = \text{Rock strength (newton meters)} \times \text{Area of fault that moved during the earthquake (square meters)} \times \text{Average slip (meters)}$$

TABLE 11.3	Some Notable Earthquakes		
Date	Place	Moment Magnitude	Fatalities
December 16, 1811	New Madrid, MO	8.1	
January 23, 1812	New Madrid, MO	7.8	
February 7, 1812	New Madrid, MO	8.0	
April 18, 1906	San Francisco, CA	7.8	700–800
November 4, 1952	Kamchatka, Russia	9.0	
May 22, 1960	Chile	9.5	5,700
March 28, 1964	Prince William Sound, AK	9.2	125
July 27, 1976	Tangshan, China	7.5	255,000–655,000
December 7, 1988	Apitak, Armenia	6.8	25,000
October 17, 1989	Loma Prieta, CA	6.9	63
June 28, 1992	Landers, CA	7.3	1
January 17, 1994	Northridge, CA	6.7	57
January 16, 1995	Kobe, Japan	6.9	5,502
August 17, 1999	Izmit, Turkey	7.6	15,637
September 21, 1999	Taicheng, Taiwan	7.8	2,100
December 26, 2003	Bam, Iran	6.6	>30,000
December 26, 2004	Sumatra, Indonesia	9.0	>150,000

the strength of the faulted rock. Analogous to the Richter scale, each whole-number increase in the moment magnitude represents a tenfold increase in the seismic moment.

Table 11.3 lists some notable earthquakes within historic time. Earthquakes with magnitudes below M4 rarely cause significant damage, whereas magnitudes between M5 and M7 typically represent large earthquakes, and earthquakes with magnitudes above M8 are catastrophic if the epicenter is in a populated area. Although no magnitude scale has an upper limit, rocks are probably not strong enough to store up the elastic strain energy to result in earthquakes with moment magnitudes larger than M10.

The magnitude scales are numerical values that give an idea of the size of an earthquake, but how do you compare a M5 earthquake to a M7 earthquake? The M7 quake is 100 times larger in seismic moment than the M5 event, but how much more energetic is the M7 earthquake? For every one unit of magnitude increase, the energy increases 32-fold. This means that an M7 earthquake is 32 times more energetic than a M6 event and 1024 times more energetic than a M5 earthquake (32 times 32 equals 1024). **Figure 11.36** graphically illustrates the energy released for earthquakes of different magnitudes and compares this to the energy of explosives and to events other than earthquakes.

Putting It Together–*How Are Earthquakes Measured?*

■ Earthquake intensity, or violence, is estimated from assessment of damage and eyewitness accounts. Many values of intensity are recorded for a single earthquake, and values are usually highest closest to the epicenter.

■ Earthquake magnitude relates to the energy released by an earthquake. In general, each whole-number increase in magnitude is a 32-fold increase in energy released.

■ There is only one magnitude value, on a particular magnitude scale, for an earthquake. Some magnitude scales rely on the response of seismometers to the passage of earthquake waves. The more recently used moment-magnitude scale is based on the area of the fault plane that ruptured, the distance that it moved, and the strength of the rock.

EXTENSION MODULE 11.1

Calculating Magnitude and Energy Released from an Earthquake.
Learn the basic mathematics behind the magnitude scales.

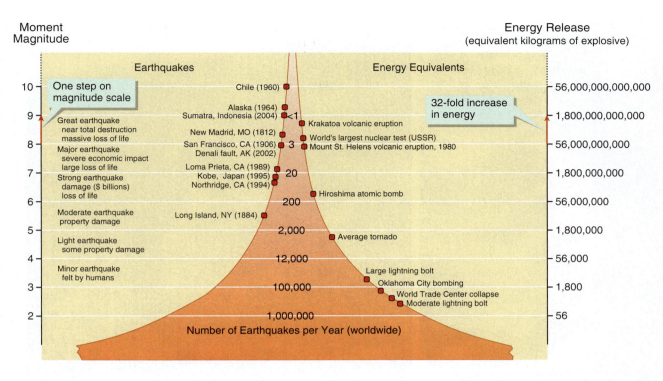

◀ **Figure 11.36** How much energy is released by an earthquake?
This diagram compares the energy released by an earthquake to the earthquake magnitude, and to the energy released by other natural and human-caused phenomena. The chart also indicates how many earthquakes of a given magnitude occur each year around the world. For instance, the 2002 Denali fault, Alaska, earthquake had a magnitude of 7.9, and about three earthquakes of this magnitude occur each year. This earthquake released energy equivalent to detonating 56 billion kilograms of explosive and about 1000 times the energy released by the atomic bomb that destroyed Hiroshima, Japan, in 1945.

11.9 Why Are Earthquakes Destructive?

The destructive effects of earthquakes are not limited to the ground rupture along the fault. Energy released by large earthquakes has far-reaching effects that can lead to catastrophic losses in urban areas, as shown in **Figure 11.37**.

Ground Shaking

The passage of surface waves radiating outward from the epicenter causes the ground to shake up and down and side to side (see Section 8.2). This ground motion exerts stress on buildings, bridges, dams, and other structures. The stress may exceed the strength of the beams or masonry, causing them to break and crumble. In some cases, pillars between floors of multistory buildings fail, causing the floors to crash down on one another (Figure 11.37c). Buildings sway back and forth and hit each other as the earthquake waves pass along the surface. Ground shaking also ruptures natural-gas lines, triggering fires that commonly destroy more property than the ground shaking itself.

Ground shaking also triggers mass movements. Figure 11.37b illustrates a landslide shaken loose by a M7.6 earthquake in El Salvador in 2001. Surface waves stressed the volcanic rocks on the hillside beyond their strength limit. The landslide buried 268 homes and killed 700 people.

Ground-shaking damage varies from locality to locality because the amplitude of seismic surface waves depends not only on proximity to the epicenter but also on the characteristics of the materials that the waves pass through. Surface-wave amplitude is greater in loose sediment than in hard rock.

Figure 11.38 illustrates how geological materials partly determine the effects of an earthquake. The intensity map for the 1989 Loma Prieta, California, earthquake reveals that the highest intensity (IX) was experienced far from the epicenter. This area of intense, deadly destruction included the collapse of a double-deck freeway bridge in Oakland, 100 kilometers from the epicenter. The segment of the bridge that collapsed was built on mud. Measurement of surface-wave amplitudes in mud, stream sand and gravel, and rock during aftershocks shows that the three materials respond very differently to the passing seismic waves. The higher-amplitude ground motion, where the surface waves

(a) Bridge collapse. Kobe, Japan, 1995. Moment magnitude 6.9.

(b) Earthquake-triggered landslide. Santa Tecla, El Salvador, 2001. Moment magnitude 7.6.

(c) Building collapse. Yulin, Taiwan, 1999. Moment magnitude 7.8.

(d) Ground-shaking damage, University of California-Northridge bookstore, 1994. Moment magnitude 6.7.

▲ **Figure 11.37 How earthquakes cause damage.**
Earthquake surface waves cause large vertical and horizontal shaking motions that topple buildings and bridges, trigger landslides, and throw objects from shelves.

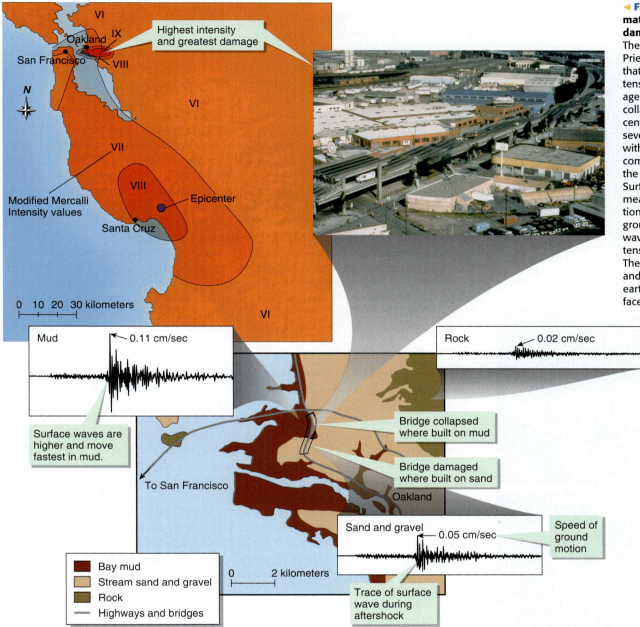

◄ **Figure 11.38 How geologic materials determine earthquake damage.**
The intensity map of the 1989 Loma Prieta, California, earthquake shows that the greatest ground shaking (intensity IX), where the deadliest damage involved a highway bridge collapse, occurred far from the epicenter. The geologic map shows that severity of bridge damage coincided with different geologic materials, with complete collapse occurring where the bridge was built on bay mud. Surface-wave heights and speeds measured at three nearby seismic stations during an aftershock show that ground shaking is most intense where waves pass through mud and least intense where waves pass through rock. The high-intensity ground shaking and great damage during the main earthquake are linked to the way surface waves pass through mud.

passed through the mud, caused more intense shaking and failure of the bridge supports.

Where ground water is close to the surface, ground shaking causes loose sediment to behave like liquid. **Liquefaction** occurs in water-saturated sediment when grains settle during ground shaking, which displaces the water upward in the intervening pore spaces such that the water pressure moves the grains apart and the whole sediment-water mixture loses strength. The wet ground turns to a fluid slurry as the surface waves pass and then settles unevenly, resulting in severe damage to buildings and other structures. Liquefaction of wet mud contributed to the bridge failure illustrated in Figure 11.38.

Tsunami

Tsunami, or seismic sea waves, caused by sudden vertical displacement of the seafloor, produce significant damage, as shown in **Figure 11.39**. These waves are sometimes erroneously called "tidal waves," although they are not caused by

Liquefaction: *See how earthquake ground shaking causes liquefaction.*

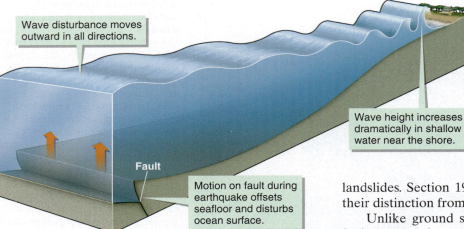

Wave disturbance moves outward in all directions.

Wave height increases dramatically in shallow water near the shore.

Fault

Motion on fault during earthquake offsets seafloor and disturbs ocean surface.

Motion on a submarine fault displaces the ocean floor and the ocean surface to produce a tsunami. Wave energy is confined into a progressively smaller space as the wave moves into shallow water near shore, which causes the wave to build up to a height that commonly exceeds 10 meters.

Brown, sediment-laden water surges inland more than 500 meters.

Tsunami ravaged Indian Ocean shorelines following a magnitude 9.0 earthquake near Sumatra in December 2004. A satellite image shows inundation of coastal Sri Lanka, more than 1600 km from the epicenter, about one hour after the first wave arrived. The waves carried boats inland over two-story buildings.

▲ **Figure 11.39** **Why tsunami are destructive.**

Tsunami: *See how an earthquake causes a tsunami.*

the gravitational pull of the Sun and Moon. Fault displacement of the seafloor during an earthquake also disturbs the ocean surface. The resulting water wave is usually less than a meter high and rarely noticed in the open sea, but the wave height increases dramatically to heights exceeding 10 meters in shallow water near land (Figure 11.39). "Tsunami" (spelled the same singular and plural) comes from a Japanese term meaning "harbor wave" and refers to the sudden buildup of high waves in shallow water. Tsunami form not only during earthquakes, but also as a result of underwater volcanic eruptions and submarine landslides. Section 19.2 presents additional insights to tsunami dynamics and their distinction from normal ocean waves.

Unlike ground shaking that affects a relatively small region around the fault-rupture site, catastrophically damaging tsunami may strike coastal areas where the earthquake was not even felt, sometimes more than 1000 kilometers from the earthquake epicenter. The Hawaiian Islands, in the middle of the Pacific Ocean, have been severely damaged by tsunami originating from distant subduction-zone earthquakes along continental margins in Alaska and South America.

Risk Assessment Tools

Geologists apply research on earthquakes and their effects to help save lives and property. Seismic-hazard mapping involves examining the geology of the area, mapping locations of the potentially active faults, and reviewing any past earthquake activity along the faults. The maps highlight the locations of geologic materials that are especially susceptible to failure by intense ground shaking or liquefaction. This information is used to map out the earthquake risk, which is the probable building damage and number of expected casualties based on the population in particular areas. Risk maps generally focus on the likely maximum intensity of ground shaking.

Figure 11.40 is a seismic-risk map of potential ground shaking for the continental United States. Notice that the higher risk zones are near active plate boundaries along the west coast of the United States (compare to Figure 1.10). In addition, there are seismically active regions in the continental interior, such as the New Madrid, Missouri, seismic zone, where large earthquakes have occurred during historic time (Figure 11.33).

Geologists also routinely prepare risk maps at the scale of states, counties, and cities. Land-use planners consult these maps when making land-use decisions, such as placing golf courses and parks, rather than dense concentrations of buildings, along faults and in areas subject to intense ground shaking. The maps also guide building codes and engineering designs, so that buildings, bridges, and dams are constructed to withstand expected ground shaking and liquefaction.

Earthquake Prediction

Can earthquakes be predicted? At present, seismologists cannot reliably predict the specific time, place, and magnitude of an earthquake. In general terms, seismologists can anticipate that there is a 67 percent chance that there will be a major (M > 6.7) earthquake in the San Francisco Bay area within the next 30

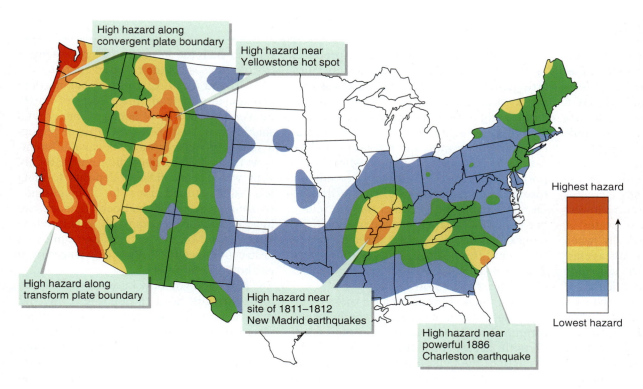

◄ **Figure 11.40 Seismic hazard map.**
This seismic hazard map of the contiguous United States uses different colors to represent the expected hazard of damage by ground shaking during an earthquake. Hazards are greatest where earthquakes frequently occur near plate boundaries along the west coast and near a hot spot below Yellowstone, Wyoming. Hazards are also somewhat higher than average near sites of powerful historic earthquakes in Missouri and South Carolina.

High hazard along convergent plate boundary

High hazard near Yellowstone hot spot

High hazard along transform plate boundary

High hazard near site of 1811–1812 New Madrid earthquakes

High hazard near powerful 1886 Charleston earthquake

Highest hazard

Lowest hazard

years, and a 60 percent chance in the Los Angeles area. These general forecasts are based on the frequency of earthquakes in the past and the measured buildup of strain in the rocks.

General forecast of a big earthquake over a 30-year period may not seem very useful to you. More specific prediction approaches are being studied. Most of these studies focus on detecting phenomena that reliably occur before a large earthquake. The establishment of reliable precursors has been elusive, and different precursor phenomena are likely to be associated with different faults, so knowledge gained in one location may not apply in another. Nevertheless, seismologists keep investigating, seeking ever more information that could help develop models for prediction.

EXTENSION MODULE 11.2

Mitigating and Forecasting Earthquake Hazards.
Learn how ground shaking, liquefaction, tsunami, landslides, and fault rupture hazards.

Putting It Together–*Why Are Earthquakes Destructive?*

■ Earthquakes cause damage by fault rupture, ground shaking, landslides, and tsunami.

■ Seismic surface waves cause ground shaking. The amplitude of the waves, and hence the damage that results, varies depending on the intensity of the earthquake and the nature of geologic materials that the waves travel through.

■ Earthquake rupture of the seafloor generates tsunami that cause destruction both close to the epicenter and as far as thousands of kilometers away.

■ Predictions for specific earthquakes are still not possible, but risk maps identify hazards for consideration in land-use planning, building codes, and engineering designs that diminish the destructive effects of earthquakes.

Where Are You and Where Are You Going?

You now know how Earth's crust deforms and how to recognize the deformation. Deformation, or strain, results from stresses that break or otherwise change rock. Geologic structures, such as folds and faults, are examples of strain resulting from tectonic stresses that exceeded the rock strength.

Rock strength, in turn, depends on temperature, pressure, and composition of the rock. Increasing pressure generally increases strength, but increasing temperature lowers strength, as does the addition of water. A rock can initially behave elastically when stressed, absorbing the applied stress as elastic strain energy and returning to its original shape and size once the stress releases. If stress increases instead of releasing, the rock undergoes permanent deformation when stress exceeds the yield strength. This deformation is brittle breakage or plastic flow. Brittle failure occurs near the surface at low temperature and pressure and produces faults and joints. Plastic deformation typically happens at higher-temperature and higher-pressure conditions deep below the surface and causes metamorphic rocks to fold and flow like viscous fluids.

Folding commonly affects sedimentary rocks close to the surface. Sedimentary rocks bend and slip past each other along bedding planes to a substantial extent before breaking along faults.

Oceanic lithosphere is stronger than continental lithosphere. The lower continental crust is probably weaker than both the overlying upper crust and the lithospheric mantle below, so that the lower crust is analogous to the soft middle of a jelly sandwich.

Deformation of Earth's crust produces both hazards and benefits for society. Hazards include destructive earthquakes as rock breaks and moves when stress exceeds the rock strength. Benefits include mineral and energy deposits. Structurally deformed areas commonly determine the formation and locations of economic deposits. Mineral deposits rich in metallic elements commonly form in fractured rocks where metal-rich fluids migrate into the fractures and precipitate ore minerals. Relatively low density oil and natural gas move upward through porous rocks until trapped below less porous layers. These traps commonly occur along the axes of anticlines and below faults.

Earthquakes occur because rocks on either side of a fault store strain energy by deforming near locked faults, and the energy abruptly releases once the stress exceeds the strength of the fault. Earthquakes can occur only in rocks that undergo brittle failure, so most foci are restricted to the upper crust, where temperature and pressure are relatively low.

Intensity of ground motion and the magnitude, or energy released, are two ways of measuring the size of an earthquake. The Modified Mercalli Intensity Scale assigns values to different locations based on severity of damage and eyewitness accounts. The magnitude is calculated either from measurement of the response of seismographs to the passage of seismic waves, or from calculations involving the area of the fault plane that ruptured, how far the rocks were offset, and the strength of the rocks. Each whole-number increase in magnitude represents a 32-fold increase in the released energy.

Ground rupture, ground shaking, and tsunami are some of the effects of earthquakes. The severity of damage from ground shaking varies with different types of geologic materials. Seismic waves behave differently as they travel through different materials, with wet, unconsolidated deposits forming less stable foundations for human-made structures than solid bedrock. Tsunami are huge ocean waves, commonly generated by fault rupture on the seafloor, that can cause destruction locally or at great distances from earthquake epicenters.

Although you now know how rocks deform, you still need to determine why this deformation occurs. The structural geology observations of faults and folds, along with locations of earthquakes, provide indications of stresses in local areas. Now, you are prepared to examine the bigger regional and global scale of stress and strain encompassed in the discipline of tectonics. What are the origins of the compressional, extensional, and shear stresses that produce the wide variety of geologic structures visible at the surface? Ultimately, convection processes you learned about in Chapter 10 produce most of these stresses within Earth. In the next chapter, you will learn how to explain global tectonic processes in terms of convective motion within Earth. You will also learn how these processes produce the most intense deformation that is concentrated along narrow zones on the planet—the plate boundaries.

 # Active Art

Folding Rock. See how rocks fold.
Fault Motions. See how motion along normal, reverse, and strike-slip faults displaces rocks.
Graphing Stress and Strain. See how to interpret graphs of stress and strain in rocks.

Elastic Rebound. See how elastic strain deforms rocks along a fault before an earthquake and what happens during an earthquake.

Liquefaction. See how earthquake ground shaking causes liquefaction.
Tsunami. See how an earthquake causes a tsunami.

Extension Modules

Extension Module 11.1: Calculating Magnitude and Energy Released from an Earthquake. Learn the basic mathematics behind the magnitude scales.

Extension Module 11.2: Mitigating and Forecasting Earthquake Hazards. Learn how ground shaking, liquefaction, tsunami, landslides, and fault rupture hazards.

Confirm Your Knowledge

1. Why is it important to study deformed rocks?

2. What are strike and dip?

3. What is the difference between a plunging fold and a nonplunging fold?

4. Draw cross sections of an eroded anticline and an eroded syncline; label the youngest and the oldest rocks in each drawing. What is the difference between an anticline and a syncline?

5. What is the difference between a fault and a joint?

6. Explain the difference between footwall and hanging wall.

7. Describe the types of dip-slip and strike-slip faults.

8. Explain the difference between stress and strain.

9. List and describe the three basic types of stress that can be applied to rocks and the resulting strains.

10. What type of deformation occurs when a rock undergoes stress less than its yield strength? What happens when stress exceeds the yield strength?

11. What factors determine rock strength?

12. For a depth of 40 km, why is the mantle beneath oceanic crust stronger than the mantle beneath continental crust?

13. Provide examples of geologic structures resulting from tension, compression and shear.

14. For each of the following geologic structures decide if the stress that caused the structure was compression, tension or shear. Normal fault, reverse fault, thrust fault, strike-slip fault, anticline fold, syncline fold.

15. Define fault scarp. What do scarps reveal about fault motion in the past?

16. Explain surveying data that support the elastic rebound theory.

17. How are earthquakes measured?

18. Explain the hazards associated with earthquakes?

Confirm Your Understanding

1. Write out an answer for each question in the Chapter Outline for the chapter sections assigned by your instructor.

2. If you were exploring for natural resources, such as metallic minerals, oil or gas, explain what geologic structures you would look for and why.

3. Explain the effects of composition, pressure, temperature and water content on rock strength.

4. For each of the localities described, determine if the dominant deformation is due to tension, compression or shear.

   • The fold and thrust belt in Wyoming

   • The normal-faulted basin and range region in Nevada

   • The right-lateral San Andreas fault in California

5. How do we know that earthquakes are the result of movement along faults?

Key Terms

aftershocks (p. 294)
anticlines (p. 273)
brittle (p. 284)
compression (p. 280)
dip (p. 272)
dip-slip fault (p. 274)
elastic (p. 283)
elastic rebound theory (p. 293)
elongation (p. 280)
fault (p. 274)
fault scarp (p. 291)
footwall (p. 274)

foreshocks (p. 294)
geologic cross section (p. 275)
geologic map (p. 275)
grabens (p. 287)
hanging wall (p. 274)
hinge line (p. 273)
horsts (p. 287)
intensity (p. 295)
joint (p. 274)
left-lateral strike-slip fault
   (p. 275)
limbs (p. 273)

liquefaction (p. 301)
magnitude (p. 296)
mainshock (p. 294)
normal fault (p. 274)
oblique-slip faults (p. 275)
plastic (p. 284)
plunging folds (p. 273)
reverse fault (p. 275)
right-lateral strike-slip fault
   (p. 275)
seismic moment (p. 297)
shear (p. 280)

shortening (p. 280)
strain (p. 280)
strength (p. 280)
stress (p. 280)
strike (p. 272)
strike-slip fault (p. 275)
synclines (p. 273)
tension (p. 280)
thrust fault (p. 275)
tsunami (p. 301)
yield strength (p. 283)

12 Global Tectonics:
Plates and Plumes

Chapter Outline

Why Study Plates and Their Motions?

EARTH MOVES BENEATH YOUR FEET. IF YOU ARE ABOUT 20 YEARS OLD and born in North America, then your birthplace is almost half a meter farther from Europe than it was on the day you were born. Perhaps that distance does not seem impressive; after all, it implies a speed similar to the rate of fingernail growth. When, however, you take into the account the humongous weight of lithosphere plates, even this sluggish speed indicates there are some gargantuan forces acting at plate boundaries. These forces account for those catastrophic events you recognize instantly in the news, earthquakes and volcanoes, as well as the patient construction of mountain ranges,

the buildup of oil and mineral resources, and the formation of entire ocean basins over the vastness of geologic time.

This chapter will both broaden your understanding of plate tectonics theory and serve as a checkpoint to integrate the diverse concepts introduced in previous chapters. Indeed, the power of plate tectonics lies in the ability of this all-encompassing theory to explain many geologic observations, most of which do not seem obviously linked at first glance.

The most basic aspect of the theory, presented in Chapter 1, is that the lithosphere, consisting of the crust and uppermost mantle, is divided into plates that primarily deform at

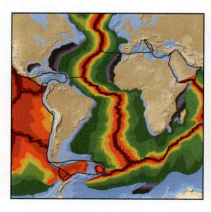

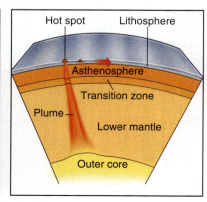
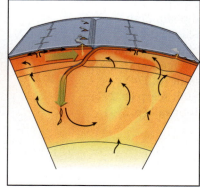

The earthquake that destroyed this city in Turkey in 1999 is evidence of the deformation caused by moving lithospheric plates. ▶

their edges by moving toward, away from, or past one another. Processes at plate boundaries explain the formation of diverse magmas and volcanoes, conditions of metamorphism, the occurrence of earthquakes and rock deformation, and the formation of mountains and basins. Plate tectonics theory, together with evolving models of convection and hot spots, describes motion at the surface that is a part of convection.

Plate tectonics theory, the blockbuster that dramatically advanced geologic science, developed quickly after a slow start. The theory is rooted in the early-twentieth-century concept of continental drift—the hypothesis that continents move about on Earth's surface. Not all geologists adopted continental drift, because there were insufficient data to demonstrate that continents actually did move or to explain how they could move. Cold War friction between countries after World War II led to the establishment of worldwide seismograph networks to monitor underground nuclear testing. These networks also established the global occurrence of earthquakes. Large sums of money funded projects to survey the seafloor to find ways of safe navigation and detection of submarines. These surveys also established the age, structure, and thickness of the virtually unknown oceanic crust. Together, all of these data were largely assembled at oceanographic and geophysical institutes in the United States and United Kingdom and unexpectedly provided the basis for quickly developing the plate tectonics theory during the 1960s.

Your knowledge of Earth materials, internal structure, and surface deformation enables you to examine plate tec-tonics as an integrating theory that explains these phenomena. The chapter objectives are

✔ To learn how geoscientists know that lithospheric plates move

✔ To explain why plates move

✔ To know what happens at plate boundaries and explain why it happens

✔ To evaluate the implications of plate motion for Earth's history

✔ To understand the origin of geologic features not explained by plate tectonics

These are the essential questions that you will answer:

12.1 How does continental drift relate to plate tectonics?

12.2 What is the evidence that plates are rigid?

12.3 What is the evidence that plates move apart at divergent plate boundaries?

12.4 What is the evidence that subduction occurs at convergent plate boundaries?

12.5 What is the evidence that plates slide past one another at transform plate boundaries?

12.6 What does the mantle-plume hypothesis explain that plate tectonics cannot explain?

12.7 *How do we know* ... that plates move in real time?

12.8 What forces cause plate motions and plumes?

12.9 What were the consequences of plate motion over geologic time?

In the LAB

You arrive in the geology lab to find the broken fragments of an intricately patterned dinner plate spread on the tabletop, as shown in **Figure 12.1**. A geologist challenges you to assemble the pieces to reconstruct the plate. You successfully fit fragments together by matching edges of similar shape and by assuring that the ornamental pattern passes continuously from one fragment to the next. You reconstruct the plate without gaps between pieces and without overlapping pieces. What does this exercise have to do with geology?

To answer this question, the geologist hands you two cardboard pieces cut in the shapes of Africa and South America (see Figure 12.1). The puzzle-piece look of eastern South America and western Africa tempt you to place the continents against one another, for the moment ignoring the presence of the intervening South Atlantic Ocean. The notion that the two continents were once joined originated in the late sixteenth century, as explorers first mapped the coastlines of the continents. Is it just a coincidence that the outlines of the two continents match, or were they really joined in the past?

You match patterns between the cardboard continents as well as their outlines, just as you matched patterns on the plate. Geologic information ornaments the cardboard pieces. Areas of different-age crust, patterns of rock deformation, and extent of ancient glaciers all terminate abruptly at the edge of each continent (Figure 12.1). When you match the outlines of the continents, you see a continuous distribution of rock and structure from one landmass to the other. Clearly, the match-up of the continents is not a coincidence. The visiting geologist tells you that German meteorologist Alfred Wegener did not think it a coincidence either, and the matching coastlines and geology of Africa and South America was a small, but crucial, part of his continental drift idea presented in the early twentieth century.

If South America and Africa were once joined and then separated, then what does that mean for the origin of the South Atlantic Ocean? You consult a

Figure 12.1 Putting a puzzle together. ▶
Africa and South America match up like pieces of a broken plate. Not only are the continent outlines similar, but the patterns of geologic features, like rocks of different ages, faults, and old glacial deposits, match from one continent to the other when the continents are placed together.

Lab notes on continental drift exercise

The pieces of this broken plate fit back together by matching both the shapes and the patterns from one piece to the next.

These cardboard models of South America and Africa match up using the same approach used to reconstruct the broken plate — matching shape and pattern.

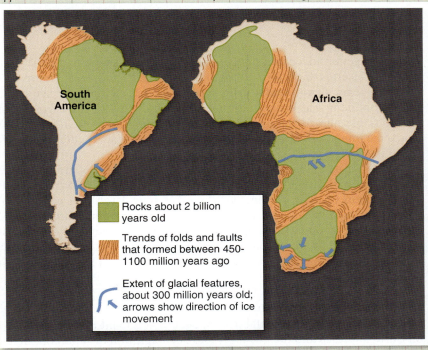

South America

Africa

Rocks about 2 billion years old

Trends of folds and faults that formed between 450-1100 million years ago

Extent of glacial features, about 300 million years old; arrows show direction of ice movement

The matching of the 2 continents suggests that they were once joined together and later drifted apart.

Both the shapes of continents and the map patterns of geologic features match up.

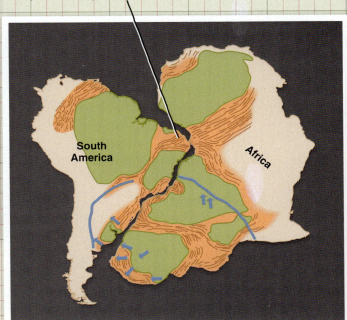

South America

Africa

North America

Europe

Mid Atlantic Ridge

Africa

South America

This downloaded map shows the Atlantic Ocean between South America and Africa. If the continents were once together, then how did the Atlantic Ocean form between them? Also, why does the Mid-Atlantic Ridge, on the seafloor, have the same outline as the coast of Africa?

map online and are impressed that the Mid-Atlantic Ridge, running like a spine down the center of the ocean, mimics the curves of the adjacent continent shorelines. You print this map and place it in your notebook (Figure 12.1) to remind you of the curious observation.

The cardboard models stimulate a number of questions. If Africa and South America moved away from one another, how did the oceanic lithosphere form between them? Unless Earth expands like an inflating balloon, formation of oceanic lithosphere in one location must be balanced by destruction of lithosphere somewhere else; does this happen? How fast did Africa and South America move away from one another, and are they still moving? When did they separate? What forces cause the continents to move? The theory of plate tectonics provides answers to all of these questions.

Scientific theories are not speculative ideas. Theories are rigorously tested concepts established as the best explanations for observed natural phenomena known at the time and are widely accepted in the scientific community. When learning any theory, it is appropriate to ask to see the supporting evidence. Throughout this chapter, you will explore the observed phenomena and tests that support plate tectonics as a valid explanation of a wide variety of geologic features and processes.

12.1 How Does Continental Drift Relate to Plate Tectonics?

Alfred Wegener advocated matching continents by shape and geological similarities in a series of books published between 1912 and 1928. **Figure 12.2** highlights Wegener's **continental drift** hypothesis: All continents once joined in a single continent that broke into pieces, which drifted to the current continent positions. He named the ancient supercontinent *Pangea* (derived from Greek words meaning "all of Earth"), and he called the seas surrounding it *Panthalassa* (meaning, "all ocean").

Wegener assembled impressive data to match the modern continents into Pangea but had difficulty explaining how continents moved from the Pangea configuration to their present positions. He speculated that continental crust plowed through oceanic crust, like an icebreaker crashing through an icy sea. Most geologists found this concept impossible to believe, and it is certainly incompatible with our current understanding that continental lithosphere is weaker than oceanic lithosphere (explained in Section 11.6). Wegener's inability to explain how continents moved across Earth's surface was an insurmountable obstacle to widespread acceptance of continental drift.

Continental drift was the unsuccessful forerunner of plate tectonics. Both plate tectonics and continental drift appeal to abundant evidence of movement on Earth's surface and the assembly and disintegration of supercontinents. Plate tectonics, however, is not only armed with more supporting data than were available to Wegener but also offers a different explanation for the appearance of drifting continents. In the plate tectonics theory, the continents are simply passengers on raftlike plates composed mostly of oceanic lithosphere—continents and oceans within the same plate move together. This is different and more supportable than the untenable idea that continents plow through oceans, and you will see why as we go through the chapter.

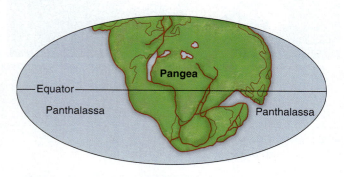

▲ **Figure 12.2 Fitting continents together.**
Alfred Wegener pieced together all of the continents using their outlines and geologic patterns. The assembled continental puzzle is consistent with his hypothesis of the existence of a late Paleozoic supercontinent, which he named Pangea, surrounded by the seas of Panthalassa.

Putting It Together—How Does Continental Drift Relate to Plate Tectonics?

■ The continental drift hypothesis, the forerunner to plate tectonics, proposed that all landmasses were joined in a supercontinent, called Pangea. Pangea broke into continental fragments that drifted to their current position.

■ The continental drift hypothesis was not universally accepted because it lacked a reasonable explanation for why continents moved. In the plate tectonics theory, continents are passengers on the plates.

12.2 What Is the Evidence That Plates Are Rigid?

Plate tectonics theory states that the strong lithosphere consists of pieces that move about on the weak asthenosphere. These pieces are the tectonic plates depicted in **Figure 12.3**. Deformation occurs at or near the edges of plates where they interact with neighbors, and the plate interiors are, by comparison, relatively undeformed. **Figure 12.4** shows that movement of blocks with deformation only at their mutual edges describes rigid plate motion.

Shifting sea ice provides an analogy for rigid plate motion, as demonstrated in **Figure 12.5**. Ice floats on the water and separates into slabs with rough edges that move toward, away from, or alongside one another, but the interiors of the slabs remain smooth, or undeformed. Is this really how the lithosphere deforms, or does it more closely resemble the semirigid or weak boundary interactions described in Figure 12.4?

What Surface Deformation Reveals

One test for rigid plates is to see if deformation preferentially occurs at the edges of these plates. **Figure 12.6** utilizes the global distribution of earthquakes between 1960 and 2000 to investigate where deformation happens. Earthquake epicenters are not uniformly spaced across Earth but are overwhelmingly concentrated into narrow zones that define plate boundaries (compare Figures 12.3 and 12.6).

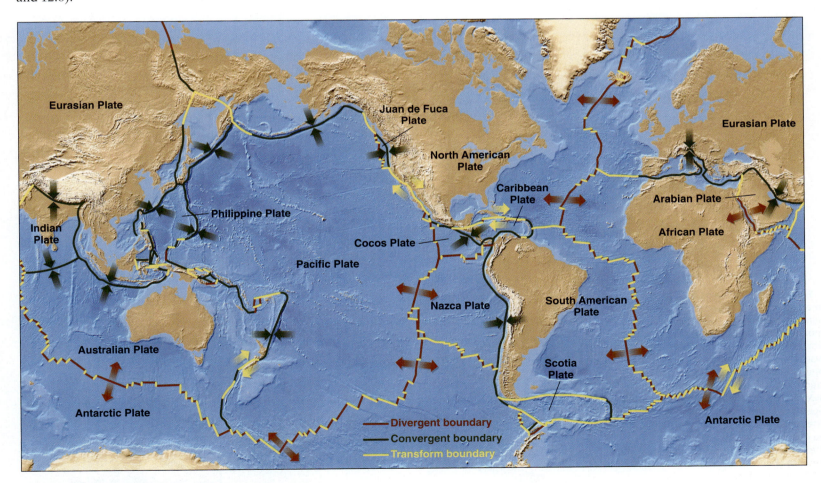

▲ **Figure 12.3** Where the plates are located.
Plate tectonics theory describes the motion of strong plates of lithosphere. Each plate is named for a prominent geographic feature on the plate. The boundaries between plates are described as

- Divergent, where plates move away from one another, as shown by red arrows
- Convergent, where plates move toward one another, as shown by dark green arrows
- Transform, where plates slide past one another, as shown by yellow arrows

Motion at Plate Boundaries: *See how plates move along their boundaries.*

▶ **Figure 12.4 What is meant by a rigid plate?**
Interacting rigid plates deform only at their edges and without any deformation in the interior. Weak plates deform throughout when they interact with neighbors, whereas semi-rigid materials exhibit intermediate behavior. Plate tectonics theory states that lithospheric plates are rigid blocks while acknowledging that some plate-boundary zones indicate semi-rigid behavior.

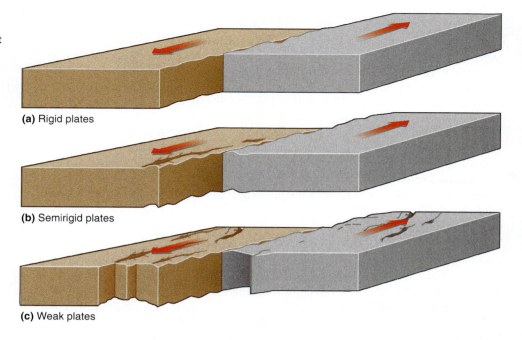

(a) Rigid plates

(b) Semirigid plates

(c) Weak plates

Although earthquake epicenters trace narrow bands along the divergent and transform boundaries within ocean basins, they are much more scattered within continents (Figure 12.6). Some plate boundaries, especially on continents, seem to be broad, diffuse zones rather than narrow, sharp lines. This is particularly true near convergent boundaries. Some earthquakes also occur within plates, far from plate boundaries. Part of the explanation for more diffuse deformation in continents is that continental lithosphere is not as strong as oceanic lithosphere (Section 11.6).

Taking all of the evidence from earthquakes, it seems that some plates are rigid, whereas others are only semirigid, in the sense illustrated in Figure 12.4, so that some plate boundaries are actually zones hundreds of kilometers wide rather than rigid edges. Recent calculations, based on earthquake data from throughout the world, suggest that more than 80 percent of lithosphere deformation occurs at narrow plate boundaries. This means that it is reasonable to assume the plates are basically rigid while acknowledging semirigid behavior along some plate boundaries.

▶ **Figure 12.5 Sea ice as an analogy for plate tectonics.**
Some aspects of lithospheric plate motion resemble rigid slabs of frozen sea ice moved about by waves and currents. (a) Leads open in sea ice where the ice cracks and adjacent slabs move away from one another to expose open water. Notice how the outlines of the broken ice slabs can be matched together, similar to the outlines of Africa and South America (Figure 12.1). The interior part of each slab is unchanged by breaking at the margin. (b) Where slabs of sea ice converge toward one another, they buckle up into the air but the deformation occurs only in a narrow zone where the slabs collide.

(a)

(b)

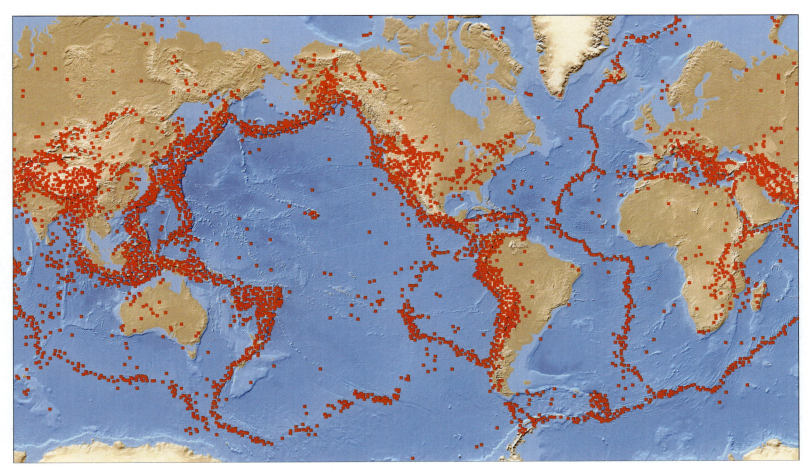

▲ **Figure 12.6 Earthquakes outline plate boundaries.**
This map shows the location of all earthquake epicenters between 1960 and 2000, for earthquakes with magnitudes greater than 4.0. Most earthquakes occur in relatively narrow bands that mark the plate boundaries labeled in Figure 12.3. Within-plate earthquakes are most notable within North America and eastern Asia.

What the Locations of Volcanoes Reveal

Figure 12.7 plots the volcanoes that erupted over the last 10,000 years. The distribution of recently active volcanoes is similar to that of earthquakes and, for the most part, delineates narrow, linear, and curving-arc-shaped belts, which are along or near plate boundaries. The low number of volcanoes mapped along divergent boundaries does not mean that there are few volcanoes along mid-ocean ridges. Geologists simply do not know the locations of many recently active, submerged seafloor volcanoes.

The presence of volcanoes near plate boundaries suggests a strong link between plate motion and locations where magma forms, where magma can reach the surface, or both. (See Section 4.6 to review how magma generation is related to plate tectonics.) At divergent boundaries, the asthenosphere rises and melts by decompression. At convergent boundaries, metamorphic reactions in subducting crust release water into the overlying asthenosphere, which decreases the melting temperature and triggers magma generation.

Not explained by plate tectonics are the large volcanoes scattered within plates, far from plate boundaries (Figure 12.7). Examples of these include the active-shield volcanoes of Hawaii near the center of the Pacific plate, many volcanoes scattered across Africa, and Yellowstone National Park in the heart of the North American plate. These hot spots require a separate and complementary hypothesis. They account for very few volcanoes and do not weaken the power of plate tectonics to explain most global volcanism.

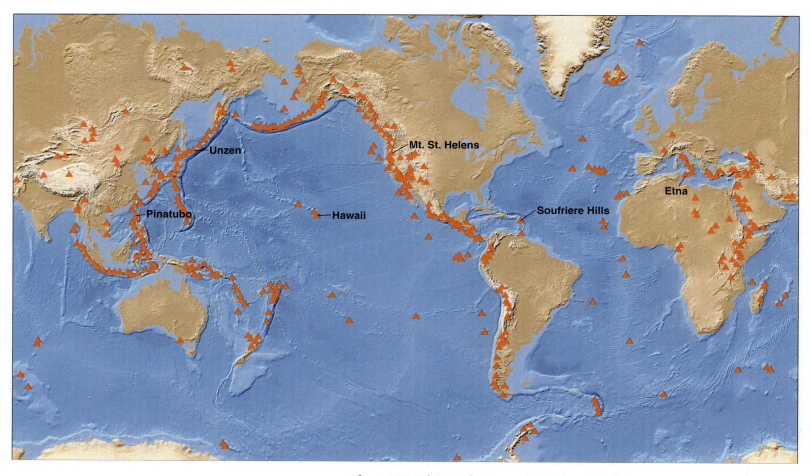

▲ **Figure 12.7 Volcanoes form at and near plate boundaries.**
This map shows the location of all volcanoes that are known to have erupted within the last 10,000 years. Compare this map to Figures 12.3 and 12.6 and notice that most volcanoes erupt along the same narrow zones where earthquakes occur at plate boundaries. Some volcanoes, like those in Hawaii, occur distant from plate boundaries and are not explained by plate interactions. The locations of some volcanoes discussed in Chapter 4 are indicated.

What the Distribution of Mountain Belts Suggests

If deformation occurs preferentially at plate boundaries, then mountain-building processes should be concentrated at these boundaries. **Figure 12.8** maps out active mountain belts, which are the areas displaying evidence for significant uplift and rock deformation within the last few million years. The outlines of these mountain belts are only approximate because not all areas of the globe are sufficiently well studied to know the age of mountain building in detail. Nonetheless, young mountains largely coincide with the occurrences of earthquakes and active volcanoes (compare Figure 12.8 with 12.6 and 12.7). Except in some areas of western North America and southeastern Asia, the active mountain belts are close to or along plate boundaries. Wider mountain belts in continents are yet another indication of the inherent weakness of continental lithosphere.

What about mountain belts of highly deformed rocks that do not coincide with active volcanoes and only few, if any, earthquake epicenters (see Figure 12.8)? Such inactive, ancient mountains, such as the Appalachian Mountains in eastern North America, are generally lower in elevation and more rounded by erosion than the tall, actively rising mountains. Ancient mountain belts contain

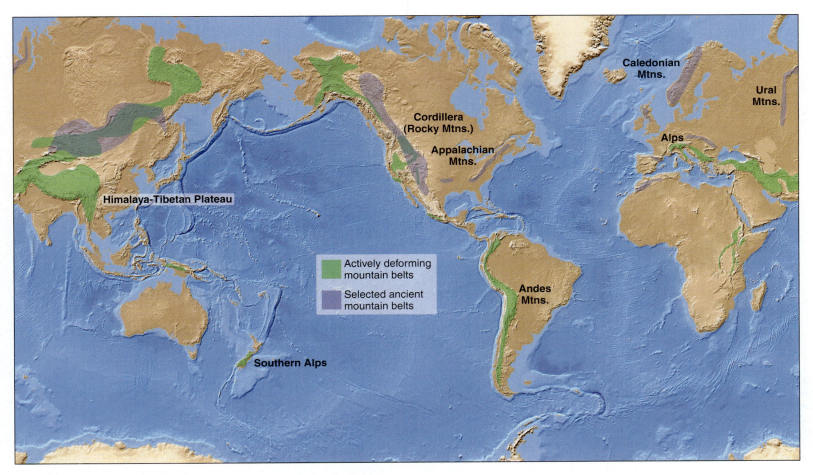

▲ **Figure 12.8** **Mountains record deformation at and near plate boundaries.**
Recently uplifted mountains on continents coincide with locations of earthquakes (Figure 12.6), which reveal that mountain building is ongoing at present. Most active mountain belts are located near plate boundaries. Some ancient mountain belts, like the Appalachian Mountains, for example, are far from plate boundaries and not associated with significant modern earthquake or volcanic activity. Geologists hypothesize that these older mountain belts formed near ancient plate boundaries.

igneous rocks like those found near modern convergent plate boundaries. You can apply uniformitarianism to hypothesize that in the geologic past, plate boundaries were in different places, near the ancient mountains. This means that plate tectonics theory not only explains active processes at present, but it also describes changing features on Earth through geologic time.

Is the Whole Lithosphere Really Strong?

Is it reasonable to hypothesize that the thickness of the whole lithosphere is strong and rigid? Rock strength decreases with increasing temperature and, according to tested data, rocks should be very weak below 15 to 40 kilometers (see Figure 11.26 to review). Yet plate tectonics theory invokes strength throughout the lithosphere, which averages about 100 kilometers thick. Without any doubt, rock strength decreases downward through the mantle part of the lithosphere. Seismic waves in the even-deeper mantle imply, however, a large and abrupt decrease in rock strength in the upper asthenosphere caused by temperatures near the melting point of peridotite (see Section 8.4). The lithosphere, therefore, is stronger and more rigid than the underlying asthenosphere.

Correlating Processes at Plate Boundaries: *See how volcanoes, earthquakes, and young mountain belts line up with plate boundaries.*

12.3 What Is the Evidence That Plates Move Apart at Divergent Boundaries?

The Mid-Atlantic Ridge forms a divergent boundary between Africa and South America (Figures 12.1 and 12.3). Matching the edges of the two continents into a single landmass requires two conditions:

1. Lithosphere containing Africa and South America existed before the ocean that now separates the continents.
2. The Atlantic Ocean grows wider through time.

A prediction of plate tectonics theory is that oceanic lithosphere forms where continental lithosphere separates into two or more continents. Is the geology of the seafloor consistent with this prediction?

What a Divergent Plate Boundary Looks Like

Plate tectonics theory incorporates the concept of **seafloor spreading**, which emerged as a successfully tested hypothesis in the mid-1960s. Seafloor spreading describes the process of oceanic lithosphere spreading apart along mid-ocean ridges, as shown in **Figure 12.9**. New lithosphere forms to fill the gap between the separating plates. Igneous rocks of the upper, crust part of the new lithosphere crystallize from magma where asthenosphere peridotite decompresses and partially melts while rising into the gap between the spreading plates. The lower, mantle part of the lithosphere is made up of unmelted residue from the original asthenosphere peridotite (see Figure 12.9). Seafloor spreading provides many testable predictions.

▼ **Figure 12.9** **What happens at a divergent plate boundary?** Divergent plate boundaries mark the separation of plates. Most divergent plate boundaries coincide with mid-ocean ridges. The gap between the separating lithospheric plates fills with up-welling asthenosphere, which partly melts to produce mafic magma that solidifies to form new oceanic crust. The lithosphere cools as it moves away from the mid-ocean ridge by continued divergent motion. As the mantle cools, the critical temperature defining the boundary between strong lithosphere and weak asthenosphere occurs at greater depth, so the lithosphere thickens as it moves away from the divergent boundary.

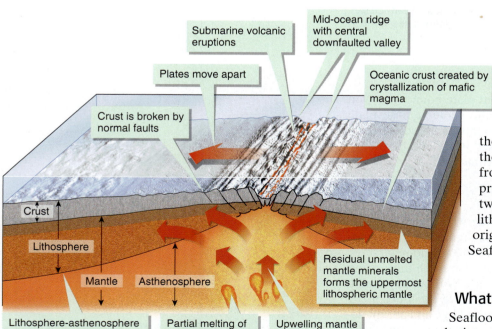

Submarine volcanic eruptions

Mid-ocean ridge with central downfaulted valley

Plates move apart

Oceanic crust created by crystallization of mafic magma

Crust is broken by normal faults

Crust

Lithosphere

Mantle   Asthenosphere

Residual unmelted mantle minerals forms the uppermost lithospheric mantle

Lithosphere-asthenosphere boundary moves down as mantle cools with increasing distance from mid-ocean ridge

Partial melting of asthenosphere by decompression produces mafic magma

Upwelling mantle

What Oceanic-Crust Composition Reveals

Seafloor spreading explains the production of oceanic crust by igneous processes at mid-ocean ridges. Laboratory experiments (see Section 4.6) indicate that partial melting of mantle peridotite produces mafic magma. Oceanic crust, therefore, should have mafic composition, and active volcanoes should occur along the mid-ocean ridges.

More than 1700 scientific drill holes into oceanic crust confirm the predicted mafic composition (see Section 8.1). **Figure 12.10** illustrates an example observation from research submarines that confirms the predicted active volcanism. Approximately 18 cubic kilometers of basalt erupt each year along the 65,000-kilometer-long mid-ocean ridge system. The high rate of volcanic activity and related high heat flow from Earth's interior readily account for extensive hydrothermal metamorphism and submarine hot springs (described and pictured in Section 6.10).

The Significance of Mid-Ocean Earthquakes and Faults

If mid-ocean ridges mark locations of plate divergence, then they should coincide with earthquakes along normal faults formed by tensional stress where plates move away from one another. Data depicted in Figure 12.6 confirm a high level of earthquake activity, and **Figure 12.11** illustrates examples of the predicted normal faults. Faults are mapped at the surface in Iceland, where the Mid-Atlantic Ridge rises above sea level (Figure 12.11a). Geologists use small research submarines to examine normal faults along the more typical submerged parts of the ridge (Figure 12.11b).

What the Age and Magnetism of Oceanic Crust Reveal

If seafloor spreading happens, then the oceanic crust is youngest along mid-ocean ridges and progressively older on either side at greater distances from the ridge crest. Measurements of the magnetic properties of seafloor basalt during the 1960s confirmed this predicted age relationship.

Shipboard measurements of the magnetic polarity of seafloor rocks document remarkable parallel stripes of normal and reverse polarity symmetrically distributed on either side of mid-ocean ridges. **Figure 12.12** illustrates an example magnetic-polarity pattern in the northeast Pacific Ocean, and explains how the magnetic data reveal the age of the crust. The measured magnetic signal originates in the mineral magnetite within the seafloor basalt. Normal-polarity basalt erupts to form oceanic crust when Earth's magnetic field is the same as it is at the present time. Reverse-polarity basalt dates to times in the past when Earth's magnetic field was reversed (see Section 10.3). The time intervals of normal and reverse magnetic fields are determined by measuring polarities in

▲ **Figure 12.10 Aftermath of a volcanic eruption along a mid-ocean ridge.**
In 1993, earthquakes along the mid-ocean ridge west of the Washington coast suggested an ongoing volcanic eruption. Within weeks, a research vessel was at the site and lowered a remotely operated, unmanned miniature submarine to see what happened. This picture of the seafloor shows part of the surface of a new, still-hot, submarine lava flow.

(a) Volcano / Fault

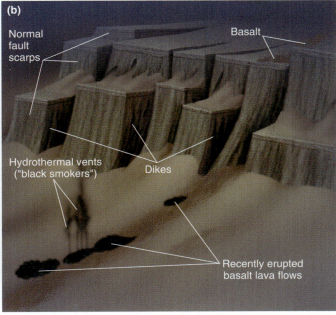

(b) Normal fault scarps / Basalt / Hydrothermal vents ("black smokers") / Dikes / Recently erupted basalt lava flows

◄ **Figure 12.11 Normal faults are present along mid-ocean ridges.**
(a) Iceland is a rare place where part of a mid-ocean ridge is exposed above sea level. Normal faults form prominent gashes across the countryside, and careful surveying shows that the island is stretching 4 mm/year. Volcanic steam issues from the faults in this picture, and a volcano forms the skyline. (b) Geologists made this sketch after observing the seafloor along the Mid-Atlantic Ridge east of Florida from a small research submarine that was used to make a geologic map of the seafloor. Normal faults form escarpments that are several hundred meters high.

▶ **Figure 12.12** **How to determine seafloor age.**

a. Shipboard measurements of seafloor-crust magnetization show a symmetrical pattern of normal and reverse magnetic polarity on either side of mid-ocean ridges.

b. Geologists studying volcanoes on land determine the eruption age and magnetic polarity for many lava flows. These data provide a magnetic polarity time scale, which depicts the times when Earth's magnetic field was either normal or reverse polarity.

c. Combining the seafloor magnetic data with the magnetic polarity time scale provides the ages of the oceanic crust. The crust is older at greater distances on either side of the mid-ocean ridge. This observation confirms the seafloor spreading hypothesis, which predicts that crust forms at mid-ocean ridges and spreads away from the ridge as new crust forms.

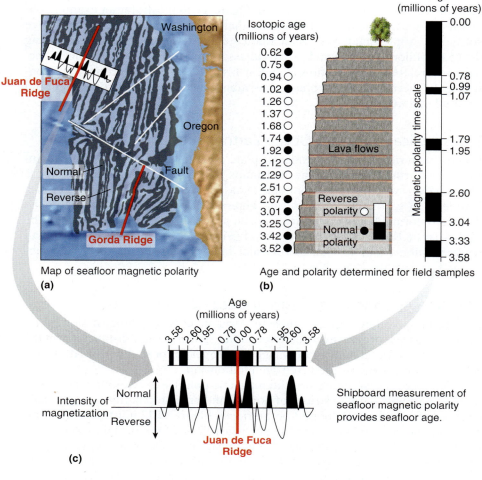

Mid-ocean ridge

Time 1

Time period of normal polarity

Magnetization recorded in newly created oceanic crust

Time 2

Time period of reverse polarity

Magnetization preserved in ancient oceanic crust

Continental lithosphere

Asthenosphere

Time period of normal polarity

Oceanic lithosphere

Time 3

Symmetrical polarity pattern

Oldest ← Youngest → Oldest

more accessible lava flows on land and dating them with radioactive-isotope methods. Geophysicists combine the data from land and sea (Figure 12.12) to determine that the seafloor is less than 780,000 years old close to the ridge crest and progressively older at greater distance on either side of the ridge.

Figure 12.13 explains how seafloor spreading forms ridge-parallel bands of basaltic crust with alternating normal and reverse magnetic polarity. As mafic magma solidifies along the spreading-ridge crest during a single magnetic polarity time interval, the new crust records that same polarity. As spreading at the divergent plate boundary continues, the original crust separates into two sections, one on either side of where still newer crust forms at the ridge crest. When the magnetic-field polarity reverses, horizontal transitions from normal-polarity crust to reverse-polarity crust can be seen on either side of the ridge crest.

Figure 12.14 portrays the age of seafloor crust based on measurements of seafloor magnetic polarity and known polarity reversals back to about 180 million years ago. Mid-ocean ridges coincide everywhere

◀ **Figure 12.13** **Seafloor spreading explains seafloor magnetization record.**
Magnetite crystals in basalt record the orientation of Earth's magnetic field at the time lava erupts and cools at the mid-ocean spreading ridge. This record documents the flip-flopping reversals of the magnetic field during the time interval when the crust forms. As the two plates separate along the spreading ridge, new seafloor forms along the ridge axis while already formed crust moves aside. The spreading process creates vertical bands of crust on either side of the ridge that have different magnetic polarity.

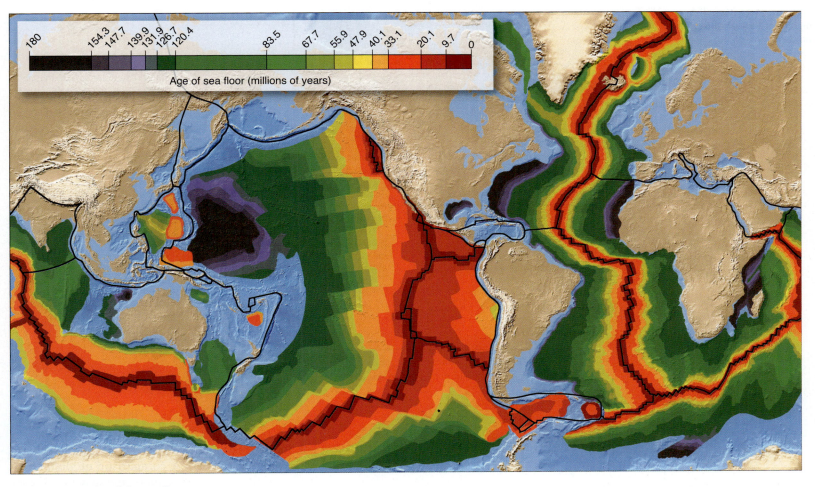

▲ **Figure 12.14 Visualizing seafloor age.**
This map of seafloor age results from shipboard measurements of magnetic polarity and assigning an age to each region of surveyed seafloor, following the approach illustrated in Figure 12.12. Black lines outline the plates. Oceanic crust is progressively older at greater distances on either side of divergent plate boundaries along mid-ocean ridges.

with the youngest crust, and they form the centerline for symmetrical patterns of seafloor age.

Examination of Figure 12.14 also answers the earlier questions about when Africa and South America separated and how fast they move apart. The oldest oceanic crust along the eastern margin of South America and western margin of Africa is 130 million years old. This was the first oceanic crust to form when the continents separated. Along the equator, the two continents are about 3900 kilometers apart. Divide the 3900 kilometers by the elapsed time of 130 million years and you have an average calculated rate of 3 cm/yr. The symmetrical pattern of crust age on either side of the Mid-Atlantic Ridge (Figure 12.14) means that South America moves 1.5 cm/yr away from Africa, while Africa moves 1.5 cm/yr away from South America. Similar calculations across other divergent plate boundaries indicate that plates move at relative rates between 1 and 10 cm/yr.

What Heat-Flow Data Show

Seafloor spreading predicts that divergent plate boundaries are unusually hot, because hot, partially melted asthenosphere rises close to Earth's surface and molten magma solidifies to make new crust (Figure 12.9). Data illustrated in **Figure 12.15** demonstrate that oceanic heat flow is highest along a mid-ocean-ridge crest. The heat flow decreases with increasing distance from the

Seafloor Spreading and Rock Magnetism: *See how seafloor spreading at divergent boundaries produces bands of crust with alternating magnetic polarities.*

▶ **Figure 12.15 Visualizing high heat flow at mid-ocean ridges.**

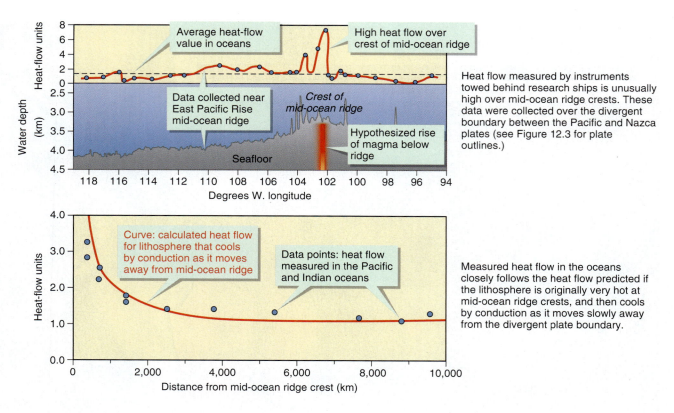

Average heat-flow value in oceans

High heat flow over crest of mid-ocean ridge

Data collected near East Pacific Rise mid-ocean ridge

Crest of mid-ocean ridge

Hypothesized rise of magma below ridge

Seafloor

Heat flow measured by instruments towed behind research ships is unusually high over mid-ocean ridge crests. These data were collected over the divergent boundary between the Pacific and Nazca plates (see Figure 12.3 for plate outlines.)

Curve: calculated heat flow for lithosphere that cools by conduction as it moves away from mid-ocean ridge

Data points: heat flow measured in the Pacific and Indian oceans

Measured heat flow in the oceans closely follows the heat flow predicted if the lithosphere is originally very hot at mid-ocean ridge crests, and then cools by conduction as it moves slowly away from the divergent plate boundary.

ridge just as expected for conductive cooling of hot, newly formed lithosphere that moves away from the ridge (Figure 12.15). A comparison of global heat flow (Figure 10.6) and the location of divergent plate boundaries (Figure 12.3) shows that all divergent boundaries are areas of high heat flow.

The decreasing elevation of the seafloor away from mid-ocean ridges relates to the cooling of the lithosphere. Lithosphere contracts while it cools down, becomes denser, and sinks down into the weak asthenosphere. **Figure 12.16** shows that the observed decrease in seafloor elevation almost exactly matches the calculated effect of simply cooling off the lithosphere as it moves away from spreading ridges.

The lithosphere also gets thicker as its ages and cools. The thickening, illustrated in Figure 12.16, occurs because rock strength is what defines the boundary between the strong lithosphere and the weak asthenosphere. Rocks are weaker at higher temperature, and the critical temperature separating strong lithosphere from weak asthenosphere slides to progressively deeper depths as lithosphere cools with increasing age and distance from the ridge (Figures 12.9 and 12.16). Hot, weak upper mantle beneath a mid-ocean ridge is part of the asthenosphere. As the upper mantle moves laterally and cools, it becomes part of the stronger lithosphere.

What Seismic Tomography Reveals

Seafloor spreading predicts that asthenosphere moves upward at mid-ocean ridges (Figure 12.9). The rising asthenosphere should be hotter than its surroundings and partly molten, so it should transmit seismic waves at slower velocities (see Section 10.2 for explanation and illustration of relationships between seismic velocity and mantle properties). **Figure 12.17** portrays seismically slow, and therefore probably warmer, upper mantle below divergent plate boundaries, as predicted. For most mid-ocean ridges, the unusually hot asthenosphere exists only down to a depth of 300 kilometers or so and not into the

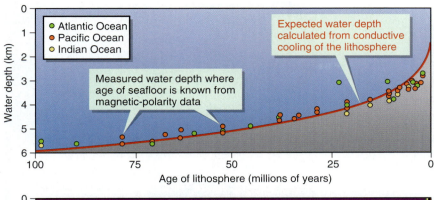

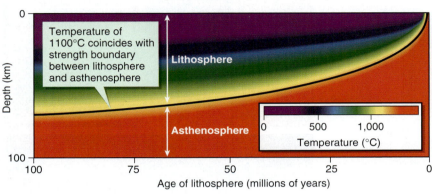

◄ **Figure 12.16** Oceanic
lithosphere cools, thickens,
and subsides with age.

Water depth is shallowest at mid-ocean ridges and increases where the seafloor is older. The increase in water depth is predicted by cooling and contraction of the lithosphere as it moves away from the ridges. Cooler, denser lithosphere sinks down into the asthenosphere, causing elevation at the surface to decrease.

Heat-flow measurements allow calculation of temperatures within lithosphere as it cools and moves away from mid-ocean ridges. A temperature of 1100°C coincides with the transition from strong lithosphere, above, to weak asthenosphere, below. Lithosphere thickens rapidly to about 60 km at an age of 50 million years, and then more slowly to 75 km by 100 million years.

deeper mantle. This shallow character of the unusually hot asthenosphere means that most divergent plate boundaries link only to processes in the upper mantle.

The Significance of Continental Rift Valleys

The separation of continents to make ocean basins, such as between Africa and South America, implies that divergent margins can originate within continents. What does a divergent plate boundary look like in the continental crust, rather than at a mid-ocean ridge? Tension within continental crust produces **rift valleys**, as explained in **Figure 12.18**. The rift valleys are graben blocks of crust that drop along normal faults and are commonly associated with volcanoes. The tensional stress eventually stretches and separates the continental lithosphere into two fragments and new oceanic lithosphere forms to fill in the gap between them (Figure 12.18).

The East African rift valleys are a region where tension may someday split the continent into two pieces, on separate plates. **Figure 12.19** illustrates the

▶ **Figure 12.17 How seismic-tomography data reflect divergent-boundary processes.**
This map shows variation in S-wave velocity between 175 and 250 kilometers below the surface. The color scale depicts how much the seismic velocity differs at each location from the average value for this mantle depth range. Notice that the upper mantle is seismically slower, and therefore probably hotter, along almost all of the mid-ocean ridges, which include divergent plate boundaries.

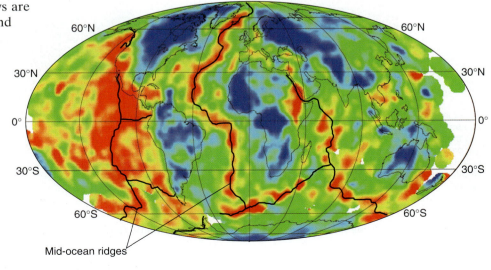

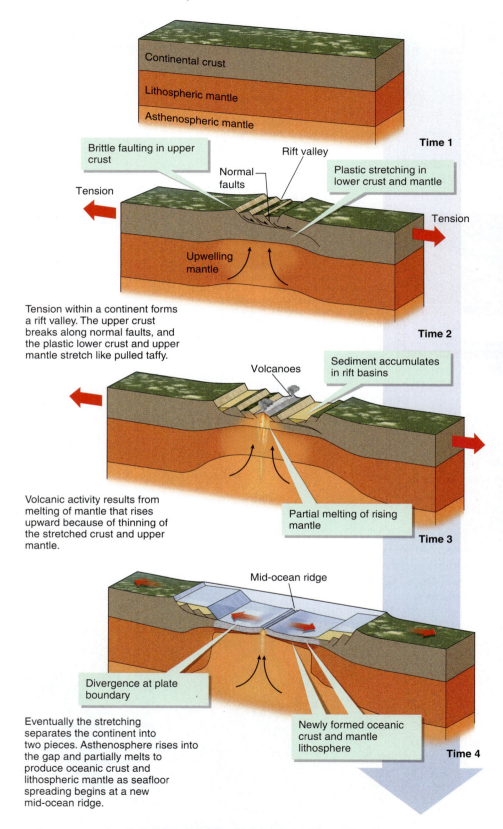

Time 1

Continental crust

Lithospheric mantle

Asthenospheric mantle

Brittle faulting in upper crust

Rift valley

Normal faults

Plastic stretching in lower crust and mantle

Tension

Tension

Upwelling mantle

Time 2

Tension within a continent forms a rift valley. The upper crust breaks along normal faults, and the plastic lower crust and upper mantle stretch like pulled taffy.

Volcanoes

Sediment accumulates in rift basins

Partial melting of rising mantle

Time 3

Volcanic activity results from melting of mantle that rises upward because of thinning of the stretched crust and upper mantle.

Mid-ocean ridge

Divergence at plate boundary

Newly formed oceanic crust and mantle lithosphere

Time 4

Eventually the stretching separates the continent into two pieces. Asthenosphere rises into the gap and partially melts to produce oceanic crust and lithospheric mantle as seafloor spreading begins at a new mid-ocean ridge.

▲ **Figure 12.18** Visualizing the origin of a divergent boundary within a continent.

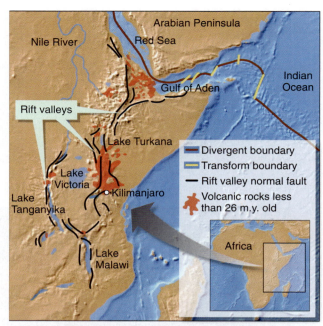

Arabian Peninsula

Nile River   Red Sea

Gulf of Aden

Indian Ocean

Rift valleys

Lake Turkana

■ Divergent boundary
■ Transform boundary
— Rift valley normal fault
🔶 Volcanic rocks less than 26 m.y. old

Lake Victoria

Kilimanjaro

Lake Tanganyika

Africa

Lake Malawi

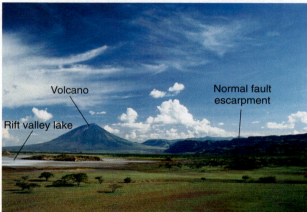

Volcano

Normal fault escarpment

Rift valley lake

▲ **Figure 12.19 Divergence in East Africa.**
Tensional stretching and thinning of continental lithosphere forms the rift valleys of East Africa. Graben blocks subside along normal faults to form elongate valleys, some of which partly fill with lakes. Upward-moving mantle melts and erupts at the surface to form volcanoes. The rifting closely relates to the divergent plate boundaries in the Red Sea and Gulf of Aden. If deformation persists, the eastern horn of Africa may separate from the rest of the continent forming a new tectonic plate. The photo shows Lake Natron and the volcano Oldoinyo Lengai in Kenya.

Active Art

Forming a Divergent Boundary: *See how a divergent boundary originates by rifting a continent.*

locations of two, parallel rift valleys, each 4000 kilometers long, and partly occupied by deep lakes. The rifts join with a divergent boundary that created the Red Sea between Africa and the Arabian Peninsula beginning about 5 million years ago.

Not all rift valleys extend sufficiently to mature into new ocean basins between fragmented continents. Some rift valleys form long tectonic gashes across continents and then the tensional stress decreases or ceases. Geologists are not sure why some rift valleys become new continental margins, whereas others fail and remain as long troughs within continents. Besides East Africa, active rifts are found in other continents, forming low valleys followed by rivers, such as the Rhine in Germany and the Rio Grande in the southwestern United States. Ancient, failed rifts are recognized in many places on all continents. The Mississippi River flows along such an ancient rift valley and the within-plate earthquakes near New Madrid, Missouri (see Figure 11.32), occur when stresses reactivate the old normal faults.

*Putting It Together–*What Is the Evidence That Plates Move Apart at Divergent Plate Boundaries?

- Plates move apart at divergent plate boundaries by the process of seafloor spreading, which creates new, mafic oceanic crust and mantle.
- Abundant earthquakes, basaltic volcanic activity, high heat flow, high seafloor elevation, and a seismically slow upper mantle confirm plate tectonics predictions for mid-ocean ridges.
- Magnetic polarity of seafloor crust, along with radioactive-isotope ages, demonstrates that oceanic crust is progressively older at greater distances from mid-ocean spreading ridges. Seafloor spreading predicts this age progression.
- Divergent plate boundaries can originate within continents to form rift valleys. Persistent extension along a rift valley causes the lithosphere to break into two continental fragments. Not all rift valleys extend sufficiently to form new continents and plate boundaries.

12.4 What Is the Evidence That Subduction Occurs at Convergent Plate Boundaries?

Convergent plate boundaries strongly contrast with divergent boundaries in three major ways.

1. Plates move toward one another at convergent boundaries and apart at divergent boundaries.
2. Subduction destroys lithosphere at convergent boundaries, whereas lithosphere is created at divergent boundaries.
3. Convergent boundaries are highly asymmetrical, with one plate angled downward beneath its neighbor. Divergent boundaries are vertical, symmetrical boundaries between plates.

What a Convergent Boundary Looks Like

Convergent plate boundaries coincide with some of the most spectacular landscapes and seascapes on Earth. A glance at southeastern Asia, shown in **Figure 12.20**, illustrates this point.

Deep-sea trenches, many kilometers deep, mark the curving line where one plate subducts beneath another in the ocean. Magma generated in the asthenosphere above the subducting plate feeds towering volcanoes on the overriding plate. The Indonesian volcanoes pictured in Figure 12.20 are just part of the great "Ring of Fire," a nearly continuous chain of volcanic islands and continental volcanoes that stand above convergent plate boundaries encircling the Pacific

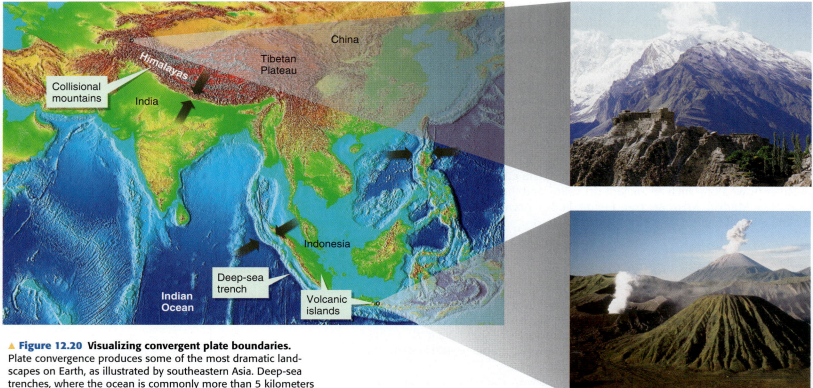

▲ **Figure 12.20** **Visualizing convergent plate boundaries.**
Plate convergence produces some of the most dramatic landscapes on Earth, as illustrated by southeastern Asia. Deep-sea trenches, where the ocean is commonly more than 5 kilometers deep, coincide with subduction of oceanic lithosphere. Volcanic island chains rise above the subducting plate, such as in Indonesia, where 76 volcanoes have erupted since 1800 in an area only about twice the size of Texas. Huge mountains, like the Himalayas, form where continental lithospheric plates collide along convergent boundaries.

Ocean (Figures 12.3 and 12.7). Some of these volcanoes erupt basalt, but most of them also erupt andesite, dacite, and occasionally rhyolite. The great diversity of magma composition suggests different processes of magma generation at convergent boundaries than occur at basaltic mid-ocean ridges.

Compressional stresses heave up great mountain ranges, especially where continents collide along convergent boundaries. The Himalayas are the most dramatic example of a modern-day convergent-margin mountain chain (Figure 12.20), but others are notable around the world (Figure 12.8). Compression forces blocks of crust tightly together along reverse and thrust faults near convergent boundaries. This tight squeezing tends to lock the faults so that huge elastic strain energy builds up before the faults fracture with large displacements over long distances to produce dramatic and devastating earthquakes. The largest historic earthquake, a M9.5 event in Chile, in 1960 (Figure 11.36), occurred along a convergent boundary. The great Denali fault, Alaska, earthquake of 2002 (Figure 11.32) and the devastating 1995 earthquake at Kobe, Japan (Figure 11.37), which took 5502 lives, are also examples of powerful earthquakes associated with convergent-boundary stresses. Convergent plate boundaries are also the only locations where earthquakes routinely occur at depths greater than 50 kilometers. The deepest recorded earthquake occurred 640 kilometers beneath Bolivia, South America, in 1994.

How a Convergent Boundary Works

Figure 12.21 illustrates how the plate tectonics theory explains all of the geologic features and phenomena that occur at convergent boundaries—places where plates collide and one descends beneath the other.

- Deep-sea trenches mark the actual plate boundary, where one plate bends and descends into the deeper asthenosphere.
- The plate collision causes compressional stress, which accounts for the folding, faulting, and associated earthquakes.

- Deep earthquakes occur within the subducted plate, which remains relatively strong at depth because it is colder than its deep mantle surroundings.
- Water-rich magma forms above subducted lithosphere and undergoes fractional crystallization to produce compositionally diverse igneous rocks.

The actual plate boundary is the narrow trench, but the convergent-boundary processes are spread out over a wide zone that contrasts with narrow divergent boundaries. Earthquakes not only occur along the trench, but also in the deeply subducting plate, and in compressed rocks of the overriding plate, so that the zone of earthquake epicenters is very wide. Unlike divergent-boundary volcanoes, which form right at the plate boundary, convergent-margin magmas rise from an area in the asthenosphere above the subducted lithosphere and form volcanoes on the overriding plate many tens or hundreds of kilometers from the plate boundary (Figure 12.21).

The convergent-plate-boundary processes depicted in Figure 12.21 summarize decades of geologic mapping, geophysical measurements, geochemical studies of igneous and metamorphic rocks, and earthquake research. These data, in turn, test the explanations that plate tectonics theory provides for features observed around the Ring of Fire and at other convergent boundaries.

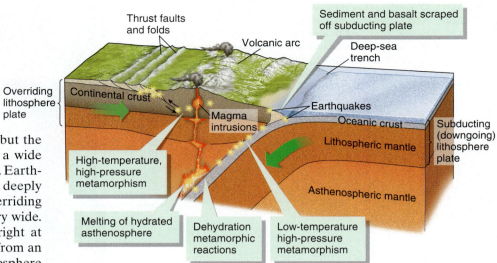

▲ **Figure 12.21 What happens at a convergent plate boundary?**
Plates move together at convergent boundaries. A deep-sea trench marks the location of subduction, where convergence occurs within or at the margin of an ocean. Rock deformation, faulting, volcanic activity, and metamorphism may extend 500 km or more from the deep-sea trench because of the inclined orientation of the subducted plate and the large compressive stresses transferred to the overriding plate. Dehydrating metamorphic reactions release water into the asthenosphere, which promotes melting. Rocks within and adjacent to the cold, subducting plate experience high-pressure, low-temperature metamorphism. High-temperature metamorphism occurs where subduction-zone magmas rise into the overriding plate.

What Earthquakes Reveal

If deep earthquakes occur in the subducted plate bending down into Earth, as shown in Figure 12.21, then earthquake foci should be deeper at greater distances from the trench. Geologists tested this prediction by carefully examining the depth of earthquake foci. **Figure 12.22** shows the depth of earthquakes occurring below central South America in relation to the distance from the subduction-zone trench. Earthquakes shallower than 50 kilometers and east of the subduction zone record shortening strain in the South American crust caused by plate convergence. The deep earthquakes occur at progressively greater depths east of the trench and define the outline of the inclined subducting plate. Earthquakes happen because of bending and downward-directed tension in the subducting plate, which occurs until the descending lithosphere becomes too hot for brittle fracture to take place. The inclined zone of earthquake foci characteristic of subduction zones is called the **Wadati-Benioff zone** in recognition of Japanese seismologist Kigoo Wadati and American seismologist Hugo Benioff who first described this feature.

What Seismic Tomography Data Reveal

Seismic tomography data, like those shown in **Figure 12.23**, also detect subducting plates. The cooler and more rigid subducting lithosphere transmits seismic waves faster than the adjacent hot, plastic asthenosphere. Tomographic studies confirm the presence of subducting plates around the world. Some regions of seismically fast mantle interpreted as subducting plates exist only in the asthenosphere, whereas others possibly penetrate the transition zone and continue to the bottom of the mantle.

What Heat-Flow Data Reveal

Plate tectonics theory predicts a complicated pattern of heat flow from the mantle to the surface near convergent plate boundaries. The subducted lithosphere is colder than the surrounding asthenosphere and draws heat from the

▶ **Figure 12.22 Earthquakes outline a subducting plate.** Earthquakes in South America deeper than about 50 km occur progressively deeper from west to east. The graph shows that the deep earthquakes outline the subducting Nazca plate, which penetrates through the asthenosphere and into the transition zone. Compression also generates shallow earthquakes in the South American plate.

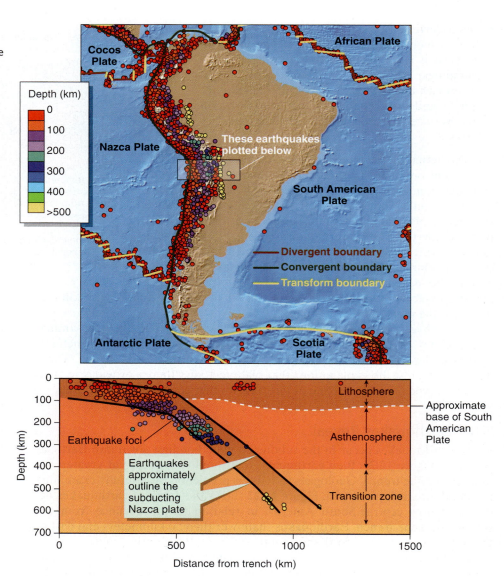

surrounding mantle by conduction. Therefore, the mantle close to the trench should be cooler than average. On the other hand, magma forms above the subducted plate and rises upward, so unusually high heat flow should occur in the overriding plate.

Figure 12.24 tests these predictions with a heat-flow profile across the convergent plate boundary in the northwestern United States. As predicted by plate tectonics, heat flow is low where the cold lithosphere subducts and is high where magma migrates upward to feed the volcanic chain. A quick review of how these thermal characteristics of convergent plate boundaries relate to igneous and metamorphic rocks that form in these settings is worthwhile (see Figure 12.21).

Magma does not form at convergent margins because of heating in the asthenosphere; indeed, the insertion of a cold subducted plate refrigerates the surrounding asthenosphere much like dropping ice cubes in water. Instead, magma forms because dehydration metamorphism of the subducted plate releases fluid into the asthenosphere (Figure 12.21). This fluid reduces the melting temperature of the peridotite so that it partially melts (see Figure 4.21b). The resulting water-rich magmas erupt explosively (Section 4.9, Figures 4.27 and 4.28) and produce economically valuable mineral deposits (Section 4.10, Figures 4.31 and 4.33).

Paired metamorphic belts, described and explained in Section 6.10, are common at convergent plate boundaries. One member of the pair is a belt of

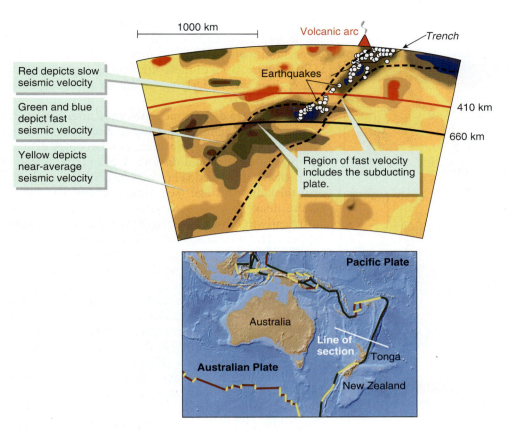

▲ **Figure 12.23** **Seismic tomography detects subducted plates.**
Seismic tomography data across the convergent plate boundary north of New Zealand reveals an inclined zone of seismically fast mantle. This fast zone is consistent with the location of relatively cold and rigid lithosphere, which is also partly outlined by earthquake foci.

high-pressure but relatively low-temperature metamorphic rocks, whose origin is readily explained by transporting rocks downward at cold subduction zones (Figure 6.29). These rocks experience increasing pressure that accounts for the metamorphic reactions as the rocks descend. The high-temperature, low-pressure belt of metamorphism, the second member of the pair, coincides with where magma rises into the lithosphere above the subducted plate (Figure 12.21).

Why Subduction Occurs

Why does subduction happen? After all, when ice floes converge in the ocean the fractured ice rises into the air, rather than descending into the water (see Figure 12.5). Ice cannot subduct into water because ice is less dense than water. Therefore, lithosphere must be denser than asthenosphere for subduction to happen. But wait—if this were generally true everywhere on Earth, then lithosphere should sink into asthenosphere everywhere, rather than only at subduction zones. Taken together, these observations imply a paradox where lithosphere is less dense than asthenosphere near mid-ocean ridges but is denser than asthenosphere near subduction zones. If this is not true, then it is impossible for plate tectonics to work. Comparing the relative densities of lithosphere and asthenosphere is, therefore, worthy of further consideration.

Figure 12.25 illustrates how lithosphere density changes through time. Lithosphere is less dense than asthenosphere when it forms at divergent boundaries and is more dense when it descends at convergent boundaries.

Lithosphere is less dense than the asthenosphere at mid-ocean ridges because these two layers have different compositions (Figure 12.25). The asthenosphere peridotite partly melts and forms two parts of the young

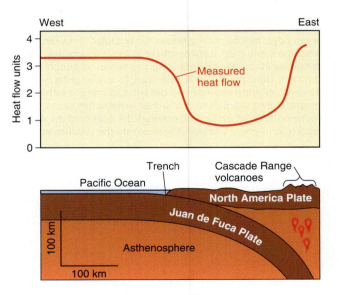

▲ **Figure 12.24** **Heat-flow data are consistent with convergent-boundary processes.**
A graph of surface heat flow across the subduction zone in the northwestern United States shows two important features: (1) Lower heat flow just east of the trench, which is consistent with the presence of relatively cold, subducted lithosphere at depth. (2) Higher heat flow in the vicinity of the Cascade Range volcanoes, which is consistent with the upward rise of hot magma generated by melting above the subducted plate.

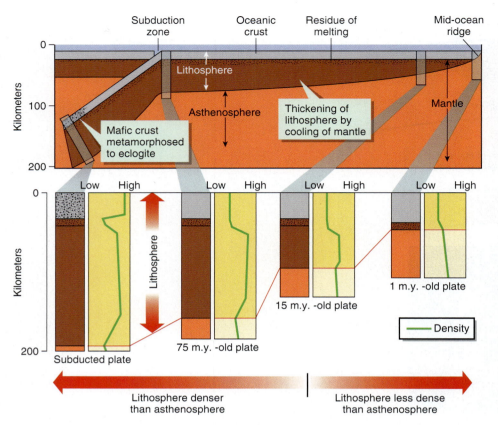

▲ Figure 12.25 Why oceanic lithosphere subducts.
This diagram illustrates how the relative densities of oceanic lithosphere and asthenosphere change while the lithosphere moves
away from the mid-ocean ridge. When first formed (right side of
diagram) the lithosphere is mostly low-density crust and the
residues of melting to make that crust; the lithosphere is significantly less dense than the asthenosphere. Farther from the mid-
ocean ridge, mantle cooling causes the boundary between strong
lithosphere and weak asthenosphere to move downward. The
thickening lithospheric mantle becomes denser as it cools and
contracts. Lithosphere that is older than 15 million years is
denser than asthenosphere and can subduct into the asthenosphere. During subduction, mafic crust metamorphoses to very
dense eclogite, which further increases the density of the subducting lithosphere and pulls the plate into the subduction zone.

lithosphere—one is the solidified basalt and gabbro of
the oceanic crust, and the other is a residue of unmelted peridotite minerals in the mantle part of the oceanic lithosphere. Both of these lithosphere layers are less
dense than the underlying unmelted asthenosphere
below the ridge.

Lithosphere density increases over time, however,
because of changing temperature and composition.
The lithosphere cools, contracts, and becomes denser
with increasing age and distance from the mid-ocean
ridge where it formed (Figure 12.25). The lithosphere
also thickens with age (see Figure 12.16). The lithosphere thickens at its base when asthenosphere cools
sufficiently to have the strength characteristics of lithosphere. The asthenosphere "added" to the base of the
cooling, thickening lithosphere has the high density of
nonmelted upper mantle, so the lithosphere density increases as it gains more and more of this dense peridotite. By the time a plate is 15 million years old, it has
gained enough dense peridotite along its thickening
base and has cooled and contracted enough so that the
overall lithosphere density is about equal to that of the
underlying asthenosphere (Figure 12.25).

Oceanic lithosphere older than 15 million years is
denser than the asthenosphere, but it does not immediately sink into the deeper asthenospheric mantle. The
strong lithosphere resists downward bending, and the
viscosity of the asthenosphere resists the sinking of the lithosphere. In addition,
any plate actively growing at a mid-ocean ridge contains some parts that are
young, warm, thin, and less dense than asthenosphere, and these less dense parts
help hold up the denser areas. An analogy would be your ability to float easily in
water if you wear a buoyant life preserver; less dense young lithosphere likewise
buoys up denser, older lithosphere.

When plates converge, and the lithosphere is denser than the asthenosphere, subduction begins with the densest plate sliding into the asthenosphere.
Where the subducting plate reaches a depth of 80–100 kilometers, the basaltic
crust metamorphoses to very dense eclogite. This converts the crust from rock
that is less dense than asthenosphere into rock that is denser than asthenosphere (Figure 12.25). The cold lithospheric mantle is also significantly denser
than the hot asthenosphere. The strong temperature contrasts and the formation of eclogite make the lithosphere substantially denser than asthenosphere.
The sinking edge of the dense subducted plate is now like a heavy anchor
pulling a chain downward through water. As more lithosphere subducts, and the
volume of the dense eclogite anchor increases, it becomes even more difficult
for less dense parts of the plate to resist being pulled down the subduction zone.
Indeed, sinking dense lithosphere is what moves the plates.

Types of Convergent Plate Boundaries

Features and processes at convergent margins vary from place to place depending on the types of plates that converge. There are two types of plate lithosphere
(continental, oceanic), so there are three types of boundaries: (1) oceanic-
oceanic, (2) continental-continental, and (3) continental-oceanic. **Figure 12.26**
illustrates how these subduction zones work.

The different characteristics of the three boundary types tie into the question of how subduction takes place to begin with—the subducted plate must
not only be denser than the asthenosphere but it must also be denser than the
overriding plate. Continental crust is much less dense than mantle peridotite.

As an outcome, when oceanic and continental plates converge, the oceanic plate always subducts (Figure 12.26).

The different "subductability" of continental and oceanic crust also explains another mystery left over from Chapters 8 and 9—the antiquity of continental crust compared to the youthfulness of oceanic crust. Plates with thin, dense oceanic crust readily subduct at convergent plate boundaries, but thick, low-density continental crust cannot subduct to any great extent. Continental crust is virtually permanent, except to the extent that it gradually weathers and erodes away as sediment. Oceanic crust is continuously destroyed by subduction at the same rate it is created at mid-ocean ridges. As a result, crust older than 1 billion years dominates continents, but there is no crust in the modern ocean basins older than about 180 million years. The ancient continental crust is scarred by faults, folds, and igneous plutons resulting from billions of years of plate tectonics, and these faults are weaknesses that account for some within-plate earthquakes.

Continent-continent convergent boundaries are extremely complex and cannot persist for long intervals of geologic time because of the buoyant nature of continental crust compared to asthenospheric mantle, as seen in Figure 12.26. Crust on both sides of the plate boundary rise because continental crust, like converging ice floes in Figure 12.5, cannot subduct. The best example of ongoing continent-continent convergence is the collision of India with Southeast Asia (Figures 12.3 and 12.20), beginning about 50 million years ago. The collision doubled up the crust, uplifted the Himalayas as the highest mountains on Earth and raised the Tibetan Plateau, an area roughly half the size of the contiguous United States, to a mean elevation of 5 kilometers above sea level. The collision is comparable to wrecking a house with a bulldozer, shattering the building and shoving it off its foundation. Likewise, eastern Asia shattered along ancient faults and slides eastward out of the way toward the Pacific Ocean. These far-flung effects of the Indian collision account for earthquakes throughout China (Figures 12.6 and 12.8).

Oceanic lithosphere subducts beneath continental lithosphere prior to continent-continent collision. Eventually, all of the oceanic lithosphere is consumed, which brings another continental block into the trench beneath the edge of the overriding continental lithosphere (Figure 12.26). Thrust-faulted slivers of oceanic crust and lithospheric mantle slide up onto land

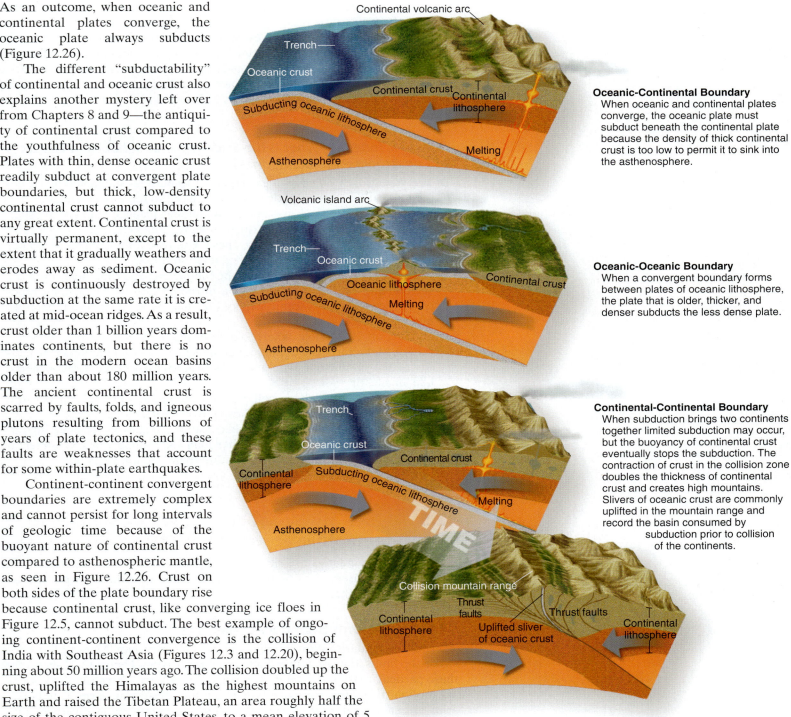

Oceanic-Continental Boundary
When oceanic and continental plates converge, the oceanic plate must subduct beneath the continental plate because the density of thick continental crust is too low to permit it to sink into the asthenosphere.

Oceanic-Oceanic Boundary
When a convergent boundary forms between plates of oceanic lithosphere, the plate that is older, thicker, and denser subducts the less dense plate.

Continental-Continental Boundary
When subduction brings two continents together limited subduction may occur, but the buoyancy of continental crust eventually stops the subduction. The contraction of crust in the collision zone doubles the thickness of continental crust and creates high mountains. Slivers of oceanic crust are commonly uplifted in the mountain range and record the basin consumed by subduction prior to collision of the continents.

▲ **Figure 12.26 Three types of convergent plate boundaries.** Geologic features of convergent plate boundaries differ depending on the types of lithosphere that meet at the boundary.

during convergence and mark the former presence of an ancient ocean intervening between continents.

All modern continents show evidence of former continent-continent collision zones, marked by highly deformed metamorphic rocks in the eroded remnants of once-tall mountains. The Appalachian Mountains, for example, mark a zone of continent-continent convergence when Pangea formed from colliding continents in the late Paleozoic (Figures 12.2 and 12.8).

Putting It Together–*What Is the Evidence That Subduction Occurs at Convergent Plate Boundaries?*

- The Wadati-Benioff zone of earthquake foci outlines subducted plates descending at an angle from deep-sea trenches into the asthenosphere.
- Seismic tomography shows subducted plates as inclined zones of unusually fast seismic velocity in the asthenosphere. Subduction zones also coincide with areas of low heat flow. These data reveal the cold, dense plates sinking into the asthenosphere at subduction zones.
- Metamorphic fluids released from the subducted plate cause melting in the surrounding asthenosphere. Magma intruding into the overriding lithosphere accounts for volcanic activity, high heat flow, and high-temperature metamorphism.
- Subduction is possible where old lithosphere is denser than asthenosphere. Metamorphism of mafic crust to eclogite further increases the density of the subducting plate and pulls it downward in the asthenosphere like an anchor.
- Where continental plates collide, high mountain ranges form because thick, low-density continental lithosphere cannot subduct.
- Oceanic lithosphere is continuously consumed at convergent plate boundaries so there is no ocean floor older than 180 million years. Continental crust cannot subduct and is mostly more than 1 billion years old.

12.5 What Is the Evidence That Plates Slide Past One Another at Transform Plate Boundaries?

Not all plate boundaries are sites of lithosphere creation or destruction. The lithosphere is conserved along transform boundaries because the plates move alongside one another.

What a Transform Boundary Looks Like

Transform boundaries show up on the plate-boundary map (Figure 12.3) where strike-slip faults join segments of other boundary types. **Figure 12.27** shows that these particular strike-slip faults are called "transforms" because plate boundary motion transforms from one type to another. Most transform plate boundaries are short and connect spreading-ridge segments along divergent plate boundaries (see Figure 12.3). Only a few transforms, including the San Andreas fault in California (Figure 12.27), form long plate boundaries. The San Andreas transform connects a divergent zone, to the south, with a convergent zone, to the north, and forms the boundary between the Pacific and North American plates (Figure 12.3).

What Fault Displacements Show

Strike-slip faults should be found at all interpreted transform boundaries. This prediction is easily tested for continental transforms by examining features displaced across the fault. **Figure 12.28** shows such a test for the San Andreas transform boundary in California. Not only do the field relationships confirm the

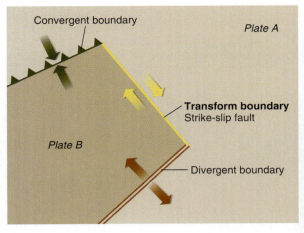

▶ **Figure 12.27 What a transform boundary looks like.**
Plates slide past one another along strike-slip faults at transform boundaries. Transform boundaries connect and transform the relative motion between typically longer divergent or convergent plate-boundary segments. Plates A and B in the diagram move in opposite directions and share a transform boundary that connects a divergent boundary with a convergent boundary. The San Andreas fault in California is part of a transform plate boundary zone separating the North America and Pacific plates.

predicted strike-slip motion, but the rate of motion along the transform, about 5.6 cm/yr, is comparable to plate velocities at mid-ocean ridges. Notice, as well, in Figure 12.28 that this plate boundary in California is a zone, with movement across many faults, rather than a simple line on the map. In stronger oceanic lithosphere, most transform boundaries are single faults.

What Earthquakes on Oceanic Transforms Show

Figure 12.29 shows how plate tectonics predicts transform faults should work where they connect mid-ocean-ridge segments. In this example, it *appears* that two mid-ocean-ridge segments are displaced by a left-lateral strike-slip fault. According to plate tectonics, however, the ridge is not an old hill displaced by a younger fault. Instead, the ridge segments and transform fault make up simultaneously active plate boundaries. Two hypotheses from plate tectonics led to this interpretation.

First hypothesis—the displacement across the example transform fault should be right lateral, not left lateral, because the plates move away from one

Fence displaced 2.6 m during 1906 earthquake

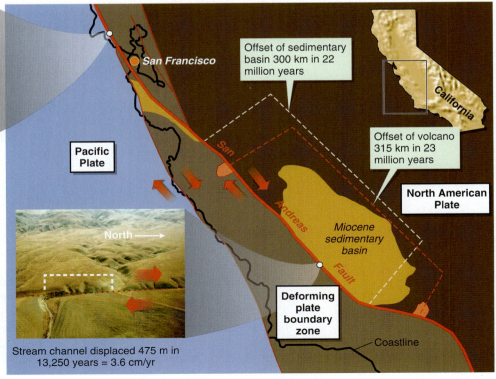

Stream channel displaced 475 m in 13,250 years = 3.6 cm/yr

◀ **Figure 12.28 Calculating plate motion along the San Andreas transform boundary.**
The Pacific plate slides northwestward past the North American plate along a transform boundary (see Figure 12.3). The San Andreas fault is the longest of several, nearly parallel faults that define the plate boundary zone. Offset of geologic features that match across the fault provides estimates of the speed of plate motion, which currently is 3.6 centimeters per year. Adding this speed to calculations for other faults in the boundary zone reveals a total movement of 5.6 centimeters per year between the Pacific and North American plates.

▶ **Figure 12.29 Determining motion on oceanic transform faults.**
Most transform boundaries connect divergent-boundary segments along mid-ocean ridges. The transform strike-slip faults reveal the plate motion between ridge segments and do not displace the ridge. The direction that each plate moves away from the nearby spreading ridges determines the movement on either side of a transform boundary. Faults exist only between spreading-ridge segments. Elsewhere, the plates move in the same direction and at the same velocity so there is no plate boundary and no fault displacement. In those areas, a tectonically quiet fracture zone separates lithosphere of different age.

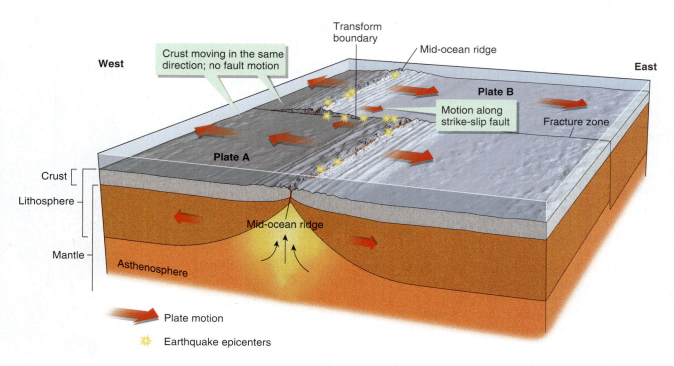

Motion at Transform Boundaries: *See how transform faults connect other plate boundaries.*

another along the segmented divergent plate boundary. To see why this is the case, notice in Figure 12.29 that the lithosphere of Plate B moves east away from the mid-ocean ridges, whereas lithosphere of Plate A moves west. These opposite directions of motion require right-lateral displacement on the transform boundary.

Second hypothesis—the transform fault is only active between the ridge segments that it connects. East and west of these ridge segments the continuation of the fault line separates lithosphere belonging to the same plate (see Figure 12.29). There should not be movement across these continuations of the fault line because the lithosphere on either side moves in the *same* direction at the *same* velocity.

Earthquake seismographs successfully tested these two hypotheses. Earthquakes only occur along the transform between the ridge segments and along the divergent boundary (Figure 12.29). The apparent fault traces that *seem* to extend farther beyond the mid-ocean ridge segments are not faults but simply **fracture zones** that separate lithosphere of different age and thickness within the same plate. Sophisticated analyses of seismograms allow geologists to determine the direction of movement along faults during earthquakes. In the Figure 12.29 example, the seismograms reveal that transform-boundary earthquakes are caused by right-lateral strike-slip motion consistent with the plate motion inferred from the divergent-boundary segments.

> *Putting It Together*—*What Is the Evidence That Plates Slide Past One Another at Transform Plate Boundaries?*
>
> ■ Transform plate boundaries are a special type of strike-slip fault along which plate motion from one boundary segment is transformed to the next segment.
>
> ■ The direction and velocity of strike-slip displacement on transform boundaries are consistent with the motions predicted by plate tectonics.
>
> ■ Most transform boundaries are short faults between mid-ocean-ridge segments. Transforms line up with tectonically inactive fracture zones that separate lithosphere of different age, thickness, and elevation but belong to the same plate and move in the same direction at the same velocity.

12.6 What Does the Mantle-Plume Hypothesis Explain that Plate Tectonics Cannot Explain?

Plate tectonics theory explains processes at, or close to, plate boundaries and does not explain active volcanoes and deformation *within* plates. The causes of these phenomena are partly speculative and less well established than plate tectonics theory. This section explains the mantle plume hypothesis, an incompletely tested companion to plate tectonics theory that offers the potential to round out a global view of tectonic processes.

The Problem of Hot Spots

Figure 12.7 reveals many volcanoes that are distant from plate boundaries and not related to plate-boundary processes. The term **hot spot** describes an area of voluminous volcanic activity not explained by melting processes at plate boundaries. **Figure 12.30** shows that there are about 40 hot spots on Earth. Curiously, and perhaps more importantly, many hot spots are within two regions of the world where seismic tomography implies warmer temperatures and corresponding upwelling in the lower mantle (Figure 12.30).

Hot spots are found in two types of locations.

1. Large volumes of young volcanic rocks in isolated locations very far from plate boundaries represent the first type of hot spot. Examples are Hawaii and Yellowstone, pictured in **Figure 12.31**. On average, about 0.1 cubic kilometer of lava erupts on the island of Hawaii each year. This is enough lava to bury San Francisco a meter deep each year and is more than 5 percent of the volume erupted along all of the world's mid-ocean ridges. This incredible eruption rate built huge shield volcanoes more than 9 kilometers above the seafloor (Figure 12.31). Yellowstone National Park marks a similarly prolific volcanic hot spot that erupted 6000 cubic kilometers of magma over the last 2 million years, mostly as rhyolitic tuff.

2. Unusually prolific volcanism along or near a divergent boundary represents the second hot-spot type. The volcanic activity is so excessive compared to

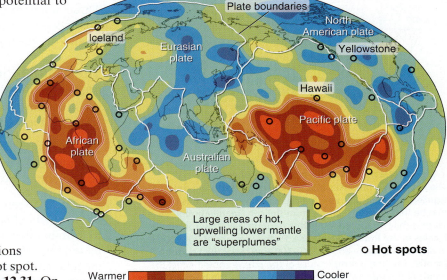

Warmer — Slower   Average   Cooler — Faster
S-wave velocity

▲ **Figure 12.30 Where hot spots are located.**
Each circle on this map is the location of a hot spot—an area of extraordinary, high-volume volcanic activity that usually is not at a plate boundary. The map also shows seismic-tomography results from the lower mantle, just above the outer core. Many hot spots coincide with areas of seismically slow, and presumed warmer, lower mantle "superplumes" centered beneath Africa and the South Pacific.

◀ **Figure 12.31 What hot spots look like.**
The "Big Island" of Hawaii (left) marks the location of a hot spot that has built volcanoes more than 9 kilometers above the seafloor in the middle of the Pacific plate. The black "fingers" visible in this Space Shuttle photograph are historically erupted basalt lava flows. The famous geysers and other thermal features of Yellowstone National Park (right) relate to a volcanic hot spot within the North American plate. The last major volcanic eruption at Yellowstone, about 620,000 years ago, blanketed most of North America in ash.

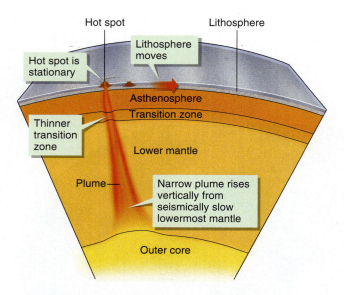

▲ **Figure 12.32 The mantle-plume hypothesis.**
The hypothesis proposes that stationary hot spots are the surface expression of plumes, which are unusually hot columns of the mantle that rise convectively from the core-mantle boundary. Plumes may be rooted in regions of unusually slow seismic velocities in the lowermost mantle. Plumes are about 200–500 km in diameter and are about 200–300°C warmer than surrounding mantle. The hotter temperature of the plume should affect the depth of the mineral transformations defining the transition zone such that this zone is thinner below hot spots than it is elsewhere.

typical mid-ocean ridges that the hot-spot volcanoes commonly build up well above sea level. Iceland is an example of this type of hot spot (see Figure 12.11a). The volcanic production is so staggeringly high at these locations that there must certainly be some mechanism other than normal divergent-margin processes to generate all of the magma.

Basalt erupted at oceanic hot spots, such as Hawaii and Iceland, differs from mid-ocean ridge basalts only in the finest details of chemical compositions. These details are important, however, because they indicate that a different and possibly deeper part of the mantle melts to produce hot-spot basalt compared to mid-ocean-ridge basalt.

The Mantle-Plume Hypothesis

The mantle-plume hypothesis emerged in the late 1960s to explain hot spots at the same time that plate tectonics theory was established. The hypothesis, illustrated by **Figure 12.32**, has two fundamental components:

1. Hot spots form where narrow columns of unusually hot mantle, called **plumes**, rise by convection from the core-mantle boundary.
2. The plume locations are stationary in the mantle.

The first component of the hypothesis explains the peculiar composition of hot-spot-island basalts. Rising asthenosphere decompresses and partially melts to form basaltic magma. If the melting asthenosphere originates deep below the plates, then the resulting basaltic magma has a different composition from the magma produced by melting near the base of the lithosphere to form mid-ocean-ridge basalt (see Figure 4.21a for a refresher on magma production at hot spots and mid-ocean ridges).

The second component of the hypothesis explains the ages of extinct volcanoes that trail away from hot spots to produce a hot-spot track. For example, **Figure 12.33** shows lines of extinct volcanic islands and submarine volcanoes, called **seamounts**, which continue for thousands of kilometers across the Pacific Ocean until terminating at an active hot spot volcano. Volcanic rocks are progressively older along each line, away from the active hot spot. Figure 12.33 shows how these data suggest a stationary plume of rising hot asthenosphere that melts to form a line of volcanoes as the lithospheric plate passes overhead. *If* this hypothesis is correct, then **Figure 12.34** shows how to determine the speed and direction of plate motion from the locations and ages of volcanic rocks along a hot-spot track. Of course, this plate "speedometer" only works if the hypothesis is correct. It is physically plausible that the hypothetical plume also moves, which negates the plate-velocity calculation unless the plume velocity is also known. Whether hot spots move or not remains unclear. Most data suggest that current hot spot movement is negligible and certainly slower than plate motion.

Testing the Plume Hypothesis

If seismic tomography data reveal the presence of narrow cylinders of lower-velocity (warmer) mantle extending from the core-mantle boundary up to hot-spot volcanoes, then the plume hypothesis would be positively tested. Figure 12.30 shows the existence of seismically slow, upwelling lower mantle around the locations of some hot spots, but not all of them. The large areas of seismically slow lower mantle rising upward beneath Africa and the southern Pacific are called "superplumes." These superplumes are not narrow plume columns connecting to distinct hot spots, like Hawaii, and are not what the plume hypothesis predicts.

These tests alone may not be adequate to test the plume hypothesis, because seismic-tomography data have limitations. Seismic-tomography data reveal many details about the upper mantle, but they still provide only a hazy view

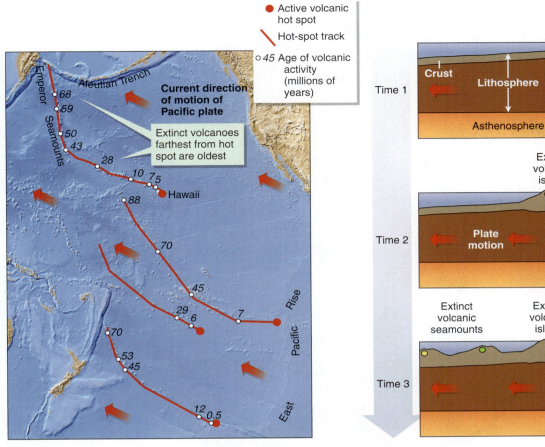

Active hot-spot volcanoes in the Pacific Ocean are present at one end of long chains of volcanic islands and submerged volcanic seamounts. Radioactive-isotope ages of volcanic rocks show that the volcanoes are progressively older at greater distances from the active hot spot.

◄ **Figure 12.33** **Hot spots leave tracks on moving plates.**

Hot-spot tracks form where lithosphere moves slowly across a nearly stationary column of rising and melting mantle. Volcanoes form at the hot spot but become extinct because plate motion carries them away from the hot spot. New volcanoes form behind the extinct volcanoes. The colored dots show shifting positions of volcanoes through time as plate motion carries the lithosphere over the hot spot.

Hot Spots and Plumes: *See how an island chain forms by plate motion across a hot spot, which might connect to a hypothesized mantle plume.*

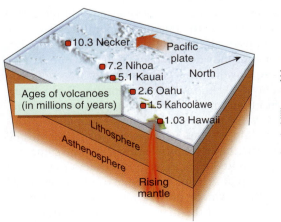

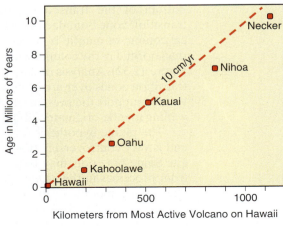

◄ **Figure 12.34** **Using hot spots to estimate plate velocity.** Pacific plate motion produced a hot-spot track of progressively older volcanoes extending northwestward from the active volcanoes on Hawaii. *If* the hot spot beneath Hawaii is stationary, then the age and distance of the older volcanoes from Hawaii indicate how fast the plate moves. On a graph of volcano age plotted against distance from Hawaii, the slope of the line is velocity. The ages and distances of the volcanoes suggest a plate velocity of about 10 centimeters per year.

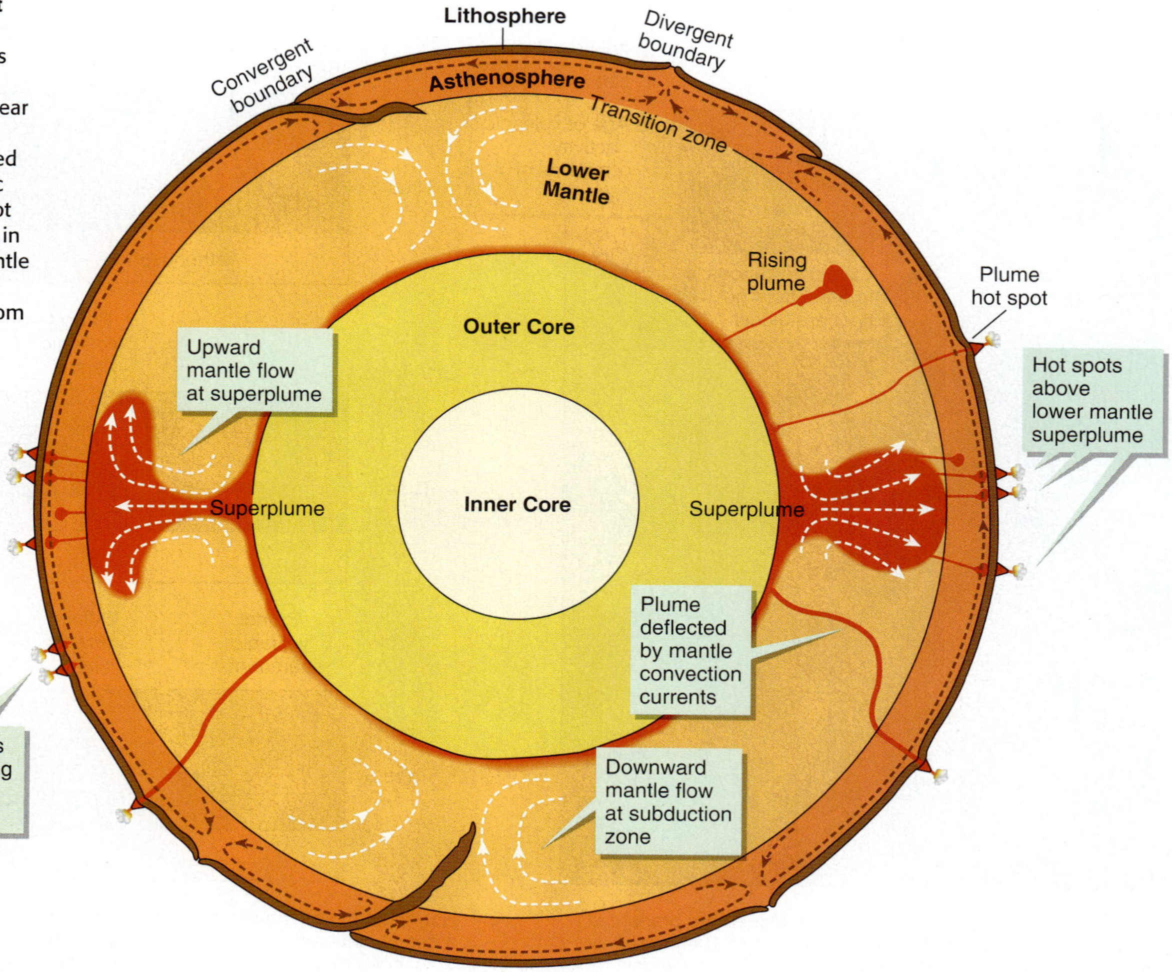

▶ **Figure 12.35 Varieties of plumes and hot spots.**
Current data suggest that only a few hot spots can be true plumes that rise from the core-mantle boundary. Some hot spots originate near the transition zone and possibly form above lower-mantle superplumes, which are centered on opposite sides of Earth beneath the Pacific Ocean and Africa (see Figure 12.30). Other hot spots originate directly below the lithosphere in the upper asthenosphere. In some cases, mantle convection currents may deflect proposed plumes to form hot spots at the surface far from where they originate in the lower mantle.

of the deepest mantle because fewer measured seismic waves pass through the lower mantle than the upper mantle. In addition, the standard seismic-tomography methods only detect large features, and plume columns less than 500 kilometers wide will not show up in images like the ones shown in Figures 10.8 and 12.17.

Geophysicists are using new instruments to gather seismic data and new methods for analyzing earthquake data that can detect the hypothesized narrow plumes. One test involves determining the exact depth of the mineral transformations that cause the seismic velocity discontinuities that define the top and bottom of the transition zone. The mineral transformations defining the transition-zone boundaries partly depend on the mantle temperature. If a vertical column of mantle is hotter than normal, then the transition zone is thinner than normal in the column with an unusually deep top and a shallower base (see Figure 12.32). Analysis of data from Hawaii and Iceland suggests unusually thin transition zones over areas only about 200 kilometers in diameter. These observations support the presence of hot upper mantle but do not prove that the upwelling mantle originates all the way down at the core-mantle boundary.

The plume hypothesis for explaining hot spots is not yet adequately tested and is very controversial. It is possible that not all hot spots have the same origin. Some hot spots may be deeply rooted plumes, whereas others may result from processes in the asthenosphere that are not yet understood. **Figure 12.35** illustrates a modified plume hypothesis where only some plumes are narrow vertical columns extending from the base of the mantle to Earth's surface. Existing data suggest, but do not prove, that there are at least six hot spots linked to such plumes, and Hawaii is a possible seventh. Perhaps the variable viscosity of the mantle and some physical boundaries, such as those defining the transition zone, also influence the shape of plumes. Some narrow plumes might rise from the transition zone above lower mantle superplumes rather than directly from the top of the core, which would explain the occurrence of many hot spots above the superplumes (Figure 12.30). At the other extreme, existing tomographic data suggest that some hot spots, Yellowstone, for example, originate within 200 kilometers of the surface, and may somehow relate to tears in the lithosphere.

Tracking narrow plumes downward through the lower mantle is challenging not only because of the limitations of seismic-tomography data but also because of mantle motion. Convection currents may deflect a vertically rising plume so that its surface expression is 1000 kilometers away from where it is hypothetically rooted in the lower mantle (Figure 12.35). This likely mantle motion also calls into question the validity of the stationary plume concept.

Putting It Together–What Does the Mantle-Plume Hypothesis Explain that Plate Tectonics Cannot Explain?

- Hot spots are areas of prolific generation of igneous rocks not explained by plate tectonics. Some hot spots are present within plates, whereas others are areas of unusually intense volcanic activity at or near divergent plate boundaries.
- Tracks of extinct volcanoes are progressively older at greater distance from active hot-spot volcanoes.
- The plume hypothesis suggests that hot spots mark places where hot mantle convectively rises to the base of the lithosphere from the core-mantle boundary.
- Conclusive evidence has not yet been found to demonstrate narrow plumes of mantle rising through the entire thickness of the mantle. Most hot spots, however, are located above large upwelling superplumes in the lower mantle.

12.7 **How Do We Know** ... That Plates Move in Real Time?

UNDERSTAND THE PROBLEM
Does the Lithosphere Move as Predicted by Plate Tectonics Theory?
Consider the possibility of unambiguously testing plate tectonics theory by measuring actual lithosphere movement. Detected movement should be consistent with plate tectonics theory, which means that rates and directions of movement of parts of the lithosphere should be different as one crosses the plate boundaries.

This very direct test of plate tectonics requires two pieces of information:

1. Theoretical calculations of predicted speed and direction of motion at different locations on Earth's surface based on plate tectonics theory.
2. A method to measure very slow plate motion, which is predicted to be only about 1–10 centimeters per year.

Comparison of predicted and measured motions completes the test.

VISUALIZING RELATIVE AND ABSOLUTE MOTION
How Do Geologists Predict Plate Motion?
Review what you do know about data that predict the speed and direction of plate motion. The arrows in Figure 12.3 show the expected directions of motion between plates where they share a boundary, based on whether that boundary is divergent, convergent, or transform. Maps of seafloor age (Figure 12.14) show

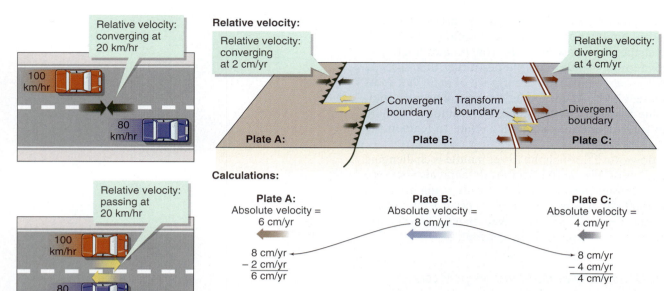

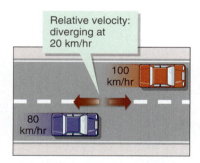

▲ **Figure 12.36 Distinguishing between absolute and relative velocities.**
Relative velocity is the difference in absolute velocity between two moving objects. Two automobiles moving in the same direction at different absolute velocities have relative velocities that show the vehicles converging, passing, and diverging. Likewise, relative velocities of plates determined by spreading rates at divergent boundaries, slip rates along transform boundaries, and convergence rates at convergent boundaries do not reveal absolute velocities. In this example, three plates have absolute velocities that move them in the same direction but at different speeds. As with the passing automobiles, this means that their relative velocities are such that they have divergent, convergent, or transform boundaries.

Relative and Absolute Motion: See how to distinguish between relative and absolute motion.

EXTENSION MODULE 12.1

Describing Plate Motion on the Surface of a Sphere.
Learn how to describe plate motion on a sphere.

the rate of plate separation by seafloor spreading along divergent boundaries. Similarly, data like those shown in Figure 12.28 provide the rate of motion between two plates along a transform boundary.

These uses of plate-boundary information reveal *relative velocities* between plates along the boundary. These data do not provide *absolute velocities*, or how fast, and in what direction, each plate moves independently of its neighbors. **Figure 12.36** illustrates the difference between relative and absolute velocities. If you catch up with and pass a car on the highway, then your speed is faster relative to the speed of the other car, but this relative speed does not specify the absolute speed of either vehicle. The absolute speeds are known from reading the speedometers in each car. Notice in Figure 12.36 that cars, and plates, can move in the same direction in an absolute sense but still be moving toward, away from, or past one another in a relative sense. This means that relative motions at plate boundaries not only fail to indicate the absolute speed of the plates but also do not indicate the actual direction that plates move. Geologic data reveal the relative speed at which two plates move toward, away from, or past one another at a shared boundary, but these data do not specify the absolute velocity of any plate, and plates do not have speedometers.

Figure 12.37 illustrates one possible pattern of absolute plate motions that satisfies what is known about relative plate motions. The speeds and direction of movement are the predicted average for the last three million years and they vary from plate to plate, which accounts for relative motion at plate boundaries. The velocities also seem to vary within plates, which may seem odd, but this is a necessary result of moving plates along the outside of the sphere-shaped Earth. There are many ways that this map could be drawn that would still satisfy the measured relative motions at plate boundaries. This particular approach is based on a physical assumption that the relative motions between boundaries that lie along a circle drawn around the Earth add up to zero. In other words, although the plates move relative to one another, there is no overall slip of the whole lithosphere in one direction or the other around the planet.

The reason why geophysicists cannot determine a single pattern of predicted absolute motion is that there is no reference frame for measuring absolute velocity. The importance of frame of reference is illustrated by an automobile race. When one car passes another, the drivers perceive the relatively slow difference in their absolute speeds so that the passing car seems to be moving very slowly compared to the car being passed. If, on the other hand, you watch the race from the stands, then you see the two cars whiz by at dizzyingly fast absolute velocities that you fully appreciate from your motionless reference frame but is less obvious to the drivers in the moving cars. What is lacking for predicting plate motion is the reference frame of a motionless spectator.

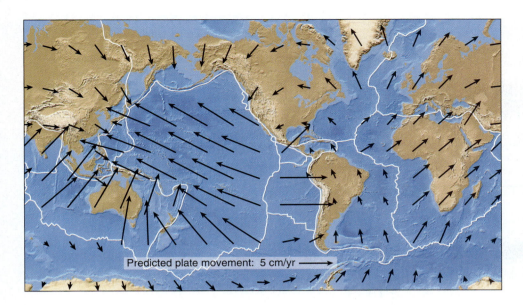

◄ **Figure 12.37** Predicting plate motion.
These arrows show a predicted pattern of global plate motions that are consistent with the relative motions observed at all plate boundaries.

TESTING THE PREDICTIONS

How Do Geologists Measure Absolute Plate Motion?

The next step is to measure the actual plate movements and compare the measurements to the predictions. Surveying locations over and over provides an indication of whether or not the locations are moving. These measurements must, however, be made from a reference frame that is not attached to the moving plates and must be able to precisely detect the very slow speeds of predicted plate motion.

The Global Positioning System, or GPS, provides the ability to make such measurements. GPS surveying, illustrated in **Figure 12.38**, uses radio signals from special satellites orbiting Earth. The positions of the satellites in space are known with very high precision and, importantly, are not moving with the plates. The time it takes for a signal to travel from a satellite to a GPS receiver on the ground determines the distance between the receiver and the satellite. The position of the receiver on Earth is accurately determined by simultaneously determining the distances from the receiver to four or more satellites.

You may be familiar with small, inexpensive, handheld GPS receivers that confidently locate hikers and boaters, or even provide navigation aids within some automobiles. Geologists use special, and more expensive, research-grade

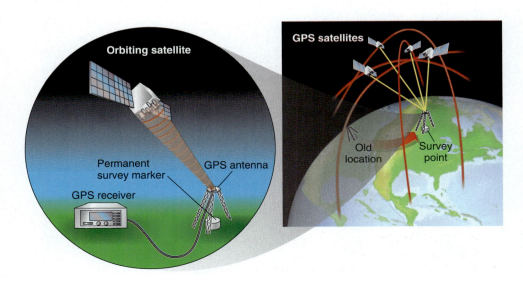

◄ **Figure 12.38** Using GPS to determine real-time plate motion.
Global Position System receivers detect radio signals from satellites located at very precisely known positions above Earth. The time required for the signal to travel from a satellite to a GPS receiver depends on the distance between them. Geologists calculate a very precise location of the receiver by using the signals simultaneously received from many satellites. If repeat measurements at the same station over a period of years yield different locations, then the station is moving.

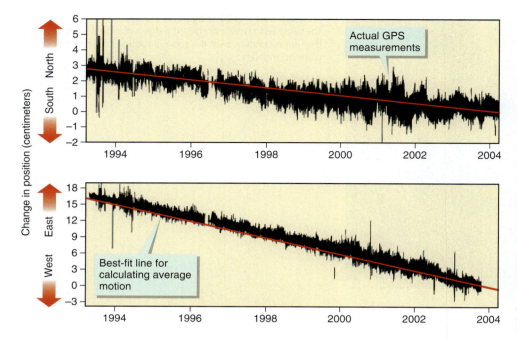

◀ **Figure 12.39 How fast does North America move?**
These graphs plot GPS data collected at a station in eastern Wisconsin. Drawing a line through the data averages out variability in the measurements. Between 1994 and 2004, this station moved southward by 2.5 cm, and westward by about 15 cm. This translates into plate motion of 1.5 cm/yr in a direction slightly south of due west.

GPS equipment to locate a point on Earth to within a few millimeters (about half the diameter of a U.S. dime). To determine plate motions, survey stations are permanently marked by a low concrete monument, and surveyed repeatedly over a number of years (Figure 12.38). **Figure 12.39** shows how the survey data determine the speed and direction of a GPS monument permanently attached to a moving plate. Data from more than one hundred GPS stations from around the world routinely monitor real-time plate motion.

COMPARING MEASUREMENTS WITH PREDICTIONS

Do Plates Move as Predicted?

Figure 12.40 shows the current ongoing lithosphere motions indicated by GPS measurements. GPS stations are only on land, so they are not widely distributed over all of the plates. Another complexity relates to the dynamic nature of Earth. The satellite data tracks locations on the surface as if they were points on a large, three-dimensional graph, with three axes projecting outward from

▼ **Figure 12.40 Map of real-time plate motion.**
The red arrows on this map depict the actual measured direction of surface movement and speed at GPS stations located around the world. These measurements correspond very closely to the predicted plate motions, which are shown by black arrows.

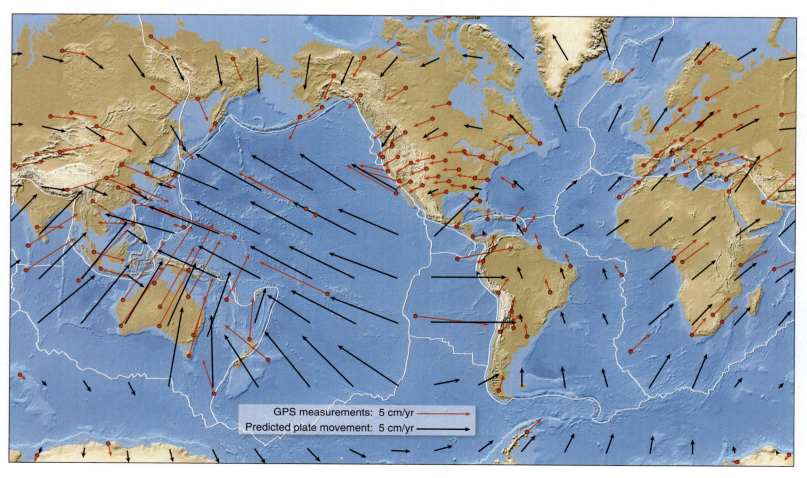

Earth's center, or more precisely from the center of mass. This seems simple at first glance—a vertical graph axis would emerge through the North Pole and the other two would occupy a horizontal plane at the equator. This hypothetical graphical reference frame is still not quite like a motionless spectator watching a car race. The rotation axis at the North Pole actually wobbles and moves relative to the surface, and the center of mass also shifts and is not exactly at the center of the planet. To accommodate these problems, everyone using GPS data to track the motion of stations on the surface adopts a frame of reference that makes the same assumption of no overall slip of the lithosphere around Earth's surface that is used in the prediction shown in Figure 12.37.

Figure 12.40 allows you to compare the predicted plate motions and those measured by GPS. The two patterns of motion match very closely. Slight differences result from the fact that the predicted velocities represent averages over the last 3 million years, whereas the GPS data depict average motion over 10 years; velocities may not be steady at the same rate over long time periods.

INSIGHTS

How Do the Results Test Plate Tectonics Theory?
Real-time measurements of absolute plate motion match the velocities and relative plate motions predicted from plate tectonics theory. This clearly is a positive test of plate tectonics. In detail, however, the measurements show that plates do not move in a perfectly rigid fashion, as predicted by the theory when it was first proposed in the late 1960s.

Figure 12.41 illustrates this nonrigid behavior in western North America. The map was made from GPS data by only looking at the movement of GPS stations relative to the North American plate. If the plates are perfectly rigid without internal deformation (see Figure 12.4), then all locations on the North American plate should appear stationary on this map, and only those locations on the adjacent Pacific plate should move. Instead, there is movement of some stations in western North America, and this movement increases toward the San Andreas transform fault. The data confirm, therefore, that the San Andreas fault does not represent all of the movement between the North American and Pacific plates (see Figure 12.28). Rather than picturing this single-fault plane as the plate boundary, instead there is a plate-boundary zone that is several hundred kilometers wide.

> ▼ **Figure 12.41** **Western North America is a plate boundary zone**
> Arrows on this map depict measured motion of GPS stations compared to the center of North America. Stations in western California move northwestward on the Pacific plate. Other stations in the western United States also move slightly northwest compared to the center of the continent. These results show that the North American plate is not perfectly rigid but deforms near its boundary with the Pacific plate.

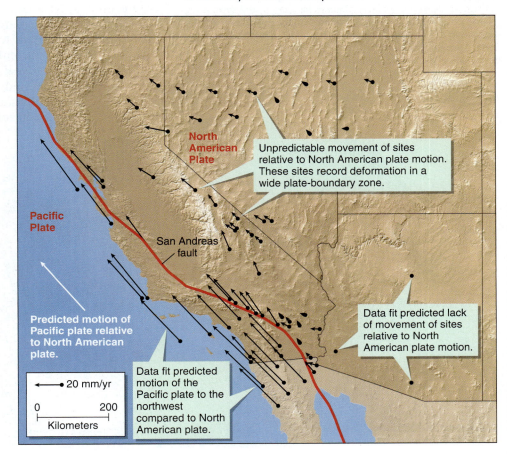

North American Plate

Pacific Plate

San Andreas fault

Unpredictable movement of sites relative to North American plate motion. These sites record deformation in a wide plate-boundary zone.

Data fit predicted lack of movement of sites relative to North American plate motion.

Data fit predicted motion of the Pacific plate to the northwest compared to North American plate.

Predicted motion of Pacific plate relative to North American plate.

← ● 20 mm/yr

0   200
├────────┤
Kilometers

Putting It Together–*How Do We Know …* *That Plates Move in Real Time?*

■ Relative motion describes the direction and speed of one plate compared to another, whereas absolute motion describes the actual velocity of a plate.

■ Global Positioning System (GPS) satellites provide a real-time test of plate motion by repeatedly surveying locations on Earth's surface to see if the locations move.

■ GPS data show that land sites move with speeds and directions that are generally compatible with plate tectonics theory but also show that the nonrigid characteristics of plates can produce wide boundary zones rather than narrow discrete boundaries as originally proposed by the theory.

12.8 What Forces Cause Plate Motion and Plumes?

When Alfred Wegener proposed continental drift, many geologists rejected his hypothesis because it lacked a plausible mechanism to explain large-scale motion on Earth's surface. Plate tectonics theory developed simultaneously with efforts to document mantle convection (Chapter 10). Linking convection with plate tectonics explains plate motion.

Plate Motion and Plumes Are Convection

The details of plate tectonics do not automatically fall out of the computer simulations of mantle convection. An essential feature of Earth's lithosphere is that it is simultaneously strong and able to move. By contrast, experiments and computer models that simulate convection assume motion of viscous fluids and do not include strong plates. The strong lithospheric plates influence mantle convection in ways that experiments or computer models cannot readily reproduce. Lithosphere capped by low-density, buoyant, continental crust cannot subduct, so the permanence of continental lithosphere is another variable not normally considered in simulated mantle convection. Nonetheless, convection explains the principal features of global tectonics.

Figure 12.42 links convection and plate motion. Convective downwelling occurs as subduction at convergent plate boundaries, and convective upwelling occurs along divergent boundaries and at hypothesized mantle plumes.

Downwelling at Convergent Boundaries

In classic convection, the upper part of the convecting system cools and sinks in long, linear downwelling zones (see Figure 10.8). In plate tectonics, lithosphere cools conductively as it moves away from a mid-ocean ridge (Figure 12.16) and becomes denser than its surroundings (Figure 12.25). The strength of the lithosphere does not permit downwelling to occur as readily as for a viscous fluid, but where plates move toward one another, the downwelling part of the convection system forms as long subduction zones (Figures 12.35 and 12.42). Geologists currently debate whether or not seismic tomography data adequately reveal subducting plates penetrating into the lower mantle (as illustrated in Figure 12.42) or if the high viscosity of the lower mantle causes sinking plates to pause or even stop in the transition zone.

Convection downwelling at subduction zones determines the direction and speed of plate motion. Gravity pulls the dense subducted slab downward and drags the rest of the plate into the trench (Figure 12.25). Geologists attribute this gravitational slab-pull force (Figure 12.42) to be the primary cause of plate motion because, as **Figure 12.43** shows, the fastest plates have the longest subduction zones along their margins. Careful calculations show that the downward pull of subducting slabs accounts for more than 90 percent of the force required to drive plate motion. Nearly all of the remaining force comes from slab suction, which is flow in the asthenosphere that draws the overriding plate toward the trench because of the downward movement of the subducted plate (Figure 12.42).

▼ **Figure 12.42 How convection explains global tectonics.** The mantle temperature profile on the right, based on Figure 10.7, implies that downwelling originates at the surface, whereas upwelling originates at the core-mantle boundary. Subduction is the downwelling part of convective circulation. The downward motion of the subducted slab exerts a slab-pull force. Subduction also forms slow currents in the asthenosphere that draw overriding plates toward the trenches; this is the slab-suction force. These two forces drive plate motion. Plumes and superplumes are candidates for the upwelling that originates in the lowermost mantle. Some upwelling currents are probably restricted to the upper mantle because of disturbances initiated by the subducting plates; downward subduction displaces deeper, warmer mantle upward in diffuse upwelling zones. These diffuse upwelling currents coincide with divergent plate boundaries, and perhaps with some hot spots that do not have roots below the asthenosphere.

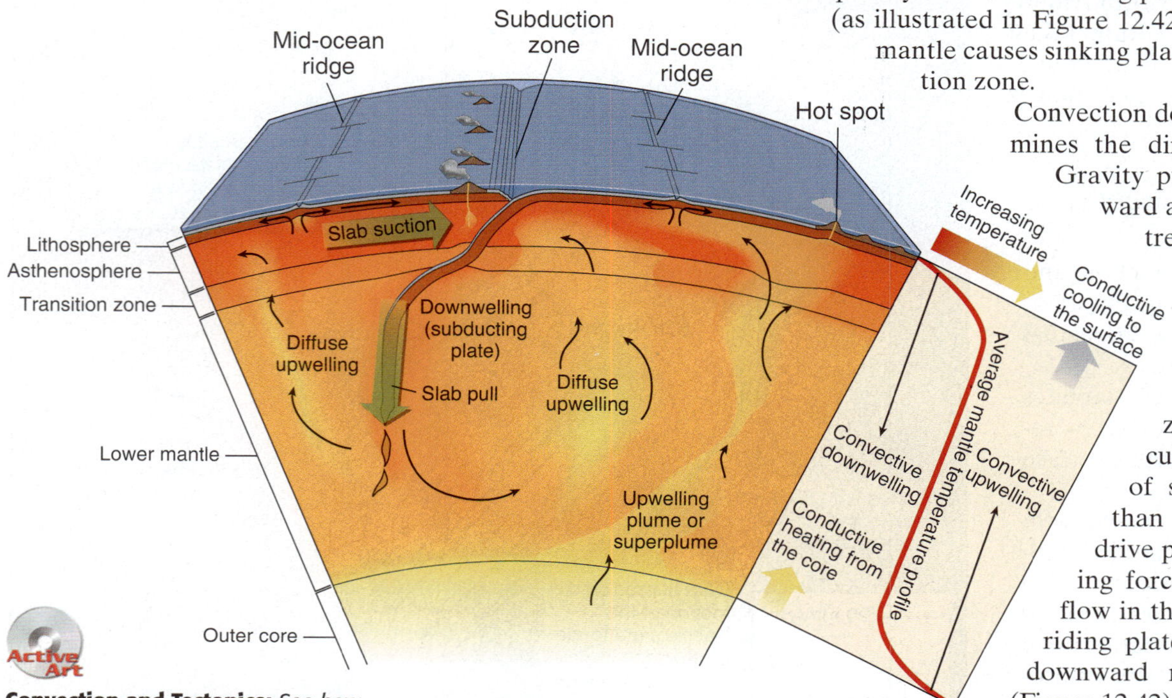

Convection and Tectonics: *See how convection motion relates to plate tectonics and plumes.*

Upwelling at Plumes, Superplumes, and Divergent Boundaries

The link between subduction and convective downwelling is well established, but the exact relationship between plate tectonics and convective upwelling is not yet resolved. Part of the uncertainty relates to the observation that upwelling, as delineated by zones of relatively slow seismic waves, tends to be more diffuse in the mantle than the narrow discrete downwelling zones characterized by fast velocities (Figure 12.42). The more diffuse nature of upwelling is consistent, however, with convection in the mantle, where viscosity varies considerably and heat is provided internally as well as from below (see Figure 10.9). Much of this diffuse, upward mantle motion can also be simply viewed as the asthenosphere displaced upward by downward injection of subducted plates, just as water level rises in a glass if you stick your finger into the water.

Divergent plate boundaries probably represent only part of the upwelling expected by mantle convection. Certainly, asthenosphere moves upward at divergent boundaries, but this upward motion only occurs because deeper mantle moves into spaces where the lithosphere stretches apart by the stresses originating as slab pull and slab suction at subduction zones. The upwelling, seismically slow mantle beneath most mid-ocean ridges only occupies shallow levels of the asthenosphere rather than being a part of convection rising through the whole thickness of the mantle.

Mantle convection should include upwelling from near the core-mantle boundary, because subduction downwelling seems to persist to this great depth (Figure 12.42). Plumes, if they exist, would be the deep upwelling component of convection. Seismically slow zones implying large masses of upwelling mantle from near the core-mantle boundary do appear in the seismic tomography data as the superplumes (Figures 12.30 and 12.35), which enclose most of the hot spots. In some cases, upwelling superplume mantle may migrate toward divergent boundaries (Figure 12.42), partly linking mid-ocean ridges to convective upwelling, and also accounting for the tendency for many hot spots to coincide with divergent boundaries.

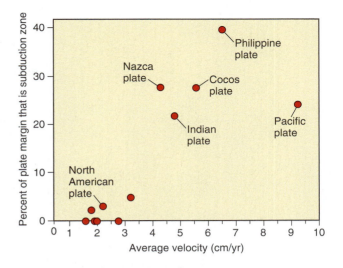

▲ **Figure 12.43** Slab pull moves plates.
Plates with long subduction zones move faster than those with short or no subduction-zone boundaries. This relationship supports the hypothesis that dense, sinking, subducting slabs provide the primary force, slab pull, for plate motion.

Putting It Together—What Forces Cause Plate Motion and Plumes?

■ Plate tectonics, mantle plumes, and superplumes are expressions of convection.

■ Subducting lithosphere represents the downwelling process of mantle convection.

■ Subducting plates provide nearly all of the force required to cause plate motion by slab pull and slab suction.

■ Convective upwelling is more diffuse and occurs in the upper mantle at divergent plate boundaries to fill a void, and in the lower mantle as superplumes, which probably connect to at least some hot spots at the surface.

12.9 What Are the Consequences of Plate Motion over Geologic Time?

Current plate boundaries and plate motions are just a snapshot in a long-running movie of Earth history. Plates move so slowly that there are no discernible changes in continental positions and sizes of ocean basins during the relatively short time of human history.

Nonetheless, it is clear that plate motions caused profound changes on the surface during the long history of Earth (Section 12.1). Although the distance a plate moves over centuries and millennia seems trivial, movement of 5 cm/yr for 10 million years, for example, adds up to 500 kilometers of travel. How has plate tectonics changed the appearance of Earth through geologic time?

Reconstructing Past Plate Motions

A variety of geologic data decipher long-term plate-motion history. Oceanic lithosphere recycles into the asthenosphere. The record of oceanic crust is, therefore, helpful only as far back as the age of crust in the modern ocean basins—about 180 million years. Over that time period the distributions of seafloor crust of different ages show where spreading ridges existed and show the direction and rate of relative motion away from those ridges.

How can we reconstruct plate motions prior to 180 million years ago? The archive of geologic history preserved on more permanent continental crust yields information that helps with this.

- Ancient igneous rocks have compositions consistent with divergent or convergent boundaries, or with hot spots, and infer the tectonic settings for magma generation. Radioactive-isotope dates on these rocks indicate when these tectonic environments existed.
- Regional metamorphism closely relates to convergent plate boundaries (Section 6.10), and it is most intense where continents collide and the partially subducted crust experiences unusually high pressure (Figure 12.26). Isotope-dating methods applied to metamorphic rocks reveal when these collisions occurred.
- Sedimentary rocks record erosion from mountains and deposition in basins shaped by plate-boundary processes. Fossils within the sedimentary strata indicate when these tectonic elements existed in the landscape, and unconformities record when tectonic forces deformed the sedimentary basins.

These are just some examples of data that geologists use to reconstruct past positions of continents as passengers on lithospheric plates.

Paleozoic Plate Tectonics

Figure 12.44 is your visual aid for fast time travel through reconstructions of Earth's past and to gain appreciation of how plate motions over geologic time account for prominent geologic features in North America and elsewhere in the world. Reconstructions for older times are more speculative than those for recent times. Older rocks are rare and thus harder to find. They are also more difficult to interpret. Old rocks get buried under younger sedimentary and volcanic rocks and metamorphosed near convergent boundaries.

Figure 12.44a–c depicts global geography at three snapshots during the Paleozoic Era (see Figure 7.9 if you need to refresh your knowledge of the geologic time scale). The continents do not have familiar outlines because current continental margins only existed after Wegener's Pangea supercontinent broke apart in the Mesozoic. Most of the land composing modern North America was near the equator and rotated almost 90 degrees clockwise from its present orientation; what is now northern Canada was on the east side of the ancient continent, and what is now the eastern United States bordered a southern ocean. The Atlantic Ocean did not start forming until the Mesozoic (Figure 12.14), so it does not appear in this Paleozoic view. Ancient oceans, whose lithosphere has long since subducted out of view, surrounded North America. The oceans to the south and east became progressively smaller during the Paleozoic as oceanic crust subducted and North America collided with neighboring continents.

The Appalachian Mountains (Figure 12.8 and partly illustrated in Figure 11.21) formed near Paleozoic subduction zones. Mountain building began in the Ordovician and was punctuated by two large continent-continent collisions. The first, in the Silurian and Devonian Periods (362 to 439 million years ago), involved collisions with continental crust now part of Western Europe and Siberia, in northeastern Asia (Figure 12.44b). The collisions created Himalaya-style mountains in what are now the northeastern United States, northern

EXTENSION MODULE 12.2

Using Paleomagnetism to Reconstruct Past Continental Positions.

Learn how paleomagnetism is used to determine past continental positions as passengers on moving plates.

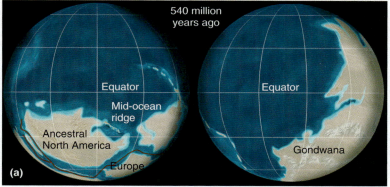

(a)

At the beginning of the Phanerozoic Eon, the continent most closely resembling modern North America was near the equator. The Gondwana continent included crust that is now found in Africa, South America, Antarctica, Australia, and India.

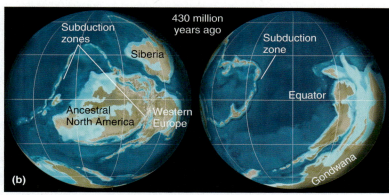

(b)

Collision of ancestral North America with continental blocks now found in Western Europe created high mountain ranges about 430 million years ago (the Caledonian Mountains shown in Figure 12.8). Sea level was high, and continents were partly submerged below shallow seas.

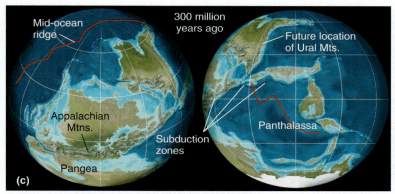

(c)

The Appalachian Mountains and other mountain belts (e.g., Ural Mountains, Figure 12.8) formed by continent-continent collisions at convergent plate boundaries when the Pangea supercontinent assembled during the late Paleozoic Era.

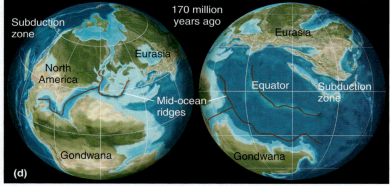

(d)

Pangea began splitting apart about 180 million years ago, as indicated by the oldest oceanic crust in the North Atlantic Ocean and elsewhere.

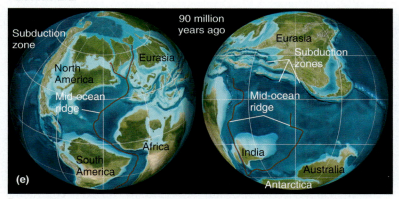

(e)

By 90 million years ago, Africa had separated from South America, and India was moving northward toward its eventual collision with Eurasia. A subduction zone was present along western North America where magma formed the batholiths of the Sierra Nevada in California, and folding and thrust faulting formed mountains throughout western North America.

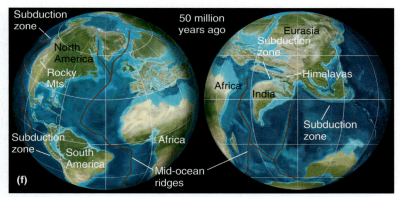

(f)

India made first contact with southern Asia about 50 million years ago, initiating uplift of the Himalayan Mountains, which continues today. The Rocky Mountains formed because of convergent-margin processes off the west coast of North America. A mid-ocean ridge was also located close to the west coast.

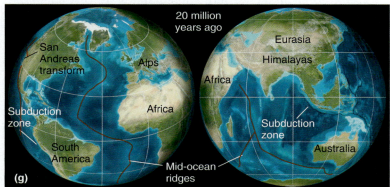

(g)

By 20 million years ago, continent-continent collision formed a nearly continuous belt of high mountain ranges from southeast Asia, through the Middle East, to the Alps of southern Europe. The earlier mid-ocean ridge off the west coast of North America began subducting beneath the continent, which changed relative plate motions and initiated the San Andreas transform boundary.

▲ **Figure 12.44 Visualizing plate motions through geologic time.**
Geology and art combine to render images of how Earth's surface changed over geologic time because of plate tectonics. Each pair of globes combines to provide a worldview for the indicated time. Only a few plate boundaries are indicated for simplicity. Continents break up and move away from divergent plate boundaries only to collide again at convergent boundaries. Oceans form in the wake of continents drifting away from mid-ocean ridges, and mountains rise near convergent boundaries, especially when continents collide.

Plate Motions through Time: *See how continents changed positions through geologic time because of plate motions.*

Canada, Greenland, the British Isles, and Scandinavia. Collision of ancestral North America with Gondwana, a continent that included modern Africa and South America, occurred during the Pennsylvanian Period as Pangea formed (Figure 12.44c). The resulting mountains stretched throughout the eastern United States, over to the present-day Ozark and Ouachita Mountains in Oklahoma and Arkansas, and continue as subtle, mostly buried features across the whole width of Texas.

Mesozoic Plate Tectonics

Plate tectonics during the Mesozoic, illustrated in Figure 12.44d–e, featured the creation of the modern ocean basins and more-familiar-looking continents. Pangea began splitting in the Triassic, and new mid-ocean ridges generated oceanic crust by the Jurassic Period (Figure 12.44d). Crust of an equatorial ocean subducted beneath southern Eurasia, bringing some fragments of old Gondwana, including India and Africa, closer to Eurasia.

Continental rifting in eastern North America during the Triassic and early Jurassic produced rift valleys from what is now Massachusetts to North Carolina. These valleys resembled the modern East Africa rift valleys. Mafic volcanic activity accompanied rifting, and numerous gabbro sills are prominent in the northeastern United States, where they form the Palisades of the Hudson River near West Point, the prominent strategic ridges of the Civil War battlefield at Gettysburg, Pennsylvania, and innumerable quarried rock resources from Connecticut to New Jersey.

A convergent plate boundary lurked off the coast of western North America from the middle part of the Paleozoic until the early Cenozoic Era. Many features of western North America formed above this subduction zone. Compression caused uplift, folding, and the formation of reverse and thrust faults from the west coast to the Rocky Mountains (see Figure 11.25). San Francisco is built on a chaotic mess of sedimentary rocks and chunks of ocean-floor basalt that were scraped off the subducting plate. Subduction-zone magmas intruded western North America and Mesozoic and early Cenozoic andesitic volcanic rocks are common in that region. The Sierra Nevada batholith, including the rocks at picturesque Yosemite National Park (Figure 4.1c), is the uplifted and exposed intrusions related to this subduction zone. The igneous rocks generated above the Mesozoic and Cenozoic subduction zones contain most of the rich mineral deposits mined in western North America.

Cenozoic Plate Tectonics

Subduction beneath southern Europe and Asia led to Tertiary continent-continent collisions, depicted in Figure 12.44f–g. The resulting mountain chains include the Himalayas, the mountainous terrain of central Asia and the Middle East, and the Alps and related mountain ranges of southern Europe.

Cenozoic tectonic events in the western United States are very complicated. A mid-ocean ridge was located off the west coast during the early Tertiary (see Figure 12.44f). The westward motion of the North American plate was, however, faster than the rate of oceanic-crust production at this mid-ocean ridge. As a result, North America ran over most of the mid-ocean ridge, beginning about 26 million years ago. The mid-ocean ridge is largely subducted out of view, but a small part remains off the Pacific Northwest coast, where the tiny Juan de Fuca plate is created, only to subduct a short distance away beneath Oregon, Washington, and northern California (Figure 12.3). Subduction of the spreading ridge places the North American plate in contact with the Pacific plate. Both plates move in westerly directions, and the Pacific plate is moving much faster than the North American plate (see Figure 12.40). The San Andreas transform boundary, therefore, replaced the convergent plate boundary that existed along the west side of North America since the middle Paleozoic.

The transition from convergence to a mostly transform boundary along the western edge of the North American plate coincided with a change from compressional to tensional stress throughout most of the western United States. The exact cause of this change in stress is the subject of vigorous ongoing research, but it at least partly relates to changing forces at the plate boundary. Tensional stress causes normal faults, which are responsible for uplift of mountain ranges and down dropping of valley floors in many western states (see Figure 11.22).

Putting It Together—What Are the Consequences of Plate Motion Over Geologic Time?

- Plate tectonics explains the history of continents and ocean basins by linking rocks and structures with plate-boundary processes.
- Paleozoic mountain ranges, such as the Appalachian Mountains, formed along convergent plate boundaries. Continent-continent collisions along several plate boundaries assembled the Pangea supercontinent by the end of the Paleozoic Era.
- Pangea rifted apart during the Mesozoic, and the modern continents dispersed to their present positions since that time. Rift valleys formed in eastern North America prior to the formation of the oldest Atlantic Ocean crust when the continents separated.
- A convergent plate boundary along the west coast of North America from the middle Paleozoic until the middle Cenozoic caused faulting and folding, intrusion and crystallization of batholiths, erupting volcanoes, and formation of valuable mineral deposits in the western United States.
- Relative plate motions changed along the western boundary of the North American plate beginning about 26 million years ago. Subduction continues offshore of southwestern Alaska, Washington, and Oregon and northern California, but transform boundaries formed elsewhere along the western edge of the plate.

Where Are You and Where Are You Going?

You now know far more about plate tectonics than the bits and pieces that revealed in Chapters 1–11. Most importantly, you know the evidence that supports plate tectonics as a viable explanation of a wide range of geological processes. Global tectonic processes, in turn, are reliably linked to convection within Earth. Applying uniformitarianism also explains the ancient tectonic history of Earth, including familiar landscapes of North America.

Relative motions of plates at divergent, convergent, and transform plate boundaries are consistent with all available geologic and geophysical data. These data include the following:

- Locations of earthquakes, volcanoes, and active mountain belts, which mostly coincide with the narrow interactive boundaries between rigid plates

- Locations of normal, reverse and thrust, and strike-slip faults, which are consistent with the stresses associated with each plate boundary
- Heat-flow and seismic-tomography data, which delineate where hot asthenosphere wells up beneath mid-ocean ridges, and where cold lithosphere sinks to the base of the mantle
- The antiquity of continental crust, which cannot be readily subducted, and the relative youth of oceanic crust, which is continuously created at divergent boundaries and consumed at convergent boundaries

Hot spots are areas of especially prolific volcanic activity not explained by plate-boundary processes. Many hot spots occur within plates, and some are found along or near divergent plate boundaries. Active hot-spot volcanoes reside at one end of a long line of extinct volcanoes that are progressively older at greater distance from the hot spot. These hot-spot tracks, and the odd composition of hot-spot basalt, imply that this volcanism results from mantle processes originating below the plates. The mantle-plume hypothesis explains hot spots as the surface expression of hot, upwelling mantle that migrates upward all of the way from the core-mantle boundary. Proving the hypothesis is difficult to do, but the coincidence of most hot spots with two broad superplume regions where hot mantle is rising from the lower mantle suggests that the hypothesis is at least partly correct.

Plate motion and mantle plumes are expressions of mantle convection. Subducting lithosphere at convergent plate boundaries records the convective downwelling of cold material. The downward pull of subducting plates provides nearly all of the driving force for global plate motion. Shallow upwelling in the asthenosphere is recorded at divergent plate boundaries. Deeper upwelling is indicated in seismic tomography by the super-plumes, which may be but are not clearly connected with the narrower plumes hypothesized to cause most hot spots.

The next chapter completes your large-scale, global view of tectonics. You will learn why continents are high-standing areas on Earth's surface, whereas oceanic crust is so much lower that it is submerged beneath the sea. You will see evidence that continents grow over time and learn how plate tectonics theory explains this growth.

 # Active Art

Motion at Plate Boundaries. See how plates move along their boundaries.

Correlating Processes at Plate Boundaries. See how volcanoes, earthquakes, and young mountain belts line up with plate boundaries.

Seafloor Spreading and Rock Magnetism. See how seafloor spreading at divergent boundaries produces bands of crust with alternating magnetic polarities.

Forming a Divergent Boundary. See how a divergent boundary originates by rifting a continent.

Motion at Transform Boundaries. See how transform faults connect other plate boundaries.

Hot Spots and Plumes. See how an island chain forms by plate motion across a hot spot, which might connect to a hypothesized mantle plume.

Relative and Absolute Motion. See how to distinguish between relative and absolute motion.

Convection and Tectonics. See how convection motion relates to plate tectonics and plumes.

Plate Motions through Time. See how continents changed positions through geologic time because of plate motions.

Extension Modules

Extension Module 12.1: Describing Plate Motion on the Surface of a Sphere. Learn how to describe plate motion on a sphere.

Extension Module 12.2: Using Paleomagnetism to Reconstruct Past Continental Positions. Learn how paleomagnetism is used to

determine past continental positions as passengers on moving plates.

Confirm Your Knowledge

1. Define the continental drift hypothesis. How does it differ from place tectonics theory?

2. Describe the behavior of plate interiors in contrast to plate boundaries.

3. How does the distribution of volcanoes and mountains support the theory of plate tectonics?

4. What is the evidence for the existence of divergent plate boundaries?

5. What is the evidence for the existence of convergent plate boundaries?

6. Why are there mountain belts on both the east and western margins of North America if only the western margin is a plate boundary?

7. Which is stronger, the lithosphere or asthenosphere?

8. On average, how fast do the plates move?

9. Where on Earth is a likely place for a new ocean basin to form in the future?

10. How does the oceanic lithosphere become denser with increasing age and distance from a mid-ocean ridge?

11. Explain the relationship between lithosphere density and subduction.

12. What is the evidence for transform faults?

13. What are the similarities and differences between the Hawaiian, Icelandic, and Yellowstone hot spots?

14. What is the difference between relative and absolute velocities?

15. Why doesn't continental crust subduct?

Confirm Your Understanding

1. Write out an answer for each question in the Chapter Outline for the chapter sections assigned by your instructor.

2. What does the word "theory" mean when used to define plate tectonic theory or the theory of evolution?

3. Although he collected impressive evidence to support his continental drift hypothesis, Alfred Wegener did not receive widespread acceptance of his idea. What obstacle prevented many scientists from accepting continental drift?

4. Use Figure 12.44 to locate the approximate location where you were born on each map. Describe where your birth location was on the globe at 540, 430, 300, 170, 90, 50, and 20 million years ago.

5. How old is the oldest crust in the oceans? Why is the oldest crust located where it is?

6. What are the dominant geologic events in North America during the Paleozoic, Mesozoic, and Cenozoic eras and how do these events relate to plate tectonics?

7. A satellite returns images of the surface of a distant planet. Conical volcanoes are clearly visible on the images. How can you determine if this planet experiences plate tectonics, plumes, or both based on the distribution of volcanoes?

Key Terms

continental drift (p. 310)
fracture zones (p. 332)

hot spot (p. 333)
plumes (p. 334)

rift valleys (p. 321)
seafloor spreading (p. 316)

seamounts (p. 334)
Wadati-Benioff zone (p. 325)

13 Tectonics and Surface Relief

Chapter Outline

Why Study Surface Relief?

EARTH IS NOT SMOOTH AND FEATURELESS BUT HAS A HIGHLY IRregular surface ranging from dramatic, towering mountains to seemingly bottomless deep-sea trenches. Beyond providing impressive scenery, the planet's varied landscapes and seascapes have played important roles in world history. Mountainous terrain impeded invading armies and effectively isolated some geographic regions, forming barriers to economic and cultural exchange until modern transportation surmounted these obstacles. Agriculture originated and still flourishes in low-lying regions where nourishing sediment washes from eroding highlands to produce fertile soil.

What causes the varied topography of Earth? Surface elevation and shape, which are the elements of topography, result from the interplay of internal and external forces. Motion driven by heat inside Earth deforms the surface, causing some areas to rise while others sink. External processes of weathering and erosion sculpt the crust into the varied landforms.

This chapter focuses on pursuing knowledge of internal processes as a prelude to understanding the external, landscape-sculpting forces in the next part of the text. Some analogies along the way will aid your understanding of internal Earth processes. Simple ideas represent general explanations to begin the discussion, leaving many natural details to be filled in. Complexity gradually increases until you have a satisfactory understanding of the whole picture. Your primary objectives in this chapter are:

✔ To understand how processes in the crust and mantle determine the varying elevations of the surface

✔ To fit in the role of mountain building to make continents

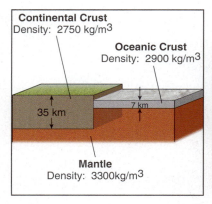
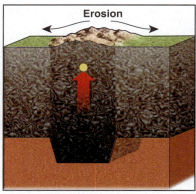

Mount Rainier, with a summit elevation 4383 meters above sea level, towers above Seattle, Washington. ▶

✔ To learn how continents grow through time

To accomplish these objectives, you will answer these questions:

13.1 Why are continents high and oceans low?

13.2 *How do we know* … that mountains have roots?

13.3 How does isostasy relate to active geologic processes?

13.4 Why does sea level change?

13.5 How and where do mountains form?

13.6 How does mountain building relate to the growth of continents?

In the FIELD

magine yourself on a vacation across the rugged terrain of the western United States. You reach the lowest elevation in the western hemisphere—Death Valley, California, portrayed in **Figure 13.1**. You stand at a point that is 85 meters below sea level. The high point in the adjacent mountains that draws your gaze upward is 3346 meters above sea level. The difference in elevation between two locations is the **relief**. Why are the mountains so high and the valley so low?

On your road trip northwest to the Pacific Ocean, you keep track of mapped elevations along the way in your field book. You use these data and maps to make a profile of the varying surface elevations across the landscape (Figure 13.1). From Death Valley, it is only 200 kilometers to the top of Mount Whitney, the highest elevation in the conterminous 48 states at 4418 meters high. Why are there such large variations in surface elevation over relatively short distances? What processes account for these differences? When you reach the shore of the Pacific Ocean, you look out across the gently undulating surface of the sea, and for a moment you are impressed by how smooth and gentle the ocean surface seems compared to the rough and rugged land surface. The next moment you realize that this is not a fair comparison—the seafloor is probably not as smooth as the sea surface and may well be just as rugged as the land you have traveled across.

You stop at a library to find some maps that depict the ocean depths. The maps allow you to continue profiling seafloor elevations west of California into the Pacific Ocean (Figure 13.1). Clearly, the seafloor also displays substantial relief.

What is the total relief on Earth's surface? On land, the highest elevation is Mount Everest, in the high Himalayas, at 8850 meters above sea level. The lowest land elevation is along the Dead Sea, which is actually a lake, at 400 meters below sea level. The deepest point in the oceans is in the Mariana Trench in the western Pacific Ocean south of Japan, at 11,040 meters below sea level. The total relief on Earth, the difference in elevation between Mount Everest and the Mariana Trench, is 19,890 meters, nearly 20 kilometers. This seems like a huge variance—compared to your height, for instance—until you remember that Earth is more than 12,740 kilometers in diameter. The relief on Earth's surface seems almost trivial when stacked up against the vast size of the entire planet. All the same, to the inhabitants of the planet, Earth's surface is rough and uneven, so why is this the case? Why are continental areas overwhelmingly elevated above sea level, whereas other regions are much lower and submerged beneath the seas? The range in elevations of Earth's surface is certainly one of the most basic features of Earth that geologists should be able to explain.

Figure 13.1 Visualizing elevation relief in California. ▶

13.1 Why Are Continents High and Oceans Low?

The phrases *above sea level* and *below sea level* probably sound familiar to you. Sea level is a convenient reference point for describing elevations of Earth's surface features, so elevations commonly are reported relative to sea level. Describing surface elevation as either above or below sea level also distinguishes the appearance of Earth as either land or ocean. What determines if regions are above or below sea level?

Elevation and Crust Type Are Related

A consideration of the total range of elevations on Earth provides some insights to this question. **Figure 13.2** depicts the percentages of Earth's surface area that are at various elevations and illustrates two important features:

1. About 29 percent of Earth's surface is land, because it is above sea level, and 71 percent is ocean.
2. Two elevations, at about 0.5 kilometer above sea level and 4.5 kilometers below sea level, are the most common elevations.

Field notes on elevation relief in California

Photo of "Badwater" in Death Valley. This is the lowest spot in the U.S., 85 meters below sea level.

The Panamint Mountains in the background have a high point that is 3346 meters above sea level, so there is a 3431 meter difference in elevation between Badwater and the mountain peak.

Downloaded map shows line of graphed elevation profile.

Pacific Ocean and Golden Gate Bridge in San Francisco

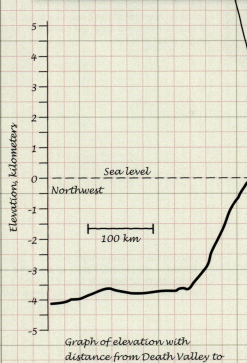

Graph of elevation with distance from Death Valley to San Francisco and into the Pacific Ocean

Mt. Whitney in the Sierra Nevada. Highest mountain in the "lower 48" United States.

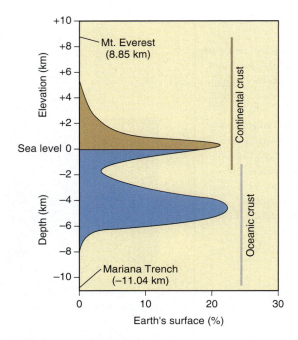

▲ Figure 13.2 Graphing global elevation.
This graph depicts the percentage of Earth's surface that is at different elevations, above sea level, and depths, below sea level. Most areas underlain by continental crust are between 1.5 km above sea level and about 0.5 km below sea level. Most areas underlain by oceanic crust are between 3 and 6 km below sea level. The two clearly defined peaks on the graph suggest that the type of crust is primarily what determines surface elevation.

The type of crust at a location seems to be the main factor that determines the elevation. Continental crust underlies most of the areas composing the graph peak at 0.5-kilometer elevation and at higher elevations. Oceanic crust underlies the lower elevations. Although sea level conveniently divides land and sea, it does not distinguish areas of continental and oceanic crust because shallow water submerges some areas of continental crust (see Figure 13.2). How does the type of crust determine surface elevation? Why are some parts of continents above sea level whereas other parts are submerged?

Isostasy Determines Elevation

The **principle of isostasy** explains variations in surface elevations in terms of the type and thickness of underlying crust. According to this principle, low-density crust floats on denser, underlying mantle so that the pressure is everywhere the same at the base of any thick vertical column through the crust and upper mantle. The word "isostasy" derives from Greek roots "iso," meaning equal, and "stasis," meaning standing. These word roots emphasize that adjacent columns of crust of different elevation are equal in terms of the pressure beneath them.

What is meant by pressure at the base of a column of crust and mantle? The pressure is simply the weight of material in the column, as illustrated in **Figure 13.3**. If pressure is the same between bases of adjacent columns, why are there differences in elevation of these columns? Different rocks have different densities, so the pressure at the base of a short column of high-density rock is the same as it is at the base of a tall column of low-density rock. You can think of high elevations as being associated with columns of lower-density material, such as granite, that composes continental crust, and low elevations with columns of denser material, such as basalt, that composes oceanic crust. However, the principle of isostasy also relates to how the crust floats on top of the mantle, which suggests that the weight measurements in Figure 13.3 only partly describe isostasy.

Weight = volume × density × acceleration of gravity

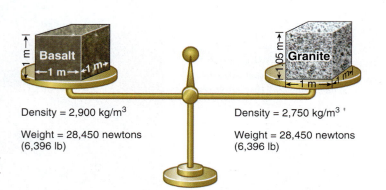

Lead is much denser than wood, so the pressure (weight) at the base of a small block of lead is the same as at the base of a much larger block of wood.

2.473 m

Wood

Density = 11,340 kg/m³

Weight = 1 m³ × 11,340 kg/m³ × 9.81 m/s² = 111,245 newtons (25,000 lb)

Density = 750 kg/m³

Weight = 15.12 m³ × 750 kg/m³ × 9.81 m/s² = 111,245 newtons (25,000 lb)

► Figure 13.3 Weight is pressure.
The pressure at the base of a block of any material is the weight of the material. Weight is equal to the volume of the block, multiplied by the density of the material and the acceleration of gravity.

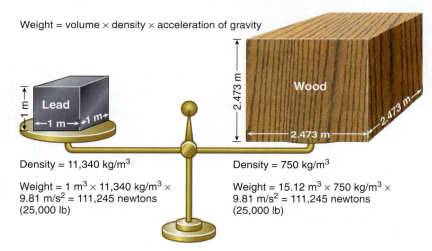

Density = 2,900 kg/m³

Weight = 28,450 newtons (6,396 lb)

Density = 2,750 kg/m³

Weight = 28,450 newtons (6,396 lb)

If you compare equal-weight blocks of two common crustal rocks, basalt and granite, then the block of basalt is smaller than the granite block, because basalt is denser than granite.

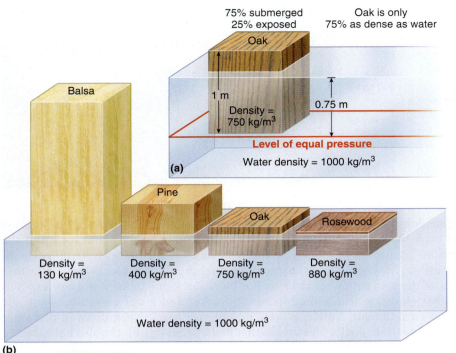

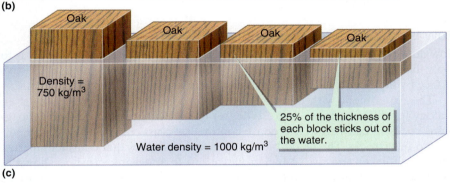

Isostasy determines the thickness of wood above and below the water line. The pressure at the base of the block must be the same as the pressure in the water at the same depth.

Wood blocks of different densities and thickness, cut to thicknesses so that they all sink to the same level in the water. Less dense blocks project higher above the water than denser blocks.

Wood blocks of the same density, but different thicknesses. The same proportion of each block is exposed as it is submerged, so thick blocks have higher surface elevations than thin blocks. The thick blocks also sink lower in the water.

▲ **Figure 13.4** **Demonstrating isostasy with wood blocks in a tub of water.**
The density of the wood and the thicknesses of the blocks determine the top-surface elevations of wood blocks floating in water.

Demonstrating Isostasy in a Tub of Water

Imagine wood blocks in a tub of water, as pictured in **Figure 13.4**a. A block of wood floats but sinks down part way in the water. What determines how much of the wood sticks up above the water line? Isostasy requires that the pressure at the base of the wood block equals the pressure in the water at the same depth as the base of the wood block. As Figure 13.4a shows, this wood block is 75 percent submerged when the pressures are equal, because the density of the wood is 75 percent of the density of water.

Blocks of different densities and sizes are shown in Figure 13.4b. The blocks are cut so that the submerged bases of the blocks are at the same level in the water. The high-density blocks almost completely submerge, whereas very little of the low-density wood sinks below the water line. Isostasy explains a higher surface elevation where the floating blocks have low density and a lower surface elevation where blocks have high density.

This consequence of isostasy also explains why most of an iceberg is submerged. Ice is only slightly less dense than water, as **Figure 13.5** shows. For the pressure to be the same at the base of the iceberg as it is below the open-ocean surface, most (about 90 percent) of the iceberg must be submerged.

Figure 13.4c illustrates another consequence of isostasy. In this bathtub experiment, all of the blocks are cut from the same wood, so they all have the

▶ **Figure 13.5** **Isostasy explains icebergs.**
Ice floats in water because ice is less dense than water. The density difference is small, so 90 percent of the ice is submerged, as predicted by the principle of isostasy. All sailors navigating icy waters know not to travel close to an iceberg. Even if an iceberg seems small above water, most of its size is submerged and hidden from view.

Density of ice:
~900 kg/m³

Density of
water:
~1000 kg/m³

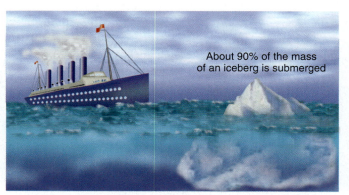

About 90% of the mass
of an iceberg is submerged

same density. The blocks are, however, cut to different thicknesses. The fraction of each block exposed above the water line must be the same for all of the blocks, because all of them have the same density. The blocks have different dimensions, however, so isostasy requires that the thicker blocks project farther out of the water than thin blocks. For blocks of equal density, isostasy explains higher elevation where the blocks are thicker and lower elevation where the blocks are thinner. Notice, as well, that thicker blocks stick down in the water more than thin blocks. This means that higher-standing blocks also have thicker "roots" below them.

How Isostasy Explains Elevations of Continents and Oceans

How do wood blocks floating in water explain elevations on Earth? The key is to apply the principle of isostasy to draw conclusions about the elevations of Earth's surface by thinking of the wood as blocks of crust and the water as the mantle.

The experiment with different types of wood blocks (Figure 13.4b) is an analogy for different types of crust: denser oceanic crust composed of mafic igneous rocks, in contrast to less dense continental crust composed of intermediate and felsic rocks. The wood-block experiment predicts that continental crust should be thicker and rise to higher elevations than oceanic crust, as shown in **Figure 13.6**a. The elevation differences between these two types of crust are as predicted, but only if both types of crust are more than 80 kilometers thick. Most oceanic crust is about 7 kilometers thick, and continental crust generally ranges between 25 and 50 kilometers (see Figure 8.7), so this is an unacceptable result. This model also does not explain differences in elevation of locations on continents underlain by more or less the same crust, such as Death Valley and Mount Whitney (Figure 13.1).

To try another approach, think back to the experiment with wood blocks of equal density but different thicknesses (Figure 13.4c). If the wood blocks represent Earth's crust, then this model assumes that all of the crust has the same density. The model then predicts that areas of thick crust have higher elevations than areas of thinner crust, as shown in Figure 13.6b. Most of the thickness difference between these modeled blocks of continental and oceanic crust appears as a deep root in the mantle, just like most of the thickness of a thick wood block, or an iceberg, is submerged in the water. The problem with this model is that it assigns the same density to continental and oceanic crust, even though rock samples and seismic data show that oceanic crust must consist of denser rock than continental crust (look back to Table 8.1 and Figure 8.2 for data about differences between continental and oceanic crust). Therefore, this model of variable crustal thickness might explain the highly varied relief on continents where the crust is of similar density, but it is unsatisfactory for explaining differences in elevation of oceanic crust versus continental crust.

Figure 13.6c shows that the best explanation for the different elevations of continental and oceanic crust combines knowledge from both experiments. Continental and oceanic crusts differ both in density *and* thickness. Continental

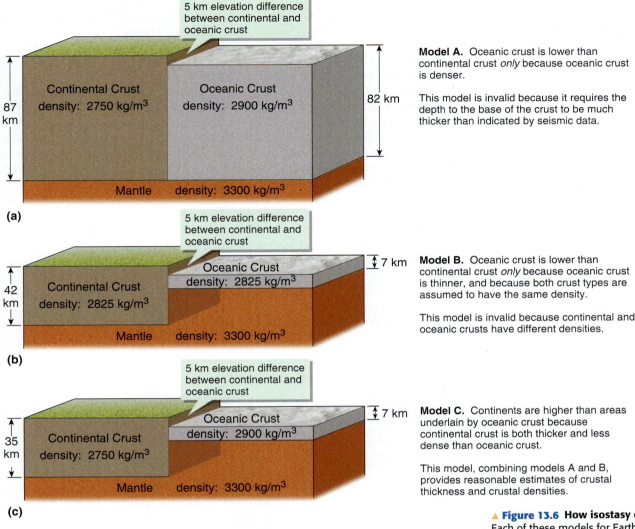

Model A. Oceanic crust is lower than continental crust *only* because oceanic crust is denser.

This model is invalid because it requires the depth to the base of the crust to be much thicker than indicated by seismic data.

Model B. Oceanic crust is lower than continental crust *only* because oceanic crust is thinner, and because both crust types are assumed to have the same density.

This model is invalid because continental and oceanic crusts have different densities.

Model C. Continents are higher than areas underlain by oceanic crust because continental crust is both thicker and less dense than oceanic crust.

This model, combining models A and B, provides reasonable estimates of crustal thickness and crustal densities.

▲ **Figure 13.6 How isostasy explains elevation.**
Each of these models for Earth elevation follows the analogy of the wood-block experiments (Figure 13.4) and data revealing a 5-kilometer average elevation difference between continents and oceans (Figure 13.2). Model A (similar to Figure 13.4b) and Model B (comparable to Figure 13.4c) do not, however, fit with other data. Only Model C, which combines aspects of the other two models, is consistent with what is known about the thickness and composition of continental and oceanic crusts.

crust is both less dense and thicker than oceanic crust, so areas of continental crust tend to stand at higher elevations above the mantle than do areas of oceanic crust.

Putting It Together—Why Are Continents High and Oceans Low?

■ The principle of isostasy states that the weight at the base of any column of crust and mantle must be the same as below every other column.
■ Isostasy explains variations in surface elevation as a result of variations in thickness *and* density of crust.
■ Areas of continental crust are higher than areas of oceanic crust because continental crust is thicker and less dense than oceanic crust.

13.2 **How Do We Know** ... That Mountains Have Roots?

PICTURE THE PROBLEM
Do Mountains Actually Have Roots?
Analogies between wood blocks in water and crustal blocks in the mantle offer an explanation for different surface elevations, but these analogies do not prove that these density and thickness differences actually exist below Earth's surface. What is the evidence that the principle of isostasy really explains surface relief?

Isostasy provides a test. High, mountainous regions should have thicker crust than lowland areas of continents, analogous to the wood blocks illustrated in Figure 13.4c. In other words, mountains should have roots.

APPRECIATE TWO HISTORICAL HYPOTHESES

How Did the Himalayas Cause Survey Errors in India?

The principle of isostasy emerged in the nineteenth century to explain puzzling surveying measurements near the highest points on Earth—the Himalayas. George Everest, the Surveyor General of India and the namesake for Earth's highest mountain, supervised an ambitious survey of India between 1840 and 1859. He discovered errors in the measured locations of some survey stations in valleys in northern India, next to the towering Himalayas. He dismissed the errors as simply inaccuracies in the surveying method.

British mathematician John Pratt proposed a different explanation for the erroneous locations in northern India. Pratt considered how some of the surveying was done using astronomical methods illustrated in **Figure 13.7**. A table is set up such that its level surface is perpendicular to a line directed toward Earth's center. Surveyors then sight a telescope on the North Star, which is almost directly above Earth's rotation axis. The angle between the line sighted to the star and a line drawn directly toward Earth's center permits calculation of latitude. A plumb weight suspended beneath the table establishes the line toward Earth's center (Figure 13.7b).

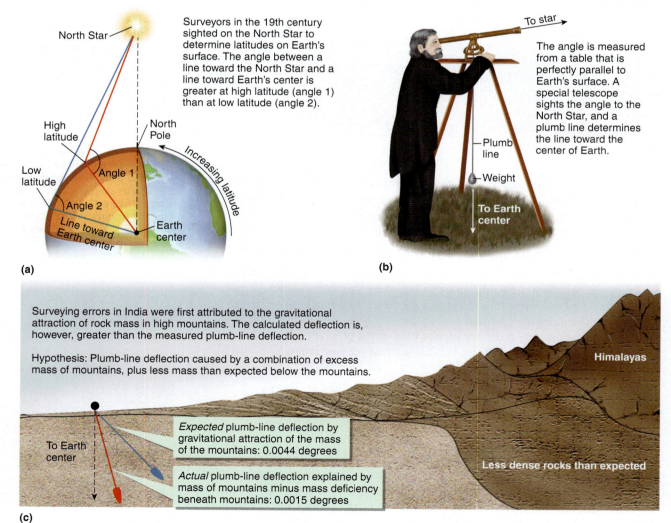

North Star

Surveyors in the 19th century sighted on the North Star to determine latitudes on Earth's surface. The angle between a line toward the North Star and a line toward Earth's center is greater at high latitude (angle 1) than at low latitude (angle 2).

High latitude

North Pole

Low latitude

Angle 1

Increasing latitude

Angle 2

Line toward Earth center

Earth center

(a)

To star

The angle is measured from a table that is perfectly parallel to Earth's surface. A special telescope sights the angle to the North Star, and a plumb line determines the line toward the center of Earth.

Plumb line

Weight

To Earth center

(b)

Surveying errors in India were first attributed to the gravitational attraction of rock mass in high mountains. The calculated deflection is, however, greater than the measured plumb-line deflection.

Hypothesis: Plumb-line deflection caused by a combination of excess mass of mountains, plus less mass than expected below the mountains.

Himalayas

To Earth center

Expected plumb-line deflection by gravitational attraction of the mass of the mountains: 0.0044 degrees

Actual plumb-line deflection explained by mass of mountains minus mass deficiency beneath mountains: 0.0015 degrees

Less dense rocks than expected

(c)

▶ **Figure 13.7 Explaining surveying errors in India.** Surveyors miscalculated latitudes in India because plumb lines on the surveyors' tables did not point directly to Earth's center as expected. This diagram explains the surveying technique and the hypothesis accounting for the plumb-line deflection.

Pratt realized that the plumb line only points to Earth's center, the assumed center of gravitational attraction, if mass is uniformly distributed within Earth and the surface is smooth. Pratt suspected that the surveying errors only occurred in the plains of northern India because of proximity to the towering Himalayas. This great mass of rock at elevations much higher than the survey stations exerts a gravitational attraction that deflects the plumb line away from the center of Earth and causes an erroneous measurement of latitude (Figure 13.7c). Pratt calculated the deflection expected from the elevation of the mountains and got a surprising result that was nearly three times larger than the actual errors in Everest's survey data.

Pratt modified his hypothesis to account for this discrepancy. He suggested that the deflection caused by the mass of the high mountains was partly offset by less mass than expected below the mountains. If the crust beneath the survey station is denser than the crust below the mountains, then the mountainous region does not exert as much gravitational attraction as it would if all of the crust had equal density. The Pratt model of isostasy accounts for different elevations with crust of different density, following model A in Figure 13.6.

British Royal Astronomer George Airy almost immediately proposed an alternative to Pratt's explanation linking isostasy and elevation. Airy agreed that the plumb-line deflections require less mass than expected below the mountains. He hypothesized that the lower mass results not from unusually low-density crust beneath the mountains but is explained instead by unusually thick crust floating on dense mantle. The Airy model of isostasy accounts for different elevations by variations in crustal thickness, similar to model B in Figure 13.6.

EVALUATING DIFFERENT HYPOTHESES

What Matters More, Crust Density or Thickness?

Which hypothesis is correct, or are neither Pratt nor Airy correct? Both hypotheses explained surface elevations equally well. The merits of each hypothesis were not tested until the last half of the twentieth century, when abundant seismic data provided the necessary knowledge of the density and thickness characteristics of crust.

When distinguishing between two hypotheses, it is important to focus on testing predictions that only support one hypothesis and not the other. Airy's hypothesis predicts that the base of the crust is deepest beneath areas of highest elevation: Mountains should have roots in the mantle. In contrast, Pratt's hypothesis calls for no change in crustal thickness between valleys and mountains. The collection of seismic data during the twentieth century provides an independent test for the presence or absence of mountain roots.

Figure 13.8 depicts the thickness of continental crust determined from seismic data below a line from the plains of northern India, over the Himalayas, and across the high Tibetan Plateau (also see Figure 12.20 for another illustration of this region). Earthquake waves move faster through mantle rocks than through the crust. By comparing the arrival times of earthquake waves at different locations, it is possible to determine the velocity of the waves at different depths, and to identify the depth of the crust-mantle boundary. These data clearly show a crustal root protruding into the mantle below the high surface elevations and support Airy's hypothesis. In the matter of the erroneous Everest survey, variations in crust thickness are to blame for the deflected plumb line and provide an isostatic explanation for surface elevation.

INSIGHTS

Does Airy Isostasy Work Everywhere?

Before leaving the impression that Airy was entirely correct and Pratt entirely wrong, take another look at Figure 13.6. Airy's hypothesis, which compares well with model B in Figure 13.6, does not explain the larger-scale differences in elevation between continents and oceans. When considering elevations over the whole globe, geologists must combine Pratt and Airy isostasy hypotheses, as in model C of Figure 13.6.

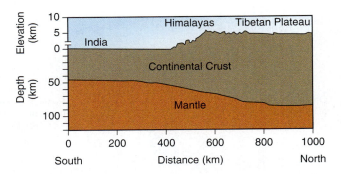

▲ **Figure 13.8** **Mountains do have roots.**
This diagram shows surface elevation and crust thickness between the low plains of India, the high Himalayas, and the Tibetan Plateau. The depth to the base of the crust is calculated from the velocity of seismic waves at different depths below the surface. The base of the crust is deeper under the higher elevations, showing that mountains have roots protruding down into the mantle.

When testing competing hypotheses, it is always important to determine if it is impossible for both ideas to be correct. Nothing about either Pratt's or Airy's hypotheses requires that the alternative view is incorrect. Indeed, it is common in science to combine hypotheses or to see that one applies best for one set of circumstances (Pratt's model is essential to explain elevation differences between oceans and continents) while the other applies best in other situations (Airy best explains elevation differences within areas of continental crust).

Think back to the field trip through California. Is the crust beneath Mount Whitney thicker than beneath Death Valley, as predicted by Airy isostasy? Seismic data suggest that the crust is only 30 kilometers thick beneath Death Valley compared to 45 kilometers thick beneath Mount Whitney. There is a problem, however. The thickness of the crust beneath Mount Whitney does not seem thick enough to explain the very high elevation. In other words, Mount Whitney has a root but it is not thick enough to support the high elevation. This means that additional hypotheses must be added to Airy and Pratt isostasy to complete your understanding of Earth's relief.

> ### Putting It Together—*How Do We Know* ... *That Mountains Have Roots?*
>
> ■ The Pratt and Airy isostasy hypotheses both successfully explain surface elevation. Although it is not possible for both hypotheses to be independently correct for all situations, combinations of each are satisfactory in most, though not all cases.
>
> ■ Seismic data demonstrate the presence of thick roots of crust projecting downward into the mantle beneath mountains as predicted by Airy's model.

13.3 How Does Isostasy Relate to Active Geologic Processes?

How do active processes, such as erosion and deformation, affect elevation and application of isostatic principles that describe stable, static conditions? If the thickness or mass in an isostatically balanced column of rock changes, then the weight at the base of that column becomes greater or less than that of neighboring columns. **Figure 13.9** illustrates this idea with another simple experiment in a water tub. If geologic processes change the thickness or mass, or both, of areas of crust, then isostasy causes vertical movements to adjust the crust to a new stable position.

Effects of Tectonic Shortening and Stretching

Compression shortens and thickens the crust by plastic flow in the lower crust and forms reverse and thrust faults in the upper crust (look back to Section 11.5 for the relationship between stress and deformation). **Figure 13.10** shows that isostasy requires the elevation of a shortened region to increase because the thickness of crust increases. Compare the change in land elevation to the elevations of thin and thick wood blocks in Figure 13.4c; the thicker block has a higher elevation.

Tension thins and stretches the lower crust by plastic flow while normal faults form in the upper crust. In Figure 13.10, you can see how elevation decreases where the crust becomes thinner. Looking back to Figure 13.4c,

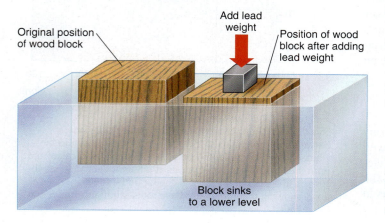

Original position of wood block

Add lead weight

Position of wood block after adding lead weight

Block sinks to a lower level

◄ **Figure 13.9 What happens when you add weight to a block?**
A wood block is initially in a stable isostatic condition. Then a lead weight is placed on the block, pushing it downward in the water to a new isostatically stable position. By analogy, changes in the distribution of weight in Earth's crust should cause some areas to subside and others to rise.

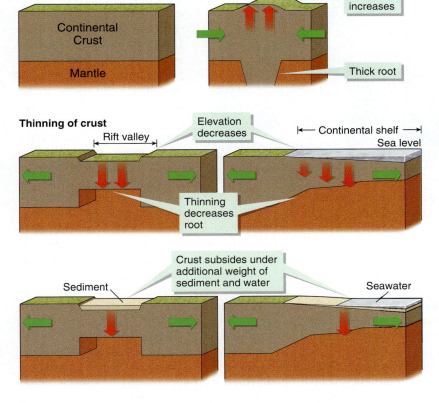

Initial crust thickness

Thickening of crust

Elevation increases

Continental Crust

Mantle

Thick root

Thinning of crust

Rift valley

Elevation decreases

Continental shelf

Sea level

Thinning decreases root

Crust subsides under additional weight of sediment and water

Sediment

Seawater

◄ **Figure 13.10** How shortening and lengthening the crust also change elevations.

Compressed continental crust shortens and thickens. Isostasy balances the thickened crust so that elevation increases above a thick root that projects into the mantle.

Extended continental crust lengthens and thins. The thinned crust subsides because of isostasy to form a rift valley. If rifting separates the continent into fragments, then the elevation of the thinned continental margin is lower than normal-thickness crust in the continental interior, forming a submerged continental shelf. The crust subsides further because of the added weight of sediment and water.

note how a thinning of the continental crust produces a block that is lower in elevation and whose root shrinks. Isostasy explains why rift valleys form where continents lengthen by tension (look at Figure 12.18). Isostasy also explains why continents have wide, submerged continental shelves. Continents separate from one another where divergent plate boundaries form. Where tension stretches the crust, the crust is also thinner and therefore has a lower elevation than the interior of the continent where the nonstretched crust is thicker. The elevation of thinned crust is lower than sea level, causing submergence of part of the continental crust.

Effects of Erosion and Deposition

Erosion and deposition of sediment redistribute mass and change the thickness of columns of crust. One more trip to the wood blocks in the tub shows how this works. The two wood blocks shown in **Figure 13.11** are analogous to columns of continental crust of the same thickness. Block A is "eroded" with a wood planer, and the shavings accumulate as "sediment" on block B.

Isostasy requires adjustments because the weights and thicknesses of the blocks changed. Block A now weighs less so it rises upward. This uplift only partially offsets the erosional loss in height of the block, so the surface elevation decreases overall. Block B subsides because of the added weight of the wood shavings. This subsidence only partly offsets the depositional gain in height of the block, so the surface elevation increases overall. You can conclude from these experiments that isostasy causes uplift where erosion occurs, even though elevation decreases, and causes subsidence where deposition occurs, even though elevation increases. Now, apply the wood-block analogy to real geology.

Mountains continue to rise long after tectonic forces cease to thicken crust and drive uplift along

▼ **Figure 13.11** Using wood blocks to model erosion and sedimentation.

These two wood blocks start out at the same thickness, and they float at the same level in the water. Block A is "eroded" with a wood planer, and the shavings are "deposited" on top of block B. Eroded block A is now thinner, so isostasy causes it to adjust to a new position with a shorter submerged root. This causes the block to move up as it erodes, although it is lower than before. Block B subsides under the added weight of the shavings, although the added thickness of the deposit causes surface elevation to increase slightly.

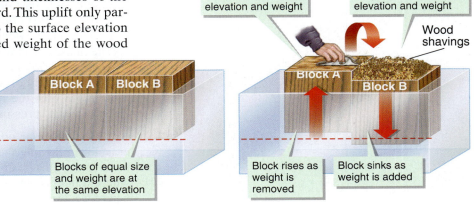

"Erosion" decreases elevation and weight

"Deposition" increases elevation and weight

Wood shavings

Block A | Block B

Block A | Block B

Blocks of equal size and weight are at the same elevation

Block rises as weight is removed

Block sinks as weight is added

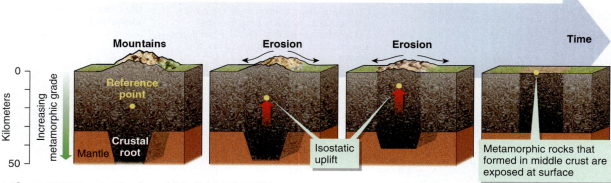

▲ **Figure 13.12 Isostasy and erosion cause rock uplift.**
Tracking the reference point in these diagrams shows isostatic adjustment causes rock uplift even while mountains erode. Erosion decreases the thickness of the crust, which means that the thickness of the root projecting into the mantle must also decrease. This causes the mountain root to move upward until erosion ceases removing rock from the surface. Although the surface elevation of the mountains gradually decreases through time, the rocks persistently rise. Erosion eventually exposes metamorphic and plutonic-igneous rocks that originally formed deep in the crust.

faults. This is because mountains result *both* from tectonic deformation and isostatic response to erosion. **Figure 13.12** illustrates how isostasy combines with erosion and causes uplift of rock, even as the surface elevations gradually decrease. As long as the crust beneath a mountain range is thicker than in adjacent areas, the mountain range stands higher than the adjacent lowlands. This observation explains why ancient mountains, such as the 300-million-year-old Appalachian Mountains, persist as topographic features long after they form. Isostasy also offers an explanation for how metamorphic rocks that form deep beneath the surface are later exposed at the surface (see Figure 13.12). Stress in the crust caused by vertical isostatic adjustments can result in movement along old faults, which is one cause of earthquakes in continents far from plate boundaries.

Sedimentary basins subside partly because of the added weight and thickness of accumulating sediment, as shown in Figure 13.10. For example, crustal thinning causes a rift valley to subside, but the total basin subsidence is greater than what thinning explains. The remainder of the subsidence occurs because the weight of accumulating sediment pushes down the basin crust. The thinned, submerged continental shelves along rifted continental margins subside under the weight of accumulating sediment delivered by rivers draining the continent and also because of the weight of overlying seawater.

Problems with Pratt and Airy Isostasy

Up to this point, combination of Pratt and Airy isostatic models explains the general features of Earth surface elevation as a result of variations in crustal density and thickness. In detail, however, both models require that crust move up and down in the mantle as readily as wood blocks move up and down in water. Is this reasonable?

Although the Pratt and Airy isostatic models explain some geologic situations in a general way (such as uplift and subsidence of thickened and thinned crust, uplift of rock in eroding mountains, and subsidence of filling sedimentary basins), they do not succeed in explaining the exact details of these phenomena, and they fail to explain a few other features at all.

There are two limitations of the wood-block analogies for crust:

1. The block models assume a very weak crust. The models also assume that faults penetrate completely through the crust so that adjacent crustal blocks readily slide past one another to adjust to changes in crustal thickness. In reality, the upper crust is strong, and the lower crust is not penetrated by faults because plastic flow occurs in the hot lower crust (look back to Section 11.6).
2. The mantle is almost entirely made up of solid rock, not liquid. Although the hot mantle flows like a very high viscosity fluid, this flow is much slower than the adjustment of water level to bobbing wood blocks. The crust moves only as quickly as the mantle very slowly flows from one place to another to equalize the pressure below regions of changing crustal weight.

To further our understanding of isostasy, it seems best to view the strong crust like a flexible sheet that is continuous over large areas rather than broken into blocks. This flexible sheet is elastic, like a rubber band. The crust depresses where weight is placed upon it, as by deposition of sediment, and it rises upward where weight erodes off.

Sitting down on a water bed might be a better analogy to how isostasy works rather than stacking weights on wood blocks. Water redistributes when your weight is added to the bed. The top of the water bed sags beneath your weight and bulges slightly next to you. When you stand up, the surface rebounds back to its original shape, showing that the deformation was elastic. Unlike a water bed, the elastic sheet of crust rests on a very high viscosity mantle. The mantle must flow away from areas of the sheet where the weight increases and rise up where the weight decreases. The sheet can only flex down or up as fast as the mantle can flow. Real isostatic adjustments are not instantaneous responses to changing distributions of mass at or near the surface. The term *flexural isostasy* describes this more complete application of the isostatic principle.

Isostatic Adjustment to Ice-Age Glaciers

The formation and subsequent melting of large continental glaciers during the last ice age provides a test of flexural isostasy. **Figure 13.13** illustrates how weight shifts on the crust when ice-age glaciers form and then melt away.

In Pratt-Airy isostasy, the addition of the thick mass of ice adds weight to the continental crust, just like placing a lead weight on the block of wood. If isostatic adjustment takes place only on local blocks, then only the crust immediately beneath the glacier subsides (Figure 13.13a). When the ice melts, the previously weighed down block of crust instantaneously pops back up to its original position.

Glacial Isostasy: *See how the crust isostatically responds to the growth and shrinkage of glaciers.*

▼ **Figure 13.13 Ice-age glaciers provide a test of isostasy.** These diagrams compare two hypotheses for isostatic adjustments resulting from the formation and melting of a thick glacier. Flexural isostasy, illustrated on the right, most satisfactorily explains observations.

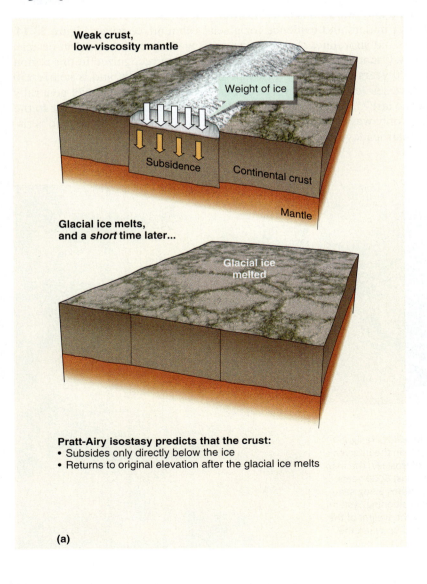

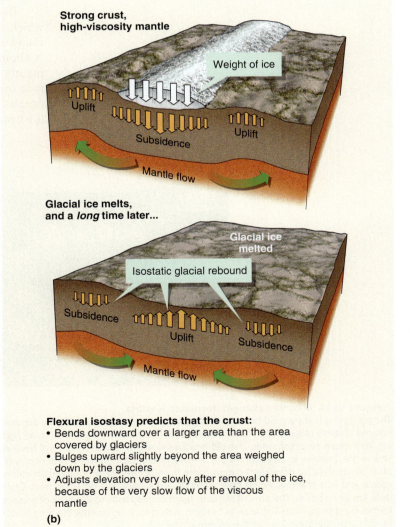

(a)

(b)

Flexural isostasy, however, predicts a different pattern (Figure 13.13b). The crust flexes down over a larger area than simply beneath the glacier, because the strong crust supports the weight over a large area. The slow flow of mantle from below the subsiding crust produces a small upward bulge adjacent to the subsiding area. After the ice melts, the crust everywhere very slowly returns to its previous elevations. The adjustments are slow because highly viscous mantle cannot flow fast enough to instantaneously even out pressure differences caused by the redistribution of weight.

Glacial ice was as much as 5 kilometers thick across large areas of northern North America and Eurasia about 21,000 years ago (the evidence appears in Chapter 18). That ice is nearly all gone now, except in Greenland, so it provides a test of block versus flexural isostasy.

If isostatic adjustment took place as quickly as wood blocks moving up and down in water, then the crust below glaciated areas should rise simultaneously with glacial melting and should have ceased rising long ago. Instead, all glaciated areas in the northern hemisphere are actively rising, and adjacent areas are sinking. This observation shows that elastic flexural isostatic adjustment is still taking place—a process called **glacial rebound**. When the ice weight was added to the crust, the underlying crust subsided and the adjacent area bulged due to corresponding movement of the mantle. Now that the ice is gone, the originally depressed areas are rising up to their original positions, and the bulges are sinking back to where they started (Figure 13.13b). The effect is like a water bed filled with a viscous liquid so that the surface does not completely recover its original shape until thousands of years after you get out of bed.

To better understand evidence for glacial rebound, consider **Figure 13.14**, which shows old shorelines that are now seen well above sea level in eastern Hudson Bay, Canada. The highest-elevation shoreline recognized in this region is about 8000 years old, based on isotope dates on seashells, and is nearly 150 meters above sea level. This is very curious, because global sea level generally rose over the last 21,000 years while the glaciers melted and added water to the oceans. The only way to explain the old, elevated beaches is that the land is rising faster than sea level is rising.

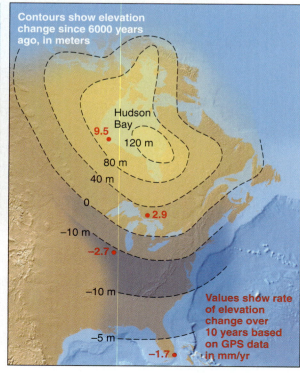

▶ **Figure 13.14 Elevation changes document isostasy.**
The photograph shows old beaches along Hudson Bay. The most recently exposed beaches are light-colored ribbons of sand. Dark green trees cover older beach ridges that stand higher than the intervening light-green swampy areas. The old beaches slowly rise above sea level because of isostasy. The map shows the changes in elevation of the land surface in eastern North America over the last 6000 years. Most of the data for drawing the map come from uplifted or submerged shoreline features along seacoasts and lake margins. Active rates of uplift and subsidence obtained from Global Positioning System data are also shown. The area of modern uplift corresponds to the area depressed by the weight of the former ice-age glaciers. The area of modern subsidence represents the area that bulged up when the flexible crust deformed under the weight of the surface ice.

Figure 13.14 shows the pattern of elevation changes over the last 6000 years in eastern North America. Uplift is greatest along Hudson Bay, which is also where the glacial ice was thickest 21,000 years ago. Land rises where the ice once existed and sinks in peripheral areas that previously bulged upward adjacent to the glaciers. Also shown on the map are current rates of uplift and subsidence detected by Global Positioning System measurements (look back to Section 12.7 for an explanation of these types of measurements). Clearly, North America is still isostatically adjusting to the melting of the ice-age glaciers, just as predicted by the flexural isostasy model (Figure13.13b). Geologists calculate that an additional 330 meters of uplift is required at Hudson Bay before isostatic stability is restored. At the current rate of rebound, this deformation will continue for 30,000–40,000 years.

Isostatic Adjustment Causes Earthquakes

Isostatic movement of crust causes stresses in the crustal rocks. What are the effects of these stresses in areas undergoing glacial rebound? The calculated stresses are less than the strength of most rocks, but the stresses are sufficient to reactivate old faults where broken rock has low strength. Scattered earthquakes, some with moment magnitudes as high as 6.0, do occur in northeastern North America, despite the great distance from active plate boundaries. Glacial rebound, rather than plate motion, is the principal source of stress to cause these earthquakes. Glacial-rebound stress shifts rocks along faults that formed long ago when plate boundaries were in this region (check Figure 12.44, for example). Even the great New Madrid, Missouri, earthquakes of 1811–1812 (revisit Figure 11.33 and Table 11.3) may partly relate to glacial-rebound stresses, because there is no geologic evidence of Cenozoic fault movement prior to the ice age. The New Madrid area isostatically rose adjacent to the ice-age glaciers and is now actively sinking back to its former elevation.

Relatively small stresses can cause earthquakes along old, weak faults. With isostatic adjustment, it does not take much weight to cause earthquakes. Even the weight of water filling reservoirs behind new dams commonly triggers hundreds of small earthquakes, and a few of these cause minor damage. Water moving into cracks and along old faults exerts pressure that separates rock materials along fractures, which decreases strength and increases the likelihood of movement that causes earthquakes.

*Putting It Together–**How Does Isostasy Relate to Active Geologic Processes?***

- Isostatic adjustments to changes in crustal weight and thickness cause vertical motion of the crust.
- Compressive stress that shortens the crust also thickens it, causing isostatic uplift. Tensional stress thins crust, causing isostatic subsidence of sedimentary basins.
- Isostasy causes crust beneath mountains to rise slowly even though the overall elevations decrease, because the weight of the crust decreases as the mountains erode. The eroded sediment accumulates in sedimentary basins, causing them to subside.
- Flexural isostasy, or bending of a strong, elastic crust above a highly viscous mantle, more accurately describes isostasy than do the Pratt and Airy models, which call for nearly instantaneous isostatic adjustments between vertical blocks of crust separated by weak boundaries.
- Slow flexural isostatic adjustment to changing weight is well documented by patterns of uplift and subsidence in eastern North America that are attributed to glacial rebound.
- Isostatic movement of the crust causes movement along old faults, thereby accounting for many earthquakes that occur far from plate boundaries.

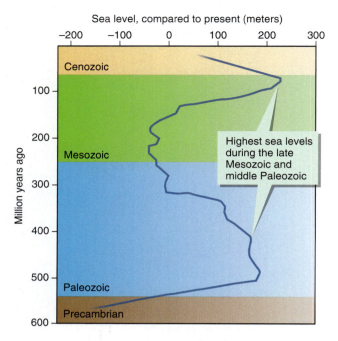

Sea level, compared to present (meters)

▲ **Figure 13.15 Sea-level changes through time.**
The graph plots estimated variations in sea level, compared to present sea level, during the last 550 million years. Geologists use the distribution of marine sedimentary deposits on continents to estimate ancient sea levels. The curve does not show large fluctuations, over very short time intervals, caused by formation and melting of glaciers, especially in the late Cenozoic.

▼ **Figure 13.16 Why sea level changes.**
(a) Sea level can change because the volume of seawater changes. Sea level falls when large glaciers form on land, and then rises when the glaciers melt and the water returns to the ocean. (b) Sea level will change if the volume of water remains the same but the shape of the ocean basin changes. The volume of water in these two hypothetical oceans is the same, but the water level rises onto land if the ocean basin becomes shallower.

13.4 Why Does Sea Level Change?

Sea level is a convenient reference position for comparing elevations, but it is not always at the *same* level. The edges of continents submerged today were dry land at times in the past. On the other hand, widespread marine sedimentary rocks in continental interiors, far from present shorelines, indicate times in the geologic past when whole continents were largely submerged. **Figure 13.15** charts the estimated sea-level changes for the last 550 million years, and the maps in Figure 12.44 illustrate varying levels of continental submergence through time. Sea level during the late Mesozoic and most of the Paleozoic was more than 200 meters higher than today (Figure 13.15).

Why does sea level change? If you think of the world ocean as a large container of water, you can change the water level by either changing the volume of water within the container or by changing the shape of the container that holds the water.

Change the Volume of Water in the Container

An easy explanation for changing sea level is that the amount of water in the oceans changes through time. **Figure 13.16**a explains sea-level fluctuations caused by the growth and demise of glaciers. At the peak of the last ice age (21,000 years ago), sea level was more than 100 meters lower than it is today simply because large volumes of water were stored on land as glacial ice. When the glaciers melted, the melt water flowed off into the oceans and sea level rose. If the remaining glaciers melt (mostly in Greenland and Antarctica), then global sea level will rise an additional 80 meters.

There are sedimentary deposits to support the presence of glaciers at earlier times. The sedimentary record of marine strata on continents indicates, however, that past sea level changed by 200 meters or more at times when there is no evidence of glaciers. Glacial oscillations cannot be the only explanation for sea-level change.

Change the Shape of the Container

Another hypothesis for explaining fluctuating sea levels states that the shape of ocean basins changes through time. This hypothesis is necessary because there are times in Earth history when shoreline sedimentary deposits show changes in sea level without any evidence for the presence of glaciers. Figure 13.16b shows that if ocean-basin depths decrease, then the water spreads out onto previous

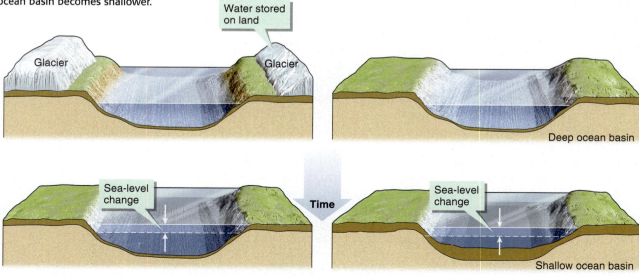

(a) Volume of seawater increases after an ice age when ice melts and adds water to oceans

(b) Same volume of water covers a larger surface area if ocean basins become shallower

dry land without changing the amount of water in the oceans. The inundation of the continents records a rise in sea level.

Plate tectonics potentially causes these changes in the shape of ocean basins. The key is the proposal that the elevation of sea floor relates to its age (look back to Figure 12.16 to see a graph of this relationship). To understand how plate tectonics affects global sea level, you need to consider what causes greater water depth above old lithosphere and shallower water above young lithosphere.

How Isostasy Explains Water Depth in Oceans

Figure 13.17 shows how isostasy explains seafloor relief. When the oceanic lithosphere forms at a mid-ocean ridge it is hot, which also means that it is expanded and has a relatively low density. This lithosphere cools, contracts, thickens, and becomes denser as seafloor spreading moves it away from the ridge (see Section 12.3). Most of the upper 125 kilometers of mantle at a mid-ocean ridge is asthenosphere. Older seafloor at greater distance from the ridge is underlain by progressively thicker, older, cooler, denser lithospheric mantle. Isostasy requires that the weight at the base of a column of crust and low-density asthenosphere at the mid-ocean ridge must equal the weight at the base of a column composed of crust and both low-density asthenosphere and higher-density lithosphere far from the mid-ocean ridge. If both columns weigh the same (meaning that they exert the same pressure), then the column with high-density mantle is shorter than the column with mostly low-density mantle. There is room for a greater thickness of seawater above the short column than above the tall column. Therefore, ocean depth increases as the age of lithosphere increases and the depths are shallowest over mid-ocean ridges.

Changes in lithosphere age, therefore, change ocean depth. If oceanic lithosphere is mostly young and warm, then isostasy keeps the lithosphere riding high in the asthenosphere, oceans are shallow, and sea level rises onto continents. If oceanic lithosphere is mostly old and cold, then it rides lower in the asthenosphere, oceans are deep, and continents stand higher above sea level. A change from deep oceans to shallow oceans is a change in the shape of the ocean container; there is less room for water in the oceans and sea level rises onto the continents.

Sea Level Changes when Average Lithosphere Age Changes

Some geologists use the connection between lithosphere age and water depth to formulate a hypothesis for the fluctuating sea levels depicted in Figure 13.15. The average age of all oceanic lithosphere on Earth depends on the amount of new lithosphere created at divergent plate boundaries and the age of the lithosphere destroyed at convergent boundaries during a particular time interval. On average the world oceans are shallower if the average lithosphere age decreases. This might happen because new mid-ocean ridges form, for instance. The seawater will not all fit into the shallower oceans and will spill over onto the continents. If the average lithosphere age later

▼ **Figure 13.17 Isostasy explains water depth in the ocean.** This diagram shows that water depth increases away from a mid-ocean ridge because the lithosphere cools and thickens over time. The weight of rock and water is the same for each of the schematic columns. At the mid-ocean ridge (far right) most of the upper mantle is low-density asthenosphere. At greater distances from the mid-ocean ridge (toward the left) an increasing thickness of denser lithosphere partly replaces the asthenosphere. This means that rock columns of equal weight are taller at the mid-ocean ridge (young lithosphere) and shorter away from the ridge (old lithosphere); therefore, more room exists for water in ocean basins when the average seafloor age is older than when it is younger.

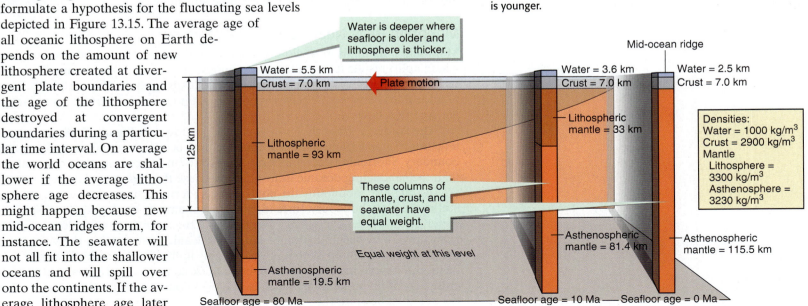

Water is deeper where seafloor is older and lithosphere is thicker.

Water = 5.5 km
Crust = 7.0 km
Plate motion

Mid-ocean ridge

Water = 3.6 km
Crust = 7.0 km

Water = 2.5 km
Crust = 7.0 km

Lithospheric mantle = 93 km

Lithospheric mantle = 33 km

125 km

These columns of mantle, crust, and seawater have equal weight.

Equal weight at this level

Asthenospheric mantle = 19.5 km

Asthenospheric mantle = 81.4 km

Asthenospheric mantle = 115.5 km

Densities:
Water = 1000 kg/m³
Crust = 2900 kg/m³
Mantle
  Lithosphere = 3300 kg/m³
  Asthenosphere = 3230 kg/m³

Seafloor age = 80 Ma

Seafloor age = 10 Ma

Seafloor age = 0 Ma

increases, then the lithosphere isostatically sinks lower in the asthenosphere, ocean basins will become deeper, and sea level will fall.

A glance back to Figure 12.44 shows that this hypothesis might explain sea-level change since the middle of the Mesozoic Era. When the supercontinent of Pangea rifted apart, starting about 180 million years ago, new mid-ocean ridges appeared in the Atlantic and Indian Oceans (Figures 12.44d and 12.44e). These spreading ridges added substantially to the areas where new lithosphere formed while at the same time old lithosphere subducted at convergent boundaries. The overall effect was to decrease average oceanic lithosphere age, which caused seafloor elevations to rise and the seas flooded onto low-lying continental areas. As the Atlantic and Indian Oceans widened, the proportion of older lithosphere beneath these oceans increased in comparison to the narrow swatch of new lithosphere forming along the mid-ocean ridges. As average lithosphere age increased, sea level fell because the ocean basins became deep enough to hold more of the seawater. Similar processes may have affected sea level earlier in geologic history.

Putting It Together—*Why Does Sea Level Change?*

■ Isostasy explains why ocean depth correlates with seafloor age: The older the lithosphere, the greater its weight, and the lower its surface elevation.

■ Sea-level changes are explained either by changing the volume of seawater (as by forming and melting ice-age glaciers) or by changing the shape of ocean basins that contain the water.

■ Plate tectonics affects global sea level through changes in oceanic lithosphere age. When the average lithosphere age decreases, the average seafloor elevation increases, which causes sea level to rise so that seawater floods onto continents.

13.5 How and Where Do Mountains Form?

Vertical movement of Earth's surface depends on isostatic adjustments. The major features of Earth's surface are related to plate tectonics, and plate motions are mostly horizontal. So how do these horizontal and vertical motions relate to one another? The most fascinating and impressive variations in elevation are continental mountains. Most of these mountain belts relate to processes at plate margins. To explain mountains, geologists link plate-margin processes and motion in the mantle with vertical isostatic adjustments.

Mountain Belts Near Continental Margins

Mountain building is most closely associated with convergent plate boundaries where uplift results from crustal thickening. Compressional shortening thickens the crust (Figure 11.13). Thickening, in turn, leads to uplift (Figure 13.10). The relationships between processes at convergent margins and surface topography are complex, however.

The Andes in western South America, portrayed in **Figure 13.18**, are an interesting example of mountains formed near an oceanic-continental convergent boundary. Shortening deformation concentrates in the continental crust, rather than the subducting oceanic lithosphere, because the continental crust is weaker. The highest mountains do not, however, form right at the trench, which marks the plate boundary. Instead, the high mountains are found 200–400 kilometers away on the overriding plate. Some of the mountains are volcanoes that grow upward by accumulation of erupted volcanic materials, but the volcanoes themselves are built on a foundation of highly uplifted older rocks. The crust beneath the mountains is 60–70 kilometers thick compared to typical continental crust east of the mountains, which is only 35 kilometers thick. What causes the thicker crust beneath the Andes?

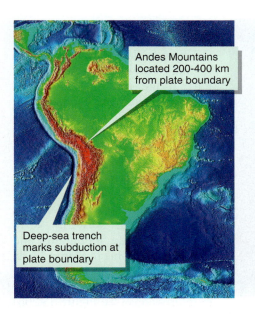

◄ **Figure 13.18 The Andes are an active mountain belt.**
The shaded-relief map of South America shows that the high Andes Mountains, like other mountains near oceanic-continental convergent boundaries, are located several hundred kilometers from the plate boundary. The photo illustrates 6914-meter-high Aconcagua, in Argentina. The highest mountain in South America is an eroded volcano perched on highly deformed and uplifted rocks.

Figure 13.19 shows two processes of crustal thickening near continental margins at convergent plate boundaries.

1. Volume is added to the crust as plutonic rocks solidify from igneous intrusions within and at the base of the crust (revisit Section 4.10 for explanations of why most magma solidifies in the crust). The volume of plutonic rocks is probably at least 10 times greater than the volume of volcanic rocks that also add an upper veneer to the total crustal thickness.
2. Crust thickens during shortening by plastic flow of lower crust caused by compression at the plate boundary. Plastic flow and thickening are greatest where magmas pass through the crust because this area is hotter and hot rocks flow more easily.

The process of crustal thickening by magmatic additions and plastic flow, and resulting isostatic adjustments, explains why high mountain belts rise skyward on the overriding plate rather than adjacent to the trench at convergent plate boundaries.

The compressional stresses also form thrust faults that shove rocks away from the volcanic chain toward the center of the continent (see right side of Figure 13.19). These thrust faults thicken the crust only slightly, so the resulting mountains are not as high as those closer to the continental margin. As the

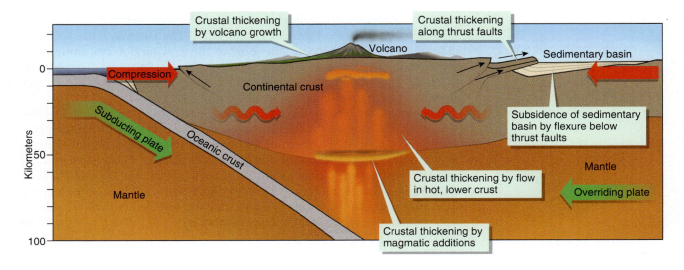

◄ **Figure 13.19 How continental-margin mountain belts form.**
Thick crust underlies high mountains. Crustal thickening occurs because of addition of magma from below that solidifies as plutonic rocks and because of plastic flow of weak, hot compressed crust. Thrust faults shove rocks farther away from the plate boundary across the overriding plate. The added weight of the overthrust rocks flexes the crust to form a basin that accumulates sediment eroded from the mountain belt.

▲ **Figure 13.20 Thrust faults make mountains and basins.**
(a) This view of Banff National Park in Alberta, Canada, is typical of the Rocky Mountains of Montana and western Canada, where sedimentary rocks were shoved eastward along thrust faults during the Mesozoic, when North America was near a convergent plate boundary. Erosion exposes the folded and thrust-faulted Paleozoic rocks that were uplifted in the Rocky Mountain thrust belt. (b) This view in western Montana shows the plains east of the Rocky Mountains, which form the distant skyline. Sedimentary rocks are as much as 6 km thick beneath the plains and accumulated in the basin that flexed down under the weight of the overthrusted rocks. These sedimentary rocks were also folded by the compression and contain rich resources of oil, natural gas, and coal.

thrust faults stack up rocks on the continent, the added weight causes the crust to flex down over a wide region, as predicted from isostasy. Part of the resulting depression fills with sediment eroded from the thrust-uplifted rocks and the more distant higher mountains. The weight of the accumulating sediment also contributes to additional subsidence of this sedimentary basin.

Mesozoic and early Cenozoic thrust belts in western North America formed mountains and deep basins, visible in **Figure 13.20**. The weight of the thrust-faulted rocks formed adjacent sedimentary depressions (see Figure 13.20b) that contain more than 6 kilometers of sedimentary rock in some places. The depressions were never 6-km-deep holes. Rather, the sediment filled in the basins as they slowly subsided under the weight of the thrust-faulted rock and accumulating sediment. The sedimentary rocks in these basins, and in parts of the thrust belt, are the source for prolific supplies of oil, natural gas, and coal throughout the Rocky Mountain region of the United States and Canada. Similar sedimentary basins adjacent to the Paleozoic Appalachian Mountains also host huge coal deposits.

Mountain Belts where Continents Collide

The mountain building where continents collide at convergent plate boundaries (Figure 12.26) is slightly different. **Figure 13.21** summarizes the features of collisional mountain belts. Crustal thickening by igneous-rock additions and shortening by plastic flow are similar to continental-margin mountain belts. In collisional mountain belts, however, the shortening takes place in both plates, because the low-density continental crust cannot readily subduct. As a result, the mountain belt has a more symmetrical cross section than in the case of oceanic-continental convergence (compare Figures 13.19 and 13.21). Huge thrust faults shove slivers of crust many kilometers thick out of the collision zone. These thrust faults cut deeply through the crust and effectively double up the thickness of continental crust (see Figure 13.21). The overthrust rocks add considerable weight to crust on both sides of the mountain belt, depressing basins that fill with 10 kilometers or more of sediment.

The extraordinary thickening of crust in collisional mountain belts forms the highest mountains (see Figures 1.2a and 12.20). The Himalayas are the highest mountains on Earth and are still rising as the slow-motion collision, which started 50 million years ago, continues between India and Asia. The Alps of southern Europe formed by collision of small continental blocks with the larger European continent beginning about 40 million years ago.

High mountains have thick roots, according to the principle of isostasy (Figure 13.8). The crust in the deep roots experiences extremely high temperature and pressure, causing high-grade metamorphism and even melting. The resulting granitic magma rises to solidify in the middle crust. This crustal melting forms large volumes of felsic magma and igneous rocks that are important to the formation of continental crust (look back in Section 9.3). The high-grade metamorphic rocks in the lowermost crust may be sufficiently dense to detach from the remainder of the crust and subduct with the underlying mantle lithosphere (Figure 13.21).

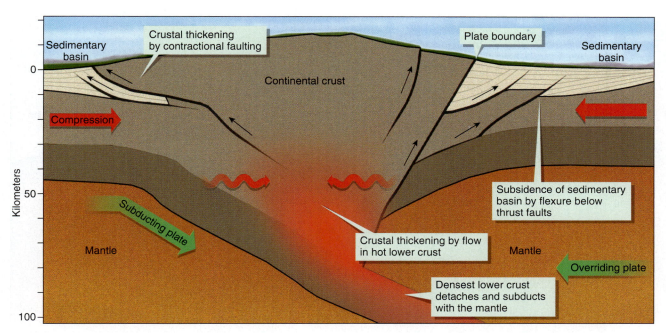

◄ **Figure 13.21**
Characteristics of a collisional mountain belt.
The highest mountains form where continental plates collide, because the crust becomes very thick. Most of the crust on the subducting plate is too buoyant to subduct and is shoved up on contractional faults and thickens by plastic flow at depth. Some of the densest lower crust and lithospheric mantle detaches from the low-density upper crust of the subducting continent and sinks into the mantle. Thrust faults shorten and thicken the upper crust on both sides of the collision zone, forming deep sedimentary basins.

What Goes Up, Comes Down

There is a limit to how high mountains grow. Section 11.6 noted that quartz-rich rocks composing continental crust are not particularly strong, especially when raised to high temperatures typical of depths in excess of 15 kilometers. The weak lower crust cannot support an infinite increase in the elevation of the surface above it. Eventually, the lower crust begins to flow horizontally away from the region of greatest crustal thickening. **Figure 13.22** shows how the spreading of crust under its own weight is similar to the spreading and sinking of a high pile of bread dough.

The horizontal flow in the lower crust not only limits the extent of crustal thickening that causes isostatic uplift, it also stretches the brittle upper crust. It is very common to find normal faults that are simultaneously active with reverse and thrust faults in high-standing mountain ranges. The horizontal extension in the area of thickest crust probably provides some of the force to drive compression in adjacent thrust belts (see Figure 13.22).

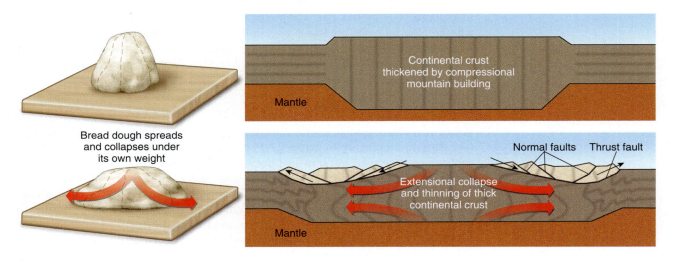

◄ **Figure 13.22 Mountains spread under their own weight.**
Thick bread dough slowly spreads out under its own weight when you pile it on a countertop, eventually covering more area while decreasing in height. Continental crust also spreads under its own weight when it becomes thicker than about 60 km. The hot, plastic lower crust cannot support the weight of overlying rock and spreads laterally, causing faulting in the brittle upper crust.

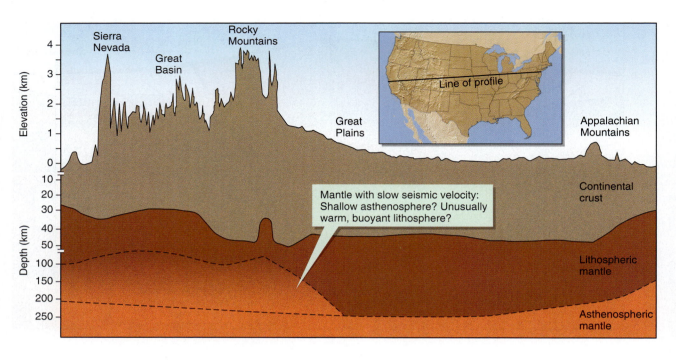

The Role of the Mantle in Topography

The high elevations of the Rocky Mountains, Sierra Nevada, and many other mountain ranges define the Western Cordillera of North America (outlined in Figure 12.8). Widespread shortening deformation and uplift occurred in western North America during the Mesozoic and early Cenozoic when there was a convergent margin along the entire west coast (described in Section 12.9).

Data illustrated in **Figure 13.23** do not show, however, a systematic relationship between surface elevation and crust thickness in the western United States. The crust below the high elevations of the Sierra Nevada (including Mount Whitney, Figure 13.1) is actually thinner than the crust of the western Great Plains, which is 3 kilometers lower in elevation. The mountains do not have the predicted thick crustal root. How can this be?

The mantle is key to explaining the high elevation of the western United States. Seismic data indicate a large region of the western United States where relatively low density mantle is unusually close to the surface. The lower-density buoyant mantle rises and flexes the surface upward to support mountainous elevations even where the crust is not very thick. The situation is somewhat similar to what you learned in Section 13.4 about the high elevation of mid-ocean ridges, in which surface elevations are higher above areas where the mantle is less dense. Geologists and geophysicists do not yet understand for certain whether the buoyant mantle is unusually low-density lithospheric mantle or if, instead, the lithosphere is unusually thin and the seismically slow mantle is part of the asthenosphere. Either way, isostasy and buoyant mantle explain high elevations in this region.

How Fast Do Mountains Rise?

All of this discussion of mountain-building processes begs the question: How fast does mountain uplift happen? Plates move horizontally at about 1 to 10 centimeters per year, but how fast does Earth's surface move vertically? Nearly 15 meters of uplift occurred on a thrust fault along the convergent boundary in Alaska during a single earthquake in 1899. Many decades to centuries of almost no uplift intervene between such great earthquakes, so these single spasms do not provide a good measure of long-term mountain uplift.

To measure uplift rates, geologists must know a vertical uplift distance and the time during which that uplift took place. One example is coral reefs near a convergent boundary in the southwest Pacific Ocean that are now 400 meters above sea level. Radioactive-isotope ages measured from the calcite in the coral indicate that the reefs were thriving below sea level 120,000 years ago. These two pieces of data indicate that the old seafloor rose at least 400 meters in no less than 120,000 years; this means that the average uplift rate was at least 3.3 millimeters per year. Another example comes from the Andes of South America, where metamorphic rocks exposed at the surface contain minerals indicating a temperature and pressure of metamorphism equivalent to about 10 kilometers depth. Radioactive-isotope ages on these minerals show that they were at this depth only 2 million years ago, which means an average rock-uplift rate of 5 millimeters per year and an equivalent erosion rate to remove the 10 kilometers of overlying rock.

Analyses of this sort in active tectonic settings around the globe show that the crust beneath mountains rises at rates of about 3–10 millimeters per year, when averaged over long time periods. The actual rates of surface elevation increase also depend on how fast erosion removes the uplifted rocks. These long-term uplift rates are not very different from measured ongoing rates of glacial rebound in eastern North America (Figure 13.14), but glacial rebound uplift and subsidence only last for tens of thousands of years, whereas mountain uplift persists for millions of years. Nonetheless, on average, plate motion moves Earth's surface at speeds about 10 times faster than isostasy and mantle processes cause vertical surface displacement.

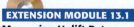

EXTENSION MODULE 13.1
Measuring Uplift Rates.
Learn how geologists measure mountain-uplift rates.

Putting It Together—*How and Where Do Mountains Form?*

■ Most mountain belts form near convergent plate boundaries where crust thickens by compressional shortening and by intrusion and crystallization of magma.

■ The hot lower-crustal root beneath very high mountain ranges is too weak to support the thick column of overlying crust. The crust spreads horizontally under its own weight and produces normal faults in parts of the mountain belt, while contraction occurs elsewhere at the same time.

■ The high elevations in the western United States do not have a thick crustal root but are held up, instead, by unusually low-density mantle.

■ Even as crustal thickening causes uplift of mountains, thrust faults weigh down adjacent lowlands to form deep basins. These basins fill with sediment eroded from the mountains and contain rich resources of oil, natural gas, and coal.

■ Mountains rise vertically at rates of 3–10 millimeters per year, or only about one-tenth the speed of horizontal plate motion.

13.6 How Does Mountain Building Relate to the Growth of Continents?

Mountain building is essential to making continents because mountain building thickens the crust. The thickness of continental crust, not just its low density, explains why continents are higher than oceans (Figure 13.6, model C). After all, where continental crust is thin, surface elevations are below sea level (Figures 13.1, 13.2, and 13.10). The crust also must be thicker than about 30 kilometers in order to stand higher than current sea level. Crust achieves this thickness by mountain building, even in areas where those mountains long since eroded away.

Another critical observation, illustrated in **Figure 13.24**, is that the edge of western North America at the end of the Precambrian (about 540 million years ago) was more than 500 kilometers east of where it is today. A large part of the Western Cordillera records westward growth of North America when the region was along a convergent plate boundary. This observation further suggests an important role of mountain building in the formation and growth of continents.

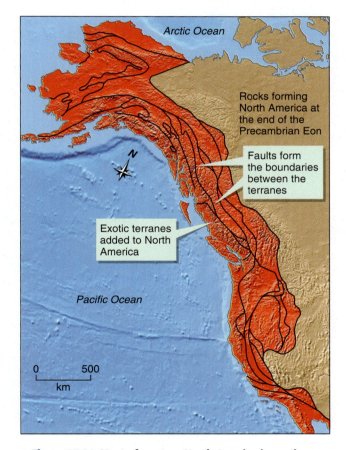

▲ **Figure 13.24 Most of western North America is exotic.** Shoreline sedimentary deposits dating from 540 million years ago indicate that the western edge of North America was more than 500 km farther east than it is today. Crust that originated outside of North America underlies the entire red region. Tectonic collisions at convergent plate boundaries added the exotic crust to the continent. The added-on crust includes many crustal blocks, outlined by faults, which have different geologic histories than neighboring blocks or North America.

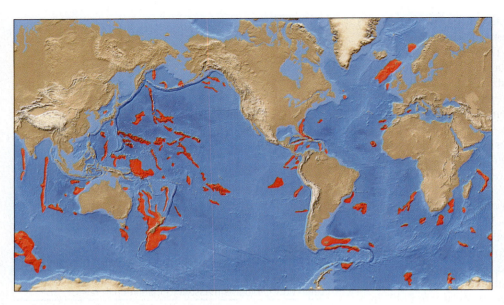

▲ Figure 13.25 Exotic terranes of the future?
The map shows the distribution of crust within modern ocean basins that is too thick to subduct. Each red area is a potential exotic terrane of the future, if it enters a subduction zone and accretes onto a continent.

Assembly at Subduction Zones

Geologic maps show that far-western North America consists of a variety of crustal blocks, each separated from its neighbors by faults (Figure 13.24). The oldest rocks exposed in each block do not resemble rocks of the same age in adjacent blocks or in the rest of North America. Each block of crust, therefore, appears to originate hundreds or thousands of kilometers from North America and was later added onto the continent. These blocks are called **exotic terranes** or **accreted terranes**. These terms emphasize regions where the crust is exotic to North America and accreted (added on) to the continent.

Crust accretes to a continental edge along a subduction zone because not all of the crust entering the zone can subduct. Continental crust resists subduction because it has a low density (Section 12.4). Crust formed at volcanic arcs near convergent plate boundaries is commonly greater than 20 kilometers thick and contains abundant relatively low-density intermediate to felsic igneous rocks that also do not readily subduct into the asthenosphere. Unusually thick oceanic crust, typical of locations experiencing excessive hot-spot volcanism, is also too thick and buoyant to subduct at convergent plate boundaries. **Figure 13.25** shows large, mostly submerged areas between the seven continents on Earth that consist of crust that is too thick to subduct. If any of these areas reach convergent plate boundaries adjoining continents, then they will accrete onto the continent.

Figure 13.26 illustrates two scenarios for how accretion and continental growth happens.

1. Where any piece of "nonsubductable" continental crust, unusually thick oceanic crust, or volcanic-arc crust enters a subduction zone at the edge of a continent, it thrusts onto the edge of the continent (Figure 13.26a).
2. If a continent follows subducting oceanic lithosphere into a trench, the buoyant continent cannot subduct and instead collides with the arc on the overriding plate. A new subduction zone forms that faces in the opposite direction (Figure 13.26b). The arc transfers from one plate to another and becomes part of the continent.

The collision of buoyant crustal blocks imparts unusually large compressive stress across the convergent-boundary continental margin. The large stresses cause considerable crustal thickening and down-bending of continental margins. Accretion of exotic terranes is, therefore, an important part of mountain building. All mountain belts contain accreted terranes, which illustrate the growth of continents by the assembly of pieces of "nonsubductable" crust at convergent plate boundaries. **Figure 13.27** illustrates examples of exotic-terrane landscape within North American mountain ranges.

Making the Ancient Continental Centers

To complete the connection of mountain building to the growth of continents, consider the origin of continental crust in the low-lying interiors of continents, far from recent, or even recognizable, older mountains. Although broken by faults in places, these interior regions of continents, called **cratons**, have been tectonically stable compared to continental margins for more than 500 million years. These low-elevation areas are typically 0.5 kilometer or less above sea level, and they expose either very ancient Precambrian rocks, or relatively thin coverings of mostly flat-lying, undeformed sedimentary rocks resting unconformably on older Precambrian rocks. Landscapes of the North American

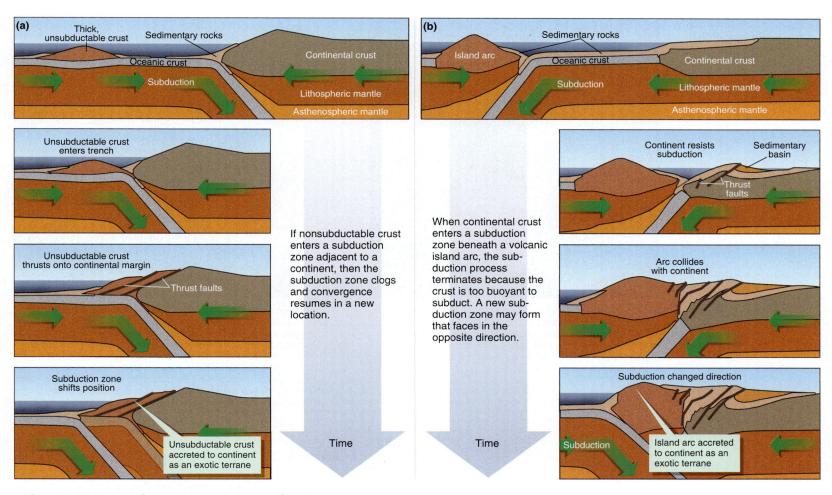

(a)
Thick, unsubductable crust
Sedimentary rocks
Oceanic crust
Continental crust
Subduction
Lithospheric mantle
Asthenospheric mantle

Unsubductable crust enters trench

Unsubductable crust thrusts onto continental margin
Thrust faults

Subduction zone shifts position
Unsubductable crust accreted to continent as an exotic terrane

If nonsubductable crust enters a subduction zone adjacent to a continent, then the subduction zone clogs and convergence resumes in a new location.

Time

(b)
Sedimentary rocks
Island arc
Oceanic crust
Continental crust
Subduction
Lithospheric mantle
Asthenospheric mantle

Continent resists subduction
Sedimentary basin
Thrust faults

Arc collides with continent

Subduction changed direction
Subduction
Island arc accreted to continent as an exotic terrane

When continental crust enters a subduction zone beneath a volcanic island arc, the subduction process terminates because the crust is too buoyant to subduct. A new subduction zone may form that faces in the opposite direction.

Time

▲ **Figure 13.26 How exotic terranes accrete to a continent.**
These diagrams show how continents grow along convergent plate boundaries by the addition of crustal blocks that originated in far distant places.

(a)

The Coast Range of Oregon and Washington consists of seamount and island volcanoes, similar to modern Hawaii, which accreted to North America about 45 million years ago because the volcanic crust was too thick to subduct beneath the continent.

(b)

The crust underlying this area of the Appalachian Mountains in central Massachusetts formed as a volcanic island arc that was shoved onto North America about 450 million years ago.

◀ **Figure 13.27 Exotic crust in North America.**

▶ **Figure 13.28 Tectonically stable landscapes of the North American craton.**

Superbly exposed rocks in northwestern Canada are highly metamorphosed Precambrian rocks typical of the craton. The low topographic relief reveals the current tectonic stability of the continental interior, although the metamorphism attests to intense deformation during Precambrian time.

Deep erosion in the Grand Canyon, Arizona, exposes the unconformity between Precambrian metamorphic and igneous rocks, recording mountain-building events that constructed North America, and much younger horizontal sedimentary rock. The undeformed sedimentary rocks are typical of the tectonically stable craton. Deep erosion exposes the underlying Precambrian rocks, whereas in most of the United States the nature of older rocks is known primarily from deeply drilled oil-exploration wells.

craton are shown in **Figure 13.28**. A map of North America, presented as **Figure 13.29**, outlines the North American craton and shows that the Precambrian crust is divided into discrete blocks, called **provinces**, composed of igneous and metamorphic rocks of different ages.

The Precambrian crust, ranging in age from about 1.0 to 4.0 billion years, is metamorphic and plutonic rock (Figure 13.28). This observation implies two things:

1. The crust probably formed by mountain building near convergent plate margins, where magma intrusion and metamorphism of thickened crust are common.
2. Considerable uplift and erosion, including isostatic adjustments, happened to expose the metamorphic and plutonic rocks at the surface before they were locally buried beneath much younger sedimentary rocks.

Figure 13.30 integrates these two conclusions to show that the exposed Precambrian rocks likely formed in the middle to lower crust during thickening of the crust by mountain building. Erosion of the high mountains, with accompanying isostatic uplift, eventually removed the now missing rocks that formed the Precambrian upper crust. The mountains eventually eroded to low-relief topography with sufficient remaining crustal thickness to maintain the surface elevation at, or slightly above, fluctuating sea level. At times when sea level was unusually high (see Figure 13.15), thin layers of marine sedimentary rocks accumulated on top of the Precambrian rocks.

Closer scrutiny of the geology of the Precambrian rocks of the craton and the map pattern in Figure 13.29 reveals how North America formed by mountain building along convergent plate boundaries. **Figure 13.31** summarizes the story. Each province composed of rock older than 2.5 billion years consists of jammed-together, convergent-boundary volcanic arcs. Metamorphosed volcanic-arc rocks dominate the oldest Precambrian history on all continents, which is

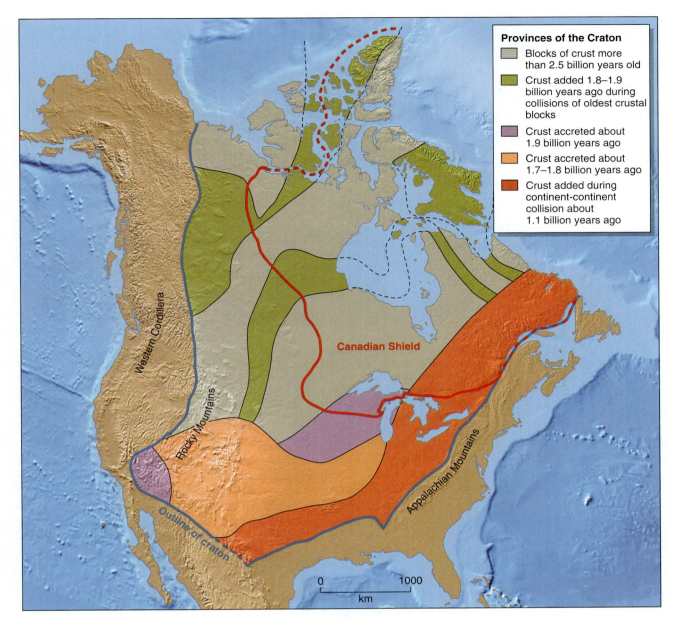

▲ **Figure 13.29 North American craton consists of Precambrian provinces.**
The North American craton is the region where ancient Precambrian rocks are exposed at the surface (mostly in the Canadian Shield) or only thinly buried by sedimentary rocks. The craton is conveniently divided into provinces with different ages of metamorphism that relate to mountain-building events that enlarged the continent. The craton has experienced very minor tectonic deformation for nearly one billion years, except in the Rocky Mountain region. The craton is partly rimmed by areas of highly deformed, in some cases metamorphosed, rocks affected by mountain building over the last 450 million years to form the Western Cordillera and the Appalachian Mountains.

one of the main reasons why convergent-margin processes were invoked in Chapter 9 to produce the first continental crust (refer back to Section 9.3 for the details). These small, embryonic continents of amalgamated arc crust then collided with each other, mostly between 1.8 and 2.0 billion years ago. Metamorphism of this age in Canada and the northern United States defines mountain belts that formed between the older provinces (see Figure 13.29). Large masses of crust were added along the southern and eastern margins of North America by major collisions with continents and volcanic island arcs between 1.1 and 1.7 billion years ago.

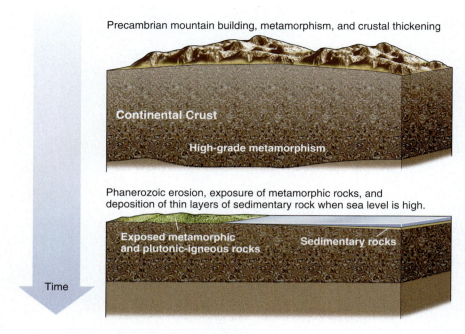

Precambrian mountain building, metamorphism, and crustal thickening

Continental Crust

High-grade metamorphism

Phanerozoic erosion, exposure of metamorphic rocks, and deposition of thin layers of sedimentary rock when sea level is high.

Exposed metamorphic and plutonic-igneous rocks

Sedimentary rocks

Time

▲ **Figure 13.30 Cratons started as mountains.**
Precambrian metamorphic and plutonic rocks formed by mountain building and crustal thickening at convergent plate boundaries. High mountains gradually wore down by erosion, combined with isostatic uplift (see Figure 13.12), until low-relief regions remained. The crustal thickness of the low regions is sufficient to keep the surface above sea level, except when sea level is exceptionally high.

Continents are the product of mountain building at ancient convergent boundaries. Where mountains stand high above the surrounding landscape, mountain-building processes are easily recognized. However, even the crust below the low-lying interior of North America is the product of mountain building that occurred as long as 4 billion years ago. The now-stable craton stands above sea level as an isostatic consequence of the development of relatively thick, low-density crust in mountain belts.

Putting It Together–How Does Mountain Building Relate to the Growth of Continents?

■ Continents grow through time by the collision and accretion of crustal fragments that cannot be subducted along convergent plate boundaries.

■ The low-elevation regions of continents are areas of long-term tectonic stability, called cratons, where Precambrian metamorphic and plutonic rocks are present at or near the surface.

■ Craton crust formed during mountain-building events between 1 and 4 billion years ago. These mountain-building events featured collisions of thick blocks of low-density, mostly igneous crust. North American crust is a collage of these crustal blocks.

■ Although the Precambrian mountains were long ago eroded down and partly buried, the elevation of the continental interior above sea level is a result of the crustal thickening that occurred during those ancient periods of mountain building.

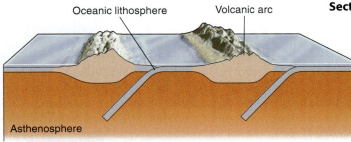
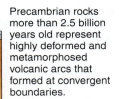

Oceanic lithosphere   Volcanic arc

Asthenosphere

Precambrian rocks more than 2.5 billion years old represent highly deformed and metamorphosed volcanic arcs that formed at convergent boundaries.

◄ **Figure 13.31 Continents are made at convergent plate boundaries.**
These highly schematic diagrams show how continents have grown through time as a result of volcanic and collisional processes at convergent plate boundaries.

Island arcs collide at subduction zones to form early continents

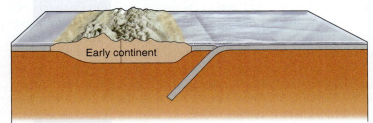

Early continent

Volcanic arcs collided in mountain-building events that formed the oldest provinces of the craton.

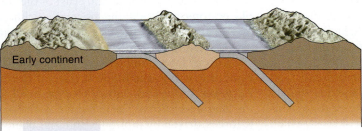

Early continent

Collisional zone of volcanic arcs between early continental blocks

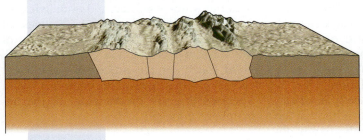

The small, early Precambrian continents collided with one another, and with intervening volcanic arcs, between 2.0 and 1.8 billion years ago to form larger continents.

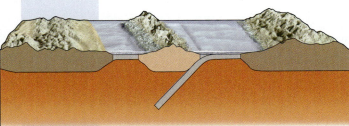

Continent-continent collision and accretion of arc and oceanic rocks

Since 1.8 billion years ago, continents grew by accretion of exotic blocks of crust, many of which originated as volcanic arcs.

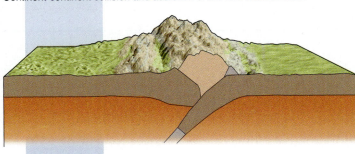

Time

Where Are You and Where Are You Going?

Isostasy explains the rough relief of Earth's surface. Areas of thick crust, or low-density crust, or both, exhibit a higher elevation than areas of thinner or denser crust. High-standing mountains, therefore, typically have a thick root of low-density rocks. Continental crust is both thicker and less dense than oceanic crust, which explains why continental areas are largely above sea level and oceanic areas are submerged beneath the sea. Crust thickens in response to compression, which causes formation of high-elevation mountains, whereas thinning occurs during tension, which produces low-elevation rift valleys and submerged continental shelves. Isostatic adjustments to changes in the distribution of mass within the crust commonly cause earthquakes within plates, far from plate boundaries.

Flexural isostasy is more appropriate for explaining surface elevations in detail than are the Pratt and Airy concepts, which are analogous to wood blocks in a bathtub. The crust is relatively strong and not broken all of the way through into discrete blocks, and the viscous, solid mantle flows very slowly. As a result, the crust bends under the added weight from deposited sediment or glacial ice and slowly rebounds when weight is removed by erosion or melting of ice. Glacial rebound of crust that flexed under the weight of the great ice-age glaciers that melted away thousands of years ago explains the slow rising and sinking of eastern North America.

Earth's mantle also plays a role in surface elevation. Where the mantle is relatively less dense, surface elevations are higher than in areas where the mantle is denser. These contrasts in mantle density are mostly determined by temperature, which in turn determines where the mantle has the properties of cold, strong lithosphere in contrast to weak, hot asthenosphere. The relatively high elevations of mid-ocean ridges, compared to surrounding seafloor, are explained by the very thin nature of the lithosphere along divergent plate boundaries. Similarly most of the mountainous region of the western United States surprisingly lacks a corresponding thick crust but is underlain instead by unusually buoyant mantle.

Mountain-building processes at convergent-plate boundaries include the accretion of blocks of crust to the edges of continents. These accreted blocks of crust, which originated far from the continent where they are now found, are too thick to subduct completely. When these blocks reach a subduction zone, they are shoved onto, or jammed beneath, the continent. Continents grow through time by the accretion of these exotic terranes at convergent boundaries.

The craton is the geologically most stable, interior part of a continent. Although the North American craton has only slightly deformed over the last half billion years, it is characterized by outcrops and shallowly buried expanses of ancient metamorphic and igneous rocks that originated near Precambrian subduction zones. Most of the continental crust seen today originated more than a billion years ago as a result of convergent-boundary magmatism and collisional accretion of nonsubductable blocks of crust. These old Precambrian rocks of the cratons were once deep below the surface in the thick crustal roots of ancient mountains. Over the course of geologic time, the mountains eroded as isostasy kept raising the crust. Eventually, continental crust of the stable cratons achieved the normal thickness and an average elevation slightly higher than average sea level.

This chapter completes your study of deformation within and at the surface of Earth. However, it does not suffice to explain all of the characteristics of Earth's surface. Tectonic forces, driven by motion within the planet, determine the general form of surface features, but water, wind, ice, and even living organisms sculpt the varied landscapes of Earth. To complete your understanding of how Earth works, you now embark on studies of Earth surface processes—where the forces of the geosphere interact with processes that are fundamental to the hydrosphere, atmosphere, and biosphere.

 ## Active Art

Glacial Isostasy. See how the crust isostatically responds to the growth and shrinkage of glaciers.

Extension Modules

Extension Module 13.1: Measuring Uplift Rates. Learn how geologists measure mountain-uplift rates.

Confirm Your Knowledge

1. What is relief? What is the total relief exhibited on Earth's surface?

2. What is the principle of isostasy?

3. What process controls elevation? How does it work?

4. Explain how icebergs are used as analogies to crustal elevations.

5. How do processes such as erosion and deformation affect isostasy?

6. How were the elevations of the Himalayas surveyed in the mid-nineteenth century? How were the anomalous data of Everest interpreted by John Pratt and George Airy?

7. Are either Pratt or Airy models correct in understanding isostasy? What are the limitations to the models?

8. What is flexural isostasy? How does it explain glacial rebound?

9. How do changes in glacier volume affect sea level?

10. How does the age of the seafloor affect sea level?

11. How are mountains formed at convergent boundaries?

12. At oceanic-continental convergent plate boundaries, why are the highest mountains and thickest crust 200 to 400 km inland of the plate boundary rather than right at the boundary?

13. Cite two examples of mountain belts formed by continental collision.

14. Explain why the mountains of the Western Cordillera of North America can exhibit high elevations but do not have deep crustal roots.

15. What is the observed range of uplift rates? How does this range compare with the speed of plate motion?

16. Continents tend to have the oldest rocks in the center, and younger material flanking the center. Explain this, with reference to cratons and exotic terranes.

Confirm Your Understanding

1. Write out an answer for each question in the Chapter Outline for the chapter sections assigned by your instructor.

2. What information do you need to calculate the pressure (weight) of a column of rock? Calculate the pressure at the base of oceanic and continental crust. Assume a column of rock 1 square meter in area.

3. Explain two lines of evidence for how we know that mountains have roots.

4. How does a mountain develop a root as well as height?

5. What are some of the problems with an isostatic model of crustal elevations using wood blocks floating in water as an analogy?

6. What evidence did the most recent ice age provide us to evaluate the merits of the Pratt-Airy isostasy and flexural isostasy models?

7. What evidence is consistent with the hypothesis that the New Madrid earthquakes of 1811–1812 were caused by isostatic adjustment from the last ice age?

8. The Appalachian Mountains formed by continent-continent collision 300 million years ago. Explain why they persist as a mountain range today despite 300 million years of erosion.

9. Explain why the two elevations, around 0.5 km above and 4.5 km below sea level, are the most common elevations on Earth?

10. Describe the "life cycle" of a mountain.

Key Terms

cratons (p. 374)
exotic terranes (accreted terranes) (p. 374)

glacial rebound (p. 364)
principle of isostasy (p. 354)

provinces (p. 376)
relief (p. 352)

14 Soil Formation and Landscape Stability

Why Study Soils?

EARTH-SURFACE PROCESSES ARE THE ACTIONS THAT TAKE PLACE at the interface of the geosphere with the hydrosphere, atmosphere, and biosphere. Nowhere are interactions of all components of the Earth system more apparent than in the formation of soil. Soils are the weathering products that remain in place at Earth's surface. All other weathering products erode or dissolve away, usually to be recast elsewhere as sediment and sedimentary rock. Soil characteristics depend on a mixture of geologic, hydrologic, atmospheric, and biologic factors. These include the materials that weather to form the soil, the climate and vegetation of the weathering environment, the relief of the landscape where weathering occurs, and the duration of soil formation. The many varieties of resulting soils are fascinating mixes to dig into, to build on, to farm, or to use to grow a garden.

Soil is essential to our survival. Soil supports the plants we eat and the crops fed to livestock. Quality of soil determines the varieties of plant ecosystems it can support and the capacity of the land to support animal life and human communities. Degradation of soil is a universal problem, from contamination to soil erosion. Soil is a nonrenewable resource, and, as such, the study of soil and its conservation will always be important.

The value of understanding soil formation lies in the many essential functions that soils perform.

- Soil is the medium for plant growth, the anchor for roots, and the source of nutrients. Agricultural economies depend on fertile soils for success, and they fail when the nutrients are exhausted or the soil erodes away.

- Soil properties determine the fate of rainfall and snowmelt. Thick, porous soil allows moisture to infiltrate the ground, where it benefits plants and soil animals and microbes, or passes farther downward to replenish ground

This view in Lancaster County, Pennsylvania, emphasizes that soil covers most of Earth's land surface and is the most essential ingredient to successful agriculture. ▶

water. Thin or nonporous soils do not absorb moisture, so that water runs off the surface and enters streams.

■ Soil organisms recycle decaying plant matter into the nutrients used by the next plant generation.

■ Soil forms the mostly loose coating on land surfaces where many buildings and roadways are constructed. Engineers are conscious of the strength of soil, just as they are mindful of the strength of rock, to assure that the natural soil foundation will support the weight and function of what is built.

Geologists also study soils as keys to the history of landscapes. Soils mature as their properties slowly change through time. Erosion or sediment deposition changes landscapes faster than soils mature. Mature soils, therefore, only cover stable landscapes that neither erode nor accumulate sediment. Landscapes may destabilize because of human activities that cause erosion and loss of soil that could otherwise benefit crops.

You may take soil for granted. Language offers quite a few disparaging words about "dirt." Soil, however, is not just "dirt." It is a valuable asset to human existence and a key to understanding the history of Earth's diverse landscapes. Your primary objectives for this chapter are as follows:

✔ To learn the variable characteristics of soil

✔ To explain the factors controlling soil formation

✔ To understand how soils relate to landscape stability

To accomplish these goals, you will answer the following questions:

14.1 What is soil?

14.2 What distinguishes soil horizons?

14.3 How do soils form?

14.4 What factors determine soil characteristics?

14.5 What are the types of soils?

14.6 *How do we know* ... that soils include atmospheric additions?

14.7 How do human activities affect soils?

In the FIELD

While driving in Ohio, you stop to look at rocks exposed in an old quarry. Hard rock does not go all the way to the top of the quarry wall. Instead, a loose mixture of sand and pebble-size rock fragments along with fine-grained clay and decomposing plant material occurs just below the surface, as shown in **Figure 14.1**a. This is what you have always referred to as dirt, but it has a number of interesting features. There are distinct color bands in the dirt, with gray shades near the top, and orange to red colors below. Trees and shrubs grow in the dirt, and accumulations of dead leaves, twigs, and small roots seem to account for the gray color close to the surface. While poking around with the tip of your pocketknife, the dirt falls apart in clods, loosely held together clumps of mineral grains and organic material. Where moist, the loose dirt is sticky and plastic, and you can roll it between your palms into balls. The sticky, clumping characteristics suggest the presence of clay minerals.

This unconsolidated material is soil. It is the stuff you enjoyed digging in as a child, that is stirred up in the preparation of planting a garden or a farm field, and that piles up around excavations for building foundations or highways. You see soil almost everywhere at the ground surface, but you may not have given much thought to how it came to be there.

On another occasion, you examine soil in a sand pit in the arid southwestern United States, which is illustrated in Figure 14.1b. This soil overlies loose, stream-deposited gravel and sand, with distinct bedding that disappears upward. There is some reddish color in the nonbedded zone, similar to but not as red as the red band at the quarry, and the desert soil lacks the gray band with abundant organic matter at the top (Figure 14.1b).

Hard, white, mineral nodules are a conspicuous feature of the desert soil (Figure 14.1b). You easily scratch the white nodules with your knife but not with your fingernail. The hardness is appropriate for calcite. Fizzing from a drop of weak acid on a nodule confirms the presence of calcite.

Your observations lead to a number of questions. Soils link to weathering processes, but how does weathering account for the features of the soils? These features include the color bands seen in both soils, the tendency of the soil to fall apart in clods, and the presence of calcite in the desert soil. How does soil form and what does it form from? How do soils relate to minerals and rocks and to the landscapes the soils cover? Why do the two soils illustrated in Figure 14.1 look different? The eastern soil is redder and has a darker gray hue near the top. The southwestern soil is lighter in color and contains calcite. If soils differ in physical and mineral properties, do these properties relate to their suitability to support growing crops? What is the significance of soil to understanding geological processes? Why is soil, rather than bare rock, the most common material found right at the land surface?

Figure 14.1 **What soil looks like.** ▶

Field notes about soils
A. Photos of rock quarry with soil, Ohio

Red soil beneath trees and on top
of loose rock rubble

About
1.5 m

Color bands in soil, gray at
top, orange-red at bottom.

B. Photos of soil in sand pit, New Mexico

Notice that bedding layers
in sand and gravel are not
present at top of pit.

White powder on pebbles
and sand grains is calcite.

Close-up view of soil color
bands, slightly reddish at
top, white lower down.

About
50 cm

14.1 What Is Soil?

Soil means different things to different people. Engineers typically define soil to include any surface material that is not solid rock. In this usage, soil includes recently deposited sediment that has not yet consolidated into rock, as well as loose material resulting from rock weathering. Soil scientists evaluate soil usefulness for different agricultural functions, so they commonly define soil as the medium for plant growth. The term "topsoil" commonly refers to just that upper part of the soil disturbed by crop cultivation or shoveling in a backyard garden. Clearly, there is no single, everyday definition of soil.

A Geologic Definition of Soil

Geologists commonly define **soil** as a layered mixture of loose mineral and organic constituents that have different physical or compositional properties (or both) than the original, nonweathered material. Chemical and physical weathering processes convert preexisting materials into soil. Plants that take root in the soil and the animals that live within it commonly influence the chemical and physical weathering processes. Soil formation, therefore, typically occurs close to the surface and links to the climatic variables of moisture and temperature, which determine both chemical-weathering reactions and the type of vegetation living on the surface.

Layers, such as the color bands visible in the field examples (Figure 14.1), are an essential component of the soil definition. These **soil horizons** are distinguished from one another by different particle sizes and mineral compositions. Horizons are not like sedimentary beds that accumulate one on top of the other. Instead, horizons form in place and reveal different physical and chemical processes at different depths below the ground surface.

Plants and soils are linked together. Plants grow upward from the surface and their roots penetrate downward into the soil. Fallen leaves and dead stems accumulate on the soil surface and are mixed into the soil by burrowing insects and other animals, whose remains also accumulate on the surface and at shallow depth in the soil. Air and water mixed into the soil aid in plant growth.

Biologic activities affects the entire thickness of the soil, even if decaying plant organic matter is mostly just at the top. Every anthill you see is a reminder that animals living in soil, especially insects and worms, persistently move soil particles. Growth of plant roots also moves soil. Biologic activity constantly churn the soil and affect its physical properties. The disruption of the original arrangement of sediment grains by growing plant roots and burrowing animals explains the lack of bedding where soil formed in the sedimentary deposit, illustrated in Figure 14.1b.

Soil Forms from Parent Material

What is soil made from? At the quarry, soil seemingly formed in fragmented rock, whereas the desert soil related to loose sediment (Figure 14.1). More generally, you observe that geologic materials at Earth's surface are either solid rock or loose unconsolidated material that typically overlies solid rock. **Bedrock** describes large, continuous occurrences of solid rock, whereas the term "rock" by itself could describe a small fragment that you pick up from the ground. Unconsolidated remains of weathered bedrock, rather than bedrock itself, cover most of Earth's surface. All unconsolidated deposits overlying bedrock comprise **regolith**, a term derived from the Greek "rhegos," meaning "blanket," and "lithos," meaning "stone." Some regolith is simply fragments from underlying or nearby outcrops of rock that were dislodged by physical weathering but have not moved, or at least have not moved very far; this type of regolith overlies the quarry bedrock in Figure 14.1a. Other regolith is sediment, the remains of weathered rock transported from their place of origin, as seen in the desert excavation in Figure 14.1b. The regolith, or less commonly, the bedrock, from which a soil forms is the **parent material** for the soil.

Soil is that part of the regolith that contains distinct soil horizons. Soil also exhibits clear evidence of biologic activity and supports plant life, which is not true of all regolith. **Figure 14.2** illustrates the differences among bedrock, regolith, and soil. The upper boundary of a soil is the atmosphere, but the lower boundary is sometimes indistinct. In most cases, the downward decrease in intensity of weathering and plant activity is very gradual. Soil scientists commonly assign a maximum thickness of two meters to a soil, and this assigned lower boundary usually encloses most plant roots and the most active weathering processes.

Putting It Together—*What Is Soil?*

- Soil is naturally occurring horizons of mostly unconsolidated mineral and organic constituents that differ from parent material in physical and compositional properties.
- The parent material is usually regolith rather than hard bedrock. Regolith is either loose sediment transported from elsewhere, or fragments generated by physical weathering of underlying rock. Soil is the part of the regolith that has horizons, has obvious biologic activity, and supports plant life.
- Soil horizons differ from one another and from parent material by contrasting grain size, mineral composition, or both.

14.2 What Distinguishes Soil Horizons?

Horizons are essential to the definition of soil, so it is important to understand why horizons exist and how they relate to soil formation processes. The first step is to determine the characteristics that distinguish each soil horizon from its neighbors. Explanations for how the horizons form must be consistent with these characteristics.

Even brief field examination (Figure 14.1) suggests at least three soil horizons:

1. A top horizon that is sometimes gray because it contains organic matter.
2. A middle horizon that can be reddish in color and contain clay.
3. A bottom horizon that is not obviously weathered.

These are the three most common horizons, simply lettered A, B, and C, from top to bottom. Some soils contain additional horizons, labeled O and E. **Figure 14.3** illustrates and summarizes the key features of the five soil horizons, and **Figure 14.4** identifies horizons in contrasting forest and desert soils.

▲ **Figure 14.2 Visualizing soil and parent material.**
This labeled photograph shows the components of the near-surface environment that are important to understanding soil formation. Regolith is unconsolidated rock and organic material that overlies bedrock. Soil is that part of the regolith that is strongly modified by chemical weathering and biological activity. The parent material is what weathered to produce the soil. Soil commonly exhibits horizons of different physical and compositional properties, which usually appear as color differences, too.

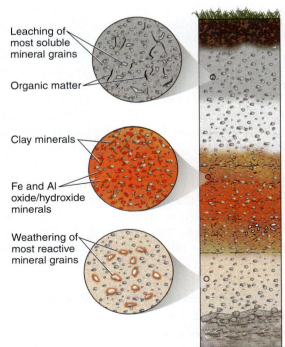

O horizon
Surface layer composed of organic residues of dead plants and animals in varying stages of decomposition. Color is dark in shades of gray and brown.

A horizon
The uppermost horizon composed of minerals; may be the surface horizon. Contains decomposed organic matter, especially in wetter climates, which may darken the horizon compared to lower horizons. Readily dissolved minerals are less abundant than in lower horizons.

E horizon
Light-colored horizon lacking clay, organic matter, and easily weathered minerals.

B horizon
The horizon of maximum accumulation of clay minerals, and iron and aluminum oxides and hydroxides. In dry climates, calcite, gypsum, and halite may accumulate in this horizon, too. Compaction and shrinking and swelling of clay minerals leads to aggregates of soil minerals into clods. Iron and aluminum oxides and hydroxides typically impart shades of yellow, orange, or red, depending on mineral type and abundance. Calcite, gypsum, and halite form white nodules and layers.

C horizon
The regolith below A and B horizons that exhibits little if any evidence of weathering; commonly the parent material for the soil. The color of this horizon is determined by the colors of the original minerals; oxidation may produce pale yellow and orange coloration.

Bedrock

Leaching of most soluble mineral grains

Organic matter

Clay minerals

Fe and Al oxide/hydroxide minerals

Weathering of most reactive mineral grains

▶ **Figure 14.3 Describing soil horizons.**

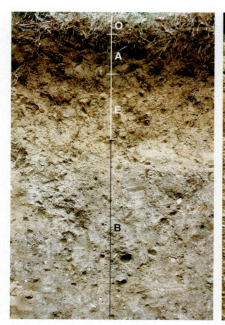

Calcite grain coatings and nodules

▲ **Figure 14.4** **What soil horizons look like.**
These photographs contrast the horizons of a forest soil (left) with a desert soil (right). The forest soil has an O horizon of decaying leaves that overlies the A horizon. There is also a light-colored, highly leached E horizon. The B horizon has yellow and orange colors representing the accumulation of iron-hydroxide minerals and clay. This desert soil has a thin, light-colored A horizon without very much organic matter. The B horizon is brown at the top, because of accumulation of iron-hydroxide and oxide minerals and clay, and has a lower, light-colored zone containing white calcite.

Horizons Close to the Surface: A, O, and E

Organic matter is most abundant near the top of the soil where biologic activity is highest and dead leaves and branches accumulate beneath the plants that live in the soil. The presence of organic matter mixed in with mineral and rock fragments defines the **A horizon**. Organic matter produces the distinctive "earthy" odor of moist soil. Where plant growth is dense, as in a forest, for example, there may be so much litter of decaying leaves, conifer needles, and wood on the ground that the surface horizon consists only of organic matter without minerals and forms an **O horizon** ("O" for "organic"). Compared to the parent material, A horizons also lack the minerals most susceptible to chemical weathering (such as the iron and magnesium silicates, micas, and soluble minerals like calcite).

E horizons appear below A horizons, if they are present at all (Figures 14.3 and 14.4). The absence of organic matter, easily weathered minerals, and weathering products like clay, oxide, and hydroxide minerals are key features of the **E horizon**. The E horizon receives its label from the word "eluviation," which comes from the Latin word roots "e" and "lavere," which mean "to wash out." Some E horizons contain nothing but quartz, which is the most weathering-resistant, common mineral in rocks. E horizons are always very pale and may be completely white, because there are no colored minerals or organic fragments.

The Middle Horizon of Mineral Accumulation: B

B horizons are most notable for colors and textures indicating accumulation of minerals that are not present in the parent material. In the field examples (Figures 14.1 and 14.4), the B horizons are reddish in color, sometimes contain calcite, and are sticky when wet. Stickiness suggests the presence of clay minerals that attach to one another because of stray electrical charges around the grain boundaries.

The accumulation of minerals in the middle part of the soil defines the B horizon (Figure 14.3). The presence of hydroxide and oxide minerals containing iron, aluminum, or both accounts for the distinctive red, orange, and yellow hues of B horizons. Clay minerals are abundant in many B horizons, even where clay is absent from the parent material. Although soluble minerals such as calcite (or even more soluble but less common gypsum and halite) are rare or absent in the A horizon, they can be more abundant in the B horizon than in the underlying parent material, but only where the climate is dry (Figure 14.4). In microscopic view (Figure 14.3), all of these added minerals—oxides, hydroxides, clays, and calcite—form coatings on the other mineral and rock fragments in the soil.

The Bottom Horizon of the Soil: C

The **C horizon**, at the bottom of the soil, is the least weathered horizon and commonly is the parent material for soil formation. The C horizon is regolith that lacks the organic matter found in the A horizon and contains little if any of the red colors, clay content, or calcite occurrences that define the B horizon (unless these are properties of the parent material).

Soil Horizons Result from Mineral Additions, Subtractions, and Transformations

Figure 14.5 provides a graphical summary of distinguishing soil-horizon characteristics. Assuming that the composition of the C horizon approximates the composition of the parent material, then comparisons of the O through B hori-

Forest soil

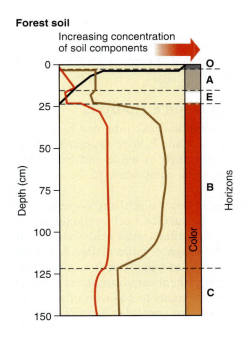

Desert soil

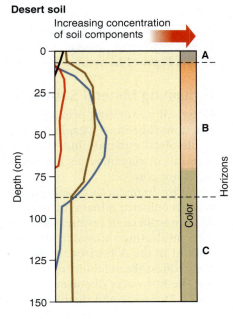

◄ **Figure 14.5 Comparing the composition of soil horizons.**
This diagram compares and contrasts the distribution of various components in a humid-region forest soil and an arid-region desert soil. Compared to the parent material in the C horizon, it is apparent that mineral components are subtracted from the A and E horizons and added in the B horizon.

Organic matter concentrates in the surface horizons, including an **O** horizon in the forest soil.

A and **E** horizons have lower mineral abundances than the **C** horizon, indicating subtraction of components from the upper horizons.

The **B** horizon contains the greatest abundances of iron oxides and hydroxides, clay minerals, and calcite (in the desert soil).

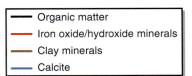

— Organic matter
— Iron oxide/hydroxide minerals
— Clay minerals
— Calcite

zons to the C horizon reveal soil-forming processes. *Added* components or those *transformed* from the parent material are more abundant in the upper horizons compared to the C horizon. *Subtracted* components are less abundant in the upper horizons compared to the C horizon. Formation of soil horizons, therefore, relates to chemical and biologic processes that remove components from some levels in the soil and add components or transform them into new minerals in other levels. The defining characteristics of the soil horizons, therefore, provide important insights into soil formation:

- Some horizon colors imply addition of constituents not present in the parent material. The gray color of the O and A horizons, the reddish colors lower down, and the white calcite nodules in the desert soil are characteristics of the soil that do not appear in the parent material.
- Organic matter is added to make O and A horizons.
- The most readily weathered minerals are subtracted from A and E horizons.
- Oxide, hydroxide, and clay minerals and, in dry regions, soluble minerals such as calcite appear in B horizons by addition or transformation during weathering reactions, or both.

Putting It Together—What Distinguishes Soil Horizons?

- Soil horizons are distinct because of differences in organic matter and mineral content that result from additions, subtractions, and transformations of minerals, and organic matter resulting from chemical weathering and biologic activity.
- A and O horizons consist of added organic matter. A horizons also show evidence of mineral dissolution.
- The E horizon, when present, is a nearly white horizon defined by an absence of organic matter, clay, colorful oxide and hydroxide minerals, or easily weathered minerals present in the parent material.
- The B horizon is notable for additions of colorful oxide and hydroxide minerals and clays. Calcite is present in arid-region B horizons.
- The C horizon is weakly weathered, unconsolidated regolith below the zone of accumulated minerals defining the B horizon.

14.3 How Do Soils Form?

The material additions, subtractions, and transformations that define soil horizons imply active physical and chemical processes that convert parent material to soil. **Figure 14.6** summarizes these processes.

Explaining Mineral Subtractions from A and E Horizons

Chemical weathering primarily involves dissolution, oxidation, and hydrolysis reactions (Section 5.1 explains these reactions). A and E horizons contain partly dissolved minerals, but weathering products, such as oxide and hydroxide minerals or clay minerals, are not very abundant in comparison to the underlying B horizon.

Why do minerals dissolve near the surface? Natural acid and organic compounds enhance mineral dissolution close to the surface. Carbon dioxide from the atmosphere, or respired through plant roots, makes a weak acid when mixed with infiltrating rainwater and snowmelt. Although organic matter is always present in the A horizon, it also readily dissolves as quickly as it accumulates. You witness the solubility of organic compounds whenever you make tea or coffee; the hot water dissolves some of the organic solids to color and flavor the water. Decaying and dissolving organic matter also generates additional acids and other organic compounds that enhance mineral dissolution. Minerals containing calcium, sodium, magnesium, and potassium held in crystal structures by relatively weak ionic bonds are most susceptible to dissolution by these acids that form in the A horizon.

The evidence of mineral dissolution in surface horizons but the low abundances of weathering products mean that two things happen:

1. Weathering reactions chemically remove soluble chemical components from soil minerals, and infiltrating water then flushes the dissolved ions downward.
2. Solid weathering products of oxidation and hydrolysis reactions physically wash downward with infiltrating water, so that only the most weathering-resistant minerals and persistently accumulating organic matter remain.

E horizons represent the most intense flushing out of reactive components and weathering products. Water-soluble organic compounds in the E horizon leach aluminum and iron from crystal structures, which even breaks down relatively resistant clay minerals. As a result, E horizons are clay poor and have less aluminum and iron than the parent material (see Figure 14.5). Even organic matter dissolves almost completely from E horizons, and only the most strongly weathering-resistant minerals remain. The intensity of chemical and physical leaching observed in E horizons requires considerable throughput of water and lots of reactive organic acids. These requirements are usually only met in humid forests, which accounts for the relatively restricted occurrence of E horizons to forest soils (Figure 14.3).

Explaining Soil Fertility and Infertility

Mineral dissolution in the A and E horizons also plays a critical role in soil fertility. Natural soil nutrients such as potassium, phosphorus, iron, and trace metals are released from minerals during weathering. Plants absorb these nutrients where they dissolve in water or bond to organic molecules. Plants rarely are able to draw the elements directly from mineral structures. The parent material must weather to be fertile. Too much weathering, however, leaches the nutrient elements from the soil and it becomes infertile, unless the plant tissues that absorb nutrients return these components to the soil through decay.

▼ **Figure 14.6 Explaining soil horizons.**
Weathering is most intense near the surface because of moisture availability and natural acidity. Minerals are subtracted from the A and E horizons by weathering dissolution and physical transport of small particles by downward-infiltrating water. Minerals are added to the B horizon by a combination of chemical and physical processes.

O horizon
Decomposing organic matter

A horizon
Mixture of weathered mineral grains and organic matter

E horizon
Maximum leaching and physical removal of mineral grains and organic matter

B horizon
Accumulation of clay minerals, Fe and Al oxide/hydroxides, calcite, and other soluble minerals. Accumulation occurs by combination of physical transport from upper horizons, precipitation from water infiltrating from upper horizons, and in-place weathering transformation.

C horizon
Unweathered or slightly weathered parent material.

Bedrock

Mineral subtraction

Mineral addition

Soils become infertile if the mineral nutrients extracted by weathering and plant growth are not returned as recycled organic matter. Naturally, infertility results from long periods of intense weathering that not only leach all of the nutrients from the minerals but also oxidize and dissolve all of the organic matter. Intense agricultural activity hastens extraction of nourishing elements from the soil, and then most of the crops are removed from the land, so the nutrients are not returned to the soil by plant decay. As a result, even the richest soils gradually lose their fertility and require addition of fertilizer to sustain agricultural productivity.

Explaining Mineral Additions and Transformations in the B Horizon

Although water leaches the most soluble mineral components from the whole thickness of a soil, the B horizon is primarily the site of mineral addition and mineral transformation, rather than subtraction (Figures 14.5 and 14.6). Water infiltrating down to the B horizon brings large quantities of dissolved ions and fine clays resulting from mineral weathering in the A and E horizons. There are limits to the ion-carrying capacity of the water, so minerals begin to precipitate from solution in the B horizon, especially when the water evaporates. Additional weathering in moist B horizons transforms parent-material minerals into weathering products in place.

Three processes, therefore, account for accumulation of colorful iron oxide and hydroxide minerals and clay minerals in B horizons:

1. Minerals precipitate in the B horizon from water that contains large concentrations of the ions that dissolved from minerals in the A and E horizons.
2. Hydrolysis and oxidation of feldspars and iron- and magnesium-bearing silicate minerals form clay, oxide, and hydroxide minerals in the moist, oxidizing environment. Infiltrating water carries away only the most soluble ions.
3. Tiny clay grains drain downward with infiltrating water from the overlying horizons.

The abundance of clay minerals in B horizons explains particle aggregates, such as the soil "clods" that you observed in the field. **Figure 14.7** illustrates cracks in the B horizon that allow the soil to be pried apart into aggregates, several centimeters across. The soil aggregates are a mild form of consolidation (discussed in Section 5.5) caused by the electrical attraction of clay minerals, which commonly have stray electrical charges around the periphery of the crystals. When clay minerals press close to one another, they tend to adhere and stick together.

The particle aggregates and the cracks between them are important features of B horizons. The cracks between soil aggregates are important pathways for downward transport of water, clay particles, and organic matter from above. Washed-in clay particles coat the surfaces of the cracks, and roots preferentially follow the cracks.

Explaining Calcite in Desert Soil

Calcite is a common ingredient of arid-region B horizons (see Figures 14.4 and 14.5), and **Figure 14.8** shows how common calcite is in the dry areas of the western United States. The presence of calcite and other water-soluble minerals, such as gypsum and halite, indicates that downward-percolating soil water cannot flush out all of the soluble ions. Calcite, gypsum, and halite dissolve when water moves through the upper soil horizons and only precipitate when ion concentrations in the water are very high. These high ion concentrations only occur because of evaporation, or nearly complete withdrawal of soil moisture by plants.

These observations explain why calcite, gypsum, and halite most commonly accumulate in B horizons that form in dry climates. In humid regions, there is almost always sufficient water draining through the soil to prohibit precipitation

▲ **Figure 14.7 Soil particles combine to form aggregates.** Soil falls apart into clods that are loosely held together by fine roots and cohesion between clay minerals. Open cracks separate the soil aggregates.

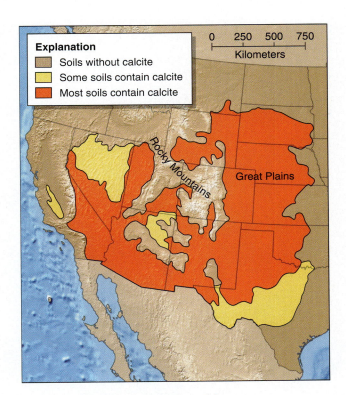

▲ **Figure 14.8 Where soil contains calcite.** This map shows where calcite accumulates in soils. Soils where calcite accumulates in the B horizons are found in the western Great Plains and in lower-elevation areas of the southwestern United States. Soil calcite only accumulates where the climate is both dry and seasonally hot. Infiltrating moisture leaches calcite from the soil where precipitation is abundant or where generally cool temperatures diminish evaporation even where rainfall is relatively sparse.

"Caliche"

▲ **Figure 14.9 Calcite cements soil into rock.**
Over time, enough calcite accumulates in desert soils to cement
the soil particles into solid rock, which is popularly called
"caliche." Calcite cemented nearly all of the B horizon in this soil
in southern Texas.

of these minerals, and the ions are flushed into deeper ground water. This is why calcite is absent from the B horizons of the example forest soils in Figures 14.1 and 14.4.

In some cases, there is sufficient precipitation of calcite to cement the soil particles into hard rock, as illustrated in **Figure 14.9**. In these circumstances, the soil is not easily excavated for foundation construction, plants have difficulty extending deep roots, and water cannot readily percolate downward. Some geologists consider this white, hard, rocky part of the soil a different horizon, called a calcic horizon, but soil scientists do not distinguish a separate horizon. The term "caliche" is commonly used to refer to well-cemented calcic B horizons, but this term is not formally used by geologists or soil scientists. Calcite is simply another example of mineral accumulation in B horizons, but soil calcite is restricted in occurrence by the availability of moisture.

Putting It Together–How Do Soils Form?

■ Chemical-weathering processes form soils, as indicated by the mineral additions, subtractions, and transformations that define soil horizons.

■ Mineral dissolution and downward flushing of dissolved ions and fine-grained weathering products are characteristic of A and E horizons, where water first enters the soil from rain or snowmelt and biologic activity forms acids and organic compounds that enhance mineral weathering.

■ The B horizon contains colorful oxide and hydroxide minerals and clays that physically wash down from overlying horizons, precipitate from ions dissolved in the overlying horizons, or form in place by weathering reactions. Calcite precipitates in arid-region B horizons when water evaporates.

14.4 What Factors Determine Soil Characteristics?

What factors account for the overall differences noted in the forest and desert soils illustrated in Figure 14.1? The grayer color of the A horizon above the quarry, compared to the desert soil, is consistent with greater accumulation of plant organic matter because of greater plant abundance in the more humid climate. The presence of calcite in the desert soil, but not in the forest soil, is consistent with accumulation of soil calcite in arid regions but not in humid regions. Vegetation and climate, then, are factors that affect soil characteristics. Geologists and soil scientists identify additional factors, including the nature of the parent material, location on an irregular landscape, and the length of time that the soil has been forming. **Figure 14.10** summarizes the relationships between these soil-forming factors and soil characteristics.

The Role of Parent Material

Soils result from weathering, and different rocks weather to produce different soil characteristics. Minerals have varying susceptibilities to chemical weathering and decompose into different weathering products. For these reasons, the parent material from which soil forms must affect the composition of soil.

For example, clay minerals form relatively quickly in soils whose parent material contains abundant feldspar, mica, or iron- and magnesium-bearing silicate minerals. Clay is already abundant in fine-grained sedimentary deposits or shale that forms the parent material for other soils. On the other hand, parent material composed mostly of quartz sand will not weather to produce a thick, clay-rich B horizon.

Another example is volcanic ash, which forms the parent material over large areas near active volcanoes. The glassy ash particles dissolve rapidly in contact with water to provide essential elements for vegetation growth. Plants extract their nutrients from water that passes through the soil and weathers the

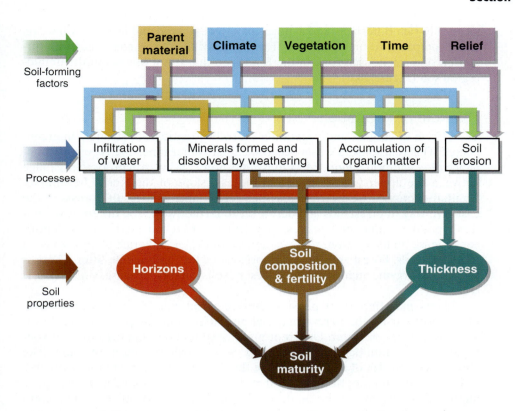

◄ **Figure 14.10** **Links between soil-forming factors, processes, and soil properties.**
Follow the arrows to see how the processes that account for soil properties relate to the soil-forming factors.

ash. This explains why agriculture in tropical, volcanically active regions, such as Central and South America, Indonesia, and the Philippines, focuses close to the volcanoes.

Physical characteristics of the parent material also influence soil characteristics. If the parent material is porous or has deeply penetrating cracks, then water readily infiltrates downward in contrast to less porous parent material. These parent-material properties that affect water movement determine the depth of mineral subtractions and additions and the extent of dissolution weathering reactions.

The thickest fertile soils develop on sedimentary deposits. In contrast, by the time weathering forms thick regolith on bedrock, most of the mineral nutrients are already leached out of the soil. Fertile soils of the upper Midwest and Northeast United States developed in widespread sedimentary deposits formed during the last ice age, which ended about 10,000 years ago. The glaciers left behind a blanket of crushed rock that was widely redistributed by rivers and wind. These deposits are sufficiently young to retain mineral nutrients but old enough to have well-developed horizons.

The Role of Climate

Quantity of precipitation and temperature are key variables in soil formation. Moisture is essential for weathering reactions to take place, and the volume of infiltrating water determines the depth of the boundaries between horizons of subtraction and addition. Chemical-weathering reactions are enhanced by high temperatures in the hot tropics, whereas the freeze-thaw process in colder regions enhance physical weathering to break rock into regolith and produce pathways for water.

If all other variables affecting soil formation are the same in humid and arid regions, then you can predict important differences in the soils formed in these regions. These differences are seen in the field sites (Figure 14.1) and also portrayed in Figure 14.5. There is more mineral dissolution from near-surface horizons in the humid region and greater accumulation of clay, oxide, and hydroxide minerals in the B horizon. As a result, the humid-region soil has a redder and more clay-rich B horizon, and a more heavily weathered and leached A horizon when compared to the desert soil. The aridity of the desert soil fosters the formation of a calcite-rich zone in the B horizon that is not present in the humid-region soil (Figure 14.8).

The Role of Vegetation

The effects of climate and vegetation on soil formation are strongly interconnected. This relationship exists because climate determines the type and abundance of vegetation. **Figure 14.11** illustrates how soil-horizon properties change between the upper Great Lakes and the east slope of the Rocky Mountains. Rainfall decreases westward across this region and, as a result, forests in the east give way to grasslands and desert scrub farther west. The different vegetation types determine the thickness and richness of organic matter in the A horizon. Dense mats of fine roots generate thick, organic-rich A horizons in grassland soils. Forest soils have O horizons of leaf and needle litter. There is very little organic matter in most desert soils, simply because vegetation is scarce.

Plants play important physical and chemical roles in soil development. Plant roots stir up the parent material and affect soil compaction and downward movement of water. Different plant communities support different soil-organism communities of insects and worms, which churn the soil. The abundance and size of roots determine the amount of organic matter provided to the soil and the depth to which it is distributed. The abundance of organic matter also determines the amount of organic acid and other organic compounds produced during organic decay. Greater organic matter content promotes faster weathering reactions and faster soil formation.

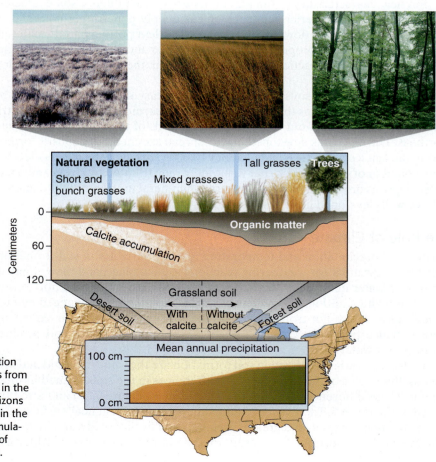

▶ **Figure 14.11 How vegetation and climate affect soil horizons.** Soil characteristics change along with changes in vegetation and precipitation along a line from Wyoming to Wisconsin. Precipitation gradually decreases from east to west, so vegetation changes from forest in Wisconsin, to grassland in the plains states, to desert scrub in northeastern Wyoming. Organic-rich A horizons are thickest below the tall grass prairie of Minnesota. Calcite accumulates in the B horizon only in the western, dryer region. The depth to the calcite accumulation increases as precipitation increases, because the greater the amount of water moving through the soil, the deeper the calcite is carried in solution.

The Role of Time

Up to this point, you have examined and interpreted soil characteristics at single snapshots in time, associated with particular characteristics of parent material, climate, and vegetation. The properties of soils change through time, however, so even when the other soil-forming variables remain the same, the soil changes as chemical weathering progresses. Soils mature through time.

Excluding those soil properties affected by agricultural activity, most natural changes in soil properties are not recognizable on human time scales. It is challenging, therefore, to understand the role of time in determining soil properties. To understand how soils mature, geologists compare properties of soils that started forming at different times and are, therefore, of different ages. For example, soil does not form on sedimentary deposits until after sediment deposition, so comparison of soils formed on deposits of different ages reveals how soils change over time.

Soils of different ages differ mostly in the thickness and mineral content of horizons. How the horizons change through time mostly depends on climate and vegetation. **Figure 14.12** illustrates changing soil characteristics over time for grassland, forest, and desert environments. Very young, immature soils exhibit very little chemical weathering and contain only a thin B horizon if any; there may simply be an A horizon resting on a C horizon. Old, mature soils have well-developed horizons, with removal of all easily weathered minerals from upper horizons, and extensive mineral additions to a thick B horizon.

Soil thickness generally increases through time. The longer that water moves through the regolith, the deeper the effects of chemical weathering. If weathering does not increase the thickness of regolith below the B horizon as quickly as mineral additions and transformations build the B horizon, then the soil may eventually extend through the C horizon to develop directly in bedrock.

Weathering effects add up through time, leading to increasing modification of the parent material. The abundance of easily weathered minerals decreases in progressively older soils, and the clay content increases in B horizons. Increasing clay content usually slows down water infiltration so that water remains in the soil longer to benefit plants that require abundant water. Calcite accumulation, in dryer climates, gradually cements the lower B horizon into rock.

How long does it take to form soil? There is no straightforward answer to this commonly asked question. The rates of soil formation strongly link to parent material, climate, and vegetation. This means that soils with very similar physical and mineral properties may be of different ages in different places. The formation of an A horizon on recently exposed regolith may require only a few decades. Soils containing B horizons more than a meter thick may represent tens of thousands of years of weathering and mineral additions. On average, however, mature soils require at least several thousand years to form. This means that in terms of human lifetimes, soils that erode away are nonrenewable resources.

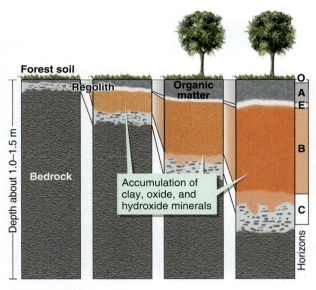

An immature forest soil consists of an **A** horizon developed on unweathered regolith. Over time, the **B** horizon thickens and an **E** horizon develops where organic acids infiltrated from thick **O** and **A** horizons. Weathering converts more and more of the original rock to regolith, which is further modified into soil.

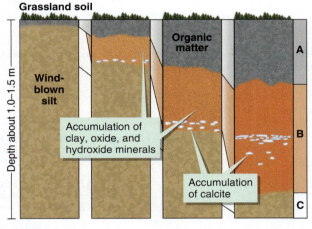

Grassland soils gradually develop a thick, organic-rich **A** horizon. In dry regions, calcite accumulates in the **B** horizon, and the depth to calcite accumulation increases through time.

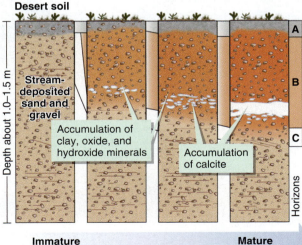

Desert soil **A** horizons are thin and low in organic content throughout their development. Calcite increases in abundance through time and eventually cements part of the soil into hard rock. Weathering and stirring of sediment by biologic activity gradually obliterates original bedding in unconsolidated sediment.

▲ **Figure 14.12 How soils mature through time.**
The diagrams show changes in soil thickness and properties through time for three different climatic and vegetation situations. Different starting materials are also depicted to show how soil formation modifies parent material. Notice the increased reddening of the B horizon caused by accumulation of iron and aluminum oxide and hydroxide minerals.

The Role of Landscape Relief

The irregular topography of Earth's surface plays an important role in soil development. Mature soils require adequate time for soil processes to generate thick B horizons and extensively weathered minerals. Mature soils, therefore, require land surfaces that are undisturbed, neither eroding nor being buried under new sediment layers, for many millennia.

Figure 14.13 shows how landscape relief affects soil development. Thick soil cannot form on steep slopes, because erosion removes regolith before chemical weathering substantially modifies it into soil. The material eroded from the hillside accumulates at the base of the slope. This means that regolith is thicker at the base of the hill than along the hillside, but the persistent deposition of material at the bottom of the slope continually buries the land surface and interrupts soil formation. Figure 14.13 shows that the most mature soil forms on flat surfaces that are stable because of lack of erosion or deposition.

Flat areas near rivers tend to lack mature soils, however, because new increments of sediment bury the soil each time the river floods (Figure 14.13). Soil formation starts over at the upper surface of the newly deposited sediment, only to stop and start over when the next flood occurs. On a positive note, the newly deposited sediment replenishes the surface with the soluble nutrient elements that provide high soil fertility.

▼ **Figure 14.13 How soil formation varies across an uneven landscape.** Mature soils form where the landscape is most stable.

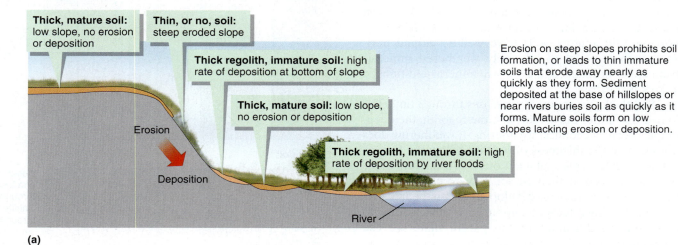

(a)

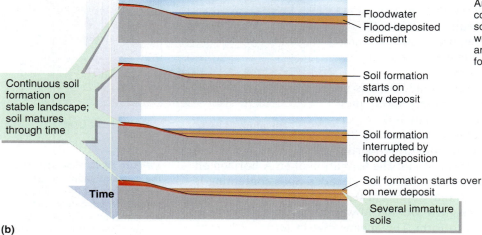

(b)

Geologists use soil maturity as a gauge of landscape stability. Where the land surface is underlain by a mature soil, the landscape is stable because there has been neither erosion nor deposition for a long time. Where soil is thin or immature, a geologist infers that the landscape has changed at some point in the recent past, because the current land surface has not existed long enough for mature soil to form. This observation suggests high rates of erosion or deposition that may make a location unsuitable for buildings, highways, or other structures.

Putting It Together—*What Factors Determine Soil Characteristics?*

- The primary factors determining soil characteristics are parent material, climate, vegetation, time, and landscape relief and location.
- Parent material determines what minerals dissolve and which precipitate during chemical weathering.
- Climate and vegetation are closely linked variables in soil formation. Vegetation is denser in moister climates, and the combination of greater moisture and greater development of organic acids from decaying vegetation leads to soils with the most distinct horizons.
- Soils mature through time, because the effects of chemical weathering add up over time. Soils generally become thicker and contain more clay, or calcite, or both in their B horizons with increasing age.
- Stable or very slowly changing landscapes have mature soils. Unstable landscapes, where rates of erosion or deposition are faster than rates of soil formation, are marked by immature soils, or no soil at all.

14.5 What Are the Types of Soils?

Variations in soil-forming factors, including parent material, climate, vegetation, relief, and time, create different soil types. The properties of different soils link to agricultural uses and productivity. Most countries where agriculture forms a major part of the economy make use of classification schemes based on soil characteristics that are affected by the soil-forming factors. The classifications vary from country to country because of regional differences in climate and parent materials. **Table 14.1** presents a simple, informal classification of soil, and **Figure 14.14** illustrates the distributions of these soil types in North America. This scheme is based on the classification used by soil scientists in the United States and many other countries. Terms used in the more complete classification appear in Table 14.1 for reference. Brief consideration of soil types integrates the knowledge of soil horizons and soil-forming factors in previous sections.

Forest Soils

Soils in temperate-zone forests, such as those found in eastern North America and parts of the mountain West (Figure 14.14), share several common ingredients (Table 14.1 and Figures 14.1, 14.4, and 14.5). Forest soils commonly have thin A horizons; instead they have thick O horizons composed of decaying leaves and conifer needles. B horizons are notable for considerable clay accumulation.

Figure 14.15 illustrates how the degree of weathering and leaching varies according to soil moisture. Very moist forest soils in New England, the upper Great Lakes, and eastern Canada form in cool, wet climates and have distinct E horizons. Forest soils in the warm, rainy southeastern United States are highly oxidizing and strongly leached of soluble mineral components. These soils are notable for red, clay-rich B horizons with abundant iron and aluminum oxide and hydroxide minerals. Forest soils in the Midwest, Mississippi Valley, and mountains of the western United States form where precipitation is not as

TABLE 14.1	Characteristics and Fertility of Major Soil Types		
Soil Type	Characteristics	Fertility	U.S. Department of Agriculture Soil Order Names
Forest soils	Characteristics vary depending on moisture and vegetation type, but typically include an O horizon, a thin A horizon, and a clay-rich B horizon. An E horizon is present where the weathering by organic acid is intense.	Fertility depends on the extent of weathering to dissolve and remove mineral nutrients. Soils that form under the wettest, warmest, and most acidic conditions are the least fertile.	Alfisols—moderately leached, fertile soil; forms in humid climate. Ultisols—highly leached, low-fertility soil; forms in humid to tropical climates. Spodosols—highly leached, low fertility soil with E horizon; forms in cool, humid forests.
Rainforest soils	Highly weathered soils that contain very little organic matter, and consist largely or entirely of iron and aluminum oxides and hydroxides with clay minerals.	Soils are infertile because mineral nutrients are dissolved and removed and organic matter is not retained in the soil.	Oxisols
Grassland soils	Thick, dark, organic-rich, crumbly A horizons are typical. B horizons vary considerably in character and may or may not contain calcite.	Soils are very fertile primarily because of retention of organic-matter nutrients.	Mollisols
Desert soils	Light-colored, organic-poor A horizons are typical. B horizons vary considerably in clay content and color, but typically contain calcite (and less commonly contain gypsum or halite). Degree of mineral weathering is less than for most other soil types.	Soils are commonly fertile when irrigated but otherwise are found in areas where the climate is typically too dry for productive agriculture. Although organic content is low, the soils contain mineral nutrients that are not readily dissolved in the dry conditions.	Aridisols
Wetland soils	Very dark, organic-rich soils that commonly contain more than 20% plant organic matter. The soil is remarkably lightweight when dry, because of the abundance of low-density organic particles.	Soils are very nutrient rich, but may be difficult to farm, and unsuitable for most crops, because of water-logged conditions.	Histosols
Soils with weakly developed horizons	Soils with weakly developed horizons. Weak horizon development usually results from limited weathering, either because soil is very young or forms in a cold climate where weathering rates are very slow. Some soils lack obvious horizons because soil constituents are mixed so that horizon boundaries are destroyed.	Soils are commonly fertile and support productive agriculture with adequate moisture and growing season, because mineral nutrients have not been removed by weathering.	Entisols—very slightly weathered to unweathered soil in very recently deposited or rapidly eroded parent material. Inceptisols—slightly weathered soil with incipient formation of a slightly reddish, clay-poor B horizon. Andisols—weakly weathered young soil formed on recently deposited volcanic ash. Gelisols—weakly weathered soil in very cold climates where soil is frozen at least part of the year. Vertisols—soils formed in clay-rich parent material that shrinks and swells by wetting and drying; shrink and swell physically mixes horizons as organic material falls deep into the soil shrinkage cracks.

great, so the soils are less strongly leached than those in wetter climates. These less leached and relatively fertile forest soils in the Midwest were extensively cleared of tree cover during the nineteenth century to permit farming.

Rainforest Soils

The most intense chemical weathering occurs in hot, wet, tropical rainforests. In the United States, these conditions occur only in small areas of Hawaii and Puerto Rico, but rainforest soils are common in South America and Africa, as shown in **Figure 14.16**. The large volume of water that passes through the soil, along with the high acidity provided by the decay of abundant plant debris, strongly weathers almost all minerals, including quartz. In the most extreme cases, only iron and aluminum oxides and hydroxides remain, with varying proportions of clay minerals. These oxide and hydroxide minerals may be sufficiently abundant to cement the soil into rock.

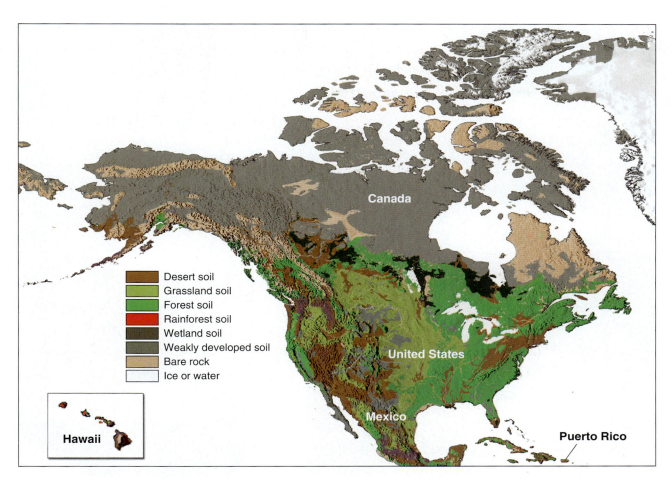

◄ **Figure 14.14** **Distribution of soil types in North America.**
Soil types in North America correlate closely to climate and vegetation, and to the maturity of soil development. Immature soils form in cold areas (where weathering is very slow), on steep mountainous terrain (where erosion removes regolith before it is deeply weathered), and along rivers (where soil formation is interrupted by deposition during floods).

Desert soil
Grassland soil
Forest soil
Rainforest soil
Wetland soil
Weakly developed soil
Bare rock
Ice or water

Canada
United States
Mexico
Hawaii
Puerto Rico

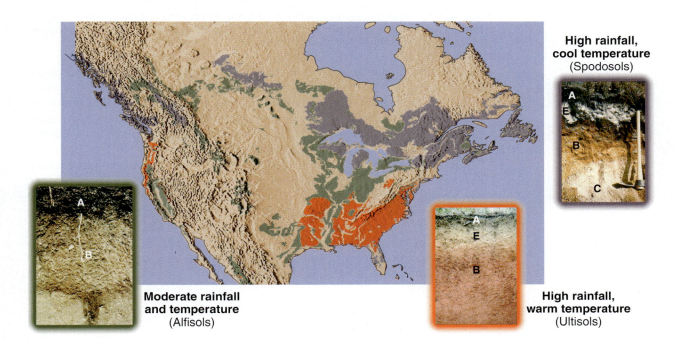

High rainfall, cool temperature
(Spodosols)

Moderate rainfall and temperature
(Alfisols)

High rainfall, warm temperature
(Ultisols)

◄ **Figure 14.15** **Characteristics of forest soils vary with climate.**
This map shows the distribution of three forest soil types in North America. In areas where soil moisture is highest, soils have the reddest B horizons and best developed E horizons, which indicate the most intense leaching of minerals near the surface and accumulation of minerals at depth. Warmer temperature promotes more development of red iron oxide minerals.

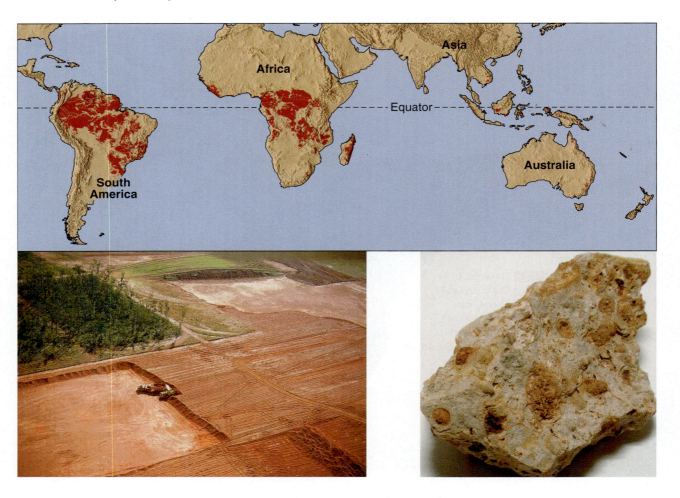

▲ **Figure 14.16 Rainforest soils are the most weathered soils.**
The map shows the distribution of the most strongly weathered soils on Earth. These soils primarily form in the near-equatorial rainforests of Africa and South America where rainfall is heavy and temperatures are very warm year-round. Nearly complete dissolution of most parent-material minerals and organic matter leaves a soil that is very rich in brightly colored iron and aluminum hydroxide and oxide minerals. Modern and ancient rainforest soils are mined for the aluminum ore bauxite, composed primarily of aluminum hydroxide minerals. The photos show a bauxite mine in Australia and a sample of bauxite.

Rainforest soils are very infertile, because the intense leaching and oxidation destroy most of the organic matter along with mineral nutrients. Agriculture in the tropics commonly involves clearing a region that is only suitable for farming for a few years. This necessitates moving on to clear new sections of rainforest at frequent intervals. This agricultural practice plays an important role in deforestation, especially in South America.

Rainforest soils can form mineral deposits. Soils rich in aluminum hydroxide minerals are the primary sources of aluminum, an economically important, lightweight, flexible, corrosion-resistant metal. The aluminum ore is called **bauxite**, a multicolored mixture of aluminum hydroxide (Figure 14.16). Bauxite mining occurs in the modern tropics, in South America, Africa, and Jamaica, and also in areas covered by rainforest in the recent geologic past. Both the world's largest bauxite deposits, in northern Australia, and smaller deposits in Arkansas, are ancient rainforest soils. Some iron-oxide-rich tropical soils are mined as iron ore.

Grassland Soils

Grassland soils exist throughout the center of the United States (Figure 14.14) and are especially fertile for growing wheat, corn, and soybeans. Grasslands form in climate zones that are drier than forested regions, so grassland soils are not strongly leached of soluble mineral components compared to forest soils. The closely spaced network of thin roots, along with fast-decaying small grass blades, leads to the formation of thick, dark A horizons as seen in **Figure 14.17** (also see Figure 14.12). The abundant organic matter provides much of the nutrient base for successful agriculture. Grassland soils in the United States straddle the boundary between climate regimes that are appropriate for, or inappropriate for, the formation of calcite-rich B horizons, so some grassland soils contain calcite and others do not (see Figures 14.8 and 14.11).

Desert Soils

Desert soils are common in the western United States (Figure 14.14), where they form in a dry climate with very sparse vegetation. As a result, the soil (a) contains very little organic matter in the A horizon, (b) has very thin and clay-poor B horizons unless the soil is very old or developed in clay-rich parent material, and (c) commonly contains calcite in the B horizon. Figures 14.1, 14.4, 14.5, 14.9, and 14.12 illustrate these features. Limited moisture means that mineral nutrients are not strongly leached from desert soils, but nutritious organic content is very low. Agricultural production depends on irrigation, because of aridity, and fertilizers, because of low soil fertility. Although some of the most productive farmland in the world is irrigated desert soil, the soil quality degrades by the accumulation of evaporite minerals, including halite, from evaporation of irrigation water in the hot, dry climate.

Wetland Soils

Marshy wetlands are at the opposite climate and vegetation extreme from desert soils. Wetland soils commonly consist entirely of O and A horizons. The soil is water saturated so there is no place for soil water and dissolved ions to percolate toward to form B horizons. The soil contains more than 20 percent organic matter, primarily as poorly decomposed plant remains. The abundance of organic carbon prevents much oxidation in the water. Not only does this mean that there are no oxidizing weathering reactions to create oxide and hydroxide minerals, but the lack of oxygen prohibits further decay of plant residues. North America's wetland soils are mostly in Canada and New England, and in small areas of the western mountains, where precipitation is abundant and cool temperatures inhibit evaporation. Wetland soils also are present locally as coastal marshes in the southeastern United States and alongside streams and lakes. The largest areas of wetland soils visible in Figure 14.14 stretch discontinuously across Canada in areas that were covered by lakes near the end of the last ice age, about 10,000 years ago.

The black, organic-rich soil, like that shown in **Figure 14.18**, may contain more plant residue than mineral grains, in which case it is called **peat** (or peat moss). When buried and subjected to temperatures of 80–100 degrees centigrade, peat transforms to coal (also see Section 5.4). Peat holds as much as four times its weight in moisture, which is 10 times greater than the wettest mineral soils. This extraordinary capacity to hold water is why peat is added to potting soils for houseplants and compressed into peat pots for germinating plant seeds.

Soils with Weakly Developed Horizons

Large areas of North America are covered by soils with weakly developed horizons, regardless of climate and vegetation type (Figure 14.14). Weakly formed horizons, such as seen in **Figure 14.19**, exhibit only subtle color contrasts, little or no evidence of mineral additions and subtractions, and easily weathered minerals persist in the A horizon. In most cases these are immature soils present on unstable landscapes and not old enough to exhibit strongly developed horizons (Figure 14.13).

▲ **Figure 14.17 Grassland soils have dark and light horizons.** This excavated soil from Kansas shows the dark, organic-rich A horizon that is typical in grasslands, where dense networks of fine roots and abundant decaying organic material contribute substantially to the composition and texture of the soil. Limited weathering in a relatively dry climate leaves a light-colored B horizon that includes streaks of white calcite with only minor accumulation of iron oxides or hydroxides.

▲ **Figure 14.18 Wetland soils are mostly organic matter.** Large amounts of plant material accumulate along with only minor, if any, mineral sediment in marshy wetlands and swamps. The soil is saturated in water and contains very little oxygen, so the organic matter does not readily decay. The abundant organic matter accounts for the very dark color of this illustrated wetland soil. Notice the water ponded on the surface, which demonstrates that the soil is saturated like a sponge.

▲ **Figure 14.19 Immature soils lack well-developed soil horizons.**
This excavation exposes an immature soil where weathering has not modified the parent material. Organic material forms a thin, dark A horizon at the surface, but the soil lacks horizons with different mineral contents.

Regions experiencing frequent ash deposition downwind of active volcanoes also have soils without distinct horizons, because soil formation starts over every time a new layer of thick ash accumulates. These soils are common in Hawaii and in the northwest United States near the active volcanoes of the Cascade Range. Volcanic soils are typically fertile because elements are leached much more rapidly from glassy volcanic ash than from crystalline minerals. Some of these elements, especially phosphorus, are essential for plant growth. Volcanic soils support robust agriculture all around the Pacific Ocean "Ring of Fire."

The widespread presence of immature soil in Alaska and northern Canada (Figure 14.14) reflects the very slow rates of mineral weathering and organic matter decay in very cold climates. In many places, water is permanently frozen in these soils, which greatly decreases the rates of weathering reactions.

Putting It Together—*What Are the Types of Soils?*

■ Soil types relate to the different soil-forming factors in different locations.
■ Forest soils usually contain less organic matter but more clay-enriched B horizons than do grassland soils. Desert soils contain the least organic matter of all soils, and clay accumulates slowly in the B horizon because of limited chemical weathering. Desert soils and grassland soils in relatively dry areas contain calcite in the B horizon.
■ Rainforest soils are very strongly weathered and usually infertile because of a nearly complete dissolution of organic matter and mineral nutrients. Highly weathered tropical soils are the source of bauxite, which is mined to produce aluminum.
■ Wetland soils are composed primarily of organic matter.
■ Immature soils, with only faintly developed horizons, form where there has been insufficient time for significant weathering. Most immature soils indicate unstable landscapes or areas too cold for weathering reactions to occur effectively.

14.6 How Do We Know . . . That Soils Include Atmospheric Additions?

PICTURE THE PROBLEM
How Can Geologists Explain Calcite-Rich Soil?
A problem emerges when examining a thick zone of calcite in a desert soil like that seen in Figure 14.9. For each molecule of calcite in the soil, there must have been a molecule of calcium in the parent material. The calcium dissolves from the parent material and is then incorporated in the precipitated calcite. In many cases, however, tests show that the parent material contains very little calcium, so it is difficult to account for the amount of calcite present in the soil.

Figure 14.20 illustrates the problem of calcium-rich soils in the western United States that formed in calcium-poor, igneous-rock regolith. Weathering would need to dissolve as much as 100 meters of rock to account for the mass of calcite in the soil. Other observations summarized in Figure 14.20 argue against such extensive chemical weathering, however. For instance, the presence of vesicular, broken up rock typical of the top of a lava flow rules out the possibility that a great thickness of the lava flow weathered away. In addition, although calcium dissolves from silicate minerals in the rock, the much more abundant silicon and other elements should remain in relatively dry desert soil. If 100 meters of rock weathered to release the required amount of calcium, then there should be an immense, thick residue of insoluble minerals at the surface, and this is not the case. These observations indicate that the composition of the parent material cannot closely match the composition of the regolith in the C hori-

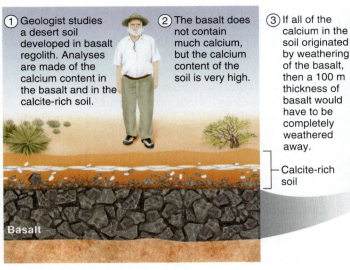

① Geologist studies a desert soil developed in basalt regolith. Analyses are made of the calcium content in the basalt and in the calcite-rich soil.

② The basalt does not contain much calcium, but the calcium content of the soil is very high.

③ If all of the calcium in the soil originated by weathering of the basalt, then a 100 m thickness of basalt would have to be completely weathered away.

Calcite-rich soil

Basalt

Original thickness of basalt implied by the analyses

Calcite-rich soil

Current thickness

◄ **Figure 14.20 Why calcite in soil presents a problem.**
This diagram walks you through the dilemma of many calcite-rich soils, if you assume that the soil formed by the weathering of only the underlying rock. To account for the amount of calcite in the soil requires an unrealistic amount of rock weathering and would leave behind in the soil a far greater abundance of relatively insoluble elements than actually exist.

④ **Problem 1:** Top of basalt is not very weathered and has a vesicular, rubbly appearance as the top of a lava flow should look.

⑤ **Problem 2:** If the calcium is from weathering 100 m of basalt, then there should be a large volume of less-soluble elements left behind in the soil in addition to the calcium.

No evidence that 100 m of weathering occurred.

Less-soluble elements measured in the soil: No evidence that 100 m of weathering occurred.

Less-soluble elements expected from weathering 100 m of basalt.

zon or the underlying bedrock. How, then, could the soil form? Calcite-rich B horizons are common over large areas of the western United States (Figure 14.8), so this is an important problem to solve.

STATING THE HYPOTHESIS
Is Calcium Delivered in Dust and Rainfall?
Any visitor to an arid region is immediately impressed with, and possibly distressed by, the abundance of windblown dust. Deserts are dusty because the surface is dry, and there is very little vegetation to protect soil particles from blowing away or to blunt the force of strong wind blowing across the surface.

Relative to the problem of calcite in soil, scientists hypothesized that accumulating dust provides continuous additions of calcium, and other ions, to the surface soil horizons. The windblown dust may originate far from where the soil develops and contain minerals different from those of the underlying parent material. There was also speculation that rainfall delivers dissolved calcium ions. Infiltrating water then dissolves calcium delivered by dust and rainfall from the surface A horizon and carries it downward to precipitate as calcite in the B horizon. The parent material composition, in other words, is not constant but receives new components from the atmosphere while the soil forms.

COLLECTING THE DATA
How Much Calcium Is Present in Dust and Rainfall?
To test the hypothesis, Leland Gile and Robert Grossman, of the U.S. Soil Conservation Service (now the Natural Resources Conservation Service), conducted dust-sampling experiments over a 10-year period in the 1960s. Eight trays

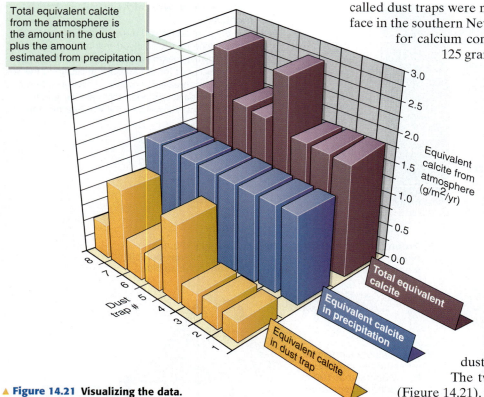

Total equivalent calcite from the atmosphere is the amount in the dust plus the amount estimated from precipitation

▲ **Figure 14.21 Visualizing the data.**
Each dust trap yielded a different rate of accumulation of calcite-equivalent calcium, indicated by the orange columns in the graph. Regional data on the amount of calcium dissolved in rainfall permits calculation of the contribution of calcite-equivalent calcium from precipitation, which the blue columns represent as the same value at each location. The total calcite contribution from atmospheric sources, shown by the purple columns, is the sum of the calcium in the dust and precipitation.

called dust traps were mounted no more than one meter above the ground surface in the southern New Mexico desert to collect dust, which was then analyzed for calcium content. Gile and Grossman showed that between 10 and 125 grams of dust fall on each square meter of the desert surface each year after taking into account the size of the pans in the traps and the period when dust was collected.

The geology of the experiment area consists of felsic igneous rocks and sediment eroded from those rocks. These parent materials contain very little calcium, and what little exists is found in low-solubility silicate minerals. This means that dust with calcium did not simply blow up into the traps from the nearby surface, but instead blew in from some distant location where regolith contains dust-sized grains of water-soluble calcium minerals, such as calcite or gypsum.

All of the dust samples contain calcium that dissolves in water, and **Figure 14.21** graphs the results. To allow easier comparison to the amounts of calcite in the nearby soils, the calcium contents were calculated as the equivalent amount of calcite that could precipitate from the calcium supplied in the dust. Six of the eight traps yielded similar amounts of equivalent calcite, between 0.35 and 0.55 g/m^2/year. The other two dust traps yielded much higher values near 1.3 g/m^2/year.

The two very high values are clear outliers on the data plot (Figure 14.21). Outlier values are common in scientific data collection, and it is important to know whether they are measurement error or are naturally explained. Gile and Grossman determined that the data were correct but that these two traps are in unusual locations. One trap was very close to small, blowing sand dunes, and accumulated two to ten times as much dust as the other seven traps. The other outlier data came from a trap located where eroded soil with calcite was exposed at the surface, so this trap had a potential calcium source that the other traps did not. Given the unusual characteristics of where these two traps were located, Gile and Grossman felt it best not to include these large calcite contents in their data analysis. About 0.5 g/m^2/year of calcite accumulates, on average, at the remaining six trap sites.

Next, Gile and Grossman considered how much calcium, as a building block for calcite, might be delivered by rain. Rainwater contains ions dissolved from dust in the atmosphere (the primary source of calcium), volcanic gas, and air pollution. Measurements show that rainwater in southern New Mexico contains about three thousandths of a gram of dissolved calcium ions in each liter of water. This is an extremely small concentration, about equal to dissolving one teaspoon of calcium in 400 gallons of water. Nonetheless, by using the measured rainfall data, Gile and Grossman calculated that rainwater annually delivers the equivalent of about 1.5 grams of calcite to each square meter of the soil. Adding this amount to each of the dust-trap abundances, and excluding the two data outliers, indicates that the total equivalent calcite added by atmospheric dust and precipitation is about 2.0 g/m^2/year (see Figure 14.21).

USING THE DATA

Can Atmospheric Calcium Additions Account for Soil Calcite?
The data show that wind and rain supply calcium to soils as hypothesized, but is the modern atmospheric calcium addition sufficient to account for the amount of calcite seen in the local desert soils? **Figure 14.22** illustrates how Gile and Grossman continued their test of the hypothesis. They excavated all of the soil below a 1-meter-by-1-meter square on the surface. The calcite content was measured in the laboratory and found to be about 12 kilograms. If the hypothesis is correct, then nearly all of the 12 kilograms of calcite should result from atmospheric addition of calcium ions over the history of soil formation.

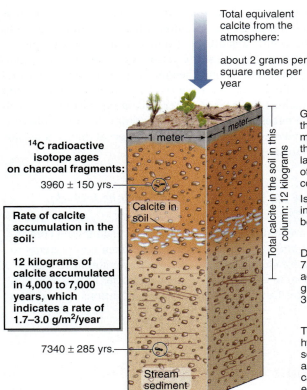

◄ **Figure 14.22 A simple calculation tests the hypothesis.**

Total equivalent calcite from the atmosphere:

about 2 grams per square meter per year

←1 meter→

←1 meter→

Total calcite in the soil in this column: 12 kilograms

^{14}C radioactive isotope ages on charcoal fragments:

3960 ± 150 yrs.

Calcite in soil

Rate of calcite accumulation in the soil:

12 kilograms of calcite accumulated in 4,000 to 7,000 years, which indicates a rate of 1.7–3.0 g/m²/year

7340 ± 285 yrs.

Stream sediment

Gile and Grossman excavated all of the soil material below a 1 square meter plot of ground and measured the total calcite content determined by laboratory analyses. Twelve kilograms of calcite are present in the soil column.

Isotope ages of charcoal fragments indicate that the soil started forming between 4,000 and 7,000 years ago.

Dividing 12 kilograms by 4,000 and 7,000 years indicates that calcite accumulated at an average rate greater than 1.7 g/m²/yr and less than 3.0 g/m²/yr.

These values are consistent with the hypothesis that all of the calcite in the soil is explained by the present-day accumulation of about 2 grams of calcite-equivalent dust that falls on each square meter each year.

To determine the duration of soil formation, the soil scientists separated charcoal fragments from the soil and underlying stream sediment comprising the parent material. The ^{14}C radioactive-isotope dating method provided ages of about 4000 years and 7000 years for these charcoal samples (Figure 14.22). The older charcoal was deposited in the stream sediment *before* the soil formed, and the younger charcoal may have been mixed into the soil *while* the soil formed. This suggests that the soil formed over at least 4000 years but no more than 7000 years. To account for the 12 kilograms of calcite over these time intervals requires addition of somewhere between 1.7 and 3.0 grams per year, on average, into the square meter of surface where the soil was excavated.

The dust trap and rainfall data match well with the observed calcite content of the soil. The measured atmospheric addition of 2.0 g/m²/yr compares closely to the average rate of calcite accumulation in the soil, 1.7 to 3.0 g/m²/yr. Gile and Grossman made similar measurements of many soils and obtained similar results.

INSIGHTS

Do Soils Form at a Steady Rate?

The dust-trap data match up reasonably well with the amount of calcite in the soils to support the hypothesis that explains calcite accumulation where parent material appears to be calcium deficient. You can ask, however, if it is valid to compare 10 years of dust-trap data with the amount of calcite that accumulates in a desert soil over thousands of years. This problem frequently arises in geologic studies where measurements cannot be made over the same time spans as geologic processes lasting thousands or even millions of years.

Is the rate of calcite accumulation and, therefore, the rate of soil formation constant over thousands of years? If not, then do the dust-trap data adequately test the hypothesis? There is no reason to expect that calcite accumulates in southern New Mexico soils at a constant rate of 2 grams over each square meter each and every year for thousands of years. Climate changes over short time intervals likely change the rate of calcium addition to the soil. The amount of accumulating dust and calcium is probably greater during dry times. During wetter times, the amount of calcium delivered in dust probably decreases, but more is delivered by rainwater, which contains much more calcium than does the dust. On the other hand, not all of the calcium in the rainwater soaks into the soil, because some of the water runs off the ground surface into streams.

The important thing to remember in a study of this type is not that the numbers add up exactly. What is important is that the amount of calcium in dust and rainwater is simply "in the ballpark" of that required to account for the amount of calcite in the soil. Inasmuch as the parent material contains virtually no calcite, the hypothesis that the necessary calcium arrives from the atmosphere is reasonably supported by the data collected over a time interval that is necessarily much shorter than the natural process.

Putting It Together–*How Do We Know . . . That Soils Include Atmospheric Additions?*

■ Soil parent material is not simply rock and regolith of unchanging composition. Windblown dust and rainfall deliver new chemical components to the surface, where they are chemically moved into lower soil horizons.

■ Measurements of calcium in dust and rainfall support the hypothesis that calcite in most desert soil forms from atmospheric additions rather than the original parent material.

14.7 How Do Human Activities Affect Soils?

Soil fertility and conservation are essential to successfully raising crops. Soils, however, form within natural ecosystems, rather than managed pastures and fields of specialized crops. Agricultural activities also disturb soil horizons by plowing and cultivating, and, in some cases, reshape the land surface, which changes patterns of water runoff and infiltration into the soil.

Soil erosion is the greatest threat to future agricultural production. Each year about 4 billion metric tons of soil erodes away in the United States. Flowing water causes about two-thirds of this erosion, and the remainder results from blowing wind. The problem for agriculture is that more than half of the soil erosion in the United States is from croplands. Even where some soil remains, the upper horizons containing the nutrient-rich organic matter are commonly lost.

To what extent is soil erosion in agricultural regions the natural process of regolith erosion, and to what extent is the erosion enhanced by human activity? Landscape stability is important for soil formation (see Sections 14.3 and 14.4). Soil erosion reflects landscape instability and the removal of weathered regolith from where it accumulated. To evaluate soil erosion, it is important to understand how stable landscapes of slow soil formation become unstable landscapes of soil removal.

Soil Erosion by Flowing Water

Agricultural activities destabilize landscapes by increasing water runoff and erosion potential. Erosion in natural forests and grasslands is relatively small compared to soil loss in the same regions after clearing of natural vegetation and conversion to crop production. Rainfall generally infiltrates more efficiently into natural soils covered in thick vegetation and decaying organic matter, because plants and dead plant matter slow down water flowing across the surface and enhance the opportunities for it to soak into the soil. In contrast, many farm fields lie bare and unused for significant parts of a year, or are only sparsely covered by plants sown in rows and then tilled to remove weeds. In addition, heavy farm implements moving through a field compact the soil and decrease its porosity. Infiltration is lower on bare, compacted ground than on loose vegetated soil covered in plant debris.

Lower infiltration means greater runoff from rainfall and snowmelt, which translates to increased soil erosion. Water moving across bare ground picks up mineral and organic grains from the soil and carries them away. The water may flow in shallow continuous sheets across the fields, or it may focus into small channels where deeper flow enhances erosion. A **rill** is a tiny channel that is sufficiently shallow to be removed by tilling the ground with a plow. A **gully** is a deeper channel that erodes too deep to be smoothed over by tillage. **Figure 14.23** shows how the flow of water through rills and gullies, as well as in shallow sheets, removes nutrient-rich upper soil horizons from fields.

Soil Erosion by Blowing Wind

Wind erosion is most common in dry regions where agricultural production may already be marginal. Wind most effectively erodes sand- and silt-size particles from the soil where the soil is dry and there is scant vegetation. Vegetation decreases the effectiveness of wind erosion by binding soil with roots and by providing obstacles that substantially decrease wind velocity. Wind erosion of soil generally happens during periods of drought when natural or crop vegetation dies off over large areas. The loss of vegetation destabilizes the landscape. The decreased soil moisture further enhances the likelihood of wind erosion because dry soil particles do not clump as well as moist ones.

Droughts are natural events, but agricultural activities may enhance wind erosion. In some locations, water-demanding crops replace native drought-tolerant grasses. Parts of the landscape covered by living native grasses erode

▲ **Figure 14.23 Soil erodes from bare fields.**
(a) Muddy water rushes from a farm field during a heavy rainstorm. Soil erosion occurs where the water flows in sheets, in narrow shallow channels called rills, and in deeper gullies.
(b) Rills formed in this field because of heavy rainfall prior to crop planting in the spring. After crops are well established, the surface-water runoff moves more slowly and is not as erosive. Although plowing smoothes over the rills, the eroded topsoil is permanently gone. (c) Runoff from heavy rainfall on this overgrazed pasture eroded rills and deep gullies. The depth of soil removal is sufficient to undermine the roots of trees, causing them to topple.

▲ **Figure 14.24** **Soil erosion during the Dust Bowl of the 1930s.**
(a) Vegetation died over large areas of the Great Plains during the 1930s drought, turning the region into a "Dust Bowl" of blowing soil. (b) During the largest dust storms, dark dust clouds towered more than 100 m over the surface and obscured sunlight to cause total darkness.

less strongly than cultivated areas where crops die for lack of moisture. In other cases, native grasses die or are considerably weakened by livestock grazing, which decreases plant cover during drought even among drought-tolerant plants.

Figure 14.24 shows dramatic examples of soil erosion by wind. These photographs show dust storms in the western plains of the central United States during a protracted period of drought in the 1930s. The combination of low rainfall and soil loss by wind erosion turned the region into what came to be known as the Dust Bowl. One sky-darkening windstorm stripped 350 million metric tons of soil (enough to fill more than 4 million railroad cars) from the Great Plains and carried it eastward where it fell like fine snow for two days along the Atlantic coast.

Figure 14.25 shows how to easily recognize soil loss by wind erosion. Plant roots bind soil particles and protect soil from erosion. In a sparsely vegetated landscape, therefore, the soil preferentially blows away in areas between plants, leaving the plants on noneroded pedestals.

Preventing Soil Erosion

Slope is an important factor in soil erosion. Water flows faster down steep slopes and causes the most erosion on the steepest slopes. Where possible, it is best to leave steep slopes undisturbed. Where necessary to farm in steep terrain, it is best to modify the slopes by making terraces, as shown in **Figure 14.26**. The flat terrace surfaces not only slow down the water and decrease erosion but also allow the water to soak into the soil to support plant growth rather than flowing downhill to streams.

Soil-erosion damage occurs even on gentle slopes of rolling hills. **Figure 14.27** shows how crop-planting practices can diminish this erosion. If crop rows run downhill, the water follows the rows and causes erosion. Corn rows shown in Figure 14.27 follow along the contours of the slopes rather than down the hillside. This planting practice allows the crops to interrupt downslope water flow to simultaneously decrease erosion and increase infiltration. The farmers also alternate the planting of corn with hay or grains, which, like natural grasses, is not grown in wide rows but forms a continuous cover of vegetation. Water runoff is generally slow and nonerosive through the close-spaced grasses. Ero-

▲ **Figure 14.25** **Evidence of soil erosion.**
The soil scientist points to the position of the ground surface at the time the grasses started to grow. Wind erosion subsequently lowered the land surface to the level at his feet. All of the soil eroded away. To the left of the soil scientist is a light-colored sand dune. Strong winds frequently move the sand so that no vegetation is reestablished and there is no stable land surface for new soil development.

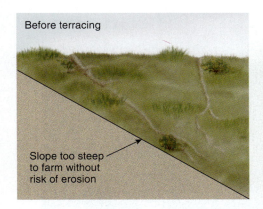

Before terracing

Slope too steep
to farm without
risk of erosion

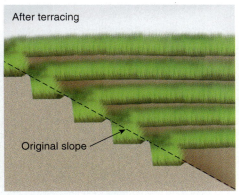

After terracing

Original slope

◀ **Figure 14.26** Terraces diminish soil erosion on steep slopes.
The diagram shows how disturbance of vegetation and soil on a steep slope increases the risk of erosion by surface water runoff. A solution, shown on the right, is to cut terraces on the slope. Flat terrace surfaces slow down the runoff so that water infiltrates to support crop growth, rather than rushing downslope and eroding the soil. The photo shows a California vineyard established on a terraced hillside.

sion in the row-crop part of the field can be further diminished during the winter months by leaving the unutilized crop residues in the field rather than plowing them under. This "no-till" farming practice decreases erosion because the plant remains slow down runoff rather than leaving bare soil exposed to be washed or blown away. The plant remains also add nutrients back to the soil.

Windbreaks, like the rows of trees shown in **Figure 14.28**, diminish the erosive force of wind. The trees form obstacles to wind that both slow down the wind speed and force the blowing air up and away from the land surface. By keeping the wind from blowing at high velocity across large distances of dusty fields, there is very little wind erosion of the soil, and soil particles are not transported out of the field.

Crops planted
in strips along
contours of hillside

Corn

Hay

Grass planted
along drainage

◀ **Figure 14.27** Planting crops to decrease water erosion.
The pattern of crop planting diminishes soil erosion on this hilly Iowa farm. Crop rows parallel the contours of the hillsides, rather than running straight downhill where they would convey rapid and erosive water runoff. The alternation of corn, planted in spaced-out rows, with hay, which forms a dense cover of grass, also diminishes erosion because water flows very slowly through the hay. Grass planted along steeper drainage channels keeps water from flowing fast enough to erode gullies.

▶ **Figure 14.28 Windbreaks diminish wind erosion.**
Rows of trees form windbreaks between these North Dakota wheat fields. The potential for water erosion is minimal, because the land is nearly flat. The soil is very susceptible to wind erosion when the wheat fields are bare between harvest and new growth. The windbreaks slow down the wind blowing along the ground surface and deflect the wind upward away from the surface; both of these effects decrease wind erosion of fine soil particles.

Putting It Together–How Do Human Activities Affect Soils?

■ Plowing and cultivating disturb soil horizons. Agricultural activities may also reshape the land surface, which increases water runoff and decreases infiltration into the soil.

■ Bare areas, rows of sown crops, and compacted land surfaces leave significant parts of the field exposed to water runoff and soil erosion.

■ The loss of vegetation, especially during drought, promotes soil erosion by wind. Dry conditions and the lack of vegetation decrease soil moisture and enhance the ability of wind to pick up and move soil particles.

■ Terracing steep slopes and planting parallel to contours diminishes water runoff. Planting continuous covers of vegetation and trees serves as a windbreak and helps retain soil moisture. Practicing "no-till" methods also decreases erosion by inhibiting water runoff and restores nutrients to the soil.

Where Are You and Where Are You Going?

Soils are not just dirt—they are mixtures of organic matter and that part of surface regolith that has weathered but not been transported away as sediment. Soil is not only important for growing productive crops but is also the source of aluminum ore and peat, the forerunner of coal. Soil characteristics also allow geologists to estimate landscape stability.

Soil formation is a complex series of processes that act on parent material.

• *Mineral changes:* Chemical reactions, enhanced by the breakdown of organic matter, modify or destroy some minerals and provide the chemical building blocks to form new minerals.

• *Movement of soil components:* Infiltrating water and, to a lesser extent, burrowing organisms and plant roots move materials from one horizon to another, usually from near-surface A, O, or E horizons into the B horizon.

• *Gain of materials:* Organic matter from decaying plants, dust from the atmosphere, and ions dissolved in rainwater are added into the soil as it forms.

• *Loss of materials:* Leaching of the most soluble components into ground water and surface erosion by water and wind remove material from the soil.

The variations in the physical and compositional characteristics of soils from place to place are linked to critical soil-forming factors. These factors are

• The parent material that weathers to form each soil.

• The climate and vegetation that determine the chemical-weathering reactions that form the soil.

• The duration of the soil-forming processes that determine maturity and strength of horizon development.

- The relief and stability of the landscape, which determine how long a soil forms before burial or erosion. Mature soils are associated with stable landscapes, because well-developed soils will not form on landscapes that are actively eroding or being buried beneath sediment.

Agricultural activities may destabilize the landscape and cause soil erosion, which removes soil and nutrients. Planting agricultural crops in place of native vegetation, along with soil compaction by farm implements, commonly leads to greater water runoff, which erodes soil. Wind also removes the finer particles of soil, especially in dry seasons or during droughts in arid regions.

In the coming chapters you will learn about different processes that sculpt Earth's surface and reflect interactions among the geosphere, hydrosphere, atmosphere, and biosphere. These processes include the failure of hillsides by mass movements, the transport of water and sediment in rivers, the dynamics of glaciers, the action of waves on shorelines, and the movement of surface sediment by wind. In all of these situations, soil development plays an important role in understanding the landscape history, because soils record the stability of the landscape and the environmental conditions that have existed during the evolution of the landscape.

Confirm Your Knowledge

1. What is the geologic definition of soil? How does it differ from other definitions?

2. What are the layers of soil called? How are they formed?

3. What is the difference between the formation of soil horizons and sedimentary layers?

4. What are the principle soil horizons? What processes dominate in each horizon?

5. Why is soil development critical to a soil's ability to support plant growth?

6. Identify the factors that affect the overall characteristics of a soil.

7. Which soil horizon contains an abundance of soil aggregates? How do they form? Why are they not as abundant in the other soil horizons? Why are they important?

8. Why is it that rocks containing calcite occur all over the United States yet soils containing calcite are restricted to the southwestern region?

9. Explain the development of soils over time.

10. Figure 14.12 illustrates how soils mature over time. What other factors that affect the overall characteristics of a soil are illustrated in the figure?

11. How does landscape affect soil development?

12. Describe the main types of soils and the processes and features that distinguish them from one another?

13. Describe how soil formation can result in an economically significant mineral deposit.

14. Why are large areas of North America covered with immature soils characterized by weakly developed horizons?

15. How can calcium-rich B horizons form in soils on top of calcium-poor rock in hot and arid regions?

16. Erosion is a major threat to soils. Explain.

17. Why did the Dust Bowl happen?

18. Explain some ways to reduce soil erosion.

Confirm Your Understanding

1. Write out an answer for each question in the Chapter Outline for the chapter sections assigned by your instructor.

2. Find a soil profile exposed somewhere in your community. Identify, describe, and measure the soil horizons present in your profile. (Note: if conditions prevent you from visiting an actual soil profile, work from a photograph and omit the measurements if a scale is not provided).

3. Why do E horizons not occur everywhere?

4. You discover an ancient reddish soil deposit buried beneath younger sediment and the modern soil. What horizon does it likely represent? What would you look for in this ancient soil to determine the ancient climate?

5. Plants are important in producing soils. How do plants at the surface affect the B horizon?

6. If you lived in the tropics, what environmental factors would you look for in choosing a site for farming?

7. Explain the apparent discrepancy that lush tropical rainforests characteristically grow on very infertile soils.

Key Terms

A horizon (p. 388)
B horizon (p. 388)
bauxite (p. 400)
bedrock (p. 386)

C horizon (p. 388)
E horizon (p. 388)
gully (p. 407)
O horizon (p. 388)

parent material (p. 386)
peat (p. 401)
regolith (p. 386)

rill (p. 407)
soil (p. 386)
soil horizon (p. 386)

15 Mass Movements:
Landscapes in Motion

15.1 What are the characteristics of mass movements?

15.2 What causes mass movements?

15.3 What factors determine slope stability?

15.4 When do mass movements occur?

15.5 How do we know . . . how to map mass-movement hazards?

15.6 How do mass movements sculpt the landscape?

Why Study Landslides?

MANY GEOLOGIC PROCESSES MOVE FRAGMENTED ROCK DOWN-slope. Processes including the work of streams, wind, and glaciers shape landscapes. This chapter focuses on how rock and regolith move downslope because of the downward pull of gravity alone. Falling, sliding, and flowing masses of rock and regolith are found almost everywhere on Earth, even underwater. Later chapters evaluate the contrasting erosion and transport of rock and regolith by running water, flowing glaciers, and blowing wind.

"Landslide" is a generic and popular term for the gravity-driven, downslope movement of rock and regolith, and the facing page illustrates a dramatic example. Geologists and engineers prefer, however, to use terms that describe and define the processes taking place rather than just referring to "land that slides." You are about to embark on an investigation of these processes, and you will learn precise terms that more accurately describe types of landslides.

Truly catastrophic landslides are infrequent, especially in human lifetimes. As a result, people in communities at risk to landslides can have a false sense of security. Added up, however, small and catastrophic landslides cause staggering numbers of casualties and substantial economic losses—about 1 to 2 billion dollars annually in the United States. Increased urbanization in areas prone to slope failure along with deforestation and the local increase in precipitation caused by changing climate patterns all contribute to increasing socioeconomic landslide losses. Landslides destroy or damage residential and industrial developments, wipe out productive agricultural and forest lands, and kill people and livestock. By understanding how landslides happen and why they occur, geologists help detect, predict, and mitigate potential hazards.

The objectives of this chapter are as follows:

✔ To understand the causes of gravity-driven, downslope movement of rock and regolith

✔ To examine how these processes modify the landscape

✔ To learn how to recognize areas where potential hazards exist

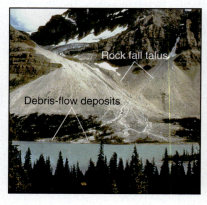

Rock fall talus

Debris-flow deposits

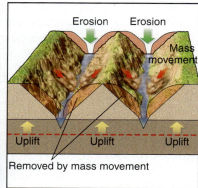

Erosion   Erosion

Mass movement

Uplift   Uplift   Uplift

Removed by mass movement

A landslide triggered by an earthquake killed more than 1000 people in this El Salvador city in 2001. ▶

To accomplish these goals, you will consider the following questions:

15.1 What are the characteristics of mass movements?

15.2 What causes mass movements?

15.3 What factors determine slope stability?

15.4 When do mass movements occur?

15.5 *How do we know* ... how to map mass-movement hazards?

15.6 How do mass movements sculpt the landscape?

In the FIELD

What do landslides look like? To answer that question, you will virtually witness four events and examine the different processes involved. Notice what processes the events have in common, how are they different, and what they might indicate about how and why rock and regolith move downslope.

Your first stop is Yosemite National Park, California, on July 10, 1996. It is the height of the tourist season; the campgrounds are full, and hikers crowd the trails. At 6:52 P.M., without warning, a giant block of granite separates from a cliff high above Yosemite Valley. The result is shown in **Figure 15.1**a. The granite block has a volume of about 30,000 cubic meters (equivalent to filling 250 eighteen-wheeler semitrailers with rock). The rock slides down the slope in contact with the rock wall for 167 meters and then free-falls 550 meters to the valley below. The impact of the rock on the valley floor is strong enough to show up on seismometers 200 kilometers away. The air displaced from beneath the falling rock produces an air blast that wipes out 1000 trees in the forest covering an area of 0.13 square kilometers (equivalent to the area of 24 football fields). The combination of falling rock and blasting air damages a nature center and several bridges, destroys a snack bar, kills one person, and seriously injures several others. The entire event is over in seconds.

The next field-trip stop is Madison Canyon, Montana, on August 17, 1959. Near midnight, an earthquake with a moment magnitude of 7.3 shakes southwestern Montana. Hundreds of vacationers are violently awakened in their tents and camp trailers along the Madison River Canyon, northwest of Yellowstone National Park. The ground heaves beneath campers' sleeping bags, and their tents shudder. At the peak of the ground shaking, a huge mass of rock and loose regolith rips from the mountainside and slides down the canyon wall. A wave of broken rock, soil, and trees descends through the darkness. The rock moves so fast that it travels partway up the opposite side of the steep canyon. Campers, trailers, and cars are buried within seconds—28 people die.

The earthquake shaking dislodged 30 million cubic meters of rock and regolith on the steep canyon slope. This landslide, illustrated in Figure 15.1b, was 1000 times larger in mass than the block of rock that later fell in Yosemite Valley. The Madison Canyon landslide destroyed part of a campground, buried the only highway in the valley, and dammed the Madison River to form Earthquake Lake.

The next field excursion is a real-time trip to Slumgullion Creek near Lake City in southwest Colorado. Figure 15.1c shows a scarred hillside in the San Juan Mountains where regolith has slowly and continuously crept downhill for 2000 years. More than 170 million cubic meters of regolith and forest, along a 6.8-kilometer-long hillside, have moved hundreds of meters. The outline of the moving regolith in Figure 15.1c resembles an oozing, pasty flow of earthen debris. The surface of the displaced regolith is cracked and heaved in different directions. About 700 years ago, the slowly moving mass dammed the Lake Fork of the Gunnison River to form Lake San Cristobal. Most landslide events happen quickly and are studied after the fact, but the Slumgullion slide moves so slowly that it can be studied in real time. You can go out and actually observe the movement in action at varying slow rates between about 0.01 and 6 meters per year.

Your last field-trip excursion is to the San Bernardino Mountains, east of Los Angeles, California. Wildfires burned more than 350 square kilometers of steep, forested mountainsides in October 2003, removing vegetation and leaving the ground covered with ash and charred wood. Heavy winter storms strike on Christmas Day, dropping as much as 20 centimeters of rain in a 24-hour period, the most severe rainfall in 20 years. Steep hillsides, covered in fire debris and loose soil, seem to melt into flowing masses of regolith and water. The sound of the moving mass is like thundering freight trains crashing through narrow mountain canyons. The saturated debris plows through a church camp at 50 kilometers per hour and resembles a flow of wet concrete more than four meters thick. The flow knocks buildings from their foundations, tears down walls, and ultimately kills half of a group of 30 people gathered for the holiday. Figure 15.1d shows the devastation. A few hours later, the event repeats in a neighboring canyon, scattering recreation vehicles in a campground as occupants sit down for Christmas dinner—two more lives are lost.

Figure 15.1 **What mass movements look like.** ▶

A. Field notes, Yosemite National Park, California 7/10/96

Photos taken just after large granite blocks broke loose from cliff and dropped to valley floor

Blocks detached from here.

Photo from helicopter showing the results

Shattered granite at base of cliff

Trees blown down by air blast

Damaged building

Dust cloud formed when blocks shattered on the valley floor.

B. Field notes, Madison Canyon, Montana, August 1959

Scar left by giant landslide

Slide debris traveled part way up opposite side of the canyon

Lake formed when slide dammed the Madison River

Field notes, Madison Canyon, Montana, August 2000

Effects of the 1959 landslide are still very evident, 41 years later. Very little vegetation grows on the slide scar. Notice dead tree snags in Earthquake Lake.

C. Field notes, near Lake City, Colorado, August 2004

Giant "tongue" of slowly flowing regolith almost 7 km long. Notice the highway for scale.

Lake formed when slow-moving landslide dammed a river.

D. Field notes, San Bernardino Mtns., California, December 26, 2003

Boulders, sand, mud, water, and trees flowed rapidly through the canyon the previous evening.

Flowing debris demolished buildings and buried some of the victims.

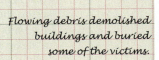

15.1 What Are the Characteristics of Mass Movements?

The real events described above are examples of **mass movement**—gravity-driven downslope motion of rock and regolith. Although water and ice commonly aid mass movements of rock and regolith, the motion occurs solely because of the force of gravity pulling on the solids and not because of flowing water.

Criteria for a Meaningful Classification

How would you describe the processes involved in the four events that you "witnessed" in the previous section? All of these events are mass movements, but what distinguishes each one from the others? Answers to these questions provide insights into how mass movements work and also form the basis for a useful, process-based classification.

The transported materials differ between the events—solid rock at Yosemite, mostly regolith at Slumgullion, and a mixture of materials at the Madison Canyon and San Bernardino Mountain sites. Water played an important role in the San Bernardino Mountain events but seems to have been of negligible, if any, consequence in the others.

The material also moves in different ways. The Yosemite event mostly featured free-falling rock. Rock and regolith at Madison Canyon primarily moved with a sliding motion. The masses moving at Slumgullion and in southern California, although consisting mostly of solid objects, seem to flow like a highly viscous fluid.

Speed also distinguishes the different mass movements. The falling and sliding rock events at Yosemite and Madison Canyon happened quickly and ended abruptly. In contrast, the flow of regolith along Slumgullion Creek is slow and ongoing.

Based on these observations, it is easy to see why three factors—(1) the nature of the mixture of solids (rock or regolith), (2) the type of motion (sliding, falling, flowing), and (3) the velocity of motion (fast or slow)—describe the variety of mass-movement processes. These characteristics are the criteria for the classification scheme of mass movements illustrated in **Figure 15.2**.

Distinguishing the Materials in Motion

Rock and regolith comprise the two types of moving material. The key factor in classification is noting what the material was prior to its descent, since originally solid rock breaks up during movement and may resemble regolith when the event is over. In the classification, regolith only refers to the weathered loose blanket of debris on the original hillslope. Debris and earth are categories of regolith. *Debris* is coarse grained; 20 to 80 percent of the fragments are larger than 2 millimeters in size. In the category *earth*, 80 percent or more of the particles are less than 2 millimeters in size. The terms "rock," "debris," and "earth" completely describe the range of material that moves downslope and allow for specific classification. For example, rock fall and debris fall share the same process of movement but involve the transport of different types of material.

Mass Movement by Free Fall

During a **fall**, material detaches from a steep slope and then free falls through the air, or bounces and rolls downslope (Figure 15.2). The key feature is that the moving mass loses contact with the surface. Rock falls are common where the rock is highly jointed and on a steep slope, such as shown in **Figure 15.3**, and describe the observed event at Yosemite Valley (Figure 15.1a).

Falls account for piles of loose rock, debris, and earth that pile up at the base of steep slopes, as seen in **Figure 15.4**. These accumulations are called **talus**. The lack of vegetation or soil on talus indicates that material is actively accumulating.

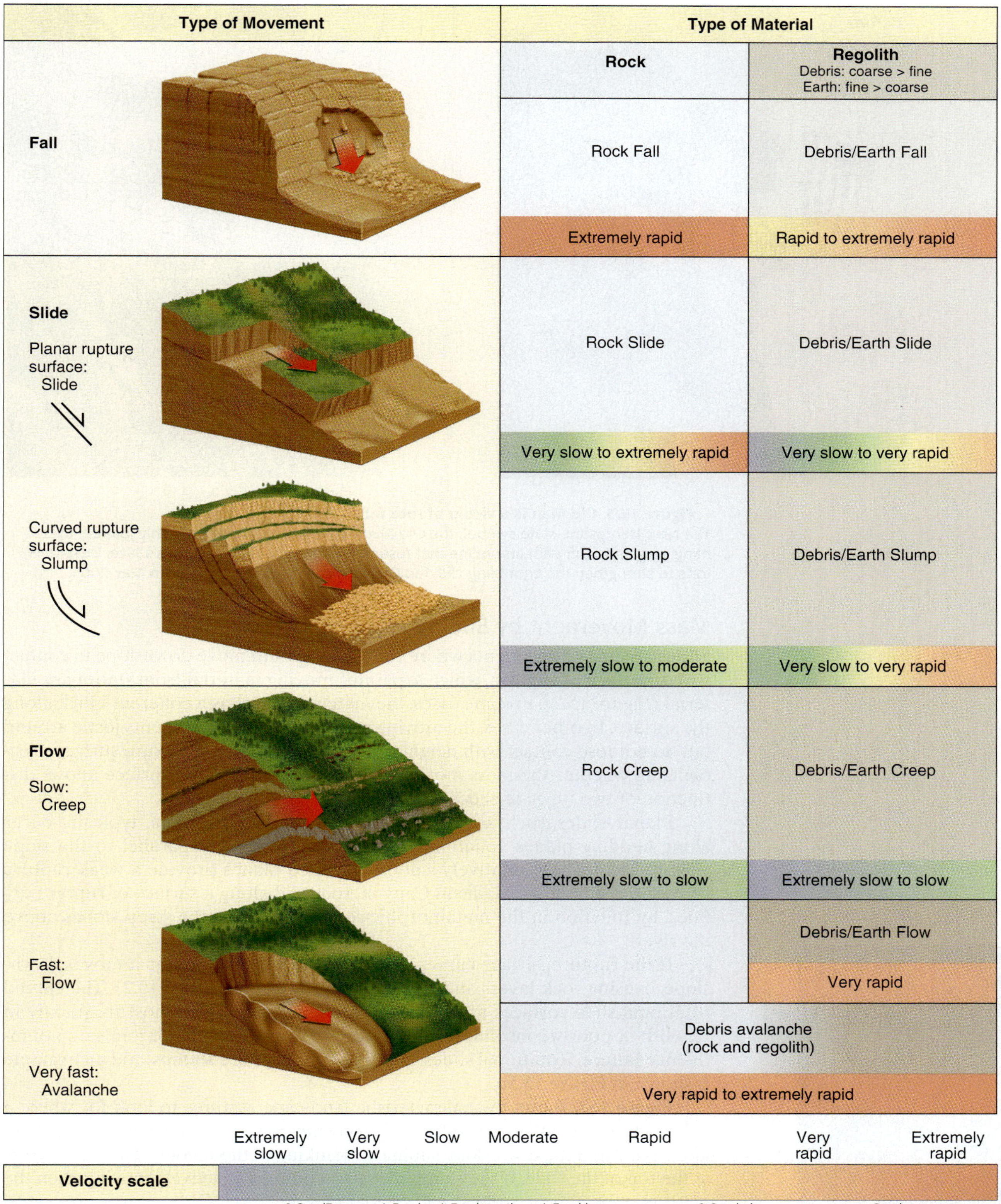

Type of Movement		Type of Material	
		Rock	**Regolith** Debris: coarse > fine Earth: fine > coarse
Fall		Rock Fall	Debris/Earth Fall
		Extremely rapid	Rapid to extremely rapid
Slide Planar rupture surface: Slide		Rock Slide	Debris/Earth Slide
		Very slow to extremely rapid	Very slow to very rapid
Curved rupture surface: Slump		Rock Slump	Debris/Earth Slump
		Extremely slow to moderate	Very slow to very rapid
Flow Slow: Creep		Rock Creep	Debris/Earth Creep
		Extremely slow to slow	Extremely slow to slow
			Debris/Earth Flow
			Very rapid
Fast: Flow Very fast: Avalanche		Debris avalanche (rock and regolith)	
		Very rapid to extremely rapid	

	Extremely slow	Very slow	Slow	Moderate	Rapid	Very rapid	Extremely rapid
Velocity scale							
	0.3 m/5yrs	1.5 m/yr	1.5 m/month	1.5 m/day	0.3 m/min	3 m/sec	

▲ **Figure 15.2 How to classify mass movements.**
Mass-movement classification relies mostly on identifying the type of material that moves and the type of movement. The velocity of the movement is characteristic of some processes but ranges considerably for others.

Mass Movements: *See how mass movements work.*

▲ **Figure 15.3 Old Man is a victim of rock fall.**
The New Hampshire state symbol, the Old Man of the Mountain, was a highly jointed, over-hanging granite cliff with an outline that resembled the silhouette of a man's face. Despite ef-forts to strengthen the crumbling cliff, the Old Man collapsed in a rock fall in May 2003.

Mass Movement by Sliding along a Surface

Slides are mass movements where rock and regolith move downslope in contact with a **surface of rupture**, which separates moving material from stationary ma-terial (Figure 15.2). In some cases, the material moves as a coherent block along the surface. In other cases, the moving rock and regolith fragments jostle around but do not lose contact with neighboring fragments or the rupture surface. Rec-ognizing whether the mass moves along a planar or curved surface allows dis-tinction of two types of slides.

Planar slides, such as the one illustrated in **Figure 15.5**a, typically occur along bedding planes, foliation, or joint planes oriented parallel to the slope (Figure 15.2). These relatively smooth, inclined planes provide a weak rupture surface for sliding. At Madison Canyon, rock slid along a surface of rupture de-fined by foliation in the metamorphic rocks underlying the steep slopes above the river.

If the rupture surface curves, then the slide mass rotates as it moves down-slope, causing rock layers and surface features to tilt (Figure 15.2). The curved rotational slide surfaces are scallop shaped, and they form most frequently in regolith or poorly consolidated or weak rock where bedding or joints do not in-fluence failure. Rotational slides on curved surfaces are **slumps**, and an example is shown in Figure 15.5b.

Figure 15.6 shows the characteristic landscape features to look for where a debris slump recently occurred or is actively moving. The surface of rupture looks like a fault, and the downslope sliding and sinking of the regolith produces scarps at the top of the slide. If the slump moved recently or is actively moving, then the scarp may be almost vertical and devoid of vegetation. Older scarps erode and are less obvious. Sliding regolith conceals the rest of the surface of rupture, but its curved shape is revealed by the rotation of the slumping mass, which tilts the land surface back into the hillside (see Figures 15.5b and 15.6). This tilting may disrupt downslope runoff of water to form swampy areas or ponds along the hillslope. Open cracks form in the land surface and are especially evident where road pave-ment breaks and tilts. Where the rupture surface curves back up at the bottom of the slump, the ground bulges unevenly into very irregular and bumpy topography.

▲ **Figure 15.4 Talus slopes, indicators of falls.**
Rocks falling from these steep sedimentary-rock cliffs in the Canadian Rockies produce towering cone-shaped slopes of talus. The lack of soil or vegetation on the talus indicates that rockfalls occur frequently and that the slopes are very unstable. Water from heavy rainfall and melting snow occasionally moves some of the talus farther downslope in debris flows.

◀ **Figure 15.5 What slides and slumps look like.**

(a)

This highway clings to an unstable slope south of San Francisco, California. Steeply dipping sedimentary rocks slide along bedding planes into the ocean, leaving a barren, unvegetated slope. Rock slides frequently close the road.

Dipping sedimentary beds

(b)

This earth slump along the Genesee River in New York shows the curving, scallop-shaped rupture scarp typical of a slump. Also notice that the broken pavement and trees tilt back into the slumping hillside, which indicates rotational movement along a curved rupture surface.

Mass Movement by Flow Resembling a Liquid

Flows are continuous movement of rock, regolith, or both that behave like a high-viscosity liquid (Figure 15.2). **Figure 15.7** illustrates the liquid appearance of moving flows and the nature of the deposits. Check back to Figure 15.4 to see debris flows formed by remobilizing rock-fall talus.

You were a virtual witness to two flows at the beginning of the chapter. Based on fragment size, the Slumgullion example is an earth flow (Figure

▶ **Figure 15.6 Recognizing a slump.**
This illustration shows characteristic surface features produced by an active or recently formed slump (compare to Figure 15.5b). Curving scarps mark the upslope end of the rupture surface, and bumpy ridges show the upward bulging of displaced regolith where the rupture surface curves back up to the ground surface. Buildings and pavement crack as the slope material moves. Rotation along the curved, concave-up rupture surface causes trees and utility poles to lean back toward the hillside, and disrupts drainage to produce swampy areas.

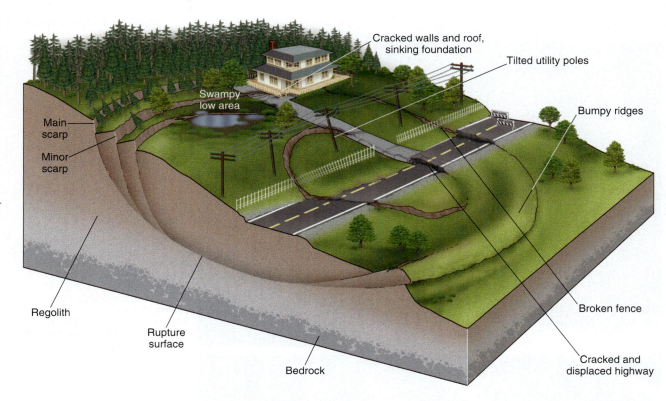

15.1c), whereas the San Bernardino Mountain events were debris flows (Figure 15.1d). The fluidlike characteristics of most debris flows partly result from water included in the flows, but solid particles constitute more than three-quarters of the mass of these flows. Debris flows and earth flows leave deposits of chaotically mixed fragments of different sizes, with very irregular, bumpy upper surfaces (Figures 15.1c, 15.1d, 15.7). Contrasting the Slumgullion and San Bernardino Mountain events also illustrates a wide range in flow velocity.

Figure 15.8 shows how very slow flows, called **creep**, are detected only by dislocation or bending of features at the surface. The bending of the rock layers in Figure 15.8a indicates that the surface regolith and part of the rock flow

(a)

(b)

▶ **Figure 15.7 What debris flows look like.**

This photo of a moving debris flow in the Philippines shows the flowing-concrete appearance that is typical of debris flows. Although water-saturated and moving like a fluid, the debris flow consists mostly of sand and gravel. The wave in the foreground is about 1 meter high.

This coarse, bouldery debris flow damaged homes near Durango, Colorado. Notice that the deposit consists of a very poorly sorted mixture of boulders, sand, mud, and broken trees.

These nearly vertical sedimentary rock layers bend downslope, to the right, because of rock creep.

Bent tree trunks are an indication of earth creep on this steep hillside. The bending results from simultaneous vertical growth of the tree and downslope tilting caused by creep of near-surface regolith above the main root zone of the tree.

downslope without evidence of a rupture surface that would define a slide. The tree roots anchor the trees in stable or only slightly moving regolith at depth while surface regolith slowly moves downslope. The movement tilts the trees but they adjust their growth back to a vertical position, causing a downslope bend in the tree trunks that becomes more obvious over time (see Figure 15.8b).

The term "avalanche" also describes some flows. Most familiar might be snow and ice avalanches, which are mass movements that sometimes bury skiers and mountain towns. **Debris avalanches** are very rapid flows of rock, regolith, vegetation, and sometimes ice. Notice that in this definition of a debris avalanche, the moving material is not restricted to regolith of debris size, despite the name. Some debris avalanches are gigantic in volume and involve the failure of significant parts of whole mountains, as depicted in **Figure 15.9**. Despite very rapid flowing motion, debris avalanches contain very little water.

▲ **Figure 15.9** **Destructive debris avalanche.**
This view from an airplane shows where a debris avalanche flowed down a mountainside in Nicaragua during torrential hurricane rainfall. The avalanche buried two villages and killed 2000 people. The smaller bare spots and stripes on the mountainside represent smaller-volume debris slides and debris flows that occurred during the same storm.

▶ **Figure 15.10 Single event, multiple processes.**
The drawing illustrates some of the morphological parts of a complex mass movement that is both a slide and a flow. These features are visible in the 1995 photograph of La Conchita, California. Another slide and flow near this location destroyed thirteen homes and killed 10 people in January 2005. The scarp is the surface along which the mass detaches as a slide, and commonly several scarps are present. The toe marks the farthest advancement of the displaced material. The mass movement is a slide near the top, as indicated by the curved surface of rupture. The irregular bumpy topography at the base indicates the part of the mass movement that flowed beyond the rupture surface.

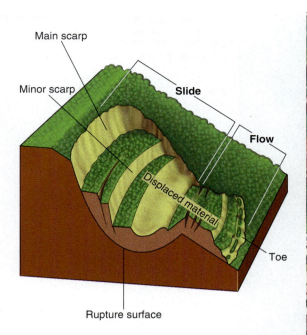

Single Events—Multiple Processes

Many mass-movement events involve combinations of processes. **Figure 15.10** shows how movement commonly originates at the upslope end as a slide, but then the material breaks up and flows beyond the surface of rupture. The displaced rock mass at Yosemite slid along a rupture surface before free falling to the valley floor (Figure 15.1a). The steep scars at the top of the Slumgullion earth flow (Figure 15.1c) and the debris avalanche illustrated in Figure 15.9 are evidence of a slide at the upslope end of the mass movement. The detached regolith and rock then continue downslope beyond the rupture surface as a flow.

Putting It Together—*What Are the Characteristics of Mass Movements?*

■ Mass movements are downslope movement of rock, regolith, or both, due solely to the downward pull of the force of gravity.

■ The three characteristics used to describe mass movements are (1) the material that moves (rock, debris, earth), (2) the movement process (fall, slide, flow), and (3) the velocity of movement (fast, slow).

■ Falls are material falling free from a steep slope. Slides move along planar or curved surfaces of rupture. Flows move like viscous liquids despite being composed primarily of solid particles.

■ Landscape features such as scarps, surface cracks, tilted or bent trees, bumpy topography, and talus provide clues to the occurrence and type of mass movement affecting a slope.

■ Some mass-movement events involve more than one process. Rock and regolith initially separate along a rupture surface, but the slide-displaced material continues downslope as a fall or flow.

15.2 What Causes Mass Movements?

Field observations provide descriptions to classify the different types of mass movements, but what causes mass movements? Gravity pulls mass downhill, but not all hillslopes seem primed to fall, slide, or flow away. Mass movements occur only where slopes are unstable and susceptible to failure. The next objective is to determine why slopes are unstable or become unstable. This problem requires examining the driving forces that pull material downslope, in contrast to resisting forces that tend to keep material where it is. Mass movement occurs when the driving forces exceed the resisting forces.

Using an Analogy of Mass Movement

To help you understand the factors that enhance or impede movement, consider the simple analogy of a brick on an inclined board that is illustrated in **Figure 15.11**. Place the brick on the board and then gradually raise one end of the board. Eventually the board rises to a critical slope angle where the brick moves. What can you conclude?

- The first conclusion—increasing slope favors motion and slope instability. Next, sand the board to make it smoother, and repeat the experiment. This time, the brick begins to move at a lower angle of tilt (Figure 15.11b).
- The second conclusion—smoother surfaces favor motion and slope instability. Finally, wet the sanded board with a film of water and repeat the experiment one more time. This time the brick begins to move at an even lower tilt angle (Figure 15.11c).
- The third conclusion—the presence of water favors motion and slope instability. This simple series of experiments suggests that slope, surface roughness, and water are factors affecting the balance between driving and resisting forces.

Gravity Is the Driving Force

Gravity, expressed as the weight of an object pulling straight down toward the center of Earth, is the driving force that causes downslope movement. **Figure 15.12** illustrates how gravity holds a boulder stationary against a horizontal surface. In contrast, along a vertical surface, gravity pulls the boulder downward.

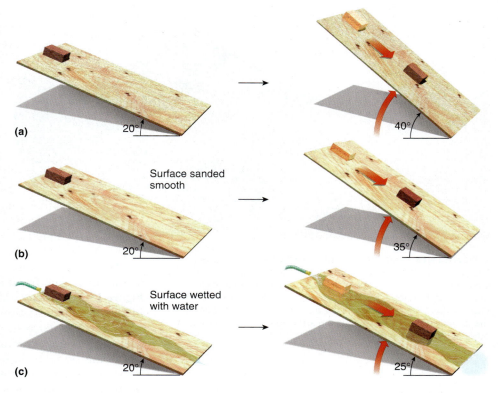

▲ **Figure 15.11 Understanding causes of mass movement by analogy.**
(a) A brick slides down the inclined slope of a wood board when the board tilts past a critically steep angle. (b) The critical angle for the onset of movement decreases if the board is first sanded smooth. (c) The angle decreases even further if the board is wetted with water.

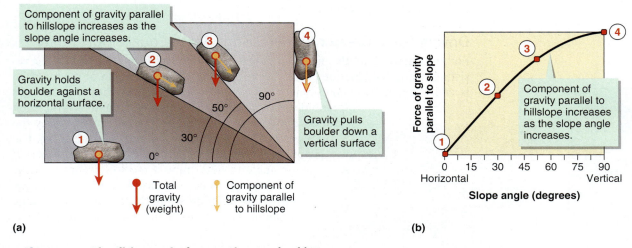

▲ **Figure 15.12 Visualizing gravity forces acting on a boulder.**
a. On a flat surface (position 1), the total gravity force, or weight, of the boulder pulls the boulder down and the boulder is stable. Gravity also pulls downward parallel to a vertical surface (position 4), which likely means that the boulder is unstable and moves. On an inclined surface (such as positions 2 and 3), a component of gravity is directed parallel to the surface (follow the yellow arrows).
b. The graph shows how the gravity force parallel to the ground surface increases with increasing steepness. The labeled positions on the curve correspond to the four scenarios illustrated in panel (a). The larger the slope angle, the larger the gravitational driving force for motion along the surface.

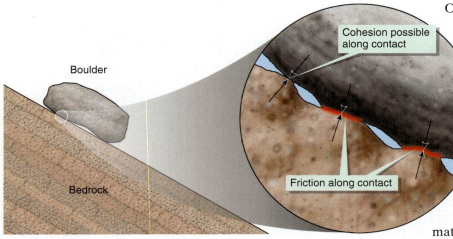

On a slope in between horizontal and vertical, there is a component of gravity parallel to the sloping surface, and this gravity component increases as the slope angle increases (Figure 15.12a). The driving force for mass movement, therefore, is greater on steep slopes than on gentle slopes. This explains why the brick moved down the board after the board was tilted to a steeper angle.

Friction and Cohesion Are the Resisting Forces

Forces acting between the boulder and the underlying material resist movement of the boulder down the inclined surface. The resisting forces are friction and cohesion, which add together as the *resisting strength* of the material. Stronger materials resist mass movements better than weak materials.

▲ **Figure 15.13 Resisting strength is friction plus cohesion.** Friction relates to the smoothness of the hillslope surface and the smoothness of the surface of the boulder. The friction force also depends on the slope, because at lower slope the weight of the boulder more effectively forces the boulder into contact against the underlying rock. Cohesion is the attraction of particles to each other at the atomic level.

Friction is the force opposing motion between two objects in contact with one another. The friction increases as the roughness of the surfaces increases. **Figure 15.13** illustrates how friction affects the stability of a boulder on a slope. In this example, friction depends both on the smoothness of the surface on which the boulder rests and the smoothness of the surface of the boulder. The unsanded board experiment (Figure 15.11) demonstrated this same effect, where the slope angle for initial sliding of the brick is steeper than when the board was sanded smooth. If many objects are present, then the friction along the boundaries of each object also must be considered. Friction also varies with slope angle because on lower slopes the weight of fragments pulls them down in stronger contact with underlying material. This means that friction is higher on gentle slopes compared to steep slopes.

Cohesion, the other component of resisting strength, is the attraction of particles to each other at the atomic level (Figure 15.13). Cohesion results from opposite electrostatic charges on adjacent particles, and it is particularly important in clay minerals (Section 5.4 and 14.2). Materials such as loose sand and gravel have very little cohesion between grains. The addition of clay to these materials raises the cohesion by effectively increasing the "stickiness." An increase in cohesion, especially by adding clay particles, helps material stay intact and resist mass movement.

Driving Force Greater than Resisting Force Equals Movement

For motion to occur, the gravity force parallel to the surface must exceed the resisting strength of friction plus cohesion. If the resisting strength of the material is small, then motion takes place on a very low slope, such as when a pencil rolls off a nonlevel tabletop. If the resisting strength is large, then motion only occurs on a steep slope, as illustrated by the brick resting on a rough, dry board (Figure 15.11a). The graph in **Figure 15.14** emphasizes the importance of slope angle to determine if motion occurs, because the driving force increases with slope angle, whereas resisting strength decreases with slope angle. For any particular circumstance, there is

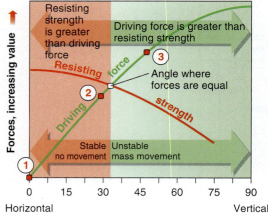

(a)

(b)

▲ **Figure 15.14 The onset of mass movement.** Mass movement occurs when the driving force of gravity exceeds the resisting strength of the rock or regolith. Boulders resting on slopes of different angles, labeled positions 1, 2, and 3, experience different forces, as shown in the graph. Driving force increases and the resisting strength decreases as the slope angle increases. For the illustrated examples, boulders in positions 1 and 2 are stable whereas the boulder in position 3 moves because it is unstable.

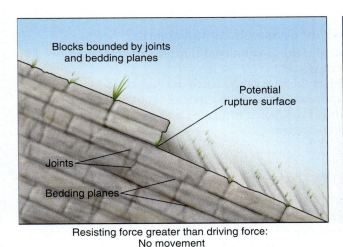

Resisting force greater than driving force:
No movement

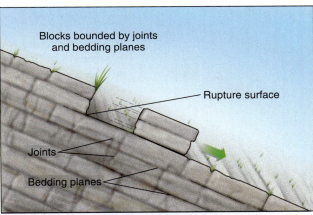

Resisting force less than driving force:
Movement

◄ **Figure 15.15 Resisting strength in fractured rock.** So-called solid rock is usually not very solid at all but is broken along joints and commonly separated into layers by sedimentary bedding planes or metamorphic foliation. Fractured rock is stable if the friction along joints and bedding planes and cohesion between blocks are greater than the pull of gravity on the rock mass parallel to the slope. Mass movement, as a slide or fall, occurs by loss of frictional and cohesive resistance along these preexisting breaks in the rock.

a critical angle below which there is no movement and above which movement takes place. This angle depends on the cohesive and frictional characteristics of the materials.

Figure 15.15 and **Figure 15.16** extend the simple case of a boulder on a slope (Figures 15.12–15.14) to other mass-movement scenarios. In the case of a fractured rock outcrop (Figure 15.15), the friction and cohesion between the interfaces of adjacent blocks of rock determine the resisting strength. The frictional resistance to sliding along a bedding plane, as a potential rupture surface, is greater if the weight of the rock above the surface is greater and if the surface is rough. This is analogous to the unsanded board in Figure 15.11. The cohesion is mostly determined by the mineral composition of the rock, and it is usually low between blocks of rock. For the case of potential movement of regolith (Figure 15.16), the resisting strength includes friction and cohesion between regolith particles and between the interfaces of the regolith and the bedrock.

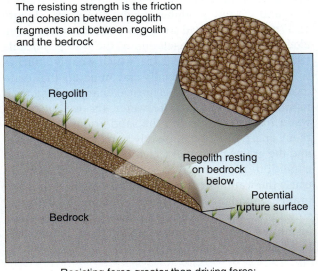

Resisting force greater than driving force:
No movement

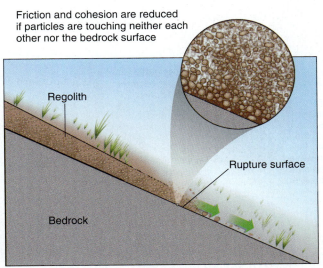

Resisting force less than driving force:
Movement

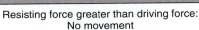

▲ **Figure 15.16 Resisting strength in regolith.**
Regolith stability is determined by friction and cohesion between regolith particles *and* between the particles and the underlying rock. Movement commonly occurs because particles lose contact with one another and with the underlying rock, which decreases the friction and cohesion forces that resist motion.

15.3 What Factors Determine Slope Stability?

Rock and regolith do not continually fall, slide, and flow down hillsides, which suggests that mass movement occurs by increase in the magnitude of the driving force, decrease in the resisting strength, or both. What processes change the slope, friction, or cohesion of a mass of rock or regolith so that the slope fails? The initial field observations at the beginning of the chapter, along with the brick and board experiment, provide some insights.

The Role of Water

Water commonly contributes to slope instability. You can quickly see how significant water is for causing mass movement by noting how easily the brick slides on a wet board (Figure 15.11c) and by thinking of the debris flows triggered by heavy rainfall in California (Figure 15.1d).

Figure 15.17 provides a familiar analogy of playground or beach sand to show how water influences slope stability. Adding small amounts of water to sand initially increases the cohesion between particles, but too much water leads to failure.

The **angle of repose** is the maximum angle of a stable slope determined by friction, cohesion, and the shapes of the particles. Figure 15.17a shows the loose dry particles with an angle of repose. Adding a small amount of water to a pile of sand, silt, or clay grains produces thin films of water between particles. The thin water films increase the cohesion between particles and permit a higher angle of repose (Figure 15.17b). The increased cohesion is caused by the weak electrical charges on each end of the water molecule (Section 2.4 and Figure 2.15), which weakly attaches to adjacent grains, especially clay minerals that commonly have unbalanced electrical charges along their margins. As you add more water, the weight of the water exerts pressure on the sand grains and moves them apart. Water completely surrounds each grain and the grains no longer touch (Figure 15.17c). In this condition, the friction is substantially reduced because the water partly supports the weight of the sand grains. Abundant water, therefore, exerts a pore pressure that reduces resisting strength and causes slope instability and movement along grain boundaries, bedding planes, and joint surfaces.

The Role of Geologic Materials

Field and lab studies show that the composition and grain size of the slope material also influence stabili-

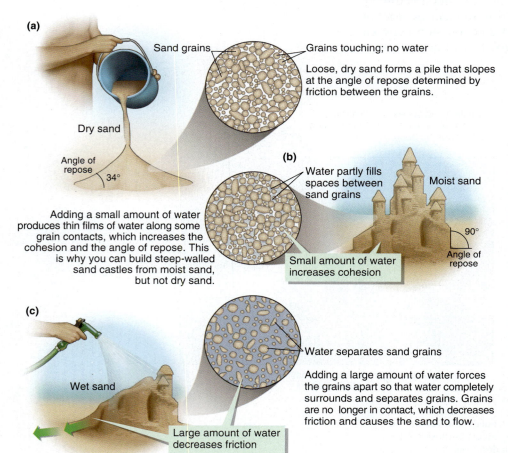

(a) Sand grains — Grains touching; no water

Loose, dry sand forms a pile that slopes at the angle of repose determined by friction between the grains.

Dry sand

Angle of repose 34°

Adding a small amount of water produces thin films of water along some grain contacts, which increases the cohesion and the angle of repose. This is why you can build steep-walled sand castles from moist sand, but not dry sand.

(b) Water partly fills spaces between sand grains   Moist sand

90°

Angle of repose

Small amount of water increases cohesion

(c) Water separates sand grains

Wet sand

Adding a large amount of water forces the grains apart so that water completely surrounds and separates grains. Grains are no longer in contact, which decreases friction and causes the sand to flow.

Large amount of water decreases friction

◄ **Figure 15.17** Visualizing the effect of water on the angle of repose of loose sand.

ty. Cohesion depends on the mineral composition of the rock or regolith. Clay minerals have high cohesive properties. This cohesion is greatly reduced, however, when clay saturates with water. Water saturation separates the mineral grains and may cause the crystal structure to expand, which turns the clay-rich regolith into a soupy, swelling muck. Cohesion increases, however, where mineral cements, such as calcite, are present between fragments.

The Role of Slope

The importance of slope steepness to cause mass movements is clear from the brick and board experiments illustrated in Figure 15.11 and from the analysis of the gravity force increasing with steeper slope (Figure 15.12). **Figure 15.18** illustrates examples of natural and artificial processes that increase slope angles to enhance mass movement. Slopes increase along stream channels because many streams erode valleys downward into the landscape. As a stream cuts down, the adjacent slopes become steeper (Figure 15.18a) and may reach the critical angle where driving force exceeds resisting strength. Stream bank erosion caused the slump illustrated in Figure 15.5b. Excavation and artificial filling on hillsides also steepens slopes so as to increase the likelihood of mass movement (Figure 15.18b).

The Role of Geologic Structures

Joints, bedding planes, foliation, and faults are planar features that interrupt cohesion otherwise provided by interlocking mineral grains (Figure 15.15). These planes may be smooth enough to offer minimal frictional resistance to movement. If smooth bedding or foliation planes parallel the slope, then movement is more likely to occur than if the planes dip into the slope, as illustrated in **Figure 15.19**. Joints, bedding planes, foliation, and faults also provide avenues for water to percolate downward from the surface. Water along the planes further reduces friction and cohesion to favor mass movement.

The Role of Weathering

Chemical and physical weathering processes produce unconsolidated regolith and determine its resisting

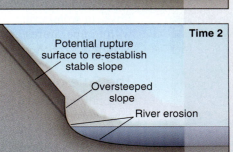
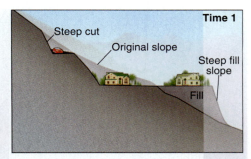

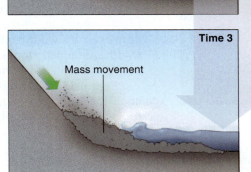

(a)

(b)

▲ **Figure 15.18 Changing slope influences slope stability.**
Natural processes and human activities increase slope angles and increase the likelihood of mass movements. (a) Vertical and horizontal river erosion forms steep unstable slopes. (b) Hillside construction can make slopes more susceptible to mass movement, as does using unconsolidated fill with steep slopes as a building foundation.

▶ **Figure 15.19 Local geology influences slope stability.**
The bedding planes of the rock on the left side of the valley dip parallel to the hillslope. The addition of water along joints adds weight and causes loss of cohesion in the clay-rich shale layer. The decrease in friction and cohesion results in a rock slide. By comparison, the situation on the steeper right side of the valley shows the bedding planes oriented in the opposite direction of the slope. Although rock falls may occur on this steep slip, the orientation of the bedding planes means that slides are highly unlikely.

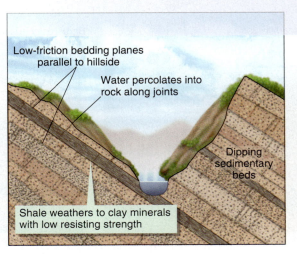
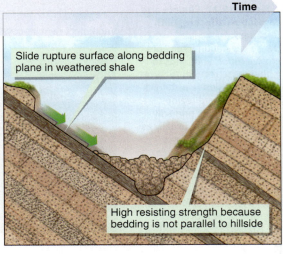

◀ **Figure 15.20** **Mass movements follow wildfires.**
This photo from western Colorado shows a common occurrence of debris flows caused by destruction of regolith-stabilizing vegetation by wildfire. Heavy rain causes debris slides and earth flows that carry loose regolith, ash, and charred wood into canyon bottoms, where the debris mixes with stream water to form debris flows.

strength. Freezing and thawing of water in cracks or shrinking and swelling of clay minerals by alternating wetting and drying decreases friction and cohesiveness to promote slope failure. The rock-fall destruction of the Old Man of the Mountain in New Hampshire, shown in Figure 15.3, illustrates these effects. Physical weathering weakened the highly jointed rock. Following several days with temperatures appropriate for freezing and thawing to take place, the Old Man fell from the steep cliff.

The Role of Vegetation

Vegetation is a very important factor in slope stability. Roots penetrate and bind together regolith and soil and absorb water from precipitation. Removal of vegetation, especially the loss of deep tree roots, takes away this added cohesion and frictional resistance to movement. Vegetation loss comes into play where trees are removed from hillslopes during logging and to clear land for agricultural and urban expansion, where livestock overgraze hillslopes, and where vegetation is destroyed by wildfire. **Figure 15.20** shows debris flows following rainfall on wildfire-denuded slopes. Large, fire-related debris flows caused the Christmas 2003 disaster in southern California (Figure 15.1d).

Figure 15.21 illustrates the role of vegetation, along with climate, to determine the thickness of regolith on hillslopes. Where plants are abundant, the roots bind the regolith that forms by weathering on the hillslopes so that runoff

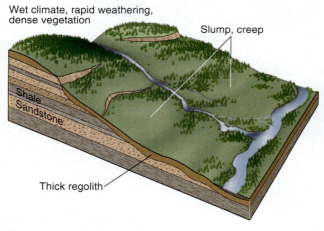

Wet climate, rapid weathering, dense vegetation

Slump, creep

Shale
Sandstone

Thick regolith

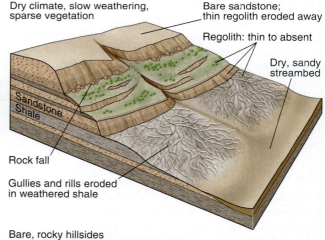

Dry climate, slow weathering, sparse vegetation

Bare sandstone; thin regolith eroded away

Regolith: thin to absent

Dry, sandy streambed

Sandstone
Shale

Rock fall

Gullies and rills eroded in weathered shale

▶ **Figure 15.21** **Climate and vegetation affect regolith thickness.**
Dense vegetation intercepts erosive runoff and binds together loose regolith to form rounded hillslopes with thick regolith in humid regions. Mass movements primarily happen in regolith, usually as creep, slumps, and flows. In contrast, arid regions without abundant vegetation lack thick regolith because weathered fragments wash away soon after they separate from the rock. Mass movements are typically rock falls and rock slides.

Vegetated, regolith-covered hillsides

Great Smoky Mountains National Park, Tennessee

Bare, rocky hillsides

Arches National Park, Utah

erodes only a small volume of the potential sediment. In arid regions with low vegetation abundance and slow rates of weathering, surface-water runoff removes most regolith particles almost as quickly as they loosen from bedrock. Hillslopes in the relatively humid eastern United States and moist regions farther west have thick coatings of regolith that mostly fail in debris or earth flows, slides, slumps, and creep. In arid parts of the western United States, however, regolith is thin or absent and slopes mostly fail by rock falls and slides.

*Putting It Together–**What Factors Determine Slope Stability?***

■ Factors determining slope stability are the abundance of water, the composition and texture of material, presence and orientation of planar features in the rock that may form rupture surfaces, the steepness of the slope, the amount of weathering, and vegetation.

15.4 When Do Mass Movements Occur?

We have identified the factors that change driving and resisting forces to permit mass movement, but what determines *when* the movement occurs? The relative magnitudes of the driving and resisting forces must reach a threshold of imbalance where cohesion and friction are insufficient to offset the downslope pull of gravity. In some situations, the changing force magnitudes are gradual and unsuspected, leading to a sudden and unexpected slope failure, such as the rock falls at Yosemite (Figure 15.1a) and the Old Man of the Mountain (Figure 15.3). In other cases, there is a sudden stimulus that causes a rapid change in the slope, resisting strength, or both that crosses the threshold for movement on what was presumed to be a stable slope. To the extent that these stimuli, called *triggers,* can be anticipated in terms of either location or timing, or both, it is possible for geologists to assess the potential hazard from mass movements.

Rainfall and Snowmelt Triggers

Addition of water to hillslope materials increases the likelihood of mass movement, so it is unsurprising that water can trigger failures. This was the case in the San Bernardino Mountains field example, where extremely heavy rainfall in a short period of time caused damaging and deadly debris flows (Figure 15.1d). Melting snow not only adds lubricating moisture but weight as well. Weight of snow on the hillslope increases the magnitude of the gravity driving force for slope failure. These weather-related triggers of mass movement occur on steep hillsides across the United States and are linked to extreme weather conditions associated with unusually heavy rain, hurricanes, and rapid spring warming. When weather forecasts suggest that these conditions may occur, warnings are issued for areas having steep slopes that may be most susceptible to slope failure. Particular attention is given to hillsides recently stripped of vegetation by wildfire or human activity.

Earthquake Triggers

Strong earthquake ground shaking triggers mass movements. The most common effects are rock and debris falls that occur because the seismic surface waves literally lift rock fragments off the ground and away from neighboring fragments. This reduces the frictional resistance to movement and the fragments bounce and roll downhill. Dust clouds from innumerable rock falls fill the air after large earthquakes. Vibrations from a magnitude-7.9 earthquake triggered a debris avalanche on Nevado Huascaran, Peru, in May of 1970 that killed 18,000 people. The avalanche buried the entire town of Yungay and part of another. Damage from this devastating avalanche is shown in **Figure 15.22**. The Madison Canyon rock slide (Figure 15.1b) is a smaller-volume example of slope failure triggered by an earthquake.

Before the earthquake

After the earthquake

▲ **Figure 15.22 Earthquake-triggered debris avalanche destroyed Peruvian village.**
Prior to 1970, the tall peaks of Nevado Huascaran loomed above the village of Yungay, Peru. A 1970 earthquake shook loose rock, regolith, and ice near the top of the mountain to form a gigantic debris avalanche that buried Yungay and 18,000 of its residents.

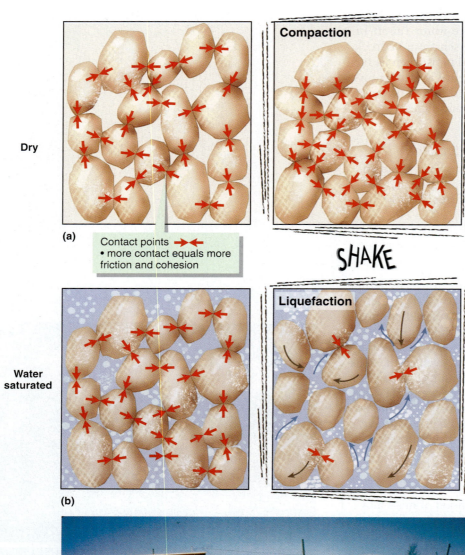

Contact points ➤◄
• more contact equals more friction and cohesion

(a)

Compaction

Dry

SHAKE

Water saturated

Liquefaction

(b)

(c)

◄ **Figure 15.23** **Earthquake-induced liquefaction causes mass movements.**
(a) During earthquake ground shaking, dry silt and sand grains compact, and the number of contact points between grains increase. (b) In water-saturated silts and sands, the grains attempt to compact during ground shaking, but the water cannot compact and instead is forced out of the pore spaces, which causes the grains to move apart. Friction decreases as the number of grain contacts decreases, and the mass behaves like a viscous fluid. (c) Liquefaction during the Great Alaskan Earthquake of 1964 caused mass movements that destroyed these homes in Anchorage.

You learned in Chapter 11 about liquefaction, the process that underlies mass movement of unconsolidated, water-saturated regolith that is shaken by strong earthquake waves (see Section 11.9). Earthquake-wave vibrations cause grains in the regolith to compact closer together in the same way that you shake a box to settle its contents, as illustrated in **Figure 15.23**a. When there is water between the grains, however, the compacting grains displace water that forces other grains apart, which decreases cohesion and friction, as illustrated in Figure 15.23b. The water-rich layer of particles behaves like a viscous fluid and flows downslope. Damage from an earthquake-induced liquefaction and mass movement is shown in Figure 15.23c.

Volcanic Eruption Triggers

The largest observed debris flows, rock slides, and debris avalanches that have originated on the slopes of volcanoes. Mass-movement hazards are anticipated in valleys adjacent to steep volcanoes at the beginning of any eruption activity and commonly lead to evacuations.

A combination of circumstances cause volcanic debris flows, also known as lahars (see Section 4.9). First, explosive volcanic eruptions deposit thick accumulations of loose, unconsolidated volcanic ash and pumice lapilli on steep mountain slopes. Second, eruptions typically destroy extensive areas of slope-stabilizing vegetation. Third, snowmelt generated by lava flows and pyroclastic flows, or heavy rainfall following eruptions, causes rapid erosion and sliding of the loose ash, which mixes with water in stream channels to make huge debris flows. In short, almost all of the ingredients for mass movement are typically present after a volcanic eruption, and effects on human populations can persist for many years.

For example, heavy rain triggered many debris flows during and for years following the June 1991 eruption of Mount Pinatubo in the Philippines (see the beginning of Chapter 4). Over the next three years more than 1.9 cubic kilometers of volcanic ash, pumice, and rock washed and slid from steep hillsides and debris-choked valleys and then was transported mostly as lahars that flowed as far as 50 kilometers from the volcano. To get an idea of the volume of these lahar deposits, imagine debris covering all of Washington, D.C., up to three stories high. The lahars tore out bridges, buried entire cities, and filled in river valleys, causing streams to change course and destroy villages and farm-

Liquefaction: *See how earthquake ground shaking causes liquefaction.*

land. **Figure 15.24** shows some of the damage caused by these events. Similarly, tragic lahars generated from melting snow by an eruption of Nevada del Ruiz destroyed the city of Armero, Colombia, in 1985 (see Section 4.9) and killed 25,000 people.

Many volcanoes collapse in huge rock slides and debris avalanches. An excellent example of this phenomenon is the largest mass movement ever witnessed by humans. When Mount St. Helens, Washington, erupted in 1980, intruding magma swelled the flank of the volcano. The swelling steepened the volcano slope and stressed the rock to the point where it began to crack and fragment. These changes simultaneously reduced the resisting strength of the rock and increased the slope and gravitational driving force for slope instability. The threshold for failure was reached, possibly enhanced by ground shaking from an earthquake caused by magma movement. The result, illustrated in **Figure 15.25**, was a gigantic, 2.8-cubic-kilometer rock slide and debris avalanche that quickly descended the north slope of the volcano and traveled 22 kilometers down an adjacent river valley.

Figure 15.26 summarizes and integrates the processes and triggers that cause hillslopes to fail by mass movements. If something happens to increase slope angle, decrease cohesion, or decrease frictional resistance, then slope stability decreases.

▲ **Figure 15.24** Debris flows following a volcanic eruption. This view of a city street in the Philippines shows the buildings buried almost to the top of the first story by debris-flow deposits. The debris flows, also called lahars, were triggered by heavy rainfall or should be on ash and pumice deposited on the steep slope of Pinatubo volcano by an explosive eruption in 1991.

*Putting It Together–**When Do Mass Movements Occur?***

- Triggers are stimuli that abruptly imbalance driving and resisting forces governing the occurrence of mass movements.
- The most common triggers are addition of water by rainfall or snowmelt, ground shaking during earthquakes, and slope failures during volcanic eruptions.
- Mass movements do not require a specific trigger but instead occur when a threshold is gradually reached between forces that favor or resist slope failure.

15.5 **How Do We Know** ... How to Map Mass-Movement Hazards?

PICTURE THE GOAL
How Do Geologists Make Maps of Areas That Are Prone to Future Mass Movements?

Mass movements are clearly hazardous, and it is desirable to recognize areas of potential risk when making land-use and construction decisions. The goal is to identify where a hazard exists and to depict the extent of the hazard on maps consulted by planners and developers.

The hazard assessment must include a detailed inventory of slope-instability factors across the landscape of interest. These factors are the characteristics that influence slope stability—slope angle, type of rock or regolith, orientation of bedding and foliation planes, extent of breakage by joints and faults, extent of weathering, and so forth. The likelihood of mass movement at any particular location is, however, a complicated combination of all of these factors. To make a map that shows the level of hazard risk in different locations, it is necessary to combine all of these data together in a meaningful way.

To show how the processes works, you will consider the example of a slide-susceptibility map created by the U.S. Geological Survey for a part of San Mateo County south of San Francisco, California. The motivation for the study is rapid urban growth that pushes the boundaries of residential development away from flat valleys and coastal areas and into rural areas with steep hillslopes.

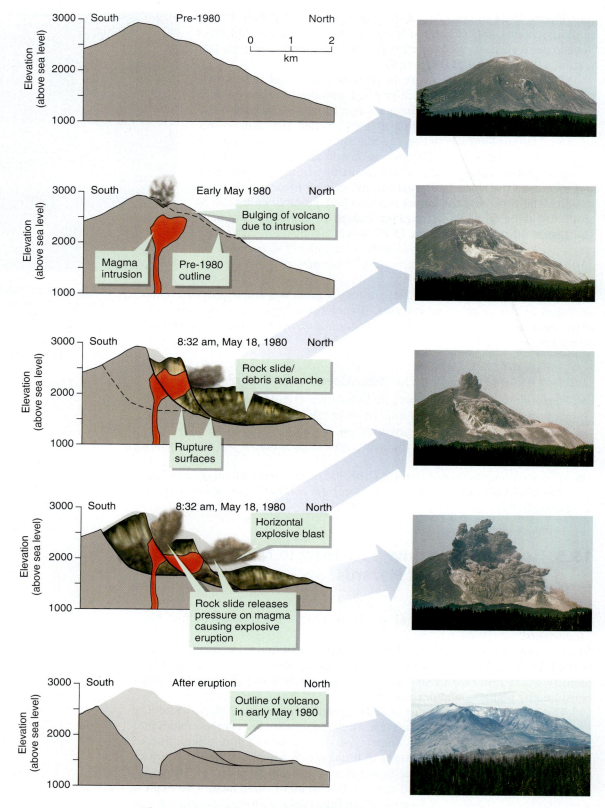

▲ **Figure 15.25** **Debris avalanche from Mount St. Helens.**
The diagram and accompanying photographs show the largest mass movement ever witnessed by humans. The rock slide and debris avalanche decapitated approximately 400 m from the top of the mountain and removed its entire north flank (compare the top and bottom photographs). The magma inside the mountain was simultaneously erupted as pyroclastic debris that not only exploded vertically but also blasted horizontally out of the failed flank of the volcano.

Active Art

Collapse of Mt. St. Helens: *See how the Mt. St. Helens debris avalanche happened.*

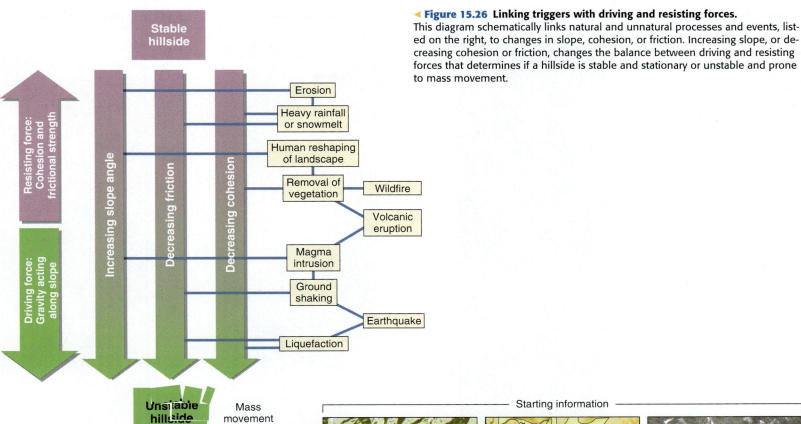

◄ **Figure 15.26 Linking triggers with driving and resisting forces.**
This diagram schematically links natural and unnatural processes and events, listed on the right, to changes in slope, cohesion, or friction. Increasing slope, or decreasing cohesion or friction, changes the balance between driving and resisting forces that determines if a hillside is stable and stationary or unstable and prone to mass movement.

ASSEMBLING THE DATA

What Types of Maps Are Needed?

Figure 15.27 summarizes the data used to develop a hazard map. Judging from the factors that cause unstable hillslopes, the critical information includes steepness of slopes, types of rocks and regolith, and orientation of bedding planes in sedimentary rocks. Distribution of past mass movements is also useful to identify where unstable hillslopes exist.

It would be quite confusing to illustrate all of this information on a single map, so separate maps, illustrated in **Figure 15.28**, are made for each data type. The topographic-elevation map is used to calculate the slope angle everywhere in the area of interest so that areas of very steep and potentially unstable slopes are recognized. A part of the resulting slope map is shown in Figure 15.28a. The geologic map, illustrated in Figure 15.28b, shows the

▶ **Figure 15.27 Flowchart for making a hazard map.**
Slope angles, geologic material properties, and locations of past mass movements are essential data for constructing a hazard map and form three data layers in the geographic information system (GIS). Slope angles are calculated from digital versions of topographic-elevation maps, and past slides are inventoried from aerial photographs. A geologist compares and analyzes the data layers to produce a map that provides a visual guide depicting areas where future hazards are greatest.

▶ **Figure 15.28 GIS data layers used to make a hazard susceptibility map.**
Here are parts of the slope, rock-type, and mass-movement-inventory maps that form the data layers for GIS analysis of hazards in part of San Mateo County, California.

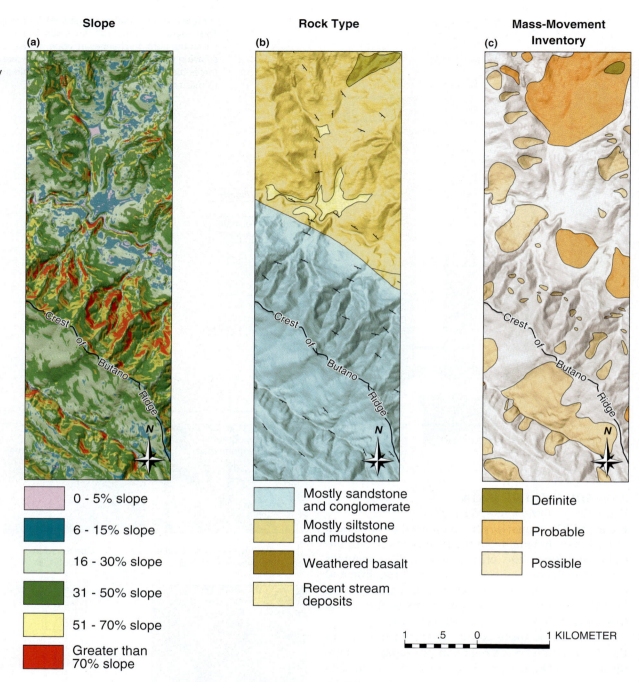

Slope

(a)

- 0 - 5% slope
- 6 - 15% slope
- 16 - 30% slope
- 31 - 50% slope
- 51 - 70% slope
- Greater than 70% slope

Rock Type

(b)

- Mostly sandstone and conglomerate
- Mostly siltstone and mudstone
- Weathered basalt
- Recent stream deposits

Mass-Movement Inventory

(c)

- Definite
- Probable
- Possible

distribution of rock types and also reveals where rocks with relatively low resisting strength are located. Topographic and geologic maps are routinely prepared and are available for many parts of the United States, so these data are typically already on hand.

Another necessary map shows the locations of past landslides because these definitively indicate the locations of unstable slopes. This map typically has to be prepared for the specific study area. The cost and time requirements of having a geologist walk over all of the landscape for a large study area are prohibitive for most hazard-assessment projects. Instead, most of the effort to recognize past mass movements focuses on careful examination of aerial photographs. Slide scarps, talus slopes, irregular ground surfaces underlain by material that has moved downslope, stream valleys blocked with slide debris, and even areas of vegetation disturbed by movement on hillsides are visible on the photographs. Geologists transfer the locations of past movements from the photographs to maps. These characteristics of landscapes affected by mass movement (see Figure 15.6) are most obvious for events that happened recently. Older mass movements are more difficult to recognize after the deposits and scars erode, overgrow with vegetation, or are modified by construction. As a result, there is some uncertainty in the recognition of past mass movements from aerial photographs, so geologists classify the past slides as definite, probable, or possible, with decreasing confidence of recognition, as shown in Figure 15.28c.

APPLYING THE TOOL

How Do Geologists Bring the Data Together?
Geologists must integrate the information on all of these maps along with interpretation of risk in order to determine the mass-movement hazard at any one location. Hazard maps are usually built by using **geographic information systems** (GIS). A GIS is a powerful set of computer-software tools that collects, stores, retrieves, analyzes, and displays data that pertain to particular locations. The GIS manipulates each component of the database in distinct data layers. For instance, each of the three maps of San Mateo County represent a data layer in the GIS (see Figure 15.27).

GIS allows comparison of the information in each database layer at each location. If geologists determine a relationship of mass-movement hazard to slope and rock type, then analysis of the slope and rock-type map layers reveals slide susceptibility at every location. The slide-susceptibility determination forms a new layer in the GIS and is printed out as a separate hazard map (Figure 15.27).

PICTURE THE RESULTS

What Locations Are Most Susceptible to Future Slides?
Figure 15.29 illustrates the interpreted slide-hazard susceptibility for the same area depicted by data in Figure 15.28. How did the geologists interpret different susceptibilities for different locations?

The geologists used three interpretations to determine the varying levels of susceptibility:

1. Areas of possible, probable, or definite mass movement in the past are unstable areas where future movement is most likely. These locations are assigned the highest susceptibility (compare Figures 15.28c and 15.29).

2. Hillslopes underlain by some rock types are more susceptible to failure than other slopes associated with other rocks. The data layer depicting past slides was superimposed on the geologic data layer within the GIS to see if some rock types were more likely than others to be associated with mass movement. The area of each rock type affected by slope failures was estimated using GIS. By surveying all of the area in rural San Mateo County, by just the areas shown in Figures 15.29, about half of the hillsides underlain by mostly clay-rich siltstone and mudstone or weathered basalt have possibly, probably, or definitely slid in the past. In contrast, only about one-third of the area underlain by mostly sandstone and conglomerate is affected by past slides. Therefore, interpreted susceptibility is lower for the areas underlain by sandstone and conglomerate compared to the other rock types (compare Figures 15.28b and 15.29).

3. Steeper slopes are more likely to fail in future slides than are gentler slopes. The slope data layer was superimposed on the past-slide data layer. Not surprisingly, slides are more common on steeper slopes. As a result, within a

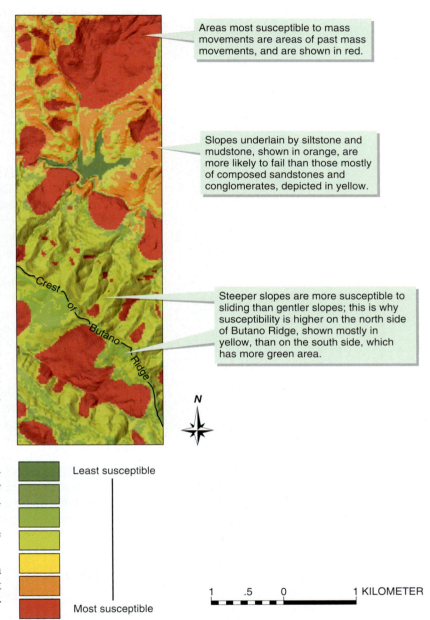

Areas most susceptible to mass movements are areas of past mass movements, and are shown in red.

Slopes underlain by siltstone and mudstone, shown in orange, are more likely to fail than those mostly of composed sandstones and conglomerates, depicted in yellow.

Steeper slopes are more susceptible to sliding than gentler slopes; this is why susceptibility is higher on the north side of Butano Ridge, shown mostly in yellow, than on the south side, which has more green area.

Least susceptible

Most susceptible

1   .5   0   1 KILOMETER

▲ **Figure 15.29 Rock and debris-slide susceptibility map.** The hazard map results from analysis of the three data-layer maps shown in Figure 15.28.

particular rock type, higher slide susceptibility occurs on steeper slopes than on lower slopes. Contrasting the susceptibility of the steep north side of Butano Ridge with the less steep south side best shows this interpretation, even though both are underlain by the same rock type (compare Figures 15.28 and 15.29).

INSIGHTS

How Can These Results Be Understood Scientifically?

The susceptibility map illustrates correlations between phenomena and locations, but scientists also seek to understand why these correlations exist. Comparison of the GIS data layers indicates that hillside slope steepness and type of underlying rock correlate with the number of past slides and the areas impacted by past slides.

Why are there more slides on the steeper slopes? The relation between steeper slopes and a larger driving force for mass movement readily explains this correlation, as illustrated and developed in Figures 15.12 through 15.15.

Why are some rock types associated with more slides? The siltstone and mudstone contain abundant clay minerals. Shrinking and swelling of clay minerals during wetting and drying periods cause these rocks to deteriorate rapidly by physical and chemical weathering. This results in low rock strength and the formation of thick, clayey regolith that readily loses frictional and cohesive strength when wet. The basalt weathers to form similar weak, clay-rich regolith. The sandstone and conglomerate beds do not weather as quickly, and the resulting regolith does not contain as much clay as is found on the siltstone, mudstone, and basalt hillslopes.

> *Putting It Together–How Do We Know . . . How to Map Mass-Movement Hazards?*
>
> - Hazard maps indicate areas susceptible to future mass movements. The hazard maps combine data from other maps that depict relevant geologic information and topographic information.
> - GIS is computer software that is capable of collecting, storing, retrieving, and manipulating a large variety of data sets.
> - GIS can be used to construct hazard maps by comparing and analyzing the assembled geologic and topographic data.

15.6 How Do Mass Movements Sculpt the Landscape?

You now understand why mass movements occur and why they are hazardous, but how important are these processes for determining the appearance of landscapes? Mass movements seem infrequent, at least based on how rarely they are reported in the news media. In reality, mass movements occur widely and frequently on Earth's surface, even though most events are not reported in the news. Mass movements account for most of the rock and regolith removed from mountainous hillslopes, and these movements play an important role in landscape development.

Mass Movements and Mountain Building Go Together

Of all the factors determining slope stability, the most important is slope steepness. Steep slopes, in turn, dominate the landscape of mountains, and slopes are steepest where mountains actively rise in response to ongoing tectonic activity.

Steep mountain slopes results from simultaneous uplift and erosion, as depicted in **Figure 15.30**. Mountain uplift increases land-surface elevation. Streams flow vigorously down the steep slopes between mountains and low-

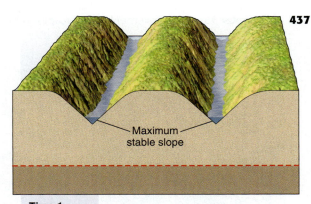

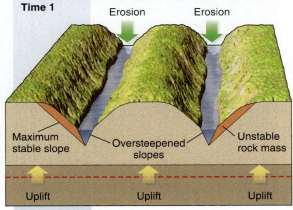

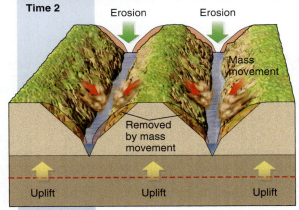

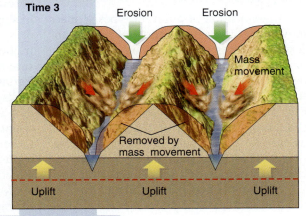

▶ **Figure 15.30 Uplift and mass movement work together.**
Uplift of mountains causes rivers to erode down into ever-deeper valleys. The erosion produces steep, unstable slopes, which promote mass movements. The landslides return the slopes to the maximum stable slope angles and the rivers transport the mass-movement debris to distant lowlands. The river responds to the continuous uplift by more erosion, followed by more mass movements. Both mass movement and river erosion contribute to the changing face of the uplifting mountain landscape, but mass movements lower the elevations more effectively than river erosion.

lands and erode deep canyons into the uplifting rock (canyon carving is further explained in Chapter 16). Stream erosion lowers elevation along valleys, which produces high relief between mountain peaks and valley bottoms. Canyon walls are very steep where rivers erode in hard rock, producing unstable slopes that are prone to mass movement. Slides, falls, and flows lower the relief, but the continued river incision into uplifting rock renews the steep slopes, which leads to more mass movement.

There are other factors besides steep slopes that make actively rising mountain hillslopes unstable:

- Tectonic deformation that drives mountain building pervasively fractures the rock, producing many planes of potential rupture.
- High-magnitude earthquakes are common in areas of active mountain building and trigger huge slides and debris avalanches, like the one illustrated in Figure 15.22.
- Mountain building brings metamorphic rocks to the surface, with closely spaced foliation planes that serve as rupture surfaces. Many metamorphic minerals are susceptible to weathering to clay minerals that further weakens the rock.
- High mountains are regions of heavy rain and snowfall. Abundant moisture contributes to slope instability by lubricating regolith fragments and rupture surfaces and perpetuating more weathering to weaken rock and generate still more loose, unstable regolith.
- Very high mountains experience the cold temperatures necessary for freeze-and-thaw processes that enhance rock-fall occurrences.

Mass Movements Move Mountains

Recent studies reveal that mass movements are the primary agents of regolith movement in mountains. As implied in Figure 15.30, the amount of rock and regolith removed by the incision of a valley by persistent flow of water in a narrow stream channel is small compared to the volume of rock and regolith that episodic falls, slides, and flows bring down the steep valley slopes.

A study in the tectonically active Southern Alps of New Zealand drives this point home. **Figure 15.31** shows the result of a recent rockslide and debris avalanche that lowered the elevation of the country's highest peak by about 10 meters. Comparison of aerial photographs taken 60 years apart showed that more than 7000 mass movements occurred in an area of about 5000 square kilometers within the Southern Alps. For comparison, this area is a little smaller than the state of Delaware. These slope failures affected areas ranging in size from as little as 100 square meters to about 1 square kilometer. Estimated volumes of rock and regolith moved by these events equate to an average lowering of elevation by more than 5 millimeters per year over the entire area. This may seem like a small erosion rate, but if such a rate persists for one million years, which is a short time for mountain building, then mass movement will remove a 5-kilometer thickness of rising crust!

There is little doubt that mass movements determine the steepness of slopes and the maximum heights of mountain peaks in regions of tectonic uplift. As tectonic forces push rocks high above the surrounding Earth surface, gravity works to pull the rocks back down in mass movements.

Uplift and Mass Movement: *See how tectonic uplift, river erosion, and mass movements combine to shape mountain landscapes.*

Mass movement
started here

Path of debris
avalanche

Debris-avalanche
deposit spread out
on top of glacier

◀ **Figure 15.31 Mass movement moves mountain.**
Just after midnight on December 14, 1991, part of New Zealand's highest mountain, Mount Cook (also known as Aoraki), collapsed. The resulting debris avalanche traveled nearly 7.5 kilometers at speeds reaching 200 kilometers per hour. The mountain was shortened by 10 meters.

Putting It Together—How Do Mass Movements Sculpt the Landscape?

■ Mass movements are the primary process of rock and regolith movement in mountainous regions and, therefore, in landscape development.

■ Active mountain uplift increases land-surface elevation and produces high relief. Stream erosion lowers elevation along valleys, promoting very steep and unstable slopes prone to mass movement.

■ In addition to steep slopes, rock fractures, earthquakes, foliation of metamorphic rocks, heavy rainfall, and freeze-thaw weathering all promote mass movements in mountains.

Where Are You and Where Are You Going?

Mass movements are gravity-driven downslope transport of rock and regolith. They occur widely over Earth's entire surface and shape landscapes. Mass movements are a very costly natural hazard, and geologists mitigate their effects by understanding the factors that cause mass movement and recognizing where those factors exist in the landscape.

Classification of different types of mass movements relies on a variety of readily observed features of active mass movements, or their resulting landforms and deposits, or both. The classification emphasizes the nature of the moving material (rock or regolith) and the type of movement (e.g., flow, slide, fall). The speed of the movement is useful for describing mass movements; speed varies from fractions of a meter per year to tens of meters per second.

Imbalances between forces that drive or resist motion cause mass movements. Mass movement occurs when the driving force exceeds the resisting strength of the material. Gravity is the driving force that pulls on rock and regolith in the downslope direction. The magnitude of the force increases with increasing slope angle. A combination of friction and cohesion defines the resisting strength that holds material in place. Friction and cohesion are properties of the material but change in response to human or natural causes so as to change the likelihood of slope failure. Friction, for example, is less on steeper slopes than on gentler ones, and friction diminishes if water exerts pore pressure along potential surfaces of movement. Variations in composition and textures of rocks, structural features of rocks (e.g., bedding and foliation planes, faults, joints), the abundance of water and vegetation (and therefore climate), and slope angles contribute to the stability of slope materials by affecting the magnitudes of the gravity driving force and the frictional and cohesive resisting strength.

Mass movements are the primary process of rock and regolith movement in steep, mountainous terrains. A complex interplay among tectonic uplift of rock, stream erosion of canyons into the rock, and mass movements that reduce the slope angles between mountaintops and canyon bottoms produces the elevation and relief of mountains.

Mass movements are only one of many processes that modify and sculpt the surface of Earth. In the upcoming chapters, you will explore erosional processes, especially processes related to movement of water in both liquid and frozen states. You learned that mass movements tear down mountain sides, but you need to understand the role of flowing streams to generate the relief exploited by slides and falls in mountains, how the streams carry away the mass-movement debris, and where the debris ends up going. In the next chapter you will learn about the significance of streams as sculptors of the landscape, how human activities influence the work of rivers, and how and why rivers affect human lives.

 **Active Art**

Mass Movements. See how mass movements work.

Liquefaction. See how earthquake ground shaking causes liquefaction.

Collapse of Mt. St. Helens. See how the Mt. St. Helens debris avalanche happened.

Uplift and Mass Movement. See how tectonic uplift, river erosion, and mass movements combine to shape mountain landscapes.

Confirm Your Knowledge

1. What is a landslide? What other processes besides landslides shape landscapes?

2. What factors lead to increased human costs due to landslides?

3. Define mass movement.

4. What three factors are used to describe and classify the variety of mass-movement processes?

5. What is the difference between a slide and a slump? What are the characteristic features they leave on the landscape?

6. What is the difference between a rock fall and a rock slide?

7. List and describe four triggers of mass movements.

8. What factors affect slope stability? Explain how each factor relates to slope stability.

9. What is the relationship between slope stability and resisting forces?

10. What is the angle of repose?

11. What combination of circumstances is needed to cause a volcanic debris flow (also called a lahar)?

12. What information is needed to create a mass-movement hazard map? How do you construct a hazard map?

13. What factors make actively rising mountain hillslopes unstable?

Confirm Your Understanding

1. Write out an answer for each question in the Chapter Outline for the chapter sections assigned by your instructor.

2. What factors affect the balance between the driving and resisting forces during mass movements? How?

3. The creation of a mass-movement hazard map will have some uncertainty associated with it. For each type of information needed to create a mass-movement hazard map identify the cause of the uncertainty.

4. What human activities associated with building a house increase the chances of mass movement?

5. A developer hires you to determine the suitability of a steep hillside for the construction of apartment buildings. What types of observations would you make to determine mass-movement risk? Explain how each observation applies to your objective.

Key Terms

angle of repose (p. 426)
cohesion (p. 424)
creep (p. 420)
debris avalanche (p. 421)
fall (p. 416)

flows (p. 419)
friction (p. 424)
geographic information systems (GIS) (p. 435)

mass movement (p. 416)
slides (p. 418)
slump (p. 418)

surface of rupture (p. 418)
talus (p. 416)

16 Streams:
Flowing Water Shapes the Landscape

Why Study Streams?

STREAMS PROVIDE WATER AND DEPOSIT FERTILE SOIL FOR AGRI-culture, and they form pathways for commerce and trade. The trade-off is that streams also flood with costly losses in lives, property, and crops. Streams erode long, narrow, curving valleys and sculpt spectacular scenic landscapes, such as the Grand Canyon and Niagara Falls. Streams carry sediment out of the mountains and deposit it in the lowlands, so they play important roles in lowering continental elevations and in moving weathering products to sites where sedimentary rocks form.

Of all the geologic processes that are active in landscape development, including tectonic uplift, soil formation, mass movement, and erosion by glaciers, waves, and wind, perhaps you are most familiar with flowing streams. Streams are present almost everywhere, and the dynamics of moving water are clearly seen, whereas most of the other processes tend to

The Rio Grande in central New Mexico carries water and sediment from highlands, such as the stream-carved mountains in the background, to the Gulf of Mexico and along the way provides irrigation water to support farming on fertile floodplain soils. ▶

affect smaller areas of Earth, operate at invisibly slow rates, or both.

Streams are an essential natural resource. Americans withdraw approximately one trillion liters of fresh water each day from rivers, streams, and lakes. A little more than half of the water (52 percent) is converted to steam for generating electricity in nuclear, coal-fired, and gas-fired power plants. About a third (32 percent) is used for irrigation and other agricultural and aquacultural uses. Only about 10 percent of this water ends up in public water supplies (for consumption at your home, school, and job site), although surface water does account for 58 percent of public and domestic water supplies (the remainder comes from ground water). Clearly, it is important to understand how much water flows in streams in order to allocate it for these many uses.

Stream channels are extensively studied, despite comprising only a few percent of continental surface area. Large streams, commonly called rivers, are used for commercial navigation, which requires a channel of sufficient depth for boat traffic. Research on natural stream channels enters into the construction of artificial irrigation canals, which are designed so that they neither choke with sediment nor erode their banks and destroy fertile land.

Extensive farmlands and large cities occupy the lowlands adjacent to river channels. These locations are also floodplains, the areas inundated when the water flow exceeds the carrying capacity of the channel. Floods are the costliest natural hazard in the United States, causing an average of 140 fatalities and $5 billion in damages each year. This serious hazard is due to the high concentration of population and agriculture along streams. In the United States, there are about 3800 towns and cities with populations exceeding 2500 inhabitants that are located on floodplains. People have a strong desire to understand why rivers flood so that they can predict floods and minimize their effects.

In this chapter, you will learn how streams work as conveyors of water and weathered rock from land to sea. The specific learning objectives are as follows:

✔ To understand how water and sediment move in channels and across floodplains

✔ To explain the diverse patterns and dimensions of stream channels and how these features relate to dynamic processes

✔ To recognize and explain the origins of landscape features produced by river erosion and deposition

✔ To analyze the causes of floods and the origin of floodplains

✔ To understand how streams change through time in response to natural processes and human activities

✔ To describe how through-flowing streams become naturally obstructed to produce lakes

To achieve the specific goals, the following questions are answered:

16.1 Where does the water come from?
16.2 Where does the sediment come from?
16.3 How do streams pick up sediment?
16.4 How do streams transport sediment?
16.5 Why do streams deposit sediment?
16.6 Why does a stream change along its course?
16.7 What factors determine the channel pattern?
16.8 How does a floodplain form?
16.9 Why do streams flood?
16.10 *How do we know* … the extent of the "100-year flood?"
16.11 How do human activities affect streams?
16.12 How do stream-formed landscapes change through geologic time?
16.13 How do lakes form?

In the FIELD

Imagine yourself in the scene depicted in **Figure 16.1**a, alongside a roaring river in the wilds of Alaska. You walk along the riverbank on smoothly polished, round cobbles and boulders and gaze out over fast-moving water. The opposite bank is more than a hundred meters away, although small islands of bare gravel interrupt the flow in many places. There is no soil or vegetation on these midstream gravel accumulations, so these must be temporary parts of the river landscape that submerge when the river is deeper. Why does the river carry more during some periods than at others?

The stream carries more than water. You hear sounds comparable to loudly grinding teeth and realize that these are the noisy collisions of rolling and bouncing of cobbles below the water surface. The rounded nature of the cobbles along the bank is testimony to the ability of these collisions to grind and smooth sediment grains while they move with the water. You look into the river to watch the moving cobbles, but all you see is muddy, murky water. A great abundance of small sediment particles suspended in the water obscures the rolling cobbles. What determines the sizes of sediment carried by the stream, and how does the stream pick up the sediment to begin with?

A natural bench rises above the margin of the valley and provides a scenic vantage point (Figure 16.1a). You observe that the bench is underlain by rounded gravel like that alongside and beneath the

Figure 16.1 Field visits to see dynamic rivers. ▶

A. Field notes, River in Alaska

Benches of rounded stream gravel occur along valley margin.

Stream channel divides and rejoins around islands of gravel.

Other channels are dry now, but presumably carry water when river is higher. Gravel islands probably submerge at higher flows, too, because they lack vegetation or soil.

River gravel is well-rounded. I can hear gravel moving in water but I cannot see the moving cobbles because the water is very murky with suspended fine-grained sediment.

B. Field notes, flight over Mississippi River, Louisiana

Water is deep enough for tugboats and barges to navigate the river.

Artificial embankments protect communities and farmland from floods.

The Mississippi River is very wide and curvy without islands of sediment.

flowing stream. Thick soil and plants cover the bench, indicating that it is far above where the river now flows, even in flood. This bench must, therefore, record the elevation occupied by the riverbed at an earlier time in its history. Why do rivers incise (cut downward) into the landscape? Do all rivers only incise deeply into Earth's surface, or are some riverbeds at stationary elevations or even rising as sediment fills up the channels?

On another occasion, you fly above the Mississippi River in Louisiana, taking in the views seen in Figure 16.1b. What a different scene from the river in Alaska. The Mississippi is nearly a kilometer across with no islands sticking up within the channel. Judging from the size of tugboats and barges navigating the river, the Mississippi seems much deeper and wider than the Alaskan stream. Although the river banks are far below you, it is clear that the sediment along the banks is much finer grained than that seen along the Alaskan river and is likely all fine sand and mud. Why are the dimensions and sediment characteristics of the two river channels so different?

From your vantage point above the Mississippi River you admire the long ribbon of water that bends back and forth in wide, horseshoe-shaped curves. Why is the channel not straighter, like the river in Alaska? Cultivated farmland extends along the river as far as you can see, interrupted here and there by towns and cities. Long, clearly artificial ridges of dirt and gravel follow alongside the curving course of the river channel. These are levees constructed to contain the water in the channel that would otherwise flood out over the farmland and towns. This powerful river provides commercial shipping and fertile farmland, but its value may be offset by the threat of floods, which you occasionally learn about in the local and national news. How do river processes affect human activities, and how do human activities affect rivers?

16.1 Where Does the Water Come From?

The field observations demonstrate that **streams** are flowing water in channels that drain water from land and simultaneously transport sediment. Streams are complex features governed by the physics of flowing water, the ease of erosion of geologic materials forming the banks, and tectonic forces that deform the landscape along the stream course. Worldwide, streams deliver about 700,000 cubic meters of water to the ocean each second. For comparison, this volume of water would fill 374 Olympic-size swimming pools. To learn about streams, there is no more important place to begin than to consider where the water comes from and what factors determine how much water flows through the channel.

The Hydrologic Cycle

Water falls from the atmosphere as rain or snow and evaporates back into the atmosphere. Careful observation shows, however, that water follows many paths on and below Earth's surface rather than just shuffling back and forth between ground and sky.

Figure 16.2 illustrates the **hydrologic cycle**, a concept describing the movement of liquid water and water vapor through all parts of the Earth system. Stream water originates from precipitation (rainfall and snow). Precipitation results from the condensation of atmospheric water vapor into water droplets and ice crystals. One important cause of condensation and precipitation is high continental elevations that force moist air to rise upward. The rising air cools, which causes the water vapor to condense. This process explains why precipitation is generally greatest in mountainous regions and along boundaries between lowlands and higher elevations.

Not all of the precipitation that reaches the land surface drains into streams. Some of the water soaks into the ground and follows different routes in the hydrologic cycle. Plants quickly take up some fraction of the infiltrated water. Some soil moisture returns to the atmosphere by evaporation or **transpiration**, the process in which plant leaves and stems release water vapor. The remaining infiltrated water becomes **ground water**, which is simply the water present below Earth's surface in pore spaces and fractures within regolith and rock. The **water table** is the undulating boundary between unsaturated regolith or rock, above, and saturated regolith and rock, below. In the saturated zone, all

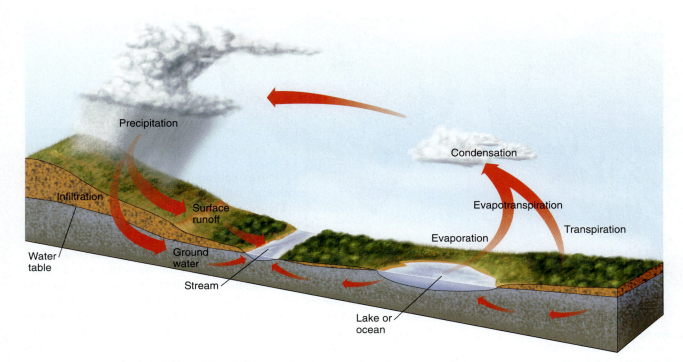

▲ **Figure 16.2** **Visualize the hydrologic cycle.**
Water moves through the Earth system along pathways in the hydrologic cycle. Evaporation and transpiration (or combined as evapotranspiration) transport water vapor into the atmosphere from Earth's surface. Water vapor condenses into liquid or frozen water that falls as precipitation. Precipitation either runs off the surface or infiltrates into the ground. Some ground water seeps back to the surface. Streams transport surface water within the hydrologic cycle.

pores and fractures are filled with water, whereas in the unsaturated zone, these open spaces are partly filled with water and partly filled with air.

Ground-water flow, covered in Chapter 17, and stream flow are connected (Figure 16.2). Most ground water eventually emerges back at the surface, especially in low-elevation valleys. This means that the total water flow in a stream is the surface runoff of rainfall and snow melt plus whatever infiltrated ground water seeps into the channel. If streams only contained surface runoff, then they would only flow when there was rainfall or melting snow to feed them. While this is true of many small streams, all large rivers and other smaller streams flow almost all of the time. Persistent flow in streams results from ground water entering the stream, and this only happens if the bottom of the stream, its bed, is lower than the water table (Figure 16.2).

Drainage Basin: The Area where the Water Comes From

How can you describe the size of a stream? There are several possibilities. The length of the river from its **headwaters**, where it begins, to its **mouth**, where it ends, is one measure of size. The measure of the width, depth, or both at different locations along the stream can also give an indication of size. River length, along with channel width and depth, figures into two other ways to measure streams—drainage basin area and discharge.

The **drainage basin**, schematically illustrated in **Figure 16.3**, describes the area from which a stream gathers water. The drainage basin of one stream is separated from adjacent drainage basins by higher ridges called **divides**. The stream headwaters are at high elevations close to a divide, whereas the mouth is the lowest elevation in the drainage basin where the stream enters another stream, a lake, or the ocean. Smaller **tributary** streams flow into larger streams. This means that the drainage basin of a large river encloses the drainage basins of all of its tributaries (Figure 16.3b).

Most rivers funnel water and sediment to the ocean. Fewer streams flow into topographically low areas surrounded by divides, in which case the river terminates in a lake, rather than at the ocean. In arid regions, the lake may be short lived because evaporation and infiltration through the lakebed occur faster than runoff delivers water to the lake. In this case, the lake dries up and

Hydrologic Cycle: *See how liquid water and water vapor move through the Earth system.*

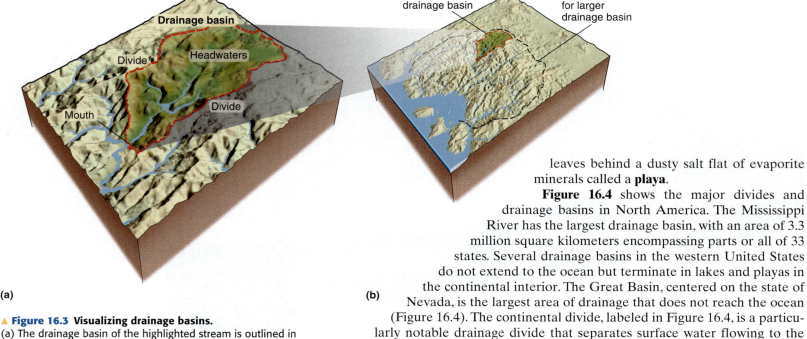

(a)

(b)

▲ **Figure 16.3 Visualizing drainage basins.**
(a) The drainage basin of the highlighted stream is outlined in red. Ridges outline the divides that separate neighboring drainage basins. All surface runoff from precipitation that falls within one drainage basin flows into the same stream. (b) The drainage basin of a large stream encloses the drainage basins of all of its tributary streams.

leaves behind a dusty salt flat of evaporite minerals called a **playa**.

Figure 16.4 shows the major divides and drainage basins in North America. The Mississippi River has the largest drainage basin, with an area of 3.3 million square kilometers encompassing parts or all of 33 states. Several drainage basins in the western United States do not extend to the ocean but terminate in lakes and playas in the continental interior. The Great Basin, centered on the state of Nevada, is the largest area of drainage that does not reach the ocean (Figure 16.4). The continental divide, labeled in Figure 16.4, is a particularly notable drainage divide that separates surface water flowing to the Atlantic and Arctic Oceans from that flowing to the Pacific Ocean.

Discharge: Amount of Water Flowing in a Stream

Another way to describe stream size is to measure how much water flows through the channel. Think back to the field observations and imagine holding a rope across the channel and then measuring the volume of water passing under that rope in each second. You are measuring the volume of water per interval of time (e.g., cubic meters per second) that passes through that part of the channel. This measurement is the **discharge**. The discharge in the Alaskan river would be much smaller than the Mississippi River observed in your field outings.

Figure 16.5 shows how to calculate the discharge by multiplying measurements of the flow velocity and the cross-section area of the flowing water. It is impractical to measure discharge everywhere along a stream. Instead, discharge is determined by measurements of water level at **stream gages** located along parts of the stream where the channel cross section is carefully surveyed. Data collected over many years permit estimation of discharge simply by knowing how deep the water is in the channel.

◀ **Figure 16.4 Outlining the drainage basins of North America.**
The orange lines on this map are the large-scale drainage divides that separate streams draining to different oceans surrounding North America. The Continental Divide separates streams flowing westward toward the Pacific Ocean from those flowing eastward and northeastward to the Atlantic and Arctic Oceans. The Great Basin is an area of internal drainage, where runoff does not drain to the ocean but accumulates and evaporates in shallow lake basins.

Geologist measuring water
depth and flow velocity

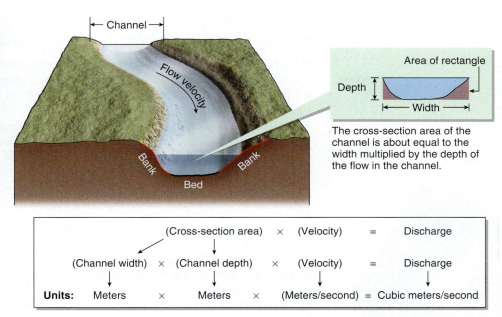

The cross-section area of the
channel is about equal to the
width multiplied by the depth of
the flow in the channel.

	(Cross-section area)	×	(Velocity)	=	Discharge	
(Channel width)	× (Channel depth)	×	(Velocity)	=	Discharge	
Units: Meters	× Meters	×	(Meters/second)	=	Cubic meters/second	

◄ **Figure 16.5 Visualize the discharge of a stream.** Discharge is the amount of water flowing in a stream. It is defined as the volume of water passing through a particular cross section of channel in an interval of time. Discharge is calculated by multiplying the cross-section area of the flow and the average flow velocity. The cross-section area of flow is the width multiplied by the depth for a rectangular channel. Real channels do not have perfectly rectangular cross sections, so careful surveys of the channel bed and banks are needed to calculate the area. The average velocity is obtained by making many measurements of velocity across the channel, because the velocity varies from place to place.

Discharge changes downstream within a drainage basin. There is very little discharge in a stream near its headwaters, but the stream gradually gains water downstream by runoff from increasingly larger areas and contributions from tributaries. Therefore, the discharge of a river usually increases downstream. Exceptions to this general rule are streams in dry regions, where stream water soaks through the streambed to become ground water. This causes the discharge to decrease downstream.

The discharge at one location along a stream varies with time. Discharge increases when runoff is high because of heavy rain or rapidly melting snow. Most of the time, however, the water does not fill a channel to the top of its banks. Stream-gage data show that the discharge necessary to fill the channel, called the bankfull discharge, usually occurs about once every two years. A **flood** occurs when the banks of the stream can no longer contain the discharge.

EXTENSION MODULE 16.1
How a Stream Gage Works.
Learn how a stream gage is constructed and how hydrologists use the gage data to determine discharge.

Putting It Together–*Where Does the Water Come From?*

■ Stream water is surface runoff from rainfall and snowmelt plus infiltrated ground water that reemerges at the surface where the water table intersects stream channels.

■ The drainage basin is the area from which water flows to a stream and divides separate adjacent drainage basins.

■ Discharge is the volume of water that passes through a cross section of a stream channel during an interval of time.

16.2 Where Does the Sediment Come From?

When geologists examine the rock and mineral fragments transported by streams, they find that the composition of the sediment corresponds to the types of rocks and soils observed in the drainage basins. Streams erode, transport, and deposit sediment liberated from rocks by weathering and mass movement. By these processes, high mountains gradually wear down and the resulting sediment accumulates in river lowlands or ocean basins. The rivers examined in the field (Figure 16.1) are as notable for the sediment they carry as for the amount of water flowing in the channel. It is worthwhile, therefore, to understand the origin of this sediment.

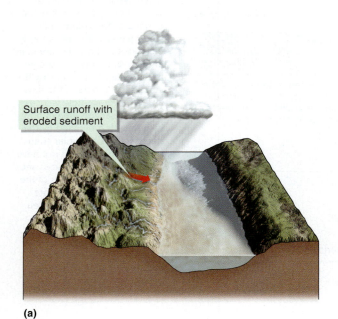

Surface runoff with eroded sediment

(a)

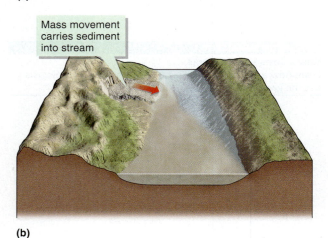

Mass movement carries sediment into stream

(b)

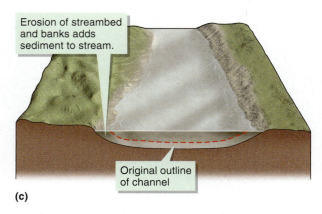

Erosion of streambed and banks adds sediment to stream.

Original outline of channel

(c)

▲ **Figure 16.6 Three ways to get sediment into flowing stream water.**
(a) Surface runoff on hillslopes above the stream washes sediment into the channel. (b) Mass movements on hillslopes carry rock and regolith into the stream channel. (c) Streams erode their beds and banks.

Sources of Sediment in Streams

Figure 16.6 shows these observations of how sediment gets to streams:

- Surface runoff picks up loose sediment and washes it into the stream through rills and gullies described in Section 14.7 (see Figure 14.23).
- Mass movements deliver regolith and rock from hillslopes directly to the stream (see Figure 15.30).
- Streams pick up sediment where they erode horizontally into the banks when the channel widens or shifts position. Streams also pick up sediment when they erode vertically downward through their beds.

How Much Sediment Do Streams Carry?

Measurements show that each year, the world's rivers transport about 14 cubic kilometers of weathered material to the oceans as either solid particles or dissolved ions. This annual volume of sediment would fill a train that encircles Earth 34 times at the equator. If spread equally over Earth's land surfaces, this volume of sediment represents 8 millimeters of elevation–lowering of the continents every century.

The particles carried by a stream constitute its **sediment load**. Streams with large drainage basins and discharges, such as the Mississippi River, carry hundreds of millions of metric tons of sediment to their mouths each year. Not all of the sediment eroded from upland areas immediately flushes out to the river mouth. This is because streams not only erode and transport sediment, they also store it to form the gravel bars and floodplains observed in the field.

To determine if some rivers transport more sediment than other rivers of similar size, it is convenient to divide the total mass of sediment transported by the area of the drainage basin. On this basis, the sediment load of major rivers graphed in **Figure 16.7** varies from only about 10 metric tons per square kilometer per year to more than 1700 metric tons per square kilometer per year. How can geologists explain these differences? Most of the extremely sediment-laden rivers are in Southeast Asia and drain the steep slopes of the Himalayas. Are these rivers carrying more sediment because steep slopes erode faster? Then again, Southeast Asia has a wet climate with a profound wet season of heavy rainfall. Does the greater precipitation increase weathering and discharge to cause the greater sediment delivery to the ocean?

Geologic data show that relief *and* climate determine the sediment load of streams. Sediment loads are largest in areas of steep relief, because river slopes are steeper and steep valley walls are more prone to mass movement (see Figure 15.30). In a general way, sediment load also increases with increasing precipitation, although the relationship is a little complicated. Higher rainfall promotes more vegetation, plants slow down surface runoff, and roots bind soil particles together. Vegetation, therefore, diminishes the erosive consequences of runoff from precipitation. Vegetation is sparse in desert regions, but there is also less rainfall to erode the sediment. More intense weathering in humid climates also generates more erodible regolith than will form during the same time in a dry climate. As a result of these factors, the amount of sediment delivered to streams from hillslopes depends on the relationship among the amount of sediment produced by weathering, the amount of precipitation to generate runoff to erode the sediment, and the effectiveness of vegetation to reduce erosion.

Dissolved Load—The Hidden Load of Streams

Weathering reactions generate ions in solution (Section 5.1). Rivers, therefore, transport ions as **dissolved load** in addition to the sediment load. Figure 16.7 shows that the dissolved load of rivers is less than the sediment load but is still substantial, especially in drainage basins that experience warm, wet climates and contain easily weathered rocks, like limestone.

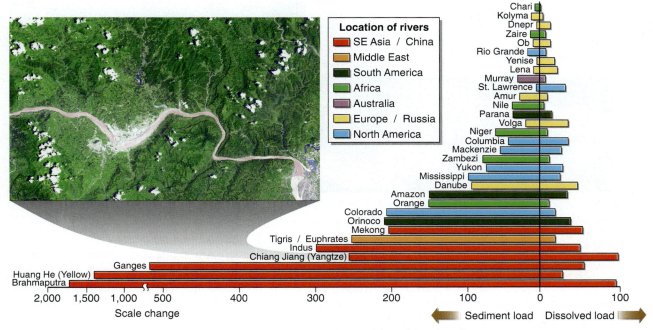

The bar chart shows:

Location of rivers
- SE Asia / China
- Middle East
- South America
- Africa
- Australia
- Europe / Russia
- North America

Rivers (top to bottom): Chari, Kolyma, Dnepr, Zaire, Ob, Rio Grande, Yenise, Lena, Murray, St. Lawrence, Amur, Nile, Parana, Volga, Niger, Columbia, Mackenzie, Zambezi, Yukon, Mississippi, Danube, Amazon, Orange, Colorado, Orinoco, Mekong, Tigris / Euphrates, Indus, Chiang Jiang (Yangtze), Ganges, Huang He (Yellow), Brahmaputra

Scale (sediment load, left): 2,000 1,500 1,000 500 | 400 300 200 100 0

Scale change

Dissolved load (right): 0 100

← Sediment load   Dissolved load →

Metric tons per square kilometer of drainage basin per year

▲ **Figure 16.7 Rivers transport weathering products.**
This bar graph illustrates the average sediment and dissolved loads for the world's major rivers. Dividing the total mass of sediment particles (sediment load) and dissolved ions (dissolved load) by the drainage basin area helps to compare rivers with different-sized drainage basins. There are large variations in the load carried by different rivers, with the largest masses of sediment carried in rivers in Southeast Asia and China. The inset photo shows the muddy Chiang Jiang River in China, as seen from the Terra Earth Observing System satellite. Dissolved load is smaller than sediment load for most rivers, with notable exceptions such as the St. Lawrence River, where weathering of limestone produces large masses of dissolved ions in comparison to particulate sediment load.

Putting It Together–Where Does the Sediment Come From?

- Streams transport the products of rock weathering, which include particles comprising the sediment load, and dissolved ions constituting the dissolved load.
- Surface runoff from hillslopes, mass movements from hillslopes, and erosion of channel bed and banks deliver sediment to flowing streams.
- Streams with larger drainage basins and higher discharges typically carry more sediment. The sediment load also is greater in regions with steeper relief and varies with the amount of precipitation and vegetation.

16.3 How Do Streams Pick Up Sediment?

A stream must pick up sediment grains before carrying them downstream. Stream channels are evidence of the erosive power of water to carve through regolith or bedrock. Why do streams erode sediment? Picking up and moving sediment grains requires work and expends energy. Stream energy mostly results from the fact that stream water flows downhill. When water enters a stream channel, it possesses the potential energy of that elevation in the landscape (Section 1.6 and Figure 1.16 explain potential and motion energy). The potential energy converts to motion energy as the water flows downslope. The motion energy is sufficient for the water to do work—it picks up and moves sediment.

Streams have energy to move sediment but how do streams actually erode their beds? How big are the sediment particles that a stream can pick up and move? To answer these questions, it is first necessary to consider, using the examples in **Figure 16.8**, that rivers either flow over loose sediment or solid rock. The loose sediment deposited by a stream is **alluvium**, and rivers that flow in alluvium are **alluvial streams**. In contrast, **bedrock streams** flow through channels cut into solid rock.

How Alluvial Streams Pick Up Loose Sediment

It might seem logical to think that faster water carries larger particles, so a stationary particle eventually moves if the flow velocity increases enough. Velocity,

▶ **Figure 16.8 Contrasting alluvial and bedrock streams.**
(a) The Arkansas River in Oklahoma is an alluvial stream flowing in a valley of stream–deposited alluvium. (b) Snow-bird Creek, in North Carolina, is a bedrock stream, flowing directly on rock.

however, does not really explain how streams erode and transport sediment. For example, alluvium is almost always much coarser grained near the headwaters than it is downstream at the mouth of the stream, despite the fact that measured flow velocity typically *increases* downstream. This implies that faster flows only transport finer sediment and leaves us looking for another explanation for why streams pick up sediment.

Force, not velocity, must be exerted to move any object. Flowing water exerts a downstream force parallel to the streambed. This force is a shear stress that works in a manner similar to shear stress applied parallel to a fault plane (see Section 11.3 for more about shear stress).

The shear stress exerted on a streambed by flowing water depends on the weight of the water and the steepness of the streambed. The greater the water depth, the greater the weight of water bearing down on the particles on the bed. Increasing the water depth (weight), the streambed slope, or both increases the shear stress.

Figure 16.9 illustrates how shear stress moves sediment. The concept is similar to evaluating the driving and resisting forces of mass movements in Chapter 15. Shear stress is a measure of the gravity driving force. The forces resisting movement are the weight of the particle plus friction and cohesion with its neighbors. Initially, stationary particles usually start to move when shear stress increases because of increasing water depth when discharge increases. Larger grains require steeper slopes to move than is true of smaller grains, because shear stress is greater on steeper slopes.

Smaller sediment grains weigh less than larger grains, so you can initially assume that small grains erode at lower shear stresses (shallower water or gentler slopes) than large grains. **Figure 16.10** shows that this is only partly true. Silt- and clay-size particles include cohesive clay minerals (see Sections 5.4 and 15.2), so relatively high shear stress is required to overcome the cohesion and erode these small grains. Once the grains move, however, they are so small and lightweight that streams easily transport silt and clay.

The overall conclusion reached from these observations is that loose sediment grains erode from the streambed when the shear stress of the flowing water is sufficient to do so. For example, gravel first moves at a higher shear stress than is required to move sand. For a stream to move gravel it must be deeper, or steeper, or both, than a stream that moves

▼ **Figure 16.9 What it takes to pick up and move alluvial sediment grains.**

A sediment grain is held down on the streambed by its weight, which is the downward pull of gravity. Flowing water, also driven by gravity, exerts a shear stress on the grain surface. For motion to occur the shear stress must exceed the weight of the grain, plus friction and cohesion at grain contacts.

Flowing stream

Shear stress of flowing water is force that drives motion

Weight of grain is a downward force that resists motion

Friction and cohesion at grain contact resists motion

(a)

Shear stress increases with increasing water depth, and eventually the stress is sufficient to move the grain.

(b) Increasing water depth = increasing shear stress

Shear stress increases with increasing channel slope, so sediment grains move most easily on steep slopes.

(c) Increasing slope = increasing shear stress

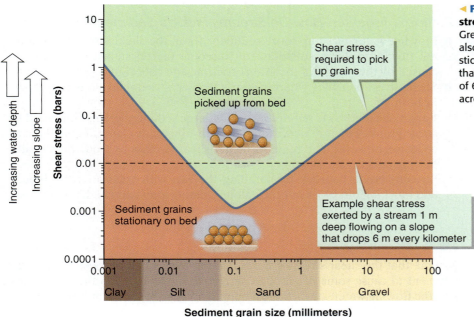

◄ **Figure 16.10 Different grain sizes move at different shear stresses.**
Greater shear stress is required to pick up gravel than sand. Silt and clay also require large shear stresses because the grains are cohesive and stick to one another. The horizontal line drawn on the graph illustrates that shear stress exerted by a 1-meter-deep stream flowing on a slope of 6/1000 is sufficient to erode grains between 0.02 mm and 1 mm across.

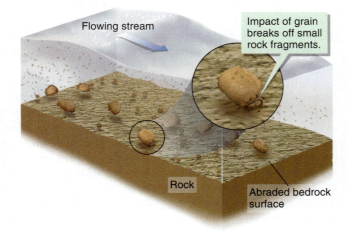

sand but does not move gravel. Cohesive clay and silt grains require relatively higher shear stresses to get picked up despite their small size.

How Bedrock Streams Break Off Pieces of Rock

Some streams flow in the bottoms of bedrock canyons that are more than 1000 meters deep. The Grand Canyon, pictured in Figure 1.2b, is a good example. Mass movements from the steep canyon walls widen the valley and bring particles to the stream, which then move the particles as alluvium. Mass movements, however, do not deepen stream channels. In order to cut a deep canyon, the stream must break up solid rock into moveable pieces.

Figures 16.11 and **16.12** illustrate two observable processes that dislodge rock fragments at the bottom of a bedrock stream:

1. Abrasion removes rock fragments from a solid rock surface by the hammer-like impact of sediment grains carried in the water.
2. Plucking gradually pries loose blocks of rock along preexisting joints and bedding planes.

▲ **Figure 16.11 Streams erode bedrock channels by abrasion.**
Loose sediment grains carried by a stream act like sandpaper to break off and grind down the underlying bedrock. The photo shows scoured and polished bedrock exposed along the Indus River in Pakistan. This outcrop is submerged and abraded when the river flows with higher discharges during floods.

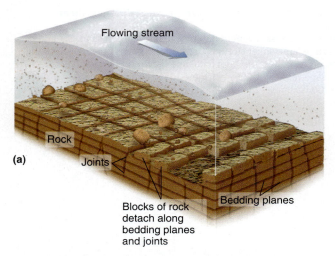

(a)

Flowing stream

Rock

Joints

Blocks of rock detach along bedding planes and joints

Bedding planes

1995

These rocks, broken loose along bedding planes and joints, are gone a year later.

1996

(b)

▲ **Figure 16.12 Streams erode bedrock channels by plucking.** The shear stress of the flowing water along with the pressure of water and sediment grains forced into cracks pluck bedrock apart along bedding planes and joints. The photos, taken 6 years apart at the same spot along a stream in Ontario, Canada, show how sedimentary rocks break loose along joints and bedding planes. The resulting rock fragments are carried away by the stream.

Putting It Together–How Do Streams Pick Up Sediment?

■ Flowing water picks up loose sediment when the shear stress exerted by the flowing water is sufficient to overcome the resistance to movement caused by the weight of the grains, friction, and cohesion.

■ Shear stress is higher where water is deeper, streambed slope is steeper, or both.

■ Larger shear stress is required to erode large sediment grains compared to small grains. Fine-grained sediment with abundant clay minerals is very cohesive, however, and requires higher shear stresses to erode than would be expected for the small grain size.

■ Streams erode bedrock by abrading the bed with transported sediment and plucking out blocks of rock along existing joints and bedding planes.

16.4 How Do Streams Transport Sediment?

Assume that the shear stress of water flowing over a bed of loose fragments is sufficient to cause some or all of the grains to move. How do the grains move with the flowing water, and is there a limit to how much sediment is transported?

Sediment Moves as Bedload and Suspended Load

From the banks of the Alaskan river in Figure 16.1a, cobbles and boulders banging into one another beneath the water surface made noises, but the moving gravel was not visible because the water was too murky to see through. A better way to understand how sediment moves in flowing water is to make laboratory observations of sediment and water moving through transparent glass-walled channels. **Figure 16.13** summarizes the observations from laboratory and natural channels. Moving sediment occupies two positions in the stream—on or near the bed, and suspended in the water.

Large grains roll, slide, and bounce along the bed and spend most of their time in contact with the bed. These particles comprise the **bedload** of the stream. The flow may mold bedload particles into submerged dunes and ripples, whose movement along the riverbed cause cross-bedding (look back to Section 5.6 and Figure 5.16 for more about dunes, ripples, and cross-bedding).

Geologists observe that all grain sizes present on the streambed typically move when the stream fills to the top of its banks. Otherwise, shear stress is low when discharge is low, because water is shallow at low discharge. Under these less–than–bankfull conditions, only part of the sediment moves and the rest remains stationary. Some sediment **bars** may be exposed above the water level and only move when the flow is deeper. The unvegetated islands viewed in the Alaska field site are bars (Figure 16.1a) and are also visible in the river pictured on the opening page of this chapter and in Figure 16.8a. The fact that bars lack vegetation and soil development indicates that they are not stable land surfaces but are submerged and moved frequently by high-discharge flows.

Small grains comprise the **suspended load**, where sediment mixes with the flowing water and is transported above the bed, which it rarely touches. The small clay and silt grains suspended in the water give the muddy appearance to many streams and may obscure the view of bedload transport along the streambed. The small suspended particles are denser than water, but swirling currents and eddies continuously stir the particles up and away from the bed so that they are not deposited.

Stream Power Puts Limits on Sediment Transport

Just because a stream exerts sufficient shear stress to move sediment of a particular size does not necessarily mean that dislodged grains travel very far. The flowing water has to do work and expend energy to erode and transport sediment. The ability of a stream to do work is the **stream power**. There are different ways to measure stream power, but a simple way is to multiply the shear stress and the average flow velocity. This calculation tells you the power of the stream to do work on and above a square meter of the streambed.

Stream power depends on both shear stress and flow velocity because shear stress determines the ability of the flow to place particles in motion, and velocity determines how quickly the particles move. The larger the volume of sediment to be transported, the more stream power is required to do the job. If the stream power along a stretch of the stream is just sufficient to transport the sediment carried in from upstream, then all of the power is consumed to continue moving that sediment, and no additional sediment erodes from the streambed. If there is excess stream power, then more sediment erodes and is transported. If the stream power is insufficient to carry what has come from upstream, then some sediment is deposited until the sediment load matches the power available to move it.

Putting It Together–*How Do Streams Transport Sediment?*

■ Sediment that rolls, slides, and bounces along the bed of the stream is bedload, whereas finer sediment intimately mixed with the water flowing above the bed is suspended load.

■ Bars are mounds of sediment that are stationary and exposed at low discharge but are submerged and transported at high discharge.

■ Stream power, which is the product of shear stress and flow velocity, describes the ability of a stream to do work. If the stream power is just appropriate for transporting the sediment delivered from upstream, then there will be no erosion or deposition.

▲ **Figure 16.13** **Visualizing sediment moving as bedload and suspended load.**
Coarse grains roll, slide, and bounce along the bed as bedload. Fine grains remain perpetually suspended by eddies in the stream current and completely mix with the flowing water. The photo shows how geologists study bedload and suspended load transport by observing artificial streams through a glass-walled channel.

16.5 Why Do Streams Deposit Sediment?

What conditions determine where a stream deposits sediment? Based on the preceding discussion, a stream must lose its ability, or power, to move sediment when it switches from erosion and transport to deposition. Stream power is the product of shear stress and velocity; so for stream power to decrease, the shear stress, velocity, or both must also decrease. You can predict locations of sediment deposition from knowing the factors that change shear stress and velocity.

Deposition Occurs when Discharge Decreases

Erosion and deposition alternate over time along a stream because discharge also varies. Consider the example of a brief increase and then decrease in stream flow following a heavy rain. When the discharge decreases, the velocity, width, and depth of flow also decrease. Decrease in water depth also decreases shear stress (Figure 16.9). Stream power also decreases because velocity and shear stress decrease, and this causes deposition of the sediment that eroded when the discharge was high.

How Streams Move Sediment: *See how flowing water moves the sediment load.*

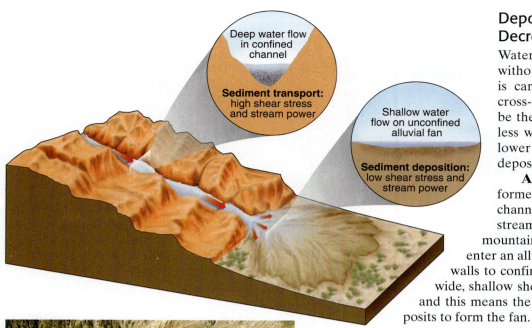

Deposition Occurs where Water Depth Decreases

Water depth, and therefore shear stress, can decrease without a change in discharge if the same discharge is carried in a wider but shallower channel. The cross-sectional area of flow and the velocity could all be the same but the decrease in water depth means less water weight on the bed, which translates to a lower shear stress, lower stream power, and sediment deposition.

Alluvial fans are fan-shaped masses of alluvium formed where flow abruptly changes from a confined channel to being unconfined. **Figure 16.14** shows how streams flow through narrow bedrock channels in a mountain range, and then form alluvial fans where they enter an alluvial valley where there are no hard-rock valley walls to confine the flow. The flow expands horizontally in a wide, shallow sheet. The shallower flow exerts less shear stress, and this means the stream power drops, so sediment abruptly deposits to form the fan.

Deposition Occurs where Slope Decreases

Shear stress is very sensitive to slope (see Figure 16.9). If the channel slope decreases without a proportional increase in water depth, then the shear stress also decreases. The slope of most stream channels decreases downstream while depth increases, as will be examined more closely in Section 16.6. The slope decrease is more substantial than the depth increase, so shear stress usually decreases downstream. The downstream decrease in shear stress means that it is increasingly difficult for the stream to erode the largest particles on the bed, so they are left behind while finer sediment continues downstream.

Deposition Occurs where Velocity Decreases

If flow velocity decreases, then stream power also decreases, which diminishes the sediment-carrying capacity of a stream. This effect is seen where streams flow into large, relatively still bodies of water such as the ocean, lakes, or artificial reservoirs. Where a river enters still water, the current velocity drops to almost nothing, and this causes water to back up at the mouth of the channel. The decrease in stream velocity where the stream approaches and enters a standing body of water causes deposition of sediment.

A **delta** is the landform produced by deposition of sediment where a stream enters a lake, reservoir, or sea. The Mississippi River forms the largest delta in North America. This delta, shown in **Figure 16.15**, covers about 28,600 square kilometers, which by comparison is a little larger than the area of Maryland. Figure 16.15 shows how a delta forms where velocity slows at the mouth of a river. Deposition of the coarsest bedload clogs the channel and causes the channel to split into multiple smaller **distributary channels** that separate around bars and vegetated islands of river-deposited sediment. The bedload accumulates along the coastline and is redistributed by waves and ocean currents. The suspended load continues out to sea, where it gradually settles to the seafloor.

▲ **Figure 16.14 How alluvial fans form.**
Alluvial fans form where water flowing in a confined channel abruptly spreads out on an unconfined valley floor. The abrupt increase in the flow width causes an equally sudden decrease in the flow depth, which decreases shear stress and stream power to cause deposition. The photograph shows an alluvial fan at the mouth of a bedrock canyon in Death Valley National Park, California.

Putting It Together–**Why Do Streams Deposit Sediment?**
■ Streams deposit sediment where the stream power decreases because the discharge decreases, the water depth decreases, the slope decreases, or the velocity decreases.

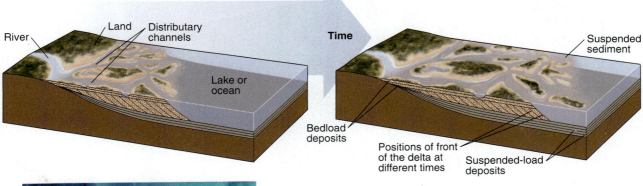

▲ **Figure 16.15 How a delta forms.**
Deltas form where streams enter the ocean or a lake. Flow velocity decreases where the stream approaches still water, which also decreases the stream power and causes sediment deposition. Bedload sediment accumulates as inclined layers that build up the delta to sea level and causes the shoreline to gradually migrate outward from land. Deposition of bedload also obstructs the channel, causing it to divide around swampy islands into distributary channels. The suspended sediment settles farther offshore as muddy-sediment layers. The satellite photo shows the Mississippi delta in Louisiana.

■ Alluvial fans form where sediment deposition results from abrupt change from a narrow, deep, confined channel to a wide, shallow, unconfined sheet.
■ Deltas result from deposition of sediment because of a sharp decrease in flow velocity where a stream enters still water.

16.6 Why Does a Stream Change along Its Course?

Figure 16.16 graphically summarizes hundreds of measurements of stream characteristics moving downstream from headwaters to mouth. The scientific challenge is to explain these measurements. Clearly, elevation must decrease because water flows downhill. It also makes sense that discharge and sediment load increase toward the mouth because of progressive inputs of water and sediment from tributaries. Although discharge increase could happen simply by increasing any one of its three components (width, depth, and velocity), all three tend to increase downstream. Why is this the case? Why do the slope angle and grain size decrease? What causes shear stress to decrease? The graph in Figure 16.16 holds the clues to understanding many dynamic aspects of water and sediment transport in streams.

Streams Adjust to Carry Available Water and Sediment

Observations show that the width, depth, and slope of a channel at any location naturally adjust to convey the amount of water and the amount of sediment delivered from immediately upstream. In particular, at bankfull discharge, a

► **Figure 16.16** **Graphing downstream changes.**
This graph depicts how the properties of a stream change from its headwaters to its mouth. The curves schematically represent the changes seen in most streams, although there are many natural variations in each of these characteristics.

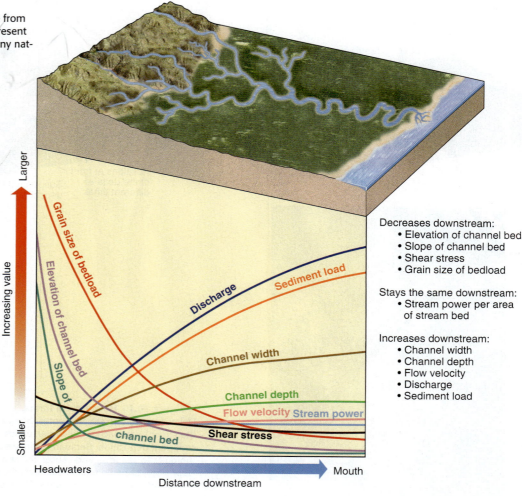

Decreases downstream:
• Elevation of channel bed
• Slope of channel bed
• Shear stress
• Grain size of bedload

Stays the same downstream:
• Stream power per area of stream bed

Increases downstream:
• Channel width
• Channel depth
• Flow velocity
• Discharge
• Sediment load

stream has the appropriate stream power to transport all of the sediment on its bed without causing any further erosion. Any significant change in the amount of water, or amount or grain size of sediment moving through the channel, causes adjustments in channel dimensions or slope to match to the new conditions. These adjustments result in deepening or widening of the channel and erosion or deposition of sediment.

Streams Become Wider, Deeper, and Faster Downstream

Discharge increases downstream, so either the cross-sectional area of the stream, or the velocity of flow, or both must also increase because discharge is the cross-sectional area of flow multiplied by velocity. Surface runoff and tributary flow to the main stream not only add water to the channel, but they also add sediment. This means that the river fine-tunes the channel width, depth, and velocity to carry the increasing amounts of both sediment and water. The channel widens so that there is more streambed for moving more sediment, just like a wide conveyor belt in a factory moves more material than a narrow one.

There is a limit to how much the stream can widen. For a particular discharge a wider channel is also shallower, and the flow is slower because of greater friction of water against the wider bed. If depth and velocity decrease as the channel widens, then the stream loses stream power and is unable to carry the increasing amount of sediment (see Section 16.5). This means that not only does the channel widen downstream but it also becomes deeper, and the flow velocity increases slightly (Figure 16.16).

Slope Decreases along the Base Level of Erosion

The downstream decrease in slope connects to the observation that the stream power per area of channel bed generally does not change along the length of a stream (Figure 16.16). Velocity increases downstream, so the shear stress must decrease downstream in order for the stream power to remain unchanged.

Figure 16.17 contrasts the slopes of a well-adjusted alluvial river and a bedrock river with waterfalls. Where the channel slope decreases downstream, a graph of elevation along the alluvial channel makes a smooth, concave-up curve. Some bedrock streams, however, lack such well-adjusted channel-slope profiles. Relatively soft rocks erode easily to form a concave-up elevation curve. Where hard, resistant rocks are present, the elevation curve has stretches of relatively low slope, alternating with steep slopes and waterfalls (also see Figure 16.8b).

Study of channel-slope profiles leads to the conclusion that a river has the ability to erode its bed down to a specific elevation, called **base level**, everywhere along its course. If the stream slope is adjusted to carry the water discharge and sediment load, then the base level along each point in the stream is the elevation along the expected concave-up elevation profile of the stream bed shown in Figure 16.17a. If the stream enters the ocean, its ultimate lowest elevation, or ultimate base level, is sea level.

Geologists assume that any deviations from a smooth concave-up profile reveal locations where the river has not eroded to its base level. For example, the bedrock stream in Figure 16.17b has not eroded to its potential base level because resistant rock perturbs the elevation profile at the waterfall.

When the elevation of a stream is at base level, the stream neither erodes nor deposits sediment. The stream simply transports the sediment that washes in from slopes or tributaries or enters the channel through mass movements. If changes in land-surface slope, water discharge, or sediment supply knock the stream out of adjustment, then it erodes or deposits sediment, or both in different places, to achieve a new base level.

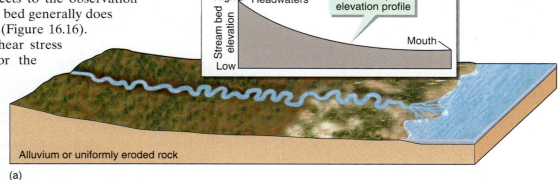

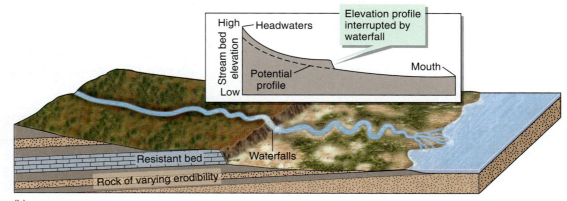

▲ **Figure 16.17 Comparing river profiles.**
(a) The elevation profile of a stream commonly follows a concave-up curve, with steeper slopes near the headwaters and gentler slopes near the mouth. This profile suggests that the stream slope is well adjusted to the sediment load and discharge and is typical of streams flowing over alluvium or uniformly eroded rock. (b) The smooth elevation profile is interrupted where a stream flows over rock that is resistant to erosion. A waterfall may form where the resistant rock forms the streambed. Further bedrock erosion takes place as the stream attempts to establish the smooth potential profile.

Sediment Size Decreases Downstream

Imagine measuring sediment grain size at many locations along the Mississippi River. Your data would show that gravel comprises 30 percent of the bedload in southern Illinois, but only finest sand, silt, and clay make it to the mouth of the river. Why does grain size decrease downstream?

Stream-transported particles are rounded, so it is reasonable to suggest that large grains grind down into small grains. After all, sediment fragments liberated from rock by weathering are angular. With increasing downstream transport, however, they become more and more rounded (also see Figure 5.10). Rounding happens as the sharp edges and corners are knocked off by grain collisions, which also diminishes the grain size.

Careful observations indicate abrupt decreases in grain size along streams that grain rounding and abrasion cannot explain. Streams clearly selectively transport smaller particles downstream and leave larger ones behind, despite downstream increase in velocity (Figure 16.16). The selective transport happens because even though water depth gradually increases downstream, the slope angle drops off dramatically so that the overall effect is decreasing shear stress. Large fragments only move when shear stress is high, so larger grains are left behind in upstream locations as shear stress decreases downstream.

> *Putting It Together–**Why Does a Stream Change along Its Course?***
>
> ■ The width, depth, and slope of a channel adjust to changing discharge or sediment load in order to maintain the appropriate stream power to transport the available water and sediment.
>
> ■ Discharge and sediment load increase downstream, causing channel width, channel depth, and flow velocity to increase downstream, too.
>
> ■ Slope decreases downstream along the base level of erosion, which is the elevation to which the stream can erode its bed at any location along its course. If a stream is everywhere at its base level, then a graph of elevation of the streambed from headwaters to mouth is a smooth, concave-up curve.
>
> ■ Sediment grain size decreases downstream. This happens mostly because decreasing slope causes diminished shear stress and partly because of abrasion of grains during transport.

16.7 What Factors Determine the Channel Pattern?

Channel width, depth, slope, and flow velocity are not obvious when standing on a stream bank. The outline of the channel itself is visible, however, and can be highly variable (see Figure 16.1). Why are most channels very curvy, like those observed for the Mississippi River, instead of straight? Why do some rivers have sediment bars within the channel, like the Alaskan river, and others do not?

Meandering and Braided Channels

The Mississippi River channel is very curvy, and sinuous like a snake. In fact, it is extremely rare to find a natural stream channel that is perfectly straight. Most sinuous channels also shift across the valley by a process called **meandering**.

Figure 16.18 illustrates an explanation for meandering in terms of velocity and shear stress variations within the channel. The flow is faster and deeper around the outside of a channel bend, the **cutbank**, causing erosion. At the same time, sediment accumulates as a **point bar**, where the flow is slower and shallower on the inside of the bend. Simultaneous erosion and deposition on opposite stream banks cause the channel to meander.

Flow in the less sinuous, gravelly Alaskan river (Figure 16.1a) separates and rejoins around bars. This channel shape is described as **braided**, because the pattern of water flow around the bars resembles braided rope or hair. The braid bars lack vegetation or soil because they submerge and move with the flow when the channel is bankfull of water.

Figure 16.19 shows the factors that geologists identified by observation to determine meandering and braided patterns in streams. Many streams are both meandering and braided, whereas others are described as primarily either braided or meandering. There is a complete gradation between, and combinations of, the attributes of these two channel patterns.

The Roles of Bedload and Discharge to Determine Channel Pattern

Braiding is more common where streams carry large volumes of bedload. This is because the channel must be wide to transport all of the sediment, but when the discharge does not completely fill the channel, some of the bedload is immobile and exposed as bars. The more variable the discharge, the greater the amount of time that the stream is unable to transport all of its bedload. This circumstance favors the formation of bars in a braided stream.

◀ **Figure 16.18 Why a sinuous stream meanders.**

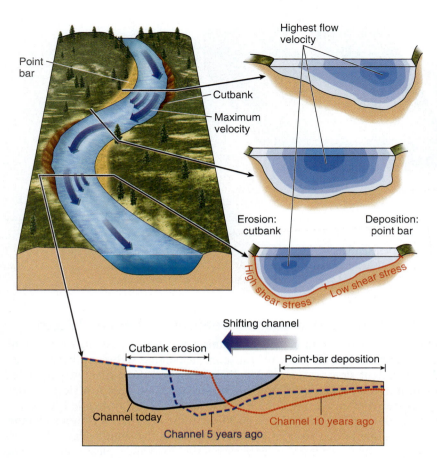

Point bar

Cutbank

Maximum velocity

Highest flow velocity

Erosion: cutbank

Deposition: point bar

High shear stress   Low shear stress

Shifting channel

Cutbank erosion

Point-bar deposition

Channel today

Channel 5 years ago

Channel 10 years ago

Flow velocity is slower close to banks and bed because of friction. When water flows around a curve in a channel, the flow is faster near the outside of the bend than it is near the inside of the curve. This velocity pattern causes erosion of the cutbank on the outside of the curve, and deposition of sediment to form a point bar on the inside. Erosion maintains a deep channel on the outside of the bend, whereas point-bar deposition causes shallow water on the inside of the curve.

Simultaneous erosion and deposition on opposite banks cause the channel to shift position through time.

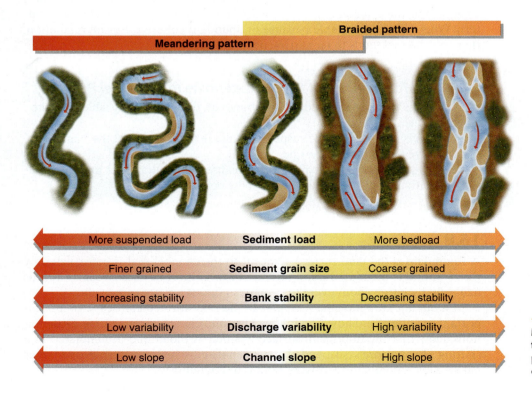

Braided pattern

Meandering pattern

More suspended load	**Sediment load**	More bedload
Finer grained	**Sediment grain size**	Coarser grained
Increasing stability	**Bank stability**	Decreasing stability
Low variability	**Discharge variability**	High variability
Low slope	**Channel slope**	High slope

◀ **Figure 16.19 Factors determining channel pattern.**
Most streams exhibit a meandering or braided channel pattern, or some combination of both patterns. The channel pattern relates to variations in the characteristics listed in the diagram.

▶ **Figure 16.20 A straight channel is steeper than a sinuous channel.**
Slope equals the elevation change along the length of the channel. The length of channel is the actual distance that water flows in the channel. A straight channel between two points is shorter than a curving channel between the same two points. The slope of a long, curvy, sinuous channel is, therefore, less than that of a short, straight channel with the same downstream drop in elevation.

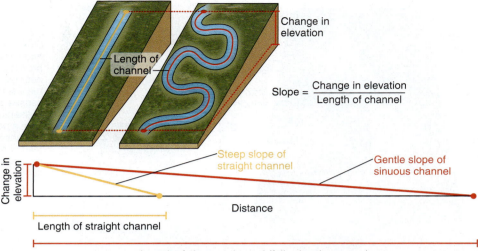

The Role of Suspended Load to Determine Channel Pattern

Meandering is more common where streams carry mostly suspended sediment. A wide channel is unnecessary to transport suspended sediment, because suspended sediment does not move on the streambed. Fine-grained suspended sediment moves at low stream power. The stream flows on a low slope so that stream power remains just adequate to transport mostly suspended load with minimal erosion of the bed. **Figure 16.20** demonstrates that a sinuous channel has a lower slope than a straight channel flowing in the same valley, which explains why streams carrying abundant suspended load are typically sinuous and meandering.

The Role of Banks to Determine Channel Pattern

Bank erodibility influences channel pattern, because a channel widens to enhance its sediment-transport capability only if the banks erode. Where banks easily erode, the channel tends to be wide and shallow, leading to a braided pattern. Where banks are stable because of abundant cohesive clay or vegetation, then the channel tends to be narrow, deep, and sinuous.

> *Putting It Together—What Factors Determine the Channel Pattern?*
> ■ Channels braid where bedload is abundant, discharge is variable, and the banks erode easily.
> ■ Channels meander where suspended load is dominant, and the banks are cohesive and difficult to erode.
> ■ Highly sinuous channels meander because of simultaneous erosion and deposition on opposite banks where the channel bends.

16.8 How Does a Floodplain Form?

A **floodplain** is the land surface adjacent to the channel that is constructed by the river and is inundated during floods. Flow in most natural streams exceeds bankfull discharge about every two years, so floods are not rare events.

Floodplains are the inhabited part of the river-formed landscape so it is important to understand how floodplains form and change through time. Observations show that the floodplain surface is either bedload that accumulated in a shifting channel or is suspended load deposited when rivers spill over their banks. Most floodplains consist of both types of deposits.

Floodplain Construction by Shifting Channels

Figure 16.21 shows how rivers widen their valley to form a floodplain. Erosion along one bank is compensated for by deposition along the opposite bank so that the channel width remains the same but shifts position. Channels typically shift at rates of about 0.5–10 m/yr, although rates as high as 200–400 m/yr have been measured along some large rivers. The floodplain is simply the part of the valley floor not occupied by the active channel, but that is underlain by older bedload deposits of the shifting stream. These former channel deposits are commonly excavated as sources of sand and gravel for road construction and a variety of industrial uses.

Figure 16.22 shows that if the meandering stream is highly sinuous, adjacent cutbanks may erode toward one another until part of the channel is cutoff from the flow. Ground water may seep into the cutoff channel segment to form a lake on the floodplain. These crescent-shaped **oxbow lakes** host ecologically diverse wetland habitats on floodplains.

Floodplain Construction by Overbank Flooding

Suspended sediment settles from flood waters that inundate floodplains. **Figure 16.23** illustrates how silt and clay deposits raise the elevation by a few centimeters to as much as a meter during each flood. These periodic overbank inundations interrupt soil development and deposit new, less weathered sediment and organic matter that renew the nutrient content of floodplain soils (see Figure 14.13). The rejuvenation of soil fertility by flooding has long drawn agricultural activity to the edges of rivers.

When river water and suspended sediment spill out of the channel and onto the floodplain, the flow spreads out as a thin sheet. The abrupt decrease in water depth leads to sharp decreases in shear stress and flow velocity that cause deposition of most of the transported sediment. More sediment accumulates next to the channel than farther away, and over time builds up a ridge parallel to the channel, called a **natural levee** (see Figure 16.23). The gradual upbuilding of the natural levee raises the elevation of the stream banks. Natural levees are different from artificial levees, which are walls or earthen embankments built along river channels to raise the banks, so that the channel holds more water at high discharges. Without the artificial levees, overbank flooding occurs more frequently and damages property.

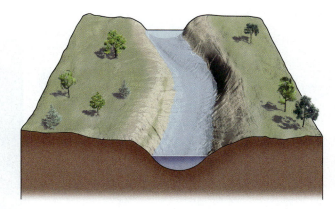

A stream channel may start out in a narrow valley, without a floodplain

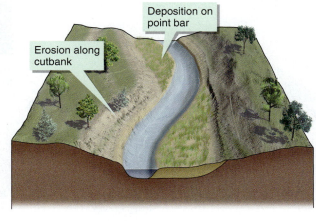

Deposition on point bar

Erosion along cutbank

Cutbank erosion and point-bar deposition cause the stream to shift horizontally.

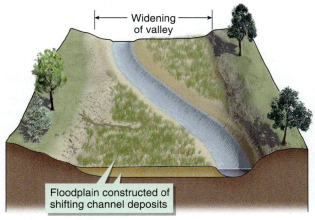

Widening of valley

Floodplain constructed of shifting channel deposits

Channel shifting, back and forth, gradually produces a wide, valley-floor floodplain that is underlain by channel-deposited sand and gravel

Time

▲ **Figure 16.21** Shifting channels form floodplains.

Putting It Together—How Does a Floodplain Form?

■ Floods occur when the discharge is too great for the flow to be contained between the stream banks.

■ Floodplains are the low-relief areas adjacent to streams that inundate during floods.

■ Rivers construct floodplains by horizontal shifting of the channel over time, by slow incremental deposition of suspended-load sediment during floods, or both.

▶ **Figure 16.22 How an oxbow lake forms.**

In a highly sinuous stream, cutbank erosion causes adjacent curves of the stream to erode toward one another.

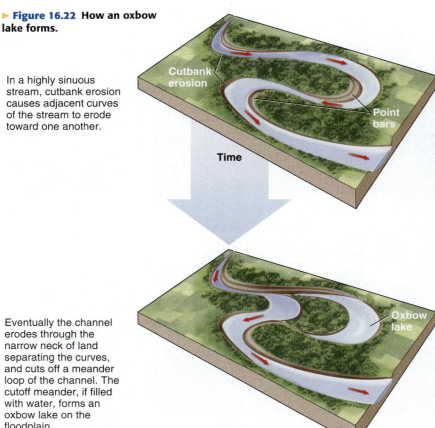

Cutbank erosion

Point bars

Time

Eventually the channel erodes through the narrow neck of land separating the curves, and cuts off a meander loop of the channel. The cutoff meander, if filled with water, forms an oxbow lake on the floodplain.

Oxbow lake

Future meander cutoff will form an oxbow lake here.

Oxbow lakes caused by recent meander cutoffs.

The photo shows recently formed, and about-to-form, oxbow lakes along a meandering stream.

Active Art

Meandering-Stream Processes: *See how meandering streams form floodplains and oxbow lakes.*

Between floods

Floodplain   Suspended load   Floodplain

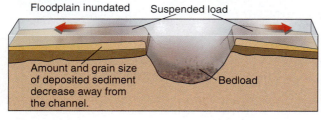

Bedload

Bedload moves at the bottom of the channel while suspended load is distributed throughout the depth of the flowing water.

During floods

Floodplain inundated   Suspended load

Amount and grain size of deposited sediment decrease away from the channel.

Bedload

When the stream floods, the bedload remains at the bottom of the channel. Some suspended load moves onto the floodplain with the flooding water. This sediment quickly deposits, with the coarser grains accumulating closer to the stream.

Between floods

Floodplain   Natural levees   Floodplain

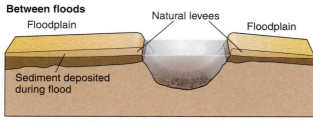

Sediment deposited during flood

Deposition of silt and clay during floods builds up the floodplain vertically, with greatest deposition adjacent to the channel to form natural levees that are higher than the rest of the floodplain.

Cleaning up mud deposited from floodwater suspension is a common scene following a flood.

▲ **Figure 16.23 Suspended-load deposits build up floodplains.**

Active Art

Flooding and the Formation of Natural Levees: *See how floods deposit sediment and form natural levees.*

16.9 Why Do Streams Flood?

What causes flood discharges that exceed the carrying capacity of the channel? Stream water originates as precipitation (Section 16.1), so unusually high discharge clearly relates to greater than normal precipitation. Floods occur when a drainage basin is incapable of soaking up all of the water from rainfall or snowmelt, such that any additional water must run off the surface. High-precipitation weather conditions lead either to short-duration flash floods that last only minutes to a few hours, or long-duration prolonged floods that persist for days to weeks.

Flash Floods

Damaging flash floods, like the one illustrated in **Figure 16.24**, usually result from high-intensity rainfall, commonly associated with spring or summer thunderstorms, in small, steep, drainage basins. The rainfall is so intense (10–20 cm/hr is not unusual) that the precipitation reaches the ground surface faster than it can infiltrate the soil. As a result, most of the water runs off the surface rather than soaking in. The problem is worse on steep slopes where soil cover is thin and water infiltrates very slowly into underlying rock. Rapid runoff from steep saturated slopes overwhelms channels very quickly. Some of the most deadly flash floods occur where the heavy runoff completely fills reservoirs behind dams, causing the dams to fail and unleash even more water downstream.

Flash floods are deepest where a narrow bedrock canyon confines the channel. Where a canyon exits the mountains onto an alluvial fan, the flow expands over the wide fan surface as a relatively thin sheet of water. Figure 16.14 shows that although the water is shallower on the alluvial fan than in the canyon, the area of inundation can be quite large, so that the entire fan surface is, in effect, a floodplain (see Figure 16.24b). Many large western cities are nestled below scenic mountain ranges, with large areas of the cities constructed on alluvial-fan surfaces. Examples include large areas of greater Los Angeles, Las Vegas, and Salt Lake City.

Flash floods are hard to predict, although weather forecasts assess the likelihood of heavy thunderstorms that may cause flash floods. People are then watchful of water levels in streams and avoid travel along stream valleys. Street flooding in cities may occur with only modest rainfall because there is no infiltration where rain falls on roofs and pavement. Once a flood begins, the floodwaters move so rapidly that it is difficult to pass along effective warnings downstream. As a result, flash floods can be deadly.

Prolonged Floods

In contrast to flash floods are the cases where water levels rise gradually along a stream over periods of days to weeks, and progressively inundate larger and larger areas of the floodplain. The worst such recent example in the United

▼ **Figure 16.24** **Flash flood!**
(a) A flash flood did this damage in Fort Collins, Colorado, in 1997. (b) This city in Venezuela was devastated in 1999 by debris flows and floods that spread out over a large alluvial fan after exiting the canyon visible in the upper left.

(a)   (b)

▶ **Figure 16.25 The Great Flood of 1993.**
Compare these two satellite images, which show the region where the Illinois and Missouri rivers join the Mississippi River near St. Louis, Missouri. The typical channel widths of the rivers are visible as dark ribbons on the left image. The image on the right, obtained near the peak of the 1993 flood, shows large areas of submerged floodplain.

States is the 1993 flood in the upper Mississippi River drainage basin, illustrated in **Figure 16.25**, which submerged 26,000 square kilometers (which roughly compares to the state of Maryland). Prolonged floods dominate the news media for weeks with stories about the destruction of entire towns, as exemplified by **Figure 16.26**, and heroic efforts by emergency workers and volunteers to add sandbags to the tops of levees.

Prolonged floods occur along rivers in large drainage basins. Unusual weather conditions lead to exceptional water runoff. Heavy rain over days to weeks is the typical trigger. Although infiltration occurs early in the rainy period, the ground eventually saturates so that additional rainfall runs off to channels. Ground water in the soggy, shallow subsurface also exits to channels at unusually high rates. The higher-than-average runoff in streams with small drainage basins may not cause severe problems. However, each small stream is a tributary to a larger stream, and each successively larger stream accumulates all of the unusually high discharges from its overloaded tributaries. These discharges add up until the larger river is overwhelmed to a degree that is only seen every few decades, or perhaps only once a century.

Prolonged floods along rivers are extremely costly because of the large urban and agricultural areas affected. The number of fatalities is small as a percentage of the number of people affected by the flood; unlike flash floods, prolonged floods rarely occur as a complete surprise, and water levels usually rise slowly enough for people to evacuate. Regional weather patterns leading to floods are commonly forecast several days to a week or more in advance.

▲ **Figure 16.26 Prolonged floods inundate large areas.**
Flooding along the Red River in the spring of 1997 submerged nearly all of the city of Grand Forks, North Dakota. Fire fighters could not respond to a fire that destroyed several buildings visible in the foreground of the photo.

Putting It Together–*Why Do Streams Flood?*

■ Floods occur because of unusually heavy rainfall or snowmelt that generates water far in excess of the volume that readily infiltrates into the ground. The resulting excess surface runoff and ground water flow to channels causes stream discharges that are too large to be confined between the stream banks.

■ Flash floods start and end abruptly and usually result from heavy thunderstorm rainfall in steep, rocky, drainage basins where infiltration is low and surface runoff is rapid.

■ Prolonged floods with slowly rising and falling discharges persist for days or weeks and result from unusually rainy conditions over days to months.

16.10 **How Do We Know . . .** the Extent of the "100-Year Flood"?

PICTURE THE PROBLEM

How Often Do Different Parts of a Floodplain Flood?

There are substantial investments and economic incentives for expansion of communities and agriculture along nearly flat, fertile, floodplains close to river channels used for navigation and irrigation. Floodplains naturally flood, but not all areas of a floodplain are at equal risk of inundation. Areas of the floodplain that are at the lowest elevations close to the stream channel flood most often and most deeply.

When hearing about big floods on television or reading about them in the newspaper, you frequently learn that a particular flood is the 50-year flood, the 100-year flood, or the 500-year flood. What do these numbers mean and how are they determined?

When evaluating acceptable risk of habitation on floodplains, the United States National Flood Insurance Program determines the probability that different parts of a floodplain flood each year. It was decided that the area with a 1 in 100, or greater, chance of flooding each year represents a significant risk. This probability of 1 in 100, or 0.01, is also described as a **recurrence interval** of 100 years, which means that on average there is an expectation of such a flood occurring once during every 100 years.

DESIGNING THE ANALYSIS

What Are the Specific Questions to Answer?

Federal law established the area inundated by a flood with a 100-year recurrence interval as the standard for planning and insuring. New development is largely excluded from the 100-year floodplain, and existing development within this floodplain is insured with federal subsidies. The law requires maps to show the extent of the 100-year flood.

Geologists must answer two questions to construct these maps:

1. How big is the flood that has a 0.01 probability of occurring each year?
2. To what elevation will the water rise during such a flood?

ANALYSIS OF DISCHARGE DATA

How Big Is the 100-Year Flood?

The potential size of the 100-year flood is determined by examining discharge data collected by stream gages over periods of several decades to perhaps as long as a century. **Figure 16.27** illustrates a 50-year flood record and shows how to do the analysis.

A measured discharge greater than or equal to 1500 cubic meters per second, shows up in the example stream-gage record five times over a period of 50 years (Figure 16.27b). You can reasonably expect, therefore, that a discharge of at least 1500 cubic meters per second will occur another 5 times during the next 50 years. On average, then, the 1500-cubic-meters-per-second flood occurs once every 10 years. Its recurrence interval is 10 years, and the probability of a flood at least this large happening during any particular year is 1 in 10, or 0.1. By comparison, when flipping a coin there is a one–in–two probability of the coin landing heads up, meaning a probability of 0.5, and a recurrence of one time in every two coin flips.

You can further analyze the data with the same method and determine the discharge of the 25-year flood, and the 50-year flood, and so forth. If, however, there is only 50 years of record, how can you determine the 100-year-flood discharge?

Figure 16.27c shows how to use the graph of the 50-year stream-gage record to estimate the discharge for the 100-year flood. Simply draw a line through the data points, extend the line out to a recurrence interval of 100 years, and the estimated discharge of 3500 cubic meters per second is read from the graph (see Figure 16.27c). This process of extrapolating from known data to a value outside

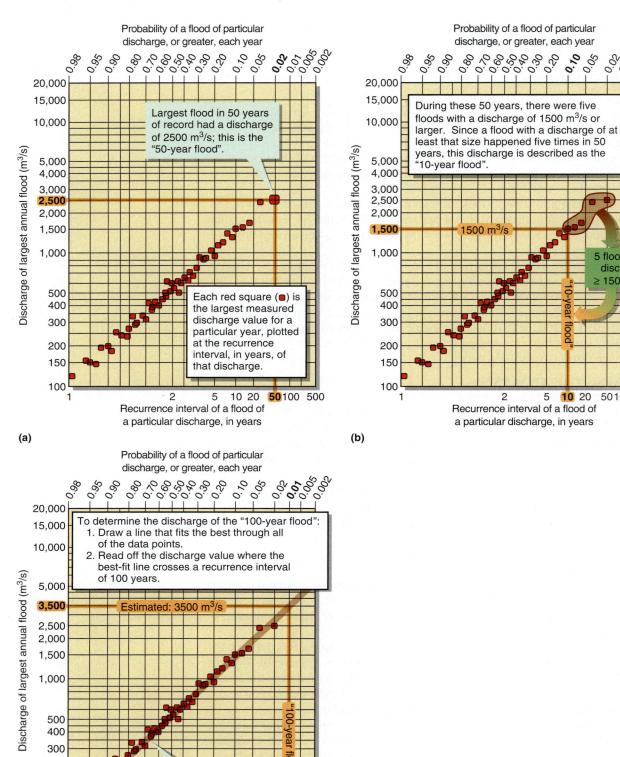

▲ **Figure 16.27 Graphing the 100-year flood.**
These graphs show how to use 50 years of stream-gage data to determine the probabilities of stream flows with different discharges. Discharges and recurrence intervals are plotted on special graphs with nonlinear axes that cause the data points to fall close to a straight line.

the data range has uncertainties that must also be considered, so a range of values for the 100-year discharge is possible.

ANALYSIS OF WATER DEPTH

How Deep Is the 100-Year Flood?

In order to determine how far up on the floodplain the water will rise during the 100-year flood, the water depth has to be known, and not simply the discharge. It is important to estimate this elevation everywhere along the floodplain, and not just where there is a stream gage whose record is used to estimate the 100-year flood discharge.

Scientists and engineers use sophisticated computer models to calculate how rapidly the flood discharge moves downstream and to estimate discharges in locations between the stream gages. Combining discharge estimates with surveyed elevation and channel-depth measurements provides estimates of the water depth of the 100-year flood discharge at every location along the valley.

The flood-depth estimates are uncertain because many factors influence the velocity of the flood water. **Figure 16.28** illustrates how the uncertainty in flood depth affects delineation of the area inundated by flood. As important as it is to delineate the 100-year floodplain for land management and insurance policies, it is impossible to mark its boundaries precisely. Just a minor uncertainty in the depth of the 100-year flood translates into a very large uncertainty in the area that floods.

Uncertainty is inherent to any scientific calculation or forecast. Everyone desires exact answers, but the very nature of scientific inquiry does not always permit high accuracy. Scientists acknowledge where the uncertainties are and how large they are so that planners recognize the uncertainties of the forecast.

INSIGHT

What Does the "100-Year Flood" Really Mean?

Some people mistakenly believe that if the 100-year flood happened last year, then such a flood will not occur again until 100 years from now. The 100-year flood is actually the flood discharge that has a 0.01 probability of occurring in *any* year, regardless of when the last flood of that size last occurred. Think again about the outcomes of coin flips. Although the probability of the coin landing heads up is 0.5, or once in every two flips, you may end up getting several heads in a row. Just because the probability is only one heads in every two flips does not mean that heads will not appear more frequently. By the same analysis, it is possible for the 0.01-probability flood to recur with a time period shorter than 100 years. Actually, this flood could happen two years in a row. It is just highly unlikely that it will.

▼ **Figure 16.28 Uncertainty of floodplain extent.**
The extent of a frequent flood with a recurrence of 10 years is well known from experience. The 100-year floodplain, however, is estimated from computer calculations. Although the uncertainty in the water depth of the 100-year flood may be small, it translates to a large area on the map, because the slope of the floodplain is very low. Properties located close to the river within the 100-year floodplain are also within the 10-year floodplain, so they have a higher flood probability than 0.01.

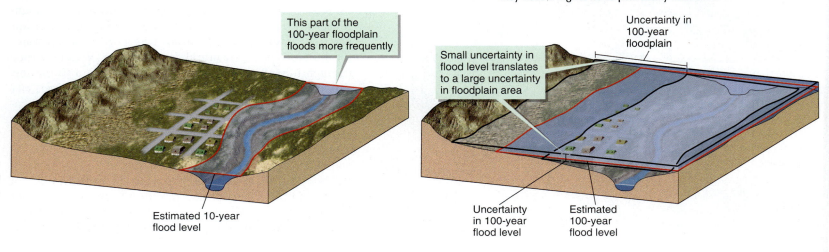

This part of the 100-year floodplain floods more frequently

Estimated 10-year flood level

Small uncertainty in flood level translates to a large uncertainty in floodplain area

Uncertainty in 100-year floodplain

Uncertainty in 100-year flood level

Estimated 100-year flood level

In other cases, property owners complain that their property on the 100-year floodplain submerges so frequently that the delineation of the floodplain must be wrong. It is essential to remember that only the highest-elevation fringe of the 100-year floodplain coincides with the estimated extent of the 100-year flood. Lower-elevation areas closer to the river have a higher probability of flood inundation with recurrence intervals much shorter than 100 years (see Figure 16.28).

EXTENSION MODULE 16.2

How to Determine Recurrence Intervals of Floods.
Learn how to use stream-gage data to calculate the recurrence times of different discharges.

EXTENSION MODULE 16.3

How to Reduce Flood Hazards.
Learn how dams, levees, and floodplain management diminish the destructive effects of floods.

Putting It Together–*How Do We Know* ... *the Extent of the "100-Year Flood"?*

■ The "100-year flood" has a 1-in-100 probability of occurring each year. The time interval between two 100-year floods may, therefore, be less than 100 years or more than 100 years.

■ Where stream gage records are less than a 100 years long, the 100-year-flood discharge is estimated by extrapolating the discharges measured for floods with shorter recurrence intervals.

■ Computer programs combine stream-gage data, the physics of flow in stream channels, and the surveyed topography along the stream to estimate the area inundated by the 100-year flood.

■ Like most scientific calculations, the discharge and depth of the 100-year flood are uncertain, which must be considered when determining the risk from the 100-year flood.

16.11 How Do Human Activities Affect Streams?

Widespread human changes of drainage-basin landscapes and stream channels commonly throw streams out of natural adjustment. Some people view modifications of streams as detrimental to the environment, whereas others emphasize the resulting increased economic productivity or public safety. These modifications are also unplanned scientific experiments. This is because the alterations of the stream channel, or changes in the supplies of water and sediment from drainage basins to the channel, provide tests and illustrations of how streams adjust to provide a balance among channel dimensions, slope, discharge, and sediment load (these variables were explored in Section 16.6). Now you can apply understanding of how streams work to explain stream responses to human changes in the natural landscape.

How Dams Affect Rivers

Dams and reservoirs interrupt nearly all large rivers and many small streams, and serve a variety of purposes. They control floods by impounding high discharges and gradually releasing the water downstream to keep the flow below the stream banks. A dam holds back some water during a rainy season or when spring snowmelt flows from mountains, and then releases it as needed during drier seasons when the river may not naturally transport sufficient water for human and agricultural consumption. Water passing through tunnels from the high side of the dam to the low side turns turbines to generate electricity. Hydroelectricity accounts for 12 percent of the total electrical generating capacity of the United States. How do these interruptions in sediment and water transport affect streams? **Figure 16.29** summarizes the observed effects.

Dams trap 90–100 percent of the sediment loads of rivers within their reservoirs. A reservoir is a still body of water, and the water-surface elevation forms a new, higher base level for the stream. As a result, the river deposits sediment to reduce the slope as flow approaches the reservoir

▼ **Figure 16.29 Dams change stream channels.**
The water elevation in a reservoir establishes a new, higher base level upstream of a dam, which causes sediment deposition in the channel and on the floor of the reservoir. Downstream of the dam, the stream is not carrying sediment, and excess stream power causes channel erosion to a deeper base level. Post-dam discharge is lower and varies less than pre-dam discharge, so the stream channel becomes narrower, and plants commonly stabilize the banks.

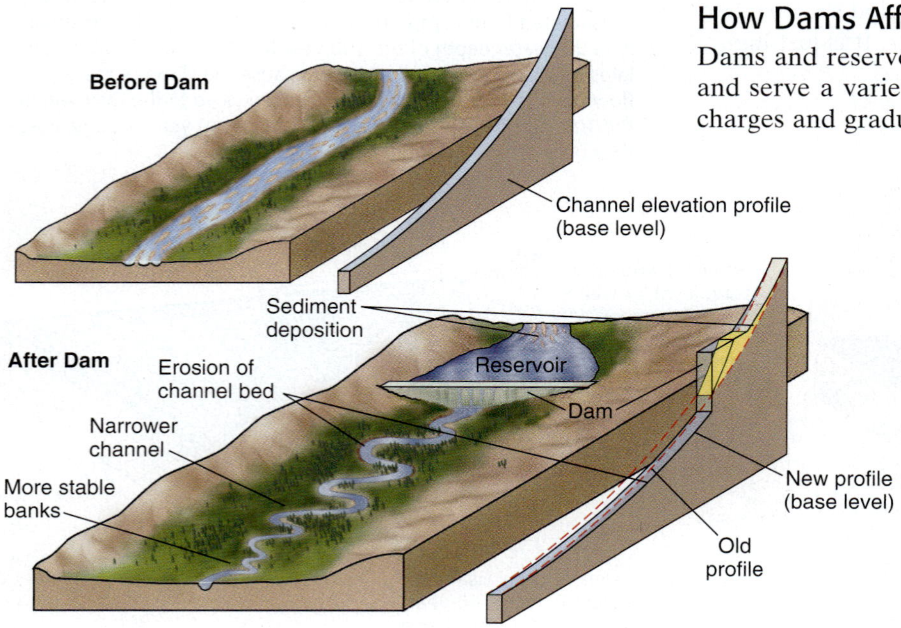

Before Dam

Channel elevation profile
(base level)

Sediment deposition

After Dam

Erosion of channel bed

Narrower channel

More stable banks

Reservoir

Dam

New profile
(base level)

Old profile

(Figure 16.29). The decrease in river velocity where flow enters the reservoir causes all of the bedload, and most or all of the suspended load, to deposit as stream power drops to zero. Over many decades or centuries, the persistent sediment deposition eventually fills in the reservoir and makes the dam obsolete.

Streambeds and banks almost always erode downstream of a dam. Water released into the channel downstream of a dam contains very little sediment, although the stream has power to carry sediment. This means that there is excess stream power for erosion and the river erodes down to a new base level. The depth of erosion, length of channel that experiences erosion, and duration of erosion vary considerably. Some of the most dramatic effects were measured after completion of Hoover Dam on the Colorado River near Las Vegas, Nevada. Erosional deepening of the riverbed by as much as 7 meters occurred along more than 100 kilometers of channel over a 14-year period.

Other, predictable changes occur in channel pattern and dimension downstream of a dam. Large discharges no longer occur because the dam impounds high flows in the reservoir. This means that the existing channel below the dam is larger than required to transport the available water, so the channel fills in along its edges and becomes narrower (see Figure 16.29). Vegetation encroaches closer to the channel when there are no floods and the plant roots decrease the erodibility of the banks, which also contributes to the narrowing of the channel. The decrease in discharge variability, decrease in coarse sediment load now trapped in the reservoir, and increase in bank stability offered by vegetation combine to enhance channel meandering instead of braiding (see Figure 16.19).

How Changing Land Use Affects Rivers

Many aspects of human land use alter the natural movement of water and sediment to a stream channel. Some scientists estimate that global sediment load in rivers has increased tenfold because of human activities. Consider these observations:

- Vegetation affects surface runoff to channels and the extent of soil erosion or mass movement to deliver sediment to channels. As pointed out in Section 14.7, replacement of natural vegetation by cropland and pasture typically increases surface runoff and causes soil erosion, especially where slopes are steep.
- Logging of forests has similar consequences as agricultural activity, and may enhance mass movement on steep slopes once the binding strength of tree roots to hold regolith in place is lost (see Section 15.3).
- Mining and large-scale construction projects that produce large piles of loose rock and soil introduce large volumes of sediment into stream channels.
- Rainfall in cities mostly falls on pavement and roofs where there is virtually no infiltration. **Figure 16.30** shows the result—surface runoff increases substantially when humans convert natural or rural landscapes to urban and suburban development.

Figure 16.31 shows that stream response to landscape changes depends on the relative importance of increases in water to do work and increases in sediment to transport. If the addition of water exceeds the addition of sediment, then there is stream power in excess of that required to transport the sediment. The stream erodes its bed and increases the width, depth, or both of its channel. This is one reason why city engineers commonly line urban stream channels with concrete, to reduce stream erosion that damages property and to keep the stream from shifting position into developed areas. If the enhanced sediment supply exceeds the ability of streams to transport it, then the channels partly or even completely fill in. The channel adjusts to carry its newly acquired sediment load by increasing channel width so that there is more room on the bed to carry the sediment. This causes erosion of banks and increases the braided character of the stream. Channel depth also decreases, which increases the risk of overbank flooding at even modest discharges.

▶ **Figure 16.30 Visualizing the effect of urbanization on discharge.**
These graphs summarize many observations of changes in stream discharge caused when urban development replaces natural landscapes. Infiltration is greatly diminished after development, so more water runs off to streams and the runoff reaches the streams faster. As a result, stream discharges in urban areas rise rapidly to flood stage.

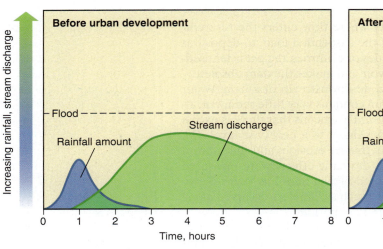
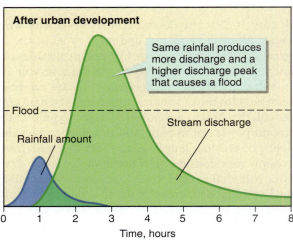

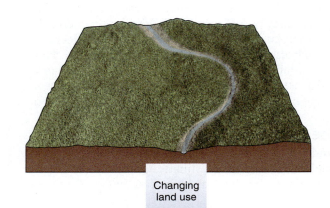

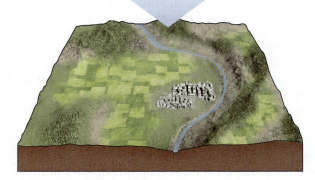

▶ **Figure 16.31 Changing land use means changing streams.**
Surface runoff and hillside erosion typically increase when natural vegetation is removed by urbanization, clearing for agriculture, and logging. Stream response to increasing runoff and sediment load depends on which factor increases the most. Channel erosion and downcutting occur if the increase in runoff is greater than the increase in sediment load. In the opposite case, the stream channel fills with sediment, widens, and floods more frequently.

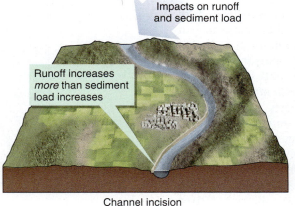

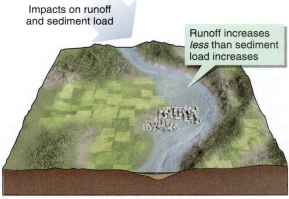

Putting It Together—How Do Human Activities Affect Streams?
- Human activities such as construction of dams and modification of naturally vegetated landscapes cause predictable adjustments in stream behavior.
- Changes in channel dimensions and shape result from modifications of the water discharge, sediment load, or both, of the stream.
- Stream adjustments may include detrimental erosion or deposition that requires lining channels with concrete to control channel location and the nature of water and sediment transport.

16.12 How Do Stream-Formed Landscapes Change through Geologic Time?

Observed base-level and channel-shape responses to human modifications of the landscape are analogies for how streams respond to naturally caused variations in stream power and sediment load over longer intervals of geologic time. Not all rivers are adjusted to their base level of erosion. Some streams are currently cutting deep canyons to reach lower elevations, whereas other channels are filling with sediment to reach higher base-level elevations. These changing characteristics of stream erosion and deposition are important aspects of changing landscapes on Earth's surface.

Terraces—Evidence of Downcutting and Filling of Valleys

Figure 16.32 shows **terraces**, which are step-and-bench landforms alongside and above a river channel. Excavation of the flat top of a terrace reveals soil developed on well-rounded gravel or sand resembling bedload deposits. These apparent river deposits are, however, high above the current streambed. It certainly seems unlikely that the river would flood tens of meters, or higher, above the bed. The presence of soil also indicates that the top of the terrace is a stable land surface, currently unaffected by stream erosion or deposition.

How do the terrace landforms and their deposits originate? The presence of bedload deposits high above the present channel records an earlier time when the river occupied a higher position in the landscape. Each terrace level records a former position of the floodplain as illustrated in **Figure 16.33**.

Terraces are evidence of the downcutting or filling of valleys over long time intervals. River downcutting and deposition indicate adjustment of the stream to offset imbalances between stream power and sediment load. Climate change, tectonic processes, and fluctuations in sea level that determine ultimate base level drive the imbalances that the river responds to over time to create terraces.

▲ **Figure 16.32 What stream terraces look like.**
The stair-step-like benches alongside this river are terraces that record former positions of the stream channel and floodplain.

Forming Stream Terraces: *See how stream terraces form over time.*

▶ **Figure 16.33 How stream terraces form.**
Stream terraces are former locations of floodplains. When a stream channel erodes downward, part of the former floodplain remains as a nearly flat terrace surface above the active channel. Soil forms on the terrace surface because it is too high above the channel for floodplain deposition to take place. Alternating time periods of downcutting and deposition eventually form multiple terrace levels. The oldest terrace has the highest elevation, and the most mature soil, because it has existed for the longest time.

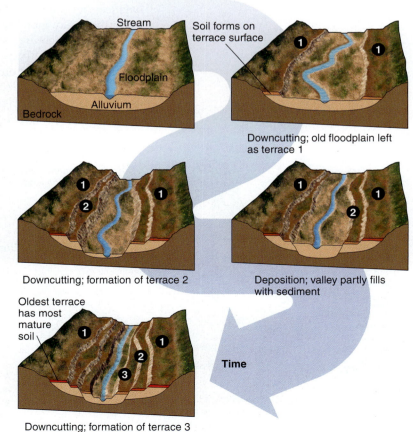

Stream / Soil forms on terrace surface / Floodplain / Alluvium / Bedrock

Downcutting; old floodplain left as terrace 1

Downcutting; formation of terrace 2

Deposition; valley partly fills with sediment

Oldest terrace has most mature soil

Downcutting; formation of terrace 3

Time

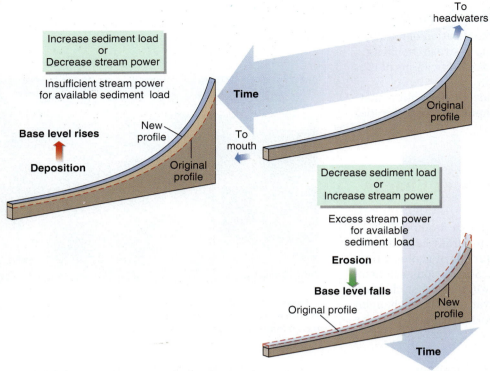

▲ **Figure 16.34 How changing sediment load and stream power change base level.**
A stream-channel profile that is well adjusted to stream power and sediment load will change if either of these variables changes. Deposition occurs and raises the base level of the stream if the sediment load increases or stream power decreases. The channel erodes to a lower base level if sediment load decreases or stream power increases.

How Climate Changes River Characteristics

The magnitudes of precipitation and temperature vary over time at any location. These climate variations are reflected in historic records for human time frames and by fossils and sediment characteristics over geologic time frames.

Rivers are very sensitive to changes in precipitation for two reasons:

1. Stream flow originates from precipitation, so changes in precipitation cause changes in discharge (Section 16.1).
2. Sediment load varies depending on the amount and intensity of rainfall and the amount of vegetation cover of the land surface (Section 16.2).

Figure 16.34 illustrates two scenarios of climate change that cause a base-level change. Alternations between intervals of erosion and deposition produce terraces like those illustrated in Figures 16.32 and 16.33. Keep in mind that the base level of the stream is a balance of available stream power to erode and transport sediment compared to the amount of sediment available to move. If one of these two variables increases while the other decreases, or if both increase or decrease but by different amounts, then the stream is out of adjustment and erodes or deposits sediment to adjust the bed to a new base level.

If sediment load increases or stream power decreases, then base level rises to a higher elevation (Figure 16.34). Deposition occurs because the stream lacks the power to transport all of the available sediment. In order to achieve a new balance between stream power and sediment load, the elevation of the streambed may increase more toward the headwaters to cause an increase in slope, which increases shear stress and velocity to move the additional sediment. Channel width also increases to facilitate transporting more bedload, and the channel may take on a more braided pattern.

If sediment load decreases or stream power increases, then base level falls to a lower elevation (Figure 16.34). The stream has excess stream power for the available sediment, so it erodes down into its bed. In order to achieve a new balance between stream power and sediment load, the overall slope of the channel decreases by having more incision near the headwaters than near the mouth. It is also typical for channel width to decrease and for the channel to become more sinuous and meandering.

How Tectonics Changes River Characteristics

Tectonic uplift and subsidence warp the elevations of the land surface. **Figure 16.35** shows how deformation changes the slope of the stream and

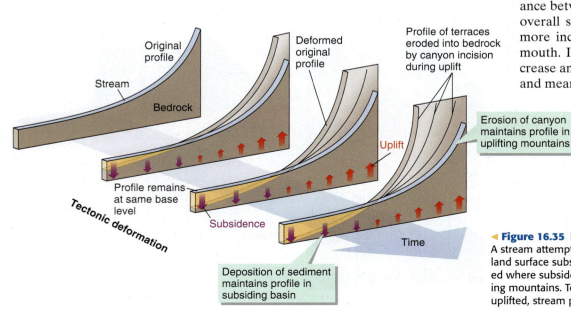

◀ **Figure 16.35 How streams respond to tectonic deformation.**
A stream attempts to maintain a steady base-level elevation as the land surface subsides or is uplifted. As a result, sediment is deposited where subsidence occurs, and canyon erosion takes place in rising mountains. Terraces may form in the canyon along former, now uplifted, stream profiles.

◄ **Figure 16.36 Streams incise canyons.**
The New River in West Virginia, on the left, and the Colorado River in Utah, on the right, eroded deep canyons into bedrock. The streams have a sinuous channel pattern, suggesting that they started out as alluvial channels at, or above, the elevations of the surrounding flat plateaus (see Figure 16.37).

the adjacent land surface. If the stream power and sediment supply are not changing, however, then the base level of the stream also does not change. This means that the channel adjusts for the warping of the surface to maintain a constant base level of erosion. The adjustments cause stream incision where uplift occurs and sediment deposition where the land subsides (Figure 16.35). Terraces form along the uplifting stretch of the river and mark former locations of the river floodplain that were abandoned as the river cut downward.

Incised Rivers—Tectonics or Climate?

Many rivers incise deep scenic canyons as seen in **Figure 16.36**, with wild rafting rapids and waterfalls. These rivers actively erode into bedrock, even where a meandering pattern suggests that the stream was once an alluvial river before cutting down into the rock, as explained in **Figure 16.37**. Until recently, most geologists interpreted incised rivers to represent areas of active tectonic uplift.

An alternative possibility, however, is that streams are responding to climate change that increased stream power and caused a downward adjustment of base level. This explanation is appealing to explain why river canyons recently formed in areas that are not tectonically active, such as the location illustrated in Figure 16.36a.

Which process causes river incision—uplift or climate change? Geologists wrestle with this question because it is very difficult to distinguish between these two causes of river downcutting. The two processes are likely intertwined and act together to at least some extent. Recall from Section 13.3 (see Figure 13.12) that erosion causes isostatic adjustment of the crust that results in uplift. The isostatic uplift increases relief, which encourages more erosion by rivers, which results in more isostatic uplift, and so on in a perpetual loop between stream erosion and isostatic adjustment of elevation, even if the stream erosion was initiated by changing climate rather than by tectonic uplift.

How Sea-Level Variation Changes River Characteristics

Sea level is the ultimate, lowest, base-level elevation that a stream draining to the ocean can achieve. A change in sea level should change the streambed elevations that determine the slope of the channel upstream of the coastline. Geologists see terraces along coastal streams that document these predicted effects.

Figure 16.38 shows how sea-level fluctuation causes downcutting or sediment deposition in channels. The elevation profile of the stream defines the slope at any location that is necessary for transporting the available sediment. The slope is always very low near the mouth of the stream where it enters the ocean.

If sea level falls, then base level falls, and the bed elevation along the stream adjusts so that the channel meets the new, lower shoreline along a very low slope. This adjustment causes incision of the lower part of the stream valley.

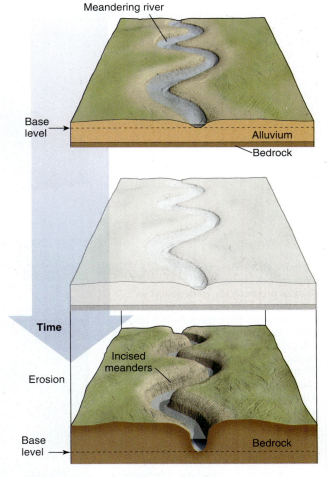

▲ **Figure 16.37 How incised meanders form.**
A meandering stream pattern forms in alluvium. If the stream channel incises down through the alluvium into bedrock, then the sinuous channel form is etched into the underlying rock.

▼ **Figure 16.38** **How streams respond to sea-level change.**
Base level rises when sea level rises, causing sediment deposition in the lower part of a stream valley near the ocean. Base level falls when sea level falls, causing erosion in the lower part of a stream valley.

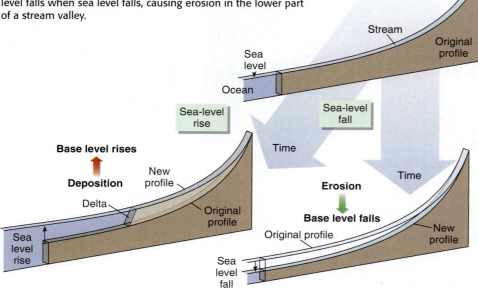

If sea level rises, then base level rises, and the bed elevation along the stream adjusts to a higher shoreline position. The adjustment causes deposition in the lower part of the stream valley. The effect is very similar to the base level adjustment where a reservoir fills behind a dam (compare Figure 16.38 with Figure 16.29).

How Bedrock Geology Affects Landscape Development

If all streams were alluvial streams, or flowed on easily weathered and eroded regolith, the material into which the stream channel erodes could be virtually ignored. The geology of the consolidated bedrock, however, imposes important controls on landscape development in two ways. First, rock types do not weather and erode equally, meaning that stream channels and valleys more easily erode in some rocks than in others. Second, the locations of rock with variable resistance to weathering and erosion determine where streams will erode most easily.

Figure 16.39 shows how dipping rock layers with different resistance to weathering and erosion determine the location of stream valleys. Large rivers with large stream power may cut across dipping layers, but smaller streams erode valleys in the most readily weathered and eroded rock types. The more resistant, inclined rock layers remain as higher ridges and drainage divides.

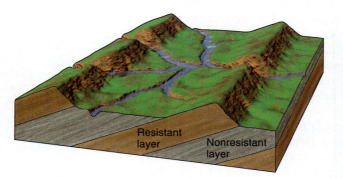

▶ **Figure 16.39** **How streams erode dipping rock layers.**
Streams preferentially erode valleys in rocks that weather into easily eroded regolith, whereas more resistant rock types remain as higher ridges. Parallel ridges and valleys form where layers dip in the same direction. The photo to the upper right illustrates such a landscape that forms Comb Ridge in Utah, where knife-edge ridges of cemented sandstone rise above valleys eroded in mudstone. The image at the bottom left, composed from satellite radar scans of Earth's surface, highlights folded sedimentary rocks of the Appalachian Mountains in the Valley and Ridge region. The structural patterns of the folded rocks are accentuated where valleys erode into less resistant rocks and more resistant rocks remain as ridges.

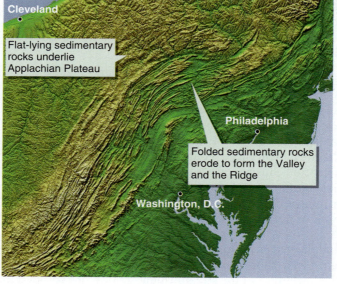

Erosion of Dipping Rocks: *See how streams erode through dipping rock layers.*

Over long erosional time periods, stream erosion etches the pattern of the geologic structure into the landscape. This effect of erosion to accentuate inclined rock layers as ridges or valleys is also evident in many photographs in Chapter 11 (see Figures 11.4, 11.8, and 11.22).

> *Putting It Together–***How Do Stream-Formed Landscapes Change through Geologic Time?**
>
> ■ Stream channels may erode their beds or deposit sediment over long geologic time intervals in response to changes in stream power and sediment load.
>
> ■ Terraces are benches alongside and higher than modern floodplains and channels. Terraces represent former positions of the floodplain and record the history of downcutting and sediment filling of stream channels.
>
> ■ The changing streambed elevations result from (a) tectonic uplift and subsidence of the land surface, (b) variations in climate that determine discharge and sediment load, and (c) fluctuations in sea level that determine ultimate base level.
>
> ■ Streams preferentially erode deeper and wider valleys where rock is most easily eroded, whereas more resistant rock forms ridges or cliffs.

16.13 How Do Lakes Form?

Of course, not all streams flow uninterrupted to the ocean. Some streams terminate in lakes. A lake may have a downslope outlet through another stream, or the water simply accumulates in a lake without an outlet and evaporates.

In simple terms, a lake accumulates water in a low part of the landscape that resembles a bowl, as shown in **Figure 16.40**. Surface runoff flows downslope into the bowl from all sides. Ground water seeps into the lake if the bottom of the bowl is lower than the water table. If the bowl completely fills, then water spills over the lowest elevation along the edge of the bowl and continues downslope in a stream channel. If water evaporates from the lake faster than it flows in, then it will not spill out. Figure 16.4 shows the outline of the Great Basin, an area where streams terminate in the continental interior rather than reaching the ocean. Stream flow in these drainage basins ends up in drying lakes (Figure 16.40).

Some lakes, such as the Great Lakes, owe their origin to scouring of the landscape by glacial ice and will be considered further in Chapter 18. This section describes lakes related to surface processes that disturb the courses of rivers.

▼ **Figure 16.40 Lakes—open and closed.**
Lakes form in low topographic bowls that collect runoff from adjacent slopes and ground water. Open-basin lakes, like Lake Tahoe, along the California-Nevada border at left, spill over into an outlet stream. If evaporation keeps the lake from filling up, then it is closed, without an outlet, as illustrated on the right by Deep Springs Lake, California.

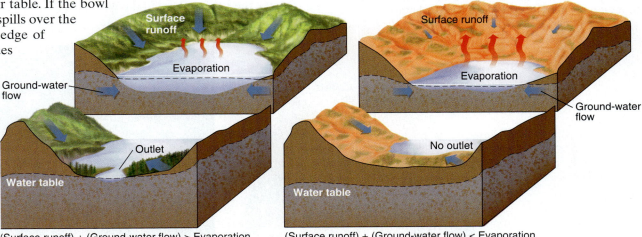

(Surface runoff) + (Ground-water flow) > Evaporation

(Surface runoff) + (Ground-water flow) < Evaporation

How to Naturally Dam a River

Many lakes form by natural blockage of a through-flowing stream, just like an artificial dam blocks a stream to form a reservoir. **Figure 16.41** illustrates the most common processes that naturally dam streams, where lava flows and mass movements (also see Figures 15.1b and 15.1c) block a stream and water pools on the upstream side of the obstacle. Eventually, the water rises to the elevation of the top of the natural dam and pours out through a newly formed outlet channel. Bedload and most suspended load sediment are deposited below the still water of the lake, just as in the case of an artificial reservoir (see Figure 16.29) so that the lake eventually fills in with sediment.

How Tectonic Activity Forms Lakes

Tectonic activity forms long-lasting lakes by dropping blocks of crust along faults so that surface runoff flows toward a central depression surrounded by higher elevations, as illustrated in **Figure 16.42**. If the down-dropped fault block subsides faster than the stream deposits sediment, then an enclosed depression collects all of the surface runoff and usually some ground-water seepage.

If more water enters the lake than is lost to evaporation, then the lake eventually fills to the lowest spot on its rim and spills over into an adjacent river valley. The lakes of the East African Rift Valleys (Figure 12.19) and Lake Baikal in Russia, the deepest lake in the world (1620 meters deep), are examples of lakes in tectonic basins with outlets.

▲ **Figure 16.41 Naturally dammed streams.**
(a) When the volcano named Cinder Cone erupted in northeastern California in 1650, a lava flow filled parts of two stream valleys, causing lakes to form on the upstream side of the lava.
(b) When an earth flow blocked the Spanish Fork River in Utah in 1983, a lake formed upstream of the blockage and completely submerged the town of Thistle along with a major highway and railroad. The cost of relocating the highway and railroad, along with property losses in Thistle, totaled $400 million, making this the most costly mass-movement event in U.S. history.

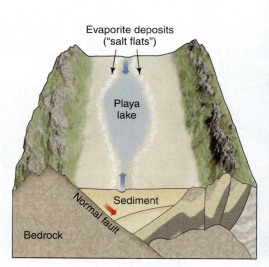

▲ **Figure 16.42 Lakes form in fault-block valleys.**
Normal faults commonly drop down blocks of crust to form valleys between uplifted mountains. Runoff from the mountains forms lakes in the valleys. In dry regions, like the Great Basin, evaporation leaves behind salty playa lakes within enclosed valleys without stream outlets, as seen in the Space Shuttle view looking southwestward across Nevada toward eastern California.

In arid regions, like the Great Basin, evaporation may keep the depression from filling with water. The resulting shallow, salty playa lakes may dry up completely during the warmest, driest times of the year (Figure 16.42). The valley floors surrounding and underlying the lake bed encrust with evaporite minerals like halite, gypsum, and less common but commercially important minerals like borax, which is used in the manufacture of detergents, glass, and ceramics. Great Salt Lake in Utah is the remains of a much larger lake that evaporated to leave surrounding salt flats. These chemical sedimentary deposits represent the dissolved load of streams that would flush out to the ocean if evaporation of the inland lakes did not occur.

Putting It Together—*How Do Lakes Form?*

- Lakes form naturally where lava flows and mass-movement deposits block a stream channel or where faults shift crustal blocks downward to produce a central depression that collects water.
- If lake water evaporates faster than water enters the lake from streams or ground water, then the lake dries up, at least in part, and precipitates evaporite minerals.

Where Are You and Where Are You Going?

Streams are an integral part of the hydrologic cycle. Stream water originates as direct surface runoff or ground water that initially reached Earth's surface as rain or snow. Some surface precipitation also infiltrates deeply to form ground-water resources, and evaporation and plant transpiration carry water back to the atmosphere. Atmospheric water vapor condenses into water droplets or ice crystals and eventually returns to the surface as precipitation, thus completing the cycle.

Streams are complicated, dynamic features on Earth's surface. Stream erosion determines most of the irregular relief on the surface by carving valleys separated by intervening drainage divides. Streams transport sediment that primarily erodes from areas of steep relief and areas where runoff from rainfall exceeds the landscape-stabilizing effect of vegetation. Shear stress of the flowing water, which strongly relates to channel slope and water depth, determines the ability of streams to pick up sediment particles. Stream power, which relates equally to the shear stress and the velocity of the flowing water, determines the ability of streams to transport the picked-up sediment. Streams deposit sediment where the shear stress or velocity or both are insufficient for continued sediment transport.

The width, depth, and slope of a stream channel adjust to establish the stream power required to transport the available sediment. Changes in discharge or sediment load cause changes in these channel characteristics, which lead to erosion or deposition to establish new, well-adjusted channels. Stream adjustments over time create terraces that show how climate change, tectonic activity, sea-level change, or human activity affected stream processes. Responses of streams to human activities that change discharge, sediment load, or modify channel characteristics are predictable because geologists understand the physical processes that determine these responses.

Stream-channel patterns include mixtures of braided and meandering behavior. High coarse-grained-bedload sediment supply, steep slope, fluctuating discharge, and easily eroded banks favor braided streams. Flowing water diverts around exposed bars in braided streams. Largely suspended load, low slope, consistent discharge, and less erodible banks favor laterally shifting, sinuous, meandering streams.

Streams form floodplains, which pose both hazards and benefits. Floodplain deposits accumulate from laterally shifting channels and by the settling out of suspended sediment from overbank floodwaters.

Agricultural productivity depends on fertile floodplain soils rejuvenated by floods and proximity to irrigation water. Large rivers also attract urbanization centered on shipping commerce in deep, wide channels. The concentration of populations close to rivers, however, increases human vulnerability to flood hazards. Geologists estimate the probability of floods inundating particular floodplain areas by combining analysis of stream-gage data, detailed knowledge of floodplain topography, and knowledge of the physics of river flow. The public learns these probabilities as recurrence intervals, although a flood with a particular recurrence interval may occur more or less frequently than the interval implies.

Natural lakes form where stream channel blockages form depressed areas where water accumulates. Processes that form lakes include eruption of lava flows or mass-movement events that naturally dam channels and tectonic subsidence of crust into enclosed depressions. Where inflow to the lake from runoff and ground water is greater than evaporation, the lake fills up to the lowest spot on its edge and spills out into a stream channel. Where evaporation exceeds water inflow, the lake dries up, at least partly, and leaves behind evaporite deposits of possible economic value.

In the next chapter, you will consider the water that is present beneath Earth's surface. Ground water and surface water are interconnected, and both originate by precipitation at the surface. Ground water accounts for 42 percent of the public water supply in the United States, so it is very important to understand the geology of this important resource. Ground water flows through minute pores and cracks in rock and sediment and does not form underground rivers or stagnant lakes. Underground flow of water is very different from the flow observed in stream channels but is important to understand in order to evaluate the extent of ground-water supplies in a region and how poor-quality water might move into drinking-water supplies. Although ground water is hidden beneath your feet, it still plays a role in the development of surface landscapes. You will soon learn how ground water makes spectacular caves, creates steep canyons, and sometimes forms holes that swallow whole city blocks.

 Active Art

Hydrologic Cycle. See how liquid water and water vapor move through the Earth system. **How Streams Move Sediment.** See how flowing water moves the sediment load. **Meandering-Stream Processes.** See how meandering streams form floodplains and oxbow lakes.

Flooding and the Formation of Natural Levees. See how floods deposit sediment and form natural levees. **Forming Stream Terraces.** See how stream terraces form over time.

Erosion of Dipping Rocks. See how streams erode through dipping rock layers.

Extension Modules

Extension Module 16.1: How a Stream Gage Works. Learn how a stream gage is constructed and how hydrologists use the gage data to determine discharge.

Extension Module 16.2: How to Determine Recurrence Intervals of Floods. Learn how to use stream-gage data to calculate the recurrence times of different discharges.

Extension Module 16.3: How to Reduce Flood Hazards. Learn how dams, levees, and floodplain management diminish the destructive effects of floods.

Confirm Your Knowledge

1. Define "stream". What role does a stream play in distributing sediment on Earth?

2. Define and describe the hydrologic cycle.

3. Streams are important natural resources. How do humans use the water withdrawn from streams on a daily basis?

4. What are the two paths that rainfall can take after it reaches the land surface?

5. What is a drainage basin? Describe how water moves through the drainage basin.

6. About how often does a natural stream overflow its banks?

7. Define discharge. How is discharge measured?

8. How does sediment get into a stream?

9. What has to happen for loose sediment grains to erode from a streambed?

10. Streams are important agents of transport for sediments. Define sediment load, and explain the link of sediment load to drainage-basin climate.

11. What is the difference between bedload and suspended load? What is the dissolved load?

12. What is the link between stream power and deposition?

13. Define and describe an alluvial fan. How does it differ from a delta?

14. Define base level. What processes cause base level to change?

15. What is a floodplain? How does it form?

16. What is the difference between a flash flood and a prolonged flood?

17. Why does flooding occur? How is the frequency of flooding determined?

18. How could a 100–year flood happen two years in a row?

19. List the positive and negative consequences of building a dam on a river.

20. What are some of the ways that human activities affect rivers?

21. What are terraces and what do they tell us about the past?

22. What processes cause rivers to erode incised valleys and canyons?

Confirm Your Understanding

1. Write out an answer for each question in the Chapter Outline for the chapter sections assigned by your instructor.

2. What is the difference between stream discharge and stream velocity? What are the units used to measure discharge and velocity? Why does stream discharge vary along the length of the stream? Why does stream velocity vary along the length of a stream?

3. What would you suspect to be the cause of dramatic differences in the sediment load per square kilometer between two streams with very similar climates? What would you suspect to be the cause of dramatic differences in the dissolved load per square kilometer between two streams with very similar climates?

4. About how often do the largest sediments on a streambed typically move?

5. What would you suspect to happen to a stream's load if its channel shape changes from being narrow and deep to being wide and shallow (assume the area of the channel and average velocity remain constant)?

6. Using Figure 16.16, list the stream properties that increase with downstream distance and those that decrease. Explain which properties are most responsible for the increase in sediment load with downstream distance and the decrease in the grain size of the sediment load with downstream distance.

7. Consider a planet with little precipitation. What would the landscape look like?

8. Using the map in Figure 16.4, determine where to the streams in your region of the country flow. Where is the local divide (hint: identify the topographic highs)?

9. Go to http://water.usgs.gov/waterwatch/ and click on a stream gage location near your home or school. Examine the discharge and gage-height graphs. Use your knowledge of how the stream may be regulated by dams, along with recent weather, to explain variations in these data over the last week and how these values compare to long-term average values.

Key Terms

alluvial fans (p. 454)
alluvial streams (p. 449)
alluvium (p. 449)
bars (p. 452)
base level (p. 457)
bedload (p. 452)
bedrock streams (p. 449)
braided (p. 458)
cutbank (p. 458)

delta (p. 454)
discharge (p. 446)
dissolved load (p. 448)
distributary channels (p. 454)
divides (p. 445)
drainage basin (p. 445)
flood (p. 447)
floodplain (p. 460)
ground water (p. 444)

headwaters (p. 445)
hydrologic cycle (p. 444)
meandering (p. 458)
mouth (p. 445)
natural levee (p. 461)
oxbow lakes (p. 461)
playa (p. 446)
point bar (p. 458)
recurrence interval (p. 465)

sediment load (p. 448)
stream (p. 444)
stream gages (p. 446)
stream power (p. 453)
suspended load (p. 452)
terraces (p. 471)
transpiration (p. 444)
tributary (p. 445)
water table (p. 444)

17 Water Flowing Underground

17.1 What is ground water and where is it found?

17.2 Why and how does ground water flow?

17.3 How do we know . . . how fast ground water moves?

17.4 What is the composition of ground water?

17.5 How does ground water shape the landscape?

Why Study Ground Water?

AMERICANS PUMP ABOUT 314 BILLION LITERS OF WATER OUT of the ground every day. This resource provides drinking water to about half of the population. It makes up 40 percent of the nation's public water supply systems and supplies individual water wells for most rural homeowners. Almost two-thirds of the ground water pumped in the United States irrigates farmland and supports livestock.

Ground water comprises approximately 98 percent of Earth's fresh water that is not frozen into glacial ice; most of the remainder is in streams and lakes. How are ground-water resources located? Are these resources adequate for the increasing demands of a growing global population? Is all ground water suitable for drinking?

Ground water usually is a reliable water source. It is available year-round, whereas some streams stop flowing during dry seasons, just when crops need water the most. In locations where surface water is not immediately available, ground water is a valued resource. Ground water is less quickly polluted than surface water by sewage, industrial wastewater, and runoff from agricultural fields that might include pesticides, herbicides, or harmful microbes.

This is not to say that ground water is pure H_2O. Perhaps you have trouble with hard water in your pipes. Earth materials dissolve in ground water. These compounds precipitate minerals that clog pipes, cement sediment grains into sedimentary rock, and also form economically valuable mineral deposits.

Underground water influences the formation of surface landscapes. Springs are locations where ground water exits to the land surface. In some cases, the perpetual wetting of the rock near springs enhances weathering and leads to mass movement on steep slopes. Ground water reacts with subterranean minerals in the rock, regolith, and soil. These reactions dissolve some rocks to form large caverns. Sinkholes form at the surface when cavern roofs collapse, and occasionally "swallow" buildings.

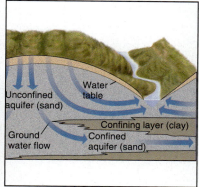

Unconfined aquifer (sand)
Water table
Confining layer (clay)
Ground water flow
Confined aquifer (sand)

Ground water spills from springs high on the wall of the Grand Canyon. ▶

Study of ground water hydrology is an important field of geology. Ground water is a critical resource for sustaining human existence on the planet. Ground water flow also modifies the surface, so the study of ground water coordinates with your understanding of landscape development and potential geologic hazards. Your specific objectives in this chapter are as follows:

✔ To learn where ground water exists below the surface and the fundamentals of how ground water flows

✔ To understand how human activities affect ground-water flow and water quality

✔ To connect ground-water chemistry with water quality, cementation of sedimentary rocks, and formation of some economic ore deposits

✔ To visualize how ground-water flow modifies surface landscapes

Along the way to accomplishing these objectives, you will answer these questions:

17.1 What is ground water and where is it found?
17.2 Why and how does ground water flow?
17.3 *How do we know* … how fast ground water moves?
17.4 What is the composition of ground water?
17.5 How does ground water shape the landscape?

In the FIELD

Think of taking a hike in the forests of southern Oklahoma. The trail climbs alongside a stream named Travertine Creek. To your surprise, the source of the stream is a natural **spring**—a place where ground water emerges onto the surface. The stream suddenly appears from the subterranean world through cracks in rock, as seen in **Figure 17.1**a. A light-colored mineral incrustation covers the banks alongside the spring. At a nearby visitor's center you see a soft-drink bottle displayed that is completely encased in the same incrustations (Figure 17.1a). The display identifies the mineral as calcite that forms a type of limestone, called travertine, for which the creek is named. Clearly, the calcite precipitates from the spring water as it flows away in the stream channel.

On another occasion, you vacation in Florida and come upon a scene illustrated in Figure 17.1b. A hole in the ground was less than a meter across when it was first noticed, but in less than a day it gradually expanded to more than 75 meters in diameter. This natural hole swallowed a house, several vehicles, and part of the community swimming pool and severely damaged two businesses. The newspaper story you read tallies the damage at more than two million dollars.

The story calls the feature a sinkhole and explains that sinkholes form where underground water dissolves cavities in limestone.

Perhaps you have never given much thought to underground water. At the Oklahoma spring, however, you see underground water emerging onto the surface, and it makes you wonder. How much water in streams generally comes from underground rather than originating as runoff from rain and snow? You know that water wells pump underground water to the surface for drinking, cleaning, crop irrigation, and industrial processes. How do you know the water is safe to drink and not contaminated? How much water really is present underground? How fast does it move and how does it move through rock? Are there underground rivers? Why does underground water emerge at the surface at springs? What is the origin of the dissolved minerals that precipitate around the spring? Why does limestone precipitate along Travertine Creek but dissolve in Florida? What landscape features other than springs and sinkholes also relate to water flowing beneath the surface? What factors determine the quality of ground-water resources used by humans?

Figure 17.1 **Evidence for ground water is visible at the surface.** ▶

17.1 What Is Ground Water and Where Is It Found?

In the hydrologic cycle, described in Section 16.1, underground water originates as rainfall and snowmelt that soaks into regolith and rock at Earth's surface. Some of this water moves short distances through shallow soil horizons and reemerges to join surface runoff in streams. The remainder of the infiltrated water percolates deeper below the surface through open spaces and fractures. **Ground water** is the term that refers to this subterranean water.

How Ground Water Is Stored in Rock and Regolith

A simple lab experience, illustrated in **Figure 17.2**, provides insights into where ground water resides. Water added to a container of loose solid particles fills in

A. Field observations of a spring: Travertine Creek, Oklahoma

Stream emerges from a spring at the base of this rock cliff.

Calcite forms a limestone encrustation, called travertine, along the stream bank.

This travertine encrusted bottle appears at the visitors' center.

B. Field observations of a sinkhole: Winter Park, Florida

This sinkhole gradually collapsed over about one day and swallowed up several buildings. Notice the car and the camper that also sank into the hole.

This newspaper picture shows the sinkhole from the air. It is about 75 meters across from left to right.

Porosity in sediment

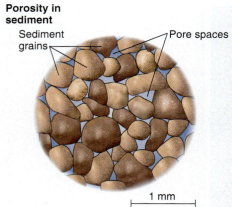

Sediment grains

Pore spaces

1 mm

Porosity in fractured rock

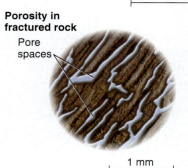

Pore spaces

1 mm

Porous sediment

Water

Approximately 50% porosity

▶ **Figure 17.2 What porosity looks like.**
Porosity is the percentage of pore spaces found between grains and fractures. Water poured into a beaker of sand fills the pores, which in this case represent about 50 percent of the volume of loose sand. The greater the porosity, the more water can be stored in sediment and rock.

the spaces between the grains. Ground water similarly occupies open spaces between particles in regolith and some sedimentary rocks, and it fills fractures that cut through all types of rocks. These open spaces are **pores**. **Porosity** is the percentage of the total volume of the regolith or rock that consists of pores. **Table 17.1** lists typical porosity values for rock and sediment, ranging from just a few percent to more than 50 percent. In other words, although some rocks are very solid, others actually have more empty space than mineral grains. Materials with greater porosity have the potential to store greater amounts of ground water.

Only very rarely are pores more than a few centimeters across, and most are less than a millimeter. This means that it is extraordinarily rare to encounter ground water flowing in underground rivers.

Locating Ground Water

How do geologists know about ground water concealed from view beneath our feet? **Figure 17.3** shows how wells are drilled to determine where ground water is present below the surface and how to extract it. At some point while drilling downward, ground water starts to seep into the bottom of the well. After drilling several wells in an area, a pattern of how the depth to water below the land surface varies from place to place is apparent.

The **water table** is the surface that marks the top of the ground water. Notice in Figure 17.3 that the water table is not flat, despite what the word "table" might imply. Instead, the water table rises and falls in elevation in a pattern that reflects variations in the surface topography. Where the land surface is high, the water table is also relatively high. Where the land surface is low, the water table is also relatively low.

The water table intersects the ground surface along a deep stream valley illustrated in Figure 17.3. At this intersection some ground water seeps

TABLE 17.1	Example Porosity Ranges for Different Earth Materials	
Material	Typical Porosity Range
Unconsolidated sediment	
Well-sorted sand and gravel	25–50%
Mixed sand and gravel	20–35%
Silt and clay	30–60%
Rock	
Shale	0–10%
Sandstone	3–30%
Limestone and dolostone	1–30%
Plutonic and metamorphic rocks	0–5%
Volcanic rocks	1–50%

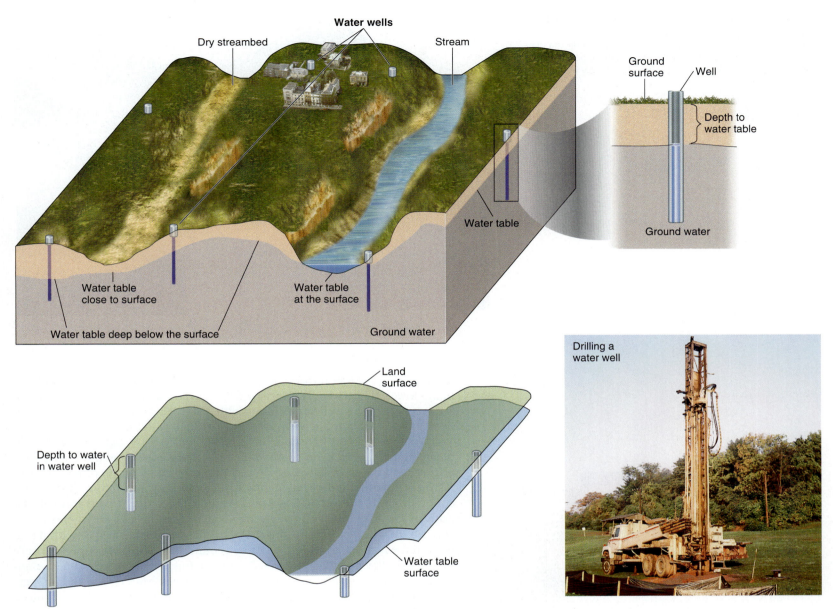

▲ **Figure 17.3 Visualizing the water table.**
Geologists and hydrologists determine the location of the water table by drilling wells. Most wells are drilled to withdraw ground water for use at the surface, although sometimes wells are drilled simply to find and monitor the elevation of the water table. The water table surface is not flat and in a very general way mimics the ground surface, which means the water table is usually low under valleys and high under hills and ridges, although local exceptions may occur. The relief of the water-table surface is typically less than the relief of the ground surface so that the water table is closer to the surface under valleys than under hills and ridges. In some cases, valley elevations are actually lower than the ground surface so that ground water emerges to form flowing streams.

through the streambed and flows away as surface water. The flow of ground water into streams explains why many streams flow every day, year-round even when it has not rained or snowed for many weeks. Water runoff to the channel from the adjacent hillsides ceases a few days after it rains. Ground water, however, seeps into the channel and sustains the persistent flow of streams with beds eroded to or below the water table. Where there is not a channel to drain the water away, natural and artificial lakes may occupy surface depressions where the water table intersects the surface; this case is shown in **Figure 17.4**.

Notice the dry streambed in Figure 17.3. This stream flows only when runoff enters the channel during and shortly after rainfall. The channel is not eroded deeply enough to intersect the water table, so it does not receive ongoing additions from seeping ground water.

Notice, too, how the water-table elevation varies over short distances in Figure 17.3. This observation, along with knowledge that the water-table surface mimics the land-surface topography, means that once the depth to the ground water is known from several wells, it is possible to predict the depth to which a new well needs to be drilled.

▶ **Figure 17.4 Ponds form where the water table intersects topographic depressions.**
Some depressions on Earth's surface are lower than the water table, so these low areas fill with water. Natural ponds form this way, and rock quarries and mines commonly fill with ground water.

Ground Water Flows

Water flows into a well from the pores or fractures in the rock or regolith. A pump lowered into a well extracts and brings the water to the surface, but the bottom part of the well keeps filling with water. This means that water flows through the rock into the well. How does ground water move?

Water flows between pore spaces through narrow gaps between sediment grains, or cracks in more solid rock. **Permeability** is the ability of a porous material to permit fluid flow through the material. A coffee maker illustrates porosity and permeability. After pouring water into a filter full of coffee grounds, the water flows into the pot. The water flowed through the pore spaces between the coffee grounds and then through tiny holes in the paper filter. The water flows into the pot almost as quickly as it is poured into the filter, which means that the coffee grounds and the filter paper are very permeable. For geologic materials, pores or fractures connect with one another so that water flows through the open spaces. Materials with high permeability have well-connected porosity. Water flows faster through more permeable materials than through less permeable ones. **Figure 17.5** illustrates how variations in sediment grain size or sorting, or variations in the abundance and spacing of rock fractures, influence both the porosity and permeability of Earth materials.

The Water Table Separates Saturated and Unsaturated Zones

Figure 17.6 provides a closer view of what happens at the water table. Water completely fills pores in the **saturated zone** below the water table, whereas air partly fills pores in the overlying **unsaturated zone**. The water table is at the interface between the unsaturated and saturated zones. Infiltrating water moves downward through the unsaturated zone to join ground water in the saturated zone. Water flows into wells only if wells are drilled into the saturated zone, where water completely fills the pores below the water table.

Higher permeability ⟷ Lower permeability

Well sorted sand

Poorly sorted sand

Ground-water flow

Uncemented sand

Cemented sandstone

Mineral cement

Connected fractures in igneous rock

Unconnected pores in igneous rock

◀ **Figure 17.5 Visualizing the factors that affect porosity and permeability.**
Variations in sediment grain size and sorting, cementation, or abundance and spacing of fractures cause variable porosity and permeability within rock and regolith. Porous rocks can still have low permeability, as shown by the examples on the right, because the pores are not well connected and water cannot easily flow from pore to pore.

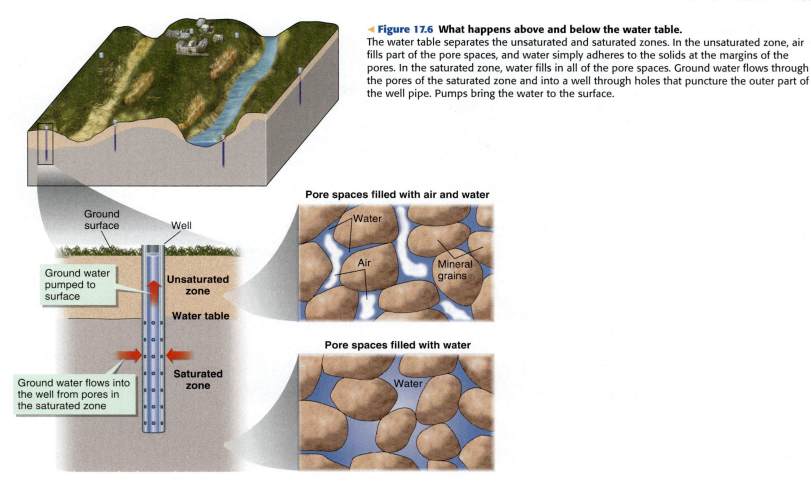

◀ **Figure 17.6 What happens above and below the water table.**
The water table separates the unsaturated and saturated zones. In the unsaturated zone, air fills part of the pore spaces, and water simply adheres to the solids at the margins of the pores. In the saturated zone, water fills in all of the pore spaces. Ground water flows through the pores of the saturated zone and into a well through holes that puncture the outer part of the well pipe. Pumps bring the water to the surface.

Water in the unsaturated zone is the primary water source for most plant roots. Soil feels moist when you dig below the surface because there are thin films of water on the particles. Electrical charges at the molecular scale attract water molecules to adhere on mineral surfaces in the soil and regolith in this zone.

Why do separate saturated and unsaturated zones exist? *If* Earth materials were uniformly highly porous and permeable, as portrayed in **Figure 17.7**a, then gravity would consistently pull water down toward the center of the planet, and it would seemingly never saturate pores close to the surface. Earth materials are *not* uniformly well suited for water storage or flow; see Figures 17.2, 17.5, and Table 17.1 for comparisons.

In general, porosity and permeability decrease downward in Earth. Why is this? Compared to the situation in shallow sediment layers, grains in deeply buried sedimentary rocks are closely compacted together, and pore spaces more likely are filled with cementing minerals. Porosity in igneous and metamorphic rocks that compose most of the crust is typically limited to fractures, which only form in the brittle upper crust. This means that there is effectively a downward limit to which water can readily move, so the pore spaces in the near-surface materials fill up to produce the saturated zone.

Why, then does the saturated zone not rise up to the surface with the addition of more water, as depicted in Figure 17.7b? In some places, this actually does happen, such as where ground water seeps into streams (Figure 17.1 and 17.3) or lakes (Figure 17.4). More generally, however, ground water does not flood out onto the surface. Earth's surface is not flat and the water table only reaches the ground surface at the low spots, whereas higher surface elevations remain above the water table. Figure 17.7c shows that removal of ground water at the low elevations in the landscape keeps the water table from rising to the surface everywhere.

► **Figure 17.7** **Understanding why the water table is not flat.**

► **Figure 17.7** **Understanding why the water table is not flat.**

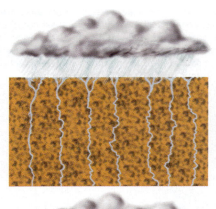

If Earth materials were uniformly porous, then water would infiltrate deep into Earth and there would be no ground water near the surface **But...**

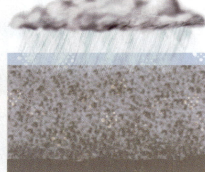

... porosity and permeability decrease downward, so infiltration is limited and pores fill with water.

If Earth's surface was flat, then water would fill up to the surface. **But...**

Rising water table

... the land surface is uneven, so ground water only rises to discharge points where it is taken away by surface flow in streams.

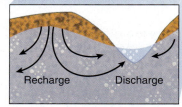

Water table highs and lows usually mimic the land surface.

Recharge   Discharge

Ground water flows from recharge areas to discharge areas.

Active Art

Why Is There a Water Table? *See why a water table exists.*

Why Water Is Sometimes Found at Surprisingly Shallow Depths

Figure 17.8 depicts the same hypothetical area as seen in Figure 17.3, with a newly drilled well encountering ground water high above the predicted water table. Clearly, this is advantageous to the landowner, because the drilling expense was less than expected, and it will be less expensive to pump the water a shorter distance to the surface. What causes this unexpectedly shallow ground-water resource?

An important clue appears on a nearby hillside, where ground water emerges at a spring far above the valley floor. Close examination of the rocks exposed near the spring shows that the water emerges where a porous and permeable sand

▶ **Figure 17.8 Visualizing perched ground water.**
Impermeable layers in the unsaturated zone may interrupt the downward infiltration of water to perch some ground water above the water table. These local areas of saturation within the otherwise unsaturated zone provide water to shallow wells. Perched ground water may also flow to the surface to produce springs on valley walls.

layer rests on top of hard, impermeable shale. The impermeable shale stops the downward-moving water and causes a local pooling, a local saturated horizon within the otherwise unsaturated zone. This is an example of **perched ground water**, where low permeability in specific spots creates a shallow saturated horizon perched above the water table.

Characteristics of Aquifers

Variations in porosity and permeability clearly affect where ground water is stored and how readily it flows. This means that not all geologic materials provide equally good supplies of ground water. An **aquifer** is a body of rock or regolith with sufficient porosity and permeability to provide water in useful quantities to wells or springs. **Confining beds** (sometimes also called aquitards) are the contrasting low-permeability materials that restrict the movement of ground water into or out of adjacent aquifers.

The term "useful" in the definition of aquifer may sound nebulous, but this is because not all aquifers are equal. A rock layer with relatively modest porosity and permeability may provide a sufficient aquifer for scattered wells supplying water to individual homes where water need is perhaps less than 100 liters per minute. Wells supplying water for a city, however, need to produce thousands of liters of water per minute and require aquifers with greater porosity and permeability.

The best aquifers are Earth materials that have both high porosity and high permeability, as shown in **Figure 17.9**. This combination of characteristics means that there are large volumes of water stored in the aquifer (because of high porosity) and the water readily flows in large volumes to pumping wells

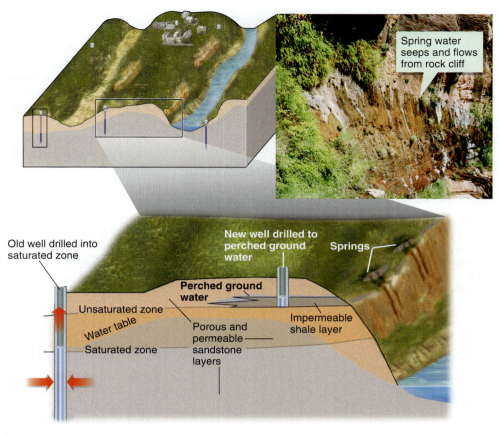

Spring water seeps and flows from rock cliff

EXTENSION MODULE 17.1
Anatomy of a Water Well.
Learn how a well is drilled and how ground water is withdrawn from a well.

Limestone aquifers

Sandstone aquifers

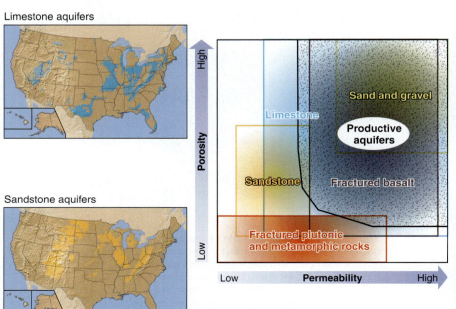

Sand and gravel aquifers

Edge of glacial deposits

Igneous/metamorphic rock aquifers

◀ **Figure 17.9 Aquifer types and properties.** Permeability and porosity vary for different rock types and are highest for loose sand and gravel. Materials with high porosity and high permeability form the most productive aquifers. Aquifer materials vary across the United States depending on the local geology.

▲ Figure 17.10 Ground water gushes from ancient lava flows. Ground water pours from a basalt aquifer at Thousand Springs, Idaho, which demonstrates the high permeability and porosity of fractured volcanic rocks. Water wells tapping this aquifer yield as much as 450 liters of water per second. This well yield would fill an Olympic-size pool in less than an hour.

(because of high permeability). In many parts of the United States, the most productive aquifers are sandy or gravelly sediment, or sandstone and conglomerate that is only slightly consolidated and cemented into rock, so that porosity and permeability remain high (Figure 17.9). Many of these aquifers are located in geologically recent deposits of streams and glaciers.

Some rocks also form good aquifers (see Figure 17.9). Ground water partially dissolves limestone (by processes explored in Section 17.4), which enlarges pores to the point of forming large caverns that fill with water and make superb aquifers. Open space along sedimentary-rock bedding planes allows rapid movement of ground water between rock layers. Volcanic rocks are usually very porous and permeable, as illustrated in **Figure 17.10**, because they have many fractures formed by breakage of the congealing lava as it flowed or cracks that formed by contraction when the rock cooled. Any rock with closely spaced fractures, including metamorphic and intrusive igneous rocks, can possess sufficient permeability to form a productive aquifer.

Water-Table Level Changes through Time

The inflow and outflow of ground water from an aquifer affect the elevation of the water table. Water exits the saturated zone where the water table intersects the land surface or where wells extract water. If more water does not enter the saturated zone to replace the water that leaves, then the volume of water in the saturated zone diminishes, and the elevation of the water table declines. Water removed from the saturated zone is discharge, and water added to the saturated zone is **recharge**.

Natural recharge occurs wherever surface water infiltrates completely through the unsaturated zone. The recharging water is rainfall and snowmelt that soak into the soil or stream water that infiltrates through the bottom of a streambed located at a higher elevation than the water table. Notice in **Figure 17.11** that some streams gain flow from ground water, whereas others lose flow to ground water. This is why some streams flow year-round, regardless of when

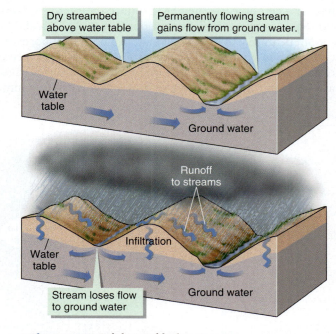

▲ Figure 17.11 Gaining and losing streams. Permanently flowing streams intersect the water table and gain discharge from ground-water flow. Streams not in contact with the water table are dry during periods of limited rainfall. When streams above the water table receive runoff, some of the flow is lost to ground water by downward infiltration through the streambed.

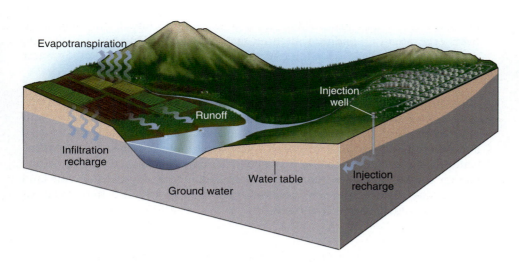

◄ **Figure 17.12 Artificial recharge.**
Humans redistribute surface water and inject treated waste-water into aquifers. Irrigation and injection wells are sources of artificial recharge. Water not evaporated or used by plants during irrigation of cropland or pastures produces runoff and infiltration to the water table.

the last precipitation fell on the surface, whereas others dry up shortly after each rainfall.

People redistribute water on the surface or inject it into the aquifer, causing artificial discharge or recharge, as depicted in **Figure 17.12**. Agricultural irrigation removes water from streams or wells and then spreads it across cropland and pasture. The water that does not evaporate, run off to streams, or get used by plants slowly seeps downward to the saturated zone. Some communities inject treated wastewater into aquifers to partly offset the amount of water discharged through wells.

Where the water table is close to the surface, its elevation may fluctuate seasonally or even daily, as **Figure 17.13** shows. Throughout the last 10,000 years, water-table elevation has decreased naturally and gradually in many regions in response to drier climates in the wake of the last ice age. Where ground water is heavily used for irrigation, industrial uses, or to supply cities,

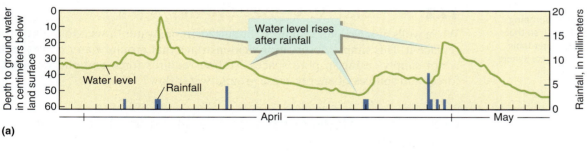

(a)

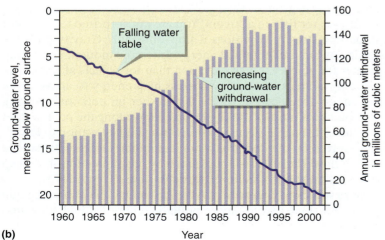

(b)

◄ **Figure 17.13 The water table changes over time.**
(a) Shallow water tables may fluctuate over hours or days in response to infiltration following precipitation. The water level in this shallow well in Nevada rises shortly following each rainfall event. (b) Water-table elevations change even more dramatically over longer time intervals where water is pumped from aquifers. This graph shows the falling water table measured in a non-pumping well in Albuquerque, New Mexico. The decline is caused by withdrawal of ground water throughout the city, for use by residents and businesses, which far exceeds the quantity of water that naturally enters the aquifer.

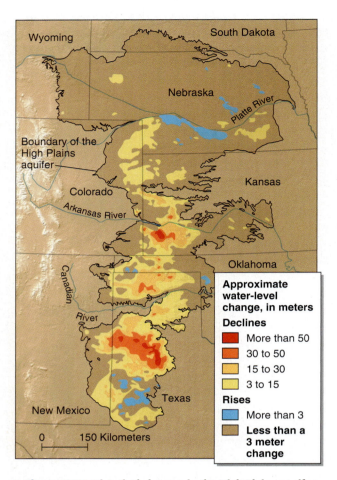

▲ **Figure 17.14 Historical changes in the High Plains aquifer.** Withdrawal of ground water for irrigation has caused widespread decline in the water table over large areas of the High Plains since the late 1800s. Just within the Texas Panhandle and adjacent New Mexico there are more than 170,000 wells pumping out a total of 6 million cubic meters of water each year, so that half of the aquifer thickness has been drained. The water table rose during this same time period in relatively small areas where stream water is diverted for irrigation and then artificially recharges the aquifer.

the discharge usually exceeds the recharge, causing decline in the water table over large areas, a condition pictured in **Figure 17.14**. This may require the drilling of new wells or deepening of existing wells to "chase" the falling water level in the aquifer.

What happens to the land surface when the water table drops? **Figure 17.15** illustrates that pumping large quantities of ground water causes the land surface to sink. Pore water in sediment layers partly supports the weight of the sediment grains. When ground-water discharge exceeds recharge, the water table lowers, and this causes the sediment grains to compact more closely together where water disappears from the pores. The compaction decreases the porosity of the aquifer and also decreases the volume of the sedimentary layers so that the land surface sinks. This subsidence, or sinking, may cause large cracks to open across roads and below building foundations.

> *Putting It Together—What Is Ground Water and Where Is It Found?*
> ■ Ground water occupies pore spaces in regolith and rock. Porosity is the percentage of the rock or regolith occupied by pores.
> ■ The water table separates the unsaturated zone (where air and water fill pores) from the saturated zone (where water completely fills pores). Ground-water wells are drilled below the water table.
> ■ Aquifers are saturated Earth materials that supply ground water. Productive aquifers have high porosity and high permeability.
> ■ Water-table elevation remains the same only if the amount of water recharging the saturated zone equals the amount of water discharging to the surface at springs, streambeds, lakes, or wells.
> ■ In most places, discharge through wells exceeds recharge, so water levels are falling. Falling water level in unconsolidated sediment may cause the sediment grains to compact more closely together, which leads to land-surface subsidence.

17.2 Why and How Does Ground Water Flow?

When wells pump ground water to the surface, more water flows into the well to replace it. This simple observation demonstrates that ground water moves and is not simply sitting in pore spaces. Motion requires force and energy. What forces and sources of energy cause ground-water flow?

Ground Water Flows from Areas of High Energy to Areas of Low Energy

To understand the energy in ground-water flow, let us first consider the more familiar case of surface-water flow in a stream. You are not surprised to see water flowing downslope in a stream. Why does it flow downhill? Gravity pulls the water molecules downward and potential energy drives the motion. The potential energy is greater for objects at high elevation than for objects at low elevation. Objects move from areas where they possess high energy to areas where they possess low energy (see Section 1.6 for a refresher on potential energy). Motion energy accounts for most of the change in energy along the path of flow. This is why water flows through a stream channel from high elevation to low elevation. Pressure exerted by the depth of water is another, more subtle source of energy for stream flow. The effect is appreciated by imagining flow in a steep channel that gets shallower and shallower through time. Eventually the flow stops because it lacks sufficient energy to move, even though the remaining thin film of water rests on a slope and has more potential energy at high elevation than at low elevation.

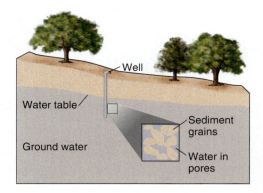

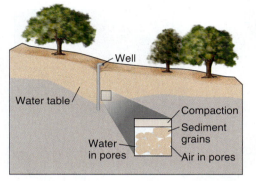

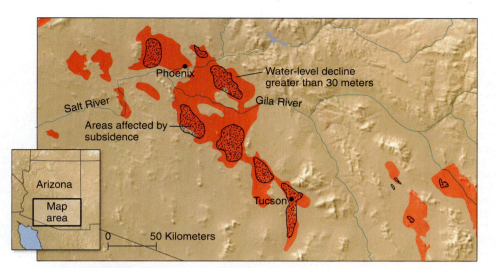

▲ **Figure 17.15 Subsidence caused by ground water withdrawal.**
Withdrawal of ground water from pore spaces in loosely consolidated sediment causes the sediment grains to compact more closely together. This compaction causes the land surface to subside. For example, ground-water withdrawal for irrigation and drinking water has caused both large declines in water levels and large areas of subsidence in southern Arizona. The photographs show open fissures caused by subsidence.

Ground water flows away from areas where the water-table elevation is high toward areas where the water table is low, because of the differences in potential energy. The slope of the water table, therefore, determines the direction of ground-water flow. This does not mean, however, that ground water simply moves along the top of the water table from high elevation to low elevation like surface water flowing along a streambed. Ground water flows through pores *everywhere* in the saturated zone, which requires taking into account energy generated by pressure.

Remember that *if* Earth materials were uniformly highly porous and permeable, then water would always flow downward (see Figure 17.7a), due to the force of gravity. Instead, aquifers effectively have a bottom, where porosity and permeability are so low that there is only a tiny amount of water present and it is barely able to move (see Figure 17.7c). The deeper the water is in the aquifer, however, the greater the pressure exerted on it by the weight of water in the

▶ **Figure 17.16**

Understanding the direction of ground-water flow.
The direction of ground-water flow between any two locations in an aquifer is determined by comparing the total energy of the water at each location. The total energy is the sum of the potential energy, determined by elevation, and the pressure energy, which relates to proportional to distance below the water table because the pressure energy is determined by the weight of overlying water in the aquifer. In this example, the overall flow pattern is downward and horizontal below the hill and upward toward the stream.

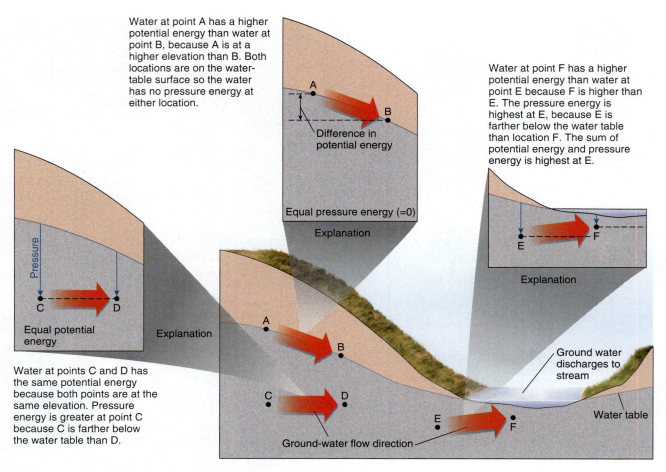

Water at point A has a higher potential energy than water at point B, because A is at a higher elevation than B. Both locations are on the water-table surface so the water has no pressure energy at either location.

Water at point F has a higher potential energy than water at point E because F is higher than E. The pressure energy is highest at E, because E is farther below the water table than location F. The sum of potential energy and pressure energy is highest at E.

Water at points C and D has the same potential energy because both points are at the same elevation. Pressure energy is greater at point C because C is farther below the water table than D.

pore spaces above. The pressure may not simply push the water deeper, it may also cause it to move horizontally or possibly even force it upward. An analogy is seen by pressing down on a saturated sponge resting on a countertop. The water cannot pass downward into the impermeable countertop so some of it flows out the sides and top of the sponge as you exert pressure.

The energy that drives ground-water motion is the sum of the potential energy, related to elevation, and pressure energy, determined by the weight of the overlying pore water. Potential energy is greater at the surface than at lower elevations below ground, but pressure energy increases with increasing depth below the water table, so ground-water flow from high energy to low energy is much more complex than water flowing in a stream.

Figure 17.16 explains ground-water flow by considering the combined effects of potential energy and pressure. Water flows from regions of high energy to low energy as determined by the combined effects of elevation and pressure. Ground-water flow, therefore, is not always downward and is commonly more or less horizontal, and is directed upward at many locations.

Ground water flows toward discharge locations (Figure 17.16). The pressure is lowest at discharge locations because water withdrawals decrease the weight of water. Where ground water discharges into low-lying valleys, the potential energy is low because the elevation is low, and the pressure energy is low because there is no overlying ground water.

Where Ground Water Flows

Figure 17.17 shows that ground water follows curving paths through the saturated zone from relatively high areas on the water-table surface to discharge locations. The curving flow paths result from the combined effects of gravity pulling the water down, pressure forcing the water laterally, and the sum of the

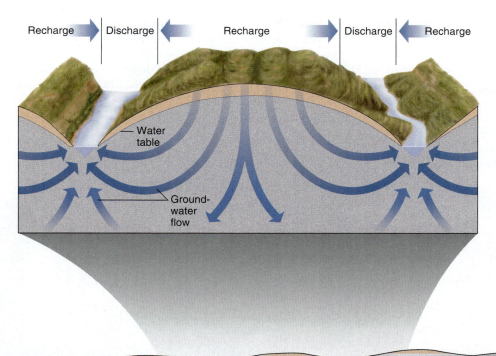

Recharge | Discharge | Recharge | Discharge | Recharge

Water table

Ground-water flow

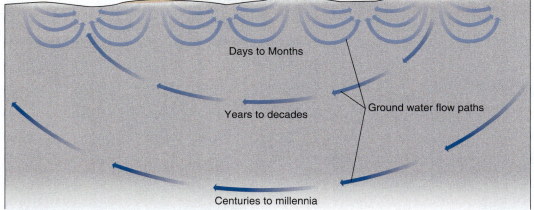

Days to Months

Years to decades

Ground water flow paths

Centuries to millennia

◀ **Figure 17.17 Ground water moves from recharge areas to discharge areas.**
Ground water follows curving paths down and away from recharge areas where water enters the aquifer and up and toward discharge areas where water leaves the aquifer. The time that the water spends in the aquifer depends partly on porosity and permeability, partly on the elevation difference of the water table between recharge and discharge areas, and mostly on the length of the flow path. Water following short flow paths close to the surface may only reside in the aquifer for days to years. Water following routes deep below the surface may not reach discharge locations for centuries or millennia.

▼ **Figure 17.18 Ground-water flow in a confined aquifer.**
Confined aquifers form where impermeable or low-permeability confining layers separate more permeable aquifer materials. Once channeled between the confining layers, the deeper flow remains mostly or entirely separate from the ground water in the shallower, unconfined aquifer.

two being lowest where water discharges. Where the distance between points of recharge and discharge is short, the curving flow path remains close to the water table. Where the distance between recharge and discharge locations is long, then the curving flow path extends deeply down into the saturated zone (Figure 17.17).

Ground-water flow is a little more complicated in sedimentary aquifers consisting of both permeable and impermeable layers. **Figure 17.18** illustrates ground-water flow in a deep, permeable sand layer between impermeable clay confining beds. This is an example of a **confined aquifer** where impermeable layers separate, or confine, the permeable aquifer layers. The confining beds keep the shallow and deeper ground water from mixing.

Deeply Flowing Ground Water Forms Hot Springs

Deep, far-traveling ground water passes through rocks that are much warmer than materials close to the surface. Heat conducts from rock into water more readily than heat conducts from water into rock. This means that ground water acquires the temperature of the warmest rock that it passes through and then cools only slightly while moving toward discharge to the surface or a well. Warm springs or hot springs (distinguished only by the relative warmth of the discharging ground water) have long been popular for relaxing and purportedly

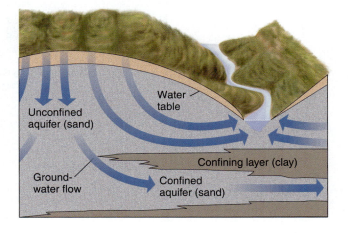

Unconfined aquifer (sand)

Water table

Ground-water flow

Confining layer (clay)

Confined aquifer (sand)

🔵 **EXTENSION MODULE 17.2**

Darcy's Law.
Learn about Henry Darcy's experiment that defined a simple mathematical formula for describing ground-water flow.

therapeutic soaking. **Figure 17.19** illustrates that the discharging ground water is warm because it followed a subsurface path through warm rocks.

Ground water near shallow magma chambers in volcanically active areas may encounter rocks heated to more than 100°C at depths less than 1 kilometer (see Figure 17.19). This means that even shallowly circulating ground water can

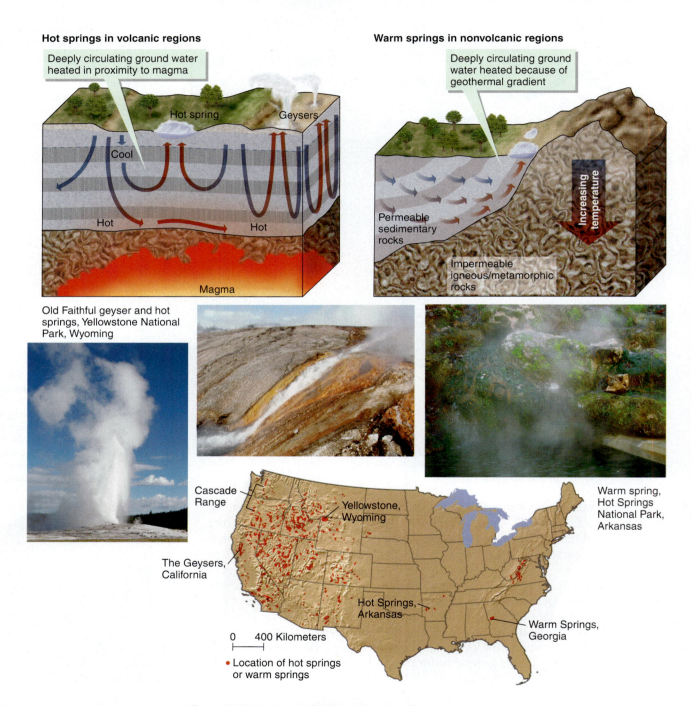

▲ **Figure 17.19** **How ground water gets warmed up.**
Hot springs and warm springs form where ground water circulates through hot rocks and then discharges at the surface. The hottest springs form where ground water heats up in close proximity to molten magma or igneous intrusions that have recently solidified and remain very hot. Many hot springs in the western United States relate to these igneous processes, including those in Yellowstone National Park, the Cascade Range in Oregon and Washington, and The Geysers north of San Francisco, California. Cooler warm springs form where deeply circulating ground water rises to the surface, usually along a boundary between permeable and impermeable rocks. The water heats up by conduction where it passes through warm rock at depth and then carries the heat to the surface.

discharge as boiling hot springs at the surface. Well-known examples attract tourists and scientists to Yellowstone National Park.

Deeply circulating ground water causes warm springs in locations remote from volcanoes, such as Warm Springs, Georgia, and Hot Springs, Arkansas (see Figure 17.19). With typical geothermal gradients, ground water that circulates to a depth of 2 kilometers is heated to about 60°–75°C, making it comfortably warm when it discharges at the surface. Warm springs in the eastern United States, and other nonvolcanic regions of the world, are always associated with upward flow and discharge of deep ground water, usually where flow in thick sedimentary-rock aquifers is diverted to the surface along steep contacts with less permeable, commonly metamorphic or igneous, rocks.

Why the Water Table Is Depressed Near a Pumping Well

The physical process of ground-water flow also explains observations of the water table near pumping wells. **Figure 17.20** shows how the shape of the water table changes near a pumping well. Monitoring of water levels in closely spaced wells shows that the water-table elevation declines as water is withdrawn, and the greatest decline occurs at the pumping well. The reshaped water table resembles an inverted cone centered on the well (Figure 17.20). This local lowering of the water-table elevation is the **cone of depression**.

Water-table elevation declines when discharge exceeds recharge (Figure 17.15). It seems reasonable, therefore, to attribute the cone of depression to pumping water from the aquifer faster than water can flow toward the well to replace what is pumped. Surprisingly, though, after a while the cone of depression no longer enlarges even though water is still pumped from the well at the same rate. To understand why this happens, remember that the well is a point of ground-water discharge, so water flows toward the well. The direction of ground-water flow parallels the slope of the water table, so the cone of depression merely establishes the water-table slope required for water to flow to the discharging well. As the cone enlarges, the well draws ground water from a larger and larger region of the aquifer. Eventually, the increased flow toward the well balances the discharge of water pumped from the well, and the water-table elevations stabilize.

The shape of the cone of depression, therefore, reflects the balance between the pumped water and ground-water flow toward the well. If the discharge increases, then the cone of depression enlarges and then stabilizes again. If the discharge decreases, then the cone of depression shrinks to a new stable configuration. As long as water is pumped from the well, there will always be a cone of depression because ground water flows toward the well. The depth of the cone of depression will never fall below the bottom of the pumping well, because if the well does not discharge water, there is no flow toward the well and there is no cone of depression. Other nearby wells that are pumped infrequently or with very small discharges compared to a neighboring heavily pumped well may, however, be adversely affected by the cone of depression that forms around the heavily pumped well (Figure 17.20).

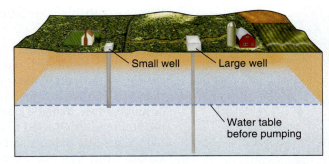

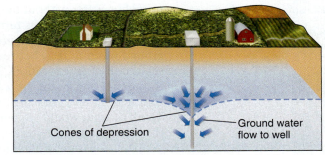

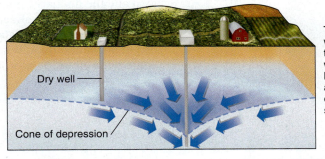

The water table always slopes inward toward a pumping well because ground water always flows parallel to the slope of the water table to a discharge point. The inward-sloping water table has the shape of an inverted cone where the water table elevation is depressed.

Each well draws ground water from within the boundary of its cone of depression, so the size of the cone is determined by the amount of water withdrawn.

The cone of depression stabilizes when the amount of water flowing to the well matches the amount of water pumped from the well. The large cone of depression around a well that pumps a large amount of water may cause adjacent smaller discharge wells to dry up.

▲ **Figure 17.20** How a cone of depression forms.

Forming a Cone of Depression: *See how a cone of depression forms.*

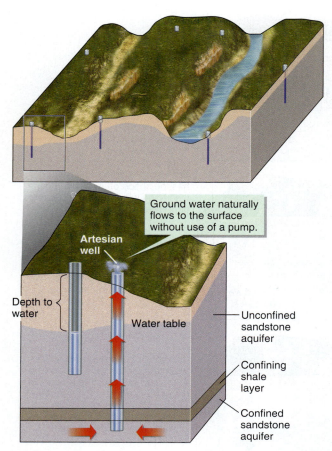

▲ **Figure 17.21 Artesian wells.**
Ground water usually rises within wells that are drilled into a confined aquifer that exists below a confining layer, like an impermeable shale bed. The water in the confined aquifer is under artesian pressure, which causes it to flow up the well without being pumped. In the case of a flowing artesian well, the ground water flows all of the way to the ground surface and may even fountain into the air. In contrast, ground water that enters a well drilled into the unconfined aquifer immediately below the water table must be pumped to the surface.

▶ **Figure 17.22 How an artesian aquifer works.**
The combined pressure and potential energy in a confined aquifer cause water to rise higher than the water table and even higher than the land surface where wells or faults penetrate the upper confining layer. The rise of water in wells can happen where the recharge area is much higher than where wells are drilled. The effect is very similar to how public water supplies are delivered from high water tanks. Water is pumped into the water tank and then flows to users. The total potential and pressure energy allows the water to rise in the plumbing nearly as high as the water level in the tank. The water-level elevation decreases with greater distance from the recharge area, for an artesian aquifer, and from the water tank, in a public water system, because energy also converts to heat by the friction of water flowing through pores spaces and pipes.

Why Ground Water Sometimes Flows Higher than the Ground Surface

Figure 17.21 shows another well drilled in the original, hypothetical study area to a greater depth than the other wells, and surprisingly the water flows from the top of the well without using a pump. The water pressure at the bottom of the well is sufficiently high to force the water up to an elevation higher than the ground surface. When water rises in a well or at a spring above the elevation of the aquifer, it is referred to as **artesian** discharge. Flowing wells and springs are the special case of artesian flows that rise on their own all of the way to the land surface. Although not common, flowing artesian wells have the advantage of providing water without the cost of pumping water from the well.

Artesian conditions form where a confined aquifer recharges at a high elevation, as explained in **Figure 17.22**. The confining impermeable layers force the ground water to flow between them rather than following curved paths back to surface-discharge locations. The upper confining layer is like a lid on this part of the aquifer and keeps the water at high pressure. When a well is drilled into the confined aquifer, the high-pressure water rapidly discharges from the well. Think of what would happen if you punched a hole in a running garden hose—the effect is similar to drilling a well into a confined aquifer.

The elevation of the recharge area and the distance between the recharge area and the discharge point determine the height to which water rises from the confined aquifer. The effect is similar to dropping a ball from a measured height above the floor. The ball starts out with the potential energy associated with the height above the floor. When the ball hits the floor, it bounces back up to nearly the same elevation as where it started. Energy cannot be gained or lost, so the

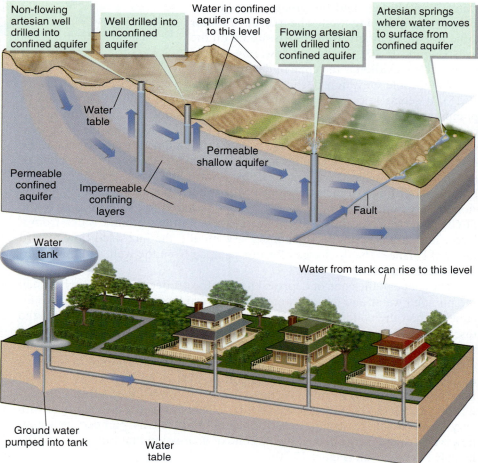

ball has to have the same energy after it bounced as before it was dropped, except for whatever energy was expended by frictional heating when the ball struck the floor. The ball rises less high on each successive bounce as more of the potential energy converts to heat by the friction of the bounce. In a similar fashion, the potential energy of the water in the confined aquifer is the energy associated with the high elevation where it entered the aquifer. The water would rise to the same elevation within the artesian well if it were not for energy expended as heat by the friction of the water flowing through the pores in the aquifer.

Water-supply systems use the same physical principles to distribute water through a community, as illustrated in the lower part of Figure 17.22. Water pumped into a high water tower has sufficient potential energy to flow through pipes in all buildings that are at a lower elevation than the tower.

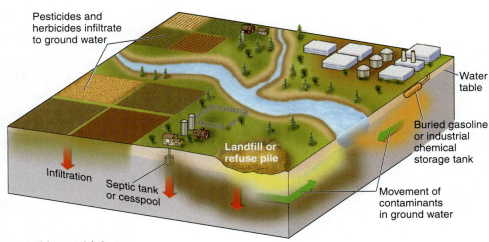

▲ **Figure 17.23 Sources of ground-water pollution.** Possible sources of pollution to ground water include infiltration of herbicides and pesticides from farmland, leaky underground storage tanks, leaky septic systems, and landfill leaching.

Pollution Also Moves in Ground Water

Understanding how ground water flows also explains risks of pollution to aquifers. Human activities introduce harmful chemicals or unhealthy microbes into ground water. Most pollutants enter the ground water with recharge, move with the ground water, and exit in discharge areas. Pollutants threaten the drinking-water supply where they enter wells. Where pollutants discharge to streams, they degrade the quality of surface water and affect stream ecology and drinking water obtained from streams.

Figure 17.23 illustrates some sources of pollution. Chemicals spilled on the surface or applied as herbicides and pesticides to farmland seep down to the saturated zone and flow in groundwater. Water soaking through landfills dissolves some materials or mixes with harmful fluids that enter the ground water. Septic systems may leak waste into ground water. Underground gasoline, fuel oil, or chemical storage tanks also may leak.

Geologists apply their understanding of the direction and path of ground-water flow to determine where pollution travels. **Figure 17.24** shows an example of ground-water contamination by water percolating through a landfill. Pollution initially threatens only one of the local ground-water supply wells, because ground water does not flow toward the other wells or springs. One of these wells pumps at an increasing rate, however, and the cone of depression in the water table threatens to cause polluted ground water to flow toward the well. Figure 17.24 also reveals that water pumped from a deeper confined aquifer is unaffected by the landfill pollution. Many

▶ **Figure 17.24 Contamination by a landfill.** Flow of liquid leachate from a landfill follows the flow directions of the ground water. Flow directions can locally reverse, however, if large cones of depression form around pumping water-supply wells. Knowledge of ground-water flow not only predicts the wells that will be affected by the pollution but also aids in cleaning up the contamination.

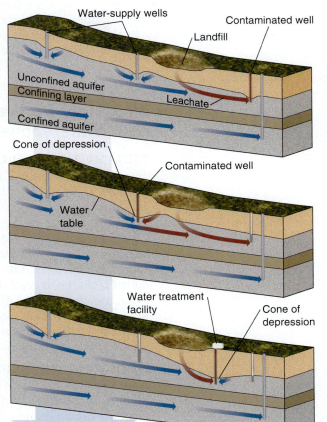

Liquid leachate from landfill infiltrates to the ground water and flows into the water-supply well. Wells upslope of the landfill are not affected because leachate does not flow toward those wells. The well drilled into the confined aquifer is unaffected because the confining layer restricts contamination to the unconfined aquifer.

Increased pumping of the initially safe water-supply well enlarges the cone of depression, which draws the leachate into the well and contaminates it.

A new well is drilled to remove and treat the polluted ground water. The well is located on the down-flow side of the landfill and is pumped heavily to create a cone of depression that draws in contaminated water and does not allow the leachate to travel further through the aquifer.

cities purposely drill to deeper confined aquifers, if they exist, to avoid near-surface pollution, even though the cost is greater to drill a deep well compared to a shallow one.

Figure 17.24 also shows how to use knowledge of ground-water flow to halt the spread of the contamination and possibly even clean up the water. A well drilled within the contaminated part of the aquifer creates a cone of depression that draws the polluted water to the well. The polluted water not only stops flowing toward the drinking-water well, but the contaminated water can be pumped from the new well, treated to remove the contaminants, and then either used at the surface or injected back into the aquifer.

Putting It Together—Why and How Does Ground Water Flow?

- Ground water flows from points of high potential and pressure energy to points of low energy, and from high elevations on the water-table surface toward the low elevations.
- Ground water flows along curving paths from recharge areas at the water table toward discharge points into surface water or wells. The farthest traveled water follows the deepest flow path through the aquifer.
- Water flowing along deep flow paths heats up because rocks are hotter at greater depth below the surface. Warm springs and hot springs form where this heated water returns to the surface.
- Artesian pressure develops where confined aquifers are recharged at significantly higher elevations than where they discharge, so that the water may rise above the land surface.
- Cones of depression in the water-table surface form where wells discharge ground water. The water-table surface slopes in all directions toward the well so that the amount of ground water flowing to the well is equal to the amount pumped from the well.
- The occurrence and travel direction of ground-water pollution is predicted by understanding the direction of ground-water flow.

17.3 How Do We Know . . . How Fast Ground Water Moves?

PICTURE THE PROBLEM
How Fast Does Ground Water Flow?
Sections 17.1 and 17.2 establish the physical processes and material characteristics that determine how and where water moves in an aquifer. Not yet addressed, however, is how fast ground water moves. It is easy to think of situations where it would be useful to know flow velocity. Consider a scenario where water levels declined in wells during a period of drought, but recently there have been several consecutive wet years. How long will it take the newly added recharge to reach the wells and potentially raise the water levels? Consider, too, the possible discovery of contamination in ground water just a few kilometers from a city water well. How long will it take the pollution to reach the water well, and how might this duration influence decisions on how to clean up the pollution or the need to drill a new water well elsewhere?

DESIGNING THE EXPERIMENT
How Can Ground-Water Velocity Be Measured?
Measuring the velocity of ground water is not as straightforward as measuring flow in a stream channel. Geologists easily know stream flow by placing meters into the water that measure how fast the water is moving (see Figure 16.5). Clearly, it is not easy to design a meter to insert within the tiny pore spaces of an aquifer to measure the velocity of the flowing water.

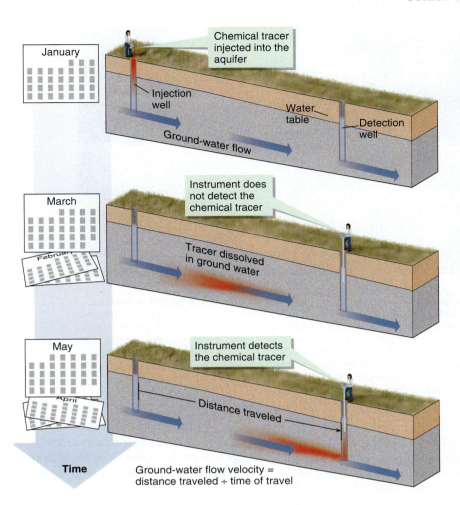

◀ **Figure 17.25 Using a tracer to measure ground-water flow velocity.** Ground-water flow velocity can be measured by injecting into the aquifer a nontoxic chemical compound that dissolves in water. The tracer is then tested for in one or more detection wells that are located along the direction of ground-water flow that is known from the slope of the water table. Water samples can also be withdrawn from the well to analyze the exact concentration of the tracer.

There is also a good reason to expect that ground water moves more slowly than surface water. Only a small part of the volume of water flowing in a stream channel slows by friction with solid banks and bed. For ground water, however, most of the water in a tiny pore or fracture is in contact with surrounding solids, so friction is more important in slowing the flow. To measure the velocity of ground water geologists need a method that detects very small flow velocities and does not require placing a device in the underground flow.

The most effective way to measure ground-water flow velocity, illustrated in **Figure 17.25**, makes use of an easily dissolved chemical substance, called a tracer, injected at a well. Water from other wells located along the ground-water-flow path are repeatedly sampled and analyzed for the presence of the tracer. The time elapsed for the tracer to travel from the injection well to a detection well reveals the average ground-water flow velocity.

Research conducted in the 1980s by the U.S. Geological Survey at Cape Cod, Massachusetts, serves as an example of such a study. The aquifer at Cape Cod is composed of unconsolidated sand and gravel. The water table is only about 5 meters below the ground surface, so it is easy and inexpensive to drill injection and detection wells into the saturated zone. Bromide ion was chosen as the injected tracer because it easily dissolves in water and is not harmful in low concentrations. The hydrologists injected 7.6 kilograms of dissolved bromide into the aquifer over 17 hours through three closely spaced wells.

Figure 17.26 shows the Cape Cod project layout. Detection wells were sited only in the direction of expected ground-water flow from the injection wells. Data from a few preliminary wells reveal that the water table slopes down toward the south. Ground water flows in the direction of decreasing water-table elevation, so the detection wells were only located south of the injection wells

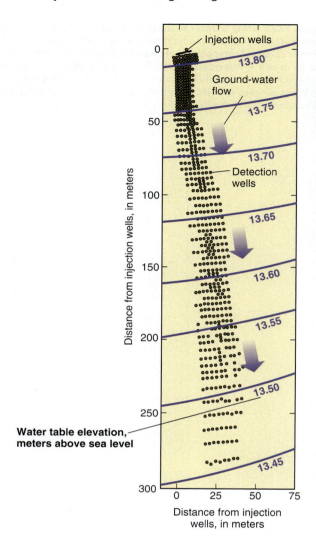

Water table elevation, meters above sea level

Distance from injection wells, in meters

Distance from injection wells, in meters

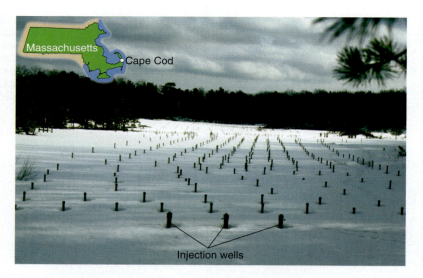

▲ **Figure 17.26 Designing a research project to measure ground-water flow.** The map shows the outline of the project area on Cape Cod. Six hundred fifty-six monitoring wells were drilled in an area extending southeast from the injection wells. The injection and detection wells stick above the snow at the research site in the photograph. The monitoring wells were drilled southeast of the injection wells because the measured water-table levels, which are contoured on the map, are high in the northwest and low in the southeast, which indicates ground-water flow to the southeast.

(Figure 17.26). Six hundred fifty-six detection wells were drilled as much as 10 meters below the water table. Water samples were extracted from some or all of the wells and analyzed for bromide concentration on 16 occasions over an 18-month period after injecting the tracer.

ANALYZING THE DATA
How Fast Did the Ground Water Flow?
Figure 17.27 maps the movement of bromide ion through the Cape Cod aquifer. Thirty-three days after injection, the highest bromide concentration was 17 meters from the injection wells. Samples collected 461 days after injection revealed bromide in many wells, with the center of the tracer located 198 meters from the injection wells. Using these distances and the elapsed times, the average velocity of the ground water is about 0.43 *meters per day*, or a distance equal to about one and a half times the height of this book page in a day. By comparison, water in a fast-flowing stream moves faster than 0.43 *meters per second*. Keep in mind, too, that permeability is high in gravel and sand so this is a rapid ground-water flow velocity. As expected, ground water flows very slowly!

The data depicted in Figure 17.27 also show other important features. First, the ground water does not flow at a uniform velocity. Four hundred sixty-one days into the experiment, some of the wells containing bromide are only 150 meters from the injection wells, whereas others are 235 meters from the injection wells. This means that some of the bromide moved in ground water at only 0.33 meters per day, whereas other water flowed at an average velocity of 0.51 meters each day. Second, the bromide concentration started out at 640 parts per million at the injection wells, but after 461 days the highest detected concentration was only 39 parts per million. The data show that the change in concentration resulted from the bromide spreading out from the small injection site to occupy about 7400 cubic meters.

Figure 17.28 illustrates how hydrologists explain why the velocity is not uniform and why the bromide spread out through the ground water. Ground water carrying the bromide tracer ions follows circuitous curving paths around the

▶ **Figure 17.27 Tracking the tracer.**
▶ **Figure 17.27 Tracking the tracer.**
The colors in this map depict measured bromide concentration in the detection wells at three stages during the experiment. The bromide was detected progressively farther to the southeast of the injection wells for more than a year after the experiment began.

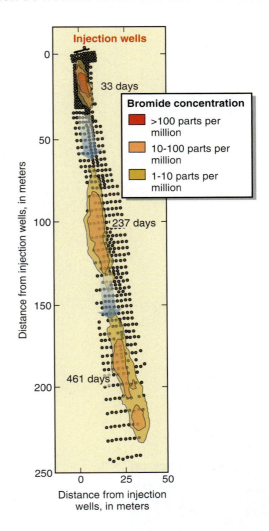

sand grains separating the pores in the aquifer. Some paths are fairly straight and the straightest paths are the fastest. Ground water following the most roundabout path around the sand grains takes longer to cover the same straight-line distance from the injection point. Every place the bromide-bearing water divides around a sand grain, the tracer spreads out over a larger area, so the concentration at any one location is less than it was at the injection point. In addition, each path through the aquifer experiences variations in permeability resulting from different sediment sizes. This means that ground water flows fastest along paths with the highest average permeability.

INSIGHT

How Is This Knowledge Used to Assess the Movement of Contaminants in Ground Water?

One motivation for understanding ground-water flow velocity is to understand how quickly polluted ground water flows toward streams or drinking-water wells. The very slow velocities revealed by the Cape Cod experiment suggest that unless a stream or well is located very close to a contamination source, and as long as the contamination is detected early, there is usually enough time to undertake steps to stop and clean up the pollution.

The way the tracer spreads through the aquifer also demonstrates that the concentration of a contaminant may naturally diminish below harmful levels at long distances from the pollution site. On the other hand, the farther the contaminant travels, the larger the area affected, because the contaminant spreads out in the aquifer rather than traveling a straight line. If a contaminant is

▼ **Figure 17.28 Contamination spreads out in an aquifer.**
This series of diagrams schematically shows the path of contamination dissolved in ground water as it moves through an aquifer. The flowing water divides around sediment grains and flows along many curving paths of different lengths. As a result, the contaminant spreads out in the aquifer both parallel to and perpendicular to the overall direction of ground-water flow. This process also dilutes the contaminant, because it gradually spreads out through a progressively larger volume of water.

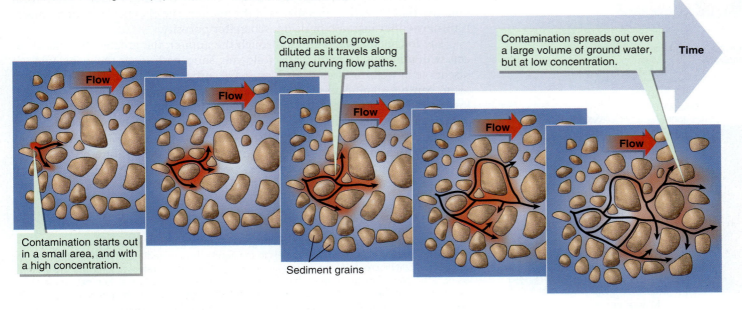

hazardous at very low concentrations, then it is also important to remember that the small amount of the contaminant that follows the shortest, fastest path through the aquifer moves faster than the average velocity and reaches wells or streams long before the more concentrated pollution that moves an average distance at an average velocity.

> *Putting It Together*–**How Do We Know . . . How Fast Ground Water Moves?**
> ■ Injecting a soluble tracer ion into the aquifer and then monitoring water composition along the flow path reveals how fast ground water flows.
> ■ Ground water flows very slowly, compared to streams, because ground water flows in tortuous paths through pore spaces and is slowed down by friction along the edges of the pores.
> ■ Not all water molecules follow the same circuitous path through aquifer pore space. Chemicals dissolved in water, therefore, spread out through the aquifer so that the concentration of the chemicals diminishes with increasing travel distance.

17.4 What Is the Composition of Ground Water?

Section 16.2 revealed that stream water is impure because it contains dissolved ions produced by weathering reactions. Precipitation of minerals around the spring you observed in the field shows that ground water contains dissolved mineral components. Where do these naturally occurring dissolved ions come from and how are they important to understanding ground-water processes? Do compounds introduced into aquifers by human activity behave similarly to natural components in ground water?

Water-Mineral Reactions in the Aquifer

Chemical reactions take place wherever water and minerals are in contact for extended periods. Most mineral grains are constantly in contact with water in the saturated zone, and the water moves very slowly (see Section 17.3), so there is plenty of time and opportunity for these reactions to occur. **Figure 17.29** shows that the chemical reactions in the aquifer dissolve some minerals, which increases the concentration of dissolved ions in the ground water. Changing composition of the water may then cause bonding of concentrated ions to form minerals that fill in the pore spaces in the aquifer. The chemical reactions, therefore, may either increase or decrease aquifer porosity.

The longer the water flows through the aquifer, the greater the time it reacts with minerals. Ground water following short paths near the water table to discharge in wells or streams may not have sufficient reaction time with minerals to cause a significant compositional change. On the other hand, water following long flow paths deep into the saturated zone typically has very high concentrations of dissolved ions because the water is in contact with reactive mineral grains for thousands or tens of thousands of years. Elevated temperature at depth also enhances chemical reactions in the deep saturated zone.

Cementation of sedimentary rocks occurs where dissolved ions in ground water precipitate in pores as minerals (Figure 17.29). Cementation decreases

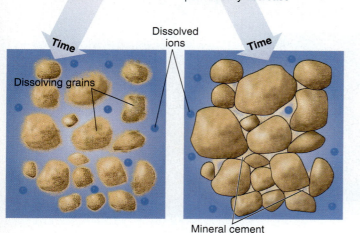

Sediment grains

Pore space

Water dissolves grains, porosity and permeability increase

Minerals precipitate from water, porosity and permeability decrease

Time

Dissolved ions

Time

Dissolving grains

Mineral cement

◀ **Figure 17.29 How chemical reactions in ground water make or destroy porosity.** If chemical reactions dissolve the minerals in the aquifer rock or sediment, then the porosity and permeability increase as the pore spaces enlarge. Dissolution releases ions into the ground water. If the ground water contains large concentrations of dissolved ions, then new mineral crystals may precipitate in the pore spaces. This mineral cement fills in the pores and closes off the connections between pores so that porosity and permeability decrease.

porosity and permeability, which slows ground-water flow. When the water slows, the water molecules spend even more time in contact with mineral grains, which promotes further dissolution of some original minerals in the rock while new cement minerals form in other places.

Some elements dissolved in ground water are unhealthy to humans. The element arsenic has attracted considerable recent attention from hydrologists, health scientists, and government decision makers. Arsenic causes a variety of health problems, ranging from skin diseases to enhanced risk of developing some cancers. **Figure 17.30** shows highly variable arsenic levels in ground water in the United States. Although some arsenic results from human pollution, the arsenic concentrations are highest where the water flows through volcanic rocks and some sedimentary rocks that contain arsenic in minerals such as pyrite. In most places, therefore, the harmful arsenic results from natural interactions of ground water with minerals. Until 2001, the acceptable level of arsenic in drinking water was 50 parts of arsenic per 1 billion parts of water, but a new standard of 10 parts per billion now has to be met by 2006.

▲ **Figure 17.30** **Arsenic threatens American ground water.** The maximum arsenic concentration permitted in United States drinking water after 2006 is 10 parts per billion. Ground water used for drinking in many parts of the country exceeds this new requirement and will require expensive treatment.

Hard and Soft Water

Minerals precipitated from ground water not only fill aquifer pore spaces; they also can clog water-supply pipes, as illustrated in **Figure 17.31**. **Hard water** describes water with high concentrations of dissolved ions. Usually these are calcium and magnesium ions derived from dissolution of calcite and dolomite, which are present in limestone and dolostone. These ions commonly reach concentrations in the water where they precipitate and clog pipes with rings of calcite, dolomite, or other minerals. Hard water also diminishes the effectiveness of soaps, because the calcium and magnesium ions react with soap and laundry detergent to draw "soap scum" compounds out of the water so that the soap does not stay in the water as a cleanser. Hard-water ions also precipitate as minerals when water evaporates, which is why you need to periodically clean coffee makers and showerheads with weak acid to dissolve clogging mineral precipitates.

Soft water has low concentrations of dissolved ions. Most soft water is artificially produced from hard water by using water softeners, which are special filtration devices that chemically remove the calcium and magnesium ions. Soft water causes problems of its own, however, because water that lacks many dissolved ions is more reactive than water that is already carrying many dissolved constituents. As a result, soft water dissolves and corrodes pipes, especially in hot-water lines, which can add unhealthy levels of lead, copper, and other metals to drinking water.

Ground Water Forms Economic Mineral Deposits

Many important ore deposits of zinc, lead, copper, and other metals form in sedimentary rocks by deeply circulating ground water. Although metal content in rock is typically very low, many metallic elements dissolve easily in warm water. Warm ground water following deep flow paths passes through a large volume of rock over long periods of time (see Figure 17.19) and scavenges available metals.

When the ground water eventually moves toward the surface, it cools slightly, mixes with shallow water with different chemistry, and reacts with near-surface rocks. All of these changing conditions also change the composition of the original, deep-traveled ground water in ways that decrease the solubility of

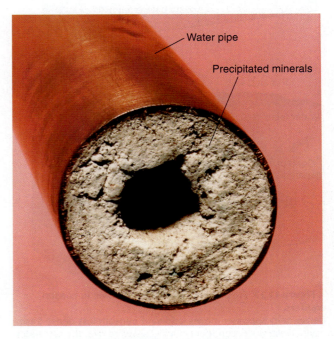

▲ **Figure 17.31** **Hard water clogs pipes with minerals.** Minerals precipitated from hard ground water have filled in most of this water pipe.

▶ **Figure 17.32** **Lead and zinc deposits formed from ground water.**
Warm, deeply circulating ground water with high concentrations of dissolved metal ions can form economically valuable mineral deposits where the water cools close to the surface and undergoes chemical reactions with near-surface ground water and rocks. Lead and zinc ore deposits in the central and eastern United States formed by this process.

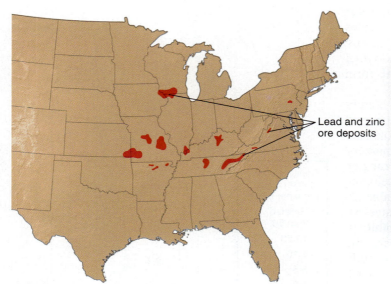

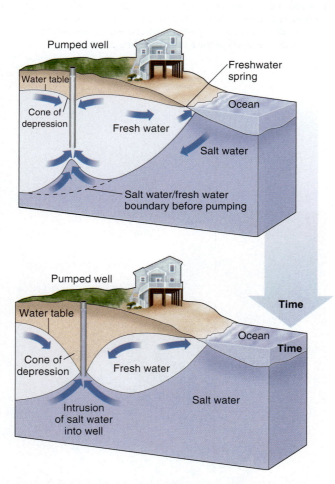

▲ **Figure 17.33** **Fresh water and salt water mix in coastal aquifers.**
Dense, salty seawater moves landward beneath less dense fresh water. Fresh water commonly moves seaward over the salt water in the aquifer and may discharge onto the seafloor at freshwater springs. Pumping of fresh water from the aquifer eventually draws the undrinkable salt water upward into the well.

the metal ions. Metallic minerals, usually sulfide minerals such as the lead ore galena, precipitate in the pore spaces of sedimentary rocks to form ore deposits.

Figure 17.32 shows that rich ore deposits of lead and zinc minerals originating by this process are scattered across the central United States. In the mid–twentieth century, nearly half of the world's supply of lead and zinc came from mines in these deposits.

Ground-Water Chemistry Near Coastlines

The presence of both fresh and salty ground water complicates aquifer chemistry near coastlines. Ground water beneath the seafloor is typically salty. This is not surprising, because seawater infiltrates underlying pore spaces. Ground water beneath the land surface is derived from freshwater precipitation. Salt water and fresh water mingle in coastal aquifers, as illustrated in **Figure 17.33**. Salt water is denser than fresh water, so the salty water invades the lower part of the aquifer while fresh water discharges onto the seafloor above the salt-water part of the aquifer. Freshwater springs are common on the seafloor near continents and islands.

Pumping of fresh ground water may draw salt water into a well. Ground water flows toward the well because it is a discharge site. Figure 17.33 shows how the flow of ground water not only forms a cone of depression in the low-density, freshwater part of the aquifer, but also draws up the high-density, salt-water part of the aquifer. There is widespread loss of water quality in coastal communities when increased pumping of fresh water causes salt water to enter the wells. Either wells must be located farther from the shoreline, with greater expense to transport water to customers, or the salt has to be removed from the water. Either option is costly.

Unnatural Additions: Ground-Water Pollution

Human activities also affect water quality. Pollution dissolved in ground water moves in the same way as tracers and naturally dissolved ions. The pollution travels the same paths and at the same velocity as the water. The concentration of the pollutant is very high where it is first introduced into the ground water, but the pollutant spreads so that the concentration decreases with distance (see Figure 17.28). Some contaminants precipitate as solids just as cement minerals precipitate in the aquifer. Therefore, a contaminant might be present at unhealthy levels close to the source, but be barely detectable at a greater distance because of spreading or precipitation.

Some harmful liquid contaminants do not mix in water but move instead as separate liquids, as shown in **Figure 17.34**. Oil and vinegar in salad dressing provide a familiar analogy for this separation behavior of some liquids. Gasoline is a common ground-water contaminant that does not mix with water. Gasoline is stored in shallowly buried underground storage tanks at fuel stations. If the tanks leak, then the fuel escapes into ground water. Gasoline is less dense than water, so it remains close to the water table rather than moving along deep flow paths into the aquifer. Technicians remove gasoline from ground water simply by pumping ground water from wells drilled to the water table at a lower elevation than the layer of contaminant. Unfortunately, a carcinogenic additive in the gasoline dissolves in ground water and is more difficult to recover. MTBE (methyl tertiary-butyl ether) was originally added to gasoline to reduce air pollution from vehicle exhaust but has now become a troublesome ground-water pollutant instead. Other harmful chemicals not only do not mix with water but they are also denser than water. These chemicals sink deep into the aquifer regardless of the ground-water flow paths, and removal is costly. Solvents used in industries and by dry cleaners are examples of these dense, insoluble liquids.

Most human-derived pollution is near the surface, so deeper wells typically avoid most contamination. Deeper wells are even less susceptible to contamination where they tap into a confined aquifer that is separated from the shallow, polluted aquifer (see Figure 17.24). Poor water quality at greater depth, however, may result from increased dissolved-ion concentrations along the deep ground-water flow paths. Many variables, both natural and human, determine where to find the best-quality ground water.

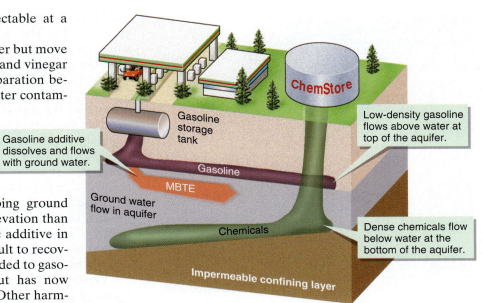

▲ **Figure 17.34** **How liquid density determines contaminant flow.**
Some liquid contaminants do not dissolve in water and their position in the aquifer is determined by comparing their density to the density of water. Dense chemicals sink to the bottom of the aquifer. Low-density liquids, like gasoline, float at the top of the aquifer. MTBE, a carcinogenic gasoline additive, dissolves in the ground water.

Putting It Together–What Is the Composition of Ground Water?

■ Slow-moving ground water reacts with the minerals in the aquifer materials. Ions release into the water where minerals dissolve.

■ Changes in ground-water chemistry, temperature, or both cause mineral precipitation that cements sedimentary rocks, clogs water pipes, and forms some economically important mineral deposits of lead and zinc.

■ Coastal aquifers contain less dense fresh water above denser salt water. An inverted cone of depression around a well may draw in the salty water.

■ Not all contaminant liquids mix in water. Gasoline is a common pollutant that is less dense than water and floats near the water table. Some solvents, however, are denser than water and sink to the bottom of the aquifer.

17.5 How Does Ground Water Shape the Landscape?

Although mostly out of sight below Earth's surface, ground water still plays an important role in forming surface landscapes.

Landscapes Molded by Ground-Water Dissolution of Rock

The property destruction you saw in the field (Figure 17.1b) resulted from a collapsing sinkhole. A **sinkhole** is a depression on Earth's surface caused by collapse of surface rock and regolith into a large underground cavity. The cavity forms where ground water dissolves rock, and then the roof of the cavity collapses.

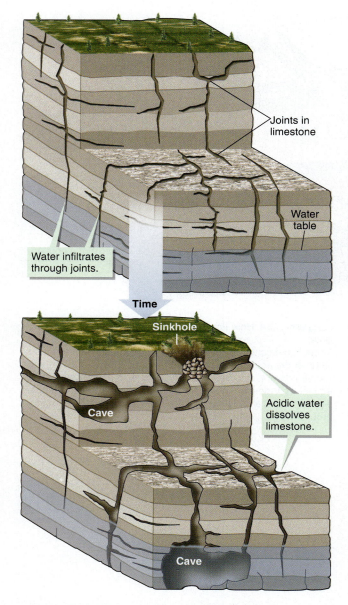

▲ Figure 17.35 **Where ground water dissolves limestone.**
Infiltrating water mixes with carbon dioxide and organic acids in soils to form a weak acid. The acidic water preferentially moves downward along preexisting joints in the rock. Limestone dissolves in contact with the weak acid, which enlarges joints and forms caves. Sinkholes form where rock and regolith collapse into near-surface cavities.

EXTENSION MODULE 17.3
The Geology of Caves.
Learn the origin of caves and the origins of the exotic rock formations commonly found in caves.

Most minerals forming chemical sedimentary rocks dissolve readily in ground water. The most soluble minerals are halite (rock salt) and gypsum, which form evaporites. Calcite, which forms very common limestone, also dissolves in ground water where large volumes of water move through the rock and the water is slightly acidic. The natural acid forms when carbon dioxide released by plant roots and by organic-matter decay in soil mixes into downward percolating water in the unsaturated zone.

The sinkhole collapse illustrated in Figure 17.1b occurred after water pumping and drought caused the water table to drop. Limestone caverns initially filled with water became open voids as the water table descended. Regolith above the limestone slid through cracks into the open cavities, forming the funnel-shaped sinkhole at the surface.

Limestone dissolves by chemical weathering where ground water moves through fractures and along bedding planes, as shown in **Figure 17.35**. Water acidity is usually greatest in the unsaturated zone and close to the water table, so most dissolution occurs under the ground but close to the surface. The dissolution process forms caverns below ground and sinkholes at the surface.

Figure 17.36 shows stages in the development of a landscape strongly affected by ground-water dissolution. Sinkholes pockmark the landscape, with as many as 2500 such depressions per square kilometer. The sinkholes start out as features as small as 10 meters or as large as 1 kilometer across but then coalesce to form valleys caused by collapse rather than stream erosion. Streams disappear into sinkholes, and springs bring ground water to the surface along fractures in the limestone enlarged by the solution. Over time, dissolution increases downward in regions of declining water-table elevation. Collapse of enlarging caverns eventually produces a highly irregular landscape of high-standing rock towers with intervening depressions and valleys.

Karst topography is the term used to describe terrains with distinctive landforms and irregular drainage patterns caused by rock dissolution. The German word "karst" originates from Italian and Slavic words that describe such an unusual landscape at the head of the Adriatic Sea in northeastern Italy and Slovenia. **Figure 17.37** shows the distribution of karst topography in the United States.

Limestone caverns, commonly just called caves, form by rock dissolution in ground water but contain many spectacular features produced by mineral precipitation. **Figure 17.38** shows how ground water causes both mineral dissolution and precipitation. Carbon dioxide gas dissolved in water produces the acid that dissolves the calcite and causes high concentrations of calcium and carbonate ions in the ground water. Where the ground water drips from the roof of a cave in the unsaturated zone, the carbon dioxide escapes from the water into the cave atmosphere, which decreases the acidity of the water and causes calcite to precipitate. Columns, pillars, and tapestry-like sheets of precipitated calcite form the common scenic highlights of famous caverns. In a similar fashion, escape of carbon dioxide from emerging spring water caused the travertine precipitation illustrated in Figure 17.1a.

Landscapes Molded by Seeping Ground Water

Figure 17.39 illustrates a scenic canyon eroded in sandstone. Although there is a stream in the bottom of the canyon, several features are uncommon for stream-eroded valleys:

- The canyon floor is very wide compared to the size of the small stream at the bottom.
- The valley width does not change downstream, whereas stream-eroded valleys generally are wider in the downstream direction.
- The main canyon and side canyons terminate upslope in vertical to overhanging alcove walls. Stream channels either are missing altogether above these alcoves or are much narrower than the width of the alcove.
- Another important observation is the appearance of dense pockets of green vegetation marking locations of springs near the base of the canyon.

Sinkholes in Indiana

Disappearing stream in Texas

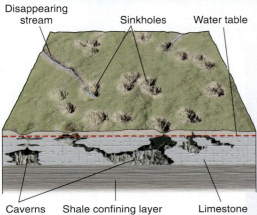

Disappearing stream   Sinkholes   Water table

Caverns   Shale confining layer   Limestone

◄ **Figure 17.36 How ground water forms karst landscapes.**
Gradual dissolution of limestone by ground water leads to the formation of caverns that collapse to form sinkholes. The pockmarked surface of collapsed sinkholes is highly irregular, and surface streams disappear into solution-enlarged joints and sinkholes. After long time periods most of the limestone may dissolve away leaving behind towering remnants of undissolved rock.

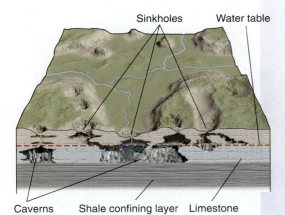

Sinkholes   Water table

Caverns   Shale confining layer   Limestone

Cave in Texas

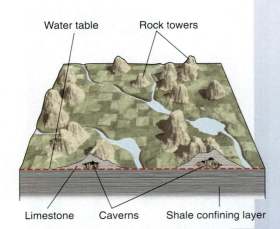

Water table   Rock towers

Limestone   Caverns   Shale confining layer

Time

Rock towers in China

► **Figure 17.37 Where ground water dissolves rocks.**
Rock dissolution by ground water is recognized over large areas of the United States. Most of these areas include caves and karst topography. Limestone dissolution accounts for most of these features, but dissolution of marble or gypsum occurs in some regions.

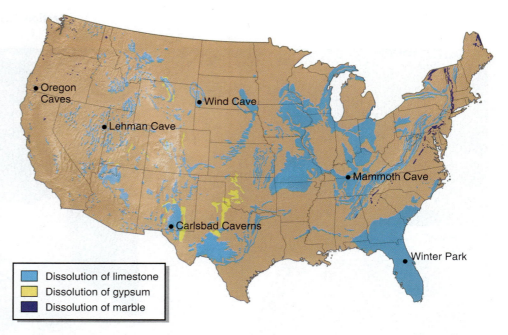

Oregon Caves

Wind Cave

Lehman Cave

Mammoth Cave

Carlsbad Caverns

Winter Park

Dissolution of limestone
Dissolution of gypsum
Dissolution of marble

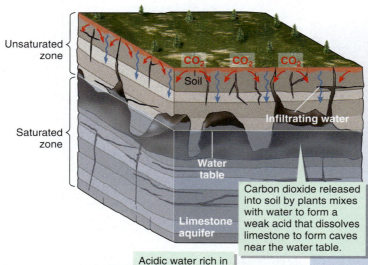

Unsaturated zone

Saturated zone

CO_2 CO_2 CO_2

Soil

Infiltrating water

Water table

Limestone aquifer

Carbon dioxide released into soil by plants mixes with water to form a weak acid that dissolves limestone to form caves near the water table.

Acidic water rich in ions from dissolved limestone and in dissolved CO_2

Time

► **Figure 17.38 How ground water forms caves.**
Calcite dissolves in limestone because the infiltrating water is weakly acidic from carbon dioxide (CO_2) that dissolves in the water as it passes through the soil. Dissolution to form caverns typically occurs mostly near the water table, and actively forming caves typically are partly submerged in ground water. When the water table falls, the cave is entirely in the unsaturated zone. CO_2 escapes into the cave atmosphere from the infiltrating ground water, which is rich in ions that resulted from limestone dissolution closer to the surface. Release of CO_2 into the cave atmosphere causes calcite precipitation to produce spectacular cave formations.

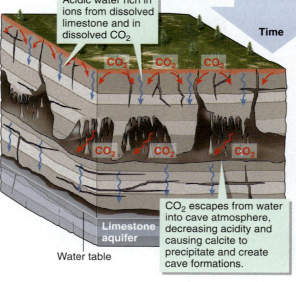

CO_2 CO_2 CO_2

CO_2 CO_2 CO_2

Limestone aquifer

Water table

CO_2 escapes from water into cave atmosphere, decreasing acidity and causing calcite to precipitate and create cave formations.

Plants grow around springs at base of overhanging alcoves.

Canyon width does not change downstream.

Small channel enters a much wider canyon.

Narrow stream channel in a very wide canyon bottom

◄ **Figure 17.39 Ground water helps form scenic canyons.**
The labels on this photograph of stream canyons in southern Utah point out features that suggest the influence of ground water on canyon formation. Perched ground water exits at springs and wet seeps near the canyon bottom. Weathering of rock around the springs forms overhanging cliffs, which fail by rock fall to produce sediment carried away by streams. The end result is wide amphitheater canyons that are too wide to be eroded by the surface streams without the assistance of ground-water weathering.

These observations are consistent with a major role of ground water in the landscape history. **Figure 17.40** illustrates how perched ground water enhances rock weathering where it emerges on a canyon wall. Wetting and drying, crystallization of salt from drying ground water, and freezing and thawing weaken and dislodge rock fragments where the water seeps out onto the outcrop face. The weathering produces an overhang above the ground-water discharge area. The overhanging cliff is unstable and fails by rock fall. The rock-fall debris weathers further into smaller fragments that are gradually carried away by flowing surface water. In this fashion, ground water plays a major role in forming wide, steep-sided canyons.

Putting It Together–*How Does Ground Water Shape the Landscape?*

■ Karst topography forms where ground water dissolves minerals in limestone and evaporite. Dissolution forms open spaces beneath the surface, including caverns, and causes the collapse of depressions, called sinkholes, that may capture surface drainage and make it flow underground.

■ Perched ground water exiting on canyon walls enhances rock weathering and mass movement. These processes form deep, wide, steep-walled canyons that have distinctly different shapes from those carved by streams alone.

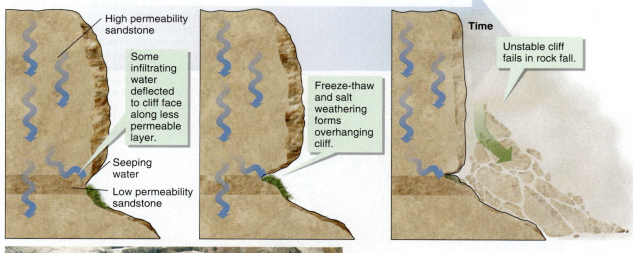

▲ **Figure 17.40** **How perched ground water contributes to mass movement.**
Infiltrating water moves to exposed cliff faces along low-permeability rock layers. The seeping water attracts plant growth and enhanced weathering by freeze and thaw, salt precipitation, and plant roots. Weathering produces an overhanging alcove that eventually fails by rock fall. The photograph shows evidence of these processes in Utah.

Where Are You and Where Are You Going?

Ground water flows through rock and regolith beneath Earth's surface. It is an essential water resource. Wells drilled below the water table, below which pore spaces are completely filled with water, extract ground water for drinking, for irrigating crops, and for use in industrial processes. The water-table elevation changes through time if there is an imbalance between the amount of ground water discharging to the surface at streams, lakes, or wells and the amount of water recharged from the surface.

Ground water flows from areas of high energy to areas of low energy where water discharges at low surface elevations or into wells. Ground water follows curving paths so that water that flows the greatest distance also travels deeper below the surface. Ground water flows much more slowly than surface water in streams, because water follows long, circuitous paths through small underground pores and is slowed by friction along pore edges.

Aquifers are Earth materials with high porosity and high permeability that provide sufficient quantities of water to wells to support the wells' intended purposes. Aquifers may be present immediately below the water table, or perched above the water table along a local impermeable horizon, or confined between impermeable layers deep in the saturated zone.

Natural and human processes influence ground-water composition. Ground water reacts with minerals. The reactions dissolve some minerals and thus concentrate dissolved ions in the water. These ions may bond to form new minerals that fill pore spaces, cement loose sediment into sedimentary rock, form valuable ore deposits, or even clog water pipes. Relatively dense salty seawater underlies less dense fresh water in coastal aquifers and encroaches into heavily pumped wells. Pollution enters ground water from many sources, including leaking fuel or chemical storage tanks, faulty septic systems, infiltration of water containing herbicides and pesticides applied to fields, and contaminated streams. If contaminants dissolve in the water, then knowledge of how ground water flows also predicts the contaminant movement. Some contaminant fluids do not mix with water; they float near the water table, if less dense than water, or sink through the aquifer, if denser than water. Pollutants typically spread out in the direction of ground-water flow, which means that the concentration decreases with increasing flow distance, although the area affected becomes larger.

Although hidden from view, ground water is important in landscape development. Highly irregular karst topography, characterized by sinkhole depressions above underground caverns and disappearing streams, results from ground-water dissolution of rocks composed of soluble minerals. Most karst topography forms in areas of limestone bedrock with ample rainfall that transmits carbon dioxide gas from the soil into shallow ground water to form weak, calcite-dissolving acid. Perched ground water seeps out along stream-canyon walls and greatly enhances rock weathering. Weathering produces rock overhangs that collapse by rock fall. These processes slowly enlarge long, wide, steep-sided canyons that are distinct from those created by stream erosion acting alone.

Next, you move from studying the processes linked to liquid water flowing on and below the land surface to consider the role of frozen water in shaping Earth landscapes. Eighty-four percent of Earth's fresh water is locked up in ice. Compared to typical silicate-rich rocks, ice is a curious mineral solid, because it plastically deforms at very low pressure. When ice builds up to only a few tens of meters thick, it actually flows under its own weight to form glaciers. Glaciers move slowly but are thick and heavy, so they exert strong erosive force on underlying rock and regolith. The importance of glaciers to sculpt Earth landscapes goes far beyond the mere 10 percent of the surface that they occupy today.

 ## Active Art

Why Is There a Water Table? See why a water table exists.

Forming a Cone of Depression. See how a cone of depression forms.

Extension Modules

Extension Module 17.1: Anatomy of a Water Well. Learn how a well is drilled and how ground water is withdrawn from a well.

Extension Module 17.2: Darcy's Law. Learn about Henry Darcy's experiment that defined a simple mathematical formula for describing ground-water flow.

Extension Module 17.3: The Geology of Caves. Learn the origin of caves and the origins of the exotic rock formations commonly found in caves.

Confirm Your Knowledge

1. What is the distribution of Earth's fresh unfrozen water?

2. What are the characteristics of ground water that make it a more valuable water source than streams in some regions?

3. Define ground water. Where is the ground water "stored"?

4. How do you find ground water?

5. What is the difference between the saturated zone and the unsaturated zone? What is the top of the saturated zone called?

6. What is permeability? How does it relate to porosity?

7. How do porosity and permeability change with depth? Which rocks have the highest porosity and permeability? The lowest?

8. What are the characteristics of an aquifer?

9. What Earth materials form the best aquifers?

10. How does a confined aquifer differ from an unconfined one?

11. Define perched ground water.

12. How does pumping water affect the water table?

13. What is natural recharge? What is artificial recharge?

14. What controls the flow of ground water? Can it flow upward, despite gravitational pull?

15. What is an artesian well? Explain how an artesian well works.

16. How do chemicals get into and pollute ground water? Give some examples.

17. Why does ground water flow more slowly than water in streams?

18. Does ground water flow at a uniform velocity through an aquifer? If not, why not?

19. What is the difference between hard water and soft water? What problems could each type cause in your home?

20. How do hot springs and warm springs form?

21. What is a sinkhole? How does it form? How do sinkholes relate to caves?

Confirm Your Understanding

1. Write out an answer for each question in the Chapter Outline for the chapter sections assigned by your instructor.

2. What is the major difference between an intermittent stream that is dry part of the year and a stream that flows year-long, even when it has not rained or snowed for several weeks?

3. How does the energy that drives ground water flow differ from the energy that drives stream flow?

4. Many brands of bottled water promote their product as artesian water. Is artesian water necessarily better than non-artesian water?

5. In the ground water study illustrated in figures 17.26 and 17.27 a chemical tracer was injected at 640 parts per million. After 461 days the highest concentration measured was only 39 parts per million. How did the chemical tracer get so diluted?

6. If you were planning to purchase a house with a well and septic system, what would you look for in regards to water availability, water quality and sinkhole risk? How would you determine where to locate your well relative to the location of your septic system?

7. Do you think there is a water table beneath the sea? Beneath lakes?

Key Terms

aquifer (p. 489)
artesian (p. 498)
cone of depression (p. 497)
confined aquifer (p. 495)
confining beds (p. 489)

ground water (p. 482)
hard water (p. 505)
karst topography (p. 508)
perched ground water (p. 489)
permeability (p. 486)

pores (p. 484)
porosity (p. 484)
recharge (p. 490)
saturated zone (p. 486)
sinkhole (p. 507)

soft water (p. 505)
spring (p. 482)
unsaturated zone (p. 486)
water table (p. 484)

18 Glaciers:
Cold-Climate Sculptors of Continents

Why Study Glaciers?

GLACIERS—FLOWING ACCUMULATIONS OF SNOW AND ICE THAT persist from year to year—currently cover about 10 percent of Earth's surface (16 million square kilometers). Nearly all of this ice is in Antarctica (13.5 million square kilometers) and Greenland (2 million square kilometers), with the remainder forming small glaciers at high latitudes, such as the Canadian Arctic Islands and Iceland, or in high mountains scattered across Earth. Glacial ice represents 84 percent of Earth's fresh water.

At times in the recent geologic past, popularly referred to as "ice ages," glaciers covered twice as much area as they do now. Data collected from temperature-sensitive fossils show that it is as warm or warmer today than it has been for 90 percent of the past 2 million years. Glaciers advanced and retreated across high-latitude and high-altitude regions many times during that time frame. The recognition of ice ages in the recent geologic past indicates that Earth's climate shifts between cold and warm extremes. The most recent glacial advance reached its peak about 21,000 years ago, and glaciers retreated to their present positions by about 6000 years ago. The glaciers seen today may seem unimportant except in remote polar regions, but glacial erosion and deposition during

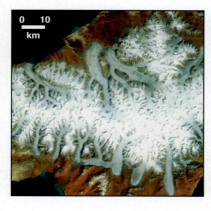

A backcountry camper enjoys an early evening view of Mendenhall Glacier in southeastern Alaska. ▶

past ice ages shaped landscapes over large areas where glaciers are no longer seen.

Glaciers actually flow. They move much more slowly than water flowing in streams, but thick, flowing ice is a more effective erosive agent. Glaciation is an important surface process, and your understanding of the origin of landscapes would be incomplete without an understanding of glaciers. Glacial processes are responsible for more than scenic, impressive alpine landscapes. Glacial deposits account for the distribution of fertile soils in the northern United States and southern Canada and the formation of many ground-water aquifers. Glaciers sculpted the Great Lakes, which together form the largest body of fresh water in the world. The Great Lakes region is key to the economy and history of central North America. Water stored in glacial ice is temporarily removed from the hydrologic cycle, so that global sea level falls during glacial advances and then rises when glaciers melt and the water returns to the ocean. The history of glaciation on land is tied to changes on coastlines that are still taking place today.

Your objectives in this chapter are as follows:

✔ To understand how glaciers work

✔ To understand how glacial erosional and depositional landforms originate

✔ To learn when, and why, glacial climates happen

The following questions are addressed to understand the work of glaciers:

18.1 What is a glacier?
18.2 How does glacial ice form?
18.3 How does ice flow?
18.4 How do glaciers erode and transport sediment?
18.5 How do glaciers deposit sediment?
18.6 What happens when glaciers reach the ocean?
18.7 How do valley glaciers modify the landscape?
18.8 How do ice sheets modify the landscape?
18.9 What did North America look like during the last ice age?
18.10 *How do we know* … how to determine when ice ages happened?
18.11 What causes ice ages?

In the FIELD

Consider a summer vacation to the Canadian Rocky Mountains. Your itinerary includes a visit to the Athabasca Glacier, in Jasper National Park, Alberta. **Figure 18.1**a shows the view from a distance. The glacier looks like a huge tongue of ice, part snow white and part dirty, like winter slush in a city street. The glacier fills the bottom of a broad, steep-sided valley that looks like it was carved with a giant ice-cream scoop. As you venture closer, you see streams emerge from beneath the glacier (Figure 18.1b); the water is milky white with suspended sediment. You encounter bare rock that is remarkably smooth but in places exhibits long, parallel scratches that look like somebody slid a giant knife across the outcrop (Figure 18.1c).

You also notice chaotic, lumpy piles and low ridges of unsorted boulders, pebbles, and sand that are scattered around the lake and the margin of the glacier (Figure 18.1d). You first wonder if these are bulldozed remains of a construction project; but a park sign describes the deposits as **moraines**, an eighteenth-century French word used by farmers to describe natural heaps of stony debris. Some of these moraine ridges coincide with signs that mark where the front of the glacier was located at various times over the last century.

You record your observations along with insights from exhibits in the visitors' center in your notebook, as seen in **Figure 18.2**. The moraines and scratched rock surfaces were beneath or alongside the glacier in the recent past, so these features probably owe their origin to glacial processes. The most impressive moraines are ridges of loose debris that parallel the two sides of the glacier and rise more than 50 meters above the icy surface. Clearly, the Athabasca Glacier is part of an actively changing landscape and not simply a static pile of ice.

You cannot resist the opportunity to take a tour to the top of the glacier. A specially equipped bus carries you and your friends onto the glacier where you set out on a short hike (Figure 18.1e). The first thing that you notice is the hard icy surface; it is slushy in a few places, but definitely not fluffy snow. The ice melts in the summer sun, and streams of water flow in shallow channels. You follow one stream to where it disappears into a deep opening in the glacier surface, as seen in Figure 18.1f. Perhaps this water plunges all of the way to the bottom of the ice and emerges in the streams that exit the front of the glacier.

Your vacation stop piques your interest in glaciers and glaciation. How do glaciers form? Does the ice, the meltwater streams, or both, transport the chaotic moraine sediment? How do glaciers modify the landscape? If the Athabasca Glacier has retreated 1.5 kilometers upvalley in the last 150 years, where was it during the ice age? What causes an ice age? When will the next ice age occur?

Figure 18.1 Visit to a glacier. ▶

A. Distant view of the glacier; notice the wide valley.

Water exits glacier here

B. The Athabaska River flows from beneath the glacier. The water looks "milky" with abundant suspended sediment.

Glacier

C. Scratched rock surface on trail to glacier. Rock is very smooth between scratches. All scratches line up in the same direction.

Glacier

Low ridge of unsorted rock debris

1992

D. Ridge of boulders, pebbles, and sand — called a "moraine." Sign points out that the front of the glacier was here in 1992.

E. The top surface of the glacier is hard ice, and it is very uneven, with bumps and cracks.

F. This is where a stream of water on top of the glacier disappears into a deep crack; cannot see the bottom.

Sketch map of features at edge of the glacier

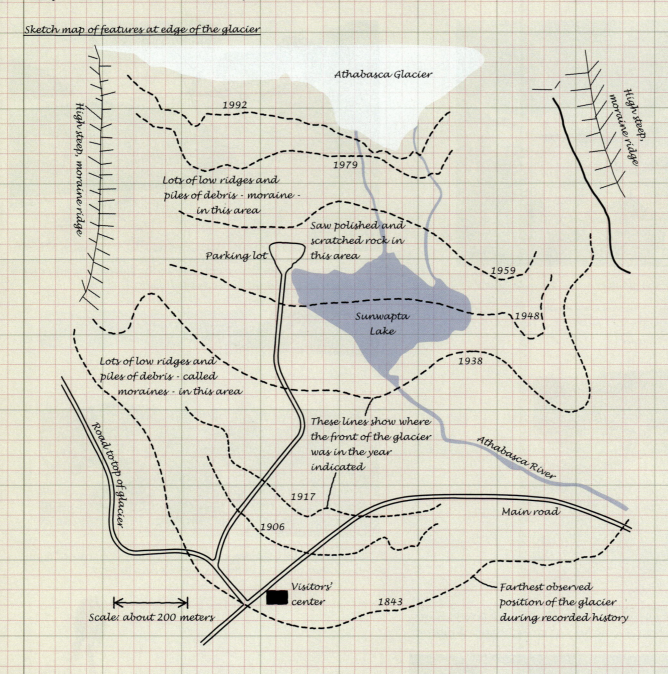

Athabasca Glacier

High steep, moraine ridge

High steep, moraine ridge

1992

1979

Lots of low ridges and
piles of debris - moraine -
in this area

Saw polished and
scratched rock in
this area

Parking lot

1959

1948

Sunwapta
Lake

1938

Lots of low ridges and
piles of debris - called
moraines - in this area

These lines show where
the front of the glacier
was in the year
indicated

Road to top of glacier

Athabasca River

1917

Main road

1906

Scale: about 200 meters

Visitors'
center

1843

Farthest observed
position of the glacier
during recorded history

1907

1998

Postcard photographs showing the retreat of the glacier

◄ **Figure 18.2 A map of glacial features.**
This map summarizes features observed in the field near the end of the Athabasca Glacier. The postcard photos at the bottom were taken from the same location at different times and show how the front of the glacier retreated upvalley during the twentieth century. The map shows that the front of the glacier retreated more than 1.5 km since 1843.

18.1 What Is a Glacier?

Geologists define a **glacier** as an accumulation of snow and ice that is thick enough to flow under its own weight. **Figure 18.3** illustrates that some glaciers occupy rock-walled valleys, whereas others are continent-scale sheets of ice. These different sizes and shapes form an easy classification.

Types of Glaciers

Valley glaciers are long and narrow glaciers confined within bedrock valleys (see Figure 18.3a). Valley glaciers, such as the Athabasca Glacier, flow from the highest elevations to the lowest elevations, just like streams.

Ice sheets are broad glaciers not confined by topography (Figure 18.3b) and are larger than 50,000 square kilometers in area. **Ice caps** are smaller unconfined glaciers (Figure 18.3c). The two ice sheets on Earth today cover most of Antarctica and Greenland (see Figure 18.3b), but you will learn in Section 18.10 that ice sheets covered large parts of North America and Europe during the last ice age. Ice sheets and caps flow radially outward from their highest ice-surface elevations, regardless of the elevation variation in the underlying bedrock.

Where Glaciers Form

Long-term observations show that glaciers form where snow persists year-round, because more snow falls during the winter than melts during the following summer. Clearly, this condition requires a cold climate, which explains why modern glaciers occur either at high latitudes or in high mountains, where year-round temperatures are chilly. If you live in western North America or have visited this mountainous part of the continent in the summer, then you have seen snow patches that remain throughout the year. A few of these snowy areas are large enough to qualify as glaciers.

Heavy winter snowfall is essential for forming glaciers. If the climate is dry, then so little snow accumulates during the winter that it all melts away even if the summer temperatures get above freezing only for a very short time. On the other hand, some snow may persist throughout the year where winter snowfall is very heavy, even in areas with relatively warm summers.

To understand where glaciers form, it is convenient to define the **snowline**, which is the elevation above which snow persists throughout the year. If a region does not have areas that rise above the snowline, then it cannot have glaciers. **Figure 18.4** shows how the snowline changes along a route drawn from the North Pole to South Pole and through the mountainous western part of North and South America. The lowest snowline elevations are at the high, cold latitudes. The snowline dips slightly near the equator because the very moist equatorial climate delivers great accumulations of snow to high mountain peaks. This is why Cotopaxi, a tall volcano located in the Andes less than one degree from the balmy equator, has glaciers. The mountain is 5900 meters high, where most of the tropical moisture falls as snow rather than rain. The snowline elevation also rises slightly around 70 degrees north latitude because the polar region is very dry.

Snowlines in the recent geologic past were much lower than today. Figure 18.4 also draws the low snowline elevations in the Americas about 21,000 years ago, during the last ice age.

► **Figure 18.3**
Visualize the types of glaciers.
Valley glaciers are confined between valley walls of rock. Ice sheets and ice caps flow radially outward across the landscape and are not confined by valleys. Ice sheets are larger than ice caps.

(a) Valley glaciers: Valley glaciers form tongues of ice descending into valleys eroded in snow-covered mountains on Bylot Island in northern Canada (left). Franz Josef Glacier in New Zealand (right) illustrates how valley glaciers are confined between bedrock valley walls.

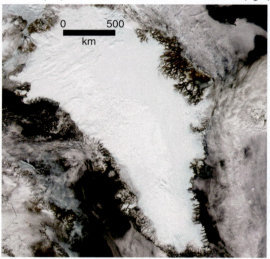

(b) Ice-sheet glaciers: This view of Greenland from space (left) shows that the large island is almost completely covered by a glacial ice sheet. The view from an airplane flying over Antarctica (right) shows that only the highest mountain peaks stick out above the kilometers-thick ice sheet that covers the continent.

(c) Ice-cap glaciers: Part of the island of Iceland is covered by an ice cap that is visible in the satellite view (left). Glaciers also descend some valleys around the periphery of the ice cap. The ice is darker in this summer view where the previous winter snow fall has melted below the snowline. The field photo (right) shows the edge of the ice cap.

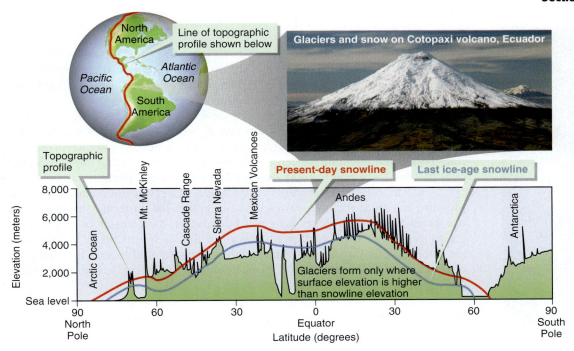

The graph shows the elevations of the present-day snowline in comparison to land-surface elevations. Only high peaks and Antarctica are above the snowline, so these are the only places where glaciers form. The snowline elevation is high near the equator but tall mountains like Cotopaxi, Ecuador, are still higher and support glaciers. The graph also shows that the snowline was much lower during the last ice age, so glaciers formed at lower elevations then than they do today.

Sizes of Glaciers

The area covered by glaciers is easily determined by looking at maps. Some valley glaciers cover less than one square kilometer, whereas the Antarctic ice sheet blankets 13.5 million square kilometers. Volume is a more complete measure of glacier size and requires knowledge of the thickness of glaciers. Drilling holes through the ice and echo-sounding methods, similar to those used to determine water depths in the ocean, provide geologists with many thickness measurements. These data show that valley glaciers are typically 50–300 meters thick, whereas large areas of the Antarctic ice sheet are more than 4 kilometers thick. The total volume of glacial ice in Antarctica is about 29 million cubic kilometers. Imagine that volume of ice spread across all of North America—it would make a layer 1.25 kilometers thick, or more than three times the height of the Empire State Building. Seventy percent of Earth's fresh water is frozen in the Antarctic ice sheet.

Putting It Together—*What Is a Glacier?*

- A glacier is a mass of snow and ice that flows under its own weight.
- Valley glaciers move downslope between the rocky walls of valleys. Ice sheets and ice caps, different only in size, are thicker ice masses that flow radially outward from a central high point.
- Glaciers form where more snow falls during the winter than melts during the summer over a prolonged period of time. Both cold temperatures and abundant winter snowfall are required to form glaciers.
- The snowline is the elevation above where snow persists all year and where glaciers form. The snowline is lowest at high latitudes, where year-round temperatures are cooler.

18.2 How Does Glacial Ice Form?

Hard, compact ice forms the interior of glaciers, as has been observed in tunnels or in icebergs, and much of the glacier surface, as you saw at Athabasca (Figure 18.1). The distinction between dense ice and fluffy snow is important for understanding glacier movement and the ability of glaciers to sculpt scenic landscapes. Where does the ice come from?

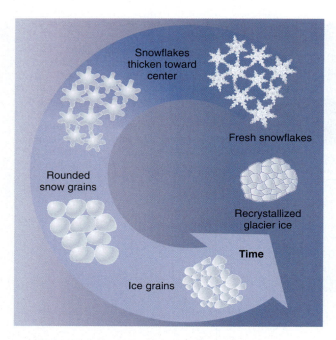

▲ **Figure 18.5 Metamorphism of snow into glacier ice.**
The porosity of fresh-fallen snow is very high because of the irregular shape of the snowflakes. As the snowflakes are buried below more snow, the ice molecules move from the edges to the center of the flakes and eventually the snow transforms into rounded grains. Minor melting and refreezing during the summer converts the snow grains into small ice grains, which pack closer together than the original snowflakes. When the ice grains are buried to a depth of a few tens of meters, the pressure causes recrystallization into nonporous glacier ice.

Snow Metamorphism

Ice is a mineral that metamorphoses at temperatures and pressures that exist close to Earth's surface. Most dense, almost nonporous glacial ice results from metamorphism of snow, as depicted in **Figure 18.5** (also see Section 6.3 and Figure 6.8 for brief coverage of snow metamorphism).

The first stage in the transition from snow to ice is the transformation of irregular snowflakes into rounded snow grains. Movement of ice molecules from the outer points of the snowflake toward its center gradually eliminates the delicate points and thickens the center of the flake. Some rounding happens as the delicate snowflake points break during movement or compaction.

Next, the rounded ice grains recrystallize. Recrystallization occurs where pressure is high at the boundaries between grains. Ice molecules transfer from one grain to another along the grain boundaries; the transfer enlarges some grains as others shrink, and eventually all of the air spaces fill with crystalline ice. The resulting interlocking ice crystals are several millimeters to a few centimeters across. The glacial ice is much denser (0.9 g/cm^3) than fresh snow (0.05 g/cm^3) but still less dense than water (1 g/cm^3).

How Much Time and Pressure Are Necessary to Make Glacial Ice?

Scientific drill holes bored into glaciers reveal the time required to convert snow to glacial ice. A hole drilled into the Greenland ice sheet encountered the transition from loose ice grains to glacial ice at a depth of 66 meters. Based on historical records of the accumulation of snow and ice, the ice at 66 meters is more than 100 years old. On the other hand, holes drilled into glaciers in Alaska reach hard ice that is only 3–5 years old at a depth of 15 meters or less. Metamorphism of snow requires the same pressure and burial depth everywhere, so these observations reveal that not all glacial ice forms by snow metamorphism.

Observations show another way of forming glacial ice. It simply forms by freezing water. Summer warmth melts the glacier surface. The meltwater soaks through the pore spaces between the snow grains and refreezes in the colder interior of the glacier. The melting-and-refreezing process fills the pore spaces with hard, crystalline ice. There is little or no surface melting in most of the interior of the Greenland and Antarctic ice sheets, even during the summer, so ice forms there only by metamorphism at depths greater than 60 meters. In warmer regions like Alaska, however, summer melting and refreezing of meltwater speed up the formation of dense, glacial ice at shallower depths.

> *Putting It Together—How Does Glacial Ice Form?*
> ■ Glacial ice forms either by the metamorphic recrystallization of snow or by freezing of meltwater that soaks into the glacier.

18.3 How Does Ice Flow?

Recall your visit to the Athabasca glacier and how you noticed that the glacier had retreated, exposing long, parallel scratches (Figure 18.2). This indicates that glaciers move and that they are not merely patches of snow that persist from year to year. The moving ice does work by eroding the underlying rock and depositing the resulting sediment.

Glacial Ice Is Weak Rock

Glacial ice flows because it is much weaker than the common rocks in Earth's crust. Ice crystals have a well-developed cleavage, similar to micas. When glacial ice is stressed, the ice crystals slide on the mineral cleavage planes and the whole mass deforms plastically. The critical pressure for plastic ice deformation is equal to a depth of only a few tens of meters, in contrast to a depth of about

10 kilometers to reach conditions for the plastic flow of granite in the continental crust. Weak ice flows under the stress exerted by its own weight. This is why glaciers form only where winter snow accumulation exceeds summer melting. Eventually the accumulating ice is thick enough to flow as a glacier.

Figure 18.6 illustrates plastic flow in the lower part of a glacier, whereas the upper brittle part breaks into large fractures. **Crevasses** are cracks in the brittle upper part of glacier caused by motion of the lower, plastically deforming part. Crevasses rarely persist below depths of about 50 meters.

Flow from Where Ice Builds Up to Where It Melts

Addition of snow and ice at high elevation causes glaciers to flow downslope because of the pull of gravity. A growing glacier eventually descends to lower, warmer elevations where summer melting exceeds winter accumulation. A glacier, therefore, consists of two zones, depicted in **Figure 18.7**:

1. The high-elevation **zone of accumulation,** where the winter snow accumulation exceeds summer melting.
2. The low-elevation **zone of wastage** (also called the zone of ablation), where summer melting exceeds the winter snow accumulation.

Over the course of a year, mass is added to the glacier in the zone of accumulation, and it is removed in the zone of wastage. Mass is also lost if icebergs form where a glacier terminates in a lake or the ocean. The snowline elevation, where there is no overall gain or loss in ice mass over the course of a year, forms the boundary between the zones of accumulation and wastage.

Ice flows from the zone of accumulation toward the zone of wastage. This is the same direction as flow from the head to the toe of a valley glacier, or from the center to the edge of an ice sheet (Figure 18.7). The ice also moves downward in the zone of accumulation, as new material is added on top each year. Ice

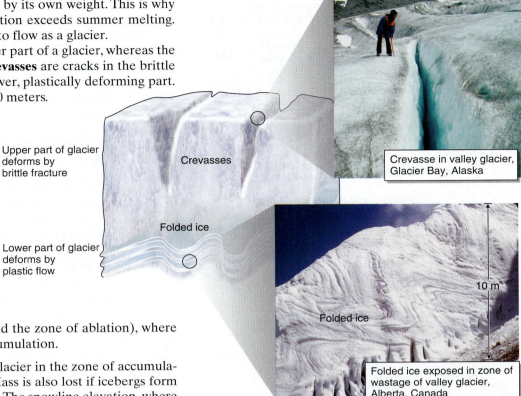

▲ **Figure 18.6 Brittle and plastic deformation of ice.** Brittle and plastic behavior of ice resembles rock deformation. Crevasses form by brittle fracture in the upper part of the glacier where pressure is low. Ice folds by plastic deformation in the lower part of the glacier where pressure is high.

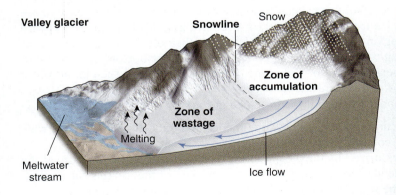

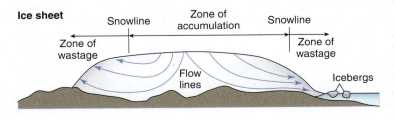

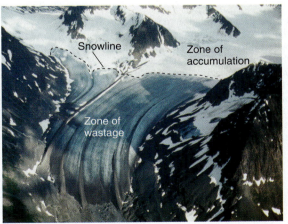

▲ **Figure 18.7 Zones of accumulation and wastage.** Zones of accumulation and wastage are separated by the snowline, where there is no overall change in ice mass throughout the year. Winter snowfall adds mass to the zone of accumulation in excess of mass lost by summer melting. Mass is lost from the zone of wastage by melting in excess of winter snowfall and by detachment of icebergs. The photo shows these two zones for a glacier in Alaska. Persistent accumulation of snow above the snowline causes the zone of accumulation to have a bright white surface, whereas summer melting exposes darker, sediment-rich ice in the zone of wastage. Ice flows from the zone of accumulation to the zone of wastage and also flows downward in the zone of accumulation and upward in the zone of wastage.

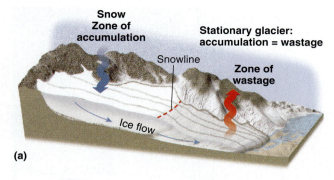

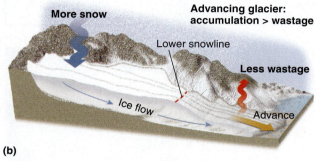

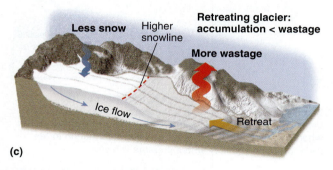

▲ **Figure 18.8** **Why glaciers advance, retreat, or remain stationary.**
Ice persistently flows from the zone of accumulation to the zone of wastage. The front of the glacier remains stationary if the annual accumulation and wastage balance one another. If wastage exceeds accumulation, then the front of the glacier retreats, even though ice flow is always toward the zone of wastage. If accumulation exceeds wastage, then the front of the glacier advances. Modern glaciers show all three behaviors, although retreating glaciers are more common (see Figure 18.2, for instance). In the Alps of southern Europe, for example, the area covered by glacier ice is now 35 percent less than in 1850.

Glacial Advance and Retreat: *See how a glacier moves when the front advances, retreats, or is stationary.*

deep in the zone of wastage moves up as surficial ice melts. These movements mean that glacial ice flows toward the underlying ground surface above the snowline and away from the ground surface below the snowline.

Figure 18.8 illustrates how the relative amounts of accumulation and wastage determine if the front of the glacier either remains stationary, advances, or retreats over time. The front of the glacier is stationary if the annual mass increase above the snowline equals the mass loss below the snowline (see Figure 18.8a). If accumulation is greater than wastage, then the front of the glacier moves downslope because more ice flows below the snowline than melts during the summer (Figure 18.8b). If accumulation is less than wastage, then the glacier toe retreats upslope (Figure 18.8c).

Look carefully at Figure 18.8 and note that the ice always flows downslope regardless of whether the glacier toe is stationary, advancing, or retreating. For example, the front of the Athabasca Glacier retreated 1.5 kilometers upvalley over the last 150 years (Figure 18.2). Survey markers placed on the glacier surface, however, always move in the down-valley direction, as illustrated in **Figure 18.9**. The fastest velocity is in the center of the glacier, because friction along the valley walls slows the flow on the sides of the glacier (Figure 18.9a).

Figure 18.9b also shows data regarding flow within the glacier interior. Velocity decreases downward in the glacier because of friction against the valley floor. The velocity is not zero at the bottom of the glacier, however, which means that the glacier slides across the valley bottom. In fact, most of the motion in the Athabasca Glacier results from this basal slip rather than by plastic flow within the ice. This is not always the case. In fact, some very slow moving glaciers in Greenland and Antarctica do not slide at the base and the measured velocity is due entirely to plastic flow within the ice.

Why It Is Important to Know the Bottom Temperature of the Glacier

It is important to know if a glacier slides at its base or if all the motion is internal plastic flow. If the glacier does not slide, then it is not effectively eroding underlying rock or regolith. Erosion from glaciers sculpts and shapes Earth's surface, and geologists want to understand how glaciers do this.

The temperature at the base of the ice distinguishes sliding and nonsliding glaciers. Ice is very adhesive when frozen to another object. This is easy to see when scraping ice from a windshield (or even in thinking about the schoolyard prank of placing one's tongue on a metal pole). If the bottom of a glacier freezes onto underlying rock, then there is no ice motion at the bottom. In contrast, if the basal temperature is at or above the melting temperature, then liquid water is present below the ice and sliding occurs.

Water enhances sliding at the base of a glacier for two reasons:

1. Water lubricates the surface between the ice and underlying regolith or rock by reducing cohesion and friction.
2. Water is pressurized by the weight of the overlying ice. Water pressure at the base of some glaciers is nearly equal to the weight of the ice; this substantially reduces the friction by lifting the glacier off of the underlying rock.

Several variables complexly interact to determine basal temperature. The distribution of nonsliding-cold-bottom and sliding-warm-bottom conditions varies within a single glacier at one point in time and through its history. As a result, it is not always appropriate to categorize a single glacier as being entirely of one type or the other. The important thing to remember is that only glaciers with warmer bottom temperatures slide at the base, and these glaciers, or parts of glaciers, do the most geologic work.

How Fast Do Glaciers Move?

Writers often describe a very slow process as occurring at glacial speed. This is because the pace of glacial motion is extremely slow. In fact, in the eighteenth

▶ **Figure 18.9 Measuring glacier flow.**
Geologists measure glacier flow using survey markers and drill pipes. The graphs show actual data collected for the Athabasca Glacier in Canada. (a) Repeated surveying of markers placed on the moving glacier surface shows that the glacier moves fastest in the center and more slowly where there is friction along valley walls. (b) Drill pipes inserted into the glacier bend because of glacier flow and show that there is slip along the base of the glacier and also internal flow, which decreases downward toward the bottom.

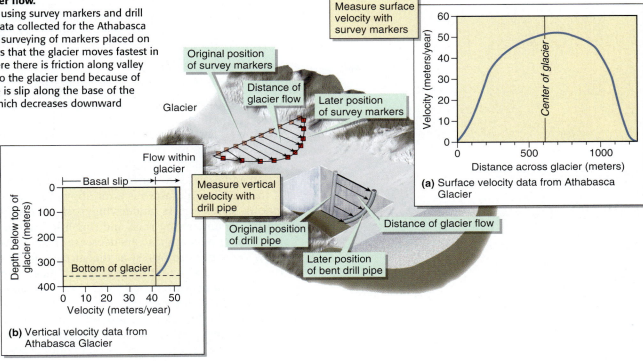

(a) Surface velocity data from Athabasca Glacier

(b) Vertical velocity data from Athabasca Glacier

century many people were skeptical of glacial erosion until experiments in the Swiss Alps proved that glaciers moved. Stakes were driven into the surface of the ice and tracked year to year as they slowly traveled downslope. Certainly, the recent surface speed of the Athabasca glacier, illustrated in Figure 18.9a, is not very impressive. If you patiently stood on the fastest part of the glacier, then you would move only half the length of a football field in one year.

Measurements reveal considerable variation in glacier velocities. Surface velocities range from less than 2 meters per year to more than 8 kilometers per year. Some glaciers rapidly surge forward over weeks or months at rates as rapid as 80 meters per day. Warm-bottom valley glaciers in southeastern Alaska and the Swiss Alps yield the fastest velocities, and cold-bottom glaciers in Greenland and Antarctica are the slowest. Furthermore, studies in tunnels underneath glaciers show that the velocity of warm-bottom glaciers is fastest when there is more water present at the base of the ice.

Figure 18.10 shows how surface features reflect the changing velocity along the length of a glacier. The ice speeds up and stretches where it moves down

▼ **Figure 18.10 How glacier surface features relate to flow velocity.**
The diagram on the left shows how flow velocity usually increases where the flow is downward and decreases where the flow is upward. Flow is also faster where the underlying slope becomes steeper and the ice slows down where the slope angle decreases. Crevasses form where the glacier speeds up and causes the ice to stretch. The ice compresses where the glacier slows down, so crevasses squeeze close and thrust faults form in the ice. The photo shows the opening and closing of crevasses in a New Zealand glacier where slope and flow velocity change.

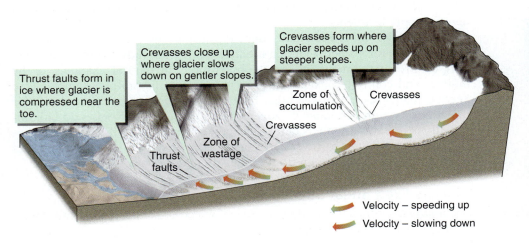

steep slopes. Tensional stress causes the upper, brittle part of the glacier to break into crevasses. Farther downslope, where the glacier slows down, the ice compresses and thickens, causing crevasses to close and, in some cases, forming thrust faults in the ice.

Putting It Together—*How Does Ice Flow?*

- Glacial ice flows under its own weight because it is a very weak rock that deforms plastically where it is more than a few tens of meters thick.
- Ice flows from the zone of accumulation to the zone of wastage, regardless of whether the toe of the glacier is stationary, advancing, or retreating. Flow is generally toward the bottom of the glacier in the zone of accumulation and toward the surface in the zone of wastage.
- If the basal temperature is at or above the melting temperature, then the glacier slides across underlying rock and regolith. Where the glacier freezes to underlying rock and regolith, it moves only by internal plastic flow. Water beneath glaciers decreases friction to enhance glacier sliding.
- Flow velocities are fastest at the surface in the center of the glacier, and range from less than 2 meters/year to more than 80 meters/day. Velocity decreases downward through the glacier and toward the valley margins (in the case of a valley glacier).
- Variations in the velocity cause different parts of the glacier to experience tensional and compressional stresses. Tension causes brittle fracturing revealed by crevasses. Compression presses crevasses closed and may cause thickening and thrust faulting in the ice.

18.4 How Do Glaciers Erode and Transport Sediment?

Geologists and naturalists have long been fascinated with the landforms of modern and ancient glaciated landscapes. Study of these landforms, and the materials that compose them, reveals the processes of glacial erosion, transport of sediment, and deposition of sediment that are complemented by observations at active glaciers. **Table 18.1** lists and explains the terms applied to many of these landforms. Most of the words probably sound unfamiliar to you and originate in Europe, where the landforms were first described. Only some of these terms are used in this chapter, although all of them appear in the geologic literature.

Comparing Glaciers and Rivers

Glaciers are sometimes described as "rivers of ice." Valley glaciers, in particular, resemble streams (see Figures 18.1a and 18.3a). Do glaciers, however, erode and transport sediment the same way streams do? It seems reasonable to expect that there are substantial differences in how solid ice and liquid water erode and transport sediment.

Observations of erosion and sediment transport are considerably more difficult to make for glaciers than for streams. Geologists study erosion and transport of stream sediment by direct observations in natural channels, or by indirect and controlled experiments in glass-walled laboratory channels (see Section 16.4). Glaciers, on the other hand, are opaque and move too slowly for direct observation of sediment movement. Nonetheless, geologists do link observations in natural tunnels below glaciers to field studies of glaciers that recently have retreated to reveal erosional features, such as the scratched and polished bedrock observed in front of the Athabasca Glacier (Figure 18.1c).

Glaciers Exert Shear Stress

Shear stress, the force exerted by a moving object, is a critical factor that determines both stream and glacial erosion (see Section 16.3 for description of shear

TABLE 18.1	Names of Glacial Landforms		
Name of Landform	Origin of Name	Formation Appearance	Illustration of Landform
Depositional landforms			
Moraine	A French word describing heaps of stony debris.	Any accumulation of nonbedded glacial sediment deposited at the base, front, or bottom of a glacier.	Figures 18.18, 18.26, 18.31
Kame	From the Scottish, *comb*, describing a steep ridge.	Outwash alluvium deposited alongside or beneath a glacier that remained as a hill or ridge after the adjacent glacial ice melted away.	Figure 18.21
Esker	From the Irish *eiscir*, meaning ridge.	Long, steep-sided curving ridge of alluvium that filled an ice tunnel beneath a glacier and was left behind when the glacier melted.	Figure 18.21
Drumlin	From the Gaelic *druim*, meaning rounded hill.	A ridge of till, or less commonly outwash, with a streamlined shape molded by flow at the base of a glacier. The slope of the ridge is steepest in the direction from which the ice approached and is more gentle and tapered in the direction of glacier flow.	Figure 18.33
Erosional landforms			
Arête	A French word for fish bone.	A narrow, knife-edge ridge separating glacially eroded valleys or cirques.	Figure 18.25
Horn	An Old German word describing the sharp, bony projection from the head of an animal.	A very steep, pointy mountain peak sculpted between three or more cirques; epitomized by the Matterhorn in the Alps.	Figure 18.25
Roche moutonée	Derives from resemblance to eighteenth-century French wigs called moutonées, after the mutton fat used to hold them in place.	An elongate bedrock ridge eroded by a glacier and parallel to the direction of glacier movement. The ridge has a gently sloping, smooth abraded slope that faces in the direction of glacier approach and a steep, plucked slope that faces in the direction of glacier flow.	Figure 18.12
Cirque	A French word meaning both circus and ring.	A steep-sided amphitheater depression or half bowl high on a mountain side at the head of a glacially eroded valley.	Figure 18.29
Tarn	Icelandic term for a small lake.	Commonly refers to a lake within a cirque where glacial erosion overdeepened the cirque floor to form an enclosed depression.	Figure 18.29

stress in a stream). Shear stress increases with the increasing weight of the flowing material. The weight of valley glaciers 50 to 300 meters thick or ice sheets more than 3 kilometers thick is huge compared to the weight of water in streams, which are usually less than ten meters deep. Glaciers, therefore, exert much stronger shear stresses on their beds than do streams, and they are also much more effective agents of lateral erosion on valley walls. For comparison, the shear stress at the bottom of the modest-size Athabasca Glacier is four times greater than the shear stress exerted by flowing water in the lower Mississippi River.

The Role of Meltwater

It is impossible to understand glacial processes without also considering the meltwater present at the bottom of the glacier. On one hand, water reduces friction, which reduces the wearing down of rock beneath the glacier. On the other hand, water enhances basal sliding, which allows the glacier to move and erode. On the whole, water increases the erosion potential of a glacier. Water flowing at the base of the ice also carries away glacial-erosion products and erodes channels into the underlying bedrock or cuts tunnels in the overriding ice.

Melting and freezing of water picks up rock debris below glaciers. Meltwater forms and then refreezes because of pressure variations at the base of the glacier, as explained in **Figure 18.11**. Experiments show that when pressure increases, the ice-melting temperature decreases, which is opposite of the effect of pressure on

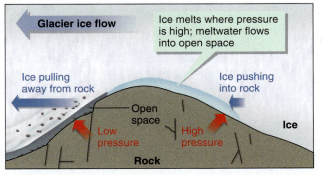

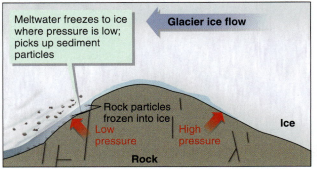

► **Figure 18.11 Melting and refreezing at the base of a glacier.**
Pressure changes at the base of a glacier cause ice to melt in some places, but then the water refreezes at other locations. Ice melts where pressure is high on the up-flow side of a rock obstacle, and the water refreezes to ice where the pressure is low on the down-flow side of a rock obstacle. Refreezing of the water traps sediment particles within the ice.

the melting temperature of silicate rocks (the temperatures and pressures of rock melting are explained in Section 4.5). This means that if ice is close to its melting temperature and pressure increases, then the ice melts. The pressure of moving ice increases where the ice pushes against a bedrock obstacle. Some ice melts because of the increase in pressure alongside the rock. The resulting melt-water then flows into the open cavity, where the ice pulls away from the down-flow side of the obstacle. The pressure is very low within this cavity so the cold water refreezes to the bottom of the glacier. Loose rock particles freeze into this newly-formed ice, providing a mechanism for glaciers to pick up rock debris.

Water flowing below a glacier erodes rock and regolith. Most water pouring from tunnels at the base of glaciers originates by melting at the glacier surface (see Figure 18.1f). The surface water spills into open crevasses where, moving rapidly and generating frictional heat, it erodes steep caverns down through the ice to the base of the glacier. The water pressure is very high where these cav-erns discharge water at the base of the glacier because the rock below and the glacier above confine the flow, much like in a garden hose. The pressurized water also injects into cracks in the underlying rock and pries it apart.

Erosion by Abrasion

Some glacial erosion occurs by abrasion (see Section 16.3 for descriptions of this process in streams). Ice has a Mohs hardness of 1.5, which is comparable to that of talc (see Figure 2.4), so ice is too soft to scratch many rocks. Glaciers, like streams, use transported rock and mineral debris as abrasive tools.

Sharp-edged rocks frozen into the bottom of a slow-moving glacier scratch and gouge the underlying bedrock. **Figure 18.12** shows examples of the parallel **striations** that demonstrate abrasion at the base of a glacier by these frozen-in rocks. Abrasion generates a huge volume of very fine-grained rock dust, called

▼ **Figure 18.12 How glaciers abrade and pluck rock.** Rock debris frozen into the bottom of a glacier scratches stria-tions and grooves into underlying rock surfaces. Ice and water dislodge rock fragments where fractured rock sticks up into the flowing ice. The asymmetric rock ridge eroded by abrasion and plucking is also called a roche moutonée (see Table 18.1).

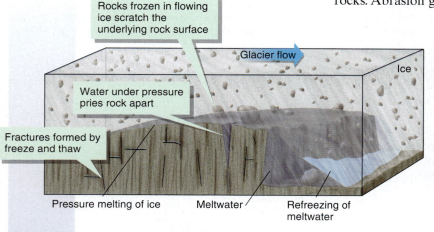

Rocks frozen in flowing ice scratch the underlying rock surface

Glacier flow

Ice

Water under pressure pries rock apart

Fractures formed by freeze and thaw

Pressure melting of ice

Meltwater

Refreezing of meltwater

Glacier flow

Glacial striations in limestone, Marblehead, Ohio

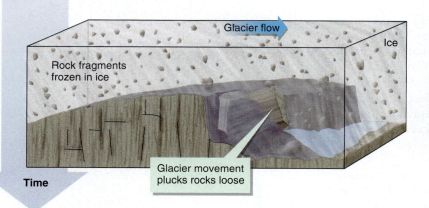

Glacier flow

Ice

Rock fragments frozen in ice

Glacier movement plucks rocks loose

Time

Jagged, plucked rock cliff

Glacier flow

Smooth, abraded rock surface

Glacially eroded rocky hill, Yosemite, California

glacial flour. If meltwater at the base of the glacier does not immediately wash away the glacial flour, then these very fine mineral particles grind along the bedrock surface and polish it smooth.

Glacial abrasion, therefore, is analogous to using sandpaper on wood. Coarse sand paper rapidly removes uneven bumps in wood, whereas fine sand paper produces a smooth polish. Rubbing coarse sand paper across a smooth, polished surface produces scratches.

Erosion by Plucking

Glaciers also pluck apart rock along preexisting joints and cracks in a very similar fashion to stream plucking (compare Figure 18.12 with Figure 16.12). The plucking process is more erosive at the bottom of a glacier than at the bottom of a stream for four reasons:

1. The higher shear stress of flowing ice more readily disaggregates fractured rock.
2. Freezing and thawing at the bottom of the glacier produces new rock fractures and further pries open existing ones.
3. The plastic ice squeezes into fractures and pries the rock apart.
4. Water pressurized at the base of the glacier by the weight of overlying ice dislodges blocks of rock.

Figure 18.12 shows how disaggregated rock blocks slide or roll off outcrops projecting into the ice. A cavity filled with refreezing water forms on the downslope side of the rocky obstacle, and the plucked fragments roll into this cavity and freeze into the ice.

Contributions from Freeze-Thaw Plus Mass Movement

Not all of the erosive action is at the bottom of the glacier. What happens on the valley walls above a valley glacier, or along mountainsides that stick up through ice caps and ice sheets? Physical weathering by freeze and thaw (see Figure 5.3a for more about physical weathering) is very intense in glacial environments, especially during seasons where temperature oscillates daily between subfreezing and above freezing. The weathered rock drops by rock falls and rock slides onto the glacier (see Figure 15.31 for a dramatic example). Glaciers are efficient conveyor belts that collect and transport debris from mass movements along the valley walls.

Where Erosion Occurs

Figure 18.13 summarizes the locations where geologists observe focused glacial erosion. Erosion is most intense where the moving ice shoves against the rock. Erosion, therefore, is characteristic of the zone of accumulation where the ice flow is downward against the underlying rock and regolith (see Figure 18.7), and also where rock projects up into the ice, in either the zone of accumulation or wastage.

Erosion is also intense along the valley walls. The cross-valley velocity variations in glacial flow (Figure 18.9) reveal strong frictional resistance between ice and rock, which enhances plucking and abrasion. Mass movements further modify the steep slopes above the glacier.

Transport of Sediment on Top, Within, and Below Glaciers

Glaciers shove and drag sediment at the base of the ice, carry it frozen within the ice, and piggyback it along the top. **Figure 18.14** illustrates these processes. Sediment eroded by abrasion and plucking freezes into the ice. Some debris remains at the base of the glacier but much of it eventually moves upward near

▼ **Figure 18.13 Where glacial erosion occurs.**
Most glacial erosion occurs where ice flow pushes downward against the underlying rock and regolith. Ice abrades the upslope sides of rocky knobs beneath the glacier and plucks the downslope side (also see Figure 18.12). Mass wasting along valley margins drops debris onto the ice surface. Deposition, rather than erosion, occurs where internal ice flow is upward away from the underlying rock in the zone of wastage. The near absence of meltwater between ice and rock along valley margins increases the friction and erosion of valley walls compared to valley floors, where meltwater slightly decreases friction.

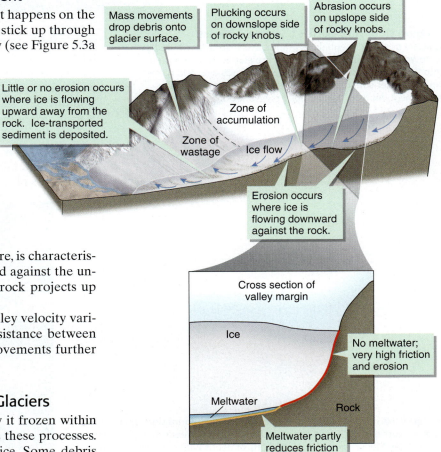

Mass movements drop debris onto glacier surface.

Plucking occurs on downslope side of rocky knobs.

Abrasion occurs on upslope side of rocky knobs.

Little or no erosion occurs where ice is flowing upward away from the rock. Ice-transported sediment is deposited.

Zone of accumulation

Zone of wastage

Ice flow

Erosion occurs where ice is flowing downward against the rock.

Cross section of valley margin

Ice

No meltwater; very high friction and erosion

Meltwater

Rock

Meltwater partly reduces friction and erosion.

Mass-movement debris carried on top of glacier.

Sediment concentrated at glacier surface by melting.

Ice flow drags and shoves sediment at base of glacier.

Zone of accumulation

Zone of wastage

Ice flow

Upward flow carries sediment to glacier surface.

Sediment frozen into the glacial ice

Bottom of glacier

▲ **Figure 18.14 How glaciers transport sediment.**
Glaciers push sediment along at the base, especially where flow is downward toward the bottom of the glacier. Sediment freezes into the ice, moves upward in the zone of wastage, and remains on the glacier surface when the surrounding ice melts. Mass-movement debris is carried along piggyback on top of the glacier.

rock obstacles and in the zone of wastage (Figures 18.10 and 18.14). When surface snow and ice melt in the zone of wastage, the sediment remains to give the glacier a dirty appearance (see Figure 18.7).

Mass-movement debris that drops onto the ice litters the tops of valley glaciers and ice sheets moving past higher mountain peaks. **Figure 18.15** shows how glaciers carry this debris as a moving ridge of boulder-rich sediment along the glacier margin, called a **lateral moraine**. Where two valley glaciers join, the lateral moraines combine to form a ribbon of sediment within the glacier, which is a **medial moraine**.

Meltwater moving beneath the glacier also transports large volumes of sediment. An example of sediment-laden meltwater occurs at the Athabasca Glacier (see Figure 18.1b). Recall that it was milky white. Why was that? The color of meltwater streams reflects the great abundance of suspended load, which is the glacial flour produced by abrasion.

Rates of Glacial Erosion

Geologists calculate the total erosive power of glaciers by measuring the amount of glacial sediment deposited over a measured time interval. These calculations show that slow-moving, cold-bottom ice-sheet glaciers in Greenland and Antarctica lower land-surface elevation only by about 0.01 millimeter per year. At the other extreme, fast-moving, warm-bottom valley glaciers in southeastern Alaska erode downward and sideways about 10–100 millimeters per year. Although this erosion rate may seem rather small, the mass of sediment liberated is gigantic. One glacier in southeastern Alaska covers an area only about twice the size of Boston, Massachusetts, but it erodes more than 200,000 metric tons of rock from every square kilometer along its base every year. If you compare an equal eroded area, then this sediment load is more than 100 times greater than that of the great rivers of southeast Asia (see Figure 16.7).

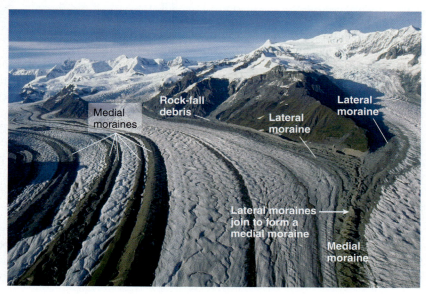

Medial moraines

Rock-fall debris

Lateral moraine

Lateral moraine

Lateral moraines join to form a medial moraine

Medial moraine

▶ **Figure 18.15 Visualizing how lateral and medial moraines form.**
Lateral moraines are ridges of accumulated rock-fall debris along the margin of a glacier. Medial moraines are ribbons of debris within the glacier. The view in this photo shows that medial moraines form where lateral moraines join at the junction of two glaciers.

Putting It Together—How Do Glaciers Erode and Transport Sediment?

■ Glacial erosion occurs by abrasion and plucking. Debris frozen into the moving ice abrades underlying rock. Plucking occurs when rocks disaggregate along pre-existing fractures, which are commonly enhanced by freeze and thaw of meltwater or by pressurized meltwater or ice that is injected into the cracks.

■ Glacial erosion is greatest where internal ice flow is downward against the underlying rock at the base of the glacier and also along the valley walls for valley glaciers.

■ Sediment is shoved along at the base of the glacier, frozen within the moving ice where it may be carried above the base of the glacier, and carried piggyback at the top of the glacier.

■ Mass movements deliver sediment to the glacier surface, where it is then transported as lateral or medial moraines on valley glaciers.

18.5 How Do Glaciers Deposit Sediment?

Geologists recognize two types of sedimentary deposits near or beneath modern glaciers, and these deposits are also widespread in previously glaciated landscapes. **Figure 18.16** illustrates these contrasting sediment types, which are described as follows:

1. Sediment deposited directly by the glacier is **till**.
2. Sediment mostly eroded by glaciers and then carried away by meltwater streams is **outwash** because it is, literally, the sediment that washes out of the glacier.

Geologists understand these depositional processes from a combination of studies around active glaciers and studies undertaken where glaciers have melted away.

Till is a very poorly sorted mixture of gravel, sand, and mud, and it typically lacks bedding (see Figure 18.16). Moraines, for example, consist of till. The rock fragments within the till erode from the entire area of glacial-ice cover. Ice sheets transport large boulders, some weighing hundreds of metric tons, more than 1000 kilometers from where they originated. **Figure 18.17** illustrates an

Till

Glacial cobble:
faceted edges, striations

Outwash

Stream cobble:
rounded, smooth

◄ **Figure 18.16 What glacial deposits look like.** Till is poorly sorted, nonbedded sediment deposited directly by a glacier. Outwash is moderately- to well-sorted, bedded alluvium deposited by meltwater streams. Stream cobbles in outwash are smooth and rounded. Glacial cobbles in till have faceted faces and edges and are commonly striated.

▲ **Figure 18.17 What a glacial erratic looks like.**
This conspicuously out-of-place boulder in Central Park, New York City, rests on very different rock with a glacier-abraded surface. The boulder is a glacial erratic eroded from a far-distant location and deposited here by an ice-age glacier.

example of these far-traveled, out-of-place rocks, called **erratics**. One piece of evidence for ice-age glaciers in the Midwestern United States is the presence of large erratics of ancient metamorphic rocks that match up with outcrops near Hudson Bay in Canada.

Outwash is much better sorted and has more distinct bedding than till. The gravelly and sandy alluvium is commonly associated with thinly bedded mud deposited in lakes that form along the front of the glacier.

Sediment Left Behind

Till deposited beneath the glacier forms a bumpy sediment sheet of irregular thickness, called **ground moraine**. Ground moraine, illustrated in **Figure 18.18**a, can originate by any of three processes:

1. A glacier drags and pushes a large volume of sediment at the bottom of the ice, which increases the friction at the base of the glacier. The frictional resistance builds up to the point where the rocky debris lodges against the ground surface and is left behind as the ice continues to flow.
2. Rock fragments frozen into the ice move upward in the zone of wastage (see Figure 18.10), but fragments dragging along the base of the glacier are left behind.
3. Melting at the bottom of the glacier releases rock fragments that previously froze into the base of the flowing ice.

Sediment Deposited at the Glacier Margin

High ridges of till flank most glacier margins. Figure 18.15 illustrates lateral moraines. The high boulder-rich ridges that tower over the Athabasca Glacier (see Figure 18.1a) are examples of lateral moraines that rise to the former surface elevation of the now shrunken glacier.

End moraines, seen in Figure 18.18b, form at the leading snout of the glacier. Some end-moraine till is unconsolidated regolith that is bulldozed along in front of the moving glacier. Most end moraines, however, are directly deposited from the glacial ice, as explained in **Figure 18.19**. Recall that ice flows even when the front of the glacier is stationary (see Figure 18.8). When melting occurs each summer, sediment melts out along the snout or accumulates on top of the glacier and then slides or washes down to the front. When the glacier is stationary for decades to millennia, the internal flow delivers sediment to the same area like a big conveyor belt and builds up the moraine ridge over time.

Ground moraine is an uneven veneer of till deposited beneath a glacier and exposed when the glacier retreats or melts away. The continuous, low ridge in the ground moraine was shaped by the glacier.

End and lateral moraines are ridges of till that mark former locations of the glacier margin. Lateral moraines only form along the margins of valley glaciers whereas valley glaciers, ice sheets, and ice caps all form end moraines.

◄ **Figure 18.18 What moraines look like.**
Moraines are landforms produced by deposition of till beneath and alongside a glacier.

18.6 What Happens when Glaciers Reach the Ocean?

Glaciers originate on land but may flow into the ocean or into large meltwater lakes. Glacial ice and water interact in complex ways because ice is less dense than water.

Formation of Tidewater Glaciers and Ice Shelves

Do glaciers float on the ocean, much like ice cubes float in a glass of water? You might think so. This conclusion is only partly true, however, because observations show that glacial ice does not immediately float right where the glacier snout moves into the water. The densities of glacial ice (0.9 g/cm^3) and water (1 g/cm^3) are close to one another in value, so water depth has to be comparable to the ice thickness—actually, more than nine-tenths of the ice thickness—before the glacier floats. **Figure 18.22** shows **tidewater glaciers**, which descend into the ocean from land and are in contact with the seafloor. Tidewater glaciers are common in southeastern Alaska, parts of northeastern Arctic Canada, and along the Greenland coast.

When a tidewater glacier moves into deeper water, it bobs up from the seafloor and floats on the water surface to form an **ice shelf. Figure 18.23** illustrates the differences between ice shelves and **sea ice**, which is simply frozen seawater. Ice shelves are much thicker than sea ice and move primarily as a result of glacial flow rather than at the whim of ocean waves and currents. Ice shelves persist throughout the year, whereas a large proportion of high-latitude sea ice forms in the winter and then melts in the summer.

Ice shelves are present along the margins of the Antarctic and Greenland ice sheets and locally in Arctic North America. Ice shelves form 44 percent of the Antarctic coastline and compose about 7 percent of the total area of the Antarctic ice sheet. On land, the predominantly cold-bottom Antarctic glaciers move very slowly, typically at rates less than 0.1 meter per year. At sea, however, the ice shelves float away from the coastline at velocities of 1–3 meters per year.

Making Icebergs

Icebergs are blocks of ice that detach from a glacier and float off into the ocean or a lake. Some icebergs break away from the front of tidewater glaciers, usually because of wave erosion that oversteepens the front of the glacier and causes mass movement (Figure 18.22). The largest icebergs break away from ice shelves. Ocean waves and currents stress the floating ice. Cracks form and eventually join up to completely separate blocks of the ice-shelf glacier, which then float free and move in directions determined by wind and ocean currents.

▼ **Figure 18.22 What a tidewater glacier looks like.**
This view from southeastern Alaska shows a tidewater glacier that originates on land and flows into the sea. The diagram shows that the ice is in contact with the seafloor because the water is too shallow for the glacier to float. Wave erosion produces a steep glacier front, which frequently collapses to form large icebergs.

Glacier front collapsing to form icebergs

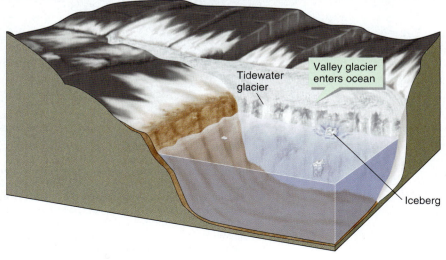

Tidewater glacier

Valley glacier enters ocean

Iceberg

▲ **Figure 18.23** **What ice shelves and sea ice look like.**
Ice shelves are thick, floating glaciers whereas sea ice is thin sheets of frozen seawater. Both of these views are near the coast of Antarctica.

Iceberg formation is an important wastage process for glaciers, especially in cold climates with minimal melting. Formation of icebergs along the Antarctic ice shelves accounts for 80 percent of the total glacier wastage for the entire continent. The largest observed iceberg-forming event occurred over 35 days in the summer of 2002, when 3250 square kilometers of an Antarctic ice shelf broke off into the South Atlantic Ocean, as shown in **Figure 18.24**.

Icebergs are very hazardous to ships. Large icebergs, some more than 50 kilometers long and 300 meters thick, float in the ocean for decades, so they cross shipping lanes far from the parent glacier. When ships collide with icebergs, the mass of the ice exerts sufficient stress against the hull to crush wood or rupture steel plate. Most of the iceberg is concealed below the water and may extend over a significantly greater area than the exposed iceberg above the water line. This is the literal meaning of the familiar expression for something more to come, "It's just the tip of the iceberg." In 1912 the *Titanic*, then the world's largest ship, sunk after colliding with an iceberg. The accident occurred south of Newfoundland, Canada, more than 2000 kilometers away from the west coast of Greenland, where the iceberg originated.

Putting It Together—What Happens when Glaciers Reach the Ocean?

■ Glacial ice is less dense than water, but glaciers float only where water depth is more than nine-tenths of the ice thickness.

■ Tidewater glaciers are in contact with the seafloor, whereas ice shelves float. Sea ice is simply thin sheets of frozen seawater and is not related to glaciers.

■ Icebergs are large blocks of floating ice that break off of tidewater glaciers and ice shelves.

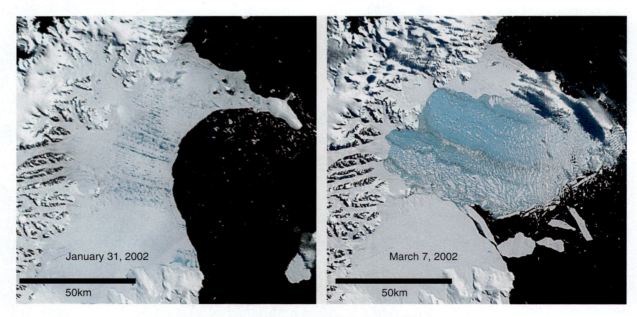

▶ **Figure 18.24** **An ice shelf breaks up to form icebergs.**
These two satellite photos provide before and after views of the break-up of the Larsen B Ice Shelf along the Antarctic Peninsula, on the Atlantic Ocean side of Antarctica. The total area of the disintegrated ice shelf is 3250 km², which you can compare to the 2717-km² area of Rhode Island. The average thickness of the ice shelf was 220 m, and the breakup liberated 720 billion metric tons of icebergs. The light blue areas between the large icebergs in the right photo are seawater littered with countless smaller icebergs.

18.7 How Do Valley Glaciers Modify the Landscape?

Glaciers play an important role in forming landscapes. Landscape modification by moving glaciers is important to understand for two reasons:

1. Glaciers are not as common as streams, but they do more erosive work. In some places, glaciers are the dominant force creating the landscape. This is especially true in high, mountainous regions, where cold temperatures spawn glaciers that erode the rock and provide for isostatic feedbacks for further uplift (to review the relationship between erosion and uplift, see Section 13.3 and Figure 13.12).

2. Ice-age glaciers sculpted the primary landscape features over large areas of North America and northern Europe that are currently ice free. These phenomena are an indication of changing global climatic conditions during Earth history.

What would you expect to see following the retreat of a valley glacier? Five landforms are particularly distinctive of the current or former action of valley glaciers: (1) U-shaped valleys, (2) knife-edge ridges and pointed peaks, (3) places where erosion overdeepened valleys to form lakes, (4) hanging valleys, and (5) moraine ridges. The first four features are erosional and the fifth is depositional. **Figure 18.25** illustrates the development of these landforms in a glaciated landscape and **Figure 18.26** points out examples at the Athabasca Glacier.

U-Shaped Valleys

Glacial valleys eroded into bedrock have a distinctive U-shaped, cross-valley profile, with very steep walls and a broad valley floor, as shown in **Figure 18.27**. This shape contrasts with bedrock stream valleys, which are usually narrow at the bottom with a V-shaped cross-valley profile (see Figure 18.25).

Fjords are U-shaped, glacier-eroded valleys along coastlines that are now partly submerged beneath the sea to form deep, elongate, steep-walled bays. *Fjord* is a Norwegian word, and the Norway coast features dozens of these glacial landforms. Fjords are also present in North America along the Atlantic, Arctic, and Pacific Ocean coasts of Canada and in southern Alaska.

Variation in the glacier velocity explains erosion of

▼ **Figure 18.25 Landscape modification by valley glaciers.** Valley glaciers modify V-shaped stream valleys into broad, steep-sided, U-shaped valleys that are separated by narrow, knife-edged ridges that rise to sharp, pointed peaks (the ridges and peaks are also sometimes called arêtes and horns; see Table 18.1). Some glacial valleys are eroded more deeply than others, leaving tributary valleys hanging in the sky with streams falling over high waterfalls.

Before glaciation

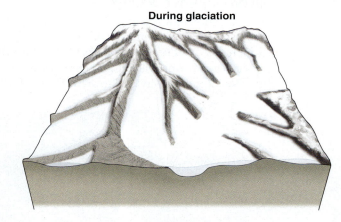

Tributary valleys

V-shaped valley

During glaciation

After glaciation

Pointed peak

Knife-edged ridge

Moraines

Cirque

Hanging valley

U-shaped valley

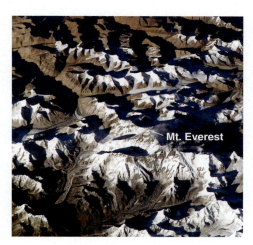

Mt. Everest

Glaciers carved the valleys around Mt. Everest, Earth's highest mountain. Notice the broad glacial valleys and narrow intervening ridges that rise to pointed peaks. This photo was taken by astronauts aboard the International Space Station.

The Grand Tetons, in Wyoming, feature the knife-edged ridges, pointed peaks, and hanging valleys of a previously glaciated mountain range. The ridge in the foreground is an end moraine.

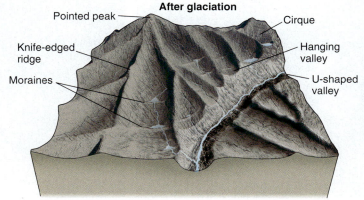

▶ **Figure 18.26 The landscape modified by the Athabasca Glacier.**
This photograph illustrates the common landscape features of valley glaciers that are seen at the Athabasca Glacier. Compare this view with the photos and map in Figures 18.1 and 18.2.

U-shaped valleys, as shown in **Figure 18.28** (also see Figure 18.9). The greatest erosion occurs where the velocity at the base of the glacier is highest. The highest sliding velocity at the base of a glacier in a V-shaped stream valley is not at the bottom of the valley but is actually partway up the valley sides (Figure 18.28). This means that erosion is stronger on the valley margins than at the valley center, so the valley floor widens into a U-shape. The zone of maximum erosion focuses along a newly widened valley bottom, so the U-shape is maintained as the valley erodes deeper (Figure 18.28).

Overdeepened Valleys

The elevation of a stream-valley floor always decreases continuously in the down-valley direction. In contrast, the long profile of glacially eroded valleys commonly includes overdeepened places where the valley floor actually slopes inward to make a bowl. Water accumulates to form lakes within these eroded bowls after the glacier melts away, as shown in **Figure 18.29** (also see Figure 18.28).

▶ **Figure 18.27 What U-shaped valleys look like.**
These photos show glaciated valleys in southeastern Alaska. The photo on the left illustrates the broad floor and steep sides that characterize glacially eroded, U-shaped valleys. The photo on the right illustrates how glaciated valleys along coastlines have submerged to form fjords as sea level rose after the ice age.

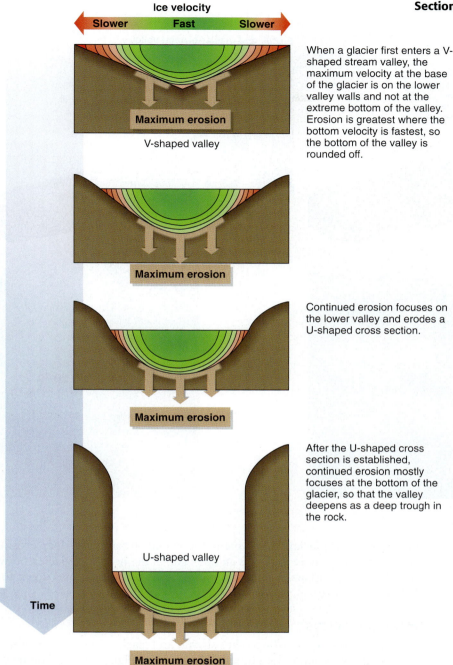

Ice velocity

Slower — Fast — Slower

Maximum erosion

V-shaped valley

When a glacier first enters a V-shaped stream valley, the maximum velocity at the base of the glacier is on the lower valley walls and not at the extreme bottom of the valley. Erosion is greatest where the bottom velocity is fastest, so the bottom of the valley is rounded off.

Maximum erosion

Continued erosion focuses on the lower valley and erodes a U-shaped cross section.

Maximum erosion

After the U-shaped cross section is established, continued erosion mostly focuses at the bottom of the glacier, so that the valley deepens as a deep trough in the rock.

U-shaped valley

Time

Maximum erosion

Glaciers erode deeply into the bedrock and overdeepen the valley floor to form these low spots. Deep erosion is especially common at the upslope end of glaciated valleys, where the steeply eroded valley walls partially enclose a natural amphitheater called a **cirque**. Some cirque depressions fill with lakes (called tarns; see Table 18.1), and they are common landforms in the Rocky Mountains (Figure 18.29).

Geologists cannot directly observe the overdeepened erosional bowls beneath active glaciers, so they are uncertain of why glaciers erode more deeply at these locations. Exceptionally deep scour sometimes coincides with softer or highly fractured rock that is more easily plucked by the glacier. In other cases, however, the rock seems uniform throughout the valley. Intense local erosion by meltwater beneath the ice may also be a cause of overdeepening, especially where crevasses permit large amounts of water to flow to the base of the glacier. The increased water pressure at the base of the glacier near these meltwater-input locations enhances fracturing in the underlying rock and permits more

▶ **Figure 18.29 What overdeepened valleys looks like.**
Lakes form in previously glaciated valleys where glaciers scour localized deep bowls into the rock. A cirque is the overdeepened head of a glacial valley that is partly encircled by an amphitheater of steep mountain ridges.

Profile of a river-eroded valley

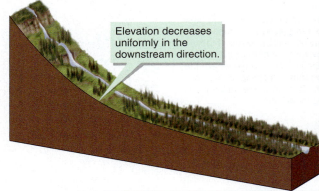

Elevation decreases uniformly in the downstream direction.

Profile of a glacier-eroded valley

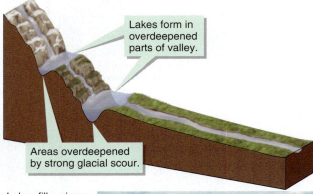

Lakes form in overdeepened parts of valley.

Areas overdeepened by strong glacial scour.

Lakes fill overdeepened parts of glaciated valley; Misty Fjords National Monument, Alaska

Lakes fill a cirque at the head of a glacial valley; Glacier National Park, Montana

plucking by the glacier. This explanation would also apply to deep erosion of cirques, because there is always an open gap at the head of a glacier, between the ice and the adjacent bedrock wall, where meltwater flows down to the base of the glacier.

Hanging Valleys

Another feature that distinguishes stream and glacial valleys is the elevation where tributary valleys join the main valley (see Figure 18.25). Streams usually join together at the same elevation—that is to say, a tributary stream neither ponds into a lake because it joins a higher-elevation main stream, nor does it cascade as a waterfall into a lower-elevation main stream. In contrast, tributary valleys in glaciated landscapes commonly terminate high above the main valley. This is not obvious while ice fills the valleys because the ice surface is uniform, and only the basal elevations are substantially different (see Figure 18.25). Retreating glaciers in side valleys typically disconnect from the glaciers in the deeper main valleys, as seen at the Athabasca Glacier (Figure 18.26). When the glacier completely melts away, the tributary-valley floor hangs hundreds of meters above the main valley, resulting in spectacular waterfalls, as illustrated in **Figure 18.30**.

Hanging valley

◀ **Figure 18.30 What a hanging valley looks like.**
Bridal Veil Falls at Yosemite National Park, California, illustrates the end of a hanging valley. Ice-age glaciers eroded the Merced River valley, in the foreground, much deeper than its tributary valleys. The mouths of tributary streams are now much higher than the main river, so the water pours over waterfalls into the Merced valley.

Hanging valleys erode because the main-valley glacier is thicker than the side-valley tributary glaciers. The main-valley glacier is thicker because it gains ice from each tributary that joins it. The shear stress is higher, and the depth of erosion is greater where the ice is thicker. This means that the thicker, main-valley glacier erodes more deeply than the thinner, side-valley glaciers.

Knife-Edged Ridges and Pointed Peaks

The ridges between widening glacial valleys become narrower, until they rise steeply to very narrow, almost knifelike ridges (also known as arêtes; see Table 18.1 and Figure 18.25). Where several valleys slope radially away from a single mountain peak, the glaciers gouge and erode narrow troughs into the central mountain. This leaves behind a very pointy pyramid (commonly called a horn). Earth's highest peak, Mount Everest, has the pointed pyramid shape that results from glacial erosion (Figure 18.25).

Lateral and End Moraines

Lateral and end moraines are recognizable alongside and in front of the Athabasca Glacier in Figure 18.26 and are also distinctive landforms in older glaciated landscapes, as seen in **Figure 18.31**. Ridges of poorly sorted till parallel to valley margins mark the edges of former valley glaciers, and the heights of lateral-moraine ridges above the valley floor provide good estimates of the thickness of the former glacier. End moraines are ridges of till that cross the valley and, along with lateral moraines or the rock walls of the valley, they may obstruct stream flow to produce a lake.

---

*Putting It Together—**How Do Valley Glaciers Modify the Landscape?***

- Five erosional and depositional landforms distinguish glacially modified mountainous landscapes from stream-eroded landscapes.
- U-shaped valleys form by glacial erosion that not only deepens but also widens valleys.
- Unusually deep erosion occurs at the cirque and at various locations along the valley. The overdeepened parts of the valley contain lakes after the glaciers melt away.
- Thick glaciers erode more deeply than their thinner tributary glaciers. After the glaciers melt, the different depths of erosion result in hanging valleys where tributary streams fall in tall waterfalls to join larger streams in the bottom of deeper valley floors.
- The widening of adjacent glaciated valleys results in the narrowing of intervening ridges and peaks. The resulting landscape consists of broad, scooped-out valleys separated by knife-edged ridges and pointed peaks.
- Lateral- and end-moraine ridges remain in glaciated valleys after glaciers melt away, and may impede streams, forming lakes on the valley floor.

18.8 How Do Ice Sheets Modify the Landscape?

Imagine a humongous retreating ice sheet—what kinds of features would you expect to see left behind? Some erosional and depositional features of huge ice-sheet glaciers simply are larger versions of those left by valley glaciers. However, since ice sheets are not confined in valleys, this means that ice sheets do not form features such as hanging valleys or lateral moraines.

Ice-sheet erosion occurs primarily beneath the zone of accumulation, where ice flow is directed downward to the glacier bed, whereas deposition takes place mostly beneath the zone of wastage. Some deposition also takes place within the original erosional zone as the ice melts in the waning stages of the ice age. Geologists' understanding of ice-sheet-modified landscapes comes largely from

▲ **Figure 18.31 Lateral and end moraines mark locations of former valley glaciers.**
Former glacial valleys include lateral and end moraines of till deposited by the glacier. The down-valley extent of a glacier in this valley in the Sierra Nevada, California, is indicated by the location of end moraines that cross the valley. The glacier was at least as thick as the top of the lateral moraines.

TABLE 18.2 Comparison of Glacial Landscape Features

Valley Glaciers	Ice Sheets
Erosional landscape features	
U-shaped valleys and hanging valleys.	Large areas of scoured, plucked, and abraded rock surfaces.
Knife-edge ridges and pointed peaks.	Streamlined ridges that parallel the direction of glacier movement.
Places where erosion overdeepened valleys to form lakes and cirques.	Lakes, scoured from rock and ranging in size from small ponds to the largest lakes on Earth.
Depositional landscape features	
Till ridges forming lateral and end moraines and typically thin ground moraine.	Large areas thickly covered with ground moraine with multiple end moraines extending for hundreds of kilometers.
Outwash-stream deposits partly fill valleys and plains downslope of the glacier.	Outwash-stream deposits partly or completely fill valleys and form widespread plains below, above, and downslope of glacial till.
	Small kettle lakes form where ice blocks melt in till and outwash.

studies of North American and European landscapes resulting from the last ice age. There are few direct observations of glacial erosion caused by the modern Greenland and Antarctic ice sheets. **Table 18.2** compares landscape features from valley and ice-sheet glaciers.

Four landscape characteristics are particularly indicative of past ice-sheet glaciation: (1) large areas of scoured, plucked, and abraded rock surfaces; (2) large regions thickly covered with till; (3) streamlined ridges that parallel the direction of glacier movement; and (4) landscapes of countless lakes, ranging from small ponds to the largest lakes on Earth, formed by both erosional and depositional processes.

Figure 18.32 illustrates landscape features related to ice-sheet glaciation. The distribution and appearance of erosional and depositional ice-age glacial landforms in northeastern North America are shown in **Figure 18.33**.

Widespread Scour of Bedrock

Ice sheets dramatically reshape the ground surface because the shear stress exerted by ice more than a kilometer thick leads to considerable erosion. Regolith is completely stripped away within the zone of accumulation, except where the ice froze to the ground surface, and the bedrock is highly abraded and plucked. Large ice sheets cover millions of square kilometers, so they erode a wide variety of rock types. Some areas are more susceptible to glacial erosion than others because of varying

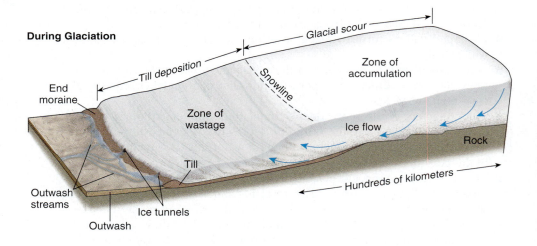

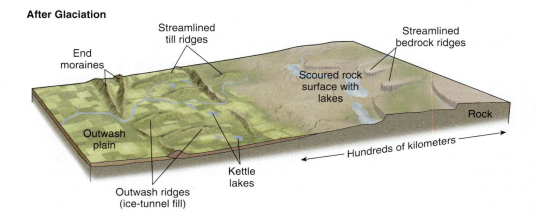

▶ **Figure 18.32 Landscape modification by ice sheets.** Moving ice sheets scour bedrock and deposit till over huge areas coinciding, respectively, with the zones of accumulation and wastage within the glacier. Outwash streams distribute sediment across vast plains beyond the ice-sheet margin. Lakes scattered across glacial landscapes formed where water fills deep scours into rock, kettle depressions left behind where ice melted in till and outwash, and low spots in irregular ground moraine and end moraine topography. Ice movement erodes bedrock and till into elongate ridges (sometimes called roche moutonée and drumlins; see Table 18.1) that are streamlined parallel to the direction of ice flow.

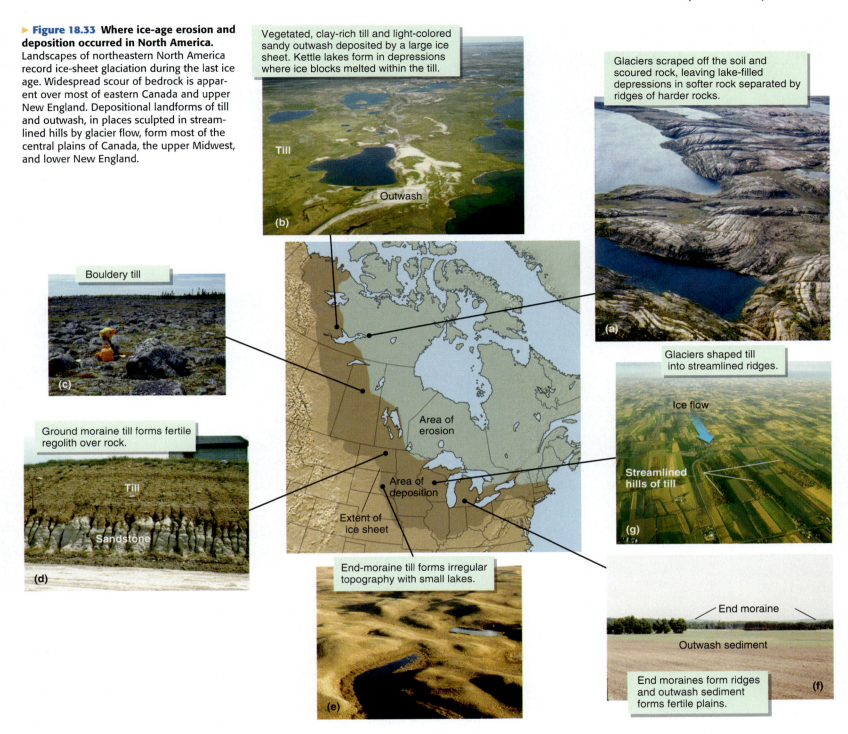

▶ **Figure 18.33 Where ice-age erosion and deposition occurred in North America.** Landscapes of northeastern North America record ice-sheet glaciation during the last ice age. Widespread scour of bedrock is apparent over most of eastern Canada and upper New England. Depositional landforms of till and outwash, in places sculpted in streamlined hills by glacier flow, form most of the central plains of Canada, the upper Midwest, and lower New England.

Vegetated, clay-rich till and light-colored sandy outwash deposited by a large ice sheet. Kettle lakes form in depressions where ice blocks melted within the till.

Till

Outwash

(b)

Glaciers scraped off the soil and scoured rock, leaving lake-filled depressions in softer rock separated by ridges of harder rocks.

(a)

Boundery till

(c)

Ground moraine till forms fertile regolith over rock.

Till

Sandstone

(d)

Area of erosion

Area of deposition

Extent of ice sheet

End-moraine till forms irregular topography with small lakes.

(e)

Glaciers shaped till into streamlined ridges.

Ice flow

Streamlined hills of till

(g)

End moraine

Outwash sediment

End moraines form ridges and outwash sediment forms fertile plains.

(f)

rock resistance to abrasion and plucking. The results, shown in Figure 18.33a, are ridges of hard rock interspersed with lake-filled depressions scoured out of softer or more fractured rock. The total depth of erosion by at least a dozen ice sheets crossing eastern Canada over the last 3 million years averages out to about 200 meters.

Thick Till and Long Moraines

In contrast to the large areas of glacial erosion are the equally impressive areas covered by unsorted glacial till (example photos shown in Figures 18.33b through 18.33e). Till in eastern North America is typically 30 to 60 meters thick and is locally more than 200 meters thick near and south of the Great Lakes. Till

▶ **Figure 18.34 Mapping the end moraines from the last ice age.**

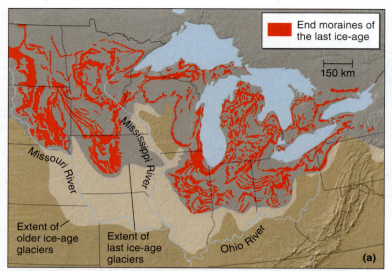

Lobate ridges winding for hundreds of kilometers across the upper Midwest are end moraines left by a retreating ice sheet during the last ice age. Notice how the moraines outline the margins of ice-sheet ice tongues that carved out the Great Lakes. Older ice ages deposited till and moraine ridges even farther south than the last ice age.

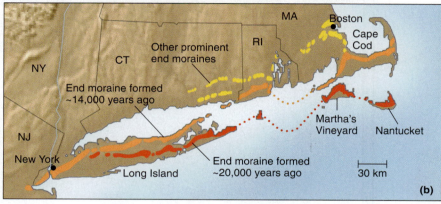

Prominent ridges along Long Island are end moraines deposited during the last ice age. The moraines are partly submerged below the Atlantic Ocean farther east, but are traced to similar moraine ridges that form islands and peninsulas in southern Massachusetts.

is so thick and widespread that the bedrock geology over tens of thousands of square kilometers is known only from rocks encountered in wells and uncommon outcrops in the bottoms of the deepest stream valleys. Landscapes are very uneven because of irregular till thickness and depressions, called **kettles**, where large blocks of ice melted within the till (or within related outwash alluvium).

End-moraine ridges continue across the landscape for hundreds of kilometers, as illustrated by **Figure 18.34**. The moraines are commonly ten or more kilometers wide and rise no more than a few tens of meters above the surrounding land surface (Figure 18.33f). Ice sheets have a curving front, and the moraines share the same curving outline (see Figure 18.34a).

Each glacier generates multiple end moraines (see Figures 18.32 and 18.34). One end moraine forms at the farthest advance of the glacier. Other moraines form during glacial retreat, where the ice front was stationary for a while or where it temporarily advanced again.

Meltwater-stream deposits are commonly found with the glacial till (Figures 18.32 and 18.33f). These include sediment deposited beneath the glacier or in ice tunnels, glacial outwash that was overrun by an advancing ice sheet, or outwash deposited on top of till by a retreating ice sheet. Outwash deposits also extend into nonglaciated regions downslope of the original ice sheet. These deposits are so voluminous that they completely fill former stream valleys more than 100 meters deep.

Streamlined Features in Rock and Till

Advancing glaciers sculpt rock, till, and outwash alluvium into furrows and ridges elongated parallel to the direction of flowing ice. These erosional effects are also seen with valley glaciers, but they are most striking in size and extent in landscapes once covered by ice sheets.

The streamlined hills of till illustrated in Figure 18.33g are tapered by the flowing ice and resemble a whale's back projecting above the water line. These ridges (sometimes called drumlins; see Table 18.1) not only provide evidence of past glaciation, they also reveal the direction of glacier movement. Furrows and ridges eroded into till may persist for several kilometers, as though a giant comb was dragged across the landscape.

Glacially Formed Lakes

Lakes are very common features in glaciated landscapes. Some mark locations where glaciers scooped out rock to leave a bowl that collected water (Figures 18.32 and 18.33a). Lakes occupy 20 percent of the land surface in the most extensively scoured areas of central Canada, a region with an average of more than 100 lakes within every 20-kilometer-by-20-kilometer square. Other lakes and ponds are kettles (Figures 18.32 and 18.33b). The irregular topography of till, especially within end moraines, commonly includes closed depressions that fill with water (Figures 18.32 and 18.33e). Minnesota's nickname, "Land of 10,000 Lakes," owes its origin to numerous lakes formed by both glacial erosion and deposition; most are kettles or correspond with the irregular till topography of moraines.

The Great Lakes are the most spectacular glacially eroded lakes on Earth. **Figure 18.35** shows these lakes, which cover an area of about 400,000 square kilometers and are as much as 400 meters deep. The lakes contain about 23,000 cubic kilometers of water—one-fifth of the world's fresh water. Many advancing ice sheets progressively scoured out the Great Lakes in locations of easily eroded rock during the last 2 million years. The glaciers carved Lakes Michigan and Huron, along with western Lake Erie, out of a dipping layer of soft evaporite sedimentary rock that encircles the lower peninsula of Michigan. Central and eastern Lake Erie and Lake Ontario were carved from easily eroded shale. Lake Superior formed where the ice sheets scoured out the sedimentary and volcanic rocks filling an ancient rift valley that is flanked by harder metamorphic rocks.

◄ **Figure 18.35 Ice sheets carved the Great Lakes.** This map consists of colored satellite radar images of North America and illustrates the irregular glacially scoured landscapes around and north of the Great Lakes in contrast to the somewhat smoother topography of glacial deposits with moraine ridges southwest of the Great Lakes. The Great Lakes are the most impressive examples of glacier scour, but also notice the countless small lakes in Canada that occupy scoured-out depressions in rock. The Finger Lakes, in New York, occupy river valleys that were substantially deepened by glacial erosion and then naturally dammed by end moraines when the ice sheet melted.

Putting It Together—How Do Ice Sheets Modify the Landscape?

■ Ice-sheet glaciation scours bedrock and deposits till across hundreds of thousands of square kilometers.

■ Rock, till, and outwash are sculpted by glacier flow into streamlined ridges and furrows that are elongate parallel to the direction of ice-sheet movement.

■ Lakes are too numerous to count in glaciated landscapes. Lakes occupy deep scours into rock and irregular depressions within glacial till, some of which result from melting of ice originally deposited with the sediment. The Great Lakes were carved out of relatively soft rock by ice-age glaciers.

18.9 What Did North America Look Like during the Last Ice Age?

The last ice age began about 120,000 years ago, late in the Pleistocene epoch (see Figure 7.9 for a refresher on the geologic time scale). Spurts of glacial advance, alternating with minor retreats, continued over the next 100,000 years. The continent-scale ice sheets reached their largest extent about 21,000 years ago. Then the glaciers melted back very rapidly compared to their long period of fitful advance, and ice sheets retreated to their present, permanent holdout positions in Antarctica and Greenland by 6000 years ago.

North America looked very different during the last ice age than it does now. Not only did ice cover huge areas, but cooler climate also affected vegetation distribution. Continent outlines were different because of lower sea level, and large lakes existed in the now arid Southwest because of diminished evaporation during this cooler time.

The Extent of Glaciers

Figure 18.36 shows the extent of Northern Hemisphere glaciers at the peak of the ice age. The outline of glaciers is determined by looking at the distribution of glacial till and where end moraines are located (see Figure 18.34, for example). There were two great ice sheets in North America—Laurentide in the east and Cordilleran in the west. At their largest extent, the two ice sheets joined together across the plains of western Canada and linked to the Greenland ice sheet. The combined area of the Laurentide and Cordilleran ice sheets was 16 million square kilometers, which is roughly two-thirds the area of North America, and the total ice volume was similar to the modern Antarctic ice sheet. Ice caps formed locally in the higher elevations of the Rocky Mountains and Sierra Nevada and fed into valley glaciers that extended to the margins of the adjacent plains. Glaciers also formed along the high volcanic peaks of the Cascade Range where smaller remnant glaciers remain today. Landscapes illustrated in Figures 18.12, 18.17, 18.26, 18.27, 18.29, 18.30, 18.31, and 18.33 owe their origin to this ice age.

◄ **Figure 18.36** **Visualizing the extent of glaciers during the last ice age.**
This map shows the maximum distribution of northern hemisphere glaciers 21,000 years ago, as seen looking down from above the North Pole. The largest ice sheets covered northern North America. Ice caps and long valley glaciers also formed in high mountains throughout Europe, Asia, and western North America. The mapped coastal outlines are different from today because sea level was substantially lower when large volumes of water were stored in the glacial ice. The Bering land bridge, connecting Asia and North America, provided a path for human migration into the Americas from Asia.

Where the Ice Sheets Came From

The Laurentide and Cordilleran ice sheets did not sweep southward from the North Pole, but they gradually grew outward from several initial ice caps in Canada (see ice-flow arrows in Figure 18.36). The far Arctic north is very dry, so the thickest glacial ice accumulated farther south where there is more snow between latitudes 55–65° North. Large areas of central and northern Alaska were ice free (Figure 18.36) because snowfall accumulation rates were too low for glaciers to form.

Geologists estimate that the Laurentide ice sheet was more than 3 kilometers thick near Hudson Bay, as shown in **Figure 18.37**. Lobes of ice extending south of the Great Lakes into the United States were only 500–1000 meters thick.

Lower Sea Levels and the Bering Land Connection

When glaciers expand on continents, sea level falls in the oceans. Ice accumulation in glaciers temporarily removes water from the water cycle. The global ice-age glacier volumes were sufficient to lower sea level by about 100–120 meters during the peak of the ice age 21,000 years ago.

The emergence of the seafloor between Asia and Alaska when ice-age sea level was low is significant for the human geography of North and South America (see Figure 18.36). This "Bering land bridge" connected the continents where the shallow waters of the Bering Sea exist today. Asiatic people crossed the land bridge and migrated to North and South America. There is no convincing evidence that Asiatic people arrived in Alaska during peak glacial conditions, about 21,000 years ago. The harshly cold climate conditions and the limited availability of game animals or edible plants were obstacles to migration into this area at the time of lowest sea level. The land bridge remained until about 12,000 years ago because sea level rose very slowly as the ice sheets melted. The first human migration into central Alaska occurred before 12,000 years ago, and migration along the coastline may have begun by 14,000 years ago.

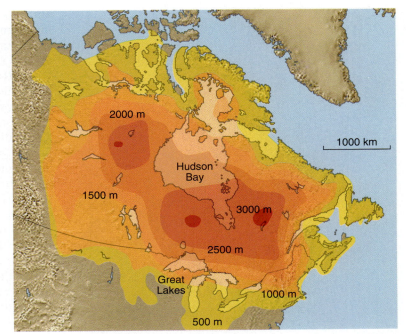

▲ **Figure 18.37** **Visualizing the thickness of an ice-age ice sheet.**
This map shows the estimated thickness of the Laurentide ice sheet. The thicknesses are calculated using equations that explain the thickness and physics of flow in the modern Antarctic and Greenland ice sheets. The ice was thickest in the areas where the ice sheet started forming, south and west of Hudson Bay.

Landscapes beyond the Glaciers

North American landscapes south of the ice sheets and lower than the glaciated mountains were also very different than today. Braided rivers choked with outwash sediment flowed southward from the ice sheets and spilled from the glaciated mountains into the surrounding plains.

Most of the meltwater from the Laurentide ice sheet entered the Mississippi River valley between 21,000 and 14,000 years ago, and the discharge in the upper Mississippi was probably 6 to 8 times larger than it is today. River valleys filled with outwash sediment in the early stages of ice-sheet melting, and then the streams eroded valleys into those deposits when the outwash-sediment supply diminished. The older braided-stream deposits underlie large areas adjacent to the Mississippi valley, as shown in **Figure 18.38**.

The sediment-laden rivers flowed southward from the Laurentide ice sheet and eastward from the Rocky Mountains across a cold, windswept, sparsely vegetated landscape. The wind blew away fine sand and silt from the floodplains

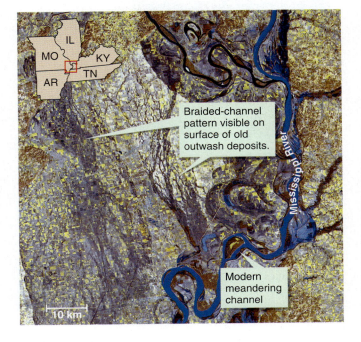

▶ **Figure 18.38** **Braided outwash rivers in the Mississippi Valley.**
This satellite image shows the meandering Mississippi River and a checkerboard pattern of farmland. The landscape west of the river is the upper surface of valley-filling outwash sediment deposited between about 8000 and 11,000 years ago. The braided pattern of the outwash is still clearly visible on this landscape.

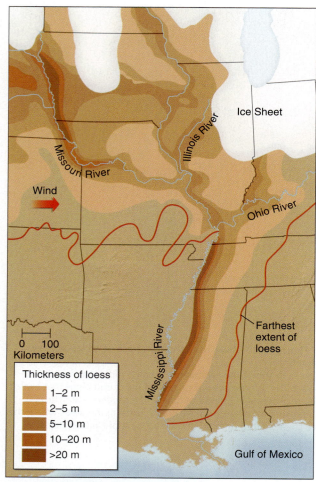

▲ **Figure 18.39** **Ice-age loess deposited downwind of outwash rivers.**
The map shows the variations in thickness of a blanket of wind-blown silt, called loess, that accumulated alongside and down-wind of the ice-age outwash rivers. The photograph shows a thick exposure of loess in northwestern Missouri.

EXTENSION MODULE 18.1

Ice Age Lakes in the Great Basin.
Learn the geologic evidence for deep lakes in the desert Great Basin during the last ice age.

and bars of the outwash streams. This sediment built up sand dunes and a frosting of fine silt that now form the parent materials for fertile soils across most of the Mississippi River drainage basin. The windblown silt deposits are called **loess**, a German term for the same type of loose silt that is common along the Rhine River in Europe. **Figure 18.39** depicts the distribution of loess in the central United States, where successive glaciations left more than 20 meters of windblown silt near major rivers.

Ice-age pollen collected from lake and floodplain deposits reveal a very different distribution of vegetation than is seen today. Cold-climate mosses and lichens present today along the Arctic Ocean coastline of Canada were the dominant vegetation much farther south in the Midwestern United States. Spruce trees now found near and north of the Great Lakes formed forests in the southeastern United States and locally persisted to the Gulf of Mexico. The southward shift of cold-climate vegetation resulted from the globally cold conditions and the effect of chilling winds that swept southward from the Laurentide ice sheet.

The deserts of the southwestern United States were awash in deep lakes during the last ice age, as shown in **Figure 18.40**. Many fault-block valleys in this region have no drainage outlets. Today, precipitation runoff flows into the valleys and then evaporates to leave salty deposits (see Section 16.13). Ancient beaches located high up on the adjacent mountainsides, and muddy lake deposits on the valley floors with remains of invertebrate animals that thrive in deep freshwater lakes reveal a different scene 21,000 years ago. Weather patterns different from today brought much more abundant winter snow and rain south of the thick northern ice sheets. Evaporation was also less, compared to today, because of the cooler temperatures of the global ice-age climate. The largest ancient Southwestern lake is Lake Bonneville, which covered 51,800 square kilometers of northwestern Utah and adjacent Nevada and Idaho to depths as great as 305 meters (see Figure 18.40). This huge lake has mostly evaporated in the warmer, drier post-ice-age climate, leaving behind the Great Salt Lake as a meager remnant along with extensive evaporites that floor the Bonneville Salt Flats. You may be familiar with the salt flats as the place where super-fast land vehicles are tested.

Landscape Changes during Glacial Retreat

Figure 18.41 maps the changing geography during the demise of the ice age. The retreat of the great ice sheets is linked to important geologic features in North America.

The northward retreat of the glaciers from the northern United States exposed freshly eroded bedrock, recently deposited till, and a new landscape for stream drainage. **Figure 18.42** shows how glacial erosion and deposition completely rearranged the drainage basins of nearly a third of North America. Glaciers are long gone from this region today, but the locations of rivers, lakes, drainage divides, and even the directions that the rivers flow are a legacy of glacial modification of the landscape.

Huge lakes, mapped on Figure 18.41, formed during melting of the Laurentide ice sheet. The initial meltwater did not readily run off in rivers to the Atlantic Ocean because the land surface was depressed by the weight of the glacial ice. The area of depressed land extended outward more than 100 kilometers beyond the terminus of the ice sheet and caused the surface to slope toward the glacier (see Section 13.3 for further description of isostatic subsidence and rebound caused by large glaciers). The big meltwater lakes have largely disappeared or are greatly reduced in size compared to 10,000–12,000 years ago (Figure 18.42) except for lakes filling deep glacier-eroded troughs in bedrock, such as the Great Lakes. Peat bogs and wetlands remain over most of the region covered by the older lakes, however. Some of the lake water drained eastward through the St. Lawrence Valley when the Laurentide ice sheet retreated to the

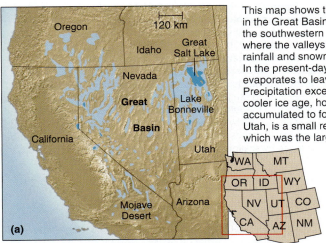

This map shows the distribution of ice-age lakes in the Great Basin and Mojave Desert regions of the southwestern United States. Lakes formed where the valleys do not have outlets, so all rainfall and snowmelt collects on the valley floors. In the present-day arid climate, this moisture evaporates to leave salt flats and playas. Precipitation exceeded evaporation during the cooler ice age, however, so the water accumulated to form deep lakes. Great Salt Lake, Utah, is a small remnant of Lake Bonneville, which was the largest ice-age lake.

◀ **Figure 18.40** Evidence for ice-age lakes in the Great Basin.

Old Lake Bonneville shorelines are visible today as beaches and wave-eroded notches along mountainsides.

Most of Lake Bonneville has dried up, leaving behind thick salt deposits that form the Bonneville Salt Flats. Notice the circled person for scale.

north, and some drained southward as the crust slowly rebounded upward in response to removal of the weight of the glacial ice. Nearly flat plains in the Upper Midwest of the United States and in southern Canada are former lake bottoms. Many large cities, including Chicago, Detroit, Toronto, and Cleveland, are built on these lake plains.

Sea level rose while the glaciers melted. The initial rate of sea-level rise was faster than the rate of isostatic-rebound uplift of the crust that had been depressed under the weight of more than 3 kilometers of glacial ice. As a result,

EXTENSION MODULE 18.2

Humongous Ice-Age Floods in the Pacific Northwest.
Learn about the incredible erosional and depositional features formed by ice-age floods more than 100 meters deep that rushed across the Northwestern United States.

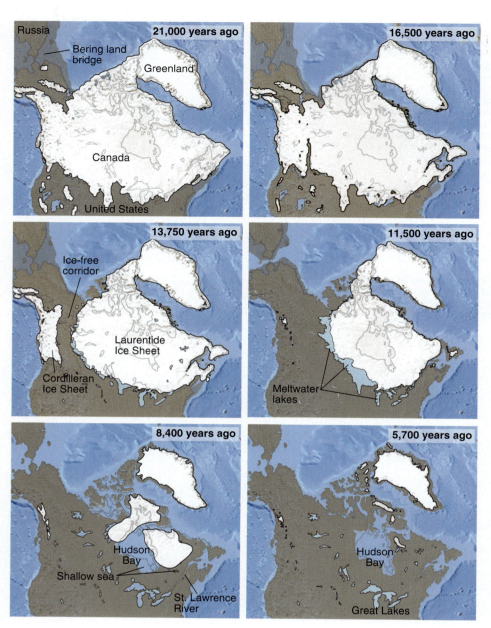

These maps show the changing distribution of glacial ice, meltwater lakes, and shoreline position since the peak of the last ice age, about 21,000 years ago. An ice-free corridor formed between the contracting Cordilleran and Laurentide ice sheets about 14,000 years ago and permitted southward migration of arctic inhabitants into the present United States. Large meltwater lakes around the periphery of the Laurentide ice sheet eventually drained to the Atlantic Ocean as the ice sheet retreated. The land was depressed by the weight of the thickest ice so that the Hudson Bay lowlands and St. Lawrence River valley were briefly submerged by shallow seas until the land gradually rose back up above sea level.

seawater flooded southwestward along the St. Lawrence River valley past Montréal, Quebec, and the Hudson Bay shoreline was as much as 250 kilometers inland of where it is today (see Figure 18.41). Eventually, the slow upward rebound of the crust raised much of this inundated area above sea level to establish the modern shoreline.

Muddy deposits of the former lakes and shallow seas present hazards and engineering challenges, especially in southern Canada. The young, poorly consolidated lake and marine clays settle unevenly under the weight of buildings and highways. The marine-clay deposits along the St. Lawrence Valley are particularly sensitive to landslide failure where wet clay slides off its foundation of glacial till or bedrock into stream valleys. The instability of these clay deposits makes the St. Lawrence Valley one of the most landslide-prone regions of the world. The most destructive of these events destroyed a Quebec village in 1971 and claimed 31 lives.

Implications for Soil Formation and Fertility

Glaciation partly explains soil characteristics in central North America. The Laurentide ice sheet scoured away the pre-existing soil and regolith over nearly half of the area that was covered by ice in Canada. Today, much of this area lacks soil or has only thin, immature soil because there has only been about 12,000 years of weathering to produce new regolith.

In contrast, very fertile soils are found in the western and southern areas of the former Laurentide ice sheet. Glacial till, locally thick outwash, and meltwater-lake sediment consist primarily of freshly ground-up rock debris eroded by the ice-age glaciers. These deposits either buried or replaced older soils resulting from tens of millions of years of weathering in a mostly humid climate. The mineral nutrients were exhausted in the old soils by this long period of weathering, and the new, unweathered, and crushed rock allowed soil formation to start over in regolith that was rich in mineral nutrients. Particularly striking differences in soil fertility are seen today within the Ohio River drainage basin, where crop productivity is very high in glaciated Illinois, Indiana, and western Ohio but markedly less productive in unglaciated Kentucky, to the south.

Fresh regolith for soil formation also characterizes most of the midcontinent plains of the United States, even south of the glaciated region. Fertile soils in this region form in thick loess deposits and valley filling outwash alluvium.

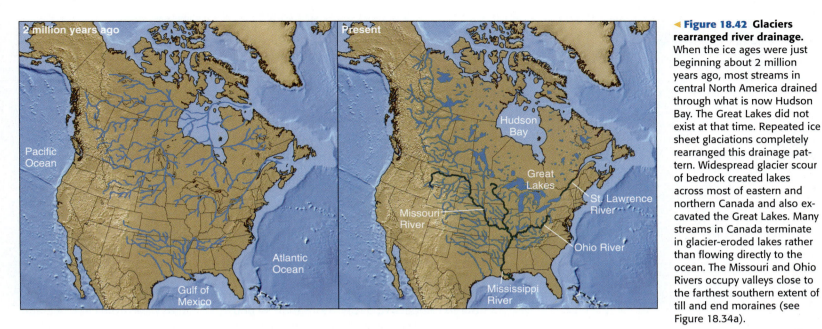

◄ **Figure 18.42** **Glaciers rearranged river drainage.** When the ice ages were just beginning about 2 million years ago, most streams in central North America drained through what is now Hudson Bay. The Great Lakes did not exist at that time. Repeated ice sheet glaciations completely rearranged this drainage pattern. Widespread glacier scour of bedrock created lakes across most of eastern and northern Canada and also excavated the Great Lakes. Many streams in Canada terminate in glacier-eroded lakes rather than flowing directly to the ocean. The Missouri and Ohio Rivers occupy valleys close to the farthest southern extent of till and end moraines (see Figure 18.34a).

Putting It Together—What Did North America Look Like during the Last Ice Age?

- During the last ice age, which peaked about 21,000 years ago, glacial ice covered about two-thirds of North America.
- Sea level was 100–120 meters lower than today, producing a land connection between Asia and Alaska along which the first humans migrated to the Americas.
- Cooler and wetter climate in the southwestern United States allowed deep lakes to form within valleys of the Great Basin, where only shallow salty lakes and salt flats are present today.
- Outwash choked river valleys, causing widespread deposition by braided streams during the early stages of glacial retreat. These deposits now form stream terraces. Wind blowing across the stream valleys swept up fine-grained sediment and deposited it in adjacent areas as loess.
- Large lakes formed on the crust depressed by the weight of the thick Laurentide ice sheet.
- Soil formation throughout the glaciated region and in areas covered by outwash and loess has been taking place for less than 21,000 years within regolith that experienced little chemical weathering and thus had lost few mineral nutrients. The most fertile soils in North America are found on deposits related to the ice age.

18.10 **How Do We Know** . . . How to Determine When Ice Ages Happened?

PICTURE THE PROBLEM
How Many Ice Ages Were There, and When Did They Happen?
The last ice age was very recent in geologic history and within human history. Clearly, a future ice age and the accompanying cold global climate would have negative impacts. It is important, therefore, to know how many ice ages occurred in the past and how often they recur.

The extent and age of the last ice age are best known because it happened most recently and its deposits are present at, or very near, the surface. In many places, however, successive layers of glacial till, separated by soils, indicate many glacial advances and retreats in the past. Each ice-age advance deposited till, and soil formed on the till after the glaciers retreated. Although the tills and soils record multiple ice ages, it is very difficult to piece together a complete history of the older glaciations because the last glacial advance substantially eroded the older deposits and highly modified the landscape. So how do geologists know? They turn to seafloor sediment, rather than glacial deposits on continents, to decipher a complete history of ice ages.

BACKGROUND TO THE METHOD

How Do Isotope Measurements of Marine Fossils Track the Timing of Ice Ages?
There are two keys to understanding how geologists determine the timing of the ice ages. The first key is a subtle change in water composition that occurs when water molecules move through the hydrologic cycle. Evidence of alternating changes in seawater composition during ice ages versus the intervening warm periods reveals a pattern. The second key is the use of chemical analyses of marine fossils collected from the seafloor to track the changes in water composition.

The first key to determining the long ice-age history rests in the behavior of two stable isotopes of oxygen within the hydrologic cycle. Both isotopes are present in water molecules. One isotope is ^{18}O (pronounced "oxygen 18") and the second one is ^{16}O. They are called stable isotopes because they do not undergo radioactive decay. More than 99 percent of all oxygen atoms are ^{16}O, and from your familiarity with atomic mass numbers, you can tell that in its nucleus ^{18}O contains two more neutrons than ^{16}O.

The abundances of the two isotopes within water, water vapor, and ice differ because water molecules containing ^{18}O are heavier than those containing ^{16}O. For example, when water partly evaporates, the water vapor has more of the lighter ^{16}O and less of the heavier ^{18}O than the remaining liquid water.

Figure 18.43 illustrates how the oxygen isotope composition of seawater changes when glaciers advance and retreat on continents. When clouds form from evaporated seawater, the water vapor preferentially includes the lighter ^{16}O. When the vapor condenses and precipitates over continents, the resulting liquid eventually finds its way back to the ocean through the hydrologic cycle, which returns the ^{16}O to the sea. As a result, the overall ratio of the two oxygen isotopes in seawater remains the same. A change in the isotope ratio occurs, however, if the precipitation in the continental interior falls as snow and remains on land as glacial ice. The snowfall in the continental interior contains more ^{16}O and less ^{18}O than seawater. This means that during an ice age the ratio of ^{18}O to ^{16}O (written as $^{18}O/^{16}O$) increases in seawater. When the ice sheets melt, this ratio in seawater decreases because the ^{16}O-rich water within the melting glacial ice returns to the ocean.

The second key to reconstructing the timing of past ice ages is analysis of marine fossils to determine how the $^{18}O/^{16}O$ ratio in seawater changed back through time. Geologists rely on the nearly microscopic shells of a group of marine organisms called *foraminifera*, an example of which is illustrated in Figure 18.43. Foraminifera secrete tiny shells composed of calcite. The $^{18}O/^{16}O$ ratio of the oxygen atoms in the calcite shells records the ratio of the seawater oxygen isotopes at the times when the foraminifera lived. Geologists extract fossil foraminifera shells from sediment layers beneath the seafloor and measure their isotope ratios in the laboratory to obtain a history of the seawater $^{18}O/^{16}O$ ratio, which simultaneously reveals the history of growth and melting of ice sheets on continents.

THE DATA

When Did the Ice Ages Take Place?
Figure 18.44 illustrates the records of oxygen-isotope ratios in foraminifera from sedimentary layers sampled from beneath the seafloor in the Atlantic and Pacific Oceans. Here are the key interpretations drawn from the data:

- The isotope ratios change in the same fashion at two widely separated locations, which indicates that the measurements track changes in global ocean composition and not local changes.

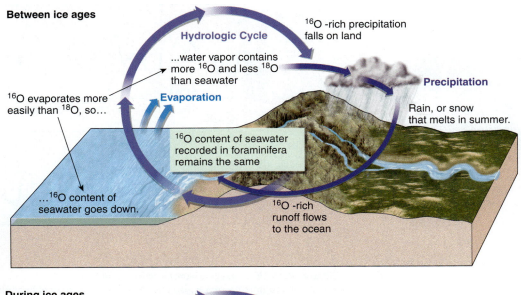

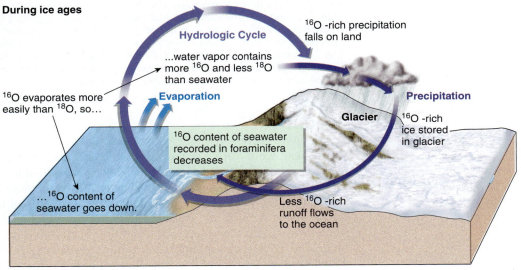

0.1 mm

Calcite shells of tiny foraminifera record oxygen-isotope content of seawater

▲ **Figure 18.43 How oxygen isotopes track ancient ice ages.**
Water molecules containing the ^{18}O and ^{16}O isotopes separate from each other within the hydrologic cycle. Water with ^{16}O more easily evaporates from the ocean, so rain and snow that fall on the continents contain more ^{16}O than seawater. During times between ice ages, like the present, the ^{16}O-rich runoff from continents returns to the ocean. During ice ages the ^{16}O-rich water is partly stored in glacial ice, so the ^{16}O content of seawater decreases. The tiny calcite shells of fossil foraminifera record these variations in the oxygen-isotope content of seawater.

- The $^{18}O/^{16}O$ ratio is very high in foraminifera deposited on the seafloor 21,000 years ago, which is consistent with the record of the last ice age on the continents. Other time periods with similar high ratios, therefore, are interpreted as older ice ages.

  Geologists have obtained several dozen similar oxygen-isotope records from locations throughout the world oceans. These records extend back more than 100 million years in a few places. They suggest that small glaciers probably existed in the Mesozoic Era, probably in Antarctica, which has been at the South Pole since Mesozoic time. The isotope data suggest that Antarctica glaciers expanded about 33 million years ago, during the middle of the Cenozoic Era. The data further indicate that significant glaciers first appeared in the northern hemisphere about 2.5 million years ago, and they have advanced and retreated dozens of times since then.

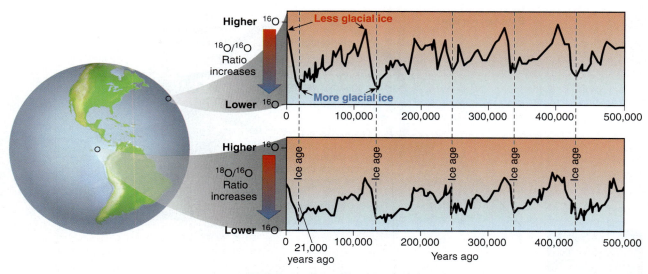

▲ **Figure 18.44 What oxygen-isotope data reveal.**
These graphs summarize oxygen-isotope measurements in fossil foraminifera from two locations. The curves are generally similar for the two locations, suggesting that they record global changes in seawater isotopic composition. Using the analysis presented in Figure 18.43, ice ages correspond to times when there is less ^{16}O in the fossil foraminifera, which also means less ^{16}O in the ocean. The occurrence of relatively low ^{16}O content 21,000 years ago, when continental deposits demonstrate the presence of large ice sheets, corroborates this interpretation and permits recognition of earlier ice ages from the isotope data.

INSIGHT

Do Ice Ages Have Rhythm?

Geologists were very excited about the oxygen-isotope records when they were first constructed during the 1960s because the data revealed something more than just when the ice ages occurred. The data plots like the ones shown in Figure 18.44 look a little bit like electronic records of a beating heart. The peaks and troughs in the isotope values are not randomly distributed through time. Instead, there seems to be a definite beat, a rhythm to the changing climate. The peaks corresponding to the warmest time periods occur about every 100,000 years. The intervening smaller peaks and troughs are also spaced fairly regularly at 23,000 years and 41,000 years. **Figure 18.45** shows how to explain the isotope curve: It is the sum of curves with rhythms of 23,000, 41,000, and 100,000 years.

The ice-age rhythm, which is explained in the next section, has turned out to be an important clue to what causes ice ages. Although not anticipated at the outset, the oxygen-isotope data collected to determine *when* ice ages occurred also opened up decades of exciting research into *why* ice ages occur.

EXTENSION MODULE 18.3
Ice Ages through Earth History.
Learn the geologic evidence for ice ages that happened hundreds of millions and billions of years ago.

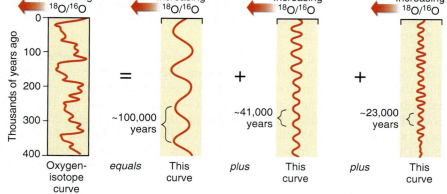

▶ **Figure 18.45 The oxygen-isotope record has rhythm.**
The pattern of peaks and troughs in the oxygen-isotope curve on the curve is simply the sum of three much more regular curves where consecutive peaks or troughs are separated by 100,000 years, 41,000 years, and 23,000 years.

Putting It Together—How Do We Know . . . How to Determine When Ice Ages Happened?

■ The ratio of two stable isotopes of oxygen present in seawater molecules changes through time as glaciers advance and retreat on continents. The variations are caused by different behaviors of the isotopes when water molecules move through the hydrologic cycle as liquid water, water vapor, and ice.

■ Marine foraminifera secrete miniature calcite shells, in which the oxygen atoms record the oxygen-isotope composition of the ocean when the foraminifera lived. Fossil foraminifera sampled from sediment layers beneath the seafloor permit reconstruction of the oxygen-isotope composition of seawater, and hence the record of glaciation on continents, back through time.

■ The oxygen-isotope ratios in foraminifera reveal dozens of ice-age glaciations in the northern hemisphere over the last 2.5 million years. These data also show rhythmic variation in glacial-ice volume that varies over time intervals of about 23,000, 41,000, and 100,000 years.

18.11 What Causes Ice Ages?

There has to be a driving force or forces that fluctuate in a regular, rhythmic fashion with the oxygen-isotope record to cause ice ages. Oceanographers and geologists recognize that the persistent beats in the isotope data at 23,000, 41,000, and 100,000 years coincide with calculated periodic variations in Earth's rotation on its axis and in its orbit around the Sun. These variations were calculated during the early twentieth century by Milutin Milankovitch and are commonly called the **Milankovitch cycles**. Milankovitch predicted that these cycles should affect the amount of solar energy that reaches Earth, but prior to measuring the oxygen isotopes in foraminifera, there was no clear evidence to support the hypothesis.

Variations in Earth's Rotation and Orbit

Astronomers recognize three climatically important effects from calculations of the gravitational pull of the Sun, Moon, and other planets on Earth's orbit and rotation. **Figure 18.46** illustrates these effects.

• Earth's rotation axis is tilted from vertical. The tilt causes seasons, because the parts of Earth's surface and atmosphere tilted toward the Sun receive more heat than areas that point away (see Figure 18.46a). The angle between vertical and the rotation axis is called the *obliquity*. The current tilt angle is 23.5°, but astronomers' calculations show that it varies from as little as 21.5° to as much as 24.5° and back about every 41,000 years.

• The shape of Earth's orbit is slightly elliptical. *Eccentricity* describes how elliptical an orbit is; a circular orbit has an eccentricity of zero. Earth's eccentricity varies over time between two very small values, 0.005 to 0.05 (Figure 18.46b). The most extreme oscillations between eccentricity values take place about every 413,000 years, but there are smaller variations that occur every 100,000 years.

• Earth wobbles as it rotates, much like a spinning top. This wobble causes the rotation axis to point in different directions over time, which causes the seasons to occur at different positions during Earth's orbit around the Sun. The orbit also wobbles, vaguely resembling a giant hula hoop as it rotates around the Sun. These wobbles are called *precession* and, acting with eccentricity variation, cause changes in the Earth-Sun distance at the seasonal solstices and equinoxes. As shown in Figure 18.46c, any particular seasonal solstice or equinox occurs at the same position in Earth's orbit around the Sun every 23,000 years.

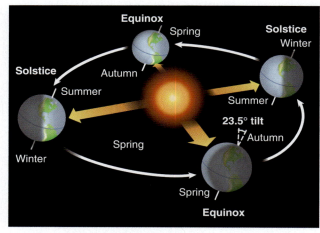

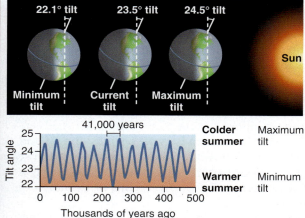

(a) Tilt of Earth's rotation axis

Earth's rotation axis is tilted at a 23.5 degree angle. The hemisphere that points toward the Sun receives more direct solar heating and is warmer than the hemisphere that points away from the Sun. The tilt, there, explains the seasons, and why northern hemisphere seasons are opposite from those in the southern hemisphere.

The tilt angle varies from low to high and back to low values every 41,000 years. Summer heating is greater and winter cooling is greater when the tilt angle is high. Summers are cooler, therefore, when the tilt angle is low. Persistence of year-round snow is favored by cooler summers.

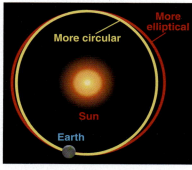

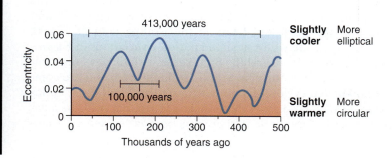

(b) Shape of Earth's orbit

Earth's orbit is slightly elliptical (or eccentric), rather than circular. When the orbit is more elliptical then year-round solar heating from the Sun is slightly less than when the orbit is more circular. Extreme values of eccentricity recur every 100,000 years, and there is another cycle that recurs every 413,000 years, but the amount of temperature change is very small.

Wobble of the rotation axis

Earth's rotation axis wobbles just like a spinning top. This means that the North Pole points to a different direction in space at different times.

The orbit around the Sun also wobbles, similar to the looping of a hula hoop around your waist.

Wobble of the orbit

Earth is at the same position in its orbit for each solstice or equinox approximately every 23,000 years. When Earth is far from the Sun on the June 21 solstice, the northern hemisphere summer is coldest and the southern hemisphere summer (December 21) is warmest.

(c) Wobble of Earth's rotation axis and orbit

▲ **Figure 18.46 How Earth's rotation and orbit vary over time.**
The Milankovitch cycles describe predictable changes in three aspects of Earth's orbit and rotation that also have predictable effects on climate.

Variations in Solar Heating of Earth's Surface

Figure 18.46 also shows how variations in tilt, elliptical orbit, and wobble affect Earth's climate.

- The poles receive the most summer sunshine and heat and the least winter warming when the tilt angle is high. As a result, the difference in summer and winter temperatures in both northern and southern hemispheres is greatest when the angle is highest. Cooler summers occur when the tilt angle is lowest (Figure 18.46a).
- The difference in Earth-Sun distance at the nearest and farthest points in the orbit from the Sun increases when eccentricity increases. This causes very slight (<0.1 percent) variations in the total solar heat received by the planet. The annual heating is slightly less when the orbit is more elliptical than when it is more circular (Figure 18.46b).
- If Earth's orbit were perfectly circular, then it would not matter where the planet was in its orbit during each season. The elliptical orbit and wobble combine, however, to produce significant climate effects. For example, if the northern hemisphere summer occurs when Earth is far from the Sun, then the summer temperatures are cool compared to when Earth is close to the Sun during the summer. Unlike obliquity, which causes warmer or cooler summers in both northern and southern hemispheres during the same years, the precession effect is the opposite in each hemisphere. This means that the coolest summers in the northern hemisphere occur in the years of the warmest summers in the southern hemisphere.

The calculated effects of the three rotational and orbital variables on climate are summed up and compared to the oxygen-isotope record of glacial ice in **Figure 18.47**. The key to making an ice age is a time period of cool summers that diminish the melting of snowfall that accumulated during the winters, especially at high latitudes that receive lots of winter precipitation. The calculated solar heating at 60 degrees north latitude varies through time with the same rhythm as the oxygen isotope curve (Figure 18.47). This similarity results from the fact that the 23,000-, 41,000-, and 100,000-year cycles in the orbital and rotational variables are also dominant in the oxygen-isotope record.

The similarities depicted in Figure 18.47 between high-latitude heating and glacial-ice volume strongly suggest that the rhythmic beat of the ice ages follows the variation in Earth's orbit and rotation, but the match is not perfect. Although the calculated swings in heating and cooling match up with the times of low and high ice volume, respectively, the magnitudes of change in glacier size do not match the magnitudes of change in solar heating. An even bigger problem is that wobble accounts for the most obvious climate swings in Figure 18.47, but while cool summers occur in the northern hemisphere, the wobble effect causes warmer summers in the southern hemisphere. The geologic record of glacial advances and retreats reveals, however, that ice ages occur simultaneously in both hemispheres. These observations suggest that the Milankovitch cycles combine to produce the rhythm of the ice ages, but something else is at work, too. Other processes must synchronize the climate effects around the globe and regulate their magnitude.

The Role of Carbon Dioxide and Ocean Currents to Cause Ice Ages

Geologists work with oceanographers and climatologists to solve the problem of how processes working on Earth interact with the variations in heat received from the Sun to cause alternating ice ages and

▼ **Figure 18.47 Comparing predicted solar heating with glacial-ice volume.**
Geologists expect glaciers to form in North America when summers are cool in central Canada. The summer temperature at 60 degrees North latitude is calculated by knowing the variations in solar heating that result from the three Milankovitch cycles. The major ice ages revealed by the oxygen-isotope curve do coincide with the predicted times of cooler summers. The predicted summer heating, however, fluctuates more dramatically than the oxygen-isotope curve. Notice that the 100,000-year eccentricity cycle should not have a significant effect on summer heating but, according to Figure 18.45, this is the dominant cycle in the oxygen-isotope curve.

Calculated summer heating at 60 degrees North Latitude for each effect

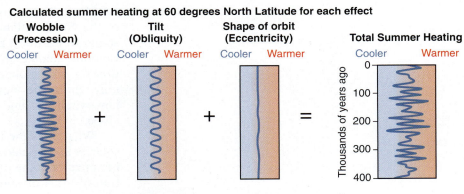

Wobble (Precession) · **Tilt (Obliquity)** · **Shape of orbit (Eccentricity)** · **Total Summer Heating**

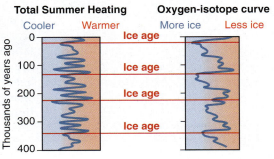

Total Summer Heating · **Oxygen-isotope curve**

warm periods. The problem is not simply academic, because it is important to understand how climate may change in the near future. We will not tackle all of the dynamics of Earth's climate, but a few observations are worth noting so that you understand where the current research focuses.

Carbon dioxide (CO_2) and global warming are hot topics of scientific, economic, and political discussions. CO_2 is one of several gases in the atmosphere that permit solar energy to reach Earth's surface but stop the heat reflected off the surface from going back out into space. The heat trapped in the lower atmosphere further warms the surface. The effect is very similar to how glass in a greenhouse raises temperature to permit year-round plant growth, so carbon dioxide is labeled a **greenhouse gas**.

Can variations in atmospheric CO_2 determine global ice ages and warm periods? Analyses of air bubbles trapped in old glacier ice in Antarctica and Greenland track atmospheric CO_2 variations over the last half-million years. Indeed, CO_2 is much lower during the ice ages and higher during the intervening warm periods. It remains unclear, however, if this greenhouse-gas variation is causing climate change or is a result of climate change. This ambiguity results because atmosphere CO_2 levels also rise and fall because of changing seawater temperature and biologic activity, which are also affected by changing climate. Recent analyses suggest that temperature started to rise at the end of the last ice age before the atmospheric CO_2 levels increased. If these results are upheld by further research, then it seems that CO_2 varied *because* of temperature changes rather than causing the temperature changes.

CO_2 variation, therefore, probably does not trigger ice ages and warm periods. It certainly must increase the extreme temperatures of both time intervals, however. For example, global increase in temperature over the last 150 years corresponds to a large increase in atmospheric CO_2 that, in turn, is attributed to the burning of fossil fuels (coal, oil, and natural gas). Herein lies a possible explanation for why the variations in solar heating and ice volume shown in Figure 18.47 do not precisely match in magnitude, because greenhouse gases also affect the overall temperatures and ice volumes.

Water holds heat much better than air or rock, so ocean currents carry heat from warm regions to cold regions in a manner that must affect global climate. Data collected by oceanographers and geologists from deep-sea sediment layers demonstrate that the global ocean circulation system works in two modes—a warm, modern mode and a cold, ice-age mode. These modes are affected not only by temperature but also by variations in the salt content of the North Atlantic Ocean. These global circulation patterns seem to respond to northern hemisphere temperature variations that are linked to the rhythmic changes in Earth's rotation and orbit. Herein lies a possible explanation for why global ice ages occur when the predicted Milankovitch effect only chills the northern hemisphere: Ocean currents communicate Milankovitch cooling or warming of the northern oceans around the entire globe.

What does the future hold in store? A glance at Figure 18.47 shows that solar heating peaked about 6000 years ago, and that another ice age seems imminent. On the other hand, burning of fossil fuels has altered the composition of the atmosphere and has contributed significantly to global temperature rise over the last 150 years. Recent data suggest that global warming is changing ocean-water temperature and saltiness (because sea-ice melting decreases salt content) in a fashion that could abruptly shift global ocean circulation into the ice-age mode. Will human-caused greenhouse warming cancel the next ice age scheduled by the orbital and rotational variations in solar heating? Or will this warming, along with possible natural warming, trigger changes in ocean circulation that will kick in the next ice age? Scientists continue to seek the answers.

Putting It Together—What Causes Ice Ages?

■ Variations in the obliquity of Earth's rotation axis, the eccentricity of Earth orbit, and the precession of the seasons around the orbit are predicted to cause rhythmic variations in the amount of solar heat received by the planet. These rhythms correspond to the rhythms in the oxygen-isotope record of glacial-ice volume, suggesting a cause and effect relationship.

■ The temperature effects predicted by the rotational and orbital variations affect the northern and southern hemispheres differently at the same time, despite the recognition that climate change is a global phenomenon. There must also, therefore, be climate-variation inputs from the Earth end of the system.

■ Variations in atmospheric carbon dioxide levels and shifts in global ocean circulation patterns appear to work with the rhythmic variations in solar heating of Earth to explain the timing of ice ages and intervening warm periods.

Where Are You and Where Are You Going?

Glaciers are slowly flowing masses of ice that cover about 10 percent of Earth's surface. The ice results from metamorphism of snow and freezing of snow meltwater. Glaciers form where temperatures are cold and precipitation is abundant so that winter snowfall exceeds summer melting. The net accumulation of snow and ice builds up until the ice flows under its own weight from high elevation to low elevation. The climate conditions for glacier formation are typically met in high mountains and at high latitudes. Valley glaciers are confined by valley walls in mountains. Ice caps and larger ice sheets are thicker glaciers that bury most or all surface topography and are not confined in valleys.

Ice always flows from the zone of accumulation to the zone of wastage regardless of whether the glacier is stationary, advancing, or retreating. The snowline defines the boundary between the two zones and marks the elevation above which winter snowfall exceeds summer melting. The relationship between the amount of accumulation above the snowline and the amount of wastage below it determines the position of the glacier front. The ice moves by internal flow and slip at the bottom of the glacier, as long as the glacier does not freeze to underlying rock and regolith. Ice deforms brittlely near the surface, to form crevasses, and plastically below about 50 meters.

Glaciers erode rock and regolith because of the high shear stress exerted by thick ice. Liquid water also plays essential roles in glacier erosion. Water at the base of warm-bottom glaciers lubricates and lifts the ice, which permits the glacier to flow faster.

Freezing and thawing of water in cracked rock and pressurized injection of water into cracks disaggregates rock that is then plucked loose by glacier flow. Freezing of water around eroded debris allows the debris to freeze into and flow with the glacier. Sediment eroded by glaciers is ultimately deposited as till left behind or melted out of the ice, or as outwash deposited by meltwater streams.

Glaciers sculpt and smooth landscapes and produce many unique landforms. Jagged peaks, knife-edge mountain ridges, broad U-shaped valleys flanked by hanging-valley waterfalls, expansive terrain pockmarked with lakes, moraine ridges, and fertile plains of glacially eroded debris are the hallmarks of glaciation. These glacial landscapes cover far more area than modern glaciers now cover, which reveals extensive modification of Earth's surface by flowing ice during past ice ages. The surficial geology of most of North America reflects glacial processes that last peaked in intensity about 21,000 years ago.

The observation that present-day landscapes are inherited from past glacial ice ages emphasizes the role of climate change in geological processes. More than a dozen ice ages occurred over the last 2 million years. The chemical compositions of tiny marine organisms reveal that the ice ages and intervening warmer climate periods, such as the present day, alternate like clockwork. The timekeepers of climate change are variations in Earth's rotation and orbit around the Sun that affect the summer heating at high northern-hemisphere latitudes where

the ice-age ice sheets are born. Natural variations in atmospheric carbon dioxide, a greenhouse gas, and abrupt shifts in ocean currents that move heat across the planet surface also play important roles in the changing climate. The onset of the next ice age is overdue, according to the astronomical timekeepers. Ironically it may be triggered by changing ocean-current patterns that some scientists predict will result from continued global warming.

You are now ready to study the important geologic boundary where land meets sea—shorelines. Water in the oceans is moved by wind and tides and not by the downward pull of gravity that governs flowing water in streams. Comprehension of how these processes move water leads to an understanding of how scenic coastal landscapes are produced, why hazardous coastal erosion occurs, and even why your favorite recreational beach exists. The highly variable characteristics of shorelines are understood by combining your acquired knowledge of rock weathering, erosion, tectonic uplift, and subsidence with new knowledge of oceanic processes.

Consideration of glacial ice-age history offers another motivation for studying shorelines. Growth of ice-age glaciers implies a simultaneous drop in global sea level. Sea level rises when glaciers melt back during warm periods. Shorelines should show a record of shifting sea level that parallels the history of glacial advances and retreats. Sea level is currently rising, with critical implications for coastal cities.

 **Active Art**

Glacial Advance and Retreat. See how a glacier moves when the front advances, retreats, or is stationary.

Glacial Processes. See the growth and movement of a glacier along with glacial erosion and deposition.

Extension Modules

Extension Module 18.1: Ice Age lakes in the Great Basin. Learn the geologic evidence for deep lakes in the desert Great Basin during the last ice age.

Extension Module 18.2: Humongous ice-age floods in the Pacific Northwest. Learn about the incredible erosional and depositional features formed by ice-age floods more than 100 meters deep that rushed across the Northwestern United States.

Extension Module 18.3: Ice Ages Through Earth History. Learn the geologic evidence for ice ages that happened hundreds of millions and billions of years ago.

Confirm Your Knowledge

---

1. What is a glacier? What features distinguishes the three types of glaciers?

2. What are the zones of accumulation and wastage? How does the snowline elevation change if climate changes?

3. What two conditions must be met for a glacier to form?

4. How does glacier ice form?

5. How does a glacier move and where does the movement take place? How quickly do glaciers move?

6. Glaciers are sometimes called "rivers of ice" but unlike rivers, direct observations of glacial flow are difficult. The ice is opaque and moves very slowly. How do geologists overcome these obstacles to studying glacier movement?

7. Explain how the front of a glacier can move upslope in a retreating valley gla-

cier if the ice is continually moving downslope.

8. Describe how glaciers erode rock and carry sediment.

9. Explain why glaciers are much more effective agents of erosion than streams, even though glaciers move much slower than streams.

10. Define "moraine". What types of moraines are there?

11. What are the two main types of sedimentary deposits formed near or beneath modern glaciers? How would you distinguish these two deposits types in an area that was glaciated in the past?

12. Which machine is more analogous to the deposition of sediment at the front of a stationary glacier, a bulldozer or a conveyor belt? Why?

13. How do icebergs form?

14. Explain the types of landscape features formed by valley-glacier erosion.

15. Describe the features left behind after an ice-sheet retreats from a region.

16. During the most recent glacial advance, which peaked at 21,000 years ago, how much of North America was covered with ice?

17. What were the consequences of ice-age climate for the parts of the continental US not covered in ice during the last ice age?

18. How do geologists determine the number of ice ages and when they happened?

Confirm Your Understanding

1. Write out an answer for each question in the Chapter Outline for the chapter sections assigned by your instructor.

2. Since glacial ice is a natural, consolidated aggregate of minerals it is considered a rock. Explain what type of rock it is, igneous, metamorphic or sedimentary?

3. Explain why a glacier carves a U-shaped valley.

4. Why are waterfalls common in a previously glaciated terrain?

5. If the conditions that existed during the last glacial maximum 21,000 years ago existed today, what would be the conditions in your hometown?

6. How would you differentiate between glacial outwash streams and braided streams associated with an alluvial fan?

7. Explain why Milankovitch cycles, by themselves, are not able to explain all aspects of the climate history of the last 400,000 years. What other factors also affect global climate change and how do they work?

Key Terms

cirque (p. 541)
crevasses (p. 525)
end moraines (p. 534)
erratics (p. 534)
fjords (p. 539)
glacial flour (p. 531)
glacier (p. 521)

greenhouse gas (p. 560)
ground moraine (p. 534)
iceberg (p. 537)
ice caps (p. 521)
ice sheets (p. 521)
ice shelf (p. 537)
kettles (p. 546)

lateral moraine (p. 532)
loess (p. 550)
medial moraine (p. 532)
Milankovitch cycles (p. 557)
moraines (p. 518)
outwash (p. 533)
sea ice (p. 537)

snowline (p. 521)
striations (p. 530)
tidewater glaciers (p. 537)
till (p. 533)
valley glaciers (p. 521)
zone of accumulation (p. 525)
zone of wastage (p. 525)

19 Shorelines: Changing Landscapes Where Land Meets Sea

19.1 What factors determine the shape of a shoreline?

19.2 How do waves form and move in water?

19.3 How do waves form shoreline landscapes?

19.4 What is the role of tides in forming coastal landscapes?

19.5 Why does shoreline location change through time?

19.6 How do we know . . . that global sea level is rising?

19.7 What are the consequences of rising sea level?

Why Study Shorelines?

HUMANS INHABIT LAND, BUT WATER COVERS **70** PERCENT OF Earth's surface, and humans have always had a practical and aesthetic relationship with the sea. Fishing has long played a crucial role as a food source, and long-distance trade has always relied on ships. People live along coasts for their livelihoods as well as to enjoy beautiful scenery, recreation, and the sound of the surf.

The total length of shorelines on Earth is about 1 million kilometers. Approximately half of the global population, or about 3 billion people, live within 60 kilometers of a shoreline. Within the United States, 14 of the 20 largest cities are located along a coast, and the counties that lie along the coastal borders contain more than half of the country's population, even though they include only 15 percent of the land area.

Most readers of this book probably live near a coast, or have at least visited a shoreline.

Along with the economic and aesthetic advantages of coastal living come the challenges of coping with active shoreline processes. On short time scales of years to decades, waves and tides erode the shore and transport huge quantities of sediment toward, away from, or along the coast. These processes, which are especially effective during windy storms, change the shape of the shoreline, damage or destroy built structures, and affect the ability of ships to access ports. On the longer time scales of centuries and millennia, changes in sea level reshape shorelines. Increases in coastal population and urbanization require an understanding of how geologic processes shape coastal landscapes.

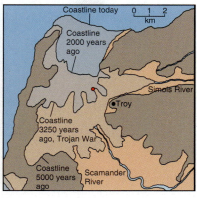

Waves shape the scenic shoreline at Big Sur, California. ▶

This chapter links the shaping and reshaping of shorelines to the processes that move water in the ocean. The focus is on ocean shorelines, although many of the concepts apply to lakeshores as well. The specific objectives are as follows:

✔ To understand the origin of waves and tides

✔ To learn how the wide variety of shoreline landscapes results from a combination of wave and tide processes, erodibility of coastal materials, sources of sediment, and sea-level change

✔ To understand how human activities modify shorelines

✔ To evaluate evidence for, and the consequences of, rising sea level

In order to accomplish these objectives you will learn answers to these questions:

19.1 What factors determine the shape of a shoreline?

19.2 How do waves form and move in water?

19.3 How do waves form shoreline landscapes?

19.4 What is the role of tides in forming coastal landscapes?

19.5 Why does shoreline location change through time?

19.6 *How do we know* … that global sea level is rising?

19.7 What are the consequences of rising sea level?

In the
FIELD

Think about taking a vacation drive along the Oregon shoreline. Perhaps you envision endless sandy beaches, but instead you see a remarkably varied landscape, as seen in **Figure 19.1**. Yes, there are beaches in some places (Figure 19.1a); there are also rocky cliffs that drop straight down into the ocean in other locations (Figure 19.1b). Steep-sided rocky islands stand a short distance off shore, suggesting that erosion separates large rock outcrops from the mainland. You wonder why the shoreline alternates between rocky cliffs and sandy beaches.

The sounds of the coast also impress you. There is the incessant crash of waves against rocky cliffs and the roar of water surging up onto a sandy beach followed by gurgling sounds, as the water pulls back into the sea. Clearly, the ocean is a powerfully moving body of water, and you wonder what processes cause the water to move.

You stroll on some beaches. Some stretches of beach are well-sorted sand, whereas others, especially near rocky sea cliffs, consist of well-rounded and polished cobbles (Figure 19.1c). When the waves rush up onto the gravel beaches, you hear the clatter of the cobbles banging together, and you know right away why they are so well rounded. Many broken seashells litter the beach, apparently transported in from offshore by waves. The waves clearly move sediment, and presumably erode rock and regolith, but how do these processes work? How might waves account for the varied shoreline scenery?

The coastal highway crosses many streams flowing to the ocean. These streams enter the sea almost always at wide, sandy beaches. Towns located where the largest rivers meet the coast are home to commercial fisheries and major tourist attractions. You wonder why people construct *jetties* (long, parallel rock walls) that stick out into the ocean on either side of the river mouths (Figure 19.1d). The wide valleys at the river mouths provide space for urban growth. The beach tends to wall off the seaward end of the valley, as seen in Figure 19.1d, to provide a bay where fishing fleets are moored in quiet water away from the waves that pound the beach.

You linger for a while in one of these coastal communities to do some shopping and visit tourist sites. During the several hours of your visit, you notice that the water level drops along the bayside waterfront, exposing barnacle-covered piers and bare, muddy banks (Figure 19.1e). You realize that the changing water level records a falling tide. What causes tides, and do they play a role in shaping shoreline landscapes?

Shorelines are dynamic places with varied landforms. It is important to understand wave and tide processes in order to understand how coastal landscapes develop. You also need to understand how these processes erode and transport sediment and where the sediment comes from. If you live in a coastal community, then you want to know how waves and tides cause changes in your familiar landscape that result in destruction of property. People build structures, like the jetties seen in Figure 19.1d, in order to control the effects of waves and tides. These engineering feats are yet another reason to study waves and tides so that you understand the consequences and effectiveness of these structures.

Figure 19.1 Views of shoreline landscapes along the Oregon Coast. ▶

Field photos of shoreline features along the Oregon Coast

A. Sandy beach with incoming waves near Newport, Oregon.

B. Steep rocky cliff of basalt drops directly into the ocean without a beach; Cape Foulweather, OR

C. Smooth, rounded pebbles and cobbles of basalt found on a small beach near where photo B was taken.

Crab, for scale; about 4 cm across

Rock outcrop rises from the seafloor; did this rock start out as part of the sea cliff, but is now eroded into an isolated island?

Water line at high tide.

Jetties stick out to sea at mouth of river.

Waves crash on sandy beach

High tide

Barnacles attached to pier are submerged at high tide.

Low tide

E. Photos taken about 6 hours apart show the different levels of high and low tides on a pier in the bay.

D. Postcard view of Yaquina Bay where the Yaquina River enters the Pacific Ocean at Newport, Oregon. The quiet water of the bay contrasts with the waves crashing on the beach on either side of the river mouth.

19.1 What Factors Determine the Shape of a Shoreline?

Shorelines are the interface between land and sea. The features illustrated in Figure 19.1 demonstrate that shorelines are not perfectly straight all along the edge of a continent. As illustrated in **Figure 19.2**, there are places where land juts out into the sea to form a **headland** (Figure 19.1b, for example). Headlands commonly have geographic names that include the words "cape" or "head." In contrast, other parts of the shoreline are deep recesses, called **bays** (see Figure 19.1d). Land may rise abruptly above the water in steep cliffs, whereas in most places there are beaches, which are low-sloping landforms composed of unconsolidated sediment moved by waves and tides (contrast Figures 19.1a and b).

Shoreline materials erode in different ways and to different extents, which contributes to irregularities in coastal outlines. The degree of consolidation, extent of jointing and faulting, and presence or absence of bedding or foliation planes are all rock characteristics that determine strength and erodibility. Strong rocks compose many headlands (Figure 19.1b), whereas weak rock or regolith may form the shoreline along adjacent bays (Figure 19.2).

Stream erosion commonly contributes to the formation of bays. Stream valleys partially submerge beneath the sea if the land subsides or sea level rises. **Estuaries** are submerged parts of stream valleys where freshwater and seawater mix. The bay seen in Figure 19.1d is an estuary at a river mouth. Along submerging coastlines the bays may simply be drowned river valleys, and the headlands are the ridges that divide adjacent valleys. In this case it is previous stream erosion rather than material properties that accounts for the irregular coastline.

The availability of loose sediment also influences the shoreline shape. Some beaches simply consist of local, wave-eroded sediment as in the case where a gravelly beach alongside a headland contains only the rock types found on the headland (see Figure 19.1c, for example). In most locations, however, rivers de-

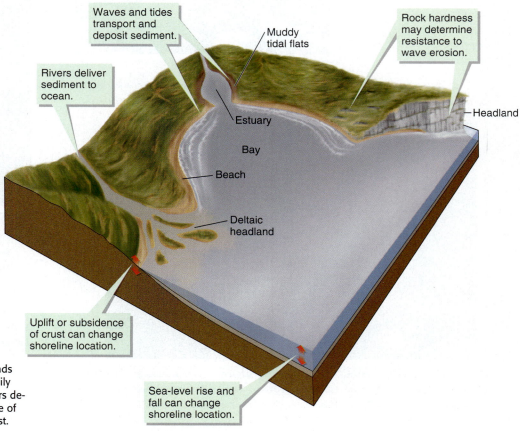

▶ **Figure 19.2 Factors that influence the appearance of coastal landscapes.**
Shorelines are rarely straight but alternate between headlands and bays. Variations in coastal landscapes relate to how easily waves and tides erode coastal rock and regolith, where rivers deliver sediment, and how shoreline position changes because of sea-level change or because of uplift and subsidence of crust.

liver sediment to the ocean where waves and tides then spread the sediment along the coastline to form beaches. Deltaic headlands (see Section 16.5) form where rivers build out the coastline by depositing more sediment than waves and tides can erode and redistribute (Figure 19.2).

Shoreline shape and position change over time (see Figure 19.2). Coastal erosion causes shorelines to retreat in the landward direction, whereas sediment deposition causes shorelines to advance seaward. Rise and fall in sea level, caused by changes in ocean-basin shape (see Section 13.4) or volume of water in the oceans (see Section 18.9) cause shorelines to shift landward or seaward. Shoreline shape also changes with uplift of the crust where former seafloor lifts above sea level or subsidence where land sinks beneath the sea.

Putting It Together–**What Factors Determine the Shape of a Shoreline?**

- Irregular shoreline shape results from variations in several factors: erodibility of rock, sediment delivery from rivers, sediment movement by waves and tides, changes in sea level, and uplift or subsidence of the crust.
- Streams may form estuaries or deltas where they meet the ocean. Estuaries are previously eroded stream valleys that submerge because of land subsidence or sea-level rise. Deltaic headlands form by sediment deposited at the mouth of a stream.

19.2 How Do Waves Form and Move in Water?

It is hard to imagine a view of the ocean without seeing waves. The origin of ocean waves and the ability of waves to erode and transport sediment are essential to understanding shoreline landscapes.

What Are Waves?

Section 8.2 introduced the general idea that waves are disturbances that transport energy through a medium like rock, air, or water. A more detailed look at the shape of waves and some of their properties is necessary to understand how waves move through water.

Figure 19.3 shows how you can form waves by shaking the end of a rope up and down. There are various features that serve to describe the wave. Low troughs separate the high crests on the wave. The **wave height** is the vertical distance between adjacent crest and trough, and the **wavelength** is the distance between successive crests, or successive troughs. The elapsed time between the passage of successive crests, or troughs, past a stationary point is the **wave period**. Typical ocean waves have wave heights of 0.5–20 meters, wavelengths of 10–200 meters, and periods of 5–20 seconds.

A wave moves through a medium, but particles in the medium return to the same location after the wave passes. For example, the wave moves from one end of the rope to the other, but no part of the rope travels from one end to the other (Figure 19.3). A typical ocean wave moves across 1000 kilometers of water in about a day, but it is the wave, and not the water, that travels that distance. Just as seismic waves represent a pulse of energy moving through Earth, ocean waves are disturbances in the water surface caused by the transfer of energy in the direction of wave movement.

How Water Waves Form

Coastal observers long ago realized that waves are highest during windy weather, which indicates that wind blowing across water produces waves. Friction at the boundary between moving air and stationary water transfers some of the

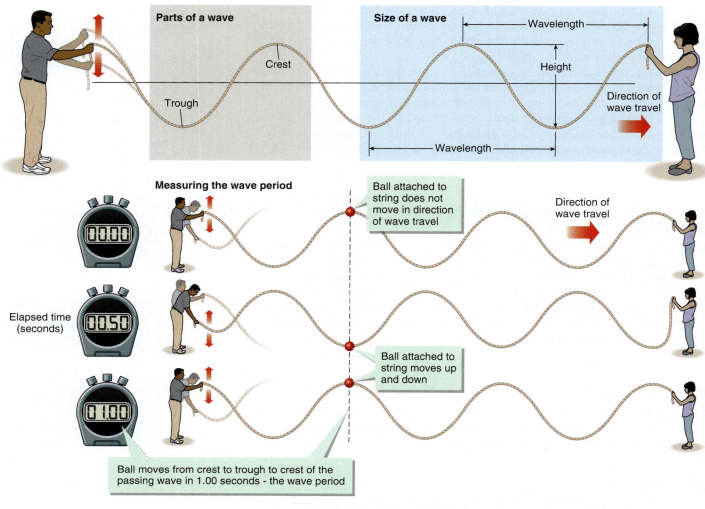

▲ **Figure 19.3 Describing the properties of a wave.**
Shaking a rope produces waves that compare well with natural waves. High and low points on the wave are called the crest and trough, respectively, and the vertical distance between them is the wave height. The wavelength is the distance between adjacent crests or troughs. Waves move through material without the material itself moving in the direction of wave travel, as illustrated by the ball attached to the rope. The wave period is the time elapsed between the passage of successive crests or troughs past a point.

Properties of Waves: *See how to recognize the parts of a wave and to measure the wave period.*

energy and momentum of the moving air to the water and places the water surface in motion. The physics of energy transfer from wind to water is a perplexing problem, but several key relationships are clear from observations.

Observations show that increases in the velocity of the wind, the duration of the wind, and the distance that the wind is in contact with the water all cause increases in wave height, wavelength, and wave period. Wind speed and duration stir the highest waves during prolonged windy storms, whereas fair weather allows the low waves of a calm sea. The importance of the distance that the wind blows over the water explains why even very high winds do not produce large waves on swimming pools or small ponds, even though hurricane winds blowing over the ocean generate waves that are more than 20 meters high.

Energy transfers from storm wind to the ocean surface, then moves across the ocean as waves, and finally arrives at the shoreline. When waves reach the shoreline, the energy performs work to move sediment and rock, and part of the energy converts to sound waves that you hear as the "roar" of the surf.

How Waves Move Water

For seismic waves, or rope waves, the energy is introduced at a single point. Wind-generated water waves are different because wind imparts energy to the ocean across the entire water surface. The amount of wave motion then decreases downward from the surface.

Figure 19.4 summarizes observations of how water moves when a wave passes. An object floating on the water surface moves up and forward as the crest of the wave goes by, and then it moves down and backward when the

▼ **Figure 19.4** **Visualizing water motion caused by a passing wave.**

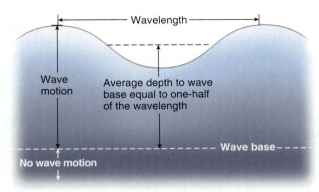

(a) Waves cause water motion only down to wave base.

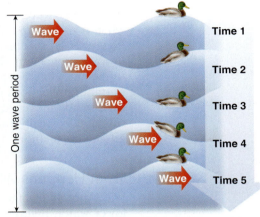

(b) Objects on the water's surface move up and down when the wave passes, but the objects stay close to the same location after each wave period.

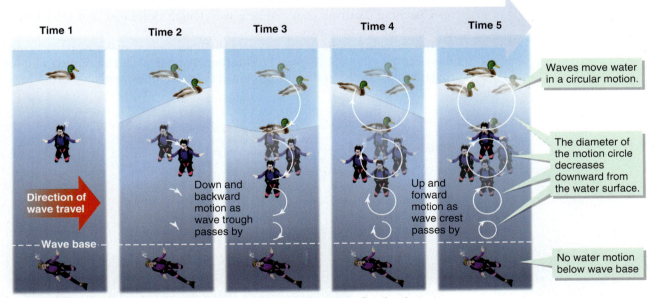

(c) These diagrams show wave movement of objects in water through the five time instants shown in (b).

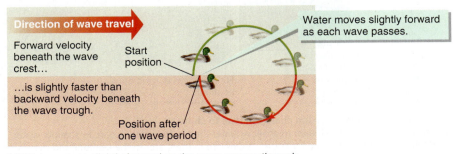

(d) Water moves slightly forward as the wave passes through.

trough of the wave passes. The floating object, therefore, traces a circular path during one wave period. Particles suspended in the water move through circular orbits of decreasing diameter at deeper depths. There is no motion below the **wave base**, the place where the depth is equal to one-half the wavelength of the waves (see Figure 19.4). The higher the wave is, the greater the wavelength and the deeper the wave base. The lack of motion below wave base explains why submerged divers and submarines are unaffected by the waves that cause seasickness on surface vessels.

The circular wave motion of water imparts a small amount of forward movement of the water that is not seen in the rope wave. The motion beneath the passing wave decreases downward, so a particle of water moves farther when it moves up and forward than when it moves down and backward (see Figure 19.4). This means that water actually does move slowly in the direction of wave movement. When the wave reaches shore, it not only transmits energy to the shoreline but also piles a mass of water against the coast, which has important implications for erosion and sediment movement. High-velocity hurricane winds move large masses of water to the shore, contributing to storm surges that locally raise sea level by several meters and cause flooding many kilometers inland.

What Happens When Waves Approach the Shoreline?

Wave motion changes where waves approach shore through progressively shallower water. **Figure 19.5** illustrates that the changes start to take place where water depth equals the wave-base depth. When waves continue into progressively shallower water, a decreasing volume of water must transport the same amount of energy. This causes many observed changes: Wave height increases, crests become narrower than troughs, wavelength decreases, and the forward velocity of waves decreases.

The changing shape of the waves in shallower water make the particle motions become elliptical (Figure 19.5) rather than circular. The motion ellipses are flatter downward so that the motion at the seafloor is simply back-and-forth

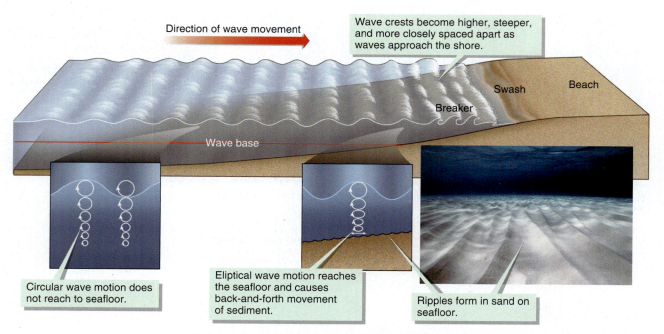

Direction of wave movement

Wave crests become higher, steeper, and more closely spaced apart as waves approach the shore.

Swash

Beach

Breaker

Wave base

Circular wave motion does not reach to seafloor.

Eliptical wave motion reaches the seafloor and causes back-and-forth movement of sediment.

Ripples form in sand on seafloor.

▲ **Figure 19.5 How wave motion changes close to shore.**
Wave crests are higher and steeper in shallow water than in the open ocean. In shallow water, near to shore, the wave motion continues downward in the water to the seafloor and moves loose sediment.

(Figure 19.5). The constant to-and-fro motion on the seafloor shapes loose sand into ripples that symmetrically slope at the same angle toward shore and away from shore (see photograph in Figure 19.5).

In still shallower water the waves become even steeper, and the forward wave velocity slows down until the water at the crest moves landward faster than the wave itself. At this point, the wave forms a **breaker** where water in the wave crest collapses down and toward the beach, as seen in **Figure 19.6**. In most cases, the wave breaks into a noisy, foaming mass of water that surges up onto the beach (Figure 19.6a). Where the steepness of the seafloor and the wave height are just right, however, the top of the wave curls over and plunges downward in the classic surfing wave (Figure 19.6b). When the wave breaks, the energy transfers from the water wave into sound waves and mechanical energy that moves sediment. The remaining momentum carries the water up the beach until the pull of gravity slows it to a stop, and then the water flows back into the ocean, as illustrated in **Figure 19.7**a. Where steep sea cliffs drop directly into deep water, the wave does not slow down sufficiently to break, so the full energy of the wave arrives with a deafening crash against the rock, as seen in Figure 19.7b.

Waves always seem to come directly toward the beach, despite the fact that waves in the open ocean move in the direction that the wind blows. Given the variety of orientations of the shoreline, how is it possible for the wave crests to be always parallel, or nearly parallel, to the shore? Refraction, the change in direction of wave motion caused by a change in wave velocity, provides the answer to this question (check back to Section 8.2 for a refresher on refraction).

(a)

New South Wales, Australia

(a)

(b)

Cape Arago, Oregon

(b)

▲ **Figure 19.6 What breakers look like.**
Wave motion in shallow water forces water up into the wave crests. Water in the steepening wave crest moves landward faster than water in the adjacent, landward trough so that the top of the wave collapses forward as a breaker. The breakers shown in (a) simply spill forward onto the beach. Excellent surfing waves, like the one shown in (b), are much higher and break with an overhanging curl.

▲ **Figure 19.7 How waves end.**
Waves end on the shore with water (a) swashing onto a beach after the wave breaks or (b) crashing into headlands where waves meet the shore before breaking.

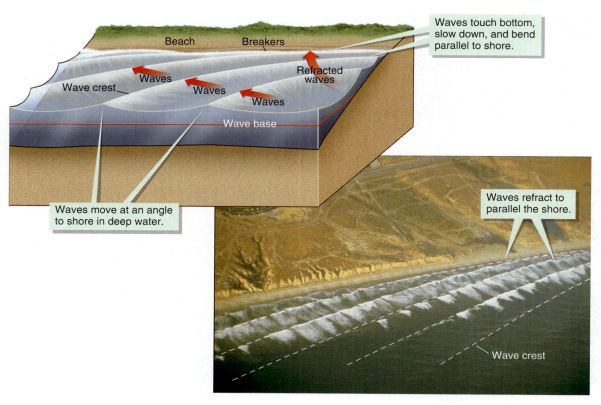

▲ **Figure 19.8 Water waves refract.**
Waves refract in shallow water to approach with wave crests parallel to the beach. Waves bend parallel to the coast because the shallow part of each wave slows down where it makes contact with the seafloor close to shore.

Water Wave Motion and Refraction: *See how waves move in water and refract along a shoreline.*

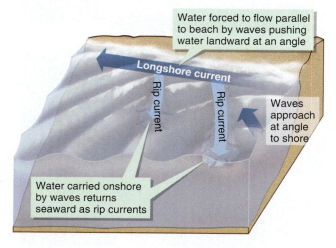

▲ **Figure 19.9 How longshore and rip currents form.**
Waves approaching the shore at an angle force water to move along the shore in the direction of wave. This along-shore water movement forms longshore currents. Some water that waves carry against the beach also moves offshore in rip currents.

Figure 19.8 illustrates how and why waves refract in water. Where a wave moves in a direction somewhat parallel to the shore, the part of the wave that is closest to land moves through shallower water than does the offshore part of the wave. The nearshore part of the wave touches bottom, steepens, and slows down. The progressive shoreward decrease in wave velocity causes each wave to bend toward the shore so that the wave crests are parallel, or nearly parallel, to the shoreline when the waves break.

How Waves Generate Currents

Wave motion is forward and backward, but observations show that waves also generate water currents that move in just one direction. These currents result from along-shore and offshore movement of the mass of water that waves move toward land. **Longshore currents** move parallel to shore, whereas **rip currents** move water away from the coast.

Figure 19.9 explains the formation of longshore and rip currents. Where waves meet the shoreline at an angle, the water carried against the land moves parallel to the coast and away from where the waves converge onto the shore. The flow of water parallel to the shore defines the longshore current. Of course, each arriving wave contributes more water into the longshore current, and this water cannot simply pile up along the shore. As a result, part of the current flows back off shore as a rip current. Longshore currents may persist for tens of kilometers unless deflected offshore by a headland or pier. Rip currents are hazardous when the current speed heading offshore is too great to swim against; swimmers encountering a rip current can be carried out to sea and drown. The safest thing to do when encountering a rip current is to swim parallel to the shore for a few tens of meters to escape the rip current and then swim toward the shore.

How Tsunami Waves Are Different

All that you have just learned relates to the normal wind-driven waves on the surface of oceans and lakes, but different processes must account for tsunami waves. Monstrous tsunami are far more deadly than storm waves, as was tragically demonstrated by the more than 150,000 lives lost in eleven nations surrounding the Indian Ocean on December 26, 2004. You learned a little bit about tsunami and their relationship to earthquakes in Section 11.9, and it is important in this section to understand how and why they are different from regular waves.

Some simple observations illustrate key differences between tsunami and regular sea waves.

- Although tall storm waves commonly arrive on distant, sunny coastlines, they rarely arrive without warning. Instead, the wave height builds gradually over many hours or days. Tsunami, in contrast, commonly arrive without warning as a series of several waves of incredible magnitude that can be more than ten meters high, as illustrated in **Figure 19.10**.

◄ **Figure 19.10** **Tsunami meets the shore.**
A towering tsunami heads toward a beach in Thailand following a giant earthquake more than 1000 kilometers away near the Indonesian island of Sumatra. The rocky seafloor is normally submerged but was exposed as the tsunami wave trough reached the shoreline ahead of the wave crest in the background. Those people curious to examine the oddly exposed seafloor, but also lacking knowledge about tsunami, then had to run for their lives as the giant wave crest rushed in.

- Instead of breaking on the beach, a single tsunami wave comes onto land as an ever-rising surge of water that floods inland for more than a kilometer on gently sloping shorelines over a period of tens of minutes. The flood wave then quickly withdraws back to the sea only to be followed by the next wave several minutes to an hour later.

The suddenness, great height, longevity of the surging wave onto land, and the long time between waves are characteristic of tsunami.

Quantitative measurements also bear out the differences between tsunami and everyday waves as indicated by the following comparisons:

- Wind-generated waves typically have a period of 5 to 20 seconds, whereas tsunami periods are typically ten minutes to two hours.
- Wind-generated waves have a wavelength of 10 to 200 meters, whereas most tsunami wavelengths are in the range of 100 to 500 kilometers in the open ocean.
- Wind-generated waves rarely move faster than about 50 kilometers per hour, whereas tsunami reach peak speeds greater than 800 kilometers per hour (comparable to a commercial jet airplane).

The differences between normal waves and tsunami relate to how they form. The fast-moving, long-wavelength tsunami require huge amounts of energy to place such a large volume of water in motion. Wind speeds would need to exceed 10,000 kilometers per hour to transfer the requisite energy that fuels a large tsunami. The long wavelengths also indicate that the entire thickness of the ocean water column, not simply a surface zone dragged along by wind, is in motion. Clearly, wind does not generate tsunami any more than wind is likely to blow water out of a swimming pool. To make a big wave that sloshes out of a swimming pool at a pool party, you need to get lots of people to jump into the pool at once. The resulting energetic wave results from displacement of water out of the pool.

Similarly, tsunami represent the sudden displacement of water over the entire depth of the ocean. Fault movement during earthquakes is the most common generating mechanism, but landslides under water or into the sea from land, the formation of calderas on volcanic islands, and the ocean impact of large meteors from space (not witnessed by humans but recorded in ancient sedimentary rocks) also cause tsunami. The connection between tsunami and

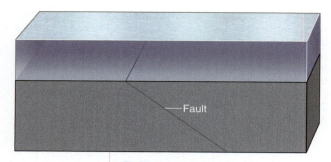

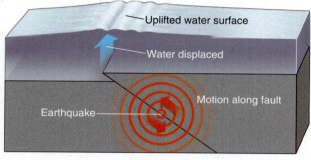

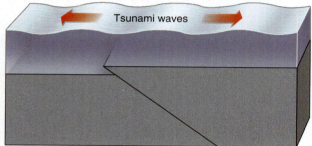

◄ **Figure 19.11** **The relationship between earthquakes and tsunami.**
Motion along a fault that ruptures the seafloor during an earthquake also displaces the overlying water and raises the ocean surface above sea level. The elevated volume of water spreads horizontally to return to equilibrium, which generates a tsunami. The longer the fault and the larger the displacement, then the larger the resulting tsunami.

sudden geologic events has been evident for centuries as residents of coastal communities, still dazed by earthquake shaking or from witnessing a huge landslide or cataclysmic volcanic explosion were then swamped by giant waves that seemingly raced onto land from nowhere.

Figure 19.11 illustrates how earthquakes generate tsunami. Although rock movement is not directly observed during submarine earthquakes, we know that it must happen by drawing comparison to displacement of rocks along faults during earthquakes on land (Figure 11.28 illustrates examples). Motion along a fault that ruptures the seafloor also moves the overlying water. Upward displacement of the ocean conveys potential energy to the water column, and the water flows away from the uplift so that the ocean surface returns to its equilibrium, minimum-energy condition. Fault motion during the December 26, 2004, Sumatra earthquake, for example, raised the seafloor and ocean surface by about 5 meters within an area approximately 400 kilometers long and 50 kilometers wide. This motion lifted more than 100 billion metric tons of water above sea level over a matter of a few minutes.

Highly energetic waves generated by an impulse, like fault motion, have very large wavelengths and low wave heights compared to waves resulting from persistent input of modest surface energy by wind. Tsunami are notoriously difficult to recognize in the open ocean because in deep water the wave may only be 20 centimeters high and more than 100 kilometers across, as depicted in **Figure 19.12**. The wave is so broad compared to its height that it is virtually invisible. Remember, too, as with any wave, that although the wave energy moves at jet-aircraft speed, the actual forward movement of the water is negligible in

Tsunami wavelength and velocity decrease when waves move from deep water to shallow water. Wave height increases dramatically. Wave period remains the same, which means that the water remains above normal sea-level for as long as an hour as each wave crest moves inland. The height, width, and velocity of tsunami causes severe destruction along low-lying coastlines, as illustrated by these satellite images obtained before and after the December 2004 tsunami struck Lhoknga, Indonesia.

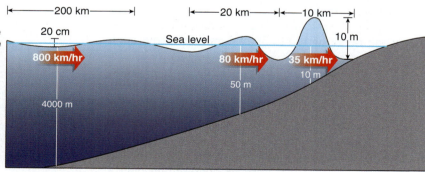

Tsunami: *See how an earthquake forms a tsunami and how a tsunami travels across the ocean.*

▶ **Figure 19.12** **Visualizing why tsunami are destructive.**

the open sea, which further decreases the likelihood of detection. The trough of the wave commonly reaches the shoreline first, so one sign of an approaching large tsunami can be the sudden withdrawal of water from the coast well beyond the low-tide mark (see Figure 19.10).

Like regular waves, however, tsunami slow down in shallow water close to shore. The immense energy condenses into a smaller volume of water, causing the ocean surface to rise as the wavelength shortens and the volume of water beneath each wave crest compresses into a smaller area above the seafloor. As Figure 19.12 shows, the fast-moving, low waves in the deep ocean build into roughly 10-meter high, 5-kilometer wide surges of water that although decelerated from their peak speed still move too fast to outrun. The actual dimensions of tsunami are hard to predict along coastlines because the height and wavelength strongly depend on the shape of the shoreline and the slope of the submerged seafloor in addition to the total tsunami energy.

Putting It Together–*How Do Waves Form and Move in Water?*

- Wind blowing across water produces waves. High waves require strong winds, blowing for long time periods, across long distances of water.
- Wave motion decreases downward in the water to wave base, the depth below the water surface to which wave motion takes place. There is a small amount of overall water motion in the direction of wave movement.
- Wave height increases and wave velocity decreases where waves move into water that is shallower than wave base. Eventually, the wave crests move forward faster than the slowing waves, causing the waves to break.
- Waves refract where they simultaneously travel slowly in shallow water and more rapidly in deep water. Refracted waves turn parallel, or nearly parallel, to the shoreline.
- Waves move water shoreward, parallel to the coast as longshore currents, then back to sea as rip currents.
- Displacement of the ocean water, usually by fault movement or landsliding generates deadly tsunami. Tsunami have a longer wavelength, larger period, and faster velocity than much less energetic wind-generated waves.

19.3 How Do Waves Form Shoreline Landscapes?

With this fundamental understanding of what waves are and how they form, it is now possible to move on to consider how waves shape the shoreline. Waves are the primary process that form and modify landscapes in most coastal regions.

How Waves Erode, Transport, and Deposit Sediment

The power of waves to do work in moving materials on a shoreline is most impressive where waves crash into rocky headlands (see Figure 19.7b). Storm waves exert incredible force against rocks, some equivalent to 10,000 kilograms weighing down on each square meter. These forces pluck off large blocks of rock that were already separated by fractures and joints or that have fallen at the base of the cliff after mass movement. Erosion is most severe near the base of a cliff and may leave an overhang that is unstable and prone to failure by mass movement, as seen in **Figure 19.13**.

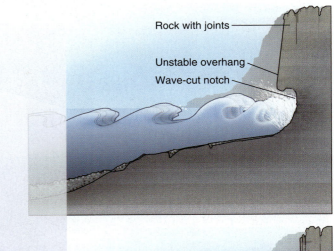

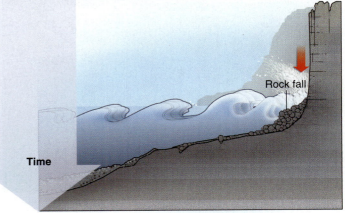

▲ **Figure 19.13 Why mass movements occur along coasts.** Waves erode steep cliffs along headlands. Wave erosion focuses at the base of cliffs to form a notch and an unstable overhang that eventually fails by rock fall.

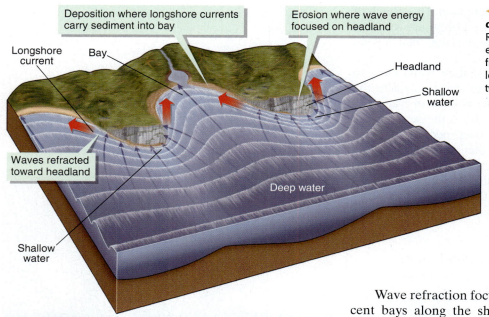

◄ Figure 19.14 Refraction causes erosion on headlands and deposition in bays.
Refraction bends waves toward headlands, which focuses wave energy on the headland. Longshore currents carry sediment eroded from headlands into the adjacent bays, where wave heights are low. Along irregular shorelines, therefore, beaches occur in bays between eroding headlands.

Wave refraction focuses erosion at headlands and favors deposition in adjacent bays along the shore, as shown in **Figure 19.14**. Incoming waves first encounter shallow water near headlands along highly irregular coasts. The waves refract and focus toward the front and sides of the protruding land. Refraction directs more water toward the headland than toward the bays, so wave height is higher along the headland than in the bays. Longshore currents carry sediment eroded from the headland toward the adjacent bays where wave height is much lower, and deposition occurs on the beach.

Waves both erode and deposit sediment along a beach and on the adjacent shallow seafloor, as illustrated in **Figure 19.15**. The to-and-fro motion of the water exerts shear stresses on the seafloor where the water is shallower than wave base. The sediment grains on the seafloor move when the shear stress is sufficient (see Figure 19.5). Sediment grains suspended in the water move landward because of the small, landward component of water transport by the waves (see Figure 19.4).

The greatest seafloor erosion occurs where waves break and plunge downward onto unconsolidated sediment (Figure 19.15). The force of the water slamming down on the seafloor suspends the sediment and carries it forward as the wave swashes up onto the beach.

There are two reasons why the sediment-carrying capacity of the water diminishes as it swashes up the beach (Figure 19.15):

1. The velocity of the water decreases because it moves upslope against the force of gravity.
2. Some water percolates into the pore spaces between the sand or gravel particles on the beach, which decreases the amount of moving water available to carry sediment.

The water rises onto the beach, slows to a brief standstill, and then retreats back down to the sea. The backwash is less erosive than the initially energetic landward swash, but it still picks up the smallest and lowest-density grains and carries them back out to sea. In this fashion, waves tend to move the coarser sediment grains landward and deposit them on the beach while the finer particles move back offshore. This explains why clastic marine sediment closest to a shoreline is coarser grained than the sediment deposited below wave-base depths (for example, see Section 5.6 and Figure 5.19).

If you watch carefully, then you will see that the swash and backwash of water on a beach concentrates high-density minerals. **Figure 19.16** shows that small grains of high-density minerals swash onto the beach with ease equal to that of large grains of lower density, such as quartz, feldspar, or calcite shells.

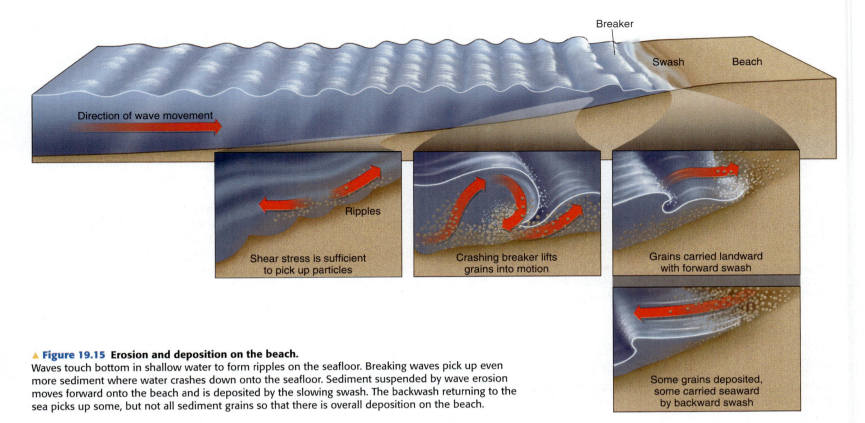

▲ Figure 19.15 Erosion and deposition on the beach.
Waves touch bottom in shallow water to form ripples on the seafloor. Breaking waves pick up even more sediment where water crashes down onto the seafloor. Sediment suspended by wave erosion moves forward onto the beach and is deposited by the slowing swash. The backwash returning to the sea picks up some, but not all sediment grains so that there is overall deposition on the beach.

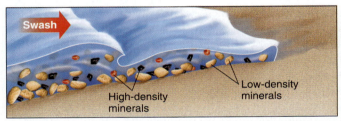

Forward swash carries sand grains up and onto the beach. Water transports small, high-density mineral grains that are equal in mass to larger, low-density mineral grains.

Sand grains settle onto beach surface as swash slows down. Small, high-density grains settle into spaces between larger, low-density grains.

Backwash preferentially picks up larger, low-density sand grains that project up into the flowing water. Most high-density grains are sheltered between larger grains and remain on the beach.

Dense sand grains from a Brazilian beach. Black grains are mostly titanium-oxide minerals.

This dredge and processing plant separates titanium-oxide mineral grains from Florida beach sand.

◄ Figure 19.16 Waves concentrate dense minerals on the beach.
Small, high-density minerals concentrate in beach deposits because they are more easily carried onto the beach by forward swash than they are carried seaward by backwash. The dense minerals are concentrated by the wave action to form dark layers in beach sand. Some economically important minerals are mined from such layers, including titanium oxide, which is a source of titanium for metal alloys and white pigments for paint and paper.

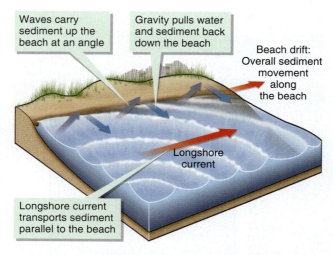

Waves carry sediment up the beach at an angle

Gravity pulls water and sediment back down the beach

Beach drift: Overall sediment movement along the beach

Longshore current

Longshore current transports sediment parallel to the beach

▲ **Figure 19.17 How sediment moves along the shoreline.** Longshore currents and beach drift transport sediment along the shoreline where waves approach the beach at an angle.

Beach Drift and Longshore Current: *See how beach drift and longshore current move sediment along the shoreline.*

The small, dense grains settle into spaces between the larger grains as the swash slows down. Here, they are protected from the backwash flow and remain on the beach. Most high-density minerals are also dark colored and produce distinctive dark layers or patches in beach sand.

Some high-density minerals that wash up on beaches are economically important. Concentrations of valuable minerals are inexpensively mined, as seen in Figure 19.16, by digging up beach sand, sorting out the high-density minerals, and returning the more abundant low-density minerals to the beach. Beach sand is a particularly important source of titanium-oxide minerals. Titanium is valuable as the principal source for white pigments in paint and paper, and it is used to strengthen lightweight steel for many uses, including aircraft engines.

At places where waves approach the beach at an angle, sediment moves parallel to the shoreline, as illustrated in **Figure 19.17**. Sediment stirred up by the breaking wave moves up the beach perpendicular to the direction of wave movement, which is at an angle to the shoreline. The backwash, however, flows directly down the beach slope into the water because of the pull of gravity. The different paths of sediment grains during transport by swash and backwash causes **beach drift,** a zigzag movement of sediment along the beach. Some of the seafloor sediment stirred up by the waves is transported parallel to shore by longshore currents (see Figure 19.8) rather than up onto the beach.

How Waves Shape Rocky Shorelines

Most rocky shorelines slowly erode in the landward direction. **Figure 19.18** illustrates how refraction focuses wave energy to erode headlands of strong rock. Wave erosion hollows out caves at the base of the headland. The caves may connect through a narrow headland to form an arch. As the arch widens, the unsupported rock roof of the arch collapses, which isolates the outermost part of the headland from the shoreline to produce a **sea stack**. This landform is simply a small rocky island that is located close to the shore (pictured in Figure 19.18; also see Figure 19.1b).

Wave-cut platforms are nearly horizontal benches of rock eroded along a rocky coast. The platform, commonly submerged and wave swept at high tide, is exposed to view when the tide is low (see Figure 19.18). Waves break up rock and carry away rock fragments loosened by biologic and chemical processes that erode the platform. **Figure 19.19** illustrates weathering and erosion processes that affect coastal rock exposures. Sea urchins, sea anemones, clams, sponges, and other marine animals bore holes into rock in order to improve their own resistance to being moved by waves. Other animals, such as snails and chitons, abrade rocks like miniature files as they graze on algae attached to the rock surface. Shallow pools of water left on the platform at low tide enhance rock weathering through growth of salt crystals in cracks by evaporation, or dissolution of soluble rocks such as limestone (see Section 5.1 for more about salt weathering). Sediment grains carried back and forth across the platform by waves also abrade the underlying rock.

How Waves Make Beaches

The definition of a beach is pretty clear to anyone who has been to a shore. A **beach** is the nonvegetated area of unconsolidated sediment that extends from the low-tide line to a landward line defined by a cliff, sand dunes, or permanent vegetation. **Figure 19.20** shows that a typical beach has a relatively steep part close to the water, called the **beach face**, and a flatter part, the **berm**. The **berm crest** marks the boundary between the beach face and the berm. Sunbathers prefer the berm part of the beach because it is beyond the reach of most fair-weather waves, is nearly flat, and is likely composed of fine sand.

Measurements by geologists show that the height of incoming waves and the size of the beach sediment determine the height of the berm crest and the

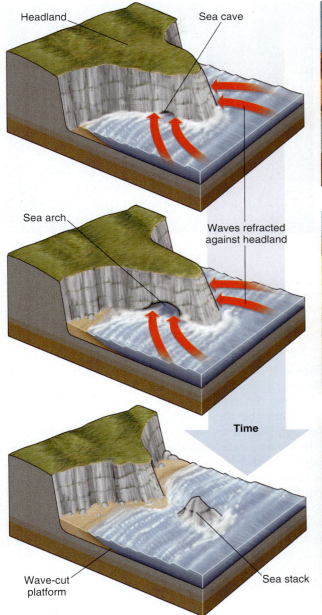

Headland

Sea cave

Waves refracted against headland

Time

Sea arch

Wave-cut platform

Sea stack

Sea arch

Sea stacks

Wave-cut platform (exposed at low tide)

◀ **Figure 19.18 What rocky coastlines look like.**
Wave erosion of rocky headlands produces caves near the waterline that may connect beneath the headland to form a sea arch. Collapse of the sea arch then separates a sea stack from the mainland. Erosion at wave base cuts a platform in the rock that is commonly exposed at low tide.

Sea urchins bore holes into rock

(a)

Salt weathering produces pits in rock

(b)

◀ **Figure 19.19 How rock weathers at the shoreline.**
(a) Biologic processes cause substantial rock weathering along the shoreline. Some animals, like sea urchins, bore holes into rock. Others, like starfish and some snails, abrade rock surfaces as they scrape off algae and other organisms for food. (b) Evaporation of salty sea spray and tide pools exposed at low tide causes precipitation of salt crystals that disaggregate rocks. Waves erode the weathered rock fragments and leave behind pitted surfaces where water accumulates and accentuates the salt-weathering process.

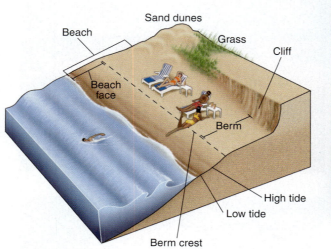

▲ **Figure 19.20 The shape of a beach.**
The beach is a bare area of unconsolidated sediment that extends from the low-tide line to a landward boundary marked by a sea cliff, vegetation, or wind-blown sand dunes. The beach includes the seaward-inclined beach face and the near horizontal to slightly landward inclined berm. The berm crest separates the beach face and the berm.

slope of the beach face. **Figure 19.21**a shows sediment deposited by the swash and backwash all along the beach-face up to the berm crest. Sediment deposited on the beach face builds the beach out toward the ocean and up to the berm-crest elevation. The higher the waves, the greater the swash, and the higher the berm crest. Where the beach sediment is coarse grained and permeable, however, more of the swash soaks into the beach. As a result, gravel beaches tend to have steeper beach faces and lower berm heights than sand beaches experiencing the same incoming wave heights. You get a feeling for this difference by comparing the beach-face slopes in the pictures within Figures 19.20 and 19.21.

The shape of a beach changes through the year because of seasonal variations in wave height (Figure 19.21b). High storm waves (usually during the winter) erode the lower part of the beach face and build a new berm at a higher elevation than the fair-weather berm. Fair-weather conditions (usually during the summer) lead to more beach deposition that builds out a new beach below the former storm berm, so that a single beach may exhibit more than one berm.

How Longshore Currents Shape Shorelines

Erosion and deposition of sediment by longshore currents produce distinctive landforms. **Figure 19.22** shows how longshore currents build **spits**, which are elongate sediment ridges that project from headlands in the direction of the longshore current. Wave refraction around the end of the spit causes sediment deposition along the tip and landward side of the spit; this may form a prominent hook-shaped beach, like that seen at Cape Cod, Massachusetts (Figure 19.22). **Figure 19.23** illustrates how longshore currents may build **baymouth bars**, which are spits that link one headland to the next and close off the intervening bay from the ocean.

Sediment erosion and transport by longshore currents is dramatically illustrated where structures are built perpendicular to the shoreline. **Figure 19.24** shows how **groins**, which are walls built perpendicular to the shoreline, trap sediment on the upcurrent side to widen a beach. The longshore current erodes the beach, however, on the downcurrent part of the shore.

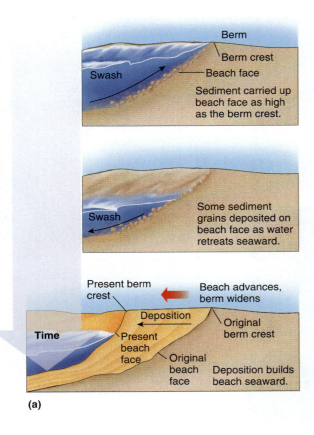

(a)

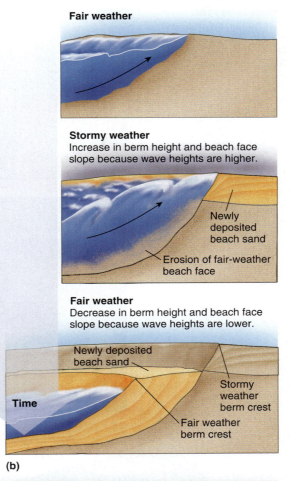

(b)

(c)

▲ **Figure 19.21 Visualizing how a berm forms and changes.**
(a) The berm crest is the height to which wave swash carries sediment up the beach face at high tide. Sediment accumulates on the beach face so that the beach builds seaward over time to produce a wide berm. (b) The berm-crest height increases when high storm waves pound the beach. Under stormy conditions the beach face erodes, and deposition builds up a higher berm. When fair-weather conditions return, waves build a new beach with a lower berm. (c) This photo of a gravel beach in England shows multiple berms and beach faces associated with different wave conditions.

Sediment Sources for Beaches

Field observations reveal many sources for beach sediment.

- Waves erode some beach sediment from shoreline rock and regolith.
- Streams deliver large volumes of sediment to the ocean, and waves carry some of this sediment onto beaches.
- Waves erode sediment from the seafloor as indicated by seashells on beaches.

Figure 19.25 illustrates a variety of beach deposits. Many white-sand beaches contain mostly quartz (Figure 19.25a). Quartz is abundant in river sediment delivered to the sea because it is the weathering-resistant mineral that is most abundant in continental rocks (see Section 5.2). Quartz also lacks cleavage, so continuous movement and abrasion by waves do not easily break it. Black-sand beaches are common on basaltic volcanic islands (Figure 19.25b). Basalt does

▶ **Figure 19.22 How longshore currents build spits.**
The longshore current carries sediment off the end of a head-land to build a spit. Wave refraction around the end of the spit shapes the sediment into a hook. Cape Cod, Massachusetts, as seen from the International Space Station, is an example of a hooked spit.

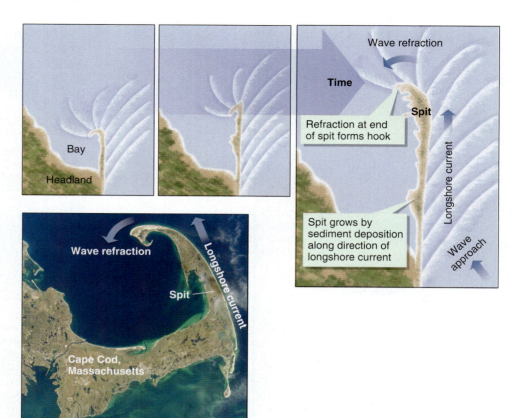

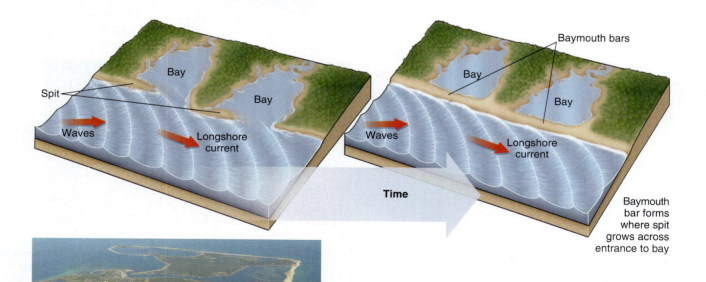

▶ **Figure 19.23 How longshore currents build baymouth bars.**
Baymouth bars form where longshore currents build spits completely across narrow bays.

Baymouth bars near Edgartown, Massachusetts

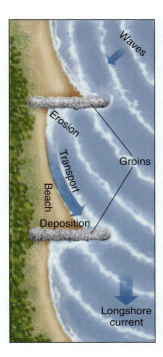

◄ **Figure 19.24 How groins modify beaches.**
Groins project into the longshore current and interrupt sediment transport along the beach. Deposition widens the beach on the upcurrent side of the groin whereas the current erodes sand from the other side of the groin. The photo shows sediment deposition alongside groins on Lake Michigan in Chicago, Illinois. Wave processes in large lakes are the same as wave processes in the ocean.

Beach eroded on downcurrent side of groin

Wide beach deposited on upcurrent side of groin

Effects of Groins and Jetties: *See how groins and jetties influence shoreline deposition and erosion.*

Quartz-sand beach, North Carolina

(a) Sand weathered and eroded from continents contains abundant quartz, because quartz resists weathering and does not abrade or break easily during transport by streams.

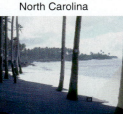

Basalt-sand beach, Hawaii

(b) Black sand beaches on volcanic islands consist of wave-eroded fragments of basalt lava.

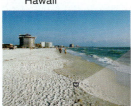

Shell-sand beach, Florida

(c) Shells eroded from the seafloor by waves compose nearly all sediment on beaches where there is little or no sediment supplied by streams or erosion of coastal rock and regolith.

◄ **Figure 19.25 What beach sediment looks like.**

not contain quartz to produce white sand, but the fine-grained igneous rock is crushed into small sand-size grains by wave erosion along the shore. Some coastlines lack major rivers to introduce sediment to the shore, and wave energy may be too weak to significantly erode rock or regolith exposed at the shore. In these cases, broken seashells carried landward by the waves compose the beach (Figure 19.25c). All beach-sediment grains are well rounded because of constant wave agitation of the grains against one another.

Sediment Budget Determines Beach Growth or Loss

Deposition increases beach width whereas erosion decreases it. Whether a beach grows, shrinks, or is relatively unchanged over time is an important aspect of human interactions with dynamic shorelines. The beach-sediment budget determines how beaches change through time.

Figure 19.26 illustrates the budget of sediment gains and losses along a coastline. Natural sediment gains to a section of shoreline include

- The amount of sediment delivered by longshore currents from the adjacent shoreline
- The amount of sediment eroded from local sea cliffs and headlands
- The amount of sediment delivered to the shore by rivers
- The amount of sediment eroded by waves offshore of the beach and then transported onshore

Rivers and coastal erosion are usually the largest sediment sources, including sediment spread out along the shore by longshore currents. Sediment gains, therefore, are largest near the mouths of major rivers and where easily eroded rock or regolith forms the shoreline. For example, erosion of steep exposures of thick, unconsolidated ice-age glacial sediments supplies abundant sand and gravel to the beaches of New England and Washington.

Natural sediment losses to a section of shoreline include

- The amount of sediment carried farther along the shore by longshore currents
- The amount of beach sediment eroded by waves and carried offshore by rip currents or backflow of storm surges during storms
- The amount of sand that wind blows off of the beach and onto adjacent coastal sand dunes

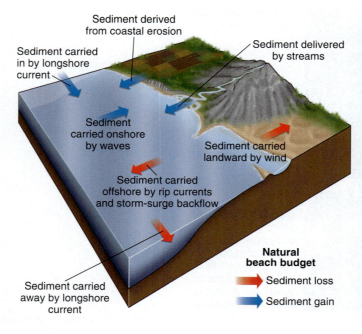
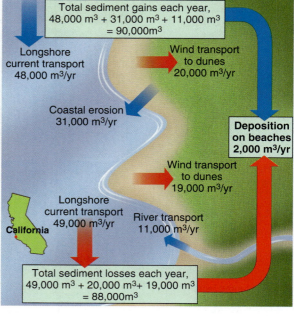

▶ **Figure 19.26 Visualizing the beach-sediment budget.** The beach along a section of shoreline grows or shrinks through time depending on the relative amounts of sediment carried to or eroded away from the beach. The example beach budget, on the left, shows that large volumes of sediment move along, toward, and away from shorelines. The actual budget illustrated on the right shows that a volume of sediment that would fill almost 500 railroad boxcars is added to this 20-kilometer-long stretch of California coastline each year.

Sediment derived from coastal erosion

Sediment delivered by streams

Sediment carried in by longshore current

Sediment carried onshore by waves

Sediment carried landward by wind

Sediment carried offshore by rip currents and storm-surge backflow

Sediment carried away by longshore current

Natural beach budget
→ Sediment loss
→ Sediment gain

Total sediment gains each year,
48,000 m³ + 31,000 m³ + 11,000 m³
= 90,000m³

Longshore current transport
48,000 m³/yr

Wind transport to dunes
20,000 m³/yr

Coastal erosion
31,000 m³/yr

**Deposition on beaches
2,000 m³/yr**

Wind transport to dunes
19,000 m³/yr

California

Longshore current transport
49,000 m³/yr

River transport
11,000 m³/yr

Total sediment losses each year,
49,000 m³ + 20,000 m³+ 19,000 m³
= 88,000m³

Coastal sand dunes, explored further in Chapter 20, form by the persistent movement of wind across loose sand on the beach. Sand dunes are common adjacent to beaches and, in many places, dunes form the highest elevations along the coast.

Comparing the sediment gains and losses reveals whether or not a beach grows or shrinks. If more sediment is gained than lost, then there is overall deposition, which causes the beach to widen toward the sea. If the losses are larger than the gains, then there is overall erosion and the beach becomes narrower.

Figure 19.27 illustrates evidence for growing and shrinking beaches. Where beaches grow seaward, subtle ridges of sand that coincide with high berm crests show older beach positions. Rows of beach ridges demonstrate that the shoreline advanced toward the sea. Where there are no beach ridges, the beach is either stationary or retreating toward land.

Human activities affect the sediment budget. Groins cause sediment surpluses in upcurrent areas and deficits in down current areas of beach (Figure 19.24). Pumping or dredging sand from offshore and adding it to the beach artificially nourishes some beaches. Along some coasts, beach sand and gravel are excavated for construction purposes, leading to narrower beaches, whereas in other cases beaches are widened by dumping of sediment dredged from adjacent harbors to deepen them for entry of large ships.

Human activities also change the sediment supply delivered by streams to the coast (see Section 16.11 for more information on human impacts on streams). Changes in land use from natural vegetation to agricultural or urban development and many mining practices add sediment to streams, whereas dams trap sediment in reservoirs that would otherwise be transported to the ocean and contribute to beach formation. **Figure 19.28** illustrates twentieth-century changes in sediment delivery to the United States Atlantic coast. Urbanization increased sediment supply to beaches in the Northeast. Construction of dams and development of agricultural practices that decrease soil erosion led to substantial decreases in sediment load south of Chesapeake Bay. Similar historic changes in sediment supply occurred on the Pacific Coast, where dam building in California reduced sediment delivery to 23 percent of the state's beaches by an average of 25 percent.

How Barrier Islands and Tidal Inlets Form

Along many coastlines, wave energy is expended along **barrier islands**, which are long, narrow ridges of land that form parallel to, but separate from, the mainland coast. Barrier islands form 13 percent of the world's coastlines, and there are about 300 such islands along the Atlantic and Gulf of Mexico coasts of the United States. Some of the most heavily developed and most valuable real estate in the country is located on barrier islands, in places like Atlantic City, New Jersey; Miami Beach, Florida; and Galveston, Texas.

Figure 19.29 illustrates the general characteristics of barrier-island coastlines. Each barrier island is typically 10 to 100 kilometers long but usually less than 5 kilometers wide. Barrier coastlines typically consist of many islands that resemble beads on a necklace. Narrow **tidal inlets** separate the islands and focus tide-produced currents between the lagoon and open ocean. The highest elevations on barrier islands are usually sand dunes constructed by wind blowing across the beach on the seaward side of the island. Barrier islands consist of sediment rather than rock.

The landward side of the barrier island faces a lagoon or marsh and is largely unaffected by waves but may exhibit features caused by tides, which are discussed in the next section. **Lagoons** are shallow, calm-water bodies of water located landward of obstacles that absorb wave energy directed toward the shoreline. Example obstacles are barrier islands, spits, baymouth bars, and offshore coral reefs. High storm waves readily cross the narrow, low barrier islands. Storm-wave erosion may breach the island to form a new tidal inlet and wash beach and dune sand into the quiet lagoon (see Figure 19.29).

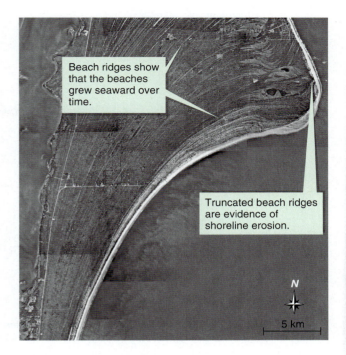

Beach ridges show that the beaches grew seaward over time.

Truncated beach ridges are evidence of shoreline erosion.

N

5 km

▲ **Figure 19.27 How to determine if a beach is growing or shrinking.**
An aerial view of Cape Canaveral, Florida (before construction of space-flight facilities), shows subtle, low-elevation beach ridges that mark former beach positions. Ridges parallel to the beach on the south side of the spit show that the shoreline has grown southward over time. In contrast, there are no ridges parallel to the east-side beach, where older beach ridges abruptly end at the modern beach. These relationships indicate beach erosion on the east side of the spit and beach growth on the south side.

Barrier islands and intervening tidal inlets shift in the direction of the long-shore current, as shown in **Figure 19.30**. Wave refraction and longshore currents erode sediment from the upcurrent end of the island and deposit it on the downcurrent end. The process of erosion on one end of the island and deposition on the other end causes both the inlet and the island to migrate.

Many inlets serve as shipping lanes between the open ocean and mainland ports in lagoons and estuaries. It is important to maintain these inlets in a stable position by building **jetties**, which are longer versions of groins that are built adjacent to an inlet. Jetties, visible in Figure 19.1d, are walls that keep the inlet from shifting with the longshore current. As illustrated in Figure 19.30, a jetty traps sediment on the upcurrent side, which keeps the barrier island from shifting into the tidal inlet. The trapping of sediment by the jetty may starve the next barrier island in the chain from sediment, causing it to retreat landward under the effects of wave erosion.

Barrier islands are common landforms, but why do these features form offshore of continents? The exact processes that form the islands remain unclear and may differ from place to place along the coast. Barrier islands form along

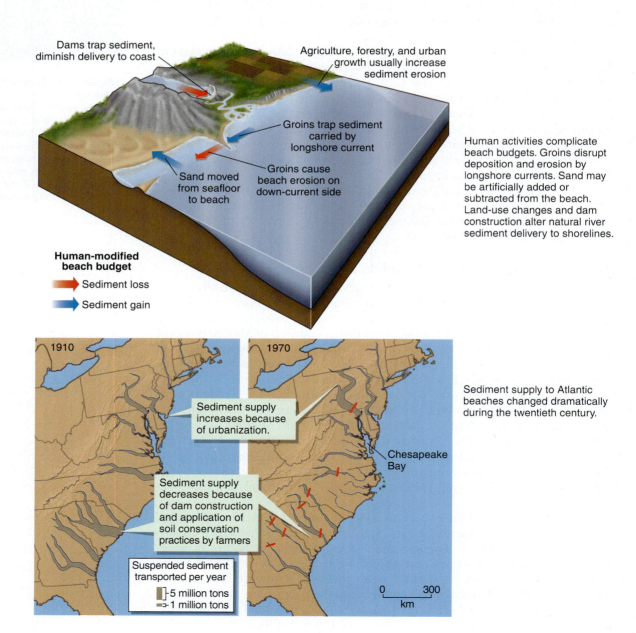

Human activities complicate beach budgets. Groins disrupt deposition and erosion by longshore currents. Sand may be artificially added or subtracted from the beach. Land-use changes and dam construction alter natural river sediment delivery to shorelines.

Sediment supply to Atlantic beaches changed dramatically during the twentieth century.

▶ **Figure 19.28** **Visualizing human modifications of the beach budget.**

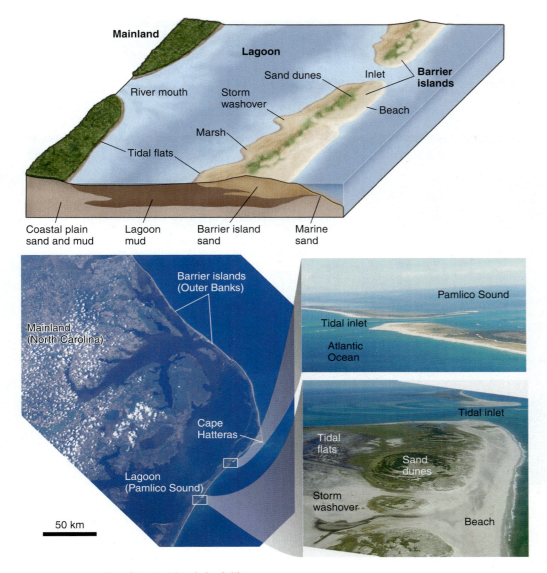

▲ **Figure 19.29 What barrier islands look like.**
Barrier islands are long, low islands that are separated from the mainland by a quiet-water lagoon. Waves form a beach on the seaward side of the island, whereas tidal mud flats and marshes are present on the lagoon side. The highest elevations along the spine of the island are sand dunes formed by wind blowing across the beach. High storm waves wash across the islands and carry beach and dune sand into the lagoon. Tidal inlets separate barrier islands. Barrier islands form long sections of the Atlantic Coast of the United States, including the pictured locations in North Carolina.

gently sloping, sandy shorelines when sea level rises. **Figure 19.31** illustrates two possible scenarios proposed by coastal geologists and oceanographers:

1. In the first scenario (Figure 19.31a), sea-level rise isolates spits and bay-mouth bars from the mainland. Storm waves erode breaches in the spits and bars to form barrier islands and intervening tidal inlets.
2. In the second scenario (Figure 19.31b), waves transport sand landward along a wide, recently submerged zone of sediment deposited by streams, waves, and wind when sea level was lower. Waves transport the large volume of sand landward as the rising ocean advances across these older deposits. Eventually, the sheer volume of the sand, much of it in dunes that rise many meters above sea level, cannot be moved landward by the waves and currents as fast as the land itself is submerged by rising sea level, and the beaches and dunes are gradually separated from the mainland to form a barrier island.

▶ **Figure 19.30 Why islands and tidal inlets migrate.**
Longshore current erosion and deposition cause barrier islands and intervening inlets to migrate in the direction of longshore transport. Erosion occurs on the up-current ends of islands and deposition occurs on the downcurrent ends of islands. Erosion and deposition on opposite sides of an inlet cause the islands and inlet to migrate together without changing the inlet width. Engineers construct jetties to keep inlets open to ships. The jetty traps longshore-transported sediment, which causes growth of one island at the expense of the other.

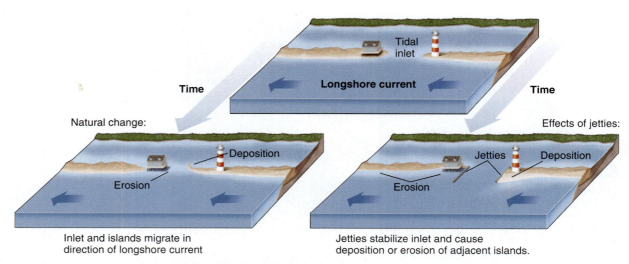

▶ **Figure 19.31 How barrier islands form.**
These diagrams show two scenarios for the formation of barrier islands during a period of sea-level rise on a gently sloping coastline with abundant sand supply.

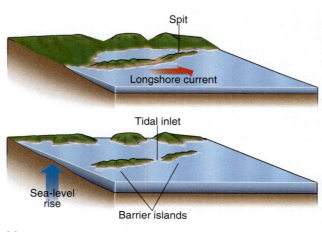

(a) Some barrier islands originate as spits that become isolated from the shoreline when sea level rises. Storm waves breach the former spit to form tidal inlets.

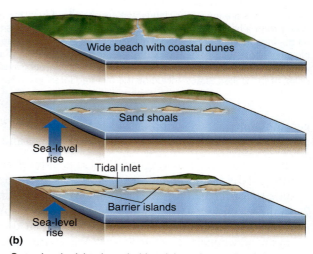

(b) Some barrier islands probably originate by sea level rise along coastlines with wide beaches and coastal sand dunes. Waves in the landward-encroaching ocean cannot transport the large volume of sand as the shoreline retreats. Instead, waves heap up the sand into shoals. Some sediment is exposed above sea level long enough for wind to construct higher sand dunes, and a barrier island forms.

Putting It Together–How Do Waves Form Shoreline Landscapes?

■ Wave motion touches the seafloor in shallow water, erodes loose sediment, and carries it landward where deposition occurs on beaches.

■ Longshore currents carry sediment parallel to shore to produce spits and baymouth bars.

■ Refracted-wave attack on rocky headlands gradually straightens the shoreline. Sea stacks and wave-cut platforms are common landforms on eroded rocky coasts.

■ Beach sediment comes from sediment carried to the ocean by rivers and from wave erosion of the seafloor and coast.

■ Overall beach growth or shrinkage depends on long-term imbalance between the volume of sediment added or subtracted from the beach. Human activities affect the beach-sediment budget.

■ Barrier islands form offshore where waves deposit sand on gently sloping seafloor during a period of sea-level rise. Barrier islands are separated by tidal inlets and migrate along the coastline in the direction of longshore currents.

19.4 What Is the Role of Tides in Forming Coastal Landscapes?

Rising and falling tides are an essential process of dynamic, changing shoreline landscapes, as is apparent by your observations on the Oregon Coast in Figure 19.1e. The **tide** is the slow, up-and-down movement of sea level that occurs each day because of gravitational interactions of the Moon and Sun with Earth.

The change in sea level between low and high tide, called the **tidal range**, is less than 3 meters along most modern shorelines, but in some places this range is greater than 15 meters. Tides typically move water at slow velocities compared to waves. Exceptions to this generalization occur where the mass of rising and falling water is constricted in inlets between islands or forced in and out of funnel-shaped estuaries. These tide-generated currents move at several meters per second.

Why Tides Exist

The attracting force of gravity that operates between Earth, Moon, and Sun explains tides. Physical measurements of the gravity force, dating back to Isaac Newton in the seventeenth century, indicate that the force exerted on one body of matter by another is larger when the masses of the objects are larger and when they are closer together. The Sun is the largest mass in our solar system, although it is far away from Earth. The Moon has a relatively small mass, but it is close to Earth and thus exerts a significant gravitational attraction. The Sun and Moon are the most important planetary bodies for calculating the gravity force that causes tides; all other planets are too small or too distant to have much effect.

Figure 19.32 diagrammatically shows how gravitational attraction produces ocean tides. All points on and within Earth are pulled toward the Moon by gravity (and the same force pulls the Moon toward Earth). For simplicity, think of Earth as a rigid sphere and the magnitude of the gravitational attraction of the Moon on Earth is simply a force acting from the center of Earth toward the center of the Moon. Also for simplicity, imagine a uniform layer of ocean water covering Earth's surface (see Figure 19.32). Points on the ocean surface are attracted toward the Moon's center by varying amounts, with water on the side closest to the Moon tugged a little bit more than water on the far side, because the water closest to the Moon experiences a larger gravitational attraction than the water on the far side of Earth.

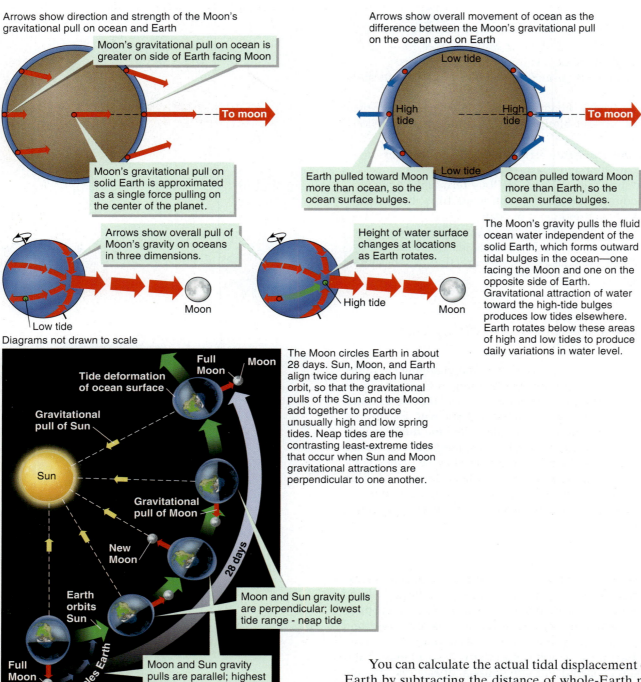

Arrows show direction and strength of the Moon's gravitational pull on ocean and Earth

Moon's gravitational pull on ocean is greater on side of Earth facing Moon

To moon

Moon's gravitational pull on solid Earth is approximated as a single force pulling on the center of the planet.

Arrows show overall pull of Moon's gravity on oceans in three dimensions.

Moon

Low tide

Diagrams not drawn to scale

Arrows show overall movement of ocean as the difference between the Moon's gravitational pull on the ocean and on Earth

Low tide

High tide

High tide

To moon

Low tide

Earth pulled toward Moon more than ocean, so the ocean surface bulges.

Ocean pulled toward Moon more than Earth, so the ocean surface bulges.

Height of water surface changes at locations as Earth rotates.

High tide

Moon

The Moon's gravity pulls the fluid ocean water independent of the solid Earth, which forms outward tidal bulges in the ocean—one facing the Moon and one on the opposite side of Earth. Gravitational attraction of water toward the high-tide bulges produces low tides elsewhere. Earth rotates below these areas of high and low tides to produce daily variations in water level.

Full Moon   Moon

Tide deformation of ocean surface

Gravitational pull of Sun

Sun

Gravitational pull of Moon

New Moon

Earth orbits Sun

28 days

Full Moon

Moon circles Earth

Moon and Sun gravity pulls are perpendicular; lowest tide range - neap tide

Moon and Sun gravity pulls are parallel; highest tide range - spring tide

The Moon circles Earth in about 28 days. Sun, Moon, and Earth align twice during each lunar orbit, so that the gravitational pulls of the Sun and the Moon add together to produce unusually high and low spring tides. Neap tides are the contrasting least-extreme tides that occur when Sun and Moon gravitational attractions are perpendicular to one another.

▲ **Figure 19.32 How gravity forces cause tides.**
Gravitational pull of the Moon and the Sun cause ocean tides on Earth.

Active Art

How Tides Work: *See how the positions of Earth, Moon, and Sun determine the rise and fall of the tides.*

You can calculate the actual tidal displacement of the ocean relative to rigid Earth by subtracting the distance of whole-Earth movement from the amount of ocean movement at the different points on the planet surface. As shown in Figure 19.32, the resulting displacement of the ocean relative to rigid Earth causes two tidal bulges, one on the side of Earth facing the Moon and the other on the side opposite the Moon. On the side facing the Moon, the ocean water is pulled farther toward the Moon than Earth is pulled, so the water surface rises. On the side opposite the Moon, Earth is pulled farther toward the Moon than the water surface is pulled, so the water surface rises relative to the rigid Earth surface. Water moves toward these rising tidal bulges in the ocean from elsewhere on the planet, where the tide is falling. Earth rotates under these two bulges in the water surface so that every location on the surface experiences two high tides, and two intervening low tides, each day.

Figure 19.32 also illustrates the gravitational effect of the Sun, which either adds to or subtracts from the tidal force exerted by the Moon. The high tidal range of **spring tides** occurs when Earth, Moon, and Sun are all aligned, so that the gravitational attraction of the Sun on Earth adds to the attraction of the Moon. The lower tidal range of **neap tides** occurs when the Moon's gravitational attraction is oriented perpendicular to that of the Sun.

Many factors complicate this simple view of tidal generation. The rotation of Earth, variations in ocean depth, and the location of land distort water movement toward the tidal bulges. These factors cause some areas to experience only one high and one low tide each day, rather than two of each, and the magnitude of each successive high or low tide can be different.

Where Tides Affect the Shoreline

Tides rise and fall along every coast, but they have the greatest effect where tidal range is high and wave energy is low. **Figure 19.33** illustrates that high and low tides are observed on waveswept beaches, but tidal changes are more obvious on shorelines where waves are small and wide beaches do not exist. Gently sloping, muddy **tidal flats**, which are marshy or barren areas of land submerged at high tide and exposed at low tide (see Figure 19.33), are common in these areas.

Locations with tidal flats protected from wave action include lagoons and estuaries. Lagoons are separated from the open ocean by barrier islands (Figure 19.29) or offshore coral reefs that absorb the incoming wave energy. Wind moving across the lagoon creates waves, but these are very small because the wind is in contact with the water over a very short distance, usually only a few kilometers. Estuaries are very deep embayments in a coastline where the ocean extends landward into a river valley (Figure 19.2). Wave height decreases inland within an estuary because wave energy focuses on headlands flanking the estuary mouth and in shallow water along the estuary shoreline.

Tides add to wave effects on the shoreline. The location of the breaker zone and beach face change through the day because of fluctuations in the overall sea level determined by the tides. If high storm waves coincide with high tide, especially the very high spring tide, then intense wave action comes closer into the shore than usual.

How Tidal Flats Form

The rising tide carries sediment landward and then leaves it behind on the tidal flat when the tide falls, as shown in **Figure 19.34**. The advancing tide slows down as it moves up the gentle slope of the tidal flat and approaches the high-tide mark, where the velocity decreases to zero before water retreats back across the flat to the low-tide mark. Sediment transported by the rising water is deposited when the rising tide slows down. When the tide falls, the initial current velocity draining off of the tidal flat is too slow to pick up all of the sediment that settled out at high tide. These changes in the velocity and direction of flowing water cause more sediment to be carried landward during the rising tide than is carried seaward during the falling tide, so that each tidal cycle leaves a veneer of newly deposited sediment. The sediment is also coarser grained on the seaward side of the flat than on the landward side, because tidal currents transport the sediment from sea toward land, rather than from land toward sea.

Water draining seaward during the falling tide locally erodes channels in the tidal-flat surface. These **tidal channels** (sometimes called tidal creeks) resemble those caused by drainage of water off of hillslopes on land. Like stream channels, tidal channels have tributaries and usually have very sinuous channel patterns consistent with the cohesive, muddy character of the sediment. Water may flow in tidal channels after the tide falls below the adjacent tidal-flat surface because shallow ground water slowly seeps out of the saturated muddy flat sediment and into the channels.

▲ **Figure 19.33** **Recognizing tides at the shore.**
The appearance of rising and falling tides is different on beach shorelines with active wave processes compared to tidal flats where waves are absent. The width of exposed beach is the only visible difference on this Mexico beach between low and high tides. Tidal flats, like this one in eastern Canada, include vegetated areas that are at or above the high tide level. Intervening unvegetated areas are submerged most of the time and when exposed at low tide are muddy surfaces crossed by small channels that drain pore water out of the exposed sediment.

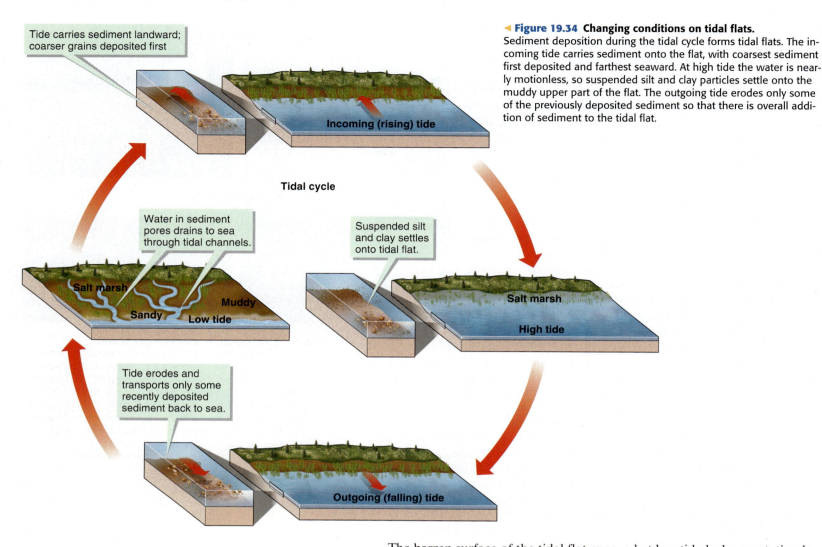

Tide carries sediment landward; coarser grains deposited first

Tidal cycle

Incoming (rising) tide

Water in sediment pores drains to sea through tidal channels.

Suspended silt and clay settles onto tidal flat.

Salt marsh

Muddy

Sandy   Low tide

Salt marsh

High tide

Tide erodes and transports only some recently deposited sediment back to sea.

Outgoing (falling) tide

◀ **Figure 19.34 Changing conditions on tidal flats.**
Sediment deposition during the tidal cycle forms tidal flats. The incoming tide carries sediment onto the flat, with coarsest sediment first deposited and farthest seaward. At high tide the water is nearly motionless, so suspended silt and clay particles settle onto the muddy upper part of the flat. The outgoing tide erodes only some of the previously deposited sediment so that there is overall addition of sediment to the tidal flat.

▲ **Figure 19.35 Chemical sedimentation occurs along desert coastlines.**
The photo shows cracked mineral crusts of salt, gypsum, and dolomite formed on tidal flats by evaporation of seawater at low tide along the Persian Gulf in the country of Qatar. The hot, dry climate favors evaporite mineral crystallization. There is very little clastic mud or sand on the tidal flat because there are no rivers to deliver sediment to the ocean.

The barren surface of the tidal flat exposed at low tide lacks vegetation because few plants are tolerant of submergence in salt water. Grasses and low shrubs with limited tolerance of salt water may form salt marshes along the landward side of the tidal flat, which submerges only during spring tides. Mangrove trees form dense forests near the high-tide mark along some coasts.

Chemical sediment covers tidal flats in desert regions where there are no streams to bring clastic sediment to low-relief, wave-protected shorelines. An example is shown in **Figure 19.35**. Seawater evaporation on the tidal flat causes calcite precipitation, partly by biologic processes in microbial mats that thrive in hot, damp conditions. Evaporation draws seawater to the surface above the high-tide mark, where it evaporates to form stark, barren deposits of evaporite minerals, like gypsum and halite.

Tidal Processes at Inlets

Tides are important along barrier-island shorelines. Not only do tidal flats form on the lagoon side of a barrier (Figure 19.29), but **Figure 19.36** shows how tidal currents move sediment back and forth through the inlets between adjacent islands. Tidal currents are much stronger in the tidal inlets than on the adjacent tidal flats because the rising and falling tide funnels back and forth through narrow constrictions between the open ocean and lagoon. The rising tide carries sediment suspended by waves and longshore currents through the tidal inlet and deposits the sediment in the quiet-water lagoon. The falling tide carries sediment back out to sea where it is deposited beyond the beach, or it may be eroded by waves and carried back onto the beach or carried away by longshore currents.

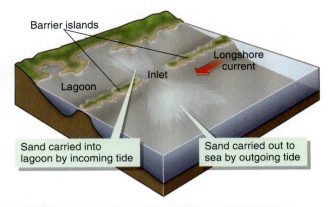

Tides as an Energy Source

The persistent rise and fall of tides is used in some places to generate electricity. Tidal power plants are constructed as dams across estuaries or inlets. The dam focuses the tidal currents to flow rapidly through a narrow artificial channel, where the water turns turbines to generate power in the same fashion that hydroelectric dams function along rivers. The rising tide generates power as it flows through the dam to enter the lagoon or estuary on the landward side. Rather than letting the water return through the dam at low tide, it is retained behind the dam until it can be released to generate electricity when power demand is high. The higher the tidal range, the more water can be held behind the dam to generate electricity by return flow through the turbines.

> ### Putting It Together–*What Is the Role of Tides in Forming Coastal Landscapes?*
>
> - Tides are periodic variations in sea-surface elevation caused by gravitational interactions of the Moon and Sun with Earth.
> - Tidal influences on shorelines are most visible where wave energy is small.
> - Tidal flats are gently sloping, muddy surfaces that are continually submerged at high tide and exposed at low tide. Some sediment carried landward by the rising tide is left behind when the tide goes out so that tidal flats gradually build upward and seaward over time. Evaporite minerals precipitate on tidal flats along arid coastlines.
> - Tide currents funnel back and forth through tidal inlets between open ocean and lagoons on barrier-island coastlines. Tidal currents at inlets carry sediment both landward into the lagoon and seaward, where longshore currents may redistribute it onto barrier beaches. Tidal currents in inlets also power some electricity generation.

▲ **Figure 19.36** **Recognizing tidal processes at tidal inlets.** Tidal currents enter and exit lagoons through the narrow tidal inlets between barrier islands. These focused currents transport sand to form submerged sand bars on either side of the inlet.

19.5 Why Does Shoreline Location Change through Time?

Clearly, shorelines are dynamic places that change in appearance on relatively short time frames. Daily tides, seasonal variations in wave energy, powerful storms, and longshore transport of sediment produce these changes. Longer-term changes are apparent over centuries or even many millennia. Successive sedimentary layers deposited in progressively shallower or deeper water record the long-term changes. (Check back to Section 5.6 and Figure 5.19 to review this sedimentary process.) These changes also determine the shape of the shoreline and occur within the time scale of human history as well as over longer intervals of geologic time.

Processes that Change Shoreline Location

Figure 19.37 illustrates the processes the geologists observe to cause changes in shoreline location. If a point on land later submerges beneath the sea, or a point on the seafloor is later located on dry land, then it is possible to assume that sea level changed. Measurement of changing sea level is made relative to either a fixed location on land or at sea. This means sea-level rise *or* land subsidence can equally explain land submergence. Likewise, emergence of the seafloor happens either by sea-level fall *or* seafloor uplift.

   Relative sea-level change describes a shift in local shoreline position caused either by global sea-level fluctuation, local uplift and subsidence of crust, or a combination of these processes. Global **absolute sea-level change** mostly results from changes in the rates of creation of seafloor at mid-ocean ridges (discussed in Section 13.4), or from changes in the volume of glacial ice on continents (see

▶ **Figure 19.37 Why shoreline positions change through time.**
Shorelines shift landward during relative sea-level rise, caused by global rise in
sea level or local subsidence of crust, or both. Shorelines shift seaward during
relative sea-level fall, which results from global sea-level fall or uplift. Erosion
and deposition also change shoreline position without any relative sea-level
change, such as when deltas build land out into the sea.

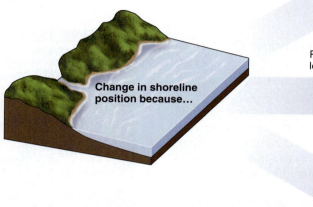
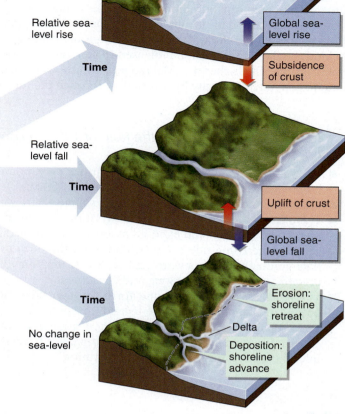

Section 18.9). Tectonic and isostatic forces (described in Chapters 11–13) cause
uplift and subsidence of the crust.

Shoreline location can change by sediment erosion and deposition without
a change in relative sea level (Figure 19.37). The largest seaward shoreline shifts
occur where deltas build out into the ocean at the mouths of large, sediment-
laden rivers (see Figure 16.15). If, on the other hand, the sediment losses out-
pace the gains in the sediment budget, and if coastal rock or regolith is easily
eroded by waves, then wave erosion causes the shoreline to retreat landward.

Submergent and Emergent Shorelines

Figure 19.38 shows that the shape of shorelines reflects relative sea-level rise or
fall. Relative sea-level rise submerges stream valleys below sea level to form a
highly irregular shoreline of drowned-valley estuaries separated by headlands
(Figure 19.38). Emergence, on the other hand, raises wave-cut platforms and old
beach deposits above sea level (Figure 19.38; also see Figure 13.14).

Global sea-level change, uplift or subsidence of crust, and deposition or
erosion of sediment may operate simultaneously to affect the position of a local
shoreline. The challenge is to separate out the role of each process. Present
shoreline configurations are unquestionably affected by 20,000 years of absolute sea-level rise resulting from melting of the last ice-age glaciers (see
Section 18.9). The evidence for absolute sea-level rise and its impacts are thoroughly explored in the next two sections. The remainder of this section demonstrates how geologists know that the other processes actively affect shorelines.

Submergent shoreline

Emergent shoreline

Present shoreline features:

Former shoreline features:

◄ **Figure 19.38** **What submergent and emergent shorelines look like.**

Sea cliff

Beach

Sea stack

Sea cliff

Beach

Sea stack

Wave-cut platform

Wave-cut platform

"Drowned" river valleys - estuaries

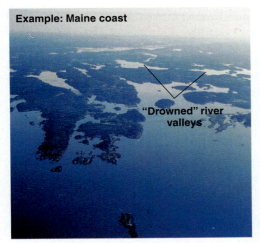

Example: Maine coast

"Drowned" river valleys

• Shoreline results from relative sea-level rise.

• Lower parts of river valleys are "drowned" below sea level to form estuaries.

• Shoreline has a highly irregular shape.

Example: California coast

Uplifted wave-cut platform

• Shoreline results from relative sea-level fall; usually resulting from uplift of crust.

• Former shoreline and seafloor features are exposed above sea level.

• Old wave-cut platforms form flat benches along relatively straight shorelines.

Evidence That Uplift and Subsidence Change Shorelines

Major earthquakes commonly result in uplift or subsidence along coastal areas, as shown in **Figure 19.39**. Over long time periods, tectonic forces have profound impacts on shorelines.

Roman ruins near Naples, Italy, clearly show signs of changing relative sea level, as shown in **Figure 19.40**. Charles Lyell illustrated these ruins at the front of his 1830 text *Principles of Geology* because it is a dramatic example of dynamic Earth processes (Section 1.4 provides more about Lyell). The illustrated Roman building was built on dry land in the second century B.C. Holes bored into the marble columns by marine mollusks are present, however, at elevations as high as 6 meters above the present water line. This means that within a scant fraction of human history, the building was submerged below sea level and then raised back up to its present elevation. The Roman city was built within a restless volcanic caldera, so scientists can say in this case that moving magma in the crust was the cause of this down-and-up movement of the land surface.

Shoreline before earthquake

Uplifted wave-cut platform

White areas-dead marine organisms

◀ **Figure 19.39 Earthquakes change shorelines.**
Shoreline modifications by tectonic uplift and subsidence are dramatically illustrated by changes resulting from a moment magnitude 9.2 earthquake in southern Alaska in 1964. The top photo shows an extensive area of uplifted seafloor, whereas the bottom photo shows a town flooded at high tide because the area subsided below sea level during the earthquake.

Uplift and subsidence affect the Oregon shoreline that you visited in the field at the beginning of this chapter. Elevations of permanently marked locations along the shore were carefully surveyed at two times, 57 years apart. **Figure 19.41** shows a graph that depicts the survey results. Some locations rose in elevation during that time, while others sank. The data indicate that some areas of the coastline are experiencing tectonic uplift while other parts are subsiding. Subduction of the Juan de Fuca plate under the North American plate (see Figure 12.3 for locations of plate boundaries) accounts for this deformation.

Evidence that Sediment Deposition and Erosion Change Shorelines

Figure 19.42 illustrates shoreline advance toward the sea where an estuary filled in with river sediment. This location, in northwestern Turkey, is significant because the geologic history is relevant to interpreting history presented in the classic epic *Iliad*, written by Greek poet Homer. *Iliad* chronicles the war fought for the conquest of Troy, around 3250 B.C. The present long distance from the shore to the presumed location of Troy is inconsistent with Homer's text, leading some archaeologists to question whether the real location of Troy was actually known.

Today

As illustrated by Lyell in 1830

Clam borings in marble columns

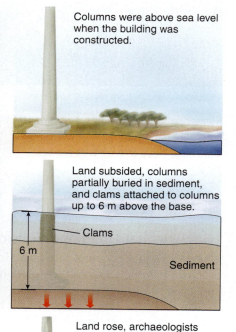

Columns were above sea level when the building was constructed.

Land subsided, columns partially buried in sediment, and clams attached to columns up to 6 m above the base.

Clams

6 m

Sediment

Land rose, archaeologists excavated sediment, and clam borings exposed in column approximately 6 m above present sea level.

Clam borings

▲ **Figure 19.40 Evidence for historic, relative sea-level change in Italy.**
Clam borings in the marble columns of a Roman ruin reveal relative sea level changes during the last 2200 years near Naples, Italy. This example of submergence and emergence of a shoreline as a result of deformation was first illustrated in Charles Lyell's geology textbook in 1830.

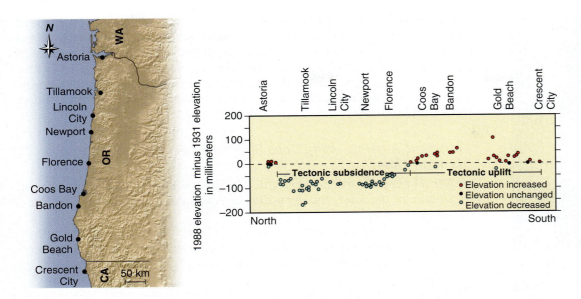

◄ **Figure 19.41 Measured uplift and subsidence along the Oregon Coast.** Repeat surveys of permanently marked locations along the Oregon Coast in 1931 and 1988 show that some parts of the coast experience uplift and other parts are subsiding. Over long time periods, these tectonic movements cause emergence where uplift occurs and submergence where subsidence occurs.

Geologists extracted sediment samples from as deep as 50 meters below the river floodplains near where Troy was thought to be and discovered marine sediment below the surface. The marine deposits are buried beneath sediment containing shells of estuarine animals, and river deposits near the surface. Geologists combined the recorded depositional environments with radioactive-isotope dates on the shells (using the ^{14}C method mentioned in Section 7.7) to determine that Troy was once located along the shore of an estuary. The estuary gradually filled in with river sediment over a period of about 8000 years. Homer's description of Troy is accurate and consistent with the location of the shoreline 5200 years ago. The confusion about the location of Troy arose because sediment deposition drastically changed the geography of the coastline.

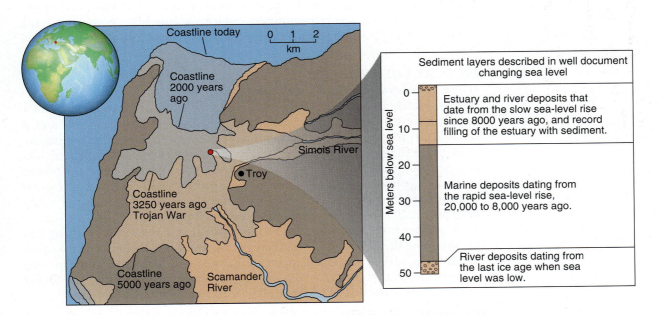

▲ **Figure 19.42 Historic shoreline change caused by sediment deposition.**
Ancient Troy was on the shoreline of an estuary 5000 years ago but is now almost 6 kilometers from the ocean. Study of sediment layers show that the estuary filled with river sediment so that the coast gradually migrated away from Troy.

▲ **Figure 19.43** **Historic shoreline change caused by sediment erosion.**
Erosion is reshaping the South Carolina shoreline, as dramatically illustrated by the Morris Island lighthouse. The lighthouse was on the beach in the 1940s but is now more than 400 meters from shore.

In contrast, **Figure 19.43** illustrates evidence of shoreline retreat because of sediment erosion. Coastal erosion occurs along 80 percent of the Atlantic and Pacific shorelines in the conterminous United States. One significant cause of the erosion is a decrease is sediment supply to the coast resulting from construction of dams along rivers (Figure 19.28). The sand deposited behind the dams never reaches the ocean, which removes sediment from the beach budget and leads to coastal erosion.

Putting It Together—Why Does Shoreline Location Change through Time?

■ Shorelines shift with time because of global sea-level change, uplift or subsidence of the coast, and deposition or erosion of sediment by waves, currents, and tides.

■ Emergent shorelines form by relative sea-level fall, which results from uplift, global sea-level fall, or both. Emergent shorelines typically have steep cliffs, and prominent flat benches that formed below sea level as wave-cut platforms.

■ Submergent shorelines form by relative sea-level rise, which results from subsidence, global sea-level rise, or both. Submergent shorelines typically have low relief and have a highly irregular outline dominated by river valleys that drowned to form estuaries.

19.6 How Do We Know . . . That Global Sea Level Is Rising?

Statements about rising sea level are commonly seen in the news. How do geologists know that sea level is rising?

We can start with data from glacial geologic studies. Observations on land reveal the extent and likely thickness of glacial ice on continents during the last ice age, which peaked about 21,000 years ago (see Figure 18.44). The calculated volume of the glacial ice requires that sea level 21,000 years ago was about 100–120 meters lower than it is today. A reasonable hypothesis, therefore, is that sea level has been rising from this time of lower sea level. Although this reasoning indicates that sea level rose after the last ice age, this does not mean that sea level is still rising now, nor does it tell us whether sea level rose at a uniform rate since the last ice age.

PICTURE THE PROBLEM
Why Is it Important to Document Changing Sea Level?
Approximately 100 million people live around the world at locations less than 1 meter above sea level, so it is essential to know if sea level is rising now and, if so, how fast it is rising. This information is necessary to determine the magnitude of the change and how to respond to it. Sea-level rise causes coastal submergence unless uplift occurs faster than sea-level rise or sediment fills in shallow, nearshore areas faster than sea-level rise. Submergence of large coastal cities (examples including New York City, and New Orleans, Louisiana) would have potentially disastrous economic and social impacts.

Consider two approaches to test the hypothesis that sea level is rising and to determine the rate of sea-level change. One approach uses historic measurements of sea level. The second approach uses geologic data to see if the historic measurements are consistent with sea-level changes measured over the longer time interval since the last ice age.

EXAMINE THE EVIDENCE FOR HISTORIC SEA-LEVEL CHANGE
What Changes in Sea Level Do Tide Gages Reveal?
Tide-gage data reveal sea-level changes at major ports, where measurements of tide elevations have been undertaken regularly for more than a century. **Figure 19.44** shows a tide-gage record for New York City. Averaging together the high-

and low-tide elevations for each day and then averaging all of these daily values determines the sea-level value for each year. Notice the large sea-level variation from one year to the next. Most of the variability relates to changes in weather conditions. Strong winds and variations in atmospheric pressure influence the water-surface elevation. Despite these year-to-year fluctuations, there is a general trend of rising sea level since 1900, at a long-term average rate of 3 millimeters per year.

The problem with the tide-gage data is that tide gages measure relative sea-level rise. The data do not exclude the possibility that sea level is stationary or actually falling while the land around the New York City tide gage is sinking. Indeed, tide-gage data from around the world show tremendous variability in the rates of relative sea-level change, and many gages reveal relative sea-level fall rather than rise. To eliminate effects of tectonic uplift and subsidence, geologists avoid using data from ports in tectonically active areas near plate boundaries.

There is still another problem, however, that relates to shifting mass on Earth's surface. During the ice age, the weight of ice pushed the continental crust down, while adjacent areas bulged slightly upward, much like the displacement observed when sitting on a water bed (Section 13.3 provides more detail on this process). Oceanic areas also changed elevation, because less water mass pushed down on the crust when sea level was lower during the ice age. Earth surface elevations are still adjusting to the removal of ice mass from some continents and the addition of water mass to the oceans. This active adjustment causes slow uplift and subsidence that affect measurements of relative sea level.

Tide gages cannot be used to separately measure change in land-surface elevation and change in water-surface elevation, so they do not measure the ongoing change in global sea level. Geologists adjust for this shortcoming by calculating the uplift and subsidence effects resulting from moving the mass of glacial ice off of the continents and adding water mass to the global ocean. In combination with the tide-gage data, these calculations reveal that global sea level is currently rising at a rate of about 1 to 2 millimeters per year.

The bottom line from this analysis is that global sea level is rising, but the rate of rise is very slow. There is also the uncertainty that results from not being able to directly measure the value of global sea-level change. The difference between estimates of 1 millimeter per year and 2 millimeters per year may seem insignificant, but the faster estimate is 100 percent larger than the slower one, so this uncertainty is not trivial. This "plus-or-minus factor" also plays a role in determining the cause of sea-level rise, which is discussed below. In the future, uncertainty will diminish as geologists use sea-surface elevation data collected from satellites orbiting Earth.

EXAMINE THE EVIDENCE FOR PREHISTORIC SEA-LEVEL CHANGE

What Changes in Sea Level Do Drowned Coral Reefs Reveal?

Geologists use corals as "dipsticks" of former sea levels. *Acropora palmata* is a Caribbean Sea coral that only grows from the water surface down to a maximum depth of 5 meters. Dead specimens of this coral species form reefs that are submerged more than 120 meters *below* current sea level. A reasonable hypothesis, illustrated in **Figure 19.45**, is that the deep, dead corals mark the locations of former reefs that grew near sea level and then drowned when sea level rose since the ice age. If this hypothesis is true, then the ages of dead corals should be progressively more recent at shallower depth, and the corals' elevations would record elevations of former ocean surfaces as sea level rose. If this is true, then the differences in depth between corals of different ages also reveals the rate of sea-level rise.

The graph in Figure 19.45 shows the age of corals at different water depths. The ages determined by use of the ^{14}C dating method on samples of submerged coral extend back to the time of the last ice age. The ages are older at greater depth, as predicted by the hypothesis. There is, however, still a problem of determining how much of the apparent sea-level rise is global sea-level rise rather than local tectonic subsidence, because the samples come from islands close to

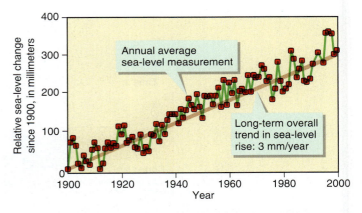

▲ **Figure 19.44** **Tide gage data reveals sea-level rise in New York City.**
Tide-gage data show that relative sea level rose about 300 millimeters (about the height of this page) during the twentieth century. Short-term variations from year to year mostly relate to weather variations. High atmospheric pressure depresses the water surface, and just a 3 percent decrease in air pressure causes the water surface to rise 300 mm. This means that the average, annual sea level is higher in years with weather conditions having generally lower air pressure and lower when there are more days with high air pressure. Long records are required, therefore, to average out these weather-related changes in sea level and isolate the actual relative sea-level change.

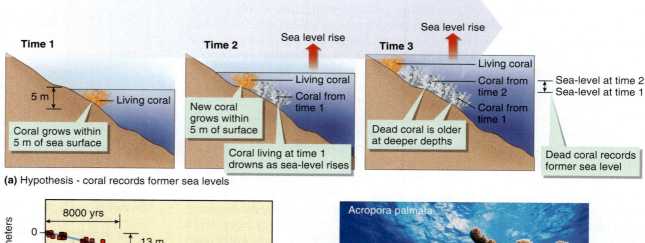

(a) Hypothesis - coral records former sea levels

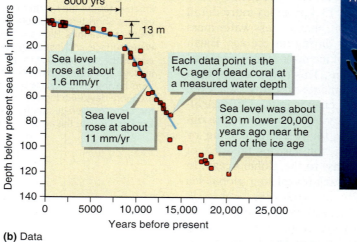

(b) Data

▲ **Figure 19.45 Ancient corals reveal sea-level rise in the Caribbean.**
(a) Researchers hypothesized that specimens of the coral *Acropora palmata* would reveal long-term sea-level change, because although living corals only live within 5 meters of the water surface, dead corals are seen at greater depths. The dead corals lived when sea level was lower and died when rising sea level submerged them deeper than 5 meters. This means that dead corals should be progressively older at greater depth. (b) Geologists used the ^{14}C radioactive-isotope dating to determine the ages of dead corals sampled from different water depths. The data confirm the hypothesis that corals are progressively older at deeper depths. The data also reveal (1) that 20,000 years ago sea level was about 120 m lower than present, and (2) the rate of sea-level rise decreased about 8000 years ago.

the convergent boundary at the eastern edge of the Caribbean plate (see Figure 12.3 for the location of the plate boundaries). In this area, however, the presence of reefs more than 100,000 years old on dry land reveals the islands are uplifted by tectonic processes, which would cause relative sea-level fall rather than rise. Clearly, the coral data require sea-level rise, and the amount of rise is adjusted in Figure 19.45 for the estimated amount of uplift determined from the older reefs exposed on land.

The Caribbean coral data also reveal changes in the rate of sea-level rise over time (Figure 19.45). Over the last 8000 years, sea level rose 13 meters, which equates to an average rise of 1.6 millimeters per year. Prior to 8000 years ago, however, the rate of sea-level rise was much faster, averaging 11 millimeters per year.

INTEGRATING THE DATA SETS
Do the Results Support One Another?
Consistency of results obtained in different ways is an important part of scientific data analysis. Scientists commonly approach the same problem with different data, collected by different methods, and used in different analyses with

different assumptions and uncertainties. If the different data and approaches yield consistent results, then the methods and assumptions employed in the investigations are strongly supported.

In the case of reconstructing sea-level change, there are three pertinent data sets to compare. Figures 19.44 and 19.45 represent two of these data sets. The third relevant data set is the extent of ice-age glaciers, which is used to calculate the total sea-level change since 21,000 years ago (see Section 18.9). The Caribbean coral data indicate 120 meters of sea-level rise since the last ice age, which is consistent with the estimates of 100–120 meters determined from calculations of ice-age glacier volume. The coral data also suggest that sea level rose at about 1.6 millimeters per year during the last 8000 years. This rate of rise agrees with the modern estimated rate of between 1 and 2 millimeters per year as calculated from tide-gage measurements and modified to account for slow adjustments of the crust to shifting masses of glacial ice and ocean water. The three data sets are, therefore, consistent with one another.

INSIGHTS

What Are the Causes and Future of Sea-Level Rise?

The data collection and analyses that establish sea-level rise as fact also lead to an evaluation of *why* sea level is rising and *how much* it might rise in the near future. It would seem simple enough to suggest that rising sea level results from melting glacial ice. There are other processes, however, that could cause global sea-level rise.

One cause of rising sea level is the expansion of ocean water as a result of global warming over the last century. Water expands when it warms up, so sea level rises as ocean temperature increases. A one-degree-Celsius increase in global water temperature produces a roughly two-centimeter increase in sea level. With current rates of global warming, this process causes sea level to rise about 1 millimeter per year.

Is it possible that human interference with the hydrologic cycle also causes rising sea level? Ground water extracted for irrigation or municipal water supplies only partly returns to aquifers. Streams probably transport most of the extracted water to the ocean. Other increases in streamflow to the oceans result from land-use changes, such as urbanization and deforestation, which decrease water infiltration and increase runoff. Other activities decrease stream discharge, however, such as evaporation of water from reservoirs behind dams and irrigated agricultural land. Some water impounded in reservoirs also infiltrates to recharge ground water rather than flowing to the ocean, and the water that is stored in reservoirs is withheld from the ocean. The magnitudes of these human impacts to increase, or decrease, the amount of water reaching the oceans are very difficult to measure or estimate. Current best estimates suggest that it is more likely that human activities decrease, rather than increase, sea level.

If the rate of sea-level rise is close to 1 millimeter per year, similar to the expectations from global warming, then it can be accounted for without including melting of glacial ice. If, however, the rate is closer to 2 millimeters a year, or even higher, then either significant glacial melting is still taking place or the estimations of direct human impact on sea level are erroneous.

How much and for how long will sea level continue to rise? After all, sea level fluctuated throughout Earth history with falling levels as well as rising ones. A clue comes from the presence of 125,000-year-old shoreline deposits exposed on land in many tectonically stable parts of the world. These ancient shorelines indicate a sea level for that time—when glacial records reveal ice sheet volume on continents that is not very different than at present—to be about 5–6 meters higher than today. One hypothesis, therefore, is that 5–6 meters of additional sea-level rise will occur before the next ice age begins. Unfortunately, this is a difficult hypothesis to test other than by waiting to see what actually happens.

Sea level rose at a rate of 1–2 millimeters per year for the last 8000 years, so a reasonable hypothesis is that it will likely continue to rise at 1–2 millimeters per year into the near future. Global sea level will likely rise 10 to 20 centimeters during the next century, and areas experiencing tectonic or isostatic subsidence can expect even more relative sea-level rise. If climate warming increases melting of ice sheets in Greenland and Antarctica, then the amount of sea-level rise will be greater. The long dimension of this page is more than 20 centimeters, so this amount of sea-level rise may seem insignificant, but it will have profound effects in many coastal areas where surface slopes are very low.

Putting It Together—How Do We Know . . . That Global Sea Level Is Rising?

■ Tide-gage records document historic relative sea-level rise and drowned corals document long-term relative sea-level rise since the last ice age.

■ Separating global sea-level rise from relative sea-level rise affected by vertical movements of the lithosphere is difficult.

■ Global sea level is currently rising at a rate of 1 to 2 millimeters per year. Expansion of seawater because of global warming explains about 1 millimeter of rise each year. Melting glacial ice is most likely responsible for any additional sea-level rise.

19.7 What Are the Consequences of Rising Sea Level?

The shape of modern shorelines reflects 21,000 years of rising sea level, and the shorelines will continue to change as sea level rises further. Some of the expected results are as follows:

- Coastal areas gradually submerge. Shoreline retreat by as much as 100 meters, roughly the length of a football field, can take place with as little as 10 centimeters of sea-level rise where the land surface is nearly flat.

- Islands gradually submerge, with decreasing area to support populations and agriculture. There are more than 1000 inhabited islands on Earth with maximum elevations less than one meter above sea level. **Figure 19.46**, for example, shows how Key West will gradually submerge as sea level rises over the next few centuries. The Florida Keys are low-elevation islands composed mostly of coral reefs that were completely submerged 125,000 years ago.

- Coastlines erode as wave energy focuses farther inland. This effect is most obvious during strong storms along shorelines composed of easily eroded material. Coastal erosion has important implications for the preservation of beaches, which are the biggest tourist attractions worldwide. Unless rivers supply sediment to shorelines in excess of the amount that is eroded, the shorelines will retreat landward.

- High tides inundate increasingly larger land areas, flooding coastal wetlands along lagoons and estuaries and killing plants that are not saltwater tolerant. Tides also extend farther up river valleys, which slows down river flow to the ocean and causes flooding along the stream banks.

- Coastal water tables rise as sea level rises and causes landward incursions of salty ground water that is undrinkable and unsuitable for irrigation.

The effects of sea-level rise on coastal landscape evolution and its effects on human structures and activities are particularly well documented for beaches and barrier islands, estuaries, deltas, and sea cliffs.

Effects of Rising Sea Level on Beaches and Barrier Islands

Sea-level rise causes erosion of beaches and landward retreat of the shoreline, as shown in **Figure 19.47**. Wave energy focuses farther inland on the beach, and more of the beach submerges at high tide. A one-centimeter sea-level rise causes the shoreline to extend inland much farther than the very short distance to

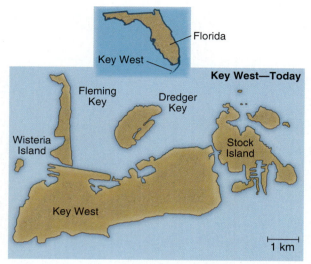

Key West—Today

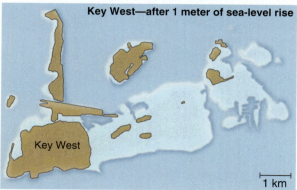

Key West—after 1 meter of sea-level rise

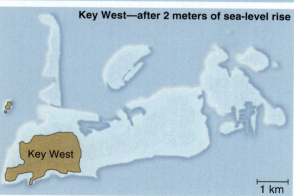

Key West—after 2 meters of sea-level rise

▲ **Figure 19.46** **Visualizing the submergence of islands.** These maps show how islands in the Florida Keys will submerge if relative sea-level rise continues. At current local rates of relative sea-level rise, a 1-meter rise will occur over the next 250 years and 2 meters of submergence will occur by 500 years from now. Large areas of these and other small islands around the world are at elevations less than 1 meter above sea level.

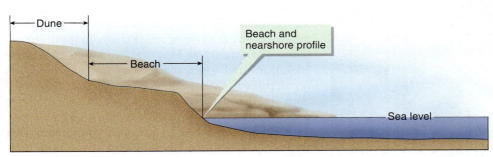

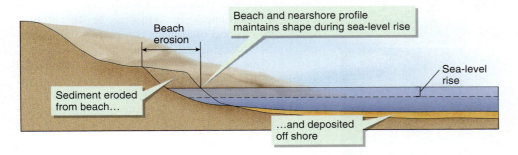

◄ **Figure 19.47 Why sea-level rise causes beach erosion.**
Even small amounts of sea-level rise cause large amounts of beach erosion and shoreline retreat. Wave height and sediment size define the profile shape of the beach and shallow seafloor. The profile shape remains the same during sea-level rise, which causes large amounts of beach erosion.

where the elevation is one centimeter higher. Wave erosion sculpts a smooth, concave-up profile along the beach and into the shallow nearshore environment. This profile shifts upward and landward as sea level rises, so that each centimeter of sea-level rise commonly causes more than 1.5 meters of shoreline retreat.

Sea-level rise causes barrier islands to migrate landward. **Figure 19.48** shows how storms erode the seaward-facing beach and wash the sediment across the island and into the landward lagoon. The barrier islands on the United States east coast have migrated for thousands of years. The sandy beach and dune deposits forming the islands rest on top of older muddy lagoon deposits that originated landward of the islands when the islands used to be farther offshore. These older lagoon deposits are detected in wells drilled on the barrier island and are locally exposed on the seaward beaches when they erode during storms (Figure 19.48).

The barrier coastlines of New Jersey, Delaware, and Maryland are heavily populated, and homes are constructed on or just landward of the beach. Relative sea-level rise will probably cause about 50 meters of shoreline erosion in this region by 2050. This amount of erosion has the potential to cause considerable damage because the beaches are generally only about 25–30 meters wide. Along the Delaware shore, the barrier islands will entirely cross the lagoons and become mainland beaches if sea level rises another six meters.

Effects of Rising Sea Level on Estuaries

Estuaries form when relative sea-level rise submerges low-gradient river valleys along coastlines. Modern estuaries, such as Chesapeake Bay, shown in **Figure 19.49**, formed during the most recent rise

▼ **Figure 19.48 Barrier islands migrate landward when sea level rises.**
Waves erode barrier-island beaches, and storms wash over the islands and carry sand into lagoons. These processes cause the islands to slowly back-peddle landward over muddy lagoon deposits. Oyster shells collected on Atlantic beaches are eroded from lagoon deposits exposed on the seafloor near the beach.

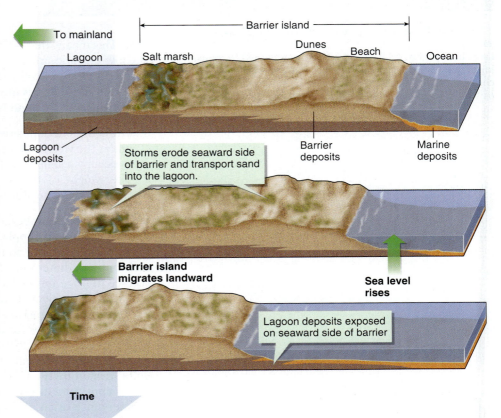

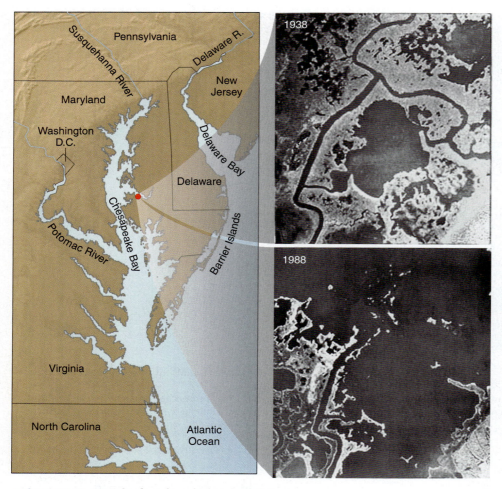

▲ **Figure 19.49 Estuaries form by submergence of river valleys.**
Large bays on the Atlantic coast are estuaries formed by drowning of river valleys beneath the rising sea. Submergence of coastal wetlands is easily recorded over short historic time scales, as seen in these two aerial photographs taken 50 years apart.

in sea level that began about 20,000 years ago. For estuaries to persist, the rate of sea-level rise to submerge the valley must exceed the rate of sediment deposition by the rivers, which tends to fill in the estuary. The estuary at Troy, for example (see Figure 19.42), formed when the rate of sea-level rise was rapid, prior to 8000 years ago (see Figure 19.45) and then filled in with sediment when the rate of rise decreased.

Chesapeake Bay is an example of an estuary that continues to enlarge by submergence as sea level rises (see Figure 19.49). The progressive submergence of the tidal flats and coastal marshes has ecological consequences in addition to threatening bayside homes and roads. In a completely natural situation, the coastal marshes would simply shift landward as sea level rose. Along heavily populated coastlines, however, there usually is only a narrow band of wetland between the high-tide mark and roads and buildings. Human-built structures restrict landward shift of the marshes, so once the wetland submerges, it is gone.

Effects of Rising Sea Level on Deltas

Deltas are headlands that build into the sea because sediment delivery by rivers outpaces shoreline retreat by sea-level rise. It might seem likely, therefore, that deltas are immune from the effects of sea-level rise, but this is not the case. Deltas cover tens of thousands of square kilometers, but sediment deposition occurs only on small parts of the delta at any one time. This means that while part of the delta builds seaward, the rest of the delta is a low-elevation plain that is susceptible to inundation by rising sea level.

Relative sea-level rise is commonly greater on delta coastlines than along the same shore that is more distant from the river mouth. Not only is global sea level rising, but the land is also sinking. Deltas slowly subside for two reasons:

1. The sediment deposited rapidly at the mouth of the river compacts over time under the weight of additional deposits. The compaction causes the sediment to occupy a smaller volume, so the land surface subsides.
2. The weight of the sediment forming the delta isostatically depresses the crust, so the land slowly subsides.

Subsidence and submergence of delta coastlines are even greater where dams diminish sediment supply from the rivers, where ground water or oil withdrawal increases sediment compaction, and where levees constructed for flood control funnel sediment out to sea rather than letting it spread out over the delta surface to fill in the subsiding areas.

Most large river deltas on Earth attracted the development of urbanized port cities and extensive agriculture to take advantage of fertile soil and readily available fresh water for irrigation. A one-meter rise in sea level will submerge about 15 percent of the densely populated Nile delta in Egypt and will submerge about 10 percent of the entire country of Bangladesh, which occupies the delta formed by the Ganges and Brahmaputra Rivers. Some shorelines on the Mississippi delta in Louisiana retreat as fast as 20 meters per year, with annual submergence of an area equal in size to Washington, D.C. The city of New Orleans has subsided below sea level and is surrounded by walls to prevent flooding. **Figure 19.50** illustrates the flooding of historic Venice, Italy, at high tide, which results from the combination of global sea-level rise and subsidence near a large river delta. A comparison of current high-tide marks on buildings along the famous Venice canals with the similar marks visible in early eighteenth-century paintings shows a relative sea-level rise of about 70 centimeters between 1727 and 2002.

Effects of Rising Sea Level on Sea Cliffs

The response of sea cliffs to sea-level rise depends on how easily waves erode the cliff-forming rock or regolith. Steep shoreline bluffs of poorly consolidated glacial deposits in the northeastern United States erode at between 10 centimeters and one meter per year. During a 1944 hurricane, a sea cliff on Long Island retreated 12 meters in a single day. Erosion rates for sea cliffs of hard granite, on the other hand, are imperceptibly slow at 1 millimeter per year, or less.

Most of the scenic, highly developed shoreline of California features homes built at the edge of sea cliffs eroded in sedimentary rocks. **Figure 19.51** illustrates damage that results from cliff erosion by storm waves. Not only is the wave energy higher during storms, but runoff from accompanying heavy rain also erodes the steep cliff faces.

Human Responses to Shoreline Erosion

There are three general responses to shoreline erosion caused by the combination of global sea-level rise, land subsidence, and alterations of shoreline sediment budgets:

1. Armor the shoreline to hold it in place where it is currently located
2. Add sediment to eroding beaches to maintain their width and location
3. Abandon the coast

Armoring usually means building **seawalls** on the beach parallel to the shoreline as shown in **Figure 19.52**. Seawalls vary in height and are constructed from wood, plastic, concrete, rock, steel, junk cars, rubber tires, or sandbags. Resistance to storm-wave erosion is greatest for high walls constructed of strong materials. Seawalls do protect coastal property but do not necessarily save the beach, as seen in Figure 19.52. Wave erosion eventually carves away the beach until the waves crash against the seawall. Longshore currents carry the eroded sand offshore or alongshore. Potential beach-sediment sources in coastal dunes

▲ **Figure 19.50 How relative rising sea level affects Venice.** Pedestrians wade in the ocean or stroll on elevated boardwalks when spring high tide floods historic Venice, Italy. Tidal flooding of Venice occurs because of the combination of global sea-level rise and subsidence of land beneath the city.

◄ **Figure 19.51 Sea-cliff erosion destroys homes.** Homes collapse onto the beach along this California shoreline because waves easily erode the soft sedimentary rock forming the sea cliff.

▶ **Figure 19.52 How seawalls protect property but not beaches.**
Wave-resistant walls stop erosional retreat of shorelines and protect coastal buildings. As sea level rises, the beach erodes away on the seaward side of the wall, which diminishes recreational value. The beach remains if it is allowed to migrate landward with rising sea level, but coastal erosion destroys unprotected buildings.

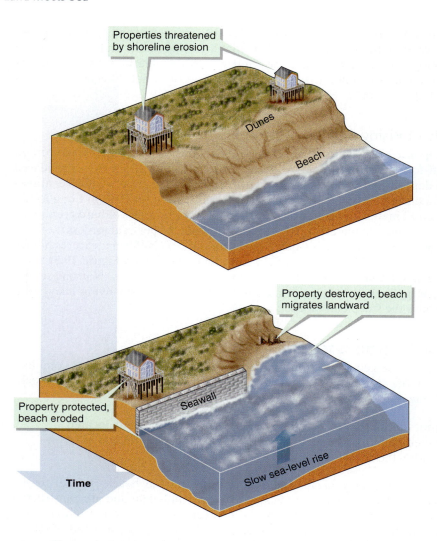

or readily eroded sea cliffs are isolated behind the seawall, which diminishes the sediment budget for the beach.

Replenishing eroded beach sand is an alternative to building seawalls. Coastal communities heavily utilize this nourishment option to restore economically important recreational beaches, as shown in **Figure 19.53**. The procedure

▶ **Figure 19.53 Replenishing Miami Beach.**
The photo on the left shows groins and seawalls protecting resort hotels on nearly beachless Miami Beach in the 1970s. The photo on the right shows a wide beach at the same location following beach replenishment efforts in the early 1980s. Millions of dollars were spent to pump wave-eroded sand onto the beach from the nearby seafloor.

also maintains beaches in front of coastal properties so that wave energy is spent on the beach and does not threaten buildings. In some places, sand eroded and transported offshore is dredged onto barges that transport the sediment back to the beach, or sand and water are pumped from the seafloor and spread onto the beach as a slurry. In other cases, beaches are replenished with sediment excavated from harbors that are filling with river deposits. Beach replenishment is very expensive and always temporary, because the conditions eroding the beach are not changed. Some resort beaches in New Jersey have been replenished more than 40 times since 1950.

Retreating in the face of rising sea level and coastal erosion is another option. Many beachfront homes are simply jacked onto trucks and moved to more inland properties. In 1999, the historic Cape Hatteras, North Carolina, lighthouse was moved 884 meters inland, as shown in **Figure 19.54**, to avoid certain destruction by coastal erosion. The lighthouse was built more than 500 meters from the beach in 1870, but storm waves lapped at its base by the early 1980s. The cost of this relocation was 12 million dollars. Wholesale movement of coastal communities and large resort hotels is clearly not economically feasible, so seawalls and beach-replenishment projects continue as costly, but still less expensive, responses to the shifting shore. If sea levels continue to rise, however, then natural coastal processes will eventually destroy coastal developments with extraordinary economic costs.

New lighthouse location

Lighthouse moving on rails

Wave erosion threatens lighthouse despite protective groin

▲ **Figure 19.54 Retreating from coastal erosion.**
The Cape Hatteras Lighthouse was moved inland in 1999 because it was at risk of being destroyed by coastal erosion. The lighthouse was more than 500 m from the shore when it was constructed in 1870.

Putting It Together—*What Are the Consequences of Rising Sea Level?*

■ Sea-level rise threatens large coastal and island populations with submergence and the hazards of coastal erosion.

■ Responses to coastal erosion on United States shorelines focus on armoring the shore against erosion or replenishing beach sand eroded by waves. These solutions are expensive and only temporary.

EXTENSION MODULE 19.1

Changing Shorelines in the Great Lakes.
Learn the causes for fluctuating water levels in the Great Lakes, and compare coastal erosion problems on Great Lakes shorelines with those that happen on ocean shores.

Where Are You and Where Are You Going?

Wind-driven waves, tides, and associated currents are agents of change at shorelines. These processes create wide sandy beaches, steep rocky headlands, broad marshy tidal flats, and long barrier islands. The diverse appearances of shoreline landscapes relate not only to waves, tides, and currents but also to the erodibility of rock and regolith at the coast, uplift and subsidence of the crust, variations in sediment supply by rivers, and absolute sea-level rise and fall.

Shoreline positions shift over short, human time scales and long, geologic time scales. Emergence of once submerged seafloor occurs when absolute sea level falls or land rises because of tectonic and isostatic processes. Submergence of land beneath the sea occurs when absolute sea level rises or the land subsides. Shorelines also extend seaward, where rivers deliver large volumes of sediment that fill in bays and nearshore shallow areas.

Sea level is currently rising at a rate of 1 to 2 millimeters per year. Most of this rise is due to the expanded volume of warming seawater resulting from globally increasing temperature. Melting of glacial ice may also contribute to sea-level rise. Some coastlines are rising because of tectonic uplift or isostatic uplift where glacial ice previously depressed the crust. Most heavily populated coastal areas and islands are, however, at risk of submergence and increasing threats of coastal erosion hazards as sea level rises.

You have now completed your study of geologic processes and landscape modification related to water—flowing in streams, flowing underground, flowing as ice, driven by wind and tides. The next chapter examines wind as a geologic process. Wind produces waves, which play the major role in shoreline geology considered in this chapter. Wind is also important on land as an agent of sediment erosion that lowers landscape elevations and as an agent of deposition to form sand dunes that cover hundreds of thousands of square kilometers in deserts. The presence of sand dunes on barrier islands indicates, however, that wind is not just a geologic agent in deserts. In fact, regardless of where you live, you only have to look at the dust that settles in your home or classroom to contemplate the importance of wind.

 **Active Art**

Properties of Waves. See how to recognize the parts of a wave and to measure the wave period.

Water Wave Motion and Refraction. See how waves move in water and refract along a shoreline.

Tsunami. See how an earthquake forms a tsunami and how a tsunami travels across the ocean.

Beach Drift and Longshore Current. See how beach drift and longshore current move sediment along the shoreline.

Effects of Groins and Jetties. See how groins and jetties influence shoreline deposition and erosion.

How Tides Work. See how the positions of Earth, Moon, and Sun determine the rise and fall of the tides.

Extension Module

Extension Module 19.1: Changing Shorelines in the Great Lakes. Learn the causes for fluctuating water levels in the Great Lakes, and compare coastal erosion problems on Great Lakes shorelines with those that happen on ocean shores.

Confirm Your Knowledge

1. Irregularities in coastal outlines are due, in part, to differences in the way shoreline materials erode. What causes these differences?

2. Where does the sand at a beach come from?

3. What factors can cause shoreline shape and position to change over time?

4. How do waves form? How do they travel? Why do they break?

5. Why do waves not form on swimming pools or small ponds?

6. Distinguish wave height from wavelength.

7. How do longshore currents cause sediment deposition or erosion along the shoreline?

8. Describe how construction of groins and jetties affect adjacent beaches.

9. If you were trying to determine is a beach was growing or shrinking, what would you look for?

10. What human activities happening tens or hundreds of kilometers from a coast can still end up affecting the shoreline sediment budget?

11. How and why do barrier islands and tidal inlets move over time?

12. Although geologists have proposed two different hypotheses for the formation of barrier islands, what is common to both ideas?

13. Which has a greater influence on Earth's ocean tides, the Sun or the Moon? Why?

14. What is the difference between a spring tide and a neap tide? Do spring tides only happen in the spring?

15. What features would you look for as evidence that a shoreline is submergent or emergent?

16. What factors control the position of the shoreline?

17. Why is it so difficult to determine how fast sea level is rising?

18. What evidence exists to support the hypothesis that global sea level will rise 5 to 6 meters higher than it is today?

19. List the expected results if sea level rises by one meter.

Confirm Your Understanding

1. Write out an answer for each question in the Chapter Outline for the chapter sections assigned by your instructor.

2. Many shoreline shapes represent a dynamic balance between sediment delivered to the shore by streams and erosion from wave action. Explain how a beach or spit might change over time because of variations in these two processes.

3. What do ocean waves and seismic waves have in common. How do they differ?

4. Explain the transfer of energy in going from wind on the open ocean to the roar of the surf.

5. Explain why a floating object moves slowly in the direction of wave movement.

6. If you lived along the shoreline, on which side of a groin, relative to the longshore current, would you prefer to locate your home? Why?

7. Barrier islands commonly contain valuable real estate that is heavily developed. What is a long-term natural problem inherent to barrier islands?

8. What would the tides on Earth be like if Earth had two moons identical in size and orbit except that they were 180 degrees apart? What if they were 90 degrees apart?

9. Are all changes in relative sea level due to changes in the volume of water in the ocean? Why or why not?

10. Some geologists hypothesize that the tug and pull of lunar tidal forces are strong enough to trigger volcanic eruptions and earthquakes. Explain how you think this might happen. How would you test your hypotheses and what data would you need?

Key Terms

absolute sea-level change (p. 595)
barrier islands (p. 587)
baymouth bars (p. 582)
bays (p. 568)
beach (p. 580)
beach drift (p. 580)
beach face (p. 580)
berm (p. 580)
berm crest (p. 580)

breaker (p. 573)
estuaries (p. 568)
groins (p. 582)
headlands (p. 568)
jetties (p. 588)
lagoons (p. 587)
longshore currents (p. 574)
neap tides (p. 593)
relative sea-level change (p. 595)

rip currents (p. 574)
sea stack (p. 580)
seawalls (p. 607)
spits (p. 582)
spring tides (p. 592)
tidal channels (p. 593)
tidal flats (p. 593)
tidal inlets (p. 587)
tidal range (p. 591)

tide (p. 591)
wave base (p. 572)
wave-cut platform (p. 582)
wave height (p. 569)
wave period (p. 569)
wavelength (p. 569)

20

Wind:
A Global Geologic Process

Why Study Wind?

ALL PROCESSES THAT ERODE AND TRANSPORT EARTH MATERIALS are geologically important and shape the landscape. Movement in the atmosphere—wind—is such a process. If you have stood on a sandy beach, or in a desert, or even near piles of dirt at a construction site on a windy day, then you have felt the wind-blown sediment stinging your skin. Clearly, wind picks up and moves material. Sand dunes are a familiar landform caused by wind transport of sediment near the surface of the ground. Just as your skin feels as though it is being sandblasted on a windy day in a sandy landscape, wind-blown particles abrade rock surfaces.

Wind-blown particles are evidence of active changes on Earth's surface, and many of these changes affect human welfare. Wind erosion removes topsoil, which diminishes agricultural productivity. Thousands of square kilometers of once vegetated land are converted into barren desert each year. Wind-blown sediment reduces visibility in populated areas

and causes highway accidents. Dust (very small sediment particles) remains aloft in the swirling atmosphere for many weeks or months, causing hazy skies. Dust particles include mineral particles along with pollen, spores, and windborne microbes that can cause many respiratory ailments. The total amount of atmospheric dust is not well known, but by some estimates more than 2 billion metric tons of fine mineral particles loft into the atmosphere from Earth's surface each year.

Wind blows everywhere. As you look back over the previous four chapters, you may not feel that all of the subject matter was relevant to where you live. Perhaps, for example, glaciers and coastal processes have no influence on your local landscape. Wind, on the other hand, is a truly global process, visible not only in the features of desert landscapes but also in other environments where you may not have thought wind to be important. Some evidence for the work of wind is subtle, and you may not have even noticed it before.

Wind moves sand along a rippled desert sand dune. ▶

Understanding how wind forms Earth's landscapes also provides insights into processes on other planets. Telescopic observations show vast dust storms that periodically obscure the surface of Mars, for example. Although water probably flowed on Mars in the ancient past, wind is now the dominant process that modifies the martian landscapes recently visited by robotic rovers.

You have four objectives in this chapter:

✔ To comprehend why and how the atmosphere moves to cause wind

✔ To describe how wind erodes, transports, and deposits sediment

✔ To explain where wind is influential in forming and modifying the landscape

✔ To understand the impact of wind-blown dust in the Earth system

You will learn answers to the following questions in pursuit of the objectives:

20.1 Why does wind blow?

20.2 Where is wind an influential process in the landscape?

20.3 What determines the locations of deserts?

20.4 How does wind pick up and transport sediment?

20.5 How does wind shape the landscape?

20.6 *How do we know* ... that wind blows dust across oceans?

In the FIELD

What is the best way to get a perspective on the geologic significance of wind? Consider two separate vacation trips to different places where wind processes are clearly at work. Your first stop is a desert in the southwestern United States. Your second trip is to the Outer Banks of North Carolina.

Figure 20.1a summarizes some important observations at Death Valley National Park, in southeastern California. From a mountain vantage point high above the valley floor, you look over a stark landscape. There is virtually no vegetation to obscure your view of steep rocky cliffs, widespread fans of stream-transported sand and gravel at the mountain bases, and glaring white salt flats where rare accumulations of surface water dried up in the past.

Similar scenes appear throughout many areas of the western United States commonly described as "deserts." You quickly note that the near absence of vegetation must relate to the lack of moisture to nourish plant growth. Indeed, your visitor's leaflet states that Death Valley receives less than 5 centimeters of rain per year, which is pretty sparse compared to a greener place like Chicago, Illinois, where 86 centimeters of precipitation falls each year.

Driving down from the mountains to the valley floor, you encounter an area of sand dunes (see Figure 20.1a). The ridges of loose sand, resembling ocean waves frozen in place, cover more than 50 square kilometers of Death Valley, and the highest dunes are 213 meters tall (or taller than a 50-story building). Smaller undulating ripples stripe the soft and sandy dune surfaces. Occasional wind gusts blow sand against your face. So much sand moves across the surface with each passing gust that for a moment the ground almost seems to flow like liquid. During more sustained gusts you notice that the ripples shift position in the direction of the wind.

Dunes form from sand blowing in the wind, but your curiosity and observations tell you there is more to the desert. Suddenly many new questions arise. Sand dunes are part of many people's images of deserts, but in reality they constitute a very small part of this one in Death Valley. In fact, sand dunes cover the surface of only about 2 percent of North American deserts. What is the relationship between deserts and dunes, and why are sand dunes only found in small areas of the American deserts? Where does the sand in the dunes come from? Looking around the valley, you see wide expanses of sand deposited by flash floods at the base of nearby mountains, and the dusty salt flat where runoff evaporated in the past. Ground surfaces are thoroughly covered by small particles that the wind blows around to make the dunes. If the wind is capable of blowing the sand that stings your face and whips across the dune surfaces, then what is the fate of the even smaller particles, the dust grains not present in this great sand pile?

Figure 20.1b summarizes your observations at Jockey's Ridge State Park, located on a barrier island on the North Carolina coast. You came here to enjoy the beach but are surprised to see that Jockey's Ridge is a sand dune. In fact, at a height of about 30 meters, Jockey's Ridge is the highest sand dune in the eastern United States.

Why do tall sand dunes form along the shoreline in the humid, rainy southeast? One reason might be that there is a lot of sand that is not covered by vegetation. The beach, especially wide at low tide, consists of loose sand that swirls around your bare legs in the wind just as it did at Death Valley. A strong wind blows, just as it does in Death Valley, making Jockey's Ridge a favorite spot for kite flying and hang gliding. The persistent winds and the soft landing spots provided by sandy beaches and dunes along the barrier islands of North Carolina are what attracted the

Figure 20.1 Wind shapes landscapes in deserts and along coasts. ▶

A. Field observations of desert dunes, Death Valley, California

Death Valley as seen from "Dante's View." The valley floor is a desolate desert, virtually devoid of vegetation and notable for broad, white salt flats where lakes evaporated in the past.

Big sand dunes cover part of the floor of Death Valley. In this view the dunes mostly bury light-colored bedrock. Notice the sand ripples in the foreground, which were observed to shift along the ground when the wind blew.

B. Field observations of coastal dunes, Jockey's Ridge, North Carolina

These big sand dunes are found along a narrow stretch of land just landward of the Atlantic Ocean beaches. This mostly barren stretch contrasts with the densely vegetated landscape more typical of the southeastern United States and present just a kilometer or so away.

The wind always seems to blow here, which attracts lots of hang-gliding and kite-flying enthusiasts.

Wright brothers to Kitty Hawk, approximately 15 kilometers from Jockey's Ridge, to successfully test their first powered airplane.

Compared to the dunes at Death Valley, the dunes at Jockey's Ridge cover a very small area. The bare, active coastal dunes form a narrow band alongside the beach and merge landward into grass-covered ridges that have the shapes of sand dunes. These ridges must be dunes that formed at an earlier time, but why did they stop moving and become stable landforms for plant growth?

These two entries into your field notebook offer a few new perspectives on sand dunes. Sand dunes are not just features of deserts, and they do not even cover very much of the desert Southwest. The common variables for sand dunes at the two places you observed, Death Valley and Jockey's Ridge, seem to be

- Blowing wind
- Lots of loose sand
- Very little vegetation, at least where the active dunes are present

Your simple list inspires new questions, however. Why does the air move to begin with? What factors of geology and climate determine the effectiveness of erosion by wind, or how loose sand for the wind to blow is even present, or how lack of vegetation commonly characterizes dune environments? What other landforms, besides sand dunes, owe their origin to blowing wind? If sand dunes are landforms that occur where wind deposits sand, then what landforms are characteristic of places where wind erodes sand? How far does wind transport sediment?

20.1 Why Does Wind Blow?

Before examining evidence of the geologic work done by wind, it is essential to understand why air moves to create wind. It is also important to understand the factors that determine how strongly the wind blows and the direction it blows from.

What Is Wind?

Wind is motion in the atmosphere. Movement of gas molecules in the atmosphere occurs for the same basic reason that motion occurs within the solid Earth (described in Chapter 10)—wind results from atmospheric convection. Denser air sinks to Earth's surface and displaces less dense air upward for the same reason that less dense mantle moves up as denser mantle sinks. This observation makes it essential for us to consider the factors that determine variations in atmosphere density.

Temperature is one variable that determines the density of the atmosphere, as illustrated in **Figure 20.2**. Heating causes the atmospheric gases to expand so that warm air is less dense than cool air. Density differences cause air to rise over regions of atmospheric heating and to sink in cool regions. Air also moves laterally along Earth's surface from the cool regions to the warm regions as dense, cool air displaces warm, less dense air. It is this lateral motion that we experience as wind.

Water vapor in the atmosphere also contributes to contrasting density between adjacent volumes of air (see Figure 20.2). Humid air is less dense than dry air at the same temperature because the water molecule has a lower mass than the nitrogen and oxygen molecules that compose most of the atmosphere (see Figure 9.16 for data on atmosphere composition). Moist, low-density air lifts from Earth's surface whereas dry air of similar temperature sinks. The temperature and moisture effects on air density commonly work together, because warm air can hold more moisture than cold air.

Convection in the atmosphere explains the variations in air pressure that figure prominently in weather forecasts. Air pressure is low where the air rises upward away from the surface and high where denser air sinks down against the surface. This means that surface winds tend to move from areas of high surface pressure toward areas of low surface pressure (see Figure 20.2). Pressure differences also cause the cir-

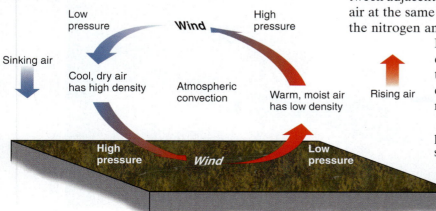

▼ **Figure 20.2 Why the atmosphere convects.**
Air density varies depending on temperature and moisture content. Warm air is less dense than cool air and moist air is less dense than dry air. This means that cool, dry air sinks while warm, moist air rises. The vertical movement of air produces regions of low and high air pressure. Air moves horizontally from regions of high pressure toward regions of low pressure. This horizontal motion is wind.

culation loop to close in the upper atmosphere. In general, air pressure decreases upward because there is progressively less overlying atmosphere weight pressing down. Sinking air reduces air pressure in the upper atmosphere even further because mass is moving downward. On the other hand, rising air compresses air mass into the upper atmosphere, which increases air pressure compared to regions of sinking air.

Convective motion in the atmosphere is much faster than convection within Earth's mantle. Primarily, this difference results from the fact that the viscosity of air is negligible compared to the viscosity of hot mantle rock. Average wind speeds are 10–20 kilometers per hour, and hurricane winds can exceed 200 kilometers per hour. These rates of convective movement hugely exceed the few-centimeters-per-year motion in the mantle driven by the same fundamental physical process.

Wind speed corresponds to the pressure difference between adjacent regions of high and low air pressure—the greater the difference in air pressure, the faster the wind speed. Windy storms and hurricanes coincide with areas of extremely low air pressure where warm, moisture-rich air rises rapidly upward. The large difference in air pressure between the storm region, which has extremely low pressure, and the adjacent areas of higher pressure is one cause of high-velocity, damaging wind.

Wind direction and speed vary from day to day in response to the shifting locations of high- and low-pressure regions in the atmosphere, but observations of how particular wind directions dominate different areas on a yearly basis reveal a global pattern of atmosphere motion. For example, in the conterminous United States the wind blows most often from the west toward the east. The direction the wind blows from describes the wind direction, so the prevailing wind in the conterminous United States is westerly. In contrast, easterly winds are most common in Hawaii. Dominant wind direction also varies with season in some locations. At Jockey's Ridge, for example, westerly winds are most common, but winter storms usually bring wind from the east or northeast. The seasonal flip-flop in wind direction moves the coastal dunes back and forth so that, instead of blowing primarily landward or seaward over time, they remain relatively permanent features. Why does wind commonly blow from different directions at different locations?

Constructing Global Wind Patterns

Figure 20.3a shows a simple view of atmosphere convection to begin your understanding of the directions of blowing wind. You will soon see that this scheme is unrealistic, but it is easiest to start with a simple idea and then add reality that is more complex. The first observation is that the equatorial region receives more direct solar heating than the poles. Therefore, it is reasonable to conclude that dense, cool air descends at the poles and moves toward the equator where less dense, warm air ascends. Air moves from surface high pressure near the poles toward a belt of surface low pressure along the equator. In this view at Earth's surface, northerly winds dominate in the northern hemisphere, southerly winds dominate in the southern hemisphere, and you do not see the expected pattern of prevailing westerly and easterly winds.

Figure 20.3b adds the effect of moisture transport on convection. Warm air rising from the low-pressure belt at the equator is also rich in water vapor, because warm temperatures favor evaporation of surface moisture and because warm air holds more water vapor than cold air. This initially warm, moist air mass cools, however, as it moves higher in the atmosphere. Cooling occurs with the upward decrease in air pressure that results from the decrease in the weight of overlying atmosphere as the air moves toward space. This cooling effect, because of pressure decrease, is the same one that occurs in convecting mantle (see Figure 10.6). The cooling air cannot hold the water vapor that was absorbed when the air was first warmer at the surface. Cooling, therefore, causes the water vapor to condense as clouds and rain. This means that the originally

Global Wind Patterns: *See the processes that determine global wind patterns.*

warm, moist, low-density air is now both cooler and drier, which are properties that increase its density.

The simple convection between equator and poles (Figure 20.3a) is cut short, therefore, by sinking of cool, dry air only 30 degrees north and south of the equator (Figure 20.3b). The descending air forms high-pressure belts at these subtropical latitudes and then flows both northward and southward. De-

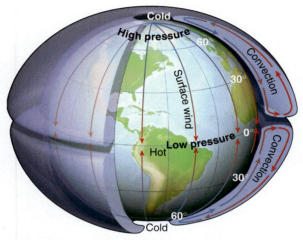

(a) Simple temperature-driven convection

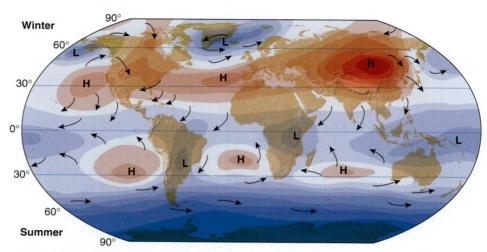

(d) Atmospheric circulation in January

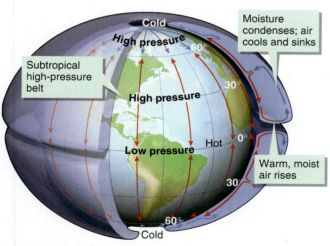

(b) Adding the effect of moisture on atmospheric circulation

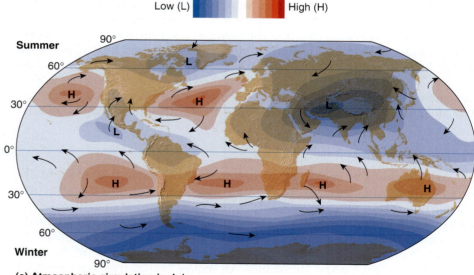

(e) Atmospheric circulation in July

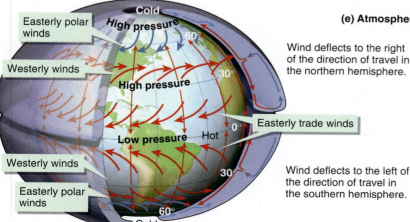

(c) Adding the Coriolis effect on atmospheric circulation

Wind deflects to the right of the direction of travel in the northern hemisphere.

Wind deflects to the left of the direction of travel in the southern hemisphere.

▲ **Figure 20.3 Understanding global wind patterns.**
If air circulation resulted only from simple convection caused by greater warming of the atmosphere at the equator, then Earth's wind pattern would resemble that seen in (a). In reality, the situation is more complicated because evaporation and condensation of water vapor also affect air density, which causes air to sink toward the surface to produce the subtropical high-pressure belts shown in (b). You must also consider Earth's rotation, which causes the Coriolis effect that deflects the air currents, as shown in (c). There are also seasonal differences in the distribution of high- and low-pressure areas on Earth's surface because solar heating of land and ocean surfaces is not equal. These seasonal variations in surface heating and air pressure cause different wind directions, which are depicted in parts (d) and (e).

scending cold air near the poles forces another convection loop at high latitudes. This pattern is more realistic than Figure 20.3a, but it still predicts only northerly and southerly winds.

The next step toward envisioning *real* global wind patterns is to add in the effect of Earth's rotation. Rotation causes all objects that move over Earth's surface or through the atmosphere to experience a horizontal drift called the **Coriolis effect** (in honor of the nineteenth-century French engineer who mathematically quantified it). The Coriolis effect shifts moving objects to the right of their initial path in the northern hemisphere and to the left in the southern hemisphere. Figure 20.3c adds Coriolis deflections to the schematic wind pattern produced by convection. For example, air moving southward from the northern hemisphere subtropical high-pressure zone deflects to the right (toward the west) to produce easterly winds, whereas air moving northward from this high-pressure belt deflects to the right to produce westerly winds. The picture is almost complete: The Coriolis effect works with convection to produce westerly and easterly winds. What remains unexplained, however, is why the wind directions do not remain the same year-round.

Figures 20.3d and 20.3e show the actual global wind directions and the locations of high- and low-pressure regions in the atmosphere during the seasonal solstices. The actual circulation differences between seasons and the general pattern predicted in Figure 20.3c relate to two factors:

1. Solar heating of the surface and atmosphere clearly differs with the seasons. This means that high- and low-pressure regions and the winds between them shift with the seasons.
2. Land and ocean surfaces heat unevenly. Water absorbs and holds solar heat much more efficiently than does rock or soil. Incoming heat is distributed through large depths of ocean water but only penetrates a short distance beneath the surface of rock and soil. This means that, compared to oceans, the land surfaces reflect more heat into the atmosphere during summer months, which produces low-pressure areas of rising warm air above the land.

The overall effect is that winter wind patterns, shown in Figures 20.3d and 20.3e, closely resemble the ideal case shown in Figure 20.3c. Subtropical high pressure forms a fairly continuous belt that encircles the globe close to 30 degrees latitude, and this belt separates zones of easterly and westerly winds. During the summer months, however, lower air pressure over the continents breaks up the continuous belt of high pressure. Very low summer air pressure over southeastern Africa, southern Asia, and northern Australia draws in moist air from adjacent oceans to cause torrential seasonal rains, called **monsoons**. Weaker summer-monsoon wind and rain also occur in the southwestern United States and Mexico.

Global wind patterns are the summed-up effect of convection and Earth's rotation, with complications caused by the seasons and uneven heating of land and sea. Hawaii and the conterminous United States are located at different latitudes and experience different prevailing wind directions. The southeast coast of the United States experiences northeasterly winds in winter (Figure 20.3d) and westerly winds in summer (Figure 20.3e) as observed at Jockey's Ridge.

Local Wind Patterns

Local surface features also cause local variations from the global wind pattern summed up in Figure 20.3. High mountains, for example, block or deflect wind near the surface. There are also small convection cells that form because of differential heating and cooling of Earth's surface. These small-scale circulation patterns produce distinctive winds in coastal and glacial regions.

In coastal regions, light winds blow landward during the day and seaward at night, especially during the summer. This air circulation between land and sea is

EXTENSION MODULE 20.1
How the Coriolis Effect Works.
Learn more about how Earth's rotation causes moving objects to follow curving, rather than straight-line, paths.

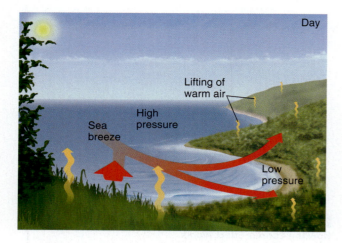

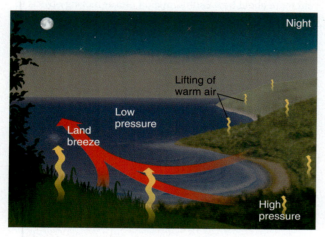

▲ **Figure 20.4 Daily changes in wind direction between land and sea.**
Warming of the air over land during the day causes air to rise, which forms low pressure over land. Sea breezes blow onshore toward the region of low pressure. At night, the land cools rapidly, but the water remains warm. This causes the region of low pressure to shift over the ocean, which results in offshore land breezes.

a miniature version of global-scale convection. The seaward and landward breezes form as illustrated in **Figure 20.4** because water absorbs heat much more efficiently than does rock or soil. The air above land is, therefore, warmer during the day than air above water. The temperature differences mean that the adjacent air masses have different densities, which result in different air pressures. Cool air moves landward from the ocean and displaces the warm air upward. Water also stores heat more effectively during nighttime hours, whereas rock and soil surfaces cool down rapidly. At night, therefore, the wind reverses and blows offshore.

Strong surface winds commonly flow away from large valley glaciers and ice sheets. Air temperatures are cold above high-altitude ice and snow. The cold, dense air flows down slope and displaces warmer air at lower elevations. Winds blowing downward and outward from ice sheets in Greenland and Antarctica are among the most persistently strong winds recorded on Earth. The rapidly moving air is not only cold but also dry, because cold air does not transport abundant water vapor.

Putting It Together—Why Does Wind Blow?

■ Wind is motion of Earth's gaseous atmosphere.
■ Convection causes atmospheric motion. Cool, dry air is denser than warm, moist air. This means that cool, dry air moves downward and displaces warm, moist air upward.
■ Global zones of easterly and westerly winds alternating with latitude result from the combination of atmosphere convection and the Coriolis effect caused by Earth's rotation. Irregular heating and cooling of continents and oceans along with seasonal temperature variations add additional complexity to global atmospheric motion.
■ Local wind patterns result from local variations in topography and from irregular heating and cooling of Earth's surface.

20.2 Where Is Wind an Influential Process in the Landscape?

Wind blows everywhere on Earth, but the impact of wind on landscapes is more obvious in places where three conditions are met:

1. Strong winds blow frequently.
2. Vegetation is sparse or absent.
3. Small, loose, dry particles are abundant on the surface.

These conditions are present at Death Valley and Jockey's Ridge and are confirmed as critical ingredients by global studies. Examples of wind erosion and sediment transport meeting these three criteria are shown in **Figure 20.5**. Wind power is the most straightforward factor; wind must blow strongly and often to be influential. The other two conditions are a little more complex and require further consideration.

Wind Is Effective Where Vegetation Is Sparse or Absent

You can observe that vegetation diminishes the geologic effectiveness of wind for two reasons:

1. Plant roots bind surface particles together so that they cannot be picked up by the wind.
2. Plants rise above the land surface and absorb and deflect the force of the wind so that surface particles are sheltered from potential wind erosion.

The presence of sand dunes in some deserts and along coastal beaches is partly explained by the lack of vegetation to hold down the sediment or to slow the near-surface winds.

The sparseness or absence of vegetation in deserts (see Figure 20.1) results from a lack of moisture. **Deserts** are arid regions where annual precipitation is less than 25 centimeters. The lack of moisture means that it is difficult for plants to grow. Plants adapted to desert climates are spaced far apart in comparison to the dense forests of wetter climate zones. Widely spaced plants do not impede the ability of wind to erode and transport surface particles. It is important to note that although deserts are commonly thought of as very hot, temperature is not part of the definition; some deserts are cold, such as in Antarctica.

Shoreline processes, rather than lack of rain, explain the sparseness of vegetation in most coastal regions, such as Jockey's Ridge (Figure 20.1b). Unlike deserts, there is no rainfall limit for formation of coastal dunes, which occur in places that may receive a meter or more of rainfall each year. Sandy beaches lack vegetation because plants do not grow within the area swept by saltwater each day by the waves and tides or seasonally by storms. In addition, lush plant growth rarely exists landward of the beach along coasts that experience strong sea breezes. The moving salty air dries out plant tissue and typically restricts plant growth to grasses that are tolerant of dry, salty conditions.

Wind Is Effective Where Surface Particles Are Small

Wind is not as effective as water in picking up surface sediment. Air density is only 1/800th the density of water, so shear stresses exerted on the surface by blowing wind are considerably less than for flowing water. Water also exerts a buoyancy effect on particles that make them easier to pick up and move. The buoyancy effect exists because water density is greater than one-third the density of most minerals, so mineral grains weigh less than two-thirds as much in water as they do in air. Grains are easier to move in water than air because the grains effectively weigh less in water than in air.

These limitations of wind as an agent of sediment erosion and transport mean that it is uncommon for mineral grains larger than about 0.5 millimeter, which is sand size, to move with the wind. If the surface has abundant particles smaller than 0.5 millimeter, then wind can pick up and move the sediment and shape the landscape. If there is no loose unconsolidated sediment or the particles are coarser than 0.5 millimeter, then the landscape is virtually unaffected by wind regardless of a lack of vegetation or the presence of strong winds. Gravel that is too coarse for wind to move covers large expanses of desert surfaces. This is why evidence for the erosion and modification of the landscape by the wind is not seen everywhere in desert climates, because many desert surface materials are too large for wind to move.

What landscapes have abundant fine-grained sediment that wind can transport? Strong winds readily move any recently deposited fine-grained material that is not covered by vegetation. Stream floodplains and bars expose sand and finer-grained sediment. Sandy beaches are another candidate for wind erosion, as seen at Jockey's Ridge. Dried-up lakebeds, which are common in the western United States, are a source of fine sediment that includes evaporite-mineral particles.

Some wind-blown sand deposits in the western United States and in the Sahara Desert of North Africa consist of grains weathered from ancient sand-dune sandstones. These sandstones are composed entirely of grains smaller than 0.5 millimeter. The grains disaggregate during weathering to produce an excellent source of sediment for today's wind to move, as shown in Figure 20.5. Ancient deposits of wind-blown sand are recycled as modern sand dunes and dust.

Dust storm

▲ **Figure 20.5 Wind erosion and sediment transport.** Dust storms occur at a variety of scales. NASA Space Shuttle astronauts took the top photo of a dust storm stretching more than 100 kilometers across Afghanistan. The bottom photo shows red haze that consists of sand and dust swept from the northern Arizona landscape on a windy spring day. The rocks in this area are 150-million-year-old sand-dune sandstones. Fine sand grains separate from the rocks during weathering and provide a bountiful source of sand for modern wind to transport. All three conditions for effective wind transport are illustrated: very little vegetation, strong winds, and and abundant fine sediment.

Putting It Together—Where Is Wind an Influential Process in the Landscape?

■ Vegetation diminishes wind erosion by absorbing wind energy that would otherwise reach the surface and by binding together loose sediment with roots.

■ Wind rarely picks up mineral grains larger than 0.5 millimeter, so wind modification of landscapes is only effective where there is abundant, loose, fine-grained sediment on the surface.

■ The combination of strong wind, sparse or absent vegetation, and abundance of fine sediment that is a necessary condition for wind erosion is primarily seen in some deserts and along some sandy coasts.

20.3 What Determines the Locations of Deserts?

Deserts are the largest regions that meet the three requirements for wind to function as an important geologic process. Although wind-blown sand deposits like the dunes in Death Valley and at Jockey's Ridge only cover about 6 percent of Earth's land surface, 97 percent of dune deposits are found in deserts.

Figure 20.6 shows the global location of deserts. Knowledge gained from Section 20.2 (Figures 20.3d and 20.3e) reveals why deserts are found in these locations. **Figure 20.7** illustrates that desert regions lack moisture because they are located where (1) dry air descends to the surface, (2) the air is too cold to hold moisture, (3) mountains block moist air from oceans, (4) moist ocean air moves away from land, or (5) some combination of these conditions exists.

Subtropical Deserts

Most of the deserts depicted on Figure 20.6 are located close to 30 degrees north and south latitude. These include the famous Sahara Desert of northern Africa, the Arabian Desert, the Kalahari Desert of southern Africa, large expanses of Australia, and the deserts straddling the border between the United States and Mexico. These deserts coincide with the subtropical high-pressure belts within Earth's atmospheric circulation (see Figure 20.3). Air descending to the surface at these latitudes is dry, having previously lost its moisture to heavy tropical rainfall near the equator.

As the descending air compresses against Earth's surface, the air also warms up (Figure 20.7a), in the reverse of the pressure-related temperature change that caused the air to cool when it first rose in the equatorial low-pressure belt. The results of this warming are staggeringly high temperatures, including the United States record of 57°C (134°F) at Death Valley and the global record of 58°C (136°F) in North Africa. The high air temperatures add to the drying effect because warm air readily absorbs moisture rather than releasing it as rain and snow.

Rain-Shadow Deserts

Deserts commonly form on the downwind side of mountain ranges, regardless of latitude (see Figure 20.7b). Mountains divert the moist air moving from oceans onto continents upward over the high topography. The lifted air expands and cools, which causes condensation of moisture as rain. The now denser cool and dry air descends the downwind side of the mountains, where it warms by compression against the surface and is depleted of moisture. The resulting deserts are called rain-shadow zones because the rain and snow falls on the upwind side of the mountain range and leaves a shadow zone deprived of precipitation on the downwind side.

The rain-shadow effect contributes to the origin of desert conditions in the western United States. High mountain ranges in the coastal states of Washington, Oregon, and California intercept moisture carried landward from the Pacific Ocean by prevailing westerly winds. Rain-shadow deserts parallel the east side of the mountain ranges through Nevada, eastern Oregon, and eastern Washington (see Figure 20.6). Rain-shadow deserts also form on the west sides

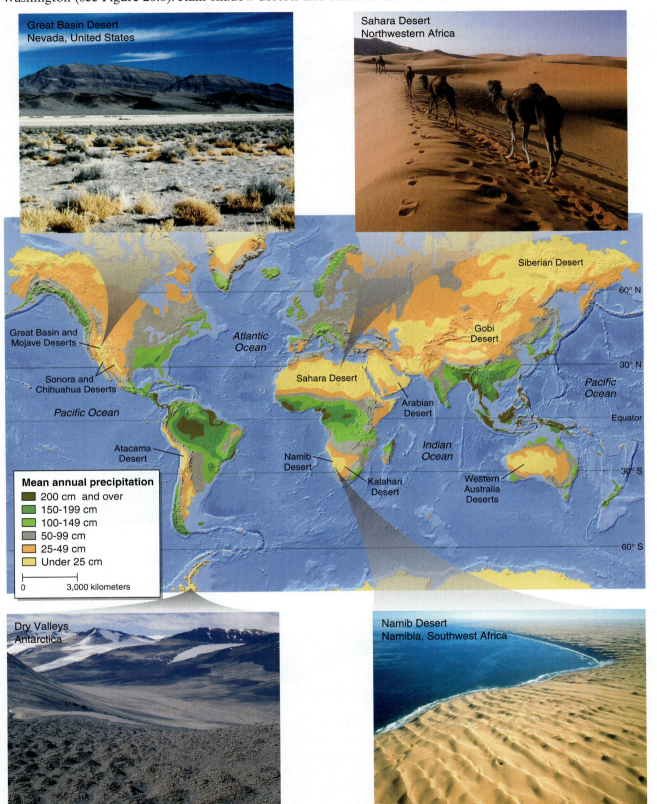

Great Basin Desert
Nevada, United States

Sahara Desert
Northwestern Africa

Siberian Desert

Great Basin and
Mojave Deserts

Atlantic
Ocean

Gobi
Desert

60° N

30° N

Sonora and
Chihuahua Deserts

Sahara Desert

Pacific
Ocean

Pacific Ocean

Arabian
Desert

Equator

Atacama
Desert

Namib
Desert

Indian
Ocean

Kalahari
Desert

Western
Australia
Deserts

30° S

Mean annual precipitation
- 200 cm and over
- 150-199 cm
- 100-149 cm
- 50-99 cm
- 25-49 cm
- Under 25 cm

0   3,000 kilometers

60° S

Dry Valleys
Antarctica

Namib Desert
Namibia, Southwest Africa

◄ **Figure 20.6 Where deserts form.**
The map of mean annual precipitation shows dramatic variability in moisture delivered to different areas of continents. Deserts are those areas that receive less than 25 centimeters of precipitation, on average, each year. The largest deserts cluster near the high-pressure belts at 30 degrees north and south latitude, in the cold and dry polar regions, within continental interiors downwind of mountains, and in narrow strips along some coastlines.

▶ **Figure 20.7 Visualizing how deserts form.**
Dry air at Earth's surface causes the formation of deserts. Dry air masses arise for
several reasons. (a) In the subtropical high-pressure belt, descending dry air creates
desert conditions. Hot wind evaporates what little moisture is present. (b) Mountains intercept moisture moving on land from oceans so that rain-shadow deserts
form downwind of the mountains. (c) Coastal deserts form where relatively dry continental air moves offshore rather than moist oceanic air moving onshore. Offshore
winds drive ocean circulation that brings cold water to the surface so that even
when sea breezes move onto land, the cold air contains very little moisture and condenses as coastal fog. (d) Cold air in polar regions holds very little moisture, so precipitation is also very low.

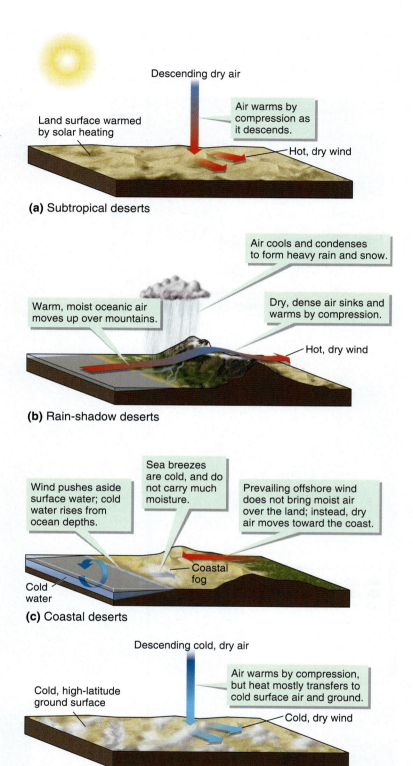

Descending dry air

Air warms by
compression as
it descends.

Land surface warmed
by solar heating

Hot, dry wind

(a) Subtropical deserts

Air cools and condenses
to form heavy rain and snow.

Warm, moist oceanic air
moves up over mountains.

Dry, dense air sinks and
warms by compression.

Hot, dry wind

(b) Rain-shadow deserts

Sea breezes
are cold, and do
not carry much
moisture.

Wind pushes aside
surface water; cold
water rises from
ocean depths.

Prevailing offshore wind
does not bring moist air
over the land; instead, dry
air moves toward the coast.

Coastal
fog

Cold
water

(c) Coastal deserts

Descending cold, dry air

Air warms by compression,
but heat mostly transfers to
cold surface air and ground.

Cold, high-latitude
ground surface

Cold, dry wind

(d) Polar deserts

of the Hawaiian Islands because the high volcanic mountains intercept the
moist tropical air delivered by prevailing easterly winds.

The great deserts of central Asia are also separated from ocean moisture by
high mountains. The problem is exaggerated here compared to North America
because of sheer distances: the interior of the huge Asian continent is so far
from oceans. The farther that moist air masses move inland from oceans, the
greater the likelihood that moisture condenses and precipitates as rain or snow,
making sure continental interiors remain very dry.

Coastal Deserts

Although the atmosphere obtains most of its moisture by evaporation over oceans, some deserts exist in coastal regions. There is a huge coastal desert in South America (Figure 20.6). The world record for the longest period without rainfall is 14 years and 4 months at a coastal village in northern Chile. Other coastal deserts include Baja California, the northwestern Sahara, the Namib Desert in southern Africa, and the northwest coast of Australia. The prevailing wind direction is from the east at each of these locations on the west edges of continents, so atmospheric circulation carries dry continental air offshore rather than bringing moist oceanic air onshore (Figure 20.7c). The offshore-directed wind also blows relatively warm surface water away from shore, so that colder water from deep in the ocean moves to the surface. The cold ocean water has a refrigerating effect on the surface air, and even when sea breezes move onshore, the air is cold and dry.

Polar Deserts

Cold deserts above 60 degrees latitude provide contrasts to the stereotypical view of scorching hot deserts. The polar regions of North America, Greenland, Siberia, and Antarctica are as dry as the Sahara and, in some places, rarely experience air temperatures above the freezing point of water. Cold air cannot hold significant moisture, so these high-latitude, high-pressure regions are dominated by the downward flow of cold, dry air (Figure 20.7d).

Putting It Together–*What Determines the Locations of Deserts?*

- Deserts form any place where atmospheric circulation deprives a region of moisture.
- Deserts form where dry air descends in high-pressure belts, where mountains intercept moisture carried landward from oceans, where wind predominantly blows offshore, and where air is extremely cold.

20.4 How Does Wind Pick Up and Transport Sediment?

You can use what you have learned about stream and glacier erosion as a starting point for understanding how wind moves sediment (look back to Sections 16.3 and 18.4 for information about stream and glacier erosion). There are both similarities and differences in how wind picks up sediment compared to erosion by flowing water and glaciers. Erosion happens because a moving mass of water, ice, or air exerts sufficient force to cause particles to move along with the current. Shear stress is the force that moves the particles. The shear stress is greatest where the slope is steeper and the thickness (and thus the weight) of ice or water is greater.

Wind also exerts shear stress, but it is not very useful to think of slope and weight as part of the process. After all, air does not weigh very much, and it is difficult to calculate what thickness of atmosphere is moving with the wind. Gravity is the force that causes movement of water, ice, and air. For water and ice, gravity pulls the moving mass down slope, so that the force is always exerted downward across the landscape. The situation is different for air. Gravity pulls denser air against Earth's surface, where it then moves toward areas of lower pressure. The direction the wind blows can be uphill or downhill, or even along flat surfaces. Geologists therefore can best estimate the shear stress of wind by measuring wind velocity at different heights above the ground surface. Wind velocity at the sediment surface is highest where the surface is smooth and there is no vegetation.

▲ **Figure 20.8 Lifting the particles off the ground.**
Fine dust is easily transported by light breezes but is not as easily picked up from the ground because these fine particles tend to be cohesive. In this photo, the truck tires rather than wind dislodged the fine particles that are then transported by the breeze. Dust is not blowing elsewhere in this view, which means that the wind is capable of transporting the dust but cannot erode the fine particles.

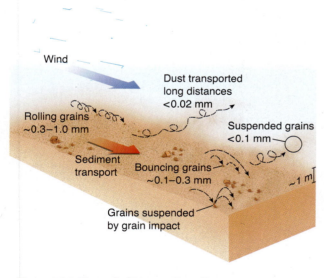

▲ **Figure 20.9 How wind transports sediment.**
Wind-transported sediment grains move by rolling, bouncing along the ground, or by being carried in suspension in whirling air currents. Bouncing grains commonly set more particles in motion when they impact the ground surface. Wind only moves sand and finer-grained sediment. The finest dust particles may travel thousands of kilometers.

How Wind Moves Sediment: *See how blowing wind moves sediment.*

Getting Particles in Motion

Fast winds are needed not only to move large sediment grains but also to move very small grains. This observation, partly intuitive and partly confusing at first glance, is based on the same principle that applies to how flowing water picks up sediment (look back to Section 16.3 for more information). Large grains weigh more than small grains, so it makes sense that large grains only move in the strongest winds. Wind is only rarely capable of moving grains larger than 0.5 millimeter across, which is within the defined range of sand (0.063–2.0 mm). Surprisingly, even smaller grains do not move more easily than sand. Measurements show that sediment grains progressively smaller than about 0.1 millimeter also need increasingly faster winds to move. Small sediment grains, especially silt and clay particles composed of clay minerals, are very cohesive because of electrostatic attractions between the small particles. In order for wind to pick up these small particles, the wind has to exert a shear stress that exceeds both the weight of the particles and the cohesion that holds them together.

Observations in the field and in wind-tunnel experiments show that many particles transported by wind are not directly picked up by wind but are instead knocked loose from the surface by the impact of other particles that are moving. Grains about 0.1 millimeter across are the easiest for wind to pick up, and their movement requires a surface wind speed of only about 5 kilometers per hour, which is equivalent to average walking speed. When these small grains bounce along the surface with the blowing wind, they knock loose larger or more cohesive grains that were motionless. Wind-tunnel experiments show that each time a bouncing sand grain hits the surface, it kicks up about 10 more sediment grains. Once propelled into the air, many of these grains are sufficiently small for the wind to carry. Larger grains fall back to the surface, where they kick loose some more grains while tiny grains drift off as dust. The end result is that most of the grains carried by wind are dislodged from the surface by impacts of a smaller proportion of bouncing wind-blown grains, rather than initially picked up by wind blowing over the ground surface.

Figure 20.8 illustrates a familiar example of how wind-induced shear stress, alone, is not enough to pick up very small sediment particles. The billowing cloud of dust that frequently follows a vehicle on a dirt road or a tractor in a field demonstrates that wind velocity is frequently sufficient to carry the small particles. The only place that these particles move, however, is where the additional shear stress exerted on the surface by rotating tires or moving farm implements is adequate to overcome the cohesive forces of the particles to resist motion.

Observations also show that dry particles are easier to erode than moist particles. Water in pore spaces exerts a cohesive force on adjacent sediment particles that resists wind erosion. This is one reason why wind erosion is most prevalent at times when ground surfaces are dry.

Transporting Particles

Whether picked up directly by wind or knocked loose by bouncing grains, the transport of sediment particles is similar to movement in flowing water. Grains move by rolling or bouncing along the ground or are completely suspended in the moving air, as shown in **Figure 20.9**. Again, field observations and wind-tunnel experiments show that small sand grains between 0.1 and 0.3 millimeter across typically bounce along the surface. Larger grains tend to roll along the ground.

What happens to particles smaller than 0.1 mm? These tiny particles are suspended in the wind. **Dust** is the term commonly applied to these suspended particles. Some dust particles settle to the surface when the wind is calm, but grains less than 0.02 millimeter across are so small that they remain suspended in even the most imperceptible breeze and may travel thousands of kilometers. Explosive volcanic eruptions demonstrate the efficiency of wind transport of tiny products, as illustrated in **Figure 20.10**. Some volcanic eruptions eject volcanic

Astronauts in the International Space Station took this picture of ash and gases erupting from Mt. Etna, on the Italian island of Sicily, in 2002. Ash from this eruption fell 560 km away in northern Africa.

Satellite instruments tracked the global spread of wind-blown condensed-gas droplets following the June 1991 eruption of Mt. Pinatubo, in the Philippines. Most of the gas droplets were sulfur dioxide, which absorbs solar heat and caused approximately 2-degree Celsius drop in average global temperature. The sulfur dioxide also mixed with water vapor to form sulfuric acid, which etched airplane windows.

Increasing condensed-gas droplets

◄ **Figure 20.10** Carrying fine volcanic particles long distances.

Before eruption

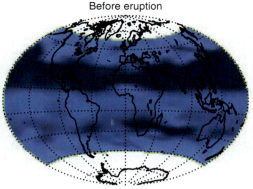

June 15 - July 25, 1991

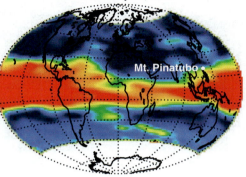

August 23 - September 30, 1991

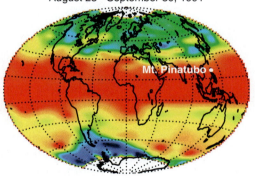

ash and gases more than 20 kilometers into the atmosphere. The gases condense into tiny liquid droplets that, along with the smallest ash particles, may travel around the world many times and remain aloft for more than a decade.

Bouncing sand grains form moving ripples and dunes (see Figure 20.1). Sand grains moved by the wind place additional grains in motion each time they bounce on the surface. All of these grains move along the surface in the direction that the wind blows. The ripple ridges, therefore, are similar to small mounds of dirt swept across the floor in front of a broom. The taller sand dunes build with variations in surface wind velocity that cause alternating areas of sediment erosion and deposition. Bouncing sand grains may come to rest against an obstacle, such as a plant or cobble, and in surface depressions sheltered from the wind. Additional deposition produces a low sand pile that becomes the nucleus of a growing dune.

Figure 20.11 shows how sand dunes move. Most sand dunes have unequal slopes. The downwind side of the dune is steeper than the upwind side. Wind erodes sand on the upwind side that faces into the current. The wind carries a sand grain to the top of the dune, where it then rolls, bounces, or settles onto the downwind side that is sheltered from the wind by the dune crest. The dune slowly migrates in the direction of the prevailing wind as erosion occurs on the

▶ **Figure 20.11** **How sand dunes migrate.**

Wind

Sand in motion

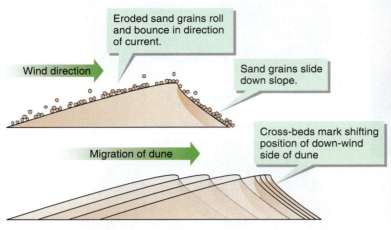

Eroded sand grains roll and bounce in direction of current.

Wind direction

Sand grains slide down slope.

Migration of dune

Cross-beds mark shifting position of down-wind side of dune

Ripples on surface of dune

Cross-bedding exposed by excavation of dune

How Sand Dunes Move: *See how sand dunes migrate over time to form cross-bedding.*

upwind side and deposition occurs on the downwind side. When you examine cross-beds in excavations, you can see the former positions of the steep, downwind side of the migrating dune (see Figure 20.11; also look back at Figure 5.16).

Putting It Together—How Does Wind Pick Up and Transport Sediment?

■ Wind erodes particles where the wind velocity exerts sufficient stress to exceed the weight and cohesion of grains, which resist movement. Cohesion results either from electrostatic attraction between clay and silt particles or from moisture.

■ Grains moved by the wind dislodge stationary particles and place them in motion, too.

■ Wind transports grains by rolling and bouncing along the ground and as suspended dust.

■ Bouncing sand grains form ripples and dunes. Dunes migrate by simultaneous erosion (on the upwind side of the dune) and deposition (on the steeper downwind side). Cross-beds show the changing position of the downwind side of the migrating dune.

20.5 How Does Wind Shape the Landscape?

Wind shapes the landscape through both erosional and depositional processes. Sand dunes are the most familiar and probably the most obvious features of wind-modified landscapes. Dunes are depositional features, and the sand has to first erode from somewhere. There are many landscape features that reveal the importance of wind erosion.

Landscapes Eroded by Wind

Observations show that wind erodes landscapes by two different processes:

1. **Deflation** (from Latin meaning "to blow away") is the process through which landscape elevations are lowered as wind removes the finer particles. Wind deflates by picking up loose particles and carrying them away.
2. **Abrasion** by wind-blown particles erodes exposed rock and regolith surfaces. This sandblasting effect reshapes rock outcrops and slowly wears away sediment grains that are too large for wind to pick up.

Figure 20.12 illustrates the features the geologists relate to deflation and abrasion. Some of these same distinctive features appear in images sent to Earth by NASA's Mars rovers, such as seen in **Figure 20.13**, which reveals the importance of wind to shape landscapes on other planets.

Deflation commonly scours out elliptical or circular areas of easily eroded sediment to leave distinctive depressions called **pans** (Figure 20.12). A low dune is typically present along the downwind side of the pan, showing that most of the deflated sediment only travels a short distance before it is deposited. In regions of fluctuating water table, pans form during dry periods and then partly fill with infiltrated ground water when the water table rises to the bottom of the pan. Some pan depressions also accumulate surface runoff. Evaporation of water deposits a crust of evaporite minerals, such as gypsum and halite, to produce a playa on the floor of the pan (go back to Section 16.13 for more information on playas).

Ventifacts are loose rocks that show evidence of abrasion by wind-blown sand (see Figures 20.12 and 20.13). The term derives from Latin roots that mean "made by the wind." Persistent sandblasting abrades smooth planar surfaces in rocks that are shaped like polished facets on a gemstone. Spots of deeper pitting form where the rock is weaker because of fractures or the presence of softer minerals. The facets commonly meet along sharp edges so that the rock resembles the outline of a brazil nut. The multiple abraded facets result from varying wind directions and shifting of the rock over time.

▼ **Figure 20.12** Features resulting from wind deflation and abrasion.

Wind erosion
Deflation
Abrasion
Pans
Yardangs
Ventifacts

Deflation pans central New Mexico

Wind erodes fine sediment and deposits it to form a dune. Deflated area, called a pan, may temporarily fill with surface runoff or ground-water seepage. Evaporite minerals crystallize when water evaporates.

Yardangs, Egypt

Blowing sand abrades soft rock or regolith and wind deflates the eroded particles. Yardangs are abraded bedrock remnants within the deflated landscape, and are elongated in the direction of the prevailing wind with pedastal-and-cap shapes resulting from greatest erosion closest to the ground.

Ventifacts, Western United States

Blowing sand abrades cobbles and boulders on the desert surface resulting in faceted rocks called ventifacts.

This football-size rock has the faceted faces and sharp edges characteristic of ventifacts.

Low sand dunes, less than 1 m tall, sweep across the floor of Endurance Crater.

◄ Figure 20.13 Recognizing wind action on Mars.
Photos returned by NASA's Mars Exploration Rovers *Spirit* and *Opportunity* in 2004 clearly show evidence of wind processes on the martian surface. Ventifacts and sand dunes on Mars are identical to features on Earth.

Deflation and abrasion work together to produce a distinctive landform called a **yardang**, which is a wind-parallel ridge of soft rock or slightly consolidated sediment that remains after surrounding material is eroded (see Figure 20.12). The term originated in Turkey, but this landform is recognized on all continents. Most yardangs are less than 5 meters high and typically no more than 10 meters long and are very steep sided. It is typical for the upwind end of the yardang to overhang a notch, because erosion is more intense close to the surface where blowing sand grains bombard the outcrop most often. Larger yardangs are more than 100 meters high, several kilometers long, and are scattered across several hundred thousand square kilometers of the great deserts of North Africa and central Asia. The material forming the yardang must be strong enough to stand up in steep ridges yet soft enough for easy abrasion by blowing sand. Weakly cemented fine-grained sedimentary rocks or cohesive mud typically provide this combination. Abrasion dislodges small pieces of the erodible outcrop, and deflation removes the pieces.

Can wind erosion substantially lower surface elevations? This process is not as clearly evident as are the erosive effects of flowing water and glaciers. From a theoretical standpoint, it seems reasonable that wind should be able to erode deeply as long as the surface materials are loose and sufficiently fine-grained for deflation. Deflation cannot continue below the water table because wind will not deflate moist sand. In desert regions, however, soft, loose, unsaturated sediment may be hundreds of meters thick. Is it possible for wind to lower landscapes by hundreds of meters? The presence of yardangs as high as 200 meters suggests that wind may erode deeply, but it is also unclear if all this erosion relates to ongoing desert sand blasting or whether it may have partly occurred by an earlier episode of stream erosion. Some geologists hypothesize that nearly 20,000 square kilometers in the Sahara Desert have depressions with elevations below sea level formed by deflation. It is difficult to prove, however, that streams or tectonic processes did not also play roles in making these large depressions. Although features such as pans, ventifacts, and yardangs result from wind erosion, the scale at which wind lowers surface elevations remains unclear.

The Variety of Sand Dunes

Actively moving wind-blown sand forms distinctive landforms of deserts and coastlines. In some places, as shown in **Figure 20.14**a, the sand simply accumu-

▼ Figure 20.14 Sand deposits at different scales.
(a) Some sand dunes are very small, such as this low dune of sand that accumulated around the base of a desert shrub.
(b) Other dunes can be hundreds of meters high and cover tens of thousands of square kilometers, as seen in this view of the Arabian Desert in Saudi Arabia.

(a)

(b)

lates in low sheets or forms piles at the base of shrubs that block the wind and cause deposition. Larger sand dunes are the most prominent and well-known deposits resulting from wind transport of sediment. Dunes in coastal regions are conspicuous for their towering height above adjacent beaches (Figure 20.1b), but they rarely cover large areas. Many deserts, on the other hand, feature vast areas of sand dunes, like the expanse pictured in Figure 20.14b, where the seemingly endless, wavelike undulations of the dune crests and troughs suggest waves on the ocean. Sand dunes are a part of everyone's first impressions of what a desert should look like, but dunes actually cover less than 20 percent of the arid-zone surface of Earth and only 2 percent of desert landscapes in the United States.

A survey of dunes would show that they appear in a variety of shapes and sizes. **Figure 20.15** illustrates and summarizes the characteristics of five common

▼ **Figure 20.15** **Visualizing types of sand dunes.**

Parabolic dunes form where wind erosion attacks a barren area in a largely vegetated landscape, commonly near a beach. A depression forms where erosion is most intense and the sand is trapped by vegetation to accumulate as a dune. The U-shape resembles the graph of a parabola with the depression in the center and the two ends of the parabola pointing upwind.

Barchan dunes are crescent shaped dunes that form on desert surfaces where gravel or rock are more abundant than wind-transported sand. The word derives from an ancient Turkish word describing sandy hills. Barchans resemble parabolic dunes except that the ends of the crescent barchans point downwind.

Transverse dunes form where sand supply is abundant and dunes continuously carpet large areas. The term "transverse" emphasizes that the curvy dune crests are mostly transverse (perpendicular) to the prevailing wind direction. These particular dunes consist of gypsum sand grains at White Sands National Monument, New Mexico.

Linear dunes are long sand ridges oriented parallel to the prevailing wind direction. Linear dunes typically form where sand supply is more limited than where transverse dunes form and where wind direction is more variable than regions characterized by barchans.

Star dunes are radiating sand ridges resulting from highly variable wind directions crossing large areas of readily eroded sand.

▶ **Figure 20.16 The factors that determine dune shape.** Availability of sand and uniformity of wind direction determine the shapes of sand dunes.

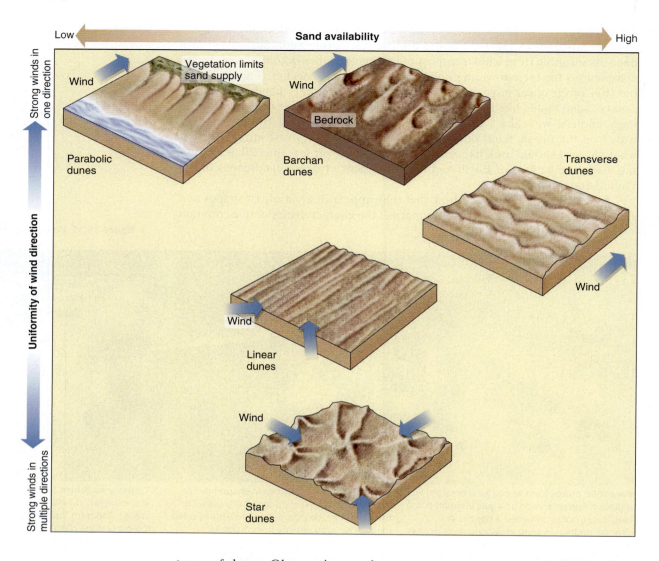

types of dunes. Observations and measurements summarized in **Figure 20.16** show that the abundance of sand for wind to move and the uniformity of wind direction are the primary factors that determine dune shape.

Blankets of Loess

What happens to the fine particles that remain suspended in the wind while sand accumulates and moves slowly in dunes? It turns out that wind-blown silt covers nearly as much of Earth's land surface as is covered by sand dunes. Although these silt deposits are tens to hundreds of meters thick, the silt does not form impressive landforms comparable to sand dunes. Instead, the silt particles accumulate gradually, mostly in vegetated landscapes where wind velocity is slow enough for small particles to settle to the ground. Surface elevations gradually rise because of sediment deposition, but the shape of the landscape is preserved as if covered by fallen snow. Geologists use a German term, **loess**, to describe on-land deposition of wind-borne sediment on land that is mostly silt size (also see Section 18.9 for more information about loess).

Figure 20.17 outlines areas on Earth where loess forms continuous blankets that are many meters thick. The loess deposits in central Asia and South America are thicker and contain more mixed-in sand in close proximity to adjacent deserts. This geographic relationship of loess and deserts confirms that silt blows out of deserts and accumulates on adjacent landscapes with greater vegetation cover and lower wind velocities. Thick loess deposits in the central and northwestern United States and in Europe tend to be thicker and sandier close

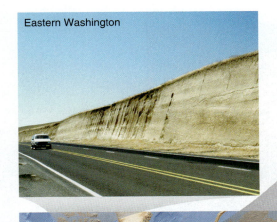

Eastern Washington

Western Iowa

China

◄ **Figure 20.17 Visualizing where loess is found.**
This map shows the global distribution of thick deposits of wind-blown silt, called loess. Thin, discontinuous loess deposits occur in other locations not shown here. The photographs show typical loess outcrops. The cohesive silt particles form deposits that erode and excavate to leave steep slopes.

to river valleys (look back at Figure 18.39, for example). Most of this loess accumulated during and shortly after the last ice age (peaking about 21,000 years ago) when cold dry winds blew off the glaciers and across river floodplains that were covered with large volumes of sediment carried by meltwater streams. During ice-age climatic conditions the river valleys were less vegetated and experienced stronger winds than are recorded today, so wind processes were more influential then than now.

Loess forms rich soil. Most surface loess accumulated within the last 20,000 years and is not extensively weathered, so mineral-nutrient content is much higher than for regions where soil formation has been ongoing for much longer. The fine grain size of the silt is ideal for retaining infiltrating ground water and allows crops to grow with little or no irrigation, even where climatic conditions border on arid. Farmers must exercise care when cultivating loess soils, however, because runoff easily erodes the silt grains where the loess is not covered with vegetation, which causes high soil-erosion rates (see Figure 20.17).

Desert Pavements: Formed by Sediment Deflation or Accumulation?

Closely spaced pieces of gravel cover barren and rocky desert floors to form a smooth surface called **desert pavement**. **Figure 20.18** illustrates typical characteristics of desert pavements, including that the pavement is usually only one or two stones thick and overlies silt. Desert pavements are an important control on wind erosion because pavement particles are too large for wind to erode and form a surface armor. On the other hand, where stream erosion or human activities remove parts of the thin gravel surface layer, wind can easily erode the exposed silt. How is desert pavement formed?

The role of wind to produce desert pavement is controversial. **Figure 20.19**a illustrates a deflation hypothesis for the origin of desert pavement. This hypothesis states that wind deflates sand and dust particles within poorly sorted surface deposits that also include gravel. The abundance of gravel fragments

Desert pavement

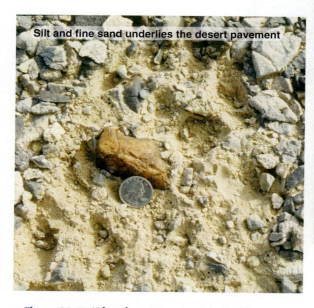

Silt and fine sand underlies the desert pavement

Pebble and cobble desert pavement on surface

Soil beneath desert pavement contains almost no pebbles or cobbles

▲ **Figure 20.18 What desert pavement looks like.**
Low-relief surfaces in deserts are commonly covered with closely spaced pebbles and cobbles that resemble a cobblestone pavement. The top-left photo shows a pavement of limestone pebbles in Nevada and the photo below it shows that the gravel is a layer that is only one-pebble thick, and rests on silt and fine sand. The arrows point to the same pebble in each photo. A deeper excavation below a pavement reveals desert soil horizons that contain almost no gravel fragments.

on the surface increases through time as finer particles blow away. Eventually the surface is completely covered with coarse particles that the wind is unable to move. The hypothesis is consistent with worldwide occurrence of pavements in deserts where wind is clearly an important geologic process. This hypothesis does not, however, readily explain two features:

1. Silty layers, in some cases tens of centimeters thick, immediately underlie the pavement armor but contain few if any gravel fragments (Figure 20.18). Deflation of the fine-grained sediment will not leave behind a layer of gravel.
2. Some geologic data reveal that all of the pavement clasts have been at the surface for the same period of time, whereas deflation requires exposure of clasts over an extended period of time as enclosing fine particles gradually blow away.

Figure 20.19b depicts an alternative hypothesis for how sediment might accumulate to form desert pavement. This hypothesis simultaneously addresses the two weak points in the deflation hypothesis. Clasts projecting above the surface of an initially coarse deposit trap wind-blown dust particles. The dust sifts into cracks between surface clasts, probably aided by infiltrating rainwater. The dust includes silt- and clay-size particles that swell and shrink during wetting and drying events. Wetting also dissolves soluble minerals in the dust, which then precipitate when the sediment dries. The shrinking and swelling, along with mineral dissolution and precipitation, shift the surface clasts around and open cracks in the accumulating fine-grained layer, and this action allows newly arriving dust particles to accumulate as well. Slow, gradual accumulation of dust progressively lifts the surface gravel fragments higher and higher above where they started. This hypothesis is more complicated than the simple deflation idea, but it is more consistent with observed data. The more likely sediment-accumulation hypothesis also reveals that development of the landscape was due to wind deposition rather than to wind erosion.

Wind-Formed Landscapes of the Recent Past

Observation of active wind erosion and deposition in deserts and along shorelines provides insights for interpreting landforms in other regions. In particular, geologists have long recognized rolling vegetated hills of sand that resemble sand dunes except for the fact that they are covered in vegetation and do not move. The conclusion, then, is that there are many places on Earth where wind shaped the landscape in the recent past when climatic conditions were more favorable than at present for producing sand dunes. Rolling hills across parts of the southeastern and central United States are inactive sand dunes where relatively recent vegetation growth has stopped sand movement and stabilized the dunes.

Inactive sand-dune deposits are common in the High Plains region. Rivers carry abundant sand eastward from the Rocky Mountains onto the semiarid, almost treeless plains where wind blows the sand out of the river valleys and across the countryside. Present climatic conditions are just sufficiently moist to support adequate grass cover to reduce significant wind erosion and transport of sand. Geologic data show that a drier climate in the recent past reduced the grass cover so that sand dunes formed.

The largest sea of dunes in this region is the Sand Hills of northwestern Nebraska, illustrated in **Figure 20.20**. The sandy hills are actually grass-covered sand dunes that cover 50,000 square kilometers and are as high as 80 meters (roughly equivalent to a 20-story building). Geologists hypothesize that these dunes formed during a past ice age when sediment supply in nearby rivers was greater than today and cold, dry winds swept southward from nearby glacial ice sheets and eastward from the cold glacial valleys in the Rocky Mountains. Careful studies show, however, that regardless of when the dunes first formed, they have been active much more recently than the demise of the ice-age glaciers. Measured wind velocities in the Sand Hills are sufficiently strong to move sand about half of the year. The prairie grasses covering the old dunes are the only factor that keeps the dunes from moving. Long periods of drought, including one episode only 200–300 years ago, reduce the distribution of grasses so that the dunes reactivate as moving features on the landscape.

(a) Forming desert pavement by wind erosion

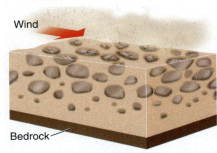

Mixture of gravel, sand, and silt.

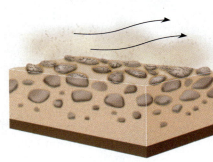

Wind blows away sand, silt, and dust, which leaves gravel behind.

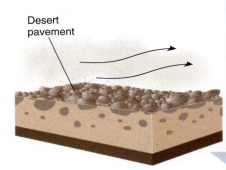

Gravel completely covers the surface producing a pavement that diminishes further wind erosion.

(b) Forming desert pavement by wind deposition

Dust accumulates on surface, especially where trapped against pebbles and cobbles.

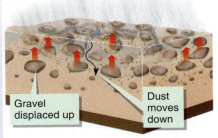

Rain washes the fine dust downward. Soluble minerals dissolve and then precipitate below the surface when the infiltrating rainwater evaporates. Accumulation of dust, shrinking and swelling of clay minerals, and precipitation of minerals lift the gravel fragments to the surface.

Continuous gravel pavement eventually accumulates at the surface above a thickening layer of finer sediment.

▲ **Figure 20.19 How to form desert pavement.**
Geologists debate the origin of desert pavement. (a) This popular hypothesis attributes pavement formation to erosion. Deflation removes fine particles and concentrates the gravel particles that are too large for wind to transport. (b) This more recent hypothesis attributes pavement formation to deposition. Deposited dust infiltrates below surface gravel and, along with shrinking and swelling and dissolution and precipitation caused by wetting and drying, the larger gravel particles are displaced to the surface. This hypothesis is consistent with observed characteristics of most desert pavements, which overlie thick deposits of fine-grained sediment without much gravel.

▶ **Figure 20.20 Sand dunes of the past.**
The photo taken by astronauts from the International Space Station shows the rolling Sand Hills of northwestern Nebraska that resemble sand dunes except that the hills are vegetated, as seen in the closer view. The Sand Hills are ancient sand dunes that are stable vegetated landforms under current climatic conditions. Wind moves the sand dunes when climate is drier and the vegetation diminishes.

Putting It Together—*How Does Wind Shape the Landscape?*

- Wind erosion, called deflation, lowers landscape elevations where fine-grained and loose sediment is abundant at the surface. Pans are low areas formed by deep deflation, and yardangs are sandblasted and streamlined ridges.

- Dunes are large sand piles moved by wind, with steep slopes that face toward the direction of movement. The variety of sand dune shapes relates to the abundance of sand, the abundance of vegetation, and consistency of wind direction.

- Deposits of wind-blown silt, called loess, blanket large areas and provide fertile soil.

- Desert pavements look like they are a covering of gravel fragments left behind by deflation of smaller particles. It is more likely, however, that these pavements form where dust accumulates beneath gravel particles that have always been at the surface.

- Some rolling-hill landscapes are ancient "seas" of sand dunes. Although the dunes are now overgrown by stabilizing vegetation, the dunes actively moved in the past when vegetation was sparse or absent because of drier climate.

20.6 How Do We Know ... That Wind Blows Dust Across Oceans?

PICTURE THE PROBLEM

What Causes the Haziest Summer Days in the Southeastern United States?
Drag your finger across any outdoor surface and you discover a layer of dust. It also finds its way indoors—dusting is a perpetual housekeeping chore. Dust is more than a simple nuisance for people who suffer from respiratory ailments such as asthma or are afflicted by dusty allergens. Dust is, therefore, monitored as a measure of air quality in the United States and other countries. Dust is also routinely monitored as part of an effort to maintain clear skies with long-distance visibility in environmentally sensitive areas, such as many national parks and wildlife refuges. Some of the tiny particles filtered from air-quality samples are airborne pollution from smokestacks and vehicles, but many are tiny mineral fragments. You have seen dust everywhere, but have you ever wondered where it comes from?

Air-quality data collected across the United States repeatedly show that the Southeast is the dustiest region during summer months. Periods of highest dust concentrations, called dust episodes, coincide with or shortly follow even more severe dusty haze conditions in the Caribbean islands south of Florida. This pattern contrasts with data collected during the spring, which shows higher dust concentrations in the western part of the country. Dust originating from the dry semiarid and arid landscapes in the West is logical, but where does dust come from to obscure summer skies in the humid Southeast?

DEVELOP A HYPOTHESIS

Does the Dust Come from Africa?

Wind transports dust. Mineral dust erodes from landscapes where wind erosion is effective, such as deserts. Consideration of wind directions and desert locations may reveal a possible source for summer dust in the Southeast.

Figures 20.3d and 20.3e show that the southeastern United States is located within a zone of slightly shifting subtropical high pressure. Most winter wind blows from the northeast, but the summer flow is from the south, as easterly winds move away from high pressure over the Atlantic Ocean and then northward toward lower pressure above the warming North America continent. This influx of warm moist air from the Caribbean keeps the southeastern United States humid, despite the fact that the region is located at the same latitude as deserts in the western United States and Africa. The easterly summer winds from Africa might carry dust from the Sahara westward into the Caribbean and then northward into the eastern United States. If this hypothesis is true, then some air-quality concerns in the Southeast relate to processes of wind erosion on another continent.

TEST THE HYPOTHESIS

How Is Dust Tracked to Its Source?

The satellite image in **Figure 20.21** clearly shows dust blowing across the Atlantic Ocean from northwestern Africa. Hazy-sky conditions in the Caribbean coincide with these dusty outbursts from the Sahara. Satellite images, therefore, provide one way of directly tracking Saharan dust by simple observation. This

Clear day in the Virgin Islands

Hazy day in the Virgin Islands

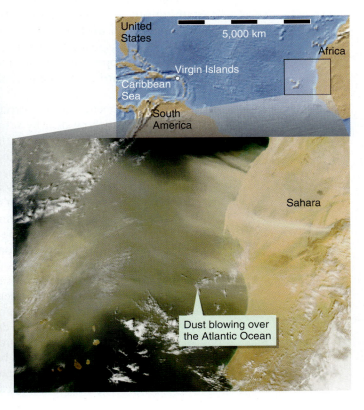
Dust blowing over the Atlantic Ocean

◀ **Figure 20.21 Dust out of Africa.**
The satellite image on the right shows Sahara Desert dust blowing eastward over the Atlantic Ocean. During these dust events, hazy skies develop in the Caribbean Sea, which is more than 5000 km downwind from Africa.

▶ **Figure 20.22** **Measuring dust concentrations in the eastern United States.**
Measurements at specially designed sampling stations track the northward movement of wind-borne mineral dust during a summer dust episode in the eastern United States.

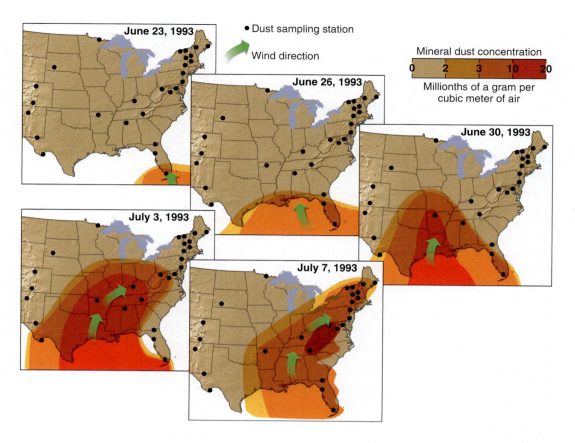

technique has limitations however, because dust concentration decreases during transport as grains slowly settle to the ocean or onto islands. When the dust is more dispersed or dilute, it is not always clear on the satellite images that the dust moves over the southeastern United States.

Figure 20.22 shows measured dust concentrations in the eastern United States during a two-week period. The maps show the airborne concentration of extremely small dust particles, less than 25/10,000ths of a millimeter across, which remain suspended in moving air for long time periods and travel long distances. The mapped changes in dust concentration clearly show that the dust moved northward at the same time that a Saharan dust episode affected the Caribbean.

Scientific conclusions are strongest when corroborated by different methods and data. Tracking dust by satellite images and measurement of dust concentration support the hypothesis that the haziest summer days in the Southeast result from wind-blown dust originating in Africa. If the dust could be "fingerprinted" to be from the Sahara, then there would be even stronger evidence to support the hypothesis. Air-quality filters in the Southeast show summer dust is reddish, because of abundant iron oxide minerals, whereas spring dust is gray. The color difference suggests a composition difference that may help track different dust sources during the two seasons.

Figure 20.23 highlights seasonal differences in dust concentration and chemical composition. Spring dust clearly originates in the dry western United States, where spring dust concentrations are highest. The dust moves with the prevailing westerly wind and affects much of the country. The composition of the fine dust particles is similar everywhere in the country. When a summer dust episode affects the Southeast, the dust composition changes in areas where the dust concentration is highest. Summer dust in the Southeast has more aluminum and less calcium than dust blowing out of the western deserts. High aluminum content in the dust reflects high clay-mineral content, whereas high calcium levels indicate abundant calcite or gypsum in the dust. Dust from the

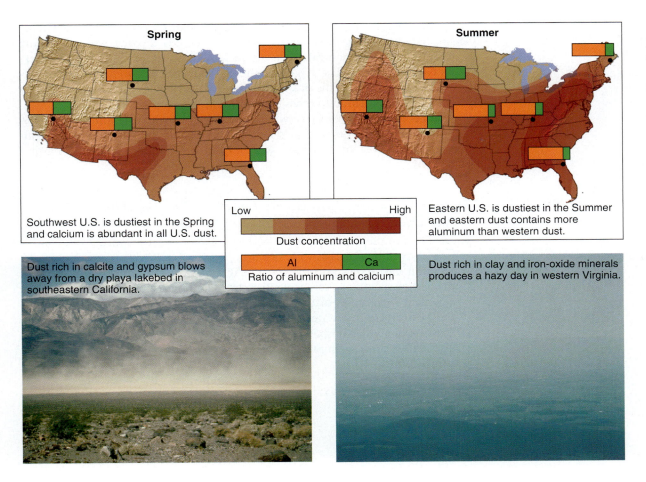

◄ **Figure 20.23** Seasonal
**differences in dust concentration
and composition.**
There are seasonal differences in
dust composition and chemical
composition across the United
States. Spring dust concentrations
are high in the southwestern United
States and summer dust concentrations are highest in the east. Comparison of aluminum and calcium
contents shows that dust from the
southwest, in both spring and summer, is higher in calcium than is the
summer dust in the southeast.
Calcium-rich dust blows off the surface of dry playa lakes in the western United States. Summer dust in
the Southeast is rich in aluminum-
rich clay minerals from Africa.

western United States deserts contains calcite and gypsum deflated from dried-
up playa lakes, as shown in Figure 20.23. Summer dust in the Southeast lacks
this calcium-rich "fingerprint" of western United States dust. Instead, the abundant clay, revealed by the very high aluminum content, indicates a desert source
that lacks numerous playa lakes. Lakes are extremely rare in the Sahara, so Saharan dust is high in aluminum and low in calcium. Saharan dust is also iron
rich, which accounts for the red color of summer dust.

INSIGHTS

What Is the Fallout of Dust from Distant Places?
By using the compositional fingerprint as a guide, scientists estimate that Saharan
dust episodes occur in the Southeast about three times each summer and, on average, last for about ten days. About 30 percent of the continental United States
is hazy with African dust during the typical dust episode. Perhaps as much as
one billion metric tons of African dust crosses the Atlantic each year.

The dust episodes not only further emphasize the importance of wind as a
geologic process, but they also explain some biologic phenomena. "Red tides" in
coastal marine waters commonly coincide with dust episodes that provide mineral nutrients for tiny plankton. Unusually large concentrations of tiny plankton
cause the red discoloration of the seawater. This plankton poisons fish and shell-
fish and causes illness in humans who eat the poisoned seafood. Red tides,
therefore, cause closure of fisheries and recreational beaches. Fortunately, the
red plankton are usually not abundant because of a lack of nutritional iron in
the seawater. However, Saharan dust delivers iron during dust episodes so that
the plankton temporarily flourish. In addition to tiny mineral grains, the African
dust includes microbes that are linked to widespread disease and death of coral
throughout the Caribbean.

Asian desert dust also enters the United States from across the Pacific Ocean. Satellite images also show large dust clouds moving with prevailing westerly winds from central Asia, across the Pacific Ocean, and completely across North America. The largest of these Asian episodes so far recorded was in April 2001; it doubled the dust in the North American sky during the six days it took for the dust cloud to cross the continent.

The geologic importance of wind seems obvious when standing among the dunes of Death Valley or Jockey's Ridge. Now you know that wind is a globally important mover of sediment, with impact on human health and natural ecosystems.

*Putting It Together–***How Do We Know** ... *That Wind Blows Dust Across Oceans?*

- Deserts are the primary source of dust, which wind transports from continent to continent across wide oceans.
- Saharan dust blows into the Caribbean and eastern United States during the summer. Atmospheric scientists track the dust clouds by satellites. A chemical fingerprint distinguishes Saharan dust from dust originating in the western United States deserts.
- Dust not only diminishes visibility and inflames respiratory ailments but also causes changes in marine ecology, such as red tides and the demise of coral reefs.

Where Are You and Where Are You Going?

Wind joins with glacial ice, water flowing in streams, and water moved by waves and tides, as an important geologic process. Like water and ice, wind erodes Earth's surface and deposits the resulting sediment to form distinctive landforms and landscapes. Wind blows everywhere; its direction and velocity are wind-determined by variations in air pressure from place to place. The varying air pressures, in turn, result from large-scale atmospheric circulation driven by a combination of convection and Earth's rotation. These air-pressure and wind patterns also determine the locations of dry regions called deserts.

Wind is most effective at modifying Earth's surface where small, loose grains are abundant on the surface, vegetation is sparse, and wind blows strongly. This combination of conditions is most commonly met in deserts. It can occur in other settings, however, especially coastal areas adjacent to wide, sandy beaches. Rolling hills in some locations are ancient sand dunes that are now stabilized by vegetation but were active during earlier drier times when plant cover was more sparse.

Sand dunes are the most recognizable landform produced by blowing wind, but most deserts contain only local regions of blowing sand. Where sand dunes are present, they reveal a wide variety of shapes determined by the abundance of sand, the sparseness of vegetation that inhibits sand movement, and the consistency of wind directions. Deflated pans, sandblasted yardangs, and abraded and pitted ventifacts reveal evidence of wind erosion, on Mars as well as on Earth. Gravel-size fragments cover large areas of desert landscapes and are too large for wind to move. Pavement-like surfaces of gravel armor over layers of fine silt that gradually accumulate beneath the larger stones.

Wind also transports and deposits fine dust. These small particles move along with the wind and may accumulate far from their original sources. Thick wind-blown silt blankets of loess weather to produce soil that is notable for high nutrient content and excellent water retention. Some loess accumulates downwind of deserts. Other loess links to river valleys, where streams deposit silt that is then picked up and moved farther across the landscape by wind. Loess is especially thick and widespread near rivers that carried large volumes of sediment and were swept by strong, cold, dry winds during and shortly after the last ice age. Even smaller dust particles, only thousandths of a millimeter across, are carried across oceans from the great deserts of North Africa and central Asia. Although springtime dust in the United States comes mostly from windy deserts and dried-up lakebeds, it is sometimes supplemented by dust that crosses the Pacific Ocean from Asia. The dustiest summer days are in the southeastern United States, where dust is delivered by wind blowing westward from the Sahara.

Movements of water, as a liquid or frozen to ice, and wind join with the direct pull of gravity on rock and regolith as the external processes shaping Earth's surface. Internal processes, ultimately linked to heat energy within Earth, cause the surface to bulge upward or sag downward, and these external processes shape the deforming surface into distinctive and scenic landforms.

 **Active Art**

Global Wind Patterns. See the processes that determine global wind patterns.

How Wind Moves Sediment. See how blowing wind moves sediment.

How Sand Dunes Move. See how sand dunes migrate over time to form cross-bedding.

Extension Module

Extension Module 20.1: How the Coriolis Effect Works. Learn more about how Earth's rotation causes moving objects to follow curving, rather than straight-line, paths.

Confirm Your Knowledge

1. What is wind?

2. What conditions are necessary for the formation of sand dunes?

3. Wind exists because of density differences in the atmosphere. What factors determine differences in atmospheric density?

4. How does air pressure affect wind direction?

5. Define Coriolis effect. How does it affect moving objects in the northern and southern hemispheres?

6. What are monsoons? Why do they occur?

7. Local winds are much more variable than predicted global patterns in Figure 20.3c. Explain the causes of these local variations.

8. What is a desert?

9. What are the factors that determine the locations of deserts?

10. Compared with water, what are the limitations wind has as an agent of erosion?

11. Why is it just as difficult for the wind to move dust as it is to move sand?

12. How does the wind move particles?

13. How would you recognize and distinguish between landscapes affected by wind erosion and deposition?

14. What is a desert pavement? How do desert pavements probably form?

15. What evidence supports the hypothesis that most of the dust that contributes to hazy summer skies in the southeastern United States originates in Africa?

Confirm Your Understanding

1. Write an answer for each question in the Chapter Outline for the chapter sections assigned by your instructor.

2. What would you expect to be the prevailing winter wind directions for the following locations?
   a. Auckland, New Zealand, Lat: 36 degrees 55 minutes S
   b. Brasília, Brazil, Lat: 15 degrees 47 minutes S
   c. Cape Town, South Africa, Lat: 33 degrees 56 minutes S
   d. Des Moines, Iowa, Lat: 40 degrees 55 minutes N
   e. Kabul, Afghanistan, Lat: 34 degrees 31 minutes N
   f. Lagos, Nigeria, Lat: 06 degrees 27 minutes N
   g. Paris, France, Lat: 48 degrees 52 minutes N

3. Although Mark Twain never actually said the "coldest winter I ever spent was a summer in San Francisco," he did say that "you can never go without a coat in the summer in the city of San Francisco." Using what you know about the general behavior of local wind patterns, explain why the summer in San Francisco is so cold compared to nearby places in California at the same latitude

4. How can some windy areas show no evidence of wind erosion?

5. If you lived adjacent to an unpaved road that generated a lot of dust in your home, what could you do to diminish the dust?

6. Dust in the southeastern United States is thought to come from Africa. How can this be, when the prevailing wind pattern at this latitude is westerly?

7. Locate your hometown or college town in Figures 20.3d and 20.3e. Explain the factors that determine the prevailing wind direction and seasonal variations in wind direction.

Key Terms

Coriolis effect (p. 619)
deflation (p. 629)
desert pavement (p. 633)

deserts (p. 621)
dust (p. 626)
loess (p. 632)

monsoons (p. 619)
pans (p. 629)
ventifacts (p. 629)

wind (p. 616)
yardangs (p. 630)

Appendix A

Periodic Table of the Elements

Key:
22
Ti
Titanium
47.88
Atomic number — Atomic weight

#	Symbol	Name	Atomic weight
1	H	Hydrogen	1.008
2	He	Helium	4.003
3	Li	Lithium	6.941
4	Be	Beryllium	9.012
5	B	Boron	10.81
6	C	Carbon	12.01
7	N	Nitrogen	14.01
8	O	Oxygen	16.00
9	F	Fluorine	19.00
10	Ne	Neon	20.18
11	Na	Sodium	22.99
12	Mg	Magnesium	24.31
13	Al	Aluminum	26.98
14	Si	Silicon	28.09
15	P	Phosphorus	30.97
16	S	Sulfur	32.07
17	Cl	Chlorine	35.45
18	Ar	Argon	39.95
19	K	Potassium	39.10
20	Ca	Calcium	40.08
21	Sc	Scandium	44.96
22	Ti	Titanium	47.88
23	V	Vanadium	50.94
24	Cr	Chromium	52.00
25	Mn	Manganese	54.94
26	Fe	Iron	55.85
27	Co	Cobalt	58.93
28	Ni	Nickel	58.69
29	Cu	Copper	63.55
30	Zn	Zinc	65.39
31	Ga	Gallium	69.72
32	Ge	Germanium	72.64
33	As	Arsenic	74.92
34	Se	Selenium	78.96
35	Br	Bromine	79.90
36	Kr	Krypton	83.80
37	Rb	Rubidium	85.47
38	Sr	Strontium	87.62
39	Y	Yttrium	88.91
40	Zr	Zirconium	91.22
41	Nb	Niobium	92.91
42	Mo	Molybdenum	95.94
43	Tc	Technetium	(98)
44	Ru	Ruthenium	101.1
45	Rh	Rhodium	102.9
46	Pd	Palladium	106.4
47	Ag	Silver	107.9
48	Cd	Cadmium	112.4
49	In	Indium	114.8
50	Sn	Tin	118.7
51	Sb	Antimony	121.8
52	Te	Tellurium	127.6
53	I	Iodine	126.9
54	Xe	Xenon	131.3
55	Cs	Cesium	132.9
56	Ba	Barium	137.3
57	La	Lanthanum	138.9
72	Hf	Hafnium	178.5
73	Ta	Tantalum	180.9
74	W	Tungsten	183.8
75	Re	Rhenium	186.2
76	Os	Osmium	190.2
77	Ir	Iridium	192.2
78	Pt	Platinum	195.1
79	Au	Gold	197.0
80	Hg	Mercury	200.6
81	Tl	Thallium	204.4
82	Pb	Lead	207.2
83	Bi	Bismuth	209.0
84	Po	Polonium	(209)
85	At	Astatine	(210)
86	Rn	Radon	(222)
87	Fr	Francium	(223)
88	Ra	Radium	(226)
89	Ac	Actinium	(227)
104	Rf	Rutherfordium	(261)
105	Db	Dubnium	(262)
106	Sg	Seaborgium	(266)
107	Bh	Bohrium	(264)
108	Hs	Hassium	(269.1)
109	Mt	Meitnerium	(268)
110	Ds	Darmstadtium	(271)
111	Uuu	Unununium	(272)
112	Uub	Ununbium	(277)
114	UUq	Ununquadium	(289)

#	Symbol	Name	Atomic weight
58	Ce	Cerium	140.1
59	Pr	Praseodymium	140.9
60	Nd	Neodymium	144.2
61	Pm	Promethium	(145)
62	Sm	Samarium	150.4
63	Eu	Europium	152.0
64	Gd	Gadolinium	157.3
65	Tb	Terbium	158.9
66	Dy	Dysprosium	162.5
67	Ho	Holmium	164.9
68	Er	Erbium	167.3
69	Tm	Thulium	168.9
70	Yb	Ytterbium	173.0
71	Lu	Lutetium	175.0
90	Th	Thorium	232.0
91	Pa	Protactinium	231
92	U	Uranium	238.0
93	Np	Neptunium	(237)
94	Pu	Plutonium	(244)
95	Am	Americium	(243)
96	Cm	Curium	(247)
97	Bk	Berkelium	(247)
98	Cf	Californium	(251)
99	Es	Einsteinium	(252)
100	Fm	Fermium	(257)
101	Md	Mendelevium	(258)
102	No	Nobelium	(259)
103	Lr	Lawrencium	(262)

Appendix B

Ions and Natural Isotopes that are Significant to Geologic Studies

Element (symbol)	Negative Ions	Native Element	Positive Ions	Isotopes (most abundant, *radioactive*)
Aluminum (Al)		Al	Al^{3+}	^{27}Al
Argon (Ar)		Ar		^{36}Ar, ^{40}Ar
Arsenic (As)	As^{3-}		As^{3+}	^{75}As
Barium (Ba)			Ba^{2+}	^{138}Ba, ^{137}Ba
Boron (B)			B^{3+}	^{11}B
Calcium (Ca)			Ca^{2+}	^{40}Ca
Carbon (C)	C^{4-}	C	C^{4+}	^{12}C, ^{13}C, ^{14}C
Chlorine (Cl)	Cl^{1-}			^{35}Cl, ^{37}Cl
Chromium (Cr)		Cr	Cr^{2+}, Cr^{3+}	^{52}Cr
Copper (Cu)		Cu	Cu^{1+}, Cu^{2+}	^{63}Cu, ^{65}Cu
Fluorine (F)	F^{1-}			^{19}F
Gold (Au)		Au	Au^{1+}	^{197}Au
Helium (He)		He		
Hydrogen (H)	H^{1-}		H^{1+}	^{1}H, ^{2}H, ^{3}H
Iridium (Ir)		Ir	Ir^{4+}	^{191}Ir, ^{193}Ir
Iron (Fe)		Fe	Fe^{2+}, Fe^{3+}	^{54}Fe, ^{56}Fe, ^{57}Fe
Lead (Pb)			Pb^{2+}, Pb^{4+}	^{204}Pb, ^{206}Pb, ^{207}Pb, ^{208}Pb, ^{210}Pb
Lithium (Li)			Li^{1+}	^{7}Li
Magnesium (Mg)			Mg^{2+}	^{24}Mg, ^{25}Mg, ^{26}Mg
Manganese (Mn)			Mn^{2+}, Mn^{3+}, Mn^{4+}	^{55}Mn
Mercury (Hg)		Hg	Hg^{1+}	^{198}Hg, ^{199}Hg, ^{200}Hg, ^{201}Hg, ^{202}Hg
Molybdenum (Mo)			Mo^{4+}, Mo^{6+}	^{92}Mo, ^{95}Mo, ^{96}Mo, ^{98}Mo, ^{100}Mo
Neodymium (Nd)			Nd^{3+}	^{142}Nd, ^{143}Nd, ^{144}Nd, ^{146}Nd
Nickel (Ni)		Ni	Ni^{2+}, Ni^{3+}	^{58}Ni, ^{60}Ni
Nitrogen (N^{3-})		N_2	N^{5+}	^{14}N, ^{15}N
Oxygen (O^{2-})		O_2		^{16}O, ^{17}O, ^{18}O
Phosphorous (P)			P^{5+}	^{31}P
Platinum (Pt)		Pt	Pt^{2+}	^{194}Pt, ^{195}Pt, ^{196}Pt
Potassium (K)			K^{1+}	^{39}K, ^{40}K, ^{41}K
Radon (Rn)		Rn		^{218}Rn, ^{219}Rn, ^{220}Rn, ^{222}Rn
Rubidium (Rb)			Rb^{1+}	^{85}Rb, ^{87}Rb
Samarium (Sm)			Sm^{3+}	^{147}Sm, ^{148}Sm, ^{149}Sm, ^{152}Sm, ^{154}Sm
Silicon (Si)			Si^{4+}	^{28}Si
Silver (Ag)		Ag	Ag^{1+}	^{107}Ag, ^{109}Ag
Strontium			Sr^{2+}	^{86}Sr, ^{87}Sr, ^{88}Sr
Sulfur (S)	S^{2-}	S	S^{+}	^{32}S, ^{33}S, ^{34}S, ^{36}S
Thorium (Th)			Th^{4+}	^{230}Th, ^{231}Th, ^{232}Th, ^{234}Th
Tin (Sn)		Sn	Sn^{2+}	^{116}Sn, ^{118}Sn, ^{120}Sn
Titanium (Ti)			Ti^{4+}	^{48}Ti
Uranium (U)			U^{4+}, U^{6+}	^{234}U, ^{235}U, ^{238}U
Zinc (Zn^{2+})				^{64}Zn, ^{66}Zn, ^{68}Zn

Appendix C

Common Conversions

Length

	Inch (in)	Feet (ft)	Mile (mi)	Centimeter (cm)	Meter (m)	Kilometer (km)
1 in =	1	0.0833	0.0000158	2.54	0.0254	0.0000254
1 ft =	12	1	0.0001894	30.48	0.3048	0.0003048
1 mi =	63,360	5,280	1	160,934.4	1,609.344	1.609344
1 cm =	0.39370	0.0328084	0.00000621	1	0.01	0.00001
1 m =	39.370	3.28084	0.000621	100	1	0.001
1 km =	39,370	3,280.84	0.621	100,000	1,000	1

Area

	Square Inch (in^2)	Square Feet (ft^2)	Square Mile (mi^2)	Square Centimeter (cm^2)	Square Meter (m^2)	Square Kilometer (km^2)
1 in^2 =	1	0.0069444	2.491×10^{-10}	6.452	0.0006452	6.452×10^{-10}
1 ft^2 =	144	1	3.587×10^{-8}	929.03	0.092903	9.2903×10^{-8}
1 mi^2 =	4.014×10^9	2.788×10^7	1	2.59×10^{10}	2.59×10^6	2.59
1 cm^2 =	0.1550	0.0010764	3.861×10^{-11}	1	0.0001	1×10^{-10}
1 m^2 =	1,550	10.764	3.861×10^{-7}	10,000	1	0.000001
1 km^2 =	1.55×10^9	1.0764×10^7	0.3861	1×10^{10}	1×10^6	1

Volume

	Cubic Inch (in^3)	Cubic Feet (ft^3)	Cubic Meter (m^3)	Quart (qt)	Liter (l)	Gallon (Gal.; U.S.)
1 in^3 =	1	0.0005787	0.0000164	0.017316	0.0163871	0.004329
1 ft^3 =	1,728	1	0.0283168	29.9220779	28.3168467	7.4805195
1 m^3 =	61,023.74	35.3147	1	1,056.6882	1,000	264.1721
1 qt =	57.75	0.0334201	0.0009464	1	0.946353	0.25
1 l =	61.02374	0.0353147	0.001	1.0566882	1	0.2641721
1 gal =	231	0.1336806	0.0037854	4	3.7854	1

Mass and Weight

	Ounce (oz)	Pound (lb)	Short Ton (T)	Gram (g)	Kilogram (kg)	Metric Ton (t)
1 oz =	1	0.0625	0.0000313	28.3	0.0283	0.0000283
1 lb =	16	1	0.0005	453.6	0.4536	0.0004536
1 T =	32,000	2,000	1	907,184.7	907.1847	0.9071847
1 g =	0.035274	0.0022046	0.00000110	1	0.001	0.000001
1 kg =	35.274	2.2046	0.00110	1,000	1	0.001
1 t =	35,274	2,204.6	1.10	1×10^6	1,000	1

Temperature

To convert from Fahrenheit (F) to Celsius (C): $°C = (°F - 32°)/1.8$
To convert from Celsius (C) to Fahrenheit (F): $°F = (°C \times 1.8) + 32°$

Fahrenheit (°F)	Celsius (°C)	
32	0	Freezing point of water at 1 atmosphere pressure
50	10	
68	20	
86	30	
104	40	
122	50	
140	60	
158	70	
176	80	
194	90	
212	100	Boiling point of water at 1 atmosphere pressure
572	300	
932	500	
1292	700	
1652	900	
1832	1000	
3632	2000	
5432	3000	
7232	4000	
9032	5000	
10832	6000	

Glossary

A horizon: A soil horizon defined by the presence of organic matter mixed in with mineral and rock fragments.

Abrasion: Erosion of rock by sediment particles carried by water, wind, or glacial ice.

Absolute age: Specific year when an event occurred or a particular number of years that have elapsed since an event occurred.

Absolute sea level change: Global change in sea level caused either by changes in the volume of water in the oceans or changes in average ocean depth.

Accreted terrane: See *exotic terrane*.

Achondrite: Stony meteorite that lacks chondrules and formed by igneous processes.

Aftershock: One of many earthquakes that follow a larger-magnitude mainshock in the same region.

Albedo: The amount of light reflected by a planetary object or particular surfaces on the object.

Alluvial fan: Fan-shaped mass of alluvium that forms where stream flow abruptly exits a confined channel onto an unconfined valley floor; commonly forms at the base of a mountain range.

Alluvial stream: Stream that flows on loose sediment (alluvium).

Alluvium: Sediment deposited by streams.

Amphibolite: Metamorphic rock composed primarily of amphibole minerals, typically with plagioclase feldspar. Usually formed by metamorphism of mafic or intermediate volcanic rocks.

Andesite: Aphanitic igneous rock solidified from intermediate-composition magma.

Angle of repose: The maximum angle of a stable slope determined by friction, cohesion, and the shapes of the particles in the mass of regolith that make up the slope.

Angular unconformity: An unconformity defined by a sharp boundary between intervals of layered rocks (either sedimentary beds or lava flows) that are inclined at different angles.

Angular velocity: The velocity of circular motion around a fixed center point, which is described as the angle of rotation divided by the time over which the rotation occurred (e.g., degrees per year).

Anthracite: The hard variety of coal produced by metamorphism of sedimentary coal.

Anticline: Arch-shaped fold where limbs dip away from the hinge line; oldest rocks are present in the center of the fold.

Aphanitic: Igneous-rock texture resulting from rapid crystallization to form very small crystals that are generally invisible to the unaided eye.

Apparent polar wander path: The apparent path of motion by a magnetic pole that is implied by paleomagnetic data collected from rocks of different ages. The word "apparent" emphasizes that it only appears like the poles are wandering, but they really are fixed and the data instead require motion of the continents where the rocks were collected.

Aquifer: A body of rock or regolith with sufficient porosity and permeability to provide water in useful quantities to wells or springs.

Arkose: Sandstone containing at least 25% feldspar in addition to abundant quartz.

Artesian: An adjective describing ground water that is confined under pressure such that the water naturally rises toward the surface in a well or spring.

Assimilation: Melting of rock adjacent to magma, which modifies the magma composition.

Asteroid: Irregularly shaped, rocky planetesimal primarily found in a belt orbiting the Sun between Mars and Jupiter.

Asthenosphere: A weak and perhaps partly molten weak layer of Earth's mantle below the lithosphere.

Atomic mass number: The sum of the number of protons and the number of neutrons in the nucleus of an atom.

Atomic mass unit: Equal to the mass of a proton, or 1.673×10^{-24} gram; abbreviated AMU.

Atomic number: The number of protons in the nucleus of an atom. Each element has a unique atomic number.

Atom: The smallest unit of matter that engages in chemical reactions and cannot be chemically broken down into simpler components.

Axial plane: An imaginary surface separating rocks on each side of a fold.

Axis: Line where the axial plane intersects an original plane in the rocks, such as a bedding plane.

B horizon: The soil horizon notable for colors and textures indicating accumulation of minerals that are not present in the parent material. Typically found below the A horizon and above the C horizon.

Bar: A body of sediment exposed above a low water level in a stream but that is submerged and moves at higher flow.

Barchan: Crescent-shaped sand dune that forms on desert surfaces where wind-transported sand is not abundant. Barchans resemble parabolic dunes except that the ends of the crescent on a barchan point downwind.

Barrier island: Long, narrow sandy ridge of land that forms parallel to, but separate from, the mainland coast.

Basalt: Aphanitic mafic igneous rock.

Base level: The elevation down to which a river has the ability to erode its bed, everywhere along its course.

Basin: Depression caused by subsidence of the crust and typically accumulating thick sediment and sedimentary rocks. May also refer to an eroded depression; also see *drainage basin*.

Batholith: Large body of intrusive igneous rocks with a total exposed area greater than 100 square kilometers.

Bauxite: Multicolored ore of aluminum hydroxide minerals that are the primary sources of commercial aluminum; commonly found in modern or ancient rainforest soils.

Baymouth bar: Spit that links one headland to the next and closes off the intervening bay from the ocean.

Bay: Deep recess along a shoreline.

Beach drift: The along-shore transport of sediment on a beach resulting from alternating oblique up-beach transport by swash and directly offshore transport by backwash.

Beach face: Relatively steep part of a beach close to the water.

Beach: Low-sloping, nonvegetated area of unconsolidated sediment moved by waves and tides that extends from the low-tide line to a landward line defined by a cliff, sand dunes, or permanent vegetation.

Bedding: Distinctive layering originating from sequential deposition of sedimentary materials at Earth's surface.

Bedload: Large grains that roll, slide, and bounce along the bottom of a stream.

Bedrock: Large, continuous occurrences of solid rock, whereas the term "rock" by itself could describe a small fragment that you pick up from the ground.

Bedrock stream: Stream that flows on solid rock.

Berm: A typically flat part of a beach formed landward of the sloping beach face.

Berm crest: Linear boundary between the relatively steep beach face and the nearly horizontal berm.

Bituminous coal: The most common lithified variety of sedimentary coal.

Body wave: Seismic (earthquake) wave that passes through the interior of Earth. See *P wave* and *S wave* as examples.

Bomb: Pyroclastic fragment more than 64 millimeters across.

Bond: The force that holds atoms together to form molecules.

Braided: Describes an alluvial stream notable for abundant bars that divide stream flow into threads that separate and rejoin around the bars; so named for the pattern of water flow around the bars that resembles braided rope or hair.

Breaker: A water-surface wave that becomes so steep that the crest outraces the rest of the wave and collapses forward.

Breccia: A clastic sedimentary rock consisting of angular fragments larger than 2 millimeters across.

Brittle: Describes deformation characterized by rock fracture. Contrast with *plastic*.

C horizon: The lowest soil horizon, which is the least weathered horizon and commonly is the parent material for soil formation.

C-type asteroid: Asteroid with low albedo and chemical composition similar to the Sun but with slightly lower abundances of gaseous elements.

Cabochon: A round or oval gem with a simple curved top, commonly used to display the beauty of a gem.

Caldera: Circular or elliptical depression at Earth's surface formed when a significant volume of magma is removed from a shallow magma chamber without immediate replenishment from deeper levels, causing the roof of the chamber to collapse.

Cementation: The filling of pore spaces with precipitated minerals that glue individual grains together; a step in lithification of sediment to form sedimentary rock.

Centrifugal force: An apparent, not real, force applied to account for the change in direction of motion of an object on the surface of a rotating object.

Chemical equilibrium: The condition in a chemical reaction when the concentrations of reactants and products no longer changes, even though the reaction continues with reactants forming products and products breaking down into reactants.

Chemical reaction: The coming together of atoms or molecular compounds that results in a change.

Chemical sediment: Sediment formed by precipitation of chemical compounds from water.

Chemical weathering: Dissolution of some minerals and formation of new minerals and dissolved ions as a result of water and oxygen reacting with minerals.

Chert: Chemical sedimentary rock composed of microscopic quartz crystals formed by chemical precipitation. Also a biologic sedimentary rock formed by accumulation of siliceous microorganisms.

Chondrite: Stony meteorite containing many tiny spheres of silicate minerals, called chondrules, with textures that indicate crystallization from molten droplets.

Cinder: Lapilli (2–64 millimeters across) of basaltic or andesitic composition. Also called scoria.

Cinder cone: A conical accumulation of loose volcanic cinder lapilli and bombs around a central crater. Also called a scoria cone.

Cirque: Steeply eroded walls at the upslope end of glaciated valleys that partially enclose a natural amphitheater, sometimes occupied by lakes.

Clastic sediment: Residue of particles that remains after rocks weather.

Clastic: Sedimentary-rock texture describing particles derived from weathering of preexisting rocks.

Cleavage: Pattern of breakage in a mineral and the shape of the resulting fragments. The separation of metamorphic rock into thin sheets.

Coal: Sedimentary rock composed almost entirely of the compacted remains of fossil plants.

Cohesion: Attraction of particles to each other at the atomic level, caused by opposite electrostatic charges on adjacent particles.

Comet: A planetismal formed mostly of ice, with subordinate rock, and that partly vaporizes to form a long gaseous tail in proximity to the sun.

Compaction: Decrease in volume caused by partial or complete elimination of pore spaces between sediment particles; a step in lithification of sediment to form sedimentary rock.

Composite volcano: Volcano with slopes generally steeper than 25 degrees and composed of interlayered lava flows and pyroclastic deposits.

Compression: A stress caused by oppositely directed forces that shorten or decrease the volume of materials.

Conduction: Process of transferring heat by contact between two surfaces of different temperatures; heat transfers without motion of matter.

Cone of depression: A cone-shaped depression in the water-table surface that forms where wells discharge ground water. The water-table surface slopes in all directions toward the well so that the amount of ground water flowing to the well is equal to the amount pumped from the well.

Confined aquifer: Ground water in a permeable layer between impermeable layers (confining beds) that prevent shallow and deeper ground waters from mixing.

Confining bed: Low-permeability material that restricts the movement of ground water into or out of adjacent aquifers.

Conglomerate: A clastic sedimentary rock composed primarily of rounded gravel-size particles (greater than 2 millimeters across).

Contact metamorphic rock: Rock changed (metamorphosed) by heat and fluid in close proximity to magma intrusions or lava flows.

Contact metamorphism: Metamorphism near the contacts of igneous intrusions or beneath erupted lava flows, produced primarily by heat and fluid flow in the region adjacent to the magma or lava.

Continental drift: Hypothesis of German meteorologist Alfred Wegener that all continents were once joined as a single continent, which he named Pangea, from where they drifted to their current positions.

Convection: Process of simultaneously transferring heat and matter by movement of fluid or plastically deforming rock because of density contrast; denser and typically colder material sinks while less dense and typically warmer material rises.

Convergent plate boundary: Curving zone where plates collide nearly head-on into one another, compressing the lithosphere and causing subduction of one plate beneath the other.

Core: The central region of Earth composed primarily of iron metal and consisting of a molten liquid outer part and a solid inner part.

Coriolis effect: The tendency of objects in motion on or above Earth's surface to appear to be deflected to the right in the northern hemisphere and to the left in the southern hemisphere because of Earth's rotation.

Correlation: Demonstrated equivalence of rocks exposed in different locations.

Covalent bond: Union formed when two or more atoms mutually share electrons.

Craton: Low-elevation, tectonically stable interior region of a continent, exposing either very ancient Precambrian rocks or relatively thin coverings of sedimentary rocks resting unconformably on older Precambrian rocks.

Creep: Very slow flow of rock or regolith detected only by dislocation or bending of features at the surface.

Crevasse: Crack in the brittle upper part of a glacier caused by motion of the lower, plastically deforming part.

Cross-bed: Sedimentary structure of inclined layers within a bed and formed by shifting dunes and ripples. Cross-beds dip in the direction of current transport.

Crust: Outermost concentric layer of Earth composed mostly of silicate minerals and containing more silicon and aluminum than the underlying mantle.

Crystal face: Smooth, flat surface with regular geometric shape that forms the outer surface of a mineral specimen.

Cutbank: A steep bank eroded on the outside of a bend in a stream channel where flow is fast and deep.

Dacite: Aphanitic igneous rock solidified from felsic magma intermediate in composition between andesite and rhyolite.

Darcy's law: The discharge of fluid through pore spaces is proportional to the area of the flow and to the hydraulic gradient. The law is an equation used to compute the quantity of water flowing through an aquifer.

Daughter isotope: An elemental isotope produced by decay of a radioactive parent isotope.

Debris avalanche: A very rapid flow of rock, regolith, vegetation, and sometimes ice.

Debris flow: A flow of regolith and water that behaves like a high-viscosity fluid.

Declination: Angle made by the two lines that connect a point on Earth's surface to the magnetic north pole and to the geographic north pole.

Decompression melting: A process of magma formation where rock partly melts when pressure decreases at a nearly constant high temperature. Usually happens by movement of mantle toward Earth's surface.

Deflation: The erosion and transport of loose particles by wind, which lowers surface elevation.

Dehydration metamorphism: High-temperature and -pressure metamorphism where water-bearing minerals react to form minerals that lack water. Water is released by these reactions.

Delta: Landform protruding outward from a coastline and produced by sediment deposition where a stream enters a lake, reservoir, or sea.

Density: A measure of how compact a substance is, and mathematically defined by the mass within a particular volume of the substance.

Desert pavement: Closely spaced gravel fragments that cover barren and rocky desert floors to form a smooth surface.

Desert: Region where annual precipitation is less than 25 centimeters.

Dike: A tabular, steeply inclined igneous intrusion that cuts across sedimentary layers, if present.

Diorite: Phaneritic igneous rock solidified from intermediate-composition magma.

Dip: Angle between an imaginary horizontal plane and the planar margin of an inclined rock layer; used with the measurement of strike to describe the orientation of any planar geologic feature such as a rock layer, fault, or margin of an intrusion.

Dip-slip fault: Fault along which rocks move parallel to the dip direction of the fault plane.

Discharge: The volume of fluid that passes a location per interval of time (e.g., cubic meters per second).

Disconformity: An unconformity defined by a sharp erosional boundary between intervals of sedimentary or volcanic rocks where layers above and below the boundary are parallel to one another.

Dissolution: Chemical reactions where bonds break between atoms or molecules that then disperse in water.

Dissolved load: The chemical ions dissolved in stream water.

Distributary channel: One of several branching streams formed by separation of flow from a single channel around bars and vegetated islands of stream-deposited sediment. Most commonly form on deltas and alluvial fans.

Divergent plate boundary: Linear or curving zones where plates move apart from one another and new lithosphere forms.

Divide: A relatively high ridge that separates the drainage basin of one stream from adjacent drainage basins.

Dolostone: Chemical sedimentary rock composed of the calcium and magnesium carbonate mineral, dolomite.

Drainage basin: Area from which a stream gathers water; can be used to describe the size of a stream.

Dune: A curving ridge of loose sediment, taller than 1 centimeter, which moves along with water or wind currents.

Dust: Wind transported particles smaller than 0.1 millimeter.

Dynamo: Device that generates electric current by rapidly rotating a large magnet inside coils of electrically conductive wire.

E horizon: The highly leached soil horizon sometimes formed below the A horizon and defined by the absence or near-absence of organic matter, easily weathered minerals, or weathering products like clay, oxide, and hydroxide minerals.

Earthquake: A nearly instantaneous release of stored energy resulting from the breaking and sudden movement of stressed rock.

Eclogite: Very high-grade metamorphic rock that lacks water-bearing minerals, is dominated by garnet and sodium-rich pyroxene (which gives the rock a blue coloration), and sometimes includes minor quartz.

Elastic: Impermanent; describes deformation in which a rock returns to its original dimensions after stress is removed, much the way a rubber band returns to its original shape after stretching and letting it go.

Elastic limit: Value of applied stress at which rock deformation is permanent and cannot be reversed; the value of increasing stress where rocks break or flow.

Elastic rebound theory: The nonpermanent bending of rock caused by strain on either side of a locked fault and which is recovered after the fault breaks during an earthquake.

Element: Chemical substance that cannot be split into simpler substances.

Elongation: The stretching strain resulting from tensional stress.

End moraine: A ridge of till deposited at the leading snout of a glacier.

Energy: A measure of the ability to do work.

Epicenter: Point on the surface directly above the focus of an earthquake.

Erratic: Rock transported by a glacier and deposited where similar rocks are not present.

Estuary: Submerged part of a coastal stream valley where freshwater and seawater mix.

Euler pole: The pole of rotation that describes the motion of an object on the surface of a sphere.

Evaporite: Chemical sedimentary rock formed by minerals, such as halite and gypsum, that crystallize when water evaporates, causing ions to bond together.

Evapotranspiration: The transfer of water vapor from Earth's surface into the atmosphere by evaporation of moisture from rock or soil and by the transpiration of moisture from plant leaves.

Exfoliation: Deformation in which rocks expand as overlying rock is eroded, forming joints parallel to the ground; in some cases, the rock displaced by expansion peels away in thin layers.

Exotic terrane: Blocks of crust added to a continent, such as most of far western North America, and that consist of rocks that do not resemble rocks of the same age in adjacent blocks or the rest of the continent.

Extrusive (or volcanic) rock: Igneous rock formed by eruption of lava flows and pyroclastic materials onto Earth's surface.

Facet: Planar surface cut on a gemstone to accentuate luster and transparency.

Fall: A mass movement where material detaches from a steep slope and then free falls through the air, or bounces and rolls downslope.

Fault: A fracture plane along which rock or regolith is displaced.

Fault scarp: Cliff or low step in the ground surface caused by displacement along a fault.

Felsic: Describes igneous rocks dominated by quartz, sodium-rich plagioclase, and potassium feldspars, and the magmas that these rocks crystallize from; derived from the words *fel*dspar and *si*lica.

Fjord: Glacier-eroded valley along a coastline that is partly submerged beneath the sea to form a long, deep, steep-walled bay.

Flood: Overflow of water beyond the banks of a stream that occurs when discharge is too large to be contained within the channel.

Flood wall: A wall typically constructed of concrete or steel along the banks of a river to confine high-discharge flows to the channel and protect the floodplain from flooding.

Floodplain: The land surface adjacent to a stream channel that is constructed by stream erosion and deposition and inundated during floods.

Floodway: An area of little or no development alongside a stream that is wide enough and low enough to carry the predicted discharge of the 100-year flood.

Flow: A mass movement of rock, regolith, or both, which behaves like a high-viscosity liquid.

Focus: Location of an earthquake within Earth (also called the hypocenter).

Foliation: Planes of minerals formed in response to stress; a feature of many metamorphic rocks.

Footwall: Rock or regolith that exist below a fault plane.

Foreshock: One of many relatively low-magnitude earthquakes that precede a larger mainshock in the same region.

Fossil: Remains of an organism preserved in rock. Fossils may consist of the original mineral matter secreted by an organism, petrified organic material, or an impression left behind after most or all of the organic material has been destroyed.

Fossil fuel: Combustible energy source such as coal, oil, or natural gas formed from organic matter buried with sediment.

Fractional crystallization: Process by which the composition of a melt changes through time because minerals that form early during crystallization differ in composition from the magma and are physically separated from the magma.

Fracture: Nonuniform breakage of a mineral to leave an uneven surface, in contrast to smooth cleavage planes. Rock breakage that characterizes brittle deformation.

Fracture zone: Linear to slightly curved boundary between oceanic lithosphere of different ages and elevations within the same lithospheric plate. Fracture zones connect to transform plate boundaries that separate mid-ocean-ridge segments.

Friction: Force that opposes motion between two objects in contact with one another.

Gabbro: Phaneritic, mafic igneous rock.

Gemstone: Mineral, rock, or organic substance that has value based on beauty, color, luster, transparency, durability, and rarity.

Geographic information system (GIS): Computer-software tool that collects, stores, retrieves, analyzes, and displays data referenced to specific locations as distinct diagrammatic layers.

Geologic cross section: Diagram showing the interpretation of subsurface geology based on surface measurements and sometimes based on rock samples recovered from wells.

Geologic map: Map that portrays the distribution and orientation of rock types, locations of faults and folds, ages of rocks, and locations and nature of contacts between rock types.

Geologic time scale: Established chronological order of time intervals in geologic history.

Geology: The science of the origin, composition, structure, and history of Earth.

Geothermal gradient: The increase in temperature with depth beneath Earth's surface.

Glacial flour: Very fine-grained particles resulting from abrasion of bedrock surfaces by rocks frozen into the bottom of glaciers; produces a distinctive milky discoloration in glacial-melt water streams.

Glacial rebound: Isostatic adjustment of surface elevation resulting from melting of glacial ice. When ice weight is added to the crust, the underlying crust subsides and the adjacent area bulges; when the overlying ice melts, the originally depressed areas rise up and the bulges sink.

Glacier: An accumulation of snow and ice that is thick enough to flow under its own weight.

Gneiss: High-grade metamorphic rock defined by foliation of parallel compositional layers of light-colored (e.g., quartz, feldspar) and dark-colored (e.g., biotite, amphibole, pyroxene, garnet) minerals.

Graded bed: A sedimentary bed defined by a gradual variation in grain size from coarse at the bottom to fine at the top.

Graben: Block of crust displaced downward along normal faults.

Granite: Phaneritic, felsic igneous rock.

Gravity: A mutually attractive force between objects that depends on the distances between objects and their masses.

Greenhouse gas: Any of several gases in the atmosphere (notably including carbon dioxide, water vapor, and methane) that permit solar energy to reach Earth's surface but stop reflected heat from going back into space; so named because these gases behave similar to glass in a greenhouse that raises temperature to permit year-round plant growth.

Greenstone: Low-grade metamorphic rock that contains abundant green minerals, usually with chlorite (iron-magnesium mica) as the primary constituent, along with green amphibole, feldspar, and quartz; typically forms by metamorphism of volcanic rocks.

Groin: A wall built perpendicular to the shoreline to trap sediment transported by longshore currents.

Ground moraine: Till deposited beneath a glacier to form a bumpy sediment sheet of irregular thickness.

Ground water: Water below Earth's surface that moves slowly through pore spaces and fractures within regolith and rock.

Gully: Channel that results from soil erosion by flowing water (and that is too deep to be smoothed over by tillage).

Half-life: Time interval required for half of the radioactive parent-isotope atoms to decay to form an equal number of daughter-isotope atoms.

Hanging wall: Rock or regolith that exist above a fault plane.

Hardness: A measure of the resistance to scratching a mineral surface.

Hard water: Water with high concentrations of dissolved ions, usually calcium and magnesium ions from dissolution of calcite and dolomite in limestone and dolostone. These ions commonly reach concentrations in the water where they precipitate and clog pipes with rings of calcite, dolomite, or other minerals.

Head: An elevation that is proportional to the total energy of a fluid. In ground-water hydrology, head is the elevation that water rises in a well and describes the potential and pressure energy at the point where the well opens to the aquifer.

Headland: Place where land juts into the sea or a lake; sometimes called a cape.

Headwaters: Source of a stream at high elevation close to a divide.

Heat flow: Total amount of heat escaping through the surface of Earth and originating within the planet, largely from radioactive decay.

High-grade metamorphism: Metamorphism occurring at temperatures in excess of 600°C.

Hinge line: The imaginary line drawn along a deformed layer where the dip direction changes (also called the axis).

Hornfels: Any very hard, nonfoliated, metamorphic rock composed mostly or entirely of microscopically small crystals.

Horst: Block of crust displaced upward along normal faults.

Hot spot: An area of intense volcanic activity not explained by plate-boundary processes. Hot spots form where asthenosphere rises beneath lithospheric plates.

Hydraulic conductivity: A quantity that is proportional to the rate at which the fluid can move through a permeable material. Hydraulic conductivity is a function of both the properties of the fluid and the material that the fluid flows through.

Hydrologic cycle: A concept describing the movement of liquid water and water vapor through all parts of the Earth system.

Hydrolysis: Chemical reaction between a solid compound and water to produce a solid compound, which contains water molecules, and dissolved ions.

Hydrothermal metamorphic rock: Rock resulting from reaction of hot fluid with a preexisting rock.

Hydrothermal metamorphism: Metamorphic process resulting from hot water circulating through pore spaces and cracks in preexisting rock.

Hypothesis: Testable prediction about a natural process that can be checked by collecting data.

Ice cap: Broad glacier not confined by topography and with an area less than 50,000 square kilometers.

Ice sheet: Broad glacier not confined by topography and with an area greater than 50,000 square kilometers. Ice sheets currently cover most of Antarctica and Greenland.

Ice shelf: The floating part of a glacier that moved from land into deep water.

Iceberg: Block of glacial ice that detaches from a glacier and floats in the ocean or a lake.

Igneous rock: Rock that crystallized from molten material originating within Earth.

Inclination: Angle between the magnetic field force line and Earth's surface.

Incompressibility: A measure of how material resists changing volume when subjected to high pressure (also known as the compressibility or the bulk modulus).

Index minerals: Minerals whose presence in metamorphic rocks allow the estimation of the pressure and temperature of rock formation.

Intensity: Measure of earthquake violence. The Mercalli Intensity Scale, denoted by Roman numerals I to XII, describes the extent to which people feel a quake, damage to structures, and secondary effects such as landslides.

Intermediate: Describes igneous rocks, and the magmas they crystallize from, that have a composition in between that of mafic and felsic rocks and magmas.

Intrusive rock: Igneous rock formed from magma injected into pre-existing rocks below the surface.

Iron meteorites: Meteorites mostly composed of iron and nickel metal.

Ionic bond: Atomic bond formed by the attraction of oppositely charged ions to one another to balance their charges.

Ion: Charged atom resulting from the gain or loss of one or more electrons so that the number of protons and electrons are unequal. A negatively charged ion is an anion and a positively charged ion is a cation.

Isograd: The boundary on a geologic map between zones defined by different metamorphic index minerals. The boundary, therefore, indicates where a metamorphic reaction took place to consume or add an index mineral in the rocks.

Isotope: One of two or more atoms of the same element that have the same number of protons but different numbers of neutrons.

Jetty: A wall built where a channel enters the sea or a lake in order to keep the channel from filling with sediment transported by longshore currents. Jetties are commonly built in pairs on either side of a harbor entrance or tidal inlet.

Joint: Fracture in rock where little or no displacement has occurred.

Karst topography: Landscape pockmarked by sinkholes or a highly irregular landscape of high-standing rock towers and intervening depressions and valleys; there may be only limited surface stream flow as a result of deranged drainage patterns caused by rock dissolution.

Kettle: Depression where large blocks of ice melted within till or outwash; typically filled with water to form a pond or lake.

Lagoon: Shallow, relatively quiet water body on the landward side of an obstacle such as a barrier island, spit, baymouth bar, or offshore coral reef that absorbs wave energy directed toward the shoreline.

Lahar: Indonesian term describing the rapid flow of water and loose debris down steep volcanic slopes.

Lapilli: Pyroclastic fragments ranging in size from 2 millimeters to 64 millimeters across (singular: lapillus).

Lateral moraine: Boulder-rich sediment that forms a ridge along the margin of a valley glacier.

Lava: Molten material that erupts from a volcano and solidifies to form extrusive igneous rock.

Lava dome: A steep-sided lava flow, commonly almost as high as it is wide, and caused by the extrusion of extremely viscous lava.

Lava flow: Molten material extruded at, and flowing away from, a volcano.

Law: Scientific description of how nature is observed to behave.

Left-lateral strike-slip fault: Strike-slip fault across which features are displaced to the left (also called a sinistral strike-slip fault).

Limb: One side of a fold where all of the rocks dip in the same direction.

Limestone: Chemical sedimentary rock composed of calcite.

Linear dune: Long, narrow sand ridge oriented parallel to the prevailing wind direction; usually formed where sand supply is limited and wind direction is variable.

Liquefaction: The transformation of loosely packed, water-saturated sediment into a fluid mass. Commonly occurs when grains settle during earthquake ground shaking, which displaces the water in the intervening pore spaces upward such that the water pressure moves the grains apart and the whole sediment-water mixture loses strength.

Lithic sandstone: Sandstone composed mostly of sand-size rock fragments.

Lithification: The process of transforming loose sediment into sedimentary rock by compaction and cementation.

Lithosphere: Outer strong shell of Earth consisting of crust and uppermost mantle.

Loess: Deposit of windblown silt.

Longshore current: A current that moves parallel to shore and forms where wave crests do not approach exactly parallel to the shoreline.

Low-grade metamorphism: Metamorphism occurring at temperatures approximately between 200°C and 400°C and at pressures less than 4 kilobars.

Low-velocity zone: The part of the upper mantle where seismic-wave velocity does not increase systematically with greater depth; the top of this zone typically defines the boundary between the lithosphere and the asthenosphere.

Luster: The nature of light reflection from mineral surfaces.

M-type asteroid: A bright asteroid consisting of metallic iron.

Mafic: Describes igneous rocks (basalt, gabbro) that have relatively high *ma*gnesium and iron (*fer*ric) contents and a relatively low silicon content.

Magma: Molten material formed and residing within Earth.

Magnetic field: A region where lines of force act on charged particles or magnets. The field is produced around a magnet or around a conductor carrying an electric current.

Magnetic reversal: An abrupt flip of Earth's magnetic poles that occurs at irregular time intervals.

Magnitude: Measure of earthquake size related to the amount of energy released by the earthquake and the amplitude of the waves recorded on a seismogram.

Mainshock: Largest of a sequence of earthquakes that is preceded by lower-magnitude foreshocks and followed by lower-magnitude aftershocks.

Mantle: The mostly solid but generally weak silicate zone of Earth below the crust and above the core.

Marble: Nonfoliated metamorphic rock composed of calcite and formed by the metamorphism of limestone.

Mass movement: Gravity-driven downslope motion of rock and regolith (also called mass wasting).

Massive: Describes rocks that lack layering or bedding.

Matter: Anything that has mass and occupies space.

Meandering: Tendency of a stream channel to gradually shift position across a valley as a result of simultaneous erosion and deposition on opposite stream banks over time.

Medial moraine: A ribbon of sediment within a glacier caused where lateral moraines combine at and downslope of the junction of two valley glaciers.

Medium-grade metamorphism: Metamorphism occurring approximately between temperatures of 400°C to 600°C.

Metallic bond: Bond formed where electrons freely roam around a number of different atoms, typically of the same element, giving the electrons a mobility that accounts for the ability of metallic substances to conduct electricity.

Metamorphic facies: An association of metamorphic minerals that are stable together over a defined range of temperature and pressure conditions.

Metamorphic rock: A rock distinguished from a preexisting rock by a change in the minerals that comprise it, or a rearrangement of the existing minerals, or both, as a result of reactions that occur at high temperature, high pressure, or in the presence of hot fluid, or a combination of all three.

Metamorphism: Processes within Earth that produce solid state mineralogical, chemical, and textural changes that alter the appearance of preexisting rocks.

Meteorite: Object from space that lands on Earth.

Meteor: Object from space that approaches Earth (called a meteorite after it lands on Earth).

Mid-ocean ridge: Long, continuous submerged volcanic mountain chain winding through Earth's oceans.

Migmatite: Rock that resembles gneiss except that light-colored bands have the igneous-crystallization texture of granite resulting from high-grade temperature, pressure, and fluid conditions that cause melting, whereas dark layers reveal metamorphic crystal growth and recrystallization.

Milankovitch cycles: Periodic variations in Earth's rotation on its axis and in its orbit around the Sun that affect the amount of solar energy that reaches Earth.

Mineral: A naturally occurring solid with a definite, only slightly variable chemical composition, and an ordered atomic structure formed mostly, but not entirely, by inorganic processes.

Mohorovičić discontinuity (Moho): The sharp discontinuity in seismic velocity where mafic to felsic igneous and metamorphosed igneous rocks of the crust are underlain by mantle peridotite. The Moho is typically encountered 5–20 km deep beneath ocean basins, is thinnest near mid-ocean ridges and is thickest near continental margins.

Mohs Hardness Scale: Relative scale for describing mineral hardness, ranging in value from 1 to 10, with the hardest mineral, diamond, having a hardness of 10.

Molecule: Substance composed of two or more atoms.

Moment magnitude: The most commonly used earthquake-magnitude scale that relies on the calculation of the seismic moment rather than the response of a seismometer to an earthquake.

Monsoon: Torrential seasonal rains caused by seasonal changes in air pressure that bring warm, moist air from oceans over land.

Moraine: Heap of stony debris deposited along the margins of, or beneath, a glacier.

Mouth: Lowest elevation along a stream where it enters another stream, a lake, or the ocean.

Mudcrack: Open crack in muddy sediment or filled crack in mudstone or shale caused by contraction of clay minerals that absorb water and swell when wet, but then lose water, shrink, and crack while drying. Many cracks form and intersect to define polygon-shaped blocks of sediment.

Mudstone: Clastic sedimentary rock composed primarily of mud (silt and clay); called shale if the rock easily splits into thin sheets.

Natural levee: A ridge of sediment alongside a stream channel produced by deposition of sediment adjacent to the channel during floods.

Neap tide: The lowest high tide or highest low tide that occurs twice during a month. The relatively low tidal range results from the subtractive effect of gravitational pull exerted by the Moon and Sun.

Nebula: Cloud of gas and dust in space.

Nonconformity: An unconformity defined by a sharp boundary between plutonic-igneous or metamorphic rocks and overlying younger sedimentary or volcanic rocks.

Normal fault: A fault formed where the hanging-wall block moves downward compared to the footwall block.

Normal-polarity interval: Time interval when Earth's magnetic field is oriented as it is today, with the field lines exiting the north magnetic pole and entering the south magnetic pole.

Normal stress: Force applied perpendicular to a surface.

Nuclear fission: A process where an atom breaks into two roughly equal atoms with energy releases far in excess of radioactive decay.

Nuclear fusion: The combining (fusing) of two atomic nuclei to form a new element; results in the release of large amounts of energy.

Nucleus (atomic): The center of the atom. The nucleus contains one or more protons, which are particles with a positive electrical charge, and usually one or more neutral neutrons.

O horizon: The surface soil horizon that consists only of organic matter without minerals.

Oblique-slip faults: Faults along which rock movement includes a component along the dip of the fault plane and a component along the strike direction of the fault plane.

Obsidian: Volcanic glass that is opaque, commonly dark gray to black or brown, and with a felsic composition.

Ore: Rock containing important metallic elements that must be extracted from the minerals by metallurgical processing that breaks the mineral bonds.

Outwash: Sediment eroded by glaciers that is carried away by meltwater streams.

Oxbow lake: A highly curved lake (resembling an oxbow on a yoke) formed on a river floodplain when erosion cuts off a meander loop of a sinuous stream channel.

Oxidation: Chemical reaction between a substance and oxygen that produces new substances.

P wave (primary wave): Seismic body wave that displaces material in the same direction that the wave is moving, which causes alternating squeezing and stretching of the material as the wave passes.

pH: A measure of the acidity or alkalinity of a solution. Explicitly, pH is the logarithm of the hydronium-ion concentration. A solution with a pH of 7 is neutral, pH values less than 7 are acidic, and pH values greater than 7 are alkaline.

Paleomagnetism: The ancient orientation of Earth's magnetic field that is recorded in rocks.

Pan: Wind-eroded depression in regolith, commonly circular or elliptical, and commonly elongate parallel to the prevailing wind direction.

Parabolic dune: Crescent-shaped dune with an overall shape that resembles the graph of a parabola with a depression ("blowout") in the center and the two ends of the parabola pointing upwind.

Parent isotope: Radioactive isotope that decays through time to a daughter isotope.

Parent material: Rock or regolith from which a soil forms.

Peat: Black, organic-rich soil that contains more plant residue than mineral grains.

Pegmatite: Atypical coarse-grained igneous rock with crystals as large as several meters.

Perched ground water: A locally saturated region above the water table formed where an impermeable layer impedes downward infiltration.

Peridotite: Phaneritic ultramafic igneous rock; uncommon at the surface but composes the upper mantle.

Period: The fundamental time interval on the geologic time scale, with boundaries defined by the presence of key fossils.

Permeability: The ability of a porous material to permit fluid flow through the material.

Phaneritic: Igneous rock texture defined by coarse, easily visible mineral crystals formed by slow crystallization.

Phyllite: Moderately high-grade metamorphic rock that forms from slate; mica grains in phyllite are coarser than in slate and generate a silky sheen from the reflection of light off the parallel mica cleavage surfaces.

Physical weathering: Disaggregation of rocks by mechanical processes.

Planetary accretion: The processs of planetary formation and growth by the gradual accumulation of small objects to make larger ones.

Planetary differentiation: Process of physical separation of relatively dense compounds toward the center of a spinning, growing planetesimal or planet, while less dense compounds remain near the surface.

Planetary embryo: A planetary object about one-hundredth to one tenth the size of Earth and formed by accretion of planetesimals.

Planetesimal: A small, solid object in space representing the first stage in accretion within the solar nebula. Planetesimals grow by cohesion between particles or by gravitational attraction between particles ranging in size from fine dust to as much as a few kilometers in diameter.

Plastic: Permanent deformation where rock flows rather than breaking

Plate: One of several discrete, rigid to semi-rigid, roughly 100-km-thick slabs that make up Earth's lithosphere.

Plate tectonics: Theory that Earth's outer shell is not seamlessly continuous but is broken into discrete pieces that move slowly relative to one another and change in size over geologic time.

Playa: A dry lake bed characterized by a dusty salt flat of evaporite minerals.

Plucking: A process of erosion by rivers or glaciers where rocks disaggregate along preexisting fractures.

Plume: One of many hypothesized narrow columns of unusually hot mantle that rise by convection from the core–mantle boundary; many hot spots may be plumes.

Plunging fold: Folded rock where the axis is not horizontal but plunges downward.

Plutonic (or intrusive) rock: Igneous rock formed where magma solidifies below Earth's surface.

Pluvial lake: A lake that exists only during a period of exceptionally high rainfall; typically refers to ice-age lakes that formed in now mostly dry enclosed drainage basins in the western United States.

Point bar: A place where sediment accumulates on the inside of a bend in a stream, where the water flow is slow and shallow.

Polarity: The orientation of Earth's magnetic field, which is described as either normal or reverse polarity.

Polymorph: One of two or more minerals with identical chemical composition but with different arrangements of atoms.

Pores: Open spaces in a rock between mineral grains or open spaces between particles in regolith.

Porosity: Percentage of the total volume of regolith or rock composed of pores.

Porphyritic: Igneous rock texture defined by some large crystals surrounded by smaller crystals.

Potential energy: Energy that an object possesses because of its elevation. The potential energy that causes objects to move is greater for objects at high elevation than for objects at low elevation. Objects move from areas where they possess high potential energy to areas where they possess low potential energy.

Principle: A guiding concept that consistently works to describe the natural world.

Principle of cross-cutting relationships: Geologic features that cut across rocks (e.g., faults, igneous intrusions) must have formed after the rocks that they cut through.

Principle of faunal succession: Fossils serve to determine the relative age of enclosing rocks because there is a consistent chronologic sequence of fossil animals through geologic time.

Principle of inclusions: Objects enclosed in rock must be older than the time of rock formation.

Principle of isostasy: Surface elevations are adjusted by uplift and subsidence to maintain a condition where low-density rock floats on denser, underlying rock so that the pressure is everywhere the same at the base of any thick vertical column through the crust and upper mantle.

Principle of lateral continuity: Sedimentary beds are continuously deposited over large areas until encountering a barrier that limits their deposition.

Principle of original horizontality: Sedimentary beds are horizontal or near horizontal when deposited.

Principle of superposition: Where rocks are found in layers, one above the other, the lowest rock formed first, with each successively higher layer being younger than the one below.

Principle of uniformitarianism: Geologic processes and natural laws now operating on and within Earth have acted throughout geologic time; the logic and method by which geologists reconstruct past events.

Protostar: An early stage of star formation where dust and gases have concentrated into a dense central cloud but there is insufficient heat to initiate hydrogen fusion as a heat source, as in a true star.

Province: A discrete block of fault-bounded Precambrian crust composed of igneous and metamorphic rocks that formed during a particular time interval.

Pumice: Lightweight, highly vesicular, felsic volcanic fragments; typically pyroclastic fragments of lapilli or bomb size.

Pyroclastic: A class of fragmental volcanic rocks formed when explosions break apart magma and shoot out fragments that quickly quench to glass and fall to the ground.

Pyroclastic-fall deposits: Deposits consisting of pyroclastic fragments ejected high above a volcano, drift downwind, and then settle to the ground. The deposits exhibit a uniform decrease in particle size with increasing distance, up to one or two thousand kilometers, from the source volcano.

Pyroclastic-flow deposits: Poorly sorted mixtures of ash, lapilli, and bombs deposited by avalanches of pumice and ash flowing down the slope of a volcano; commonly thinner and finer grained at greater distances from the source.

Quartz sandstone: Sandstone composed primarily (>90%) of quartz, suggesting an environment where chemical weathering removed or altered all of the other minerals in the source rock.

Quartzite: Nonfoliated metamorphic rock consisting of recrystallized quartz; formed by metamorphism of quartz sandstone.

Radiation: The process of energy transport in the form of waves or particles, such as light.

Radioactive decay: The process by which unstable (radioactive) isotopes transform to new elements by a change in the number of protons (and neutrons) in the nucleus.

Radioactivity: Energy released when atoms of one element are transformed into atoms of another element resulting from processes that change the number of protons and neutrons in the nucleus.

Rayleigh number: The ratio of the gravity force that drives convection to the viscous force that resists convection. Calculating the Rayleigh number for a substance experiencing a temperature gradient assesses whether heat transfers by convection or conduction.

Recharge: Water added to the saturated zone.

Recrystallization: The transfer of atoms from one part of a crystal to another part of the same crystal or to a different crystal; generally causes an increase in the size of some crystals at the expense of others and commonly changes crystal shape.

Recurrence interval: The time interval between the occurrences of a type of event. Usually used in relation to the probability that a floodplain floods each year; a recurrence interval of 100 years means that there is a one-in-one-hundred (0.01) probability of a flood with a specified discharge each year.

Reflection: The phenomenon where waves bounce off a boundary.

Refraction: The bending of a wave caused by a change in wave velocity.

Regional metamorphic rock: Rock changed by metamorphic processes affecting an entire region, such as occurs near a convergent plate boundary.

Regional metamorphism: Metamorphism over large areas not related to specific igneous intrusions or sources of hydrothermal fluid. Typically related to the formation of mountain belts near subduction zones, and involving progressively increasing temperature- and pressure-driven mineralogical and textural changes to rock.

Regolith: Unconsolidated remains of weathered rock overlying solid rock, some of it transported as sediment and some of it remaining as fragments found above or near the source bedrock.

Relative age: Ordering of objects or features from oldest to youngest; establishing the age of one thing as older or younger than another.

Relative sea level change: Shifting shoreline position caused either by absolute sea level fluctuation, uplift and subsidence of crust, or a combination of these processes.

Relief: Difference in elevation between two locations.

Reverse fault: A fault inclined at an angle steeper than 45 degrees and formed where the hanging wall rock moves upward compared to the footwall.

Reversed polarity interval: Time interval when Earth's magnetic field is oriented the opposite of the present orientation so that the field lines exit at the south magnetic pole and enter Earth at the north magnetic pole; compasses would point toward the south pole rather than toward the north pole.

Rhyolite: Aphanitic igneous rock solidified from felsic magma.

Rift valley: Long, continuous valley occupying graben blocks of crust displaced downward along normal faults.

Right-lateral strike-slip fault: Strike-slip fault across which features are displaced to the right (also called a dextral strike-slip fault).

Rigidity: A measure of how much force is needed to change the shape of a solid object without changing its volume.

Rill: Tiny channel resulting from soil erosion by water that is sufficiently shallow to be removed by tilling the ground with a plow.

Rip current: Current that moves water directly offshore and away from the coast.

Ripple: Curving ridge of loose sediment, shorter than 1 centimeter, which moves along with water or wind currents, or moves back and forth beneath oscillating water waves.

Rock cycle: A sequence of processes and products that relate each rock type to the others, and that describes rocks as continuously forming from pre-existing rocks.

Rotation: Motion where an object turns around a pivot point or rotation axis. Points on an object move along curving paths and each point moves a different distance.

Rounding: A characteristic of the shape of clastic-sediment grains that describes the extent to which the edges and corners of grains abrade during transport by wind or water.

S-type asteroid: A high-albedo asteroid rich in metallic iron along with iron- and magnesium-rich silicates.

S wave (Secondary wave): Seismic body wave that displaces material at right angles to the direction of wave motion; only travels through solids.

Sandstone: Classic sedimentary rock composed of sand grains that are 1/16th to 2 millimeters across.

Saturated zone: The area below the water table where all pores and fractures are filled with water.

Schist: A shiny, mica-rich, foliated metamorphic rock that forms from phyllite exposed to higher temperatures and pressures; consists of parallel mica crystals that are coarse enought to be seen with the naked eye.

Scientific method: Process used to systematically and objectively examine and explain a problem or observed phenomenon.

Scoria: See *cinder*.

Sea ice: Frozen seawater.

Sea stack: Small rocky island close to a shoreline and left behind by the collapse of an arch following preferential erosion of a headland by waves.

Seafloor spreading: The process whereby oceanic lithosphere is pulled apart along the crests of mid-ocean ridges as new lithosphere fills the gap where the plates separate.

Seamount: Submarine volcano that does not reach the ocean surface to form an island.

Seawall: Structure of wood, plastic, concrete, rock, steel, junk cars, rubber tires, or sandbags built on a beach parallel to the shoreline to hold the shoreline in place against wave erosion.

Sediment load: Particles carried by a stream as bedload and suspended load.

Sedimentary rock: Rock consisting of the particulate and precipitated dissolved products of the weathering of older rocks.

Sedimentary structure: A physical feature, such as cross-bedding, produced in sediment at the time it is deposited, or shortly after deposition and before lithification into rock.

Seismic moment: Measure of the amount of energy released during an earthquake based on the strength of the rock that broke during the earthquake, the area of the fault plane that ruptured, and the distance that rocks moved on either side of the fault.

Seismic tomography: Method of determining the internal structure of Earth by mapping locations where seismic waves travel slightly faster or slightly slower than in adjacent rock at the same depth.

Seismic waves: Elastic energy waves that pass through and along the surface of Earth following an earthquake or explosion.

Seismogram: Record from a seismometer.

Seismometer: An instrument that detects seismic-wave motion and amplifies the wave pattern to make a record.

Serpentinite: Metamorphic rock composed almost entirely of serpentine, a hydrous magnesium-rich silicate; typically forms by metamorphism of peridotite.

Shadow zones: Areas at Earth's surface where P, S, or both P and S waves are not recorded.

Shale: A type of mudstone in which alignment of clay minerals cause the rock to break in thin sheets.

Shear: A deformation, or strain, where adjacent parts of a rock slide parallel to their plane of contact without overall shortening or elongation.

Shear stress: Force applied parallel to a surface.

Shield volcano: Volcano consisting of very thin and widespread basaltic lava flows and with gentle slopes typically less than 15 degrees.

Shortening: The resulting strain of applied compressive stress that decreases the distance between points, decreases the volume of a body, or both.

Silica tetrahedron: The basic building block of the silicate mineral crystal structure, consisting of four oxygen atoms surrounding and bonded to a silicon atom.

Sill: A tabular, commonly horizontal or near-horizontal igneous intrusion that is commonly formed by injection of magma between sedimentary layers.

Sinkhole: A depression on Earth's surface caused by collapse of surface rock and regolith into a large underground cavity formed by ground water dissolution of rock.

Slate: A metamorphic rock consisting mostly of fine-grained mica that causes the rock to separate along parallel rock-cleavage planes; commonly produced by metamorphism of shale or other mudrock.

Slide: A mass movement where rock and regolith move downslope in contact with a surface of rupture, which separates moving material from stationary material.

Slump: A type of slide where rock or regolith move by rotation along a curved surface of rupture.

Snowline: The elevation above where snow persists throughout the year. The snowline separates zones of accumulation and wastage in a glacier.

Soft water: Water with low concentrations of dissolved ions, usually artificially produced from hard water by using water softeners—special filtration devices that chemically remove calcium and magnesium ions.

Soil: The mostly unconsolidated to loosely consolidated surface residue that results from weathering of rock as minerals interact with water and organisms. The resulting mixture of mineral and organic constituents has different physical or compositional properties, or both, than the material from which it was derived.

Soil horizons: Distinct layers of soil, each with physical or compositional properties that distinguish them from adjacent layers.

Solar nebula: A disk-shaped spinning cloud formed by the inward collapse of a region of gas and dust in space.

Solar wind: The streams of charged particles produced by fast moving protons and electrons blasting outward from the Sun.

Sorting: Describes the range in grain size of a clastic sedimentary deposit or rock, with well sorted specimens containing mostly one grain size and poorly sorted sediment containing a wide range of grain sizes.

Specific gravity: The ratio of the mass of a substance compared to the mass of the same volume of water.

Speleothem: A mineral deposit in a cave caused by mineral precipitation from water.

Spit: An elongate sediment ridge formed by longshore currents and that projects from headlands in the direction of the longshore current. Wave refraction around the end of the spit causes sediment deposition along the tip and landward side of the spit and may form a prominent hook-shaped beach.

Spring: Place where ground water emerges onto the surface.

Spring tide: The highest high tide or lowest low tide that occurs twice a month. The relatively high tidal range results from the additive effect of the gravitational pull exerted by the Moon and Sun.

Stage: The measured elevation of the water surface in a stream. The stage is defined relative to an arbitrarily selected elevation that is typically close to the elevation of the deepest part of the stream channel.

Stalactite: A conical or cylindrical speleothem developed downward from the roof of a cave by mineral precipitation from dripping water; usually composed of calcite.

Stalagmite: A conical or cylindrical speleothem developed upward from the floor of a cave by mineral precipitation from dripping water; usually composed of calcite.

Star dune: Sand dune with radiating ridges resulting from highly variable wind directions crossing large areas of readily eroded sand.

Stock: A large body of intrusive rocks with a total exposed area less than 100 square kilometers; some stocks may simply be small parts of batholiths that have not yet been completely exposed by erosion.

Stones: A class of meteorites, including chondrites and achondrites, with compositions similar to rocks and minerals found on Earth.

Stony irons: A class of meteorites that contain approximately equal volumes of metal and silicate minerals.

Strain: The measurable deformation resulting from stress.

Stratification: See *bedding*.

Streak: Color of the residue left behind from scratching a mineral on a non-glazed porcelain plate.

Stream: Flowing water that moves through a channel and simultaneously transports dissolved and particulate products of rock weathering.

Stream gage: Device that measures water level at a location in a stream where the channel cross section has been carefully surveyed; stream gage data are used to calculate discharge.

Stream power: A measure of the ability of a stream to do work. A simple way to calculate stream power averaged over a unit area of stream bed is to multiply the shear stress and average flow velocity.

Strength: Measure of the amount of stress a material can endure before it fails, either by breaking or flowing.

Stress: The magnitude of force divided by the area over which the stress is applied.

Striation: A scratch on a bedrock surface commonly created by sharp-edged rocks frozen in the bottom of a glacier as the glacier slowly moves in a straight-line path.

Strike: Compass orientation of a line produced by the intersection of an imaginary horizontal plane with an inclined plane such as a tilted, fault plane, or edge of an intrusion; used with dip to describe the orientation of any planar geologic feature.

Strike-slip fault: A fault where rocks are displaced by horizontal movement along the strike direction of the fault plane.

Structural geology: The study of deformation in rocks at a microscopic to regional (hundreds of square kilometers) scale.

Subduction: The process by which a lithospheric plate descends beneath a neighboring plate.

Surface of rupture: Planar or curved face along which moving material of a slide separates from stationary material.

Surface waves: Seismic (earthquake) waves on Earth's surface that decrease in intensity with depth.

Suspended load: The fine-grained sediment intimately mixed with the water and flowing above the bed of a stream.

Syncline: Trough-shaped fold where limbs dip toward the hinge line; youngest rocks are present in the center of the fold.

Talus: Piles of loose, fallen rock, debris, and earth found at the base of a steep slope.

Tectonics: The field of study that encompasses deformation at a regional to global scale.

Tension: A stress caused by oppositely directed forces that elongate or increase the volume of materials.

Terraces: Step-and-bench landforms, including stream terraces alongside and above a river channel.

Theory: Established, thoroughly tested, generally accepted explanation for an observed natural phenomenon that is supported by a substantial body of data.

Thrust fault: A fault dipping at an angle less than 45 degrees and formed where the hanging wall rock moves upward compared to the footwall.

Tidal channel: Channel eroded by water draining seaward during the falling tide; sometimes called a tidal creek.

Tidal creek: See *tidal channel.*

Tidal current: Fast-moving current that forms where the rising and falling tide is constricted to inlets between islands, or forced in and out of funnel-shaped estuaries.

Tidal flat: Gently sloping, muddy, marshy, or barren area of land submerged at high tide and exposed at low tide.

Tidal inlet: Narrow body of water between barrier islands that connects a lagoon with the open ocean.

Tidal range: Change in sea-level elevation between low and high tide.

Tide: The slow rhythmic, alternating rise and fall in the surface of the ocean caused by the differential gravitational pull of the Moon and Sun on the oceans and rigid Earth.

Tidewater glaciers: Glaciers that descend into the ocean from land and are in contact with the seafloor.

Till: Very poorly sorted mixture of boulders, cobbles, gravel, sand, and mud deposited directly by a glacier.

Tillite: A poorly sorted sedimentary rock interpreted to have originated as glacial till.

Tonalite: Phaneritic igneous rock solidified from felsic magma intermediate in composition between diorite and granite.

Transform plate boundary: Zones where lithospheric plates slide past one another with neither creation nor destruction of lithosphere.

Transition zone: The region of the mantle between 410 and 660 kilometers below the surface where minerals undergo changes between peridotite upper mantle and the lower mantle composed of high-pressure minerals.

Translation: Motion where the orientation of the object does not change while it moves, and all points on an object move the same distance.

Transpiration: Process whereby water in plant leaves and stems is released as water vapor.

Transverse dune: A linear or curving sand dune with a crest oriented perpendicular to the prevailing wind direction; forms where sand supply is abundant and wind direction varies only slightly.

Travertine: A porous variety of limestone that typically forms where calcite precipitates from water around springs.

Tributary: A relatively small stream that flows into a larger stream.

Tsunami: Fast moving, long-wavelength sea waves caused by sudden displacement of the seafloor, typically caused by submarine fault motion or landslide.

Turbidite: Sedimentary deposit created by a turbidity current, typically consisting of a graded bed with a sharp, eroded base and ripple-formed cross-bedding.

Turbidity current: Mixture of sediment and water that flows along the seafloor or a lake bottom.

Ultramafic: Describes igneous rocks with a relatively low silica content, dominated by olivine and pyroxene and lacking quartz or significant feldspar.

Unconformity: Breaks in the continuity of the geologic record between rock units. The absence of rocks representing some interval of time that usually results from erosion.

Unsaturated zone: The region above the water table where pores are partly filled with water and partly filled with air. Water moves downward through the unsaturated zone to join ground water in the saturated zone.

Valley glaciers: Elongate glaciers confined within bedrock valleys.

Van der Waals forces: Uneven distribution of electrical charges around a neutral molecule that exert slight attractive and repulsive forces.

Ventifact: A loose rock with faceted, planar surfaces abraded by wind-blown sand.

Vesicles: Cavities in volcanic rocks occupied by gas when the rock solidified.

Viscosity: The property of a fluid that describes its resistance to flow.

Volcanic ash: Pyroclastic fragments less than two millimeters across.

Volcanic (or extrusive) rock: Rock originating from eruption of molten material at Earth's surface, including the products of lava flows, pyroclastic falls, and pyroclastic flows.

Volcanic neck: A nearly cylindrical pipe of plutonic rock or tuff that marks the feeding conduit of a volcano and later exposed by erosion.

Volcano: Hill, ridge, or mountain formed by the accumulation of lava flows and pyroclastic deposits around the conduit, or vent, from which they were erupted.

Wadati-Benioff zone: The inclined zone of earthquakes foci characteristic of subduction zones at convergent plate boundaries.

Water table: The surface that marks the top of the ground water and forms the boundary between the saturated and unsaturated zones.

Wave: Disruption that moves through a medium as a pulse of energy with little if any overall transport of the medium in the direction of wave movement.

Wave base: The depth to which oscillatory wave motion persists downward below the water surface.

Wave height: Distance between the adjacent crest and trough of a wave.

Wave period: The time elapsed between the passage of two successive peaks or troughs of a wave past a point.

Wave-cut platform: Nearly horizontal bench eroded along a rocky coast, commonly submerged at high tide and exposed to view at low tide.

Wavelength: The distance between two successive peaks or troughs of a wave.

Weathering: Deterioration of rocks at Earth's surface, more systematically defined as the response of the geosphere at its interface with the atmosphere, hydrosphere, and biosphere to reduce rocks into loose particles while dissolving some minerals and producing new ones.

Welded tuff: A pyroclastic deposit that forms a rock because the hot pyroclastic fragments compacted and welded to one another because of the weight of overlying and rapidly accumulating pyroclastic material.

Wet melting: A process of magma generation resulting from the introduction of water into hot rock, which lowers the melting temperature of the rock.

Wind: Movement of gas molecules in the atmosphere caused by convection.

Yardang: A wind-parallel ridge of soft rock or slightly consolidated sediment that remains after surrounding material is eroded by wind abrasion.

Yield strength: The point at which rock deformation is permanent and cannot be reversed by decreasing the stress.

Zone: Used in the study of metamorphic rocks to indicate the area on a map where a particular metamorphic index mineral is present in the rocks.

Zone of accumulation: The high-elevation zone of a glacier where the winter snow accumulation exceeds summer melting.

Zone of wastage: The low-elevation zone of a glacier (also called the zone of ablation) where melting exceeds the winter snow accumulation.

Photo Credits

Chapter 1

Chapter Opener 1: NASA/Johnson Space Center *Figure 1.1a:* Peter Carsten/ National Geographic Image Collection *Figure 1.1b:* Lester Lefkowitz/ Corbis/Bettmann *Figure 1.1c:* Melvin Grubb/Grubb Photo Service, Inc. *Figure 1.2a:* Oldrich Karasek/Getty Images Inc.-Stone Allstock *Figure 1.2b:* John Wang/Getty Images, Inc.- Photodisc *Figure 1.2c:* Jim Wark/Peter Arnold, Inc. *Figure 1.6a:* Eyewire Collection/Getty Images-Photodisc *Figure 1.6b:* Colin Keates © Dorling Kindersley, Courtesy of the Natural History Museum, London *Figure 1.6c:* Albert Copley/Visuals Unlimited *Figure 1.8a:* Dr. Parvinder S. Sethi *Figure 1.8b:* Dorling Kindersley *Figure 1.9a:* Dr. Ross W. Boulanger *Figure 1.10:* Earthquake and volcano data from Smithsonian Institution, Global Volcanism Program *Figure 1.12:* James A. Sugar/Corbis Bettmann

Chapter 2

Chapter Opener 2: Erica Van Pelt/Van Pelt Photography *Figure 2.1a:* Kevin Schafer/Corbis/Bettmann *Figure 2.1b:* Professor Gary A. Smith *Figure 2.2a:* The Natural History Museum, London *Figure 2.2b:* Charles D. Winters/Photo Researchers, Inc. *Figure 2.3:* Prof. Gary A. Smith *Figure 2.5a&b:* Aurora Pun *Figure 2.6:* Chip Clark *Figure 2.7a&b:* Aurora Pun *Figure 2.8:* Aurora Pun *Figure 2.9a:* Breck P. Kent *Figure 2.9b:* E.R. Degginger *Figure 2.9c:* Mark A. Schneider/Visuals Unlimited *Figure 2.18:* Clive Streeter © Dorling Kindersley *Figure 2.23a:* Chip Clark *Figure 2.23b:* Jeffrey A. Scovil *Figure 2.25a:* Chip Clark *Figure 2.25b:* Breck P. Kent *Figure 2.26a:* Dennis Tasa, Tasa Graphic Arts, Inc. *Figure 2.26b:* Richard M. Busch *Figure 2.26c:* Harry Taylor © Dorling Kindersley *Figure 2.28:* M. Claye/Jacana Scientific Control/Photo Researchers, Inc. *Figure 2.29:* Breck P. Kent/Animals Animals/Earth Scenes *Figure 2.30a:* Chip Clark *Figure 2.30b:* Breck P. Kent *Figure 2.30c:* Photolibrary.Com *Table 2.2:* (amphibole, chalcopyrite, corundum, gold, hematite, magnetite) Harry Taylor © Dorling Kindersley; (calcite) Colin Keates (c) Dorling Kindersley, Courtesy of the Natural History Museum, London; (copper, pyrite, sphalerite) Breck P. Kent/Animals Animals/Earth Scenes; (dolomite, garnet) Jeffrey A. Scovil; (feldspar) Charles D. Winters/Photo Researchers, Inc.; (galena) Chip Clark; (gypsum) RUNK/SCHOENBERGER/Grant Heilman Photography, Inc.; (olivine) Dennis Tasa, Tasa Graphic Arts, Inc.; (pyroxene, halite) Richard M. Busch; (quartz) Photo Researchers, Inc.; (mica) Marli Miller

Chapter 3

Chapter Opener 3: Tyler Stableford/The Image Bank/Getty Images, Inc. *Figure 3.1a:* Bert Sagara/Getty Images Inc.-Stone Allstock *Figure 3.1b:* Kim Heacox/DRK Photo *Figure 3.1c, d:* Professor Gary A. Smith *Figure 3.2a:* Bruce Forster © Dorling Kindersley *Figure 3.2b:* A. Flowers & L. Newman/Photo Researchers, Inc. *Figure 3.2c:* Andrew Jaster *Figure 3.2d:* George D. Lepp/Photo Researchers, Inc. *Figure 3.3a:* Aurora Pun *Figure 3.3b:* Professor Gary A. Smith *Figure 3.4a:* Douglas Peebles Photography *Figure 3.4b:* D.A. Swanson/Hawaiian Volcano Observatory, U.S. Geological Survey *Figure 3.4c1:* Andrew Alden *Figure 3.4c2&d:* Aurora Pun *Figure 3.7a:* Professor Gary A. Smith *Figure 3.7b:* Alan Kearney/Getty Images, Inc.-Taxi *Figure 3.9a:* Steve Austin; Papilo/Corbis/Bettmann *Figure 3.9b:* Craig Aurness/Corbis/Bettmann *Figure 3.11:* U.S. Geological Survey, U.S. Department of the Interior

Chapter 4

Chapter Opener 4: Salvatore Ragonese/Agence France Presse/Getty Images *Figure 4.1a:* Peter Mouginis-Mark *Figure 4.1b:* U.S. Air Force *Figure 4.1c:* © Craig Aurness/CORBIS *Figure 4.1d:* Professor Gary A. Smith *Figure 4.2a:* Albert J. Copley/Getty Images, Inc.-Photodisc *Figure 4.2b:* Andreas Einsiedel © Dorling Kindersley *Figure 4.2c:* Aurora Pun *Figure 4.2d:* Professor Gary A. Smith *Figure 4.2e:* Grace Davies/Omni-Photo Communications, Inc *Figure 4.3b:* Breck P. Kent/Animals Animals/Earth Scenes *Figure 4.3c, f:* Breck P. Kent *Figure 4.3d:* Geoscience/Pearson Education PH College *Figure 4.3e:* Donna Tucker *Figure 4.3g:* Jeffrey A. Scovil *Figure 4.3h, i:* Harry Taylor © Dorling Kindersley *Figure 4.3j:* Charles E. Jones *Figure 4.4a:* J.P. Lockwood/U.S. Geological Survey/U.S. Department of the Interior *Figure 4.4b:* Breck P. Kent/Animals Animals/Earth Scenes *Figure 4.4c:* Professor Gary A. Smith *Figure 4.4d:* Aurora Pun/Professor Gary A. Smith *Figure 4.5:* Professor Gary A. Smith *Figure 4.7a:* Karl E. Karlstrom *Figure 4.7b:* Marli Bryant Miller *Figure 4.8a:* Adriel Heisey/Network Aspen *Figure 4.8b:* Professor Gary A. Smith *Figure 4.9a:* R.T. Holcomb/Hawaiian Volcano Observatory, U.S. Geological Survey *Figure 4.9b:* Dan Suzio/Photo Researchers, Inc. *Figure 4.9c:* J.D. Griggs/ Hawaiian Volcano Observatory, U.S. Geological Survey *Figure 4.10a:* © Michael T. Sedam/CORBIS *Figure 4.10b:* John Gerlach/Animals Animals/Earth Scenes *Figure 4.11:* Jeff Wynn/USGS/ Cascades Volcano Observatory *Figure 4.11a:* Martin Miller/Visuals Unlimited *Figure 4.11b:* David Falconer/DRK Photo *Figure 4.12:* Solarfilma ehf *Figure 4.13a, b:* Professor Gary A. Smith *Figure 4.14a:* Michael Collier *Figure 4.14b:* J.D. Griggs/Hawaiian Volcano Observatory, U.S. Geological Survey *Figure 4.14c:* © Roger Ressmeyer/CORBIS *Figure 4.14e:* Michael Clynne/U.S. Geological survey/U.S. Department of the Interior *Figure 4.15:* J.D. Griggs/ Hawaiian Volcano Observatory, U.S. Geological Survey *Figure 4.16b:* François Gohier *Figure 4.26:* Paul Chesley/Getty Images, Inc.-Stone Allstock *Figure 4.27:* Yomiuri/AP Wide World Photos *Figure 4.28:* Archives de l'Academie des Sciences *Figure 4.29:* J. LangevinCorbis/Sygma *Figure 4.31:* Kennecott Utah Copper Corporation *Figure 4.33a:* Michael N. Spilde *Figure 4.33b:* Jeffrey A. Scovil *Figure 4.33c:* Getty Images, Inc.-Photodisc

Chapter 5

Chapter Opener 5: Yva Momatiuk/John Eastcott/Minden Pictures *Figure 5.1a:* Professor Gary A. Smith *Figure 5.1b:* Tom Bean/DRK Photo *Figure 5.1c,d,e & 5.2a:* Professor Gary A. Smith *Figure 5.2b:* Marli Miller *Figure 5.3b:* Professor Gary A. Smith *Figure 5.3c:* Larry Ulrich/DRK Photo *Figure 5.4a:* Aurora Pun *Figure 5.4b:* Dan Guravich/Photo Researchers, Inc. *Figure 5.4c:* Dr. John Bloch *Figure 5.5a:* Professor Gary A. Smith *Figure 5.5b:* Jess Alfrod/Getty Images, Inc.-Photodisc *Figure 5.5c:* © David Muench/David Muench Photography Inc. *Figure 5.6a:* Colin Keates © Dorling Kindersley, Courtesy of the Natural History Museum, London *Figure 5.6b:* Colin Keates © Dorling Kindersley *Figure 5.6c:* Breck P. Kent *Figure 5.6d:* Joy Spurr/Bruce Coleman Inc. *Figure 5.7a1:* Ted Mead/Photolibrary.Com *Figure 5.7a2:* Spike Walker/Getty Images, Inc.-Stone Allstock *Figure 5.7b:* Dave Houseknecht/U.S. Geological Survey/U.S. Department of the Interior *Figure 5.9a, b:* Marli Miller *Figure 5.9c:* Professor Gary A. Smith *Figure 5.11a:* Breck P. Kent *Figure 5.11b:* Colin Keates © Dorling Kindersley, Courtesy of the Natural History Museum, London *Figure 5.11c, d:* Dr. B. Booth/GeoScience Features Picture Library *Figure 5.11e:* Wally Eberhart/Visuals Unlimited *Figure 5.12:* Richard M. Busch *Figure 5.12a:* Joyce Photographics/Photo Researchers, Inc. *Figure 5.12b&e:* Breck P. Kent *Figure 5.12d:* © Harry Taylor/Dorling Kindersley Media Library *Figure 5.12f:* Mark A. Schneider/Visuals Unlimited *Figure 5.13a:* David R. Frazier/Photo Researchers, Inc. *Figure 5.3b&d:* Professor Gary A. Smith *Figure 5.13c:* MODIS/NASA Headquarters *Figure 5.14a:* Tom Bean/Tom & Susan Bean, Inc. *Figure 5.14b & 5.15a&b:* Professor Gary A. Smith *Figure 5.16a1:* High Sitton/Getty Images, Inc.-Stone Allstock *Figure 5.16a2:* Richard Hamilton Smith/Corbis/Bettmann *Figure 5.16c1:* Marli Miller *Figure 5.16c2:* Professor Gary A. Smith *Figure 5.17a, b:* Aurora Pun *Figure 5.17c, 18:* Professor Gary A. Smith *Figure 5.20a:* Marli Miller *Figure 5.20b, c:* Professor Gary A. Smith *Figure 5.21b&c:* "Turbidity currents as a cause of graded bedding", by P.H. Kuenen and C.I. Migliorini, *The Journal of Geology*, volume 58 (1950). *Figure 5.23:* Ronald C. Blakey

Chapter 6

Chapter Opener 6: CO & 01 Muench Photography, Inc. *Figure 6.1a&b:* Dr. J. Alcock, Penn State Abington College *Figure 6.1c:* Charles E. Jones *Figure 6.9a:* Colin Keates © Dorling Kindersley, Courtesy of the Natural History Museum, London *Figure 6.9b:* Jim Wehtje/Getty Image, Inc.-Photodisc *Figure 6.10b:* Doug Martin/Photo Researchers, Inc. *Figure 6.11b:* Marli Miller

Figure 6.12b: Marli Bryant Miller/Visuals Unlimited *Figure 6.13a, b:* Aurora Pun *Figure 6.15:* Longman Scientific and Technical, Essex/Longman International Education *Figure 6.20a:* Charles E. Jones *Figure 6.20b, g&l:* Harry Taylor © Dorling Kindersley *Figure 6.20c,e, f&h:* Charles E. Jones *Figure 6.20d:* Richard M. Busch *Figure 6.20i:* E.R. Degginger/Color-Pic, Inc. *Figure 6.20j&k:* Aurora Pun *Figure 6.22:* Larry D. Fellows/Arizona Geological Survey *Figure 6.26:* Ken MacDonald/Science Photo Library/Photo Researchers, Inc.

Chapter 7

Chapter Opener 7: Marli Miller *Figure 7.2:* Professor Gary A. Smith *Figure 7.3:* Breck P. Kent *Figure 7.4a:* Professor Gary A. Smith *Figure 7.4b:* Fletcher & Baylis/Photo Researchers, Inc. *Figure 7.6a:* Tom & Susan Bean, Inc. *Figure 7.11:* Marli Miller *Figure 7.12a & 7.13a:* Professor Gary A. Smith

Chapter 8

Chapter Opener 8: The Natural History Museum, London *Figure 8.1a, b:* Patrick Lynch, PH ESM *Figure 8.3:* Professor Gary A. Smith *Figure 8.4a:* Photo courtesy of Yildirim Dilek, Miami University *Figure 8.4b:* Dr. Aaron Yoshinobu *Figure 8.6a:* Benjamin Shearn/Getty Images, Inc.-Taxi *Figure 8.8:* Russell D. Curtis/Photo Researchers, Inc.

Chapter 9

Chapter Opener 9: NASA Headquarters *Figure 9.1:* Dorling Kindersley Media Library *Figure 9.2a:* NASA/Jet Propulsion Laboratory *Figure 9.2b:* NASA/ Johnson Space Center *Figure 9.4a&b:* C.R. O'Dell/NASA Headquarters *Figure 9.6:* Photo Researchers, Inc. *Figure 9.13:* D.A. Seal/NASA Headquarters

Chapter 10

Chapter Opener 10: J. Alean, R. Carniel, M. Fulle *Figure 10.9a&b:* Photo courtesy of P. J. Tackley *Figure 10.11b:* Phil Degginger/Color-Pic, Inc. *Figure 10.13:* Richard Megna/Fundamental Photographs, NYC *Figure 10.16 & 10.17:* Gary Glatzmaier

Chapter 11

Chapter Opener 11: Martin Bond/Science Photo Library/Photo Researchers, Inc. *Figure 11.1a, b:* Professor Gary A. Smith *Figure 11.1c:* Jon Price/Nevada Bureau of Mines and Geology NBMG *Figure 11.1d:* Photo courtesy of John W. Geissman *Figure 11.2b:* Ken Hamblin *Figure 11.2c:* William E. Ferguson *Figure 11.4a:* Breck P. Kent *Figure 11.4b:* Bernhard Edmaier/SPL/Photo Researchers, Inc. *Figure 11.5:* Phil Geusebroek *Figure 11.6a:* Professor Gary A. Smith *Figure 11.6b:* Marli Miller *Figure 11.8a:* Jim Wark/AIRPHOTO *Figure 11.8b:* Marli Bryant Miller/Visuals Unlimited *Figure 11.8c&d:* Marli Miller *Figure 11.12:* B.J. Skinner *Figure 11.15a&b:* B. Willis/ U.S. Geological Survey/ Geologic Inquiries Group *Figure 11.17:* M.S. Patterson, Mt. Stromlo & Siding Spring Observatories *Figure 11.22a:* NASA/JET Propulsion Laboratory *Figure 11.22b:* Professor Gary A. Smith *Figure 11.22c:* Lloyd Cuff/Corbis/Bettmann *Figure 11.22d:* Courtesy of Paul Breeding, Maryland Geological Survey, 1988 *Figure 11.25b:* Stephen Trimble *Figure 11.24a&b:* Professor Gary A. Smith *Figure 11.27a:* Lloyd Cuff/CORBIS-NY *Figure 11.27b:* U.S. Geological Survey, Denver *Figure 11.30:* Courtesy of Thomas McGuire/American Geological Institute AGI Image Bank *Figure 11.37a:* Asahi Shimbun/SIPA Press *Figure 11.37c:* Simon Kwong/Reuters New Media Inc./Corbis/Bettmann *Figure 11.37d:* Peter W. Weigand/ Dr. Sandra Jewett *Figure 11.37a:* U.S. Geological Survey/U.S. Department of the Interior *Figure 11.37b:* Loma Prieta Collection, Earthquake Engineering Research Center, University of California, Berkeley. *Figure 11.38:* Reuters Kimimasa Mayama/Getty Images Inc.-Hulton Archive Photos *Figure 11.39b:* DigitalGlobe *Figure 11.39c:* Jimin Lai/AFP/Getty Images, Inc.-Liaison

Chapter 12

Chapter Opener 12: AP Wide World Photos *Figure 12.1a&b:* Patrick Lynch/PH ESM *Figure 12.5a:* Norbert Wu/Minden Pictures *Figure 12.5b:* Bryan & Cherry Alexander/Arcticphoto.co.uk *Figure 12.10:* Bob Embley/NOAA PMEL Vents Program *Figure 12.11a:* Bob Krist/Bob Krist Photography *Figure 12.19b:* Nicholas Parfitt/Getty Images Inc.-Stone Allstock *Figure 12.20a:* National Geophysical Data Center, NOAA *Figure 12.20b:* Alex Blackburn Clayton/Photolibrary.Com *Figure 12.20c:* Toby Fischer *Figure 12.27b:* Photo by Joseph A. Dellinger *Figure 12.28a:* G.K. Gilbert/U.S. Geological Survey, Denver *Figure 12.28b:* Lloyd Cuff/Corbis/Bettmann *Figure 12.31a:* NASA Headquarters *Figure 12.31b:* Aurora Pun *Figure 12.43a-g:* Ron Blakey

Chapter 13

Chapter Opener 13: Paul A. Souders/CORBIS-NY *Figure 13.1a:* Thomas Hallstein/ Alamy Images *Figure 13.1b:* Robert W. Christopherson *Figure 13.1c:* Bobbe Z. Christopherson *Figure 13.14:* Photo by Dr. John Riley from his book *Flora of the Hudson Bay Lowland and its Postglacial Origins*, Courtesy of NRC Research Press. *Figure 13.18a:* NOAA/National Geophysical Data Center *Figure 13.18b:* Super Stock, Inc. *Figure 13.20a & 13.21a:* EyeWire Collection/Getty Images-Photodisc *Figure 13.21b:* AIRPHOTO—Jim Wark *Figure 13.28a:* Marli Miller *Figure 13.28b:* Paul Rezendes *Figure 13.29a:* © Robert Hildebrand *Figure 13.29b:* Marli Miller

Chapter 14

Chapter Opener 14: AIRPHOTO-Jim Wark *Figure 14.1a1:* Professor Gary A. Smith *Figure 14.1.a2:* USDA/NRCS/Natural Resources Conservation Service *Figure 14.1b1 & 14.1b2:* 2 Photo courtesy of Leslie D. McFadden *Figure 14.2b & 14.4a:* USDA/NRCS/Natural Resources Conservation Service *Figure 14.7:* Raymond Weil/Professor Gary A. Smith *Figure 14.9:* U.S. Department of Agriculture *Figure 14.11b:* USDA/NRCS/Natural Resources Conservation Service *Figure 14.11c:* Jeff Vanuga/ USDA/NRCS/Natural Resources Conservation Service *Figure 14.11d:* Tom Edwards/Animals Animals/Earth Scenes *Figure 14.15b,c, & d:* USDA/NRCS/Natural Resources Conservation Service *Figure 14.16b:* Grahame McConnell/Photolibrary.Com *Figure 14.16c:* © 2004 Theodore Gray www.element-collection.com *Figure 14.17:* USDA/NRCS/Natural Resources Conservation Service *Figure 14.18 & 14.19:* Randall J. Schaetzl, Michigan State University *Figure 14.20b:* Professor Gary A. Smith *Figure 14.23a–c, 14.24a&b, 14.25, 14.26b, 14.27, and 14.28:* USDA/NRCS/ Natural Resources Conservation Service

Chapter 15

Chapter Opener 15: AP Wide World Photos *Figure 15.1a, a2:* Dr. David F. Walter, dfwalter@hotmail.com *Figure 15.1a3:* Edwin L. Harp, U.S. Geological Survey/U.S. Department of the Interior *Figure 15.1b1:* J.R. Stacy, U.S. Geological Survey/U.S. Department of the Interior *Figure 15.1b2:* Aurora Pun *Figure 15.1c:* U.S. Geological Survey/U.S. Department of the Interior *Figure 151d1&d2:* AP Wide World Photos *Figure 15.4:* U.S. Geological Survey/U.S. Department of the Interior *Figure 15.5a:* Marli Miller *Figure 15.15b1&b2:* Richard Young *Figure 15.7a:* Michael Dolan *Figure 15.7b:* U.S. Geological Survey/U.S. Department of the Interior *Figure 15.8a:* D. Bradley/National Oceanic and Atmospheric Administration *Figure 15.8b:* National Earthquake Information Center, U.S.G.S. *Figure 15.9 & 15.10b:* U.S. Geological Survey/U.S. Department of the Interior *Figure 15.20:* U.S. Forestry Service *Figure 15.21a1a2:* Terry Donnelley/Getty Images Inc.-Image Bank *Figure 15.21b2:* Professor Gary A. Smith *Figure 15.22a, b:* Lloyd S. Cluff *Figure 15.25b1,b2,b3&b4:* Gary Rosenquist *Figure 15.25b5:* Professor Gary A. Smith *Figure 15.31:* Institute of Geological & Nuclear Sciences Ltd.

Chapter 16

Chapter Opener 16: Adriel Heisey/Adriel Heisey Photography *Figure 16.1a1:* Cliff Riedinger/Alaska Stock *Figure 16.1a2:* Liz Hymans/CORBIS-NY *Figure 16.1b1:* Marli Bryant Miller *Figure 16.1b2:* U. S. Army Corps of Engineers *Figure 16.5b:* James Bartolino/U.S. Geological Survey/U.S. Department of the Interior *Figure 16.7b:* NASA and the US/Japan ASTER Team *Figure 16.8a:* Professor Gary A. Smith *Figure 16.8b:* Kevin Adams/Kevin Adams Nature Photography *Figure 16.11b:* Robert S. Anderson *Figure 16.12b1&b2:* Keith J. Tinkler *Figure 16.13b:* Gary A. Smith/Keith J. Tinkler *Figure 16.14b:* Marli Miller *Figure 16.15b:* NASA/Jet Propulsion Laboratory *Figure 16.22b:* Altitude (Yann Arthus-Bertrand)/Peter Arnold, Inc. *Figure 16.23b & 16.24a&b:* AP Wide World Photos *Figure 16.25a&b:* Earth Satellite Corp./Photo Researchers, Inc. *Figure 16.26:* St. Paul Pioneer Press/CORBIS-NY *Figure 16.32a:* Marli Miller *Figure 16.36a:* Richard T. Nowitz/CORBIS-NY *Figure 16.36b:* Grant Meyer *Figure 16.39:* Tom Bean/Tom & Susan Bean, Inc. *Figure 16.40a:* Claver Carroll/Photolibrary.Com *Figure 16.40b:* Marli Miller *Figure 16.41a:* Michael A. Clynne/Volcano Hazards Team *Figure 16.41b:* Robert L. Schuster/U.S. Geological Survey, Denver *Figure 16.42b:* NASA Headquarters

Chapter 17

Chapter Opener 17: © Tom Till *Figure 17.1a1:* Laura Wilson *Figure 17.1a2:* National Park Service *Figure 17.1b1:* Gerry Davis/Phototake NYC

Figure 17.1b2: Jim Tuten/Black Star/Stockphoto.com/Black Star *Figure 17.2:* Patrick Lynch *Figure 17.3:* Fred Habegger/Grant Heilman Photography, Inc. *Figure 17.4a:* Tony Arruza/CORBIS-NY *Figure 17.4c:* Daniel Dempster/Bruce Coleman Inc. *Figure 17.8:* Robert Shedlock/U.S. Geological Survey/U.S. Department of the Interior *Figure 17.10:* David R. Frazier/Alamy Images *Figure 17.15b1, b2:* Copyright Larry Fellows, Arizona Geological Survey. Image courtesy of American Geological Insitute, ImageBank, http://www.earthscienceworld.org/imagebank *Figure 17.19b1&b2:* Aurora Pun *Figure 17.19b3:* Photo courtesy of Richard Allen Ashmore *Figure 17.26b:* Denis R. LeBlanc/U.S. Geological Survey/U.S. Department of the Interior *Figure 17.31:* Avonsoft Water Treatment Ltd. *Figure 17.32b:* The Natural History Museum, London *Figure 17.36b:* Eve L. Kuniansky/U.S. Geological Survey/U.S. Department of the Interior *Figure 17.36c:* G.R. "Dick" Roberts/The Natural Sciences Image Library (NSIL) *Figure 17.36d:* Jon Gilhousen/U.S. Geological Survey/U.S. Department of the Interior *Figure 17.36e:* A.C. Waltham/Robert Harding World Imagery *Figure 17.38b:* Stuart Westmorland/CORBIS-NY *Figure 17.38c:* Hans Strand/Getty Images Inc.-Image Bank *Figure 17.41b:* Marli Miller

Chapter 18

Chapter Opener 18: Peter Arnold, Inc. *Figure 18.1a:* Dennis Hallinan/Alamy Images *Figure 18.1b:* Phil Degginger/Color-Pic, Inc. *Figure 18.1c–f:* Peter J. Fawcett *Figure 18.2:* Top photo, courtesy of Mary Schaffer, Whyte Museum of the Canadian Rockies, Banff. Permission obtained from the Whyte Museum. Bottom photo courtesy of Dr. Brian Luckman, University of Western Ontario. *Figure 18.3a1:* NASA Headquarters *Figure 18.3a2:* Kim Karpeles, Life Through the Lens *Figure 18.3b1:* Orbital Imaging Corporation/Photo Researchers, Inc. *Figure 18.3b2:* Gordon Wiltsie/National Geographic Image Collection *Figure 18.3c1:* Space Imaging *Figure 18.3c2:* Macduff Everton/CORBIS-NY *Figure 18.4b:* Galen Rowell/CORBIS-NY *Figure 18.6b1:* Tom & Susan Bean, Inc. *Figure 18.6b2:* Richard Kucera/John S. Shelton *Figure 18.7b:* Bruce Molnia/American Geological Institute AGI *Figure 18.10b:* Paul A. Souders/CORBIS-NY *Figure 18.12b1:* Professor Gary A. Smith *Figure 18.12c:* Tony Waltham/Tony Waltham Geophotos *Figure 18.14b:* Simon Fraser/Photo Researchers, Inc. *Figure 18.15:* Tom & Susan Bean, Inc. *Figure 18.16b1:* Marli Miller *Figure 18.16b2:* Photo courtesy of Professor Randall J. Schaetzl, Michigan State University *Figure 18.17:* Grace Davies/Grace Davies Photography *Figure 18.18a:* Photo courtesy of Nicholas Eyles *Figure 18.18b:* Tom & Susan Bean, Inc. *Figure 18.20a:* NASA Headquarters *Figure 18.20b:* Réjean Couture & Gianluca Bianchi Fasani, Courtesy of Earth Sciences Information Centre, National Earth Sciences Library. *Figure 18.21b:* Bradford Washburn *Figure 18.22a:* Grant Heilman/Grant Heilman Photography, Inc. *Figure 18.22b:* Mark Newman/Alaska Stock *Figure 18.23a&b:* British Antarctic Survey *Figure 18.24a&b & 18.25b:* NASA *Figure 18.25c:* Aurora Pun *Figure 18.27a:* Gerald & Buff Corsi/Visuals Unlimited *Figure 18.27b:* Sunset Avenue Productions/Digital Vision Ltd. *Figure 18.29b:* Neil Rabinowitz/CORBIS-NY *Figure 18.29c:* Lowell Georgia/CORBIS-NY *Figure 18.30:* Tom Bean/CORBIS-NY *Figure 18.31:* AIRPHOTO-Jim Wark *Figure 18.33a:* © Robert Hildebrand *Figure 18.33b&c:* Lynda Dredge, Courtesy of Earth Sciences Information Centre, Canada's National Earth Sciences Library. *Figure 18.33d:* John S. Shelton *Figure 18.33f:* Randall J. Schaetzl, Michigan State University *Figure 18.33g:* Tom & Susan Bean, Inc. *Figure 18.35:* JPL *Figure 18.38:* Mike Blum *Figure 18.40b:* Trent Nelson/America 24-7/Getty Images, Inc. *Figure 18.40c & 18.43b:* Professor Gary A. Smith

Chapter 19

Chapter Opener 19: Richard Price/Getty Images, Inc.-Taxi *Figure 19.1a–c:* Professor Gary A. Smith *Figure 19.1d:* Heim/U.S. Army Corps of Engineers, Headquarters *Figure 19.1e1&e2:* © BradleyIreland.com *Figure 19.5:* Bill Curtsinger/National Geographic Image Collection *Figure 19.6a:* Penny Tweedie/Getty Images Inc.-Stone Allstock *Figure 19.6b:* Warren Bolster/Getty Images Inc.-Stone Allstock *Figure 19.7a:* Nick Green/Photolibrary.Com *Figure 19.7b:* Gary Braasch/CORBIS-NY *Figure 19.8b:* John S. Shelton *Figure 19.10:* Agence France Presse/Getty Images *Figure 19.13b:* TLM Photo *Figure 19.14b:* © Dick Jones/RJ Aerial Photo *Figure 19.12b top & bottom:* Ikonos images copyright Centre for Remote Imaging, Sensing and Processing, National University of Singapore and Space Imaging/NASA. *Figure 19.16b:* Andrew Jaster *Figure 19.18d:* G.R. "Dick" Roberts/The Natural Sciences Image Library (NSIL) *Figure 19.19a:* Professor Gary A. Smith *Figure 19.19b:* Martyn F. Chillmaid/Science Photo Library/Photo Researchers, Inc. *Figure 19.20b:* Rafael Macia/Photo Researchers, Inc. *Figure 19.21b:* John S. Shelton *Figure 19.22b:* NASA Headquarters *Figure 19.23b:* Photo courtesy of Joseph R. Melanson/skypic.com *Figure 19.24b:* AIRPHOTO-Jim Wark *Figure 19.25a1:* Bob Pardue/Alamu Images *Figure 19.25a2:* Andrew Jaster *Figure 19.25b1:* Professor Gary A. Smith *Figure 19.25b2:* Aurora Pun *Figure 19.25c1:* Dave Houser/CORBIS-NY *Figure 19.25c2:* Photo courtesy of Michael Levin *Figure 19.27:* Photo courtesy of Jon Arthur of the Florida Geological Survey and the Florida Department of Transportation. *Figure 19.29b:* NASA Headquarters *Figure 19.29c, d:* FRF/U.S. Army Corps of Engineers, Headquarters *Figure 19.30b:* (c) Joseph R. Melanson of www.skypic.com *Figure 19.35a:* Joyce Photographics/Photo Researchers, Inc. *Figure 19.36b:* U.S. Department of Agriculture *Figure 19.38a2:* John S. Shelton *Figure 19.38b2:* Marli Miller/Visuals Unlimited *Figure 19.39a&b:* U.S. Geological Survey, Denver *Figure 19.40a:* Photo courtesy of Paul E. Olsen (N.G. McDonald Collection) *Figure 19.40b:* Mimmo Jodice/CORBIS-NY *Figure 19.43:* Photo courtesy Bob & Sandra Shanklin "the Lighthouse People" www.thelighthousepeople.com *Figure 19.45c:* Larry Lipsky/Tom Stack & Associates, Inc. *Figure 19.49b&c:* Leatherman/Professor Gary A. Smith *Figure 19.50:* Professor Gary A. Smith *Figure 19.51:* Royalty-Free/Corbis RF *Figure 19.53a&b:* U.S. Army Corps of Engineers, Headquarters *Figure 19.54a:* AP Wide World Photos *Figure 19.54b:* North Carolina Dept. of Transportation

Chapter 20

Chapter Opener 20: Eric Wunrow/Index Stock Imagery, Inc. *Figure 20.1a1:* Professor Gary A. Smith *Figure 20.1a2:* Marli Miller *Figure 20.1b1:* Richard T. Nowitz/CORBIS-NY *Figure 20.1b2:* David Muench/CORBIS-NY *Figure 20.5a:* NASA Headquarters *Figure 20.5b:* Professor Gary A. Smith *Figure 20.6b:* Bruno Morandi/Robert Harding World Imagery *Figure 20.6c:* Charlie Ott/Photo Researchers, Inc. *Figure 20.6d:* Bernhard Edmaier/Photo Researchers, Inc. *Figure 20.6e:* Norbert Wu 2/Peter Arnold, Inc. *Figure 20.8:* Craig Aurness/CORBIS-NY *Figure 20.10a:* NASA Headquarters *Figure 20.10b:* Courtesy of Larry W. Thomason, NASA Langley Research Center *Figure 20.11c:* John S. Shelton *Figure 20.12b:* Photo courtesy of David Love, New Mexico Bureau of Geology and Mineral Resources) *Figure 12.b2:* Professor Gary A. Smith *Figure 20.13a&b:* NASA/Jet Propulsion Laboratory *Figure 20.13c:* Marli Miller *Figure 20.14a:* Marli Miller *Figure 20.15b:* Photo courtesy of Dr. Patrick Hesp *Figure 20.15c:* Gerry Ellis/Minden Pictures *Figure 20.15d:* Pete Oxford/DRK Photo *Figure 20.15e:* Georg Gerster/Photo Researchers, Inc. *Figure 20.15f:* Michael Collier *Figure 20.17b:* Phil Schermeister/CORBIS-NY *Figure 20.17d:* Michael Andrews/Animals Animals/Earth Scenes *Figure 20.17c & 20.18a&b:* Professor Gary A. Smith *Figure 20.18c:* Soil Science Society of America, Inc. *Figure 20.20b:* NASA Headquarters *Figure 20.20c:* Tom Bean/Corbis *Figure 20.21a1,a2 & b:* Photos courtesy of Virginia Garrison, USGS; NASA/MODIS *Figure 20.22b:* Marli Miller

Index

Gary A. Smith/Aurora Pun
How Does Earth Work?
Physical Geology and the Process of Science
Active Art and Extension Modules
Instructor's Edition
0-13186746-6
© 2006 Pearson Education, Inc.
Pearson Prentice Hall
Pearson Education, Inc.
Upper Saddle River, NJ 07458
Pearson Prentice Hall™ is a trademark of Pearson Education, Inc.

SYSTEM REQUERMENTS

Windows:
266 MHz Intel Pentium processor
Windows 98/ME/2000/NT4-sp3*/XP
16 bit Sound Card
32 MB or more of available RAM
Best viewed at 1024 x 768 monitor resolution
Mouse or other pointing device
4x CD-ROM Drive

Macintosh:
PowerPC 200Mhz processor
Operating System OS9.2.1/OSX
32 MB or more of available RAM
Best viewed at 1024 x 768 monitor resolution
Mouse or other pointing device
4x CD-ROM Drive
For some early release versions of OS9 you may need to install the "Carbonlib" system file (see "Known issues" in the readme file.)

Credits

Chapter 1
Chapter Opener 1: NASA/Johnson Space Center *Figure 1.1a:* After Peter Carsten/National Geographic Image Collection *Figure 1.1b:* Lester Lefkowitz/Corbis/Bettmann *Figure 1.1c:* Melvin Grubb/Grubb Photo Service, Inc. *Figure 1.2a:* Oldrich Karasek/Getty Images Inc.-Stone Allstock *Figure 1.2b:* John Wang/Getty Images, Inc.- Photodisc *Figure 1.2c:* Jim Wark/Peter Arnold, Inc. *Figure 1.3:* After Tarbuck and Lutgens, *Earth, An Introduction to Physical Geology* 8th ed., Prentice Hall *Figure 1.5:* Based on maps from the Association of Bay Area Governments *Figure 1.6a:* Eyewire Collection/Getty Images-Photodisc *Figure 1.6b:* Colin Keates © Dorling Kindersley, Courtesy of the Natural History Museum, London *Figure 1.6c:* Albert Copley/Visuals Unlimited *Figure 1.8a:* Dr. Parvinder S. Sethi *Figure 1.8b:* Dorling Kindersley *Figure 1.9a:* Ross W. Boulanger *Figure 1.10:* Earthquake and volcano data from Smithsonian Institution, Global Volcanism Program *Figure 1.12:* James A. Sugar/Corbis Bettmann

Chapter 2
Chapter Opener 2: Erica Van Pelt/Van Pelt Photography *Figure 2.1a:* Kevin Schafer/Corbis/Bettmann *Figure 2.1b:* Gary A. Smith *Figure 2.2a:* The Natural History Museum, London *Figure 2.2b:* Charles D. Winters/Photo Researchers, Inc. *Figure 2.3:* Gary A. Smith *Figure 2.5a&b:* Aurora Pun *Figure 2.6:* Chip Clark *Figure 2.7a&b:* Aurora Pun *Figure 2.8:* Aurora Pun *Figure 2.9a:* Breck P. Kent *Figure 2.9b:* E.R. Degginger *Figure 2.9c:* Mark A. Schneider/Visuals Unlimited *Figure 2.11:* Courtesy of David Barber *Figure 2.13:* After Chernicoff and Whitney, Geology, 3rd edition-Houghton Mifflin, 2002 *Figure 2.18:* Clive Streeter © Dorling Kindersley *Figure 2.21:* After W.D. Carlson, 1980, American Mineralogist-65, pp. 1252–1262 *Figure 2.23a:* After Chip Clark *Figure 2.23b:* Jeffrey A. Scovil *Figure 2.25a:* Chip Clark *Figure 2.25b:* Breck P. Kent *Figure 2.26a:* Dennis Tasa, Tasa Graphic Arts, Inc. *Figure 2.26b:* Richard M. Busch *Figure 2.26c:* Harry Taylor © Dorling Kindersley *Figure 2.27:* Chip Clark *Figure 2.28:* M. Claye/Jacana Scientific Control/Photo Researchers, Inc. *Figure 2.29:* Breck P. Kent/Animals Animals/Earth Scenes *Figure 2.30a:* Chip Clark *Figure 2.30b:* Breck P. Kent *Figure 2.30c:* Photolibrary.Com *Table 2.2:* (amphibole, chalcopyrite, corundum, gold, hematite, magnetite) Harry Taylor © Dorling Kindersley; (calcite) Colin Keates (c) Dorling Kindersley, Courtesy of the Natural History Museum, London; (copper, pyrite, sphalerite) Breck P. Kent/Animals Animals/Earth Scenes; (dolomite, garnet) Jeffrey A. Scovil; (feldspar) Charles D. Winters/Photo Researchers, Inc.; (galena) Chip Clark; (gypsum) RUNK/SCHOENBERGER/Grant Heilman Photography, Inc.; (olivine) Dennis Tasa, Tasa Graphic Arts, Inc.; (pyroxene, halite) Richard M. Busch; (quartz) Photo Researchers, Inc.; (mica) Marli Bryant Miller

Chapter 3
Chapter Opener 3: Tyler Stableford/The Image Bank/Getty Images, Inc. *Figure 3.1a:* Bert Sagara/Getty Images Inc.-Stone Allstock *Figure 3.1b:* Kim Heacox/DRK Photo *Figure 3.1c, d:* Gary A. Smith *Figure 3.2a:* Bruce Forster © Dorling Kindersley *Figure 3.2b:* Andrew Jaster *Figure 3.2c:* Gary A. Smith *Figure 3.2d:* George D. Lepp/Photo Researchers, Inc. *Figure 3.3a:* Aurora Pun *Figure 3.3b:* Gary A. Smith *Figure 3.4a:* Douglas Peebles Photography *Figure 3.4b:* D.A. Swanson/Hawaiian Volcano Observatory, U.S. Geological Survey *Figure 3.4c1:* Andrew Alden *Figure 3.4c2&d:* Aurora Pun *Figure 3.7a:* Gary A. Smith *Figure 3.7b:* Alan Kearney/Getty Images, Inc.-Taxi *Figure 3.9a:* Steve Austin; Papilo/Corbis/Bettmann *Figure 3.9b:* Craig Aurness/Corbis/Bettmann *Figure 3.11:* U.S. Geological Survey, U.S. Department of the Interior

Chapter 4
Chapter Opener 4: Salvatore Ragonese/Agence France Presse/Getty Images *Figure 4.1a:* Peter Mouginis-Mark *Figure 4.1b:* U.S. Air Force *Figure 4.1c:* © Craig Aurness/CORBIS *Figure 4.1d:* Gary A. Smith *Figure 4.2a:* Albert J. Copley/Getty Images, Inc.-Photodisc *Figure 4.2b:* Andreas Einsiedel © Dorling Kindersley *Figure 4.2c:* Aurora Pun *Figure 4.2d:* Gary A. Smith *Figure 4.2e:* Grace Davies/Omni-Photo Communications, Inc *Figure 4.3b:* Breck P. Kent/Animals Animals/Earth Scenes *Figure 4.3c, f:* Breck P. Kent *Figure 4.3d:* Geoscience/Pearson Education PH College *Figure 4.3e:* Donna Tucker *Figure 4.3g:* Jeffrey A. Scovil *Figure 4.3h, i:* Harry Taylor © Dorling Kindersley *Figure 4.3j:* Charles E. Jones *Figure 4.4a:* J.P. Lockwood/U.S. Geological Survey/U.S. Department of the Interior *Figure 4.4b:* Breck P. Kent/Animals Animals/Earth Scenes *Figure 4.4c:* Gary A. Smith *Figure 4.4d:* Aurora Pun *Figure 4.5:* Gary A. Smith *Figure 4.6:* After Tarbuck and Lutgens, *Earth*, 7th ed, Prentice Hall, 2002 *Figure 4.7a:* Courtesy of Karl E. Karlstrom *Figure 4.7b:* Marli Bryant Miller *Figure 4.8a:* Adriel Heisey/Network Aspen *Figure 4.8b:* Gary A. Smith *Figure 4.9a:* R.T. Holcomb/Hawaiian Volcano Observatory, U.S. Geological Survey *Figure 4.9b:* Dan Suzio/Photo Researchers, Inc. *Figure 4.9c:* J.D. Griggs/ Hawaiian Volcano Observatory, U.S. Geological Survey *Figure 4.10a:* © Michael T. Sedam/CORBIS *Figure 4.10b:* John Gerlach/Animals Animals/Earth Scenes *Figure 4.11:* Jeff Wynn/USGS/Cascades Volcano Observatory *Figure 4.11a:* Marli Bryant Miller/Visuals Unlimited *Figure 4.11b:* David Falconer/DRK Photo *Figure 4.12:* Solarfilma ehf *Figure 4.13a, b:* Gary A. Smith *Figure 4.14a:* Michael Collier *Figure 4.14b:* J.D. Griggs/Hawaiian Volcano Observatory, U.S. Geological Survey *Figure 4.14c:* © Roger Ressmeyer/CORBIS *Figure 4.14d:* Michael Clynne/U.S. Geological survey/U.S. Department of the Interior *Figure 4.15:* J.D. Griggs/ Hawaiian Volcano Observatory, U.S. Geological Survey *Figure 4.16b:* François Gohier *Figure 4.18:* After Davidson, Reed, and Davis, *Exploring Earth*, 2nd ed., Prentice Hall, 2002 *Figures 4.19, 4.20:* After Tuttle and Bowen, 1958, *Geol. Soc. America Memoir* 74 *Figure 4.26:* Paul Chesley/Getty Images, Inc.-Stone Allstock *Figure 4.27:* Yomiuri/AP Wide World Photos *Figure 4.28:* Archives de l'Academie des Sciences *Figure 4.29:* J. LangevinCorbis/Sygma *Figure 4.31:* Kennecott Utah Copper Corporation *Figure 4.32:* Data from United States Department of Energy-Energy Information Agency [USDOE-EIA] *Figure 4.33a:* Courtesy of Michael N. Spilde *Figure 4.33b:* Jeffrey A. Scovil *Figure 4.33c:* Getty Images, Inc.-Photodisc

Chapter 5
Chapter Opener 5: Yva Momatiuk/John Eastcott/Minden Pictures *Figure 5.1a:* Gary A. Smith *Figure 5.1b:* Tom Bean/DRK Photo *Figure 5.1c,d,e & 5.2a:* Gary A. Smith *Figure 5.2b:* Marli Bryant Miller *Figure 5.3a2:* Gary A. Smith *Figure 5.3b2:* Larry Ulrich/DRK Photo *Figure 5.4a:* Aurora Pun *Figure 5.4b:* Dan Guravich/Photo Researchers, Inc. *Figure 5.4c:* Courtesy of John Bloch *Figure 5.5a:* Gary A. Smith *Figure 5.5b:* Jess Alfrod/Getty Images, Inc.-Photodisc *Figure 5.5c:* © David Muench/David Muench Photography Inc. *Figure 5.6a:* Colin Keates © Dorling Kindersley, Courtesy of the Natural History Museum, London *Figure 5.6b:* Colin Keates © Dorling Kindersley *Figure 5.6c:* Breck P. Kent *Figure 5.6d:* Joy Spurr/Bruce Coleman Inc. *Figure 5.7a1:* Tim Mead/Photolibrary.Com *Figure 5.7a2:* Spike Walker/Getty Images, Inc.-Stone Allstock *Figure 5.7b:* Dave Houseknecht/U.S. Geological Survey/U.S. Department of the Interior *Figure 5.9a, b:* Marli Bryant Miller *Figure 5.9c:* Gary A. Smith *Figure 5.11a:* Breck P. Kent *Figure 5.11b:* Colin Keates © Dorling Kindersley, Courtesy of the Natural History Museum, London *Figure 5.11c, d:* Dr. B. Booth/GeoScience Features Picture Library *Figure 5.11e:* Wally Eberhart/Visuals Unlimited *Figure 5.12:* Richard M. Busch *Figure 5.12a:* Joyce Photographics/Photo Researchers, Inc. *Figure 5.12b&e:* Breck P. Kent *Figure 5.12d:* © Harry Taylor/Dorling Kindersley Media Library *Figure 5.12f:* Mark A. Schneider/Visuals Unlimited *Figure 5.13a:* David R. Frazier/Photo Researchers, Inc. *Figure 5.13b&d:* Gary A. Smith *Figure 5.13c:* MODIS/NASA Headquarters *Figure 5.14a:* Tom Bean/Tom & Susan Bean, Inc. *Figure 5.15:* After Chernicoff and Whitney, *Geology* (Houghton Mifflin, 2002, 3rd ed.) *Figure 5.14b & 5.15a&b:* Gary A. Smith *Figure 5.16a1:* High Sitton/Getty Images, Inc.-Stone Allstock *Figure 5.16a2:* Richard Hamilton Smith/Corbis/Bettmann *Figure 5.16c1:* Marli Bryant Miller *Figure 5.16c2:* Gary A. Smith *Figure 5.17a, b:* Aurora Pun *Figure 5.17c, 18:* Gary A. Smith *Figure 5.19:* After Levin, *The Earth through Time*, 4th ed. Saunders *Figure 5.20a:* Marli Bryant Miller *Figure 5.20b, c:* Gary A. Smith *Figure 5.21b&c:* "Turbidity currents as a cause of graded bedding", by P.H. Kuenen and C.I. Migliorini, *The Journal of Geology*, volume 58 (1950). *Figure 5.22:* After D. J. W. Piper et al., 1988, *Geol. Soc. America Spec. Pap.* 229, pp. 77–92 *Figure 5.23:* After Levin, *The Earth through Time*, 4th ed. (Saunders); Ronald C. Blakey

Chapter 6
Chapter Opener 6: Muench Photography, Inc. *Figure 6.1a&b:* Dr. J. Alcock, Penn State Abington College *Figure 6.1c:* Charles E. Jones *Figure 6.6:* After Davidson, Reed, and

Davis, *Exploring Earth*, 1st ed, 1997, Prentice Hall *Figure 6.8:* After Duff, *Holmes' Principles of Physical Geology*, 4th ed. 1993, Chapman and Hall *Figure 6.9a:* Colin Keates © Dorling Kindersley, Courtesy of the Natural History Museum, London *Figure 6.9b:* Jim Wehtje/Getty Image, Inc.-Photodisc *Figure 6.10b:* Doug Martin/Photo Researchers, Inc. *Figure 6.11b:* Marli Bryant Miller *Figure 6.12b:* Marli Bryant Miller/Visuals Unlimited *Figure 6.13a, b:* Aurora Pun *Figure 6.15:* After W. D. Yardley, W. S. MacKenzie, and C. Guilford, *Atlas of Metamorphic Rocks and their Textures*, 1990, Longman Scientific and Technical, Essex *Figure 6.16:* After A. D. Edgar, *Experimental Petrology: Basic Principles and Techniques*, 1973, Clarendon Press *Figures 6.17, 6.18:* Data from M. J. Holdaway and B. Mukhopadhyay, 1993, *American Mineralogist*, vol. 78, p. 298 *Figure 6.20a:* Charles E. Jones *Figure 6.20b, g&l:* Harry Taylor © Dorling Kindersley *Figure 6.20c,e, f&h:* Charles E. Jones *Figure 6.20d:* Richard M. Busch *Figure 6.20i:* E.R. Degginger/Color-Pic, Inc. *Figure 6.20j&k:* Aurora Pun *Figure 6.21:* After Marshak, *Earth, Portrait of a Planet*, Norton, 2001 *Figure 6.22:* Larry D. Fellows/Arizona Geological Survey *Figure 6.24:* After Davidson, Reed, and Davis, *Exploring Earth*, 1st ed, 1997, Prentice Hall; after Nagy and Parmentier, 1982, *Earth and Planetary Science Letters*, vol. 59, pp. 1–10 *Figure 6.26:* Ken MacDonald/ Science Photo Library/Photo Researchers, Inc. *Figure 6.27:* After Humphris et al., 1995, *Nature*, vol. 377, p. 713 *Figure 6.29:* After Duff, *Holmes' Principles of Physical Geology*, 4th ed. 1993, Chapman and Hall

Chapter 7

Chapter Opener 7: Marli Bryant Miller *Figure 7.2:* Gary A. Smith *Figure 7.3:* Breck P. Kent *Figure 7.4a:* Gary A. Smith *Figure 7.4b:* Fletcher & Baylis/Photo Researchers, Inc. *Figure 7.6a:* Tom & Susan Bean, Inc. *Figure 7.8:* After C. L. and M. A. Fenton, *Giants of Geology*, Doubleday, 1952 *Figure 7.9:* After International Stratigraphic Chart, 2003, by the International Commission on Stratigraphy *Figure 7.11:* Marli Bryant Miller *Figure 7.12a & 7.13a:* Gary A. Smith *Figure 7.19:* Data in (a) from Libby, W. F., 1961, Radiocarbon Dating, *Science*, vol. 133, pp. 621–629. Data in (b) from Williams, I. S., Compston, W., and Chappell, B. W., 1983, Zircon and monazite U-Pb systems and the histories of I-type magmas, Berridale batholith, Australia, *Journal of Petrology*, vol. 21, pp. 76–97 *Figure 7.23:* Data from Beckinsale, R. D., and Gale, N. H., 1969, A reappraisal of the decay constants and branching ratio of ^{40}K, *Earth and Planetary Science Letters*, vol. 6, pp. 289–294

Chapter 8

Chapter Opener 8: The Natural History Museum, London *Figure 8.1a, b:* Patrick Lynch, PH ESM *Figure 8.3:* Gary A. Smith *Figure 8.4a:* Courtesy of Yildirim Dilek *Figure 8.4b:* Courtesy of Aaron Yoshinobu *Figure 8.5:* After J. Elder, 1976, *The Bowels of the Earth*, Oxford *Figure 8.6a:* Benjamin Shearn/Getty Images, Inc.-Taxi *Figure 8.7:* After B. A. Bolt, 1976, *Nuclear Explosions and Earthquakes*, W. H. Freeman and Co. *Figure 8.8:* Russell D. Curtis/Photo Researchers, Inc. *Figure 8.14:* After G. C. Brown and A. E. Mussett, 1981, *The Inaccessible Earth*: Allen and Unwin *Figure 8.15:* Source: After A.E. Mussett and M.A. Khan, 2000, *Looking into the Earth*: Cambridge *Figures 8.16, 8.18:* After Christensen, N. I., and Wepfer, W. W., 1989, Laboratory techniques for determining seismic velocities and attenuations, with applications to the continental lithosphere, in L. C. Pakiser and W. D. Mooney, eds., *Geophysical framework of the continental United States: Geological Society of America Memoir* 172, pp. 91–102 *Figure 8.22:* Data from A.E. Mussett and M.A. Khan, 2000, *Looking into the Earth*, Cambridge

Chapter 9

Chapter Opener 9: NASA Headquarters *Figure 9.1:* Dorling Kindersley Media Library *Figure 9.2a:* NASA/Jet Propulsion Laboratory *Figure 9.2b:* NASA/Johnson Space Center *Figure 9.4a&b:* C.R. O'Dell/NASA Headquarters *Figure 9.5:* After W. K. Hamblin and E. H. Christiansen, *Exploring the Planets*, 1990, Macmillian *Figure 9.6:* Photo Researchers, Inc. *Figure 9.9:* After M. J. Drake and K. Righter (2002), Determining the composition of the Earth, *Nature*, vol. 416, pp. 39–44 *Figure 9.10:* After G. Schubert, D. L. Turcotte, and P. Olson, 2001, *Mantle Convection in the Earth and Planets*; Cambridge University Press *Figure 9.13:* D.A. Seal/NASA Headquarters *Figure 9.14:* After M. J. Drake and K. Righter (2002), Determining the Composition of the Earth, *Nature*, vol. 416, pp. 39–44 *Figure 9.15:* After Ron Miller *Figure 9.16:* Data for volcanoes from H. Sigurdsson, editor, 2000, *Encyclopedia of Volcanoes*, Academic Press. Atmospheric data from G. Faure, 1991, *Principles and Applications of Inorganic Geochemistry*, Macmillan Publishing Company

Chapter 10

Chapter Opener 10: J. Alean, R. Carniel, M. Fulle *Figure 10.1:* Cadre Design *Figure 10.5:* Map courtesy of Henry Pollack *Figure 10.7:* After X.-D. Li and B., 1996, "Global Mantle Shear-Velocity Model Developed Using Nonlinear Asymptotic Coupling Theory", *J. Geophys.* Res., vol. 101, No. B10, 22,245-22,272, and G. Schubert, D. L. Turcotte, and P. Olson, 2001, *Mantle Convection in the Earth and planets*, Cambridge University Press *Figure 10.8:* After B. Travis, P. Olson, and G. Schubert, 1990, The transition from two-dimensional to three-dimensional planforms in infinite-Prandtl-number thermal convection: *Journal of Fluid Mechanics*, vol. 216, pp. 71–91 *Figure 10.9a&b:* Photo courtesy of P. J. Tackley, D. J. Stevenson, G. A. Glatzmaier, and G. Schubert, 1993, Effects of an endothermic phase transition at 670 km depth in a spherical model of convection in the Earth's mantle, *Nature*, v. 361, p. 699-704 *Figure 10.11b:* Phil Degginger/Color-Pic, Inc. *Figure 10.12:* Left map after *Geotimes*, May 2002, p. 10. Graph data from the National Geophysical Data Center *Figure 10.13:* Richard Megna/Fundamental Photographs, NYC *Figure 10.15:* Top diagram after W.M. Elsasser, 1958, The Earth as a dynamo, *Scientific American*. Bottom diagram after J. Bloxham and D. Gubbins, 1989, The evolution of the Earth's magnetic field, *Scientific American* *Figure 10.16 & 10.17:* After Glatzmaier, G.A., 2002, Geodynamo simulations - how realistic are they? *Annual Reviews of Earth and Planetary Science*, v. 30, p. 237-257

Chapter 11

Chapter Opener 11: Martin Bond/Science Photo Library/Photo Researchers, Inc. *Figure 11.1a, b:* Gary A. Smith *Figure 11.1c:* Jon Price/Nevada Bureau of Mines and Geology NBMG *Figure 11.1d:* Courtesy of John W. Geissman *Figure 11.2b:* William E. Ferguson *Figure 11.4a:* Breck P. Kent *Figure 11.4b:* Bernhard Edmaier/SPL/Photo Researchers, Inc. *Figure 11.5:* Phil Geusebroek *Figure 11.6a:* Gary A. Smith *Figure 11.6b:* Marli Bryant Miller *Figure 11.8a:* Jim Wark/AIRPHOTO *Figure 11.8b:* Marli Bryant Miller/Visuals Unlimited *Figure 11.8c&d:* Marli Bryant Miller *Figure 11.9:* After L. A. Woodward and R. L. Ruetschilling, 1976, *Geology of San Ysidro Quadrangle, New Mexico*, New Mexico Bureau of Mines and Mineral Resources Geologic Map 37 *Figure 11.11:* After E. J. Tarbuck and F. K. Lutgens, 2005, *Earth, An Introduction to Physical Geology*, 8th ed., Prentice Hall *Figure 11.12:* B.J. Skinner *Figure 11.14:* After G. H. Davis, 1984, *Structural Geology of Rocks and Regions*, Wiley and Sons *Figure 11.15a&b:* B. Willis/ U.S. Geological Survey/Geologic Inquiries Group *Figure 11.16:* After S. Judson and S. M. Richardson, 1995, *Earth, an Introduction to Geologic Change*, Prentice Hall *Figure 11.17:* M.S. Patterson, Mt. Stromlo & Siding Spring Observatories *Figures 11.19, 11.20:* After M. P. Billings, *Structural Geology*, 3rd edition, Prentice Hall *Figure 11.22a:* NASA/JET Propulsion Laboratory *Figure 11.22b:* Gary A. Smith *Figure 11.22c:* Lloyd Cuff/Corbis/Bettmann *Figure 11.22d:* Courtesy of Paul Breeding, Maryland Geological Survey, 1988 *Figure 11.24a&b:* Gary A. Smith *Figure 11.25b:* Stephen Trimble *Figure 11.26:* After D. L. Kohlstedt, B. Evans, and S. J. Mackwell, 1995, Strength of the lithosphere: Constraints imposed by laboratory experiments, *Journal of Geophysical Research*, vol. 100, pp. 17587–17602 *Figure 11.27:* After A. Jones, L. Siebert, P. Kimberly, and J. F. Luhr, 2000, *Earthquakes and eruptions*, Smithsonian Institution, Global Volcanism Program Digital Information Series GVP-2 *Figure 11.27a:* Lloyd Cuff/CORBIS-NY *Figure 11.27b:* U.S. Geological Survey, Denver *Figure 11.28:* USGS *Figure 11.29:* Data from the Southern California Earthquake Data Center *Figure 11.30:* Courtesy of Thomas McGuire/American Geological Institute AGI Image Bank *Figure 11.31:* Courtesy of David Schmidt, University of Oregon *Figure 11.32:* U.S. Geological Survey Fact Sheet 014-03 *Figure 11.33:* After L. Brewer, 1992, Preliminary damage and intensity survey, *Earthquakes and Volcanoes*, volume 23, pp. 219–226; U.S. Geological Survey *Figure 11.34:* After B. A. Bolt, 1993, *Earthquakes and Geological Discovery*, Scientific American Library *Figure 11.36:* After IRIS Consortium, 2002, How often do earthquakes occur? Education and Outreach Series No. 3 *Figure 11.37a:* Asahi Shimbun/SIPA Press *Figure 11.37b:* USGS *Figure 11.37c:* Simon Kwong/Reuters New Media Inc./Corbis/Bettmann *Figure 11.37d:* Peter W. Weigand/Dr. Sandra Jewett *Figure 11.38:* After G. Plafker and J. P. Galloway, 1989, Lessons Learned from the Loma Prieta, California, Earthquake of October 17, 1989, U.S. Geological Survey Circular 1045; S. E. Hough, 2002, *Earthshaking Science, What We Know (and Don't Know) about Earthquakes*, Princeton University Press, NJ; photo from Earthquake Engineering Research Center —Loma Prieta Collection *Figure 11.39:* After F. I. Gonzales, May 1999, Tsunami! *Scientific American*, pp. 58–65; photo Kimimasa Mayama/Reuters/Getty Images *Figure 11.39b:* DigitalGlobe *Figure 11.39c:* Jimin Lai/AFP/Getty Images, Inc.-Liaison *Figure 11.40:* USGS

Chapter 12

Chapter Opener 12: AP Wide World Photos *Figure 12.1a&b:* Patrick Lynch/PH ESM *Figure 12.2:* After A. Wegener, 1915, Die Entstehung der Kontinente und Ozeane, Vieweg, Braunschweig *Figure 12.3:* After A. Jones, L. Siebert, P. Kimberly, and J. F. Luhr, 2000, *Earthquakes and Eruptions*, Smithsonian Institution, Global Volcanism Program Digital Information Series, GVP-2 *Figure 12.5a:* Norbert Wu/Minden Pictures

Figure 12.5b: Bryan & Cherry Alexander/Arcticphoto.co.uk *Figure 12.6, 12.7:* After A. Jones, L. Siebert, P. Kimberly, and J. F. Luhr, 2000, *Earthquakes and Eruptions*, Smithsonian Institution, Global Volcanism Program Digital Information Series, GVP-2
Figure 12.9: After E. J. Tarbuck and F. K. Lutgens, 2002, *Earth, An Introduction to Physical Geology*, 7th ed., Prentice Hall *Figure 12.10:* Bob Embley/NOAA PMEL Vents Program *Figure 12.11a:* Bob Krist/Bob Krist Photography; *Figure 12.11b:* After J. A. Karson, and P. A. Rona, 1990, Block-tilting, transfer faults, and structural control of magmatic and hydrothermal processes in the TAG area, Mid-Atlantic Ridge 26°N., Geological Society of America, vol. 102, no. 12, pp. 1635–1645 *Figure 12.14:* After R. D. Müller, W. R. Roest, J. Y. Royer, L. M. Gahagan, and J. G. Sclater, 1997, Digital isochrons of the world's ocean floor, *Journal of Geophysical Research*, vol. 102, pp. 3211–3214
Figure 12.15: (top after) R. P. von Herzen and S. Uyeda, 1963, Heat flow through the eastern Pacific Ocean floor, *Journal of Geophysical Research*, vol. 68, pp. 4219–4250; (bottom after) W. C. Pitman, 1978, The relationship between eustasy and stratigraphic sequences of passive margins, *Geological Society of America Bulletin*, vol. 89, pp. 1389–1403 *Figure 12.16:* (top after) W. C. Pitman, 1978, The relationship between eustasy and stratigraphic sequences of passive margins, *Geological Society of America Bulletin*, vol. 89, pp. 1389–1403; (bottom after) S. Stein and C. A. Stein, 1996, Thermo-mechanical evolution of oceanic lithosphere: Implications for the subduction process and deep earthquakes, American *Geophysical Union Monograph* 96, pp. 1–18
Figure 12.17: Tomographic image courtesy of Roderick Brown, University of Glasgow *Figures 12.18:* After S. Marshak, 2001, *Earth, Portrait of a Planet*, Norton
Figure 12.19b: Nicholas Parfitt/Getty Images Inc.-Stone Allstock *Figure 12.20a:* National Geophysical Data Center, NOAA *Figure 12.20b:* Alex Blackburn Clayton/Photolibrary.Com *Figure 12.20c:* Courtesy of Tobias Fischer *Figure 12.22:* After A. Jones, L. Siebert, P. Kimberly, and J. F. Luhr, 2000, Earthquakes and eruptions, Smithsonian Institution, Global Volcanism Program Digital Information Series, GVP-2
Figure 12.23: After R. van der Hilst, 1995, Complex morphology of subducted lithosphere in the mantle beneath the Tonga trench, *Nature*, vol. 374, pp. 154–157
Figure 12.24: After D.D. Blackwell, R. G. Bowen, D. A. Hull, J. Riccio, and J. L. Steele, 1982, Heat flow, arc volcanism, and subduction in northern Oregon: *Journal of Geophysical Research*, vol. 87, pp. 8735–8754 *Figure 12.26:* After E. J. Tarbuck and F. K. Lutgens, 2002, *Earth, An Introduction to Physical Geology*, 7th ed., Prentice Hall
Figure 12.27b: Courtesy of Joseph A. Dellinger *Figure 12.28:* After R. G. Stanley, 1987, New estimates of displacement along the San Andreas fault in central California based on paleobathymetry and paleogeography, *Geology*, vol. 15, pp. 171–174
Figure 12.28a: G.K. Gilbert/U.S. Geological Survey, Denver *Figure 12.28b:* Lloyd Cuff/Corbis/Bettmann *Figure 12.30:* After V. Courtillot, A. Da-vaille, J. Besse, and J. Stock, 2003, Three distinct types of hotspots in the Earth's mantle, *Earth and Planetary Science Letters*, vol. 205, pp. 295–308 *Figure 12.31a:* NASA Headquarters
Figure 12.31b: Aurora Pun *Figure 12.35:* After V. Courtillot, A. Davaille, J. Besse, and J. Stock, 2003, Three distinct types of hotspots in the Earth's mantle, *Earth and Planetary Science Letters*, vol. 205, pp. 295–308 *Figure 12.37:* After D. F. Argus and R. G. Gordon, 1991, No-net-rotation model of current plate velocities incorporating plate motion model NUVEL-1, *Geophysical Research Letters*, vol. 18, pp. 2039–2042
Figure 12.39: After NASA Jet Propulsion Laboratory, GPS Time Series
Figure 12.40: After D. F. Argus and R. G. Gordon, 1991, No-net-rotation model of current plate velocities incorporating plate motion model NUVEL-1, *Geophysical Research Letters*, vol. 18, pp. 2039–2042; NASA Jet Propulsion Laboratory, GPS Time Series *Figure 12.41:* After R.A. Bennett, J. L. Davis and B. P. Wernicke, 1999, Present day pattern of Cordillearn deformation in the Western United States, *Geology*, vol. 27, pp. 371–374 *Figure 12.43:* After D. W. Forsyth and S. Uyeda, 1975, On the relative importance of the dirving forces of plate motion; *Geophysical Journal of the Royal Astronomical Society*, vol. 43, pp. 163–200 *Figure 12.44a-g:* Ron Blakey

Chapter 13

Chapter Opener 13: Paul A. Souders/CORBIS-NY *Figure 13.1a:* Thomas Hallstein/ Alamy Images *Figure 13.1b:* After Robert W. Christopherson *Figure 13.1c:* Bobbe Z. Christopherson *Figure 13.2:* P. R. Pinet, 1992, *Oceanography, An Introduction to the Planet Oceanus*, West Pub *Figure 13.8:* After J. Jackson, 2002, Strength of the continental lithosphere: Time to abandon the jelly sandwich? *GSA Today*, vol. 12, no, 9, pp. 4–10
Figure 13.14: Photo by Dr. John Riley from his book *Flora of the Hudson Bay Lowland and its Postglacial Origins*, Courtesy of NRC Research Press; map after A. B. Watts, 2001, *Isostasy and Flexure of the Lithosphere*, Cambridge University Press, with GPS data from NASA/Jet Propulsion Laboratory *Figure 13.15:* P. R. Vail, R. M. Mitchum, and S. Thompson, 1978, Seismic stratigraphy and global changes in sea level, *American Association of Petroleum Geologists Memoir* 26, pp. 83–97 *Figure 13.18a:* NOAA/ National Geophysical Data Center *Figure 13.18b:* Super Stock, Inc.
Figure 13.20a: EyeWire Collection/Getty Images-Photodisc

Figure 13.20b: AIRPHOTO—Jim Wark *Figure 13.22:* After P. Rey, O. Vanderhaighe, and C. Teyssier, 2001, Gravitational collapse of the continental crust; definition, regimes and modes. *Tectonophysics*, vol. 342, pp. 435–449 *Figure 13.24:* After S. Marshak, 2001, *Earth, Portrait of a Planet*, Norton *Figure 13.25:* After A. Nur and Z. Ben-Avraham, 1982, Oceanic plateaus, the fragmentation of continents and mountain building, *Journal of Geophysical Research*, vol. 87, pp. 3644–3661 *Figure 13.27a:* Marli Bryant Miller *Figure 13.27b:* Paul Rezendes *Figure 13.29a:* Robert Hildebrand *Figure 13.28b:* Marli Bryant Miller *Figure 13.29:* After K. E. Karlstrom, S. S. Harlan, M. L. Williams, J. McLelland, J. W. Geissman, and K. I. Ahall, 2001, Long-lived (1.8-0.8 Ga) Cordilleran-type orogen in southern Laurentia, its extensions to Australia and Baltica, and implications for refining Rodinia. *Precambrian Research*, vol. 111, pp. 1–30 *Figure 13.31:* After D. R. Lowe, 1992, Major events in the geological development of the Precambrian Earth, in J. W. Schopf and C. Klein, eds. The Proterozoic Biosphere: A Multidisciplinary Study, Cambridge University Press

Chapter 14

Chapter Opener 14: AIRPHOTO-Jim Wark *Figure 14.1a1:* Gary A. Smith *Figure 14.1.a2:* USDA/NRCS *Figure 14.1b1 & 14.1b2:* Photos courtesy of Leslie D. McFadden *Figure 14.2 & 14.4a:* USDA/NRCS *Figure 14.4b:* Soil Science Society of America, Inc. *Figure 14.7:* Raymond Weil *Figure 14.8:* After M. N. Machette, 1985, *Calcic Soils of the Southwestern United States*, Geological Society of America Special Paper 203, pp. 1–21 *Figure 14.9:* U.S. Department of Agriculture *Figure 14.11:* After N. C. Brady and R. R. Weil, 1996, *The Nature and Properties of Soils*, 11th ed., Prentice Hall *Figure 14.11b:* USDA/NRCS *Figure 14.11c:* Jeff Vanuga/ USDA/NRCS *Figure 14.11d:* Tom Edwards/Animals Animals/Earth Scenes *Figure 14.14:* NRCS *Figure 14.15:* USDA/NRCS *Figure 14.16:* NRCS *Figure 14.16b:* Grahame McConnell/Photolibrary.Com *Figure 14.16c:* © 2004 Theodore Gray www.element-collection.com *Figure 14.17:* USDA/NRCS *Figure 14.18 & 14.19:* Randall J. Schaetzl, Michigan State University *Figure 14.20b:* Gary A. Smith *Figure 14.21:* Data from L. H. Gile and R. B. Grossman, 1979, *The Desert Project Soil Monograph*, U.S. Dept. of Agriculture, Soil Conservation Service *Figure 14.23a–c, 14.24a&b, 14.25, 14.26b, 14.27, and 14.28:* USDA/NRCS

Chapter 15

Chapter Opener 15: AP Wide World Photos *Figure 15.1a, a2:* David F. Walter, dfwalter@hotmail.com *Figure 15.1a3:* Edwin L. Harp, U.S. Geological Survey/U.S. Department of the Interior *Figure 15.1b1:* J.R. Stacy, U.S. Geological Survey/U.S. Department of the Interior *Figure 15.1b2:* Aurora Pun *Figure 15.1c:* U.S. Geological Survey/U.S. Department of the Interior *Figure 15.1d1&d2:* AP Wide World Photos *Figure 15.2:* After D. J. Varnes, 1978, *Landslides, Analysis and Control*, Transportation Research Board Special Report 176, National Research Council, pp. 11–33
Figure 15.3: AP World Wide Photos *Figure 15.4:* U.S. Geological Survey/U.S. Department of the Interior *Figure 15.5a:* Marli Bryant Miller *Figure 15.5b1&b2:* Richard Young *Figure 15.6:* After S. Marshak, 2001, *Earth, Portrait of a Planet*, Norton
Figure 15.7a: Michael Dolan *Figure 15.7b:* Susan Cannon/U.S. Geological Survey/U.S. Department of the Interior *Figure 15.8a:* D. Bradley/National Oceanic and Atmospheric Administration *Figure 15.8b:* Colorado Geological Survey/National Geophysical Data Center *Figure 15.9 & 15.10b:* U.S. Geological Survey/U.S. Department of the Interior *Figure 15.10:* After D. J. Varnes, 1978, *Landslides, Analysis and Control*, Transportation Research Board Special Report 176, National Research Council, pp. 11–33; photo by USGS *Figure 15.20:* Andrea Holland-Sears/ USDA/Forest Service
Figure 15.21a2: Terry Donnelley/Getty Images Inc.-Image Bank *Figure 15.21b2:* Gary A. Smith *Figure 15.22a, b:* Lloyd S. Cluff *Figure 15.23:* Photo Courtesy National Information Service for Earthquake Engineering, University of California, Berkeley *Figure 15.24:* Tom Pierson, USGS *Figure 15.25b1,b2,b3&b4:* Gary Rosenquist *Figure 15.25b5:* Gary A. Smith *Figures 15.28, 15.29:* After GIS data from E. E. Brabb, S. Roberts, W. R. Cotton, A. L. Kropp, R. H. Wright, and E N. Zin, 2000, *Possible Costs Associated with Investigating and Mitigating Geologic Hazards in Rural Areas of Western San Mateo County California*, U.S. Geological Survey Open-File Report 00-127 *Figure 15.31:* Institute of Geological & Nuclear Sciences Ltd.

Chapter 16

Chapter Opener 16: Adriel Heisey/Adriel Heisey Photography *Figure 16.1a1:* Cliff Riedinger/Alaska Stock *Figure 16.1a2:* Liz Hymans/CORBIS-NY *Figure 16.1b1:* Marli Bryant Miller *Figure 16.1b2:* U. S. Army Corps of Engineers *Figure 16.5b:* James Bartolino/U.S. Geological Survey/U.S. Department of the Interior *Figure 16.7:* Diagram after M. A. Summerfield, 1991, *Global Geomorphology*, Longman Scientific and Technical; image from NASA and the U.S./Japan ASTER Science Team *Figure 16.8a:* Gary A. Smith *Figure 16.8b:* Kevin Adams/Kevin Adams Nature

Photography *Figure 16.11b:* Courtesy of Robert S. Anderson
Figure 16.12b1&2: Courtesy of Keith J. Tinkler *Figure 16.13b:* Gary A. Smith
Figure 16.14b: Marli Bryant Miller *Figure 16.15b:* NASA and the U.S./Japan ASTER Science Team *Figure 16.16:* After M. Church, 1992, Channel morphology and typology, in *The Rivers Handbook, Hydrological and Ecological Principles*, Blackwell Scientific Publications, vol. 1, pp. 126–143 *Figure 16.18:* After D. F. Ritter, R. C. Kochel, and J. R. Miller, 2002, *Process geomorphology*, 4th ed. McGraw Hill, and L. B. Leopold, 1994, *A View of the River,* Harvard University Press *Figure 16.22b:* Altitude (Yann Arthus-Bertrand)/Peter Arnold, Inc. *Figure 16.23b & 16.24a&b:* AP Wide World Photos *Figure 16.25a&b:* Earth Satellite Corp./Photo Researchers, Inc. *Figure 16.26:* St. Paul Pioneer Press/CORBIS-NY *Figure 16.32:* Marli Bryant Miller *Figure 16.36a:* Richard T. Nowitz/CORBIS-NY *Figure 16.36b:* Courtesy of Grant Meyer *Figure 16.39b:* Tom Bean/Tom & Susan Bean, Inc. *Figure 16.39c:* NASA/Jet Propulsion Laboratory
Figure 16.40a: Claver Carroll/Photolibrary.Com *Figure 16.40b:* Marli Bryant Miller
Figure 16.41a: Michael A. Clynne/Volcano Hazards Team/USGS/Department of Interior *Figure 16.41b:* Robert L. Schuster/U.S. Geological Survey, Denver
Figure 16.42b: NASA Headquarters

Chapter 17
Chapter Opener 17: Tom Till *Figure 17.1a1:* Courtesy of Laura Wilson
Figure 17.1a2: National Park Service *Figure 17.1b1:* Gerry Davis/Phototake NYC
Figure 17.1b2: Jim Tuten/Black Star/Stockphoto.com/Black Star
Figure 17.2: Patrick Lynch *Figure 17.3:* Fred Habegger/Grant Heilman Photography, Inc. *Figure 17.4a:* Tony Arruza/CORBIS-NY *Figure 17.4c:* Daniel Dempster/Bruce Coleman Inc. *Figure 17.8:* Robert Shedlock/U.S. Geological Survey/U.S. Department of the Interior *Figure 17.9:* Map by USGS *Figure 17.10:* David R. Frazier/Alamy Images
Figure 17.13: Data from USGS *Figures 17.14, 17.15:* Map by USGS
Figure 17.15b1, b2: Larry Fellows, Arizona Geological Survey. Image courtesy of American Geological Institute, ImageBank, http://www.earthscienceworld.org/imagebank *Figure 17.17:* After M. K. Hubbert, 1940, The theory of ground-water motion, *Journal of Geology*, vol. 48, pp. 785–944, and J. A. Tóth, 1963, A theoretical analysis of ground-water flow in small drainage basins, *Journal of Geophysical Research*, vol. 68, pp. 4375–4387 *Figure 17.19:* After G. A Waring, 1965, *Thermal springs of the United States and other countries of the world*; a summary, USGS Professional Paper 492 *Figure 17.19b1&b2:* Aurora Pun *Figure 17.19b3:* Courtesy of Richard Allen Ashmore *Figures 17.26a, 17.27:* After D. R. LeBlanc, S. P. Garabedian, K. M. Hess, L. W. Gelhar, R. D. Quadri, K. G. Stollenwerk, and W. W. Wood, 1991, Large-scale natural gradient tracer test in sand and gravel, Cape Cod, Massachusetts. 1. Experimental design and observed tracer movement, *Water Resources Research*, vol. 27, pp. 895–910. Photo by Denis R. LeBlanc, USGS *Figure 17.26b:* Denis R. LeBlanc/USGS/Department of Interior *Figure 17.30:* USGS *Figure 17.31:* Avonsoft Water Treatment Ltd. *Figure 17.32:* After J. R. Craig, D. J. Vaughan, and B. J. Skinner, 2001, *Resources of the Earth*, 3rd edition, Prentice Hall *Figure 17.32b:* The Natural History Museum, London *Figure 17.36:* After T. L. McKnight and D. Hess, 2004, *Physical Geography, A Landscape Appreciation*, 7th ed., Prentice Hall
Figure 17.36b: Eve L. Kuniansky/U.S. Geological Survey/U.S. Department of the Interior *Figure 17.36c:* G.R. "Dick" Roberts/The Natural Sciences Image Library (NSIL) *Figure 17.36d:* Jon Gilhousen/U.S. Geological Survey/U.S. Department of the Interior *Figure 17.36e:* A.C. Waltham/Robert Harding World Imagery
Figure 17.37: After National Cave and Karst Research Foundation, National Park Service *Figure 17.38b:* Stuart Westmorland/CORBIS-NY *Figure 17.38c:* Hans Strand/Getty Images Inc.-Image Bank *Figure 17.39:* H.E. Holt/NASA
Figure 17.40b: Marli Bryant Miller

Chapter 18
Chapter Opener 18: Peter Arnold, Inc. *Figure 18.1a:* Dennis Hallinan/Alamy Images
Figure 18.1b: Phil Degginger/Color-Pic, Inc. *Figure 18.1c–f:* Courtesy of Peter J. Fawcett *Figure 18.2:* Top photo, courtesy of Mary Schaffer, Whyte Museum of the Canadian Rockies, Banff. Permission obtained from the Whyte Museum. Bottom photo courtesy of Dr. Brian Luckman, University of Western Ontario.; After B. H. Luckman and T. A. Kavanaugh, 2003, Documenting recent environmental changes and their impacts in the Canadian Rockies. Proceedings of the Conference on Ecology and Earth Sciences in Mountain Areas, Banff Centre, pp. 101–119 *Figure 18.3a1:* NASA Headquarters *Figure 18.3a2:* Kim Karpeles, Life Through the Lens *Figure 18.3b1:* Orbital Imaging Corporation/Photo Researchers, Inc. *Figure 18.3b2:* Gordon Wiltsie/National Geographic Image Collection *Figure 18.3c1:* Space Imaging *Figure 18.3c2:* Macduff Everton/CORBIS-NY *Figure 18.4:* Diagram after W. S. Broecker and G. H. Denton, 1990, What drives glacial cycles? *Scientific American*, vol. 262, pp. 49–56
Figure 18.4b: Galen Rowell/CORBIS-NY *Figure 18.6b1:* Tom & Susan Bean, Inc.

Figure 18.6b2: Richard Kucera/John S. Shelton *Figure 18.7b:* Bruce Molnia/American Geological Institute AGI *Figure 18.9:* Diagram after I. P. Martini, M. E. Brookfield, and S. Sadura, 2001, *Principles of Glacial Geomorphology and Geology*, Prentice Hall. Graphed data from C. F. Raymond, 1971, Flow in a transverse section of Athabasca Glacier, Alberta, Canada, *Journal of Glaciology*, vol. 10, pp. 55–84
Figure 18.10b: Paul A. Souders/CORBIS-NY *Figure 18.12a2:* Gary A. Smith
Figure 18.12b2: Tony Waltham/Tony Waltham Geophotos *Figure 18.14b:* Simon Fraser/Photo Researchers, Inc. *Figure 18.15:* Tom & Susan Bean, Inc.
Figure 18.16b1: Marli Bryant Miller *Figure 18.16b2:* Photo courtesy of Professor Randall J. Schaetzl, Michigan State University *Figure 18.17:* Grace Davies/Grace Davies Photo-graphy *Figure 18.18a:* Photo courtesy of Nicholas Eyles *Figure 18.18b:* Tom & Susan Bean, Inc. *Figure 18.20a:* NASA Headquarters *Figure 18.20b:* Réjean Couture & Gianluca Bianchi Fasani, Courtesy of Earth Sciences Information Centre, National Earth Sciences Library. *Figure 18.21:* Diagram after R. D. Dallmeyer, 2000, *Introductory Physical Geology Laboratory Text and Manual*, 6th ed., Kendall/Hunt Publishing, Dubuque, Iowa *Figure 18.21b:* Bradford Washburn *Figure 18.22:* Diagram after R. D. Powell, 1981, A model for sedimentation by tidewater glaciers, *Annales Glaciology*, vol. 2, pp. 129–134 *Figure 18.22a:* Grant Heilman/Grant Heilman Photography, Inc.
Figure 18.22b: Mark Newman/Alaska Stock *Figure 18.23a&b:* British Antarctic Survey *Figure 18.24a&b & 18.25b:* NASA *Figure 18.25:* Diagram after S. Marshak, 2001, *Earth, portrait of a planet*, Norton *Figure 18.25c:* Aurora Pun
 Figure 18.26: David Barnes/Corbis *Figure 18.27a:* Gerald & Buff Corsi/Visuals Unlimited *Figure 18.27b:* Sunset Avenue Productions/Digital Vision Ltd.
Figure 18.28: After J. M. Harbor, 1992, Numerical modeling of the development of U-shaped valleys by glacial erosion, *Geological Society of America Bulletin*, vol. 104, pp. 1364–1373 *Figure 18.29b:* Neil Rabinowitz/CORBIS-NY *Figure 18.29c:* Lowell Georgia/CORBIS-NY *Figure 18.30:* Tom Bean/CORBIS-NY *Figure 18.31:* AIRPHOTO-Jim Wark *Figure 18.33a:* Robert Hildebrand *Figure 18.33b&c:* Lynda Dredge, Courtesy of Earth Sciences Information Centre, Canada's National Earth Sciences Library. *Figure 18.33d:* John S. Shelton *Figure 18.33f:* Randall J. Schaetzl, Michigan State University *Figure 18.33g:* Tom & Susan Bean, Inc. *Figure 18.34:* After E. J. Tarbuck and F. K. Lutgens, *Earth*, 8th ed., Prentice Hall *Figure 18.35:* JPL *Figure 18.36:* After B. J. Skinner, S. C. Porter, and J. Park, 2004, *Dynamic Earth*, 5th ed., John Wiley and Sons *Figure 18.37:* After N. Eyles, 2002, *Ontario Rocks: Three Billion Years of Environmental Change*, Fitzhenry and Whiteside *Figure 18.38:* NASA Landsat imagery processed courtesy of Michael D. Blum *Figure 18.39:* After B. J. Skinner, S. C. Porter, and J. Park, 2004, *Dynamic Earth*, 5th ed., John Wiley and Sons *Figure 18.40a:* After E. J. Tarbuck and F. K. Lutgens, *Earth*, 8th ed., Prentice Hall *Figure 18.40b:* Trent Nelson/America 24-7/Getty Images, Inc. *Figure 18.40c:* Gary A. Smith *Figure 18.41:* After A. S. Dyke, A. Moore, and L. Robertson, 2003, Deglaciation of North America, Geological Survey of Canada Open File 1574 *Figure 18.42:* After W. K. Hamblin and E. H. Christiansen, 2004, *Earth's Dynamic Systems*, 10th ed., Prentice Hall *Figure 18.43:* After Photo courtesy of Jerry D. Kudenov, Scanning Electron Microscope Lab, University of Alaska–Anchorage *Figure 18.44:* Graphs after J. Imbrie, E. A. Boyle, S. C. A. Duffy, W. R. Howard, G. Kukla, J. Kutzbach, D. G. Martinson, A. McIntyre, A. C. Mix, B. Molfino, J. J. Morley, L. C. Peterson, N. G. Pisias, W. L. Prell, M. E. Raymo, N. J. Shackleton, and J. R. Toggweiler, 1992, On the structure and origin of major glaciation cycles; 1, Linear responses to Milankovitch forcing, *Paleoceanography*, vol.7, no. 6, pp. 701–738
Figures 18.45, 18.46: After R. C. L Wilson, S. A. Drury, and J. L. Chapman, 2000, *The Great Ice Age; Cimate Change and Life*, Routledge, London

Chapter 19
Chapter Opener 19: Richard Price/Getty Images, Inc.-Taxi *Figure 19.1a–c:* Gary A. Smith *Figure 19.1d:* Bob Heims/U.S. Army Corps of Engineers, Headquarters
Figure 19.1e1&e2: © BradleyIreland.com *Figure 19.5:* Bill Curtsinger/National Geographic Image Collection; Diagram after F. Press, R. Siever, J. Grotzinger, and T. H. Jordan, 2004, *Understanding Earth*, 4th ed., W. H. Freeman and Co. *Figure 19.6a:* Penny Tweedie/Getty Images Inc.-Stone Allstock *Figure 19.6b:* Warren Bolster/Getty Images Inc.-Stone Allstock *Figure 19.7a:* Nick Green/Photolibrary.Com *Figure 19.7b:* Gary Braasch/CORBIS-NY *Figure 19.8b:* John S. Shelton *Figure 19.10:* Agence France Presse/Getty Images *Figure 19.12b top & bottom:* Ikonos images copyright Centre for Remote Imaging, Sensing and Processing, National University of Singapore and Space Imaging/NASA. *Figure 19.13b:* TLM Photo *Figure 19.14b:* © Dick Jones/RJ Aerial Photo *Figure 19.16b:* Andrew Jaster *Figure 19.18b:* Photo Researchers, Inc.
Figure 19.18d: G.R. "Dick" Roberts/The Natural Sciences Image Library (NSIL)
Figure 19.19a: Gary A. Smith *Figure 19.19b:* Martyn F. Chillmaid/Science Photo Library/Photo Researchers, Inc. *Figure 19.20b:* Rafael Macia/Photo Researchers, Inc.
Figure 19.21b: John S. Shelton *Figure 19.22b:* NASA Headquarters
Figure 19.23b: Joseph R. Melanson/skypic.com *Figure 19.24b:* AIRPHOTO-Jim Wark

Figure 19.25a1: Bob Pardue/Alamu Images *Figure 19.25a2:* Andrew Jaster *Figure 19.25b1:* Gary A. Smith *Figure 19.25b2:* Aurora Pun *Figure 19.25c1:* Dave Houser/CORBIS-NY *Figure 19.25c2:* Photo courtesy of Michael Levin *Figure 19.26:* After P. D.Komar, 1998, *Beach Processes and Sedimentation*, 2nd ed. Prentice Hall *Figure 19.27:* Photo courtesy of Jon Arthur of the Florida Geological Survey and the Florida Department of Transportation. *Figure 19.28:* After R. H. Meade and S. W. Trimble, 1974, Changes in sediment loads in rivers of the Atlantic drainage of the United States since 1900, International Association for the Hydrological Sciences Publication 113, pp. 99–104 *Figure 19.29b:* NASA Headquarters *Figure 19.29c, d:* FRF/U.S. Army Corps of Engineers *Figure 19.30b:* Joseph R. Melanson, www.skypic.com *Figure 19.33a1 & a2:* Thomas McGuire/AGI Image Bank *Figure 19.33b1 &b2:* Graeme Teague *Figure 19.35:* Joyce Photographics/Photo Researchers, Inc. *Figure 19.36b:* U.S. Department of Agriculture *Figure 19.38a2:* John S. Shelton *Figure 19.38b2:* Marli Bryant Miller/Visuals Unlimited *Figure 19.39a&b:* U.S. Geological Survey, Denver *Figure 19.40a:* Mimmo Jodice/CORBIS-NY *Figure 19.40b:* Photo courtesy of Paul E. Olsen (N.G. McDonald Collection) *Figure 19.41:* After P. D. Komar and S. M. Shih, 1993, Cliff erosion along the Oregon Coast: A tectonic-sea level imprint plus local controls by beach processes, *Journal of Coastal Research*, vol. 9, pp. 747–765 *Figure 19.42:* After J. C. Kraft, G. Rapp, I. Kayan, and J. V. Luce, 2003, Harbor areas at ancient Troy: Sedimentology and geomorphology complement Homer's *Iliad*, *Geology*, vol. 31, pp. 163–166 *Figure 19.43:* Photo courtesy Bob & Sandra Shanklin "the Lighthouse People" www.thelighthouse people.com *Figure 19.44:* Graphed data from Permanent Service for Mean Sea Level Data Catalogue, Proudman Oceanographic Laboratory *Figure 19.45:* Data from R. G. Lighty, I. G. Macintyre, and R. Stickenrath, 1982, *Acropora palmata* reef framework: A reliable indicator of sea level in the western Atlantic for the past 10,000 years, *Coral Reefs*, vol. 1, pp. 125–130, and R. G. Fairbanks, 1989, A 17,000-year glacio-eustatic sea level record: Influence of glacial melting rates on the Younger Dryas even and deep-ocean circulation, *Nature*, vol. 342, pp. 637–642. *Figure 19.45c:* Larry Lipsky/Tom Stack & Associates, Inc. *Figure 19.46:* After B. H. Lidz and E. A. Shinn, 1991, Paleoshorelines, reefs, and a rising sea; South Florida, U.S.A., *Journal of Coastal Research*, vol.7, pp. 203–229 *Figure 19.49b&c:* Courtesy of Stephen Leatherman, Florida International University *Figure 19.50:* Gary A. Smith *Figure 19.51:* Royalty-Free/Corbis RF *Figure 19.53a&b:* U.S. Army Corps of Engineers, Headquarters

Figure 19.54a: AP Wide World Photos *Figure 19.54b:* North Carolina Dept. of Transportation

Chapter 20

Chapter Opener 20: Eric Wunrow/Index Stock Imagery, Inc. *Figure 20.1a1:* Gary A. Smith *Figure 20.1a2:* Marli Bryant Miller *Figure 20.1b1:* Richard T. Nowitz/CORBIS-NY *Figure 20.1b2:* David Muench/CORBIS-NY *Figure 20.3d&e, Figure 20.4:* After T. L. Mc-Knight and D. Hess, 2004, *Physical Geography, a Landscape Appreciation*, 7th ed., Prentice Hall *Figure 20.5a:* NASA Headquarters *Figure 20.5b:* Gary A. Smith *Figure 20.6:* After T. L. McKnight and D. Hess, 2004, *Physical Geography, a Landscape Appreciation*, 7th ed., Prentice Hall *Figure 20.6b:* Bruno Morandi/Robert Harding World Imagery *Figure 20.6c:* Charlie Ott/Photo Researchers, Inc. *Figure 20.6d:* Bernhard Edmaier/Photo Researchers, Inc. *Figure 20.6e:* Norbert Wu 2/Peter Arnold, Inc. *Figure 20.8:* Craig Aurness/CORBIS-NY *Figure 20.10a:* NASA Headquarters *Figure 20.10b:* Courtesy of Larry W. Thomason, NASA Langley Research Center *Figure 20.11b:* Getty Images, Inc./Taxi *Figure 20.11c:* John S. Shelton *Figure 20.12b:* Courtesy of David Love, New Mexico Bureau of Geology and Mineral Resources *Figure 20.12b2:* Richard M. Busch *Figure 20.12b3:* Peter Wilson/Corbis *Figure 20.13a&b:* NASA/Jet Propulsion Laboratory/Cornell Univ. *Figure 20.14a:* Marli Bryant Miller *Figure 20.14b:* Thomas Dressler/DRK *Figure 20.15a:* Courtesy of Dr. Patrick Hesp *Figure 20.15b:* Gerry Ellis/Minden Pictures *Figure 20.15c:* Pete Oxford/DRK Photo *Figure 20.15d:* Georg Gerster/Photo Researchers, Inc. *Figure 20.15e:* Michael Collier *Figure 20.17:* After K. Pye, 1987, *Aeolian Dust and Deposits*, Academic Press, London *Figure 20.17b:* Phil Schermeister/CORBIS-NY *Figure 20.17c:* Michael Andrews/Animals/Earth Scenes *Figure 20.17a & 20.18a&b:* Gary A. Smith *Figure 20.18c:* Soil Science Society of America, Inc. *Figure 20.20a:* NASA Headquarters *Figure 20.20b:* Tom Bean/Corbis *Figure 20.21a1,a2:* Courtesy of Virginia Garrison, USGS *Figure 20.21b:* NASA/MODIS *Figure 20.22:* K. D. Perry, T. A. Cahill, R. A. Eldred, and D. D. Dutcher, 1997, Long-range transport of North African dust to the eastern United States, *Journal of Geophysical Research*, vol. 102, pp. 11225–11238 *Figure 20.23:* Maps generalized from data provided by the IMPROVE Program (Interagency Monitoring of Protecting Visual Environments); *Figure 20.23b1* Marli B. Miller *Figure 20.23b2:* National Park Service/IMPROVE

Gary A. Smith/Aurora Pun
How Does Earth Work?
Physical Geology and the Process of Science
Active Art and Extension Modules
Student Edition
0-13145782-9
© 2006 Pearson Education, Inc.
Pearson Prentice Hall
Pearson Education, Inc.
Upper Saddle River, NJ 07458
Pearson Prentice Hall™ is a trademark of Pearson Education, Inc.

SYSTEM REQUREMENTS

Windows:
266 MHz Intel Pentium processor
Windows 98/ME/2000/NT4-sp3*/XP
16 bit Sound Card
32 MB or more of available RAM
Best viewed at 1024 x 768 monitor resolution
Mouse or other pointing device
4x CD-ROM Drive

Macintosh:
PowerPC 200Mhz processor
Operating System OS9.2.1/OSX
32 MB or more of available RAM
Best viewed at 1024 x 768 monitor resolution
Mouse or other pointing device
4x CD-ROM Drive
For some early release versions of OS9 you may need
to install the "Carbonlib" system file (see "Known issues" in the readme file.)